Springer **M**onographs *in* **M**athematics

Springer
New York
Berlin
Heidelberg
Barcelona
Hong Kong
London
Milan
Paris
Singapore
Tokyo

Davar Khoshnevisan

Multiparameter Processes

An Introduction to Random Fields

Davar Khoshnevisan
Department of Mathematics
University of Utah
Salt Lake City, Ut 84112-0090
davar@math.utah.edu

With 12 illustrations.

Mathematics Subject Classification (2000): 60Gxx, 60G60

Library of Congress Cataloging-in-Publication Data
Khoshnevisan, Davar.
    Multiparameter processes : an introduction to random fields / Davar Khoshnevisan.
      p. cm. — (Springer monographs in mathematics)
    Includes bibliographical references and index
    ISBN 0-387-95459-7 (alk. paper)
     1. Random fields. I. Title. II. Series.
QA274.45 .K58 2002
519.2′3—dc21   2002022927

Printed on acid-free paper.

© 2002 Springer-Verlag New York, Inc.
All rights reserved. This work may not be translated or copied in whole or in part without the written permission of the publisher (Springer-Verlag New York, Inc., 175 Fifth Avenue, New York, NY 10010, USA), except for brief excerpts in connection with reviews or scholarly analysis. Use in connection with any form of information storage and retrieval, electronic adaptation, computer software, or by similar or dissimilar methodology now known or hereafter developed is forbidden. The use in this publication of trade names, trademarks, service marks and similar terms, even if they are not identified as such, is not to be taken as an expression of opinion as to whether or not they are subject to proprietary rights.

Manufacturing supervised by Jerome Basma.
Camera-ready copy prepared from the author's LaTeX files.
Printed and bound by Edwards Brothers, Inc., Ann Arbor, MI.
Printed in the United States of America.

9 8 7 6 5 4 3 2 1

ISBN 0-387-95459-7   SPIN 10869448

Springer-Verlag   New York Berlin Heidelberg
A member of BertelsmannSpringer Science+Business Media GmbH

Preface

This book aims to construct a general framework for the analysis of a large family of random fields, also known as multiparameter processes. The need for such a development was pointed out in Doob (1990, p. 47). Referring to the theory of one-parameter stochastic processes, Doob writes:[1]

> Our definition of a stochastic process is historically conditioned and has obvious defects. In the first place there is no mathematical reason for restricting T to be a set of real numbers, and in fact interesting work has already been done in other cases. (Of course, the interpretation of t as time must then be dropped.) In the second place there is no mathematical reason for restricting the value assumed by the x_t's to be numbers.

There are a number of compelling reasons for studying random fields, one of which is that, if and when possible, multiparameter processes are a natural extension of existing one-parameter processes. More exciting still are the various interactions between the theory of multiparameter processes and other disciplines, including probability itself. For example, in this book the reader will learn of various connections to real and functional analysis, a modicum of group theory, and analytic number theory. The multiparameter processes of this book also arise in applied contexts such as mathematical statistics (Pyke 1985), statistical mechanics (Kuroda and Manaka 1998), and brain data imaging (Cao and Worsley 1999).

---

[1] He is referring to a stochastic process of the form $(x_t;\ t \in T)$.

My writing philosophy has been to strike a balance between developing a reasonably general theory, while presenting applications and explicit calculations. This approach should set up the stage for further analysis and exploration of the subject, and make for a more lively approach.

This book is in two parts. Part I is about the discrete-time theory. It also contains results that allow for the transition from discrete-time processes to continuous-time processes. In particular, it develops abstract random variables, parts of the theory of Gaussian processes, and weak convergence for continuous stochastic processes. Part II contains the general theory of continuous-time processes. Special attention is paid to processes with continuous trajectories, but some discontinuous processes will also be studied. In this part I will also discuss subjects such as potential theory for several Markov processes, the Brownian sheet, and some Gaussian processes. Parts I and II are influenced by the fundamental works of Doob, Cairoli, and Walsh.

My goal has been to keep this book as self-contained as possible, in order to make it available to advanced graduate students in probability and analysis. To this I add that a more complete experience can only be gained through solving many of the problems that are scattered throughout the body of the text. At times, these in-text exercises ask the student to check some technical detail. At other times, the student is encouraged to apply a recently introduced idea in a different context. More challenging exercises are offered at the end of each chapter.

Many of the multiparameter results of this book do not seem to exist elsewhere in a pedagogic manner. There are also a number of new theorems that appear here for the first time. When introducing a better-known subject (e.g., martingales or Markov chains), I have strived to construct the most informative proofs, rather than the shortest.

This book would not exist had it not been for the extensive remarks, corrections, and support of R. Bass, J. Bertoin, K. Burdzy, R. Dalang, S. Ethier, L. Horváth, S. Krone, O. Lévêque, T. Lewis, G. Milton, E. Nualart, T. Mountford, J. Pitman, Z. Shi, J. Walsh, and Y. Xiao. Their efforts have led to a much cleaner product. What errors remain are my own. I have enjoyed a great deal of technical support from P. Bowman, N. Beebe, and the editorial staff of Springer. The National Science Foundation and the North Atlantic Treaty Organization have generously supported my work on random fields over the years. My sincerest gratitude goes to them all. Finally, I wish to thank my dearest friend, and my source of inspiration, Irina Gushin.

This book is dedicated to the memory of Victor Gushin, and to the recent arrival of Adrian V. Kh. Gushin.

Davar Khoshnevisan, Salt Lake City, UT.
March 2002

Contents

Preface v

List of Figures xv

General Notation xvii

I Discrete-Parameter Random Fields   1

1 Discrete-Parameter Martingales   3
- 1   One-Parameter Martingales 4
  - 1.1   Definitions 4
  - 1.2   The Optional Stopping Theorem 7
  - 1.3   A Weak (1,1) Inequality 8
  - 1.4   A Strong (p,p) Inequality 9
  - 1.5   The Case $p = 1$ 9
  - 1.6   Upcrossing Inequalities 10
  - 1.7   The Martingale Convergence Theorem 12
- 2   Orthomartingales 15
  - 2.1   Definitions and Examples 16
  - 2.2   Embedded Submartingales 18
  - 2.3   Cairoli's Strong (p,p) Inequality 19
  - 2.4   Another Maximal Inequality 20
  - 2.5   A Weak Maximal Inequality 22

		2.6	Orthohistories	22
		2.7	Convergence Notions	24
		2.8	Topological Convergence	26
		2.9	Reversed Orthomartingales	30
	3	Martingales		31
		3.1	Definitions	31
		3.2	Marginal Filtrations	31
		3.3	A Counterexample	33
		3.4	Commutation	35
		3.5	Martingales	37
		3.6	Conditional Independence	38
	4	Supplementary Exercises		40
	5	Notes on Chapter 1		44

2 Two Applications in Analysis — 47

1	Haar Systems		47
	1.1	The 1-Dimensional Haar System	48
	1.2	The N-Dimensional Haar System	51
2	Differentiation		54
	2.1	Lebesgue's Differentiation Theorem	54
	2.2	A Uniform Differentiation Theorem	58
3	Supplementary Exercises		61
4	Notes on Chapter 2		63

3 Random Walks — 65

1	One-Parameter Random Walks		66
	1.1	Transition Operators	66
	1.2	The Strong Markov Property	69
	1.3	Recurrence	70
	1.4	Classification of Recurrence	72
	1.5	Transience	74
	1.6	Recurrence of Possible Points	75
	1.7	Recurrence–Transience Dichotomy	78
2	Intersection Probabilities		80
	2.1	Intersections of Two Walks	80
	2.2	An Estimate for Two Walks	85
	2.3	Intersections of Several Walks	86
	2.4	An Estimate for N Walks	89
3	The Simple Random Walk		89
	3.1	Recurrence	90
	3.2	Intersections of Two Simple Walks	91
	3.3	Three Simple Walks	93
	3.4	Several Simple Walks	97
4	Supplementary Exercises		99
5	Notes on Chapter 3		103

4 Multiparameter Walks — 105

- 1 The Strong Law of Large Numbers 106
  - 1.1 Definitions 106
  - 1.2 Commutation 107
  - 1.3 A Reversed Orthomartingale 109
  - 1.4 Smythe's Law of Large Numbers 110
- 2 The Law of the Iterated Logarithm 112
  - 2.1 The One-Parameter Gaussian Case 113
  - 2.2 The General LIL 116
  - 2.3 Summability 117
  - 2.4 Dirichlet's Divisor Lemma 118
  - 2.5 Truncation 119
  - 2.6 Bernstein's Inequality 121
  - 2.7 Maximal Inequalities 123
  - 2.8 A Number-Theoretic Estimate 125
  - 2.9 Proof of the LIL: The Upper Bound 127
  - 2.10 A Moderate Deviations Estimate 128
  - 2.11 Proof of the LIL: The Lower Bound 130
- 3 Supplementary Exercises 132
- 4 Notes on Chapter 4 135

5 Gaussian Random Variables — 137

- 1 The Basic Construction 137
  - 1.1 Gaussian Random Vectors 137
  - 1.2 Gaussian Processes 140
  - 1.3 White Noise 142
  - 1.4 The Isonormal Process 144
  - 1.5 The Brownian Sheet 147
- 2 Regularity Theory 148
  - 2.1 Totally Bounded Pseudometric Spaces 149
  - 2.2 Modifications and Separability 153
  - 2.3 Kolmogorov's Continuity Theorem 158
  - 2.4 Chaining 160
  - 2.5 Hölder-Continuous Modifications 165
  - 2.6 The Entropy Integral 167
  - 2.7 Dudley's Theorem 170
- 3 The Standard Brownian Sheet 172
  - 3.1 Entropy Estimate 172
  - 3.2 Modulus of Continuity 173
- 4 Supplementary Exercises 175
- 5 Notes on Chapter 5 178

6 Limit Theorems — 181

- 1 Random Variables 181
  - 1.1 Definitions 182

		1.2	Distributions	183
		1.3	Uniqueness	184
	2	Weak Convergence		185
		2.1	The Portmanteau Theorem	186
		2.2	The Continuous Mapping Theorem	188
		2.3	Weak Convergence in Euclidean Space	188
		2.4	Tightness	189
		2.5	Prohorov's Theorem	190
	3	The Space C		193
		3.1	Uniform Continuity	193
		3.2	Finite-Dimensional Distributions	195
		3.3	Weak Convergence in C	196
		3.4	Continuous Functionals	199
		3.5	A Sufficient Condition for Pretightness	200
	4	Invariance Principles		201
		4.1	Preliminaries	202
		4.2	Finite-Dimensional Distributions	204
		4.3	Pretightness	207
	5	Supplementary Exercises		210
	6	Notes on Chapter 6		213

II Continuous-Parameter Random Fields   215

7 Continuous-Parameter Martingales   217

	1	One-Parameter Martingales		217
		1.1	Filtrations and Stopping Times	218
		1.2	Entrance Times	221
		1.3	Smartingales and Inequalities	222
		1.4	Regularity	223
		1.5	Measurability of Entrance Times	226
		1.6	The Optional Stopping Theorem	226
		1.7	Brownian Motion	228
		1.8	Poisson Processes	230
	2	Multiparameter Martingales		233
		2.1	Filtrations and Commutation	233
		2.2	Martingales and Histories	234
		2.3	Cairoli's Maximal Inequalities	235
		2.4	Another Look at the Brownian Sheet	236
	3	One-Parameter Stochastic Integration		239
		3.1	Unbounded Variation	239
		3.2	Quadratic Variation	242
		3.3	Local Martingales	245
		3.4	Elementary Processes	246
		3.5	Simple Processes	247

		3.6	Continuous Adapted Processes	248
		3.7	Two Approximation Theorems	250
		3.8	Itô's Formula	251
		3.9	The Burkholder–Davis–Gundy Inequality	253
	4	Stochastic Partial Differential Equations	255
		4.1	Stochastic Integration	256
		4.2	Hyperbolic SPDEs	257
		4.3	Existence and Uniqueness	260
	5	Supplementary Exercises	263
	6	Notes on Chapter 7	266

8 Constructing Markov Processes · · · 267

	1	Discrete Markov Chains	267
		1.1	Preliminaries	267
		1.2	The Strong Markov Property	272
		1.3	Killing and Absorbing	272
		1.4	Transition Operators	275
		1.5	Resolvents and λ-Potentials	277
		1.6	Distribution of Entrance Times	279
	2	Markov Semigroups	281
		2.1	Bounded Linear Operators	281
		2.2	Markov Semigroups and Resolvents	282
		2.3	Transition and Potential Densities	284
		2.4	Feller Semigroups	287
	3	Markov Processes	288
		3.1	Initial Measures	288
		3.2	Augmentation	290
		3.3	Shifts	292
	4	Feller Processes	293
		4.1	Feller Processes	294
		4.2	The Strong Markov Property	298
		4.3	Lévy Processes	303
	5	Supplementary Exercises	307
	6	Notes on Chapter 8	311

9 Generation of Markov Processes · · · 313

	1	Generation	313
		1.1	Existence	314
		1.2	The Hille–Yosida Theorem	315
		1.3	The Martingale Problem	317
	2	Explicit Computations	320
		2.1	Brownian Motion	320
		2.2	Isotropic Stable Processes	322
		2.3	The Poisson Process	325
		2.4	The Linear Uniform Motion	326

xii   Contents

	3	The Feynman–Kac Formula	326
		3.1 The Feynman–Kac Semigroup	326
		3.2 The Doob–Meyer Decomposition	328
	4	Exit Times and Brownian Motion	329
		4.1 Dimension One	330
		4.2 Some Fundamental Local Martingales	331
		4.3 The Distribution of Exit Times	335
	5	Supplementary Exercises	339
	6	Notes on Chapter 9	340

10 Probabilistic Potential Theory   343

	1	Recurrent Lévy Processes	344
		1.1 Sojourn Times	344
		1.2 Recurrence of the Origin	347
		1.3 Escape Rates	350
		1.4 Hitting Probabilities	353
	2	Hitting Probabilities for Feller Processes	360
		2.1 Strongly Symmetric Feller Processes	360
		2.2 Balayage	362
		2.3 Hitting Probabilities and Capacities	367
		2.4 Proof of Theorem 2.3.1	368
	3	Explicit Computations	373
		3.1 Brownian Motion and Capacities	373
		3.2 Stable Densities and Subordination	377
		3.3 Asymptotics for Stable Densities	380
		3.4 Stable Processes and Capacities	382
		3.5 Relation to Hausdorff Dimension	385
	4	Supplementary Exercises	386
	5	Notes on Chapter 10	388

11 Multiparameter Markov Processes   391

	1	Definitions	391
		1.1 Preliminaries	392
		1.2 Commutation and Semigroups	395
		1.3 Resolvents	397
		1.4 Strongly Symmetric Feller Processes	398
	2	Examples	401
		2.1 General Notation	401
		2.2 Product Feller Processes	402
		2.3 Additive Lévy Processes	405
		2.4 Product Process	407
	3	Potential Theory	408
		3.1 The Main Result	408
		3.2 Three Technical Estimates	410
		3.3 Proof of Theorem 3.1.1: First Half	413

	3.4	Proof of Theorem 3.1.1: Second Half	418
4	Applications		419
	4.1	Additive Stable Processes	419
	4.2	Intersections of Independent Processes	424
	4.3	Dvoretzky–Erdős–Kakutani Theorems	426
	4.4	Intersecting an Additive Stable Process	428
	4.5	The Range of a Stable Process	429
	4.6	Extension to Additive Stable Processes	433
	4.7	Stochastic Codimension	435
5	α-Regular Gaussian Random Fields		438
	5.1	Stationary Gaussian Processes	438
	5.2	α-Regular Gaussian Fields	441
	5.3	Proof of Theorem 5.2.1: First Part	443
	5.4	Proof of Theorem 5.2.1: Second Part	448
6	Supplementary Exercises		450
7	Notes on Chapter 11		453

12 The Brownian Sheet and Potential Theory — 455

1	Polar Sets for the Range of the Brownian Sheet		455
	1.1	Intersection Probabilities	456
	1.2	Proof of Theorem 1.1.1: Lower Bound	457
	1.3	Proof of Lemma 1.2.2	460
	1.4	Proof of Theorem 1.1.1: Upper Bound	468
2	The Codimension of the Level Sets		472
	2.1	The Main Calculation	473
	2.2	Proof of Theorem 2.1.1: The Lower Bound	474
	2.3	Proof of Theorem 2.1.1: The Upper Bound	476
3	Local Times as Frostman's Measures		477
	3.1	Construction	478
	3.2	Warmup: Linear Brownian Motion	480
	3.3	A Variance Estimate	485
	3.4	Proof of Theorem 3.1.1: General Case	488
4	Supplementary Exercises		491
5	Notes on Chapter 12		493

III Appendices — 497

A Kolmogorov's Consistency Theorem — 499

B Laplace Transforms — 501

1	Uniqueness and Convergence Theorems		501
	1.1	The Uniqueness Theorem	502
	1.2	The Convergence Theorem	503
	1.3	Bernstein's Theorem	505

Contents

| | 2 | A Tauberian Theorem | 506 |

C Hausdorff Dimensions and Measures — 511
- 1 Preliminaries — 511
  - 1.1 Definition — 511
  - 1.2 Hausdorff Dimension — 515
- 2 Frostman's Theorems — 517
  - 2.1 Frostman's Lemma — 517
  - 2.2 Bessel–Riesz Capacities — 520
  - 2.3 Taylor's Theorem — 523
- 3 Notes on Appendix C — 525

D Energy and Capacity — 527
- 1 Preliminaries — 527
  - 1.1 General Definitions — 527
  - 1.2 Physical Interpretations — 530
- 2 Choquet Capacities — 533
  - 2.1 Maximum Principle and Natural Capacities — 533
  - 2.2 Absolutely Continuous Capacities — 537
  - 2.3 Proper Gauge Functions and Balayage — 539
- 3 Notes on Appendix D — 540

References — 543

Name Index — 565

Subject Index — 572

List of Figures

1.1	Orthohistories	22
1.2	Orthohistories	23
1.3	Histories	32
5.1	Covering by balls	160
9.1	Gambler's ruin	335
10.1	Covering balls	345
11.1	Planar Brownian motion	431
11.2	Additive Brownian motion	434
12.1	Planar Brownian sheet (aerial)	458
12.2	Planar Brownian sheet (portrait)	458
12.3	Planar Brownian sheet (side)	458
12.4	The zero set of the Brownian sheet	473
C.1	Cantor's set	516

General Notation

While it is generally the case that the notation special to each chapter is independent of that in the remainder of the book, there is much that is common to the entire book. These items are listed below.

Euclidean Spaces, Integers, etc.

The collection of all real (nonnegative real) numbers is denoted by $\mathbb{R}$ ($\mathbb{R}_+$), positive (nonnegative) integers by $\mathbb{N}$ ($\mathbb{N}_0$ and/or $\mathbb{Z}_+$). Finally, the rationals (nonnegative rationals) are written as $\mathbb{Q}$ ($\mathbb{Q}_+$). The latter spaces are endowed with their standard Borel topologies and Borel σ-fields which will also be referred to as Borel fields. Recall that the Borel field on a topological space $\mathbb{X}$ is the σ-field generated by all open subsets of $\mathbb{X}$; that is, the smallest σ-field that contains all open subsets of $\mathbb{X}$.

The sequence space ℓ^p is the usual one: For any $p > 0$, ℓ^p designates the collection of all sequences $a = (a_k;\ k \geq 1)$ such that $\sum_k |a_k|^p < \infty$. As usual, ℓ^∞ stands for all bounded sequences. When $\infty > p \geq 1$, these ℓ^p spaces can be normed by $\|a\|_{\ell^p} = \{\sum_k |a_k|^p\}^{1/p}$. Using these norms in turn, we can norm the Euclidean space $\mathbb{R}^k$ in various ways. Throughout, we will use the following two norms: (a) the ℓ^∞ norm, which is $|x| = \max_{1 \leq j \leq k} |x^{(j)}|$, for $x \in \mathbb{R}^k$; and (b) the ℓ^2 norm, which is $\|x\| = \{\sum_{j=1}^k |x^{(j)}|^2\}^{\frac{1}{2}}$.

Product Spaces

Throughout, the "dimension" numbers d and N are reserved for the spatial and temporal dimension, respectively.

Given any two sets F and T, the set F^T is defined as the collection of all functions $f: T \mapsto F$. When F is a topological space, F^T is often endowed with the product topology; cf. Appendix D. If $m \in \mathbb{N}$, F^m is the usual product space

$$F^m = \overbrace{F \times \cdots \times F}^{m \text{ times}}.$$

This, too, is often endowed with the product topology if and when F is topological.

Throughout, the ith coordinate of any point $s \in \mathbb{R}^m$ is written as $s^{(i)}$. We need the following special order structure on the $\mathbb{R}^m$: Whenever $s, t \in \mathbb{R}^m$, we write $s \preccurlyeq t$ ($s \prec t$) when for $i = 1, \ldots, m$, $s^{(i)} \le t^{(i)}$ ($s^{(i)} < t^{(i)}$). Occasionally, we may write this as $t \succcurlyeq s$ (or $t \succ s$). Whenever $s, t \in \mathbb{R}^m$, $s \wedge t$ designates the point whose ith coordinate is $s^{(i)} \wedge t^{(i)}$ for all $i = 1, \ldots, m$.

If $s \preccurlyeq t$, then $[s,t] = \prod_{j=1}^m [s^{(j)}, t^{(j)}]$. We will refer to $[s,t]$ as a **rectangle** (or an m-dimensional rectangle). When this rectangle is of the form $[s, s+(r,\ldots,r)]$ for some $r \in \mathbb{R}_+^1$ and $s \in \mathbb{R}^m$, then it is a (hyper)**cube**. If $s \prec t$, one can similarly define $]s,t[$, $[s,t[$, and $]s,t]$. For instance, $]s,t] = \prod_{j=1}^m]s^{(j)}, t^{(j)}]$. All subsets of $\mathbb{R}^m$ automatically inherit the order structure of $\mathbb{R}^m$. This way, $s \preccurlyeq t$ makes the same sense in $\mathbb{N}^k$ as it does in $\mathbb{R}^k$, for instance.

Probability and Measure Theory

Unless it is stated to the contrary, the underlying probability space is nearly always denoted by $(\Omega, \mathcal{G}, \mathbb{P})$, where Ω is the so-called sample space, $\mathcal{G}$ is a σ-field of subsets of Ω, and $\mathbb{P}$ is a probability measure on $\mathcal{G}$. Unless it is specifically stated otherwise, the corresponding expectation operator is always denoted by $\mathbb{E}$.

While intersections of σ-fields are themselves σ-fields, their unions are not always. Thus, when $\mathcal{F}_1$ and $\mathcal{F}_2$ are two σ-fields, we write $\mathcal{F}_1 \vee \mathcal{F}_2$ to mean the smallest σ-field that contains $\mathcal{F}_1 \cup \mathcal{F}_2$. More generally, for any index set $\mathbb{A}$, we write $\vee_{\alpha \in \mathbb{A}} \mathcal{F}_\alpha$ for the smallest σ-field that contains all of the σ-fields $(\mathcal{F}_\alpha; \alpha \in \mathbb{A})$.

An important function is the indicator function (called the characteristic function in the analysis literature): In *any* space for *any* set A in that space, $\mathbf{1}_A$ denotes the function $x \mapsto \mathbf{1}_A(x)$ that is defined by $\mathbf{1}_A(x) = 1$ if $x \in A$ and $\mathbf{1}_A(x) = 0$ if $x \notin A$. In particular, if A is an event in the probability space, $\mathbf{1}_A$ is the indicator function of the event A.

Throughout, "a.s." is treated synonymously to "almost surely," "$\mathbb{P}$-almost everywhere," "almost sure," or "$\mathbb{P}$-almost sure," depending on which is more applicable.

Both "iid" and "i.i.d." stand for "independent, identically distributed."

When μ is a (nonnegative) measure on a measure space $(\Omega, \mathcal{F})$, $L^p(\mu)$ denotes the collection of all real-valued, p-times μ-integrable functions on Ω. In particular, we use this notation quite often when μ is a probability measure $\mathbb{P}$. In this case, we can interpret $L^p(\mathbb{P})$ as the collection of all random variables whose absolute value has p finite moments. (You should recall that L^p spaces are in fact spaces of equivalence relations, where we say that f and g are equivalent when they are equivalent μ-almost everywhere.)

A special case is made for the L^p spaces with respect to Lebesgue's measure (on Euclidean spaces): When $E \subset \mathbb{R}^k$ is Borel (or more generally, Lebesgue) measurable, $L^p(E)$ (or sometimes $L^p E$) denotes the collection of all p-times continuously differentiable functions that map E into $\mathbb{R}$. For instance, we write $L^p[0,1]$ and/or $L^p([0,1])$ for the collection of (equivalence classes of) all p-times integrable functions on $[0,1]$.

Depending on the point that is being made, a stochastic process $(X_t : t \in T)$ (where T is some indexing set) is identified with the "randomly chosen function" $t \mapsto X_t$.

Throughout, Leb denotes Lebesgue's measure regardless of the dimension of the underlying Euclidean space.

Part I

Discrete-Parameter Random Fields

1
Discrete-Parameter Martingales

In this chapter we develop the basic theory of multiparameter martingales indexed by a countable subset of $\mathbb{R}_+^N$, usually $\mathbb{N}^N$ or $\mathbb{N}_0^N$. As usual, $\mathbb{N} = \{1, 2, \ldots\}$, $\mathbb{N}_0 = \{0, 1, 2, \ldots\}$, and N denotes a fixed positive integer.

We will be assuming that the reader is quite familiar with the standard aspects of the theory of martingales indexed by a discrete one-parameter set. However, we have provided a primer section on this subject to reacquaint the reader with some of the one-parameter techniques and notations; see Section 1 below.

The main thrust of this chapter is its discussion of maximal inequalities, since they are at the very heart of the theory of multiparameter martingales. Even in the simple setting of one-parameter random walks, maximal inequalities are considered deep results to this day. For instance, consider n i.i.d. random variables $\xi_1, \ldots, \xi_n$ and define the corresponding random walk as $S_n = \xi_1 + \cdots + \xi_n$. Then, Kolmogorov's maximal inequality states that whenever ξ's have mean 0, for all $\lambda > 0$,

$$\mathbb{P}\left(\max_{1 \leq j \leq n} |S_j| \geq \lambda\right) \leq \frac{1}{\lambda} \mathbb{E}\{|S_n|\}.$$

On the other hand, a straightforward application of Chebyshev's inequality implies that for all $\lambda > 0$, $\mathbb{P}(|S_n| \geq \lambda) \leq \frac{1}{\lambda} \mathbb{E}\{|S_n|\}$, and this is sharp, in general. Thus, roughly speaking, Kolmogorov's maximal inequality asserts that the behavior of the entire partial sum process $j \mapsto S_j$ ($1 \leq j \leq n$) is controlled by the value of the latter process at time n.

The aforementioned maximal property leads to one of the key steps in the usual proof of Kolmogorov's strong law of large numbers and is by no means a singular property. Indeed, maximal inequalities are precisely the tools required to prove strong convergence theorems in the theory of multiparameter martingales (Cairoli's second convergence theorem, Theorem 2.8.1), as well as Lebesgue's differentiation theorem (Theorem 2.1.1, Chapter 2). They are also an essential ingredient in the analysis of the intersections of several random walks (Theorem 2.2.1, Chapter 3), the law of the iterated logarithm (Section 2.6, Chapter 4), regularity theory of general stochastic processes (Section 2.4, Chapter 5), and weak convergence of measures and processes in the space of continuous functions (Theorem 3.3.1, Chapter 6). You might have noted that we have chosen a discrete-time example for every chapter in Part I of this book! Suffice it to say that Part II also relies heavily on maximal inequalities, and we now proceed with our treatment of discrete-parameter martingales without further ado.

Let $(\Omega, \mathcal{G}, \mathbb{P})$ be a probability space. In this chapter we will discuss martingales on $(\Omega, \mathcal{G}, \mathbb{P})$ that are indexed by $\mathbb{N}_0^N$ and sometimes $\mathbb{N}^N$. Before proceeding further, the reader should become familiar with the notation described in the preamble to this book.

1 One-Parameter Martingales

This section is not a comprehensive introduction to one-parameter martingales. Rather, it serves to remind the reader of some of the key methods and concepts as well as familiarize him or her with the notation. Further details can be found in the Supplementary Exercises.

1.1 Definitions

Suppose $\mathcal{F} = (\mathcal{F}_k; k \geq 0)$ is a collection of sub–σ-fields of the underlying σ-field $\mathcal{G}$. We say that $\mathcal{F}$ is a (discrete-time, one-parameter) **filtration** if for all $k \geq 0$, $\mathcal{F}_k \subset \mathcal{F}_{k+1}$.

By a **stochastic process**, or a **random process**, we mean a collection of random variables that are indexed by a possibly arbitrary set. A stochastic process $M = (M_k; k \geq 0)$ is **adapted** to the filtration $\mathcal{F}$ if for all $k \geq 0$, M_k is $\mathcal{F}_k$-measurable.

A stochastic process M is a **submartingale** (with respect to the filtration $\mathcal{F}$) if

1. M is adapted to $\mathcal{F}$;

2. for all $k \geq 0$, $\mathbb{E}\{|M_k|\} < \infty$; that is, $M_k \in L^1(\mathbb{P})$ for all $k \geq 0$; and

3. for all $k \geq 0$, $\mathbb{E}[M_{k+1} \,|\, \mathcal{F}_k] \geq M_k$, almost surely.

We say that M is a **supermartingale** if $-M$ is a submartingale. If M is both a supermartingale and a submartingale, then it is a **martingale**. We will refer to M as a **smartingale** if it is either a sub- or a supermartingale. By Jensen's inequality, if M is a nonnegative submartingale, Ψ is convex nondecreasing on $[0, \infty[$, and if $\mathbb{E}\{|\Psi(M_k)|\} < \infty$ for all $k \geq 0$, then $(\Psi(M_k); k \geq 0)$ is also a submartingale.

***Exercise* 1.1.1 (Random Walks)** Suppose $\xi_1, \xi_2, \ldots$ are i.i.d. mean 0 random variables. Let $S_n = \sum_{j=1}^n \xi_j$ $(n = 1, 2, \ldots)$ denote the corresponding partial sum process that is also known as a **random walk**. If $S_0 = 0$ and if $\mathcal{F}_n$ denotes the σ-field generated by $\xi_1, \ldots, \xi_n$ $(n = 0, 1, 2 \ldots)$, show that $S = (S_n; n \geq 0)$ is a martingale with respect to $\mathcal{F} = (\mathcal{F}_k; k \geq 0)$. □

***Exercise* 1.1.2 (Branching Processes)** Write $X_0 = 1$ and define $X_{n+1} = \sum_{\ell=1}^{X_n} \xi_{\ell,n}$ $(n = 0, 1, \ldots)$, where the ξ's are all i.i.d., integrable random variables that take their values in $\mathbb{N}_0$. Let $\mathcal{F}_m$ denote the σ-field generated by $X_1, \ldots, X_m$ $(m = 0, 1, \ldots)$ and show that $(\mu^{-n} X_n; n \geq 0)$ is a martingale with respect to $\mathcal{F} = (\mathcal{F}_n; n \geq 0)$, where $\mu = \mathbb{E}[\xi_{1,1}]$.

The stochastic process X is a **branching process** and, quite often, arises in the modeling of biological systems. For instance, suppose each individual gene in a given generation has a chance to give birth to a random number of genes (its offspring in the following generation). If the birth mechanisms are all i.i.d., from individual to individual and from generation to generation, and if the entire population starts out with one individual in generation 0, the above X_n denotes the total number of individuals in generation n, for appropriately chosen birth numbers. □

***Exercise* 1.1.3 (Random Stick-Breaking)** Let $X_0 = 1$ denote the length of a stick at time 0. Let X_1 be a random variable picked uniformly from $[0, X_0]$ and, conditionally on $\{X_1, \ldots, X_n\}$, define X_{n+1} to be picked uniformly from $[0, X_n]$. Show that $(2^n X_n; n \geq 0)$ is a martingale with respect to $\mathcal{F} = (\mathcal{F}_n; n \geq 0)$, where $\mathcal{F}_n$ denotes the σ-field generated by $(X_1, \ldots, X_n)$. Among other things, this process models breaking the "stick" $[0, 1]$ in successive steps. □

Suppose T is a random variable that takes its values in $\mathbb{N}_0 \cup \{\infty\}$. Then, we say that T is a **stopping time** (with respect to $\mathcal{F}$) if for all $k \geq 0$, $(T > k) \in \mathcal{F}_k$. Equivalently, T is a stopping time if for all $k \geq 0$, $(T = k) \in \mathcal{F}_k$. This is a consequence of the properties of filtrations, together with the decomposition

$$(T = k) = (T > k - 1) \cap (T > k)^\complement.$$

In order to emphasize the underlying filtration $\mathcal{F}$, sometimes we say that T is an $\mathcal{F}$-stopping time.

There are many natural stopping times, as the following shows.

***Exercise* 1.1.4** Nonrandom constants are stopping times. Also, if $T_1, T_2, \ldots$ are stopping times, so are

- $\min(T_1, T_2) = T_1 \wedge T_2$;
- $\max(T_1, T_2) = T_1 \vee T_2$;
- $T_1 + T_2$;

6   1. Discrete-Parameter Martingales

- $\inf_n T_n$, $\sup_n T_n$, $\liminf_n T_n$, and $\limsup_n T_n$.

Finally, if X is an adapted process and A is a Borel subset of $\mathbb{R}$, then T is a stopping time, where $T = \inf(k \geq 0 : X_k \in A)$ and $\inf \varnothing = \infty$. □

For any stopping time T, define

$$\mathcal{F}_T = \Big(A \in \bigvee_{n=1}^{\infty} \mathcal{F}_n : A \cap (T \leq k) \in \mathcal{F}_k, \text{ for all } k \geq 0\Big).$$

Exercise 1.1.5 Show that

(a) this extends the definition of $\mathcal{F}_k$ when $T \equiv k$ for some nonrandom k; and

(b) an alternative definition of $\mathcal{F}_T$ would be

$$\mathcal{F}_T = \Big(A \in \bigvee_{n=1}^{\infty} \mathcal{F}_n : A \cap (T = k) \in \mathcal{F}_k, \text{ for all } k \geq 0\Big).$$

Also show that, in general, $\mathcal{F}_T$ is a σ-field and T is an $\mathcal{F}_T$-measurable random variable. Furthermore, show that when T is an a.s. finite $\mathcal{F}$-stopping time and when M is adapted to $\mathcal{F}$, M_T is an $\mathcal{F}_T$-measurable random variable.[1] Finally, demonstrate that whenever T_1 and T_2 are stopping times that satisfy $T_1 \leq T_2$, we have $\mathcal{F}_{T_1} \subset \mathcal{F}_{T_2}$. □

The choice of the σ-field $\mathcal{F}_T$ is made to preserve many of the fundamental properties of martingales. We conclude this subsection with an example of this phenomenon. Recall that $\mathbf{1}_A$ is the indicator function of the measurable event A.

Theorem 1.1.1 *Suppose $\mathcal{F} = (\mathcal{F}_k; k \geq 1)$ is a filtration and Y is an integrable random variable. Define $M = (M_k; k \geq 1)$ by $M_k = \mathbb{E}[Y \mid \mathcal{F}_k]$. Then, M is a martingale with respect to $\mathcal{F}$. Moreover, if T is a stopping time with respect to $\mathcal{F}$, $M_T \mathbf{1}_{(T<\infty)} = \mathbb{E}[Y \mid \mathcal{F}_T] \mathbf{1}_{(T<\infty)}$, a.s.*

Equivalently, for any $j \geq 1$, $\mathbb{E}[Y \mid \mathcal{F}_j] = \mathbb{E}[Y \mid \mathcal{F}_T]$, on $(T = j)$. It is important to note that there is something that needs to be proved here. This is *not* a mere consequence of "replacing" $\mathcal{F}_T$ by $\mathcal{F}_j$ on $(T = j)$.

Proof Clearly, M is adapted. By Jensen's inequality,

$$\sup_k \mathbb{E}\{|M_k|\} \leq \mathbb{E}\{|Y|\} < \infty.$$

---

[1] Recall that the random variable M_T is defined ω by ω as $M_T(\omega) = M_{T(\omega)}(\omega)$.

Recall the **towering property of conditional expectations**: If $\mathcal{H}_1 \subset \mathcal{H}_2$ are two σ-fields, then for any integrable random variable U,

$$\mathbb{E}[U \mid \mathcal{H}_1] = \mathbb{E}\big[\,\mathbb{E}\{U \mid \mathcal{H}_2\} \mid \mathcal{H}_1\,\big] = \mathbb{E}\big[\,\mathbb{E}\{U \mid \mathcal{H}_1\} \mid \mathcal{H}_2\,\big], \qquad (1)$$

almost surely. The martingale property of M follows from the towering property of conditional expectations, together with the fact that $\mathcal{F}_k \subset \mathcal{F}_{k+1}$.

For any $A \in \mathcal{F}_T$, $\mathbb{E}[\mathbb{E}\{Y \mid \mathcal{F}_T\}\mathbf{1}_{(T=j)}\mathbf{1}_A] = \mathbb{E}[Y \mathbf{1}_{(T=j)}\mathbf{1}_A]$, since $A \cap (T=j) \in \mathcal{F}_T$. On the other hand, $A \cap (T=j) \in \mathcal{F}_j$ implies that $\mathbb{E}[Y \mathbf{1}_{(T=j)}\mathbf{1}_A] = \mathbb{E}[\mathbb{E}\{Y \mid \mathcal{F}_j\}\mathbf{1}_{(T=j)}\mathbf{1}_A] = \mathbb{E}[M_T \mathbf{1}_{(T=j)}\mathbf{1}_A]$. Since $M_T \mathbf{1}_{(T=j)}$ and $\mathbb{E}[Y \mid \mathcal{F}_T]\mathbf{1}_{(T=j)}$ are both $\mathcal{F}_T$-measurable random variables, the result follows. □

1.2 The Optional Stopping Theorem

Perhaps the single most important fact about smartingales is Doob's optional stopping theorem;[2] see (Doob 1990; Hunt 1966).

Theorem 1.2.1 (The Optional Stopping Theorem) *Suppose M is a submartingale with respect to a filtration $\mathcal{F}$. Then, for any integer $k \geq 0$ and all $\mathcal{F}$-stopping times T_1 and T_2 with $T_1 \leq T_2 \leq k$,*

$$\mathbb{E}[M_{T_2} \mid \mathcal{F}_{T_1}] \geq M_{T_1}, \qquad a.s.$$

Proof For any $A \in \mathcal{F}_{T_1}$,

$$\mathbb{E}\big[(M_{T_2} - M_{T_1})\mathbf{1}_A\big] = \mathbb{E}\Big[\sum_{j=T_1}^{T_2-1}(M_{j+1} - M_j)\mathbf{1}_A\Big]$$

$$= \sum_{j=0}^{k}\mathbb{E}\Big[(M_{j+1} - M_j)\mathbf{1}_{A \cap (T_1 \leq j < T_2)}\Big].$$

On the other hand, for all $j \geq 0$, $A \cap (T_1 \leq j)$ and $(T_2 > j)$ are both in $\mathcal{F}_j$. Hence, so is $B = A \cap (T_1 \leq j < T_2)$. By the submartingale property, $\mathbb{E}\big[(M_{j+1} - M_j)\mathbf{1}_B\big] \geq 0$. This shows that for all $A \in \mathcal{F}_{T_1}$,

$$\mathbb{E}[M_{T_1}\mathbf{1}_A] \leq \mathbb{E}[M_{T_2}\mathbf{1}_A],$$

which is the desired conclusion. □

---

[2] In the general theory of processes, stopping times are known as optional times, whence the name.

***Exercise* 1.2.1 (Doob's Decomposition)** Suppose $X_1, X_2, \ldots$ are integrable random variables and let $\mathcal{F}_k$ denote the σ-field generated by $X_1, \ldots, X_k$ ($k = 1, 2, \ldots$). Let $X_0 = \mathbb{E}[X_1]$ and let $\mathcal{F}_0 = \{\varnothing, \Omega\}$ be the trivial σ-field. Show that $M = (M_k;\, k \geq 1)$ is a martingale with respect to $\mathcal{F} = (\mathcal{F}_k;\, k \geq 1)$, where

$$M_k = \sum_{\ell=1}^{k} \{X_\ell - \mathbb{E}[X_\ell \mid \mathcal{F}_{\ell-1}]\}, \qquad k \geq 1.$$

In particular, show that every submartingale can be written as the sum of a martingale and an increasing process. □

***Exercise* 1.2.2** Use Theorem 1.1.1 and Exercise 1.2.1, in conjunction, to give another proof of the optional stopping theorem, Theorem 1.2.1. This proof is due to G. A. Hunt; cf. (Dellacherie and Meyer 1982; Hunt 1966) for this and much more. □

The following, perhaps more standard, form of the optional stopping theorem is an immediate corollary of the above:

Corollary 1.2.1 *If M is a submartingale and T is an $\mathcal{F}$-stopping time, then $(M_{T \wedge k};\, k \geq 0)$ is a submartingale.*

Many of the important results of martingale theory are consequences of clever applications of the above. We list some of them in order.

1.3 A Weak (1,1) Inequality

Suppose M is a submartingale. A **strong (p, q) maximal inequality** is one of the form $\mathbb{E}[\max_{0 \leq i \leq k} |M_i|^p] \leq C \mathbb{E}\{|M_k|^q\}$, where C is a positive finite constant that may depend only on p and q. When $p = q = 1$, such a result typically does *not* hold. Instead, one ought to prove a **weak (1,1) inequality** that states that the tails of the distribution of the maximum of a submartingale are well controlled by the size of the expectation of the submartingale. More precisely, we have the following theorem.

Theorem 1.3.1 (Doob's Maximal Inequality) *If M is a nonnegative submartingale, $\lambda > 0$ is a real number, and $k \geq 0$ is an integer, then*

$$\mathbb{P}(M_k^* \geq \lambda) \leq \frac{1}{\lambda} \mathbb{E}[M_k \mathbf{1}_{(M_k^* \geq \lambda)}],$$

where $M_k^* = \max_{0 \leq i \leq k} M_i$.

Typically, one uses the above as follows: Under the above conditions, for all $\lambda > 0$ and $k \geq 0$,

$$\mathbb{P}(M_k^* \geq \lambda) \leq \frac{1}{\lambda} \mathbb{E}[M_k].$$

Proof Define $T = \inf(i \geq 0 : M_i \geq \lambda)$, where $\inf \varnothing = \infty$. Then, T is a $\mathcal{F}$-stopping time and $(M_k^* \geq \lambda)$ is nothing but the event $(T \leq k)$. Note that M is nonnegative. Moreover, on $(T < \infty)$, $M_T \geq \lambda$. Consequently,

$$M_{T \wedge k} \mathbf{1}_{(T \leq k)} \geq \lambda \cdot \mathbf{1}_{(T \leq k)}.$$

Since $(T \leq k) \in \mathcal{F}_{T \wedge k}$ and $T \wedge k \leq k$, we can apply Theorem 1.2.1 to the stopping times $T \wedge k$ and k to see that

$$\mathbb{E}[M_{T \wedge k} \mathbf{1}_{(T \leq k)}] \leq \mathbb{E}[M_k \mathbf{1}_{(T \leq k)}].$$

This implies the result. $\square$

1.4 A Strong (p, p) Inequality

It turns out that Theorem 1.3.1 implies a *strong* (p, p) inequality if $p > 1$.

Theorem 1.4.1 (Doob's Strong $L^p(\mathbb{P})$ Inequality) *If $p > 1$, then for any nonnegative submartingale $M = (M_k; k \geq 0)$ and all integers $k \geq 0$,*

$$\mathbb{E}\left[\max_{0 \leq i \leq k} M_i^p\right] \leq \left(\frac{p}{p-1}\right)^p \mathbb{E}[M_k^p].$$

Proof Without loss of generality, we may assume that $\mathbb{E}[M_k^p] < \infty$; otherwise, there is nothing to prove. Since $p > 1$, Jensen's inequality shows that M_k^p is also a submartingale. In particular, check that

$$\mathbb{E}\left[\max_{0 \leq i \leq k} M_i^p\right] \leq \sum_{j=1}^{k} \mathbb{E}[M_j^p] < k\mathbb{E}[M_k^p] < \infty.$$

We integrate Theorem 1.3.1 by parts and use the notation there to obtain

$$\mathbb{E}\{|M_k^*|^p\} = p\int_0^\infty \lambda^{p-1} \mathbb{P}(M_k^* \geq \lambda)\,d\lambda \leq p\mathbb{E}\left[\int_0^{M_k^*} \lambda^{p-2}\,d\lambda \cdot M_k\right].$$

We have used Fubini's theorem. Therefore,

$$\mathbb{E}\{|M_k^*|^p\} \leq \left(\frac{p}{p-1}\right)\mathbb{E}\{|M_k^*|^{p-1} \cdot M_k\} \leq \left(\frac{p}{p-1}\right)\left(\mathbb{E}\{|M_k^*|^p\}\right)^{\frac{p-1}{p}}\left(\mathbb{E}[M_k^p]\right)^{\frac{1}{p}},$$

by Hölder's inequality. The result follows readily from this. $\square$

1.5 The Case $p = 1$

Our proof of Theorem 1.4.1 breaks down when $p = 1$. However, it does show that

$$\mathbb{E}[M_k^* \mathbf{1}_{(M_k^* \geq 1)}] = \int_1^\infty \mathbb{P}(M_k^* \geq \lambda)\,d\lambda \leq \int_1^\infty \frac{1}{\lambda}\mathbb{E}[M_k \mathbf{1}_{(M_k^* \geq \lambda)}]\,d\lambda$$

$$= \mathbb{E}\left[M_k \mathbf{1}_{(M_k^* \geq 1)} \int_1^{M_k^*} \frac{1}{\lambda}\,d\lambda\right] = \mathbb{E}[M_k \ln_+ M_k^*],$$

where $\ln_+ x = \ln(x \vee 1)$. Therefore,

$$\mathbb{E}[M_k^*] \leq 1 + \mathbb{E}[M_k \ln_+ M_k^*].$$

We will have need for the following inequality: For all $w \geq 0$, $\ln w \leq \frac{1}{e} w$.[3] Apply this with $w = \frac{y}{x}$ to see that whenever $0 \leq x \leq y$, $x \ln y \leq x \ln_+ x + \frac{1}{e} y$. We have proven the following weak $(1,1)$ inequality.

Theorem 1.5.1 *If $M = (M_k; k \geq 0)$ is a nonnegative submartingale and $k \geq 0$,*

$$\mathbb{E}\left[\max_{0 \leq i \leq k} M_i\right] \leq \left(\frac{e}{e-1}\right)\left[1 + \mathbb{E}[M_k \ln_+ M_k]\right].$$

***Exercise* 1.5.1** Suppose X and Y are nonnegative random variables such that for all $\lambda > 0$,

$$\mathbb{P}(Y > \lambda) \leq \frac{1}{\lambda} \mathbb{E}[X \mathbf{1}_{(Y>\lambda)}].$$

Show that if $\mathbb{E}[X\{\ln_+ X\}^k]$ is finite for some $k \geq 1$, so is $\mathbb{E}[Y\{\ln_+ Y\}^{k-1}]$. This is from Sucheston (1983, Corollary 1.4). □

***Exercise* 1.5.2** Suppose $M^1, M^2, \ldots$ are nonnegative submartingales, and that for any $j \geq 0$, $\lim_{n \to \infty} \mathbb{E}\{\Psi(M_j^n)\} = 0$, where $\Psi(x) = x \ln_+ x$. Prove that for any integer $k \geq 0$,

$$\lim_{n \to \infty} \max_{j \leq k} M_j^n = 0,$$

where the limit takes place in $L^1(\mathbb{P})$. □

1.6 Upcrossing Inequalities

Given a submartingale M and two real numbers $a < b$, let $T_0 = \inf(\ell \geq 0 : M_\ell \leq a)$, and for all $j \geq 1$, define

$$T_j = \begin{cases} \inf(\ell > T_{j-1} : M_\ell \geq b), & \text{if } j \text{ is odd,} \\ \inf(\ell > T_{j-1} : M_\ell \leq a), & \text{if } j \text{ is even,} \end{cases}$$

with the usual stipulation that $\inf \varnothing = \infty$. These are the **upcrossing times** of the interval $[a, b]$. Furthermore, for all $j \geq 1$, $M_{T_j} \geq b$ if j is odd and $M_{T_j} \leq a$ if j is even. For all $k \geq 1$, define

$$U_k[a, b] = \sup(0 \leq j \leq k : T_{2j} < \infty),$$

---

[3] To verify this, you need only check that $g(w) = \ln w - \frac{1}{e} w$ is maximized at $w = e$ and $g(e) = 0$.

where $\sup \varnothing = 0$. In words, $U_k[a,b]$ represents the total number of upcrossings of the interval $[a,b]$ made by the (random) numerical sequence $(M_i; 0 \le i \le k)$.

Theorem 1.6.1 (Doob's Upcrossing Inequality) *If $M = (M_k; k \ge 0)$ is a submartingale, $a < b$, and $k \ge 0$, then*

$$\mathbb{E}\{U_k[a,b]\} \le \frac{|a| + \mathbb{E}\{|M_k|\}}{b-a}.$$

It is also possible to derive downcrossing inequalities for supermartingales; cf. Exercise 1.6.1 below.

Proof By considering the submartingale $(M_k - a)^+$ instead of M, we can assume without loss of generality that $a = 0$ and that M is nonnegative.

For all $j \ge 0$, define $\tau_j = T_j \wedge k$, and note that the τ_j's are nondecreasing, bounded stopping times. Clearly,

$$\sum_{j \text{ even}} [M_{\tau_{j+1}} - M_{\tau_j}] \ge b U_k[0,b]. \tag{1}$$

On the other hand, by the optional stopping theorem (Theorem 1.2.1), $\mathbb{E}[M_{\tau_j}]$ is increasing in j. Thus, since we have a finite sum, we can interchange expectations with sums to obtain

$$\mathbb{E}\Big\{\sum_{j \text{ odd}} [M_{\tau_{j+1}} - M_{\tau_j}]\Big\} = \sum_{j \text{ odd}} \mathbb{E}[M_{\tau_{j+1}} - M_{\tau_j}] \ge 0.$$

Adding this to equation (1) above, we obtain the result. (Why? That is, what if for all j, $\tau_j < k$?) $\square$

Exercise 1.6.1 Let $M = (M_k; k \ge 0)$ denote a supermartingale with respect to a filtration $\mathcal{F} = (\mathcal{F}_k; k \ge 0)$. Show that whenever $0 \le a < b$,

$$\mathbb{E}\{D_k[a,b]\} \le \frac{\mathbb{E}[X_1 \wedge b] - \mathbb{E}[X_n \wedge b]}{b-a},$$

where $D_k[a,b]$ denotes the number of downcrossings of the interval $[a,b]$ made by the sequence $(M_i; 0 \le i \le k)$, i.e., $D_k[a,b] = U_k[-b,-a]$. $\square$

Doob's upcrossing inequality contains a prefatory but still useful formulation of the weak $(1,1)$ inequality. Indeed, note that in the special case where M is a martingale,

$$\mathbb{P}\Big(\max_{0 \le j \le k} |M_j| \ge \lambda\Big) \le \mathbb{P}(|M_0| \ge \tfrac{1}{2}\lambda) + \mathbb{P}\Big(\max_{0 \le j \le k} |M_j - M_0| \ge \tfrac{1}{2}\lambda\Big).$$

But $|M_j - M_0|$ is a nonnegative submartingale. Let $U_k(\lambda)$ denote the number of upcrossings of $[0,\lambda]$ made by this submartingale before time k. Then, by first applying Chebyshev's inequality and then Doob's upcrossing inequality (Theorem 1.6.1), we obtain

$$\mathbb{P}\Big(\max_{0\le j\le k}|M_j|\ge \lambda\Big) \le \mathbb{P}(|M_0|\ge \tfrac{1}{2}\lambda) + \mathbb{P}(U_k(\tfrac{1}{2}\lambda)\ge 1)$$
$$\le \frac{2}{\lambda}\mathbb{E}\{|M_0|\} + \mathbb{E}[U_k(\tfrac{1}{2}\lambda)]$$
$$\le \frac{2}{\lambda}\mathbb{E}\{|M_0|\} + \frac{2}{\lambda}\mathbb{E}\{|M_k - M_0|\}$$
$$\le \frac{4}{\lambda}\mathbb{E}\{|M_0|\} + \frac{2}{\lambda}\mathbb{E}\{|M_k|\}.$$

Since $\mathbb{E}\{|M_0|\} \le \mathbb{E}\{|M_k|\}$, we obtain

$$\mathbb{P}\Big(\max_{0\le j\le k}|M_j|\ge \lambda\Big) \le \frac{6}{\lambda}\mathbb{E}\{|M_k|\}.$$

This is essentially the weak $(1,1)$ inequality (Theorem 1.3.1) with a slightly worse constant. (If $M_0 = 0$, check that in the above, the constant 6 can be improved to the constant 1 of Theorem 1.3.1.)

1.7 The Martingale Convergence Theorem

Recall that a stochastic process $X = (X_k;\ k\ge 0)$ is **$L^p(\mathbb{P})$-bounded** (or **bounded in $L^p(\mathbb{P})$**) for some $p > 0$ if $\sup_{k\ge 0}\mathbb{E}\{|M_k|^p\} < \infty$.

Theorem 1.7.1 (Doob's Convergence Theorem) *If M is an $L^1(\mathbb{P})$-bounded submartingale, then $M_\infty = \lim_{k\to\infty} M_k$ exists almost surely, and $M_\infty \in L^1(\mathbb{P})$. If M is uniformly integrable, then $M_k \to M_\infty$ in $L^1(\mathbb{P})$ and $M_k \le \mathbb{E}[M_\infty\,|\,\mathcal{F}_k]$, a.s. Finally, if M is $L^p(\mathbb{P})$-bounded for some $p > 1$, then $M_k \to M_\infty$, in $L^p(\mathbb{P})$, as well.*

Among other things, Chatterji (1967, 1968) has constructed a proof of Doob's convergence theorem that is based solely on Doob's maximal inequality (Theorem 1.3.1) and the optional stopping theorem (Theorem 1.2.1). This proof was later discovered independently in Lamb (1973) and is developed in Supplementary Exercise 3. The argument described below is closer to the original proof of J. L. Doob and is based on the upcrossing inequality.

Proof Note that $k \mapsto U_k[a,b]$ is an increasing map. Therefore, by Lebesgue's monotone convergence theorem, if M is bounded in $L^1(\mathbb{P})$, Doob's upcrossing inequality (Theorem 1.6.1) implies that for all $a < b$,

$$\mathbb{E}\Big\{\sup_{k\ge 0} U_k[a,b]\Big\} < \infty. \tag{1}$$

For all $a < b$, define the measurable event

$$N_{a,b} = \left(\liminf_k M_k \leq a < b \leq \limsup_k M_k\right).$$

Then, by equation (1), $\mathbb{P}(N_{a,b}) = 0$ for all $a < b$, for otherwise, with positive probability, $[a,b]$ is upcrossed infinitely many times, which would contradict equation (1). Consequently,

$$\mathbb{P}\left(\bigcup_{a,b \in \mathbb{Q}: a<b} N_{a,b}\right) = 0,$$

where, as usual, $\mathbb{Q}$ denotes the collection of all rationals. Thus, with probability one, $\limsup_k M_k = \liminf_k M_k$, which gives the desired limit M_∞. Moreover, this is finite, almost surely. In fact, applying Fatou's lemma, we arrive at $\mathbb{E}\{|M_\infty|\} < \infty$. If M_k is uniformly integrable, then we immediately have $M_k \to M_\infty$ in $L^1(\mathbb{P})$.

If M_k is also $L^p(\mathbb{P})$ bounded for some $p > 1$, then, by the strong (p,p) inequality (Theorem 1.4.1), $\sup_k |M_k|$ is in $L^p(\mathbb{P})$. In particular, $|M_k|^p$ is uniformly integrable.

It remains to prove that under uniform integrability, $M_k \leq \mathbb{E}\{M_\infty \mid \mathcal{F}_k\}$, a.s. For $k \leq \ell$, and for any $\lambda > 0$,

$$M_k \leq \mathbb{E}\{M_\ell \mid \mathcal{F}_k\} = \mathbb{E}\{M_\ell \mathbf{1}_{(|M_\ell| \leq \lambda)} \mid \mathcal{F}_k\} + \mathbb{E}\{M_\ell \mathbf{1}_{(|M_\ell|>\lambda)} \mid \mathcal{F}_k\}, \quad \text{a.s.}$$

By the bounded convergence theorem for conditional expectations, with probability one,

$$\lim_{\ell \to \infty} \mathbb{E}\{M_\ell \mathbf{1}_{(|M_\ell| \leq \lambda)} \mid \mathcal{F}_k\} = \mathbb{E}\{M_\infty \mathbf{1}_{(|M_\infty| \leq \lambda)} \mid \mathcal{F}_k\}.$$

On the other hand, $M_\infty \in L^1(\mathbb{P})$ and the dominated convergence theorem for conditional expectations together reveal that, as $\lambda \to \infty$, the right-hand side of the above display converges to $\mathbb{E}\{M_\infty \mid \mathcal{F}_k\}$, a.s. It suffices to show that a.s.,

$$\liminf_{\lambda \to \infty} \liminf_{\ell \to \infty} \mathbb{E}\{|M_\ell| \mathbf{1}_{(|M_\ell|>\lambda)} \mid \mathcal{F}_k\} = 0.$$

But this follows from the following formulation of uniform integrability: $\lim_{\lambda \to \infty} \sup_{\ell \geq 1} \mathbb{E}\{|M_\ell| \mathbf{1}_{(|M_\ell|>\lambda)}\} = 0$. □

Exercise 1.7.1 Consider the branching process of Exercise 1.1.2 and show that when $\mu < 1$, $\lim_{n \to \infty} X_n = 0$, a.s., and when $\mu > 1$, $\lim_{n \to \infty} X_n = +\infty$, a.s. That is, if the mean number of progeny is strictly less (greater) than one, then the population will eventually die out (blow up). □

Exercise 1.7.2 Consider the stick-breaking process of Exercise 1.1.3 and show that $\lim_{n \to \infty} X_n$ exists, a.s. Identify this limit. □

***Exercise* 1.7.3** Prove that every positive supermartingale converges. (HINT: Try Exercise 1.6.1 first.) □

We conclude this section with an important application of the martingale convergence theorem. The **tail σ-field** $\mathcal{T}$ corresponding to the collection of σ-fields $\mathcal{X} = (\mathcal{X}_n;\ n \geq 1)$ is defined as

$$\mathcal{T} = \bigcap_{m=1}^{\infty} \bigvee_{n=m}^{\infty} \mathcal{X}_n.$$

We say that $\mathcal{T}$ is the tail σ-field of the random variables $X_1, X_2, \ldots$ when $\mathcal{T}$ is the tail σ-field of $\sigma(X_1), \sigma(X_2), \ldots$. We now present a martingale proof of Kolmogorov's 0-1 law. In words, it states that the tail σ-field corresponding to independent random variables is trivial.

Corollary 1.7.1 (Kolmogorov's 0-1 Law) *If $\mathcal{T}$ denotes the tail σ-field corresponding to independent random variables $X_1, X_2, \ldots$, then for all events $\Lambda \in \mathcal{T}$, $\mathbb{P}(\Lambda)$ is identically 0 or 1.*

Proof Let $\mathcal{F} = (\mathcal{F}_k;\ k \geq 1)$ denote the filtration generated by the X_i's; that is, $\mathcal{F}_k = \vee_{\ell=1}^{k} \sigma(X_\ell)$. Since for all $m > n$, $\vee_{k=m}^{\infty} \sigma(X_k)$ is independent of $\mathcal{F}_n$, $\mathcal{F}_n$ and $\mathcal{T}$ are independent. In particular, for any $\Lambda \in \mathcal{T}$, $\mathbb{P}(\Lambda) = \mathbb{P}(\Lambda \mid \mathcal{F}_n)$, a.s. The latter defines a bounded martingale (indexed by n), which, by Doob's martingale convergence theorem, converges, a.s., to $\mathbb{P}(\Lambda \mid \vee_n \mathcal{F}_n)$. Since Λ is $\vee_m \mathcal{F}_m$-measurable, this almost surely equals $\mathbf{1}_\Lambda$. We have shown that with probability one, $\mathbb{P}(\Lambda) = \mathbf{1}_\Lambda$, which proves our result. □

***Exercise* 1.7.4** It is possible to devise measure-theoretic proofs of Kolmogorov's 0-1 law, as we shall find in this exercise. In the notation of our proof of Kolmogorov's 0-1 law, show that for all $\Lambda \in \mathcal{T}$ and for all $\varepsilon \in\,]0,1[$, there exists some $n \geq 1$ large enough and some event $\Lambda_n \in \vee_{i \geq n} \sigma(X_i)$ such that $\mathbb{P}(\Lambda \triangle \Lambda_n) \leq \varepsilon$, where $\triangle$ denotes the set difference operation. Now use this to verify that $\mathbb{P}(\Lambda \cap \Lambda) = \{\mathbb{P}(\Lambda)\}^2$. □

The following difficult exercise should not be neglected.

***Exercise* 1.7.5 (Hewitt–Savage 0-1 Law)** (Hard) Recall that a function $f : \mathbb{R}^m \to \mathbb{R}$ is said to be **symmetric** if for all permutations π of $\{1, \ldots, m\}$ and for all $x_1, \ldots, x_m \in \mathbb{R}$, $f(x_1, \ldots, x_m) = f(x_{\pi(1)}, \ldots, x_{\pi(m)})$. Now suppose that $X_1, X_2, \ldots$ are independent random variables—all on one probability space $(\Omega, \mathcal{G}, \mathbb{P})$—and for all $m \geq 1$, define $\mathcal{E}_{-m}$ to be the σ-field generated by all random variables of the form $f(X_1, \ldots, X_m)$, where $f : \mathbb{R}^m \to \mathbb{R}$ is measurable and symmetric. The σ-field $\mathcal{E} = \cap_m \mathcal{E}_{-m}$ is called the **exchangeable σ-field**. Our goal is to prove the Hewitt–Savage 0-1 law: "$\mathcal{E}$ *is trivial.*"

(i) If $S_n = \sum_{j=1}^{n} X_j$, show that the event $(S_n > 0,\ \text{infinitely often})$ is exchangeable. Also, prove that the tail σ-field of $X_1, X_2, \ldots$ is in $\mathcal{E}$, although the converse need not hold. Moreover, check that $(\mathcal{E}_{-m};\, m \geq 1)$ is a filtration.

(ii) Consider integers $n \geq m \geq 1$ and a bounded, symmetric function $f : \mathbb{R}^m \to \mathbb{R}$. Show that for any $1 \leq i_1, \ldots, i_m \leq n$, a.s.,
$$\mathbb{E}\big[f(X_{i_1}, \ldots, X_{i_m}) \,\big|\, \mathcal{E}_{-n}\big] = \mathbb{E}\big[f(X_1, \ldots, X_m) \,\big|\, \mathcal{E}_{-n}\big].$$

(iii) Given integers $n \geq m \geq 1$ and a bounded, symmetric $f : \mathbb{R}^m \to \mathbb{R}$, define
$$U_n = \binom{n}{m}^{-1} \sum_{C(n)} f(X_{i_1}, \ldots, X_{i_m}),$$
where $\sum_{C(n)}$ represents summation over all $\binom{n}{m}$ combinations of m distinct elements $\{i_1, \ldots, i_m\}$ from $\{1, \ldots, n\}$. Show that U_n is $\mathcal{E}_{-n}$-measurable. Conclude that as $n \to \infty$, U_n almost surely converges to $\mathbb{E}[f(X_1, \ldots, X_m) \,|\, \mathcal{E}]$.

(iv) Show that $\lim_n \binom{n}{m}^{-1} \sum_{C(n;1)} f(X_{i_1}, \ldots, X_{i_m}) = 0$, a.s., where $\sum_{C(n;1)}$ represents summation over all $\binom{n}{m}$ combinations of m distinct elements $\{i_1, \ldots, i_m\}$ from $\{1, \ldots, n\}$, such that one of the i_j's equals 1. Conclude that $\mathbb{E}[f(X_1, \ldots, X_m) \,|\, \mathcal{E}]$ is independent of X_1. Extend this argument to prove that $\mathbb{E}[f(X_1, \ldots, X_m) \,|\, \mathcal{E}]$ is independent of $(X_1, \ldots, X_m)$.

(v) Prove that $\mathcal{E}$ is independent of itself. Equivalently, for all $\Lambda \in \mathcal{E}$, $\mathbb{P}(\Lambda) \in \{0, 1\}$.

(HINT: For part (ii), compute $\mathbb{E}[f(X_{i_1}, \ldots, X_{i_m}) \cdot g(X_1, \ldots, X_n)]$; for part (iv), start by computing the cardinality of $C(n; 1)$; for part (v), use (iv) to conclude that $\mathcal{E}$ is independent of $X_1, X_2, \ldots$.) □

2 Orthomartingales: Aspects of the Cairoli–Walsh Theory

Now that the one-parameter theory is established, we are ready to tackle multiparameter smartingales. There are two sources of difficulty that need to be overcome and they are both related to the fact that $\mathbb{N}_0^N$—and more generally $\mathbb{R}^N$—cannot be well-ordered in a useful way.[4]

---

[4] The stress here is on the word *useful*, since the 1904 well-ordering theorem of E. Zermelo states that, under the influence of the axiom of choice, every set can be well-ordered. However, the structure of this well-ordering is not usually known and/or in line with the stochastic structure of the problem at hand.

The first problem is that there is no sensible way to uniquely define multiparameter stopping times. To illustrate, suppose $(M_t;\ t \in \mathbb{N}_0^N)$ is a collection of random variables. Then, it is not clear—and in general not true—that there is a uniquely defined *first time* $t \in \mathbb{N}_0^N$ such that $M_t \geq 0$, say.

The second source of difficulty with the multiparameter theory is that there are many different ways to define smartingales indexed by several parameters.

In this book we will study two such definitions, both of which are due to R. Cairoli and J. B. Walsh. This material, together with other aspects of the theory of multiparameter martingales, can be found in Cairoli (1969, 1970a, 1970b, 1971, 1979), Cairoli and Gabriel (1979), Cairoli and Walsh (1975, 1978), Ledoux (1981), and Walsh (1979, 1986b).

Later on, in Section 3, we will study multiparameter martingales. They are defined in complete analogy to one-parameter martingales based on the conditional expectation formula $s \preccurlyeq t \implies \mathbb{E}[M_t \mid \mathcal{F}_s] = M_s$, together with the obvious measurability and integrability conditions.

As we shall see throughout the remainder of this book, multiparameter smartingales are both natural and useful. Despite this, it is unfortunate that, when considered in absolute generality, multiparameter martingales have no structure to speak of. Nonetheless, it is possible to tame them when a so-called commutation hypothesis holds. To clarify these issues, we begin with a different—seemingly less natural—class of multiparameter smartingales that we call orthosmartingales. These are the second class of multiparameter smartingales encountered in this book and are the subject of this section.

2.1 *Definitions and Examples*

Let N be a positive numeral and throughout consider N (one-parameter) filtrations $\mathcal{F}^1, \ldots, \mathcal{F}^N$, where $\mathcal{F}^i = (\mathcal{F}^i_k;\ k \geq 0)$ $(1 \leq i \leq N)$. A stochastic process $M = (M_t;\ t \in \mathbb{N}_0^N)$ is an **orthosubmartingale** if for each $1 \leq i \leq N$, and all nonnegative integers $(t^{(j)};\ 1 \leq j \leq N, j \neq i)$, $t^{(i)} \mapsto M_t$ is a one-parameter submartingale with respect to the one-parameter filtration $\mathcal{F}^i$. The stochastic process M is an **orthosupermartingale** if $-M$ is an orthosubmartingale. If M is both an orthosupermartingale and an orthosubmartingale, it is then an **orthomartingale**. Finally, we say that M is an **orthosmartingale** if it is either an orthosubmartingale or an orthosupermartingale.

For example, let us consider the case $N = 2$ and write the process M as $M = (M_{i,j};\ i, j \geq 0)$. Then, M is an orthosubmartingale if

1. for all $i, j \geq 0$, $\mathbb{E}\{|M_{i,j}|\} < \infty$;

2. for all $j \geq 0$, the one-parameter process $i \mapsto M_{i,j}$ is adapted to the filtration $\mathcal{F}^1$, while for each $i \geq 0$, $j \mapsto M_{i,j}$ is adapted to $\mathcal{F}^2$; and

3. for all $i, j \geq 0$, $\mathbb{E}[M_{i+1,j} \,|\, \mathcal{F}_i^1] \geq M_{i,j}$, a.s., and $\mathbb{E}[M_{i,j+1} \,|\, \mathcal{F}_j^2] \geq M_{i,j}$, a.s.

While the above conditions may seem stringent, we will see next that in fact, many such processes exist.

Example 1 Suppose $M^\ell = (M_i^\ell;\ i \geq 0)$ $(1 \leq \ell \leq N)$ are N stochastic processes that are adapted to independent filtrations $\mathcal{M}^\ell = (\mathcal{M}_i^\ell;\ i \geq 0)$ $(1 \leq \ell \leq N)$. We then say that $M^1, \ldots, M^N$ are **independent processes**. Given that the M^ℓ's are all (super- or all sub- or all plain) martingales, we define the **additive smartingale** A as

$$A_t = \sum_{\ell=1}^N M_{t^{(\ell)}}^\ell, \qquad t \in \mathbb{N}_0^N.$$

We also define the **multiplicative smartingale** M as

$$M_t = \prod_{\ell=1}^N M_{t^{(\ell)}}^\ell, \qquad t \in \mathbb{N}_0^N.$$

Then, it is possible to see that both M and A are orthosmartingales with respect to the $\mathcal{F}^\ell$'s, where for all $1 \leq \ell \leq N$,

$$\mathcal{F}_k^\ell = \mathcal{M}_k^\ell \vee \bigvee_{\substack{i=1 \\ i \neq \ell}}^N \bigvee_{m \geq 0} \mathcal{M}_m^i, \qquad k \geq 0.$$

In words, the filtration $\mathcal{F}_k^\ell$ is the σ-field generated by $\{X_i^\ell; 0 \leq i \leq k\}$, as well as all of the variables $\{X_i^m;\ m \neq \ell,\ i \geq 0\}$. $\square$

The following should certainly be attempted.

***Exercise* 2.1.1** Check that for each $1 \leq \ell \leq N$, $\mathcal{F}^\ell$ is a one-parameter filtration. Conclude that additive smartingales and multiplicative smartingales are indeed orthosmartingales. $\square$

We mention another example, to which we will return in Chapter 4.

Example 2 Suppose $X = (X_t;\ t \in \mathbb{N}_0^N)$ are i.i.d. random variables. Define the **multiparameter random walk** $S = (S_t;\ t \in \mathbb{N}_0^N)$ based the these X's as

$$S_t = \sum_{s \preceq t} X_s, \qquad t \in \mathbb{N}_0^N.$$

Define one-parameter filtrations $\mathcal{F}^1, \ldots, \mathcal{F}^N$ as follows: For all $1 \leq i \leq N$,

$$\mathcal{F}_k^i = \sigma(X_s;\ s^{(i)} \leq k), \qquad k \geq 0,$$

18   1. Discrete-Parameter Martingales

where $\sigma(\cdots)$ denotes the σ-field generated by the random variables in the parentheses. Multiparameter random walks are related to orthomartingales in the same way that one-parameter random walks are to one-parameter martingales, as the following simple but important exercise reveals. □

***Exercise* 2.1.2** Whenever $\mathbb{E}[X_t] = 0$ for all $t \in \mathbb{N}_0^N$, S is an orthomartingale with respect to $\mathcal{F}^1, \ldots, \mathcal{F}^N$. □

We will encounter multiparameter martingales many more times in this book. For now, let us note the following consequence of Jensen's inequality. It shows us how to produce more orthosmartingales in the presence of some.

Lemma 2.1.1 *Suppose that* $M = (M_t;\ t \in \mathbb{N}_0^N)$ *is a nonnegative orthosubmartingale with respect to one-parameter filtrations* $\mathcal{F}^1, \ldots, \mathcal{F}^N$, *that* $\Psi : [0, \infty[\to [0, \infty[$ *is convex nondecreasing, and that for all* $t \in \mathbb{N}_0^N$, $\mathbb{E}[\Psi(M_t)] < \infty$. *Then,* $t \mapsto \Psi(M_t)$ *is an orthosubmartingale.*

2.2 Embedded Submartingales

We have seen many instances where stopping times play a critical role in the one-parameter theory of stochastic processes. For instance, see the described proof of Doob's weak (1,1) maximal inequality (Theorem 1.3.1). In a multiparameter setting, we can use the following idea to replace the use of stopping times, since stopping times are not well defined in the presence of many parameters.

Proposition 2.2.1 *Suppose* $M = (M_t;\ t \in \mathbb{N}_0^N)$ *is an orthosubmartingale with respect to the one-parameter filtrations* $\mathcal{F}^1, \ldots, \mathcal{F}^N$. *If* $s, t \in \mathbb{N}_0^N$ *with* $s \preccurlyeq t$, *then* $\mathbb{E}[M_s] \leq \mathbb{E}[M_t]$. *Moreover, for all* $t^{(2)}, \ldots, t^{(N)} \geq 0$, *the following is a one-parameter submartingale with respect to* $\mathcal{F}^1$:

$$\widetilde{M}_{s^{(1)}} = \max_{0 \leq s^{(2)} \leq t^{(2)}} \cdots \max_{0 \leq s^{(N)} \leq t^{(N)}} M_s, \qquad s^{(1)} \geq 0.$$

In particular, the first part says that, as a function on $\mathbb{N}_0^N$, $t \mapsto \mathbb{E}[M_t]$ is increasing in the partial order $\preccurlyeq$. Furthermore, suppose $N = 2$ and write $M = (M_{i,j};\ i, j \geq 0)$. Then, Proposition 2.2.1 states that, as a process in i, $\max_{0 \leq j \leq m} M_{i,j}$ is a submartingale with respect to $\mathcal{F}^1$. By reversing the roles of i and j, this proposition also implies that, as a process in j, $\max_{0 \leq i \leq n} M_{i,j}$ is a submartingale with respect to $\mathcal{F}^2$. Consequently, Proposition 2.2.1 extracts 2 (in general, N) one-parameter submartingales from M.

Proof We prove the first part for $N = 2$. Write $M = (M_{i,j};\ i, j \geq 0)$ and recall that $(i, j) \preccurlyeq (n, m)$ if and only if $i \leq n$ and $j \leq m$. Fix any such pair (i, j) and (n, m). Since $i \mapsto M_{i,j}$ is a one-parameter submartingale, by Jensen's inequality, $\mathbb{E}[M_{i,j}] \le \mathbb{E}[M_{n,j}]$. On the other hand, $j \mapsto M_{n,j}$ is a submartingale, too. Another application of Jensen's inequality shows that $\mathbb{E}[M_{i,j}] \le \mathbb{E}[M_{n,m}]$, as claimed. The general case (i.e., when $N \ge 2$ is arbitrary) is proved in Exercise 2.2.1 below.

For the second and main part of the proposition, note that by the definitions, for any $s^{(j)} \le t^{(j)}$ ($2 \le j \le N$), M_s is measurable with respect to $\mathcal{F}^1_{s^{(1)}}$. Therefore, $\max_{0 \le s^{(2)} \le t^{(2)}} \cdots \max_{0 \le s^{(N)} \le t^{(N)}} M_s$ is also measurable with respect to $\mathcal{F}^1_{s^{(1)}}$. Moreover,

$$\mathbb{E}\left\{\max_{0 \le s^{(2)} \le t^{(2)}} \cdots \max_{0 \le s^{(N)} \le t^{(N)}} |M_s|\right\} \le \sum_{0 \le s^{(2)} \le t^{(2)}} \cdots \sum_{0 \le s^{(N)} \le t^{(N)}} \mathbb{E}\{|M_s|\},$$

which is finite. It remains to verify the submartingale property. This can be checked directly, as we demonstrate below: Suppose $s^{(1)} = k+1$. Then,

$$\begin{aligned}\mathbb{E}\left[\max_{0 \le s^{(2)} \le t^{(2)}} \cdots \max_{0 \le s^{(N)} \le t^{(N)}} M_s \,\Big|\, \mathcal{F}^1_k\right] &\ge \max_{0 \le s^{(2)} \le t^{(2)}} \cdots \max_{0 \le s^{(N)} \le t^{(N)}} \mathbb{E}[M_s \mid \mathcal{F}^1_k] \\ &\ge \max_{0 \le r^{(2)} \le t^{(2)}} \cdots \max_{0 \le r^{(N)} \le t^{(N)}} M_r, \quad \text{a.s.},\end{aligned}$$

where $r^{(1)} = k$. We have used the orthosubmartingale property of M in the last step, and shown that

$$\mathbb{E}\left[\widetilde{M}_{k+1} \,\Big|\, \mathcal{F}^1_k\right] \ge \widetilde{M}_k, \quad \text{a.s.}$$

This is the desired result. □

Exercise **2.2.1** Prove Proposition 2.2.1 for a general integer $N \ge 2$. □

Exercise **2.2.2** Suppose $\gamma : \mathbb{N}_0^N \to \mathbb{N}_0$ is a nondecreasing function in the sense that whenever $s \preccurlyeq t$, then $\gamma(s) \le \gamma(t)$. Prove that whenever M is an N-parameter orthosubmartingale with respect to $\mathcal{F}^1, \ldots, \mathcal{F}^N$, $M \circ \gamma$ is a 1-parameter submartingale with respect to $\mathcal{F} \circ \gamma$, where $(M \circ \gamma)_t = M_{\gamma(t)}$ and $(\mathcal{F} \circ \gamma)_t = \cap_{i=1}^N \mathcal{F}^i_{\gamma^{(i)}(t)}$ ($t \in \mathbb{N}_0^N$). □

2.3 Cairoli's Strong (p,p) Inequality

We can combine Proposition 2.2.1 with Doob's strong (p,p) maximal inequality (Theorem 1.4.1) to obtain Cairoli's strong (p,p) inequality for orthosubmartingales.

Theorem 2.3.1 (Cairoli's Strong (p,p) Inequality) *Suppose that $M = (M_s; s \in \mathbb{N}_0^N)$ is a nonnegative orthosubmartingale with respect to one-parameter filtrations $\mathcal{F}^1, \ldots, \mathcal{F}^N$. Then, for all $t \in \mathbb{N}_0^N$ and $p > 1$,*

$$\mathbb{E}\left[\max_{0 \preccurlyeq s \preccurlyeq t} M_s^p\right] \le \left(\frac{p}{p-1}\right)^{Np} \mathbb{E}[M_t^p].$$

Proof Write $\max_{0 \preceq s \preceq t} M_s$ as $\max_{0 \le s^{(1)} \le t^{(1)}} \widetilde{M}_{s^{(1)}}$, where

$$\widetilde{M}_k = \max_{0 \le s^{(2)} \le t^{(2)}} \cdots \max_{0 \le s^{(N)} \le t^{(N)}} M_{k, s^{(2)}, \ldots, s^{(N)}}.$$

By Proposition 2.2.1, $\widetilde{M}$ is a (one-parameter) submartingale with respect to the one-parameter filtration $\mathcal{F}^1$. Applying Doob's strong (p,p)-inequality (Theorem 1.4.1), we see that

$$\mathbb{E}\left[\max_{0 \preceq s \preceq t} M_s^p\right]$$
$$\le \left(\frac{p}{p-1}\right)^p \mathbb{E}\{|\widetilde{M}_{t^{(1)}}|^p\}$$
$$= \left(\frac{p}{p-1}\right)^p \mathbb{E}\left\{\left|\max_{0 \le s^{(2)} \le t^{(2)}} \cdots \max_{0 \le s^{(N)} \le t^{(N)}} M_{t^{(1)}, s^{(2)}, \ldots, s^{(N)}}\right|^p\right\}.$$

Going through the above argument one more time, we arrive at

$$\mathbb{E}\left[\max_{0 \preceq s \preceq t} M_s^p\right]$$
$$\le \left(\frac{p}{p-1}\right)^{2p} \mathbb{E}\left\{\left|\max_{0 \le s^{(3)} \le t^{(3)}} \cdots \max_{0 \le s^{(N)} \le t^{(N)}} M_{t^{(1)}, t^{(2)}, s^{(3)}, \ldots, s^{(N)}}\right|^p\right\}.$$

The result follows from induction on N. □

2.4 Another Maximal Inequality

The method of Section 2.3 can be used in the $p = 1$ case. However, this time, the relevant result is the maximal inequality of Theorem 1.5.1.

Theorem 2.4.1 *Suppose $M = (M_t; t \in \mathbb{N}_0^N)$ is a nonnegative orthosubmartingale with respect to one-parameter filtrations $\mathcal{F}^1, \ldots, \mathcal{F}^N$. Then, for all $p \ge 0$ and all $t \in \mathbb{N}_0^N$,*

$$\mathbb{E}\left[\max_{0 \preceq s \preceq t} M_s (\ln_+ M_s)^p\right] \le (p+1)^N \left(\frac{e}{e-1}\right)^N \left\{N + \mathbb{E}\left[M_t (\ln_+ M_t)^{p+N}\right]\right\}.$$

Consequently, letting $p = 0$, we obtain a multiparameter extension of Theorem 1.5.1:

$$\mathbb{E}\left[\max_{0 \preceq s \preceq t} M_s\right] \le \left(\frac{e}{e-1}\right)^N \left\{N + \mathbb{E}\left[M_t (\ln_+ M_t)^N\right]\right\}.$$

Proof Without loss of generality, we can and will assume that the expectation on the right-hand side of the theorem's display is finite. Define $\Psi^p(x) = x(\ln_+ x)^p$ $(p \ge 1)$. Note that Ψ^p is both nondecreasing and convex on $]0,\infty[$. By Jensen's inequality, $\Psi^p(M) = (\Psi^p(M_t);\ t \in \mathbb{N}_0^N)$ is an ortho-submartingale. First, let us suppose that $N = 1$. We can apply Theorem 1.5.1 to see that

$$\mathbb{E}\Big[\max_{0 \leq i \leq k} \Psi^p(M_i)\Big]$$
$$\leq \Big(\frac{e}{e-1}\Big)\Big\{1 + \mathbb{E}\Big[\Psi^p(M_k)\ln_+ \Psi^p(M_k)\Big]\Big\}$$
$$\leq \Big(\frac{e}{e-1}\Big)\Big\{1 + \mathbb{E}\big[\Psi^{p+1}(M_k)\big] + p\mathbb{E}\big[\Psi^p(M_k)\ln_+ \ln_+ M_k\big]\Big\}.$$

Since $\ln_+ \ln_+ x \leq \ln_+ x$ and $p \geq 0$,

$$\mathbb{E}\Big[\max_{0 \leq i \leq k} \Psi^p(M_i)\Big] \leq \Big(\frac{e}{e-1}\Big)\Big\{1 + (p+1)\mathbb{E}[\Psi^{p+1}(M_k)]\Big\}$$
$$\leq (p+1)\Big(\frac{e}{e-1}\Big)\Big\{1 + \mathbb{E}[\Psi^{p+1}(M_k)]\Big\}.$$

This is the desired result for $N = 1$. To prove this result in general, we write

$$\max_{0 \preccurlyeq s \preccurlyeq t} \Psi^p(M_s) = \max_{0 \leq s^{(1)} \leq t^{(1)}} \Psi^p(\widetilde{M}_{s^{(1)}}),$$

where $(\widetilde{M}_k;\ k \geq 0)$ is defined in Section 2.3. By Jensen's inequality and the embedding Proposition 2.2.1, $\Psi^p(\widetilde{M})$ is a (one-parameter) submartingale with respect to the (one-parameter) filtration $\mathcal{F}^1$. The first portion of the proof, and the monotonicity of Ψ^p for all $p \geq 0$, together reveal that

$$\mathbb{E}\Big[\max_{0 \preccurlyeq s \preccurlyeq t} \Psi^p(M_s)\Big]$$
$$\leq \Big(\frac{e(p+1)}{e-1}\Big)\Big\{1 + \mathbb{E}\Big[\max_{0 \leq s^{(2)} \leq t^{(2)}} \cdots \max_{0 \leq s^{(N)} \leq t^{(N)}} \Psi^{p+1}(M_{t^{(1)},s^{(2)},\ldots,s^{(N)}})\Big]\Big\}.$$

Once again, using the first part of our proof, together with the embedding Proposition 2.2.1, we arrive at

$$\mathbb{E}\Big[\max_{0 \preccurlyeq s \preccurlyeq t} \Psi^p(M_s)\Big]$$
$$\leq (p+1)\Big(\frac{e}{e-1}\Big)\Big\{1 + (p+1)\Big(\frac{e}{e-1}\Big)$$
$$\times \Big(1 + \mathbb{E}\Big[\max_{0 \leq s^{(3)} \leq t^{(3)}} \cdots \max_{0 \leq s^{(N)} \leq t^{(N)}} \Psi^{p+2}(M_{t^{(1)},t^{(2)},s^{(3)},\ldots,s^{(N)}})\Big]\Big)\Big\}$$
$$\leq (p+1)^2\Big(\frac{e}{e-1}\Big)^2$$
$$\times \Big\{2 + \mathbb{E}\Big[\max_{0 \leq s^{(3)} \leq t^{(3)}} \cdots \max_{0 \leq s^{(N)} \leq t^{(N)}} \Psi^{p+2}(M_{t^{(1)},t^{(2)},s^{(3)},\ldots,s^{(N)}})\Big]\Big\}.$$

The result follows upon induction. □

2.5 A Weak Maximal Inequality

In Section 1 we have shown how to use a maximal probability estimate such as Theorem 1.3.1 to obtain maximal moment estimates such as the results of Sections 1.4 and 1.5. Now we go full circle and obtain a maximal N-parameter probability estimate in terms of the N-parameter moment estimates of Sections 2.3 and 2.4. In fact, we have the following *weak* $(1, L \ln^{N-1} L)$ *inequality*:

Theorem 2.5.1 *Suppose $M = (M_t; t \in \mathbb{N}_0^N)$ is a nonnegative orthosubmartingale with respect to filtrations $\mathcal{F}^1, \ldots, \mathcal{F}^N$. Then, for all $t \in \mathbb{N}_0^N$ and all real $\lambda > 0$,*

$$\mathbb{P}\Big(\max_{0 \preccurlyeq s \preccurlyeq t} M_s \geq \lambda\Big) \leq \frac{1}{\lambda}\Big(\frac{e}{e-1}\Big)^{N-1}\Big\{(N-1) + \mathbb{E}\big[M_t(\ln_+ M_t)^{N-1}\big]\Big\}.$$

Proof By the embedding Proposition 2.2.1 and by Theorem 1.3.1,

$$\mathbb{P}\Big(\max_{0 \preccurlyeq s \preccurlyeq t} M_s \geq \lambda\Big) \leq \frac{1}{\lambda}\mathbb{E}\Big[\max_{0 \leq s^{(2)} \leq t^{(2)}} \cdots \max_{0 \leq s^{(N)} \leq t^{(N)}} M_{t^{(1)}, s^{(2)}, \ldots, s^{(N)}}\Big].$$

The above maximum is over $N-1$ variables and is the maximum of an $(N-1)$-parameter orthosubmartingale; cf. the embedding Proposition 2.2.1. We obtain the result from Theorem 2.4.1. □

2.6 Orthohistories

A stochastic process $M = (M_t; t \in \mathbb{N}_0^N)$ generates N **orthohistories** $\mathcal{H}^1, \ldots, \mathcal{H}^N$ defined in the following manner: For any $1 \leq i \leq N$ and any $k \geq 0$, $\mathcal{H}_k^i$ is the smallest σ-field that makes the collection $(M_t; 0 \leq t^{(i)} \leq k)$ measurable.

For instance, if $N = 2$ and if we write $M = (M_{i,j}; i, j \geq 0)$, then $\mathcal{H}_k^1$ is the σ-field generated by $(M_{i,j}; 0 \leq i \leq k, 0 \leq j)$ and $\mathcal{H}_k^2$ is the σ-field generated by $(M_{i,j}; 0 \leq i, 0 \leq j \leq k)$.

You can (and should) think of $\mathcal{H}_k^1$ and $\mathcal{H}_k^2$ as "the information contained in the values of the process M, over the sets" shown in Figures 1.1 and 1.2, respectively. The following lemma contains some elementary properties of orthohistories.

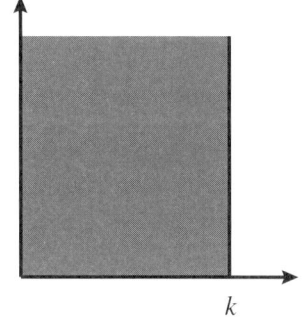

Figure 1.1: $\mathcal{H}_k^1$

Lemma 2.6.1 *Let $\mathcal{H}^1, \ldots, \mathcal{H}^N$ denote the orthohistories corresponding to the stochastic process $M = (M_t; t \in \mathbb{N}_0^N)$. Then,*

1. each of the $\mathcal{H}^i$'s forms a one-parameter filtration; and
2. $\vee_{k \geq 0} \mathcal{H}_k^1 = \cdots = \vee_{k \geq 0} \mathcal{H}_k^N$.

Proof Consider $\mathcal{H}^1$: It is an increasing family of σ-fields, i.e., a filtration. A similar argument can be applied to all the $\mathcal{H}^i$'s; this proves 1. To prove 2, note that $\vee_{i=1}^{\infty} \mathcal{H}_i^1$ is the σ-field generated by $(M_t;\ t \in \mathbb{N}_0^N)$. Since a similar remark holds for $\vee_{i=1}^{\infty} \mathcal{H}_i^2, \ldots, \vee_{i=1}^{\infty} \mathcal{H}_i^N$, our proof is complete. □

An interesting property of orthohistories is that they preserve smartingale properties.

Proposition 2.6.1 *Suppose M is an orthosubmartingale with respect to one-parameter filtrations $\mathcal{F}^1, \ldots, \mathcal{F}^N$, and let $\mathcal{H}^1, \ldots, \mathcal{H}^N$ denote the orthohistories of M. Then, M is an orthosubmartingale with respect to $\mathcal{H}^1, \ldots, \mathcal{H}^N$.*

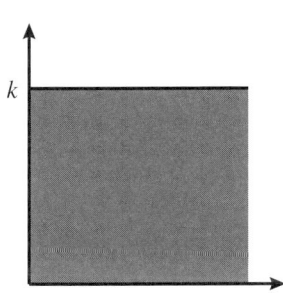

Figure 1.2: $\mathcal{H}_k^2$

Proof Recall that for all $1 \leq i \leq N$ and all $t \in \mathbb{N}_0^N$, $t^{(i)} \mapsto M_t$ is adapted to $\mathcal{F}_{t^{(i)}}^i$. Since $\mathcal{H}_{t^{(i)}}^i$ is the smallest σ-field with respect to which $t^{(i)} \mapsto M_t$ is adapted, we have shown that for all integers $k \geq 0$ and all $1 \leq i \leq N$, $\mathcal{H}_k^i \subset \mathcal{F}_k^i$. To finish our proof of the proposition, we need only check the submartingale property of M with respect to the $\mathcal{H}^i$'s. Since $\mathcal{H}^i \subset \mathcal{F}^i$, this follows from the towering property of conditional expectations; see equation (1) of Section 1.1, Chapter 1. In other words, suppose $t \in \mathbb{N}^N$ and $1 \leq i \leq N$. Then

$$\mathbb{E}[M_t \mid \mathcal{H}_{t^{(i)}-1}^i] = \mathbb{E}\Big[\mathbb{E}\{M_t \mid \mathcal{F}_{t^{(i)}-1}^i\} \Big| \mathcal{H}_{t^{(i)}-1}^i\Big], \quad \text{a.s.,}$$
$$\geq \mathbb{E}[M_s \mid \mathcal{H}_{t^{(i)}-1}^i], \quad \text{a.s.,}$$

where $s^{(j)} = t^{(j)}$ if $j \neq i$; but $s^{(i)} = t^{(i)} - 1$. As a result, M_s is measurable with respect to $\mathcal{H}_{t^{(i)}-1}^i = \mathcal{H}_{s^{(i)}}^i$. Thus,

$$\mathbb{E}[M_t \mid \mathcal{H}_{t^{(i)}-1}^i] \geq M_s, \quad \text{a.s.,}$$

for the above choice of s. This completes our proof. □

Proposition 2.6.1 frees us from sometimes having to worry about the one-parameter filtrations with respect to which M is an orthosmartingale. That is, if no such filtrations are mentioned, it is sometimes safe to assume that they are the orthohistories.

2.7 Convergence Notions

One of the highlights of the one-parameter theory of smartingales is Doob's convergence theorem (Theorem 1.7.1). In order to discuss such a result in the setting of orthosmartingales, we first need to define what we mean by "$\lim_{t\to\infty} X_t$" for a stochastic process $X = (X_t;\ t \in \mathbb{N}_0^N)$.

In this book two natural notions of convergence arise. The first is a straightforward topological one. Namely, given a finite real number L, we say that "$L = \lim_{t\to\infty} X_t$, a.s." whenever the following holds with probability one: As the distance between the point $t \in \mathbb{N}_0^N$ and the axes of $\mathbb{R}^N$ grows, X_t converges to L. That is,

$$\lim_{t\to\infty} X_t = \lim_{t^{(1)},\ldots,t^{(N)} \to \infty} X_t = L.$$

In this spirit, if $L = +\infty$ (respectively $-\infty$), "$L = \lim_{t\to\infty} X_t$, a.s." means that with probability one, for all $M > 0$, there exists $K > 0$, such that whenever $\min_{1 \le i \le N} t^{(i)} \ge K$, then $X_t \ge M$ (respectively $X_t \le -M$).

Sometimes, the almost sure existence of a topological limit is too stringent a condition. We can relax it considerably by defining *sectorial limits* as follows:

Consider a real-valued function f on $\mathbb{R}_+^N$, and let Π_N denote the collection of all permutations of $\{1, \ldots, N\}$. For any $\pi \in \Pi_N$, define

$$\pi - \lim_{t\to\infty} f(t) = \lim_{t^{(\pi(1))} \to \infty} \cdots \lim_{t^{(\pi(N))} \to \infty} f(t),$$

when it exists. Of course, the order in which the limits are taken is, in general, quite important.

We say that the function f has **sectorial limits** (at infinity) if $\pi - \lim_{t\to\infty} f(t)$ exists for all $\pi \in \Pi_N$. To illustrate, suppose $N = 2$. Then $\Pi_2 = \{\pi_1, \pi_2\}$, where $\pi_1 : (1,2) \mapsto (1,2)$ and $\pi_2 : (1,2) \mapsto (2,1)$. The two sectorial limits are

$$\pi_1 - \lim_{t\to\infty} f(t) = \lim_{t^{(1)} \to \infty} \lim_{t^{(2)} \to \infty} f(t);$$

$$\pi_2 - \lim_{t\to\infty} f(t) = \lim_{t^{(2)} \to \infty} \lim_{t^{(1)} \to \infty} f(t).$$

Simple examples show that the sectorial limits need not be equal. However, if and when they are, we write $\lim_{t \rightsquigarrow \infty} f(t)$ for their common value, which we refer to as the **sectorial limit** of f (at infinity).

Next, let us suppose that $M = (M_t;\ t \in \mathbb{N}_0^N)$ is an orthosubmartingale that is bounded in $L^1(\mathbb{P})$. That is, $\sup_{t \in \mathbb{N}_0^N} \mathbb{E}\{|M_t|\} < \infty$. We can apply the one-parameter convergence theorem (Theorem 1.7.1) to the first coordinate of M, say, to see that for all integers $t^{(2)}, \ldots, t^{(N)} \ge 0$, the following exists, almost surely:

$$M^1_{t^{(2)},\ldots,t^{(N)}} = \lim_{t^{(1)} \to \infty} M_t.$$

The next step in showing the existence of sectorial limits is to let $t^{(2)} \to \infty$ in the above. We could do this if we knew that the $(N-1)$-parameter stochastic process M^1 was an $(N-1)$-parameter orthosubmartingale. A little thought shows that for this to hold, we need more stringent conditions than mere boundedness in $L^1(\mathbb{P})$. Indeed, the following suffices:

Theorem 2.7.1 (Cairoli's First Convergence Theorem) *If M is an N-parameter uniformly integrable orthosubmartingale, the sectorial limit $M_{\infty,\ldots,\infty} = \lim_{t \rightsquigarrow \infty} M_t$ exists, almost surely and in $L^1(\mathbb{P})$. Moreover, for any $\pi \in \Pi_N$,*

$$M_t \le \mathbb{E}\left\{\cdots \mathbb{E}\left[\mathbb{E}\{M_{\infty,\ldots,\infty} \mid \mathcal{F}^{\pi(1)}_{t^{\pi(1)}}\} \mid \mathcal{F}^{\pi(2)}_{t^{\pi(2)}}\right] \cdots \mid \mathcal{F}^{\pi(N)}_{t^{\pi(N)}}\right\}, \qquad a.s.$$

To understand this theorem better, consider the case $N = 2$ and write $M = (M_{i,j};\ i,j \ge 0)$. The above result states that

$$M_{i,j} \le \min\left\{\mathbb{E}\left[\mathbb{E}\{M_{\infty,\infty} \mid \mathcal{F}^1_i\} \mid \mathcal{F}^2_j\right],\ \mathbb{E}\left[\mathbb{E}\{M_{\infty,\infty} \mid \mathcal{F}^2_j\} \mid \mathcal{F}^1_i\right]\right\}, \qquad a.s.$$

In particular, when M is a 2-parameter *orthomartingale*, then we have the following **commutation** property:

$$M_{i,j} = \mathbb{E}\left[\mathbb{E}\{(M_{\infty,\infty} \mid \mathcal{F}^1_i\} \mid \mathcal{F}^2_j\right] = \mathbb{E}\left[\mathbb{E}\{M_{\infty,\infty} \mid \mathcal{F}^2_j\} \mid \mathcal{F}^1_i\right], \qquad a.s.$$

We will explore an abstraction of this property in greater depth later on.

Proof Let us prove the result for $N = 2$. The general case is contained in Exercise 2.7.1 below.

Throughout this proof we will assume that $\mathcal{F}^1$ and $\mathcal{F}^2$ are the orthohistories of the two-parameter process M. See the discussion of Section 2.6. We will also write $M = (M_{i,j};\ i,j \ge 0)$.

For any $j \ge 0$, $i \mapsto M_{i,j}$ is a one-parameter uniformly integrable submartingale. By the one-parameter smartingale convergence theorem (Theorem 1.7.1), $M_{\infty,j} = \lim_{i \to \infty} M_{i,j}$ exists almost surely and in $L^1(\mathbb{P})$. Moreover,

$$M_{i,j} \le \mathbb{E}[M_{\infty,j} \mid \mathcal{F}^1_i].$$

Next, we need the process properties of $M_{\infty,j}$. By the asserted $L^1(\mathbb{P})$ convergence,

$$\mathbb{E}\left[M_{\infty,j+1} \mid \mathcal{F}^2_j\right] = \lim_{i \to \infty} \mathbb{E}[M_{i,j+1} \mid \mathcal{F}^2_j] \ge \lim_{i \to \infty} M_{i,j} = M_{\infty,j},$$

where the convergence takes place in $L^1(\mathbb{P})$. Hence, $j \mapsto M_{\infty,j}$ is itself a one-parameter submartingale with respect to $\mathcal{F}^2$ and

$$M_{i,j} \le \mathbb{E}[M_{\infty,j} \mid \mathcal{F}^1_i].$$

Appealing to Fatou's lemma, we see that $j \mapsto M_{\infty,j}$ is a uniformly integrable one-parameter submartingale with respect to $\mathcal{F}^2$. The one-parameter smartingale convergence theorem (Theorem 1.7.1) shows that $M^1_{\infty,\infty} = \lim_{j \to \infty} M_{\infty,j}$ exists almost surely and in $L^1(\mathbb{P})$. It may be useful to point out that in the notation of the paragraph preceding the statement of the theorem,

$$M^1_{\infty,\infty} = \pi_2 - \lim_{(i,j) \to \infty} M_{i,j} = \lim_{j \to \infty} \lim_{i \to \infty} M_{i,j}, \quad \text{a.s. and in } L^1(\mathbb{P}),$$

where $\pi_2 : (1, 2) \mapsto (2, 1)$.

Our argument, thus far, can be applied equally well to the variables in reverse order. In this way we obtain a uniformly integrable $\mathcal{F}^1$-orthomartingale $(M_{i,\infty}; i \geq 0)$ that almost surely converges to some $M^2_{\infty,\infty}$, as $i \to \infty$. That is, we have extracted $M^1_{\infty,\infty}$ and $M^2_{\infty,\infty}$ as the two possible sectorial limits of M. Now we show that with probability one, $M^1_{\infty,\infty} = M^2_{\infty,\infty}$.

For all $k \geq 0$,

$$\mathbb{E}\big[M^1_{\infty,\infty} \mid \mathcal{F}^1_k\big] = \lim_{j \to \infty} \lim_{i \to \infty} \mathbb{E}\big[M_{i,j} \mid \mathcal{F}^1_k\big] \geq \lim_{j \to \infty} \lim_{i \to \infty} M_{k,j} = M_{k,\infty}, \quad \text{a.s.}$$

Thus, $\lim_{k \to \infty} \mathbb{E}[M^1_{\infty,\infty} \mid \mathcal{F}^1_k] \geq M^2_{\infty,\infty}$, almost surely. Since $\mathcal{F}^1$ is a filtration, the one-parameter martingale convergence theorem implies that the mentioned limit is equal to $\mathbb{E}[M^1_{\infty,\infty} \mid \vee_{k \geq 0} \mathcal{F}^1_k]$, almost surely. By Lemma 2.6.1, $\vee_{k \geq 0} \mathcal{F}^1_k = \vee_{k \geq 0} \mathcal{F}^2_k$, and $M^1_{\infty,\infty}$ is measurable with respect to the latter. That is, $M^1_{\infty,\infty} \geq M^2_{\infty,\infty}$, almost surely. We now reverse the roles of the indices to deduce that with probability one, $M^1_{\infty,\infty} = M^2_{\infty,\infty} = M_{\infty,\infty}$.

We have shown the almost sure, as well as the $L^1(\mathbb{P})$, existence of the sectorial limit $\lim_{(i,j) \rightsquigarrow \infty} M_{i,j}$. That $M_{i,j}$ can be bounded by a conditional expectation of $M_{\infty,\infty}$ follows from the $L^1(\mathbb{P})$ convergence. □

Exercise 2.7.1 Prove Theorem 2.7.1 when $N \geq 2$ is arbitrary. □

2.8 Topological Convergence

In this section we will show that, at little extra cost, the sectorial convergence of Section 2.7 can be greatly strengthened:

Theorem 2.8.1 (Cairoli's Second Convergence Theorem) *Consider any orthomartingale $M = (M_t; t \in \mathbb{N}_0^N)$ that satisfies the following integrability condition:*

$$\sup_{t \in \mathbb{N}_0^N} \mathbb{E}\{|M_t|(\ln_+ |M_t|)^{N-1}\} < \infty.$$

Then, with probability one,

$$\lim_{t^{(1)} \to \infty} \sup_{t^{(2)},\ldots,t^{(N)} \geq 0} |M_t - M^1_{t^{(2)},\ldots,t^{(N)}}| = 0,$$

where $M^1_{t^{(2)},\ldots,t^{(N)}} = \lim_{t^{(1)} \to \infty} M_t$, almost surely.

In particular, by interchanging the indices we can conclude the following strengthened form of Theorem 2.7.1 for orthomartingales:

Corollary 2.8.1 *Suppose $M = (M_t; t \in \mathbb{N}_0^N)$ is an orthomartingale with the property that $\sup_{t \in \mathbb{N}_0^N} \mathbb{E}\{|M_t|(\ln_+ |M_t|)^{N-1}\} < \infty$. Then, $\lim_{t \to \infty} M_t$ exists almost surely and in $L^1(\mathbb{P})$.*

Exercise 2.8.1 Prove Corollary 2.8.1. □

Our proof of Theorem 2.8.1 requires the following technical lemma.

Lemma 2.8.1 *Suppose $M = (M_t; t \in \mathbb{N}_0^N)$ is a uniformly integrable orthosubmartingale. Suppose $\Psi : [0, \infty[\to [0, \infty[$ is convex, and that $x \mapsto x^{-1}\Psi(x)$ is finite and nondecreasing on $[0, \infty[$. If $\sup_{t \in \mathbb{N}_0^N} \mathbb{E}[\Psi(|M_t|)] < \infty$, then $M_* = \lim_{t \rightsquigarrow \infty} M_t$ exists and*

$$\lim_{t \to \infty} \mathbb{E}[\Psi(|M_t - M_*|)] = 0.$$

Proof Since M is uniformly integrable, by Theorem 2.7.1, the sectorial limit, $M_* = \lim_{t \rightsquigarrow \infty} M_t$, exists, a.s. and in $L^1(\mathbb{P})$. Moreover, Fatou's lemma shows that $\mathbb{E}[\Psi(|M_*|)] < \infty$. Observe that $t \mapsto \Psi(|M_t|)$ is itself an N-parameter orthosubmartingale and, by Theorem 2.7.1, for all $s \in \mathbb{N}_0^N$,

$$|M_s| \leq \mathbb{E}\Big(\cdots \mathbb{E}\Big[\mathbb{E}\{|M_*| \,\big|\, \mathcal{F}^1_{s^{(1)}}\}\,\Big|\,\mathcal{F}^2_{s^{(2)}}\Big] \cdots \,\Big|\, \mathcal{F}^N_{s^{(N)}}\Big),$$

almost surely. By convexity, almost surely,

$$\Psi(|M_s|) \leq \mathbb{E}\Big(\cdots \mathbb{E}\Big[\mathbb{E}\{\Psi(|M_*|) \,\big|\, \mathcal{F}^1_{s^{(1)}}\}\,\Big|\,\mathcal{F}^2_{s^{(2)}}\Big] \cdots \,\Big|\, \mathcal{F}^N_{s^{(N)}}\Big).$$

Consequently, as long as $\Lambda \in \cap_{\ell=1}^N \mathcal{F}^\ell_{s^{(\ell)}}$, the following inequality holds:

$$\mathbb{E}[\Psi(|M_s|)\mathbf{1}_\Lambda] \leq \mathbb{E}[\Psi(|M_*|)\mathbf{1}_\Lambda]. \tag{1}$$

We let $\Lambda_s = (\Psi(|M_s|) > \lambda)$ ($\lambda > 0, s \in \mathbb{N}_0^N$) and apply equation (1) with Λ replaced by Λ_s. Since $\mathbb{P}(\Lambda_s) \leq \sup_{t \in \mathbb{N}_0^N} \frac{1}{\lambda}\mathbb{E}[\Psi(|M_t|)]$, this probability (i.e., $\mathbb{P}(\Lambda_s)$) goes to 0 as $\lambda \to \infty$, uniformly in $s \in \mathbb{N}_0^N$. By equation (1), we deduce that $t \mapsto \Psi(|M_t|)$ is a *uniformly integrable*, N-parameter orthosubmartingale. Since Ψ is continuous on $[0, \infty[$, by Theorem 2.7.1,

$$\lim_{t \rightsquigarrow \infty} \mathbb{E}[\Psi(|M_t|)] = \mathbb{E}[\Psi(|M_*|)]. \tag{2}$$

We now show that the sectorial limit "$\lim_{t \rightsquigarrow \infty}$" is, in fact, a topological one. To this end, note that for all $x, y \in \mathbb{R}$,

$$\Psi(|x - y|) \leq |\Psi(x) - \Psi(y)|. \tag{3}$$

28   1. Discrete-Parameter Martingales

To verify this, write $\Psi(x) = xf(x)$ ($x \geq 0$), where f is nondecreasing, and assume without loss of generality that $x \geq y \geq 0$. Then,

$$\Psi(|x-y|) = (x-y)f(x-y) \leq xf(x) - yf(x-y).$$

Thus, if $x - y \geq y$, we obtain equation (3). If $x - y \leq y$,

$$\Psi(|x-y|) \leq (x-y)f(y) = xf(y) - yf(y) \leq xf(x) - yf(y).$$

Now that we have demonstrated equation (3), we can proceed with our proof. By Jensen's inequality, $\Psi(|M_t|)$ is an $L^1(\mathbb{P})$ bounded ortho-submartingale. Since Ψ is nondecreasing, by Proposition 2.2.1, if $s \preccurlyeq t$, then $\mathbb{E}[\Psi(|M_s|)] \leq \mathbb{E}[\Psi(|M_t|)]$. A real-variable argument shows that $\lim_{s \to \infty} \mathbb{E}[\Psi(|M_s|)]$ exists and, by our assumption on Ψ and M, is finite. Thus, by equation (2), $\lim_{t \to \infty} \mathbb{E}[\Psi(|M_s|)] = \mathbb{E}[\Psi(|M_*|)]$ and the lemma follows from equation (3). □

We can now prove Theorem 2.8.1.

Proof of Theorem 2.8.1 For the sake of illustration, we first detail our proof in the case $N = 2$ and write $M = (M_{i,j};\ i,j \geq 0)$.

Clearly, the function $\Psi(x) = x \ln_+ x$ satisfies the conditions of Lemma 2.8.1. That is, it is convex and $x^{-1}\Psi(x)$ is nondecreasing. Furthermore, by the assumption of Theorem 2.8.1, $\sup_{i,j \geq 0} \mathbb{E}[\Psi(|M_{i,j}|)]$ is finite. Define

$$M_{i,\infty} = \lim_{j \to \infty} M_{i,j}, \quad i \geq 0,$$
$$M_{\infty,j} = \lim_{i \to \infty} M_{i,j}, \quad j \geq 0.$$

On the other hand, during the course of our proof of Theorem 2.7.1, we already saw that $i \mapsto M_{i,\infty}$ (respectively $j \mapsto M_{\infty,j}$) is a uniformly integrable one-parameter martingale with respect to $\mathcal{F}^1$ (respectively $\mathcal{F}^2$) (Why?) Moreover, if $M_{\infty,\infty} = \lim_{t \leadsto \infty} M_t$ denotes the sectorial limit, then

$$M_{\infty,\infty} = \lim_{i \to \infty} M_{i,\infty} = \lim_{j \to \infty} M_{\infty,j}, \quad \text{a.s.}$$

We proceed with establishing some probability estimates. First, we note that for any fixed $m \geq 0$,

$$\sup_{j \geq m} \sup_{i \geq 0} |M_{i,j} - M_{i,\infty}| \leq 2 \sup_{i \geq 0} \sup_{j \geq m} |M_{i,j} - M_{i,m}|.$$

Hence, for any $c > 0$ and all $m \geq 0$,

$$\mathbb{P}\left(\sup_{j \geq m} \sup_{i \geq 0} |M_{i,j} - M_{i,\infty}| \geq \lambda\right) \leq \mathbb{P}\left(\sup_{i \geq 0} \sup_{j \geq m} |c(M_{i,j} - M_{i,m})| \geq \tfrac{1}{2}c\lambda\right).$$

On the other hand, for any $m \geq 0$, $(c|M_{i,j} - M_{i,m})|;\ i \geq 0, j \geq m)$ is an orthosubmartingale with respect to its orthohistories, for instance. Therefore, by the weak maximal inequality (Theorem 2.5.1 of Section 2.5),

$$\mathbb{P}\Big(\sup_{j \geq m} \sup_{i \geq 0} |M_{i,j} - M_{i,\infty}| \geq \lambda\Big)$$
$$\leq \frac{2}{c\lambda}\Big(\frac{e}{e-1}\Big)\Big\{1 + \sup_{j \geq m} \sup_{i \geq 0} \mathbb{E}\Big[\Psi\big(c|M_{i,j} - M_{i,m}|\big)\Big]\Big\}.$$

By Jensen's inequality (cf. the first part of the embedding Proposition 2.2.1),

$$\mathbb{P}\Big(\sup_{j \geq m} \sup_{i \geq 0} |M_{i,j} - M_{i,\infty}| \geq \lambda\Big)$$
$$\leq \frac{2}{c\lambda}\Big(\frac{e}{e-1}\Big)\Big\{1 + \mathbb{E}\Big[\Psi\big(c|M_{\infty,\infty} - M_{\infty,m}|\big)\Big]\Big\}.$$

Now note that for any $m \geq 0$, $i \mapsto M_{i,m}$ is a uniformly integrable one-parameter martingale with $\sup_{i \geq 0} \mathbb{E}\{|M_{i,m}| \ln_+ |M_{i,m}|\} < \infty$. In particular, by Lemma 2.8.1, $m \mapsto M_{\infty,m}$ is a uniformly integrable one-parameter martingale with $\sup_{m \geq 0} \mathbb{E}\{|M_{\infty,m}| \ln_+ |M_{\infty,m}|\} < \infty$. By another appeal to Lemma 2.8.1, for any $c > 0$, $\lim_{m \to \infty} \Psi(c|M_{\infty,\infty} - M_{\infty,m}|) = 0$. Thus, there exists $c_m > 0$ such that $\lim_m c_m = \infty$ and $\sup_{m \geq 0} \Psi(c_m|M_{\infty,\infty} - M_{\infty,m}|) \leq 1$. That is,

$$\mathbb{P}\Big(\sup_{j \geq m} \sup_{i > 0} |M_{i,j} - M_{i,\infty}| \geq \lambda\Big) \leq \frac{4}{c_m \lambda}\Big(\frac{e}{e-1}\Big).$$

Letting $m \to \infty$ and then letting $\lambda \to 0^+$ along a rational sequence, we see that
$$\limsup_{m \to \infty} \sup_{i \geq 0} |M_{i,m} - M_{i,\infty}| = 0, \quad \text{a.s.}$$

Similarly, $\limsup_{n \to \infty} \sup_{j \geq 0} |M_{n,j} - M_{\infty,j}| = 0$, a.s., which proves the result for $N = 2$. We now proceed with our proof in the general case. Write any $t \in \mathbb{N}_0^N$ as $t = (t^{(1)}, t)$, where $t' = (t^{(2)}, \ldots, t^{(N)}) \in \mathbb{N}_0^{N-1}$. The above argument extends naturally to show that for all real $c, \lambda > 0$ and all positive integers n,

$$\mathbb{P}\Big(\sup_{t' \in \mathbb{N}_0^{N-1}} \sup_{t^{(1)} \geq n} |M_t - M_{t'}^1| \geq \lambda\Big)$$
$$\leq \frac{2}{c\lambda}\Big(\frac{e}{e-1}\Big)^{N-1}\Big\{(N-1) + \mathbb{E}\Big[\Psi\big(c|M_* - M_{n,\infty,\ldots,\infty}|\big)\Big]\Big\},$$

where $M_* = \lim_{t \to \infty} M_t$ and $M_{n,\infty,\ldots,\infty} = \lim_{t' \to \infty} M_{n,t'}$ are N-parameter and $(N-1)$-parameter sectorial limits, respectively, and $\Psi'(x) = x(\ln_+ x)^{N-1}$; see Section 2.7. Hence, we can find $c_n \to \infty$ such that

$$\mathbb{P}\Big(\sup_{t' \in \mathbb{N}_0^{N-1}} \sup_{t^{(1)} \geq n} |M_t - M_{t'}^1| \geq \lambda\Big) \leq \frac{N}{c_n \lambda}\Big(\frac{e}{e-1}\Big)^{N-1}.$$

(The process M^1 is defined in the statement of Theorem 2.8.1 that we are now proving.) The rest of our proof follows as in the $N = 2$ case. □

Exercise 2.8.2 In the set-up of Theorem 2.8.1, suppose further that there exists $p > 1$ such that M is bounded in $L^p(\mathbb{P})$. That is, $\sup_{t \in \mathbb{N}_0^N} \mathbb{E}[|M_t|^p] < \infty$. Show that M_t has a limit as $t \to \infty$, a.s. and in $L^p(\mathbb{P})$. Prove a uniform version of this in the style of Cairoli's second convergence theorem. □

2.9 Reversed Orthomartingales

Roughly speaking, one-parameter reversed martingales are processes that become martingales once we reverse time. In order to be concrete, we need some filtrations. We say that a collection of σ-fields $\mathcal{F} = (\mathcal{F}_k; k \geq 0)$ is a **reversed filtration** if for all $k \geq 0$, $\mathcal{F}_{k+1} \subset \mathcal{F}_k$. That is, $\mathcal{F}$ is a collection of σ-fields that becomes a filtration, once we reverse time. Now we define a stochastic process $M = (M_k; k \geq 0)$ to be a **reversed martingale** (with respect to the reversed filtration $\mathcal{F}$) if

1. M is adapted to $\mathcal{F}$; and

2. for all $\ell \geq k \geq 0$, $\mathbb{E}[M_k \,|\, \mathcal{F}_\ell] = M_\ell$, a.s.

Exercise 2.9.1 Verify that the above Condition 2 is *equivalent* to the following: For all $k \geq 0$, $\mathbb{E}[M_k \,|\, \mathcal{F}_{k+1}] = M_{k+1}$, a.s. □

Exercise 2.9.2 Show that every one-parameter reversed martingale converges. In this regard, see also Supplementary Exercise 4.
(HINT: You can begin with by deriving an upcrossing inequality.) □

Our multiparameter discussion begins with N one-parameter reversed filtrations $\mathcal{F}^i = (\mathcal{F}_k^i; k \geq 0)$, where $1 \leq i \leq N$. A stochastic process $M = (M_t; t \in \mathbb{N}_0^N)$ is a **reversed orthomartingale** if for each integer $1 \leq i \leq N$ and all nonnegative integers $(t^{(j)}; 1 \leq j \leq N, j \neq i)$, $t^{(i)} \mapsto M_t$ is a reversed martingale with respect to the filtration $\mathcal{F}^i$. To illustrate, consider $N = 2$ and write $M_{i,j}$ ($\mathcal{F}_{i,j}$, etc.) for $M_{(i,j)}$ ($\mathcal{F}_{(i,j)}$, etc.). Then, M is a reversed orthomartingale if for all $i \geq 0$, $(M_{i,k}; k \geq 0)$ is a martingale with respect to $\mathcal{F}^2$ and $(M_{k,i}; k \geq 0)$ is a martingale with respect to $\mathcal{F}^1$.

Under a mild integrability condition, reversed orthomartingales always converge.

Theorem 2.9.1 *Suppose $M = (M_t; t \in \mathbb{N}_0^N)$ is a reversed orthomartingale with respect to one-parameter reversed filtrations $\mathcal{F}^i = (\mathcal{F}_k^i; k \geq 0)$, $1 \leq i \leq N$. If $\mathbb{E}\{|M_0|(\ln_+ |M_0|)^{N-1}\} < \infty$, then $\lim_{t \to \infty} M_t$ exists a.s. and in $L^1(\mathbb{P})$.*

Exercise 2.9.3 Prove Theorem 2.9.1. □

***Exercise* 2.9.4** Prove the following $L^p(\mathbb{P})$ variant of Theorem 2.9.1. In the same context as the latter theorem, if $\mathbb{E}\{|X_0|^p\} < \infty$ for some $p > 1$, then, as $t \to \infty$, M_t converges in $L^p(\mathbb{P})$. □

***Exercise* 2.9.5** Show that whenever M is a uniformly integrable, N-parameter reversed orthomartingale, the sectorial limit $\lim_{t \rightsquigarrow \infty} M_t$ exists and is in $L^1(\mathbb{P})$. Moreover, $\mathbb{E}\{\lim_{t \rightsquigarrow \infty} |M_t|\} \leq \mathbb{E}\{|M_0|\}$. □

3 Martingales

In this section we discuss N-parameter martingales, and more generally N-parameter smartingales. In an absolutely general setting, N-parameter smartingales do not possess much structure; cf. Section 3.3 below. On the other hand, we will see, in this section, that in some cases one can use the theory of orthomartingales to efficiently analyze N-parameter martingales.

3.1 Definitions

While the definition of orthomartingales may have seemed contrived, that of N-parameter martingales is certainly not, since N-parameter martingales are the most natural extension of 1-parameter martingales. It is the deep connections between martingales and orthomartingales that make the latter useful in the study of the former.

Suppose $\mathcal{F} = (\mathcal{F}_t; t \in \mathbb{N}_0^N)$ is a collection of sub σ-fields of $\mathcal{G}$. We say that $\mathcal{F}$ is a **filtration** if $s \preccurlyeq t$ implies that $\mathcal{F}_s \subset \mathcal{F}_t$. An N-parameter stochastic process $M = (M_t; t \in \mathbb{N}_0^N)$ is **adapted** to the filtration $\mathcal{F}$ if for all $t \in \mathbb{N}_0^N$, M_t is $\mathcal{F}_t$ measurable. The process M is an N-parameter **submartingale** (with respect to $\mathcal{F}$) if it is adapted (to $\mathcal{F}$), for all $t \in \mathbb{N}_0^N$, $\mathbb{E}\{|M_t|\} < \infty$ and for all $t \succcurlyeq s$, $\mathbb{E}[M_t | \mathcal{F}_s] \geq M_s$, a.s.

The stochastic process M is a **supermartingale** if $-M$ is a submartingale. It is a **martingale** if it is both a supermartingale and a submartingale. If M is either a sub- or a supermartingale, then it is a **smartingale**.

The following is an immediate consequence of Jensen's inequality:

Lemma 3.1.1 *Suppose $M = (M_t; t \in \mathbb{N}_0^N)$ is a nonnegative submartingale with respect to a filtration $\mathcal{F} = (\mathcal{F}_t; t \in \mathbb{N}_0^N)$ and $\Psi : [0, \infty[\to [0, \infty[$ is convex nondecreasing. If for all $t \in \mathbb{N}_0^N$, $\mathbb{E}[\Psi(M_t)] < \infty$, then $t \mapsto \Psi(M_t)$ is a submartingale with respect to $\mathcal{F}$.*

3.2 Marginal Filtrations

The collection of all multiparameter smartingales contains the class of all orthosmartingales. To be more precise, suppose $M = (M_t; t \in \mathbb{N}_0^N)$ is an

N-parameter random process that is adapted to the N-parameter filtration $\mathcal{F} = (\mathcal{F}_t; \ t \in \mathbb{N}_0^N)$. For all $1 \leq j \leq N$, define

$$\mathcal{F}_k^j = \bigvee_{\substack{t \in \mathbb{N}_0^N: \\ t^{(j)} = k}} \mathcal{F}_t, \qquad k \geq 0.$$

We define $\mathcal{F}^j = (\mathcal{F}_k^j; \ k \geq 0)$, $1 \leq j \leq N$ and refer to the σ-fields $\mathcal{F}^1, \ldots, \mathcal{F}^N$ as the **marginal filtrations** of $\mathcal{F}$.

Lemma 3.2.1 *If $\mathcal{F}$ is a filtration, then, $\mathcal{F}^1, \ldots, \mathcal{F}^N$ are one-parameter filtrations, and for all $t \in \mathbb{N}_0^N$, $\mathcal{F}_t \subseteq \cap_{\ell=1}^N \mathcal{F}_{t^{(\ell)}}^\ell$.*

Exercise **3.2.1** Prove Lemma 3.2.1. □

The marginal filtrations of an N-parameter filtration form one of the deep links between N-parameter martingales and orthomartingales, as we shall see next.

Proposition 3.2.1 *Suppose M is adapted to the N-parameter filtration $\mathcal{F}$. If M is an orthosubmartingale with respect to the marginal filtrations of $\mathcal{F}$, then M is a submartingale with respect to $\mathcal{F}$.*

Proof We present a proof that uses the towering property of conditional expectations (see equation (1) of Section 1.1), working one parameter at a time. Suppose $t \succcurlyeq s$ are both in $\mathbb{N}_0^N$. Then, with probability one,

$$\mathbb{E}[M_t \,|\, \mathcal{F}_s] = \mathbb{E}\{\,\mathbb{E}[M_t \,|\, \mathcal{F}_{s^{(1)}}^1] \,|\, \mathcal{F}_s\} \geq \mathbb{E}[M_{s^{(1)},t^{(2)},\ldots,t^{(N)}} \,|\, \mathcal{F}_s]$$
$$= \mathbb{E}\bigl[\,\mathbb{E}\{M_{s^{(1)},t^{(2)},\ldots,t^{(N)}} \,|\, \mathcal{F}_{s^{(2)}}^2\} \,|\, \mathcal{F}_s\bigr] \geq \mathbb{E}[M_{s^{(1)},s^{(2)},t^{(3)},\ldots,t^{(N)}} \,|\, \mathcal{F}_s]$$
$$\geq \cdots \geq \mathbb{E}[M_s \,|\, \mathcal{F}_s] = M_s.$$

The remaining properties are easy to check. □

To each N-parameter process M we can associate a **history** $\mathcal{H}$ as follows: For every $t \in \mathbb{N}_0^N$, let $\mathcal{H}_t$ denote the σ-field generated by $(M_s; \ s \preccurlyeq t)$. When $N = 2$, we can think of $\mathcal{H}_t$ as the information contained in the values $(M_s; \ s \preccurlyeq t)$. This is shown in Figure 1.3. You should check that the corresponding pictures for the 2 marginal filtrations $\mathcal{H}_k^1$ and $\mathcal{H}_k^2$ are given by Figures 1.1 and 1.2 of Section 2.6, respectively. It is helpful to have these pictures in mind when interpreting many of the results and arguments that involve multiparameter filtrations, processes, etc.

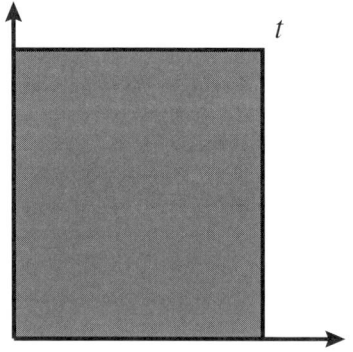

Figure 1.3: A picture of $\mathcal{H}_t$

3.3 A Counterexample of Dubins and Pitman

According to Proposition 3.2.1 and Lemma 3.2.1, there is a sense in which the class of all smartingales includes orthosmartingales. Since the latter have rich mathematical properties (Section 2), one may be tempted to think that the same is true of the former. Next, we discuss a two-parameter counterexample of Dubins and Pitman (1980); it shows that, in the absence of further features, general multiparameter smartingales do not have many regularity properties. Related results can be found in the second half of Föllmer (1984a).

Let us begin with two fixed sequences of numbers: $p_1, p_2, \ldots$, all taking values in $]0,1[$, and $m_1, m_2, \ldots$, all positive integers. Next, consider a probability space $(\Omega, \mathcal{G}, \mathbb{P})$ rich enough so that on it we can construct independent random variables $X_1, X_2, \ldots$ such that for any $k \geq 1$,

$$\mathbb{P}(X_k = 0) = p_k, \quad \text{and} \quad \mathbb{P}(X_k = j) = m_k^{-1}(1 - p_k),$$

for all $1 \leq j \leq m_k$. For every integer $i, j \geq 0$ and all numerals $k \geq 1$, define $\mathcal{F}_{i,j}^k$ to be the trivial σ-field if $i + j < k$. If $i + j = k$, define $\mathcal{F}_{i,j}^k$ to be the σ-field generated by $(X_k = 0$ or $X_k = i)$. Finally, if $i + j > k$, define $\mathcal{F}_{i,j}^k$ to be the σ-field generated by all of the X's. We can also define $\mathcal{F} = (\mathcal{F}_{i,j}; i, j \geq 0)$ by

$$\mathcal{F}_{i,j} = \bigvee_{k=1}^{\infty} \mathcal{F}_{i,j}^k, \qquad i, j \geq 0.$$

We begin with some preliminary facts.

Exercise **3.3.1** Verify the following properties:

1. for each fixed integer $k \geq 1$, $\mathcal{F}^k = (\mathcal{F}_{i,j}^k; i, j \geq 0)$ is a filtration;

2. $\mathcal{F}^1, \mathcal{F}^2, \ldots$ are independent; and

3. $\mathcal{F}$ is a filtration.

Finally, find the marginal filtrations of $\mathcal{F}$. □

Let $A = \cup_{k=1}^{\infty}(X_k = 0)$ and define

$$M_{i,j} = \mathbb{P}(A \mid \mathcal{F}_{i,j}), \qquad i, j \geq 0.$$

Clearly, $M = (M_{i,j}; i, j \geq 0)$ is a bounded two-parameter martingale. In fact, $0 \leq M_{i,j} \leq 1$ for all $i, j \geq 0$. We plan to show that it need not converge. More precisely, we have the following:

Theorem 3.3.1 *Suppose $\sum_{k=1}^{\infty} p_k < 1$ but $\lim_{k \to \infty} m_k p_k = \infty$. Then, with probability one, $\limsup_{t \in \mathbb{N}_0^N} M_t > \liminf_{t \in \mathbb{N}_0^N} M_t$.*

Remarks

(a) Once again, we identify $M_{i,j}$ with $M_{(i,j)}$ for any $i,j \geq 0$. Throughout our proof, the same remark applies to $\mathcal{F}_{i,j}$, $\mathcal{F}_{i,j}^k$, etc.

(b) The conditions of the theorem hold, for example, if $m_k = 5^k$ and $p_k = 2^{-k}$.

Proof We start with a calculation that is an exercise in conditional expectations:

$$\mathbb{P}(X_k = 0 \mid \mathcal{F}_{i,j}^k) = \begin{cases} p_k, & \text{if } i+j < k, \\ \dfrac{m_k p_k}{m_k p_k + 1 - p_k} \mathbf{1}_{(X_k=0 \text{ or } X_k=i)}, & \text{if } i+j = k, \\ \mathbf{1}_{(X_k=0)}, & \text{if } i+j > k. \end{cases} \quad (1)$$

Fix any $n, m \geq 0$ and note that for all k large enough (how large depends on n and m),

$$\sup_{(i,j) \succcurlyeq (n,m)} \mathbb{P}(X_k = 0 \mid \mathcal{F}_{i,j}^k) \geq \frac{m_k p_k}{m_k p_k + 1 - p_k}.$$

(This uses the Borel–Cantelli lemma and the fact that $\sum_k p_k < +\infty$, so that with probability one, $X_k \neq 0$ for all but finitely many k's.)

By (b) above, X_k is independent of the entire collection $(\mathcal{F}^\ell;\ \ell \neq k)$. Therefore, for all $m, n \geq 0$, and all k large enough,

$$\sup_{(i,j) \succcurlyeq (n,m)} \mathbb{P}(X_k = 0 \mid \mathcal{F}_{i,j}) \geq \frac{m_k p_k}{m_k p_k + 1 - p_k}.$$

Since $(X_k = 0) \subset A$, we obtain the following: For all $n, m \geq 0$ and all k large enough,

$$\sup_{(i,j) \succcurlyeq (n,m)} M_{i,j} \geq \frac{m_k p_k}{m_k p_k + 1 - p_k}.$$

The left-hand side is independent of all large k and bounded above by one. We can let $k \to \infty$ and use the conditions $\lim_{k\to\infty} p_k = 0$ and $\lim_{k\to\infty} m_k p_k = +\infty$ to see that with probability one, $\sup_{(i,j) \succcurlyeq (n,m)} M_{i,j} = 1$. That is,

$$\limsup_{t \in \mathbb{N}_0^N} M_t = \inf_{t \in \mathbb{N}_0^2} \sup_{s \succcurlyeq t} M_s = 1, \quad \text{a.s.} \quad (2)$$

On the other hand, $(M_{j,j}; j \geq 0)$ is a bounded, one-parameter martingale with respect to the filtration $(\mathcal{F}_{j,j}; j \geq 0)$; cf. Exercise 2.2.2. By Doob's convergence theorem, it converges almost surely and in $L^p(\mathbb{P})$ for all $p \geq 1$. That is, with probability one, $\lim_{j\to\infty} M_{j,j} = \mathbb{P}(A)$. This shows that with probability one,

$$\liminf_{t \in \mathbb{N}_0^N} M_t = \sup_{s \in \mathbb{N}_0^2} \inf_{t \succcurlyeq s} M_t \leq \mathbb{P}(A) \leq \sum_{k=1}^{\infty} \mathbb{P}(X_k = 0) = \sum_{k=1}^{\infty} p_k.$$

Since $\sum_{k=1}^{\infty} p_k < 1$, this and (2), together, have the desired effect. □

Exercise 3.3.2 Verify equation (1). □

Exercise 3.3.3 Show that Theorem 3.3.1 continues to hold, even if the condition $\sum_k p_k < 1$ is replaced by $\sum_k p_k < +\infty$.
(HINT: The Borel–Cantelli lemma.) □

Exercise 3.3.4 Construct a 2-parameter martingale that is in $L^p(\mathbb{P})$ for for all $p > 0$, and yet its maximum is a.s. unbounded. This is due to S. Janson.
(HINT: For the set-up of the Dubins–Pitman counterexample, consider $N_{i,j} = \sum_{k=1}^{\infty} \alpha_k \mathbb{P}(X_k = 0 \,|\, \mathcal{F}_{i,j})$, for a suitably chosen sequence $\alpha_1, \alpha_2, \ldots$ that goes to infinity in the limit.) □

3.4 Commutation

In Section 3.3 we saw that general smartingales can have undesirable properties; cf. Theorem 3.3.1 and Exercise 3.3.4. On the other hand, when the smartingale in question is an orthosmartingale, the theory is quite rich; see Section 2. The question arises, *when is a martingale an orthomartingale?* This brings us to the so-called commutation hypothesis.

Suppose $\mathcal{G}_1, \ldots, \mathcal{G}_N$ are sub-σ-fields of $\mathcal{G}$. We say that they **commute** if for all bounded random variables Y and all $\pi, \pi' \in \Pi_N$ (the collection of all permutations of $\{1, \ldots, N\}$),

$$\mathbb{E}\Big[\,\mathbb{E}\big[\,\mathbb{E}\{Y\,|\,\mathcal{G}_{\pi(1)}\}\,\big|\,\mathcal{G}_{\pi(2)}\big]\cdots\,\Big|\,\mathcal{G}_{\pi(N)}\Big]$$
$$= \mathbb{E}\Big[\,\mathbb{E}\big\{\,\mathbb{E}(Y\,|\,\mathcal{G}_{\pi'(1)})\,\big|\,\mathcal{G}_{\pi'(2)}\big\}\cdots\,\Big|\,\mathcal{G}_{\pi'(N)}\Big], \qquad \text{a.s.}$$

Two remarks are in order:

1. We have already encountered a weak version of such a property in Section 2.7. See the remarks after Theorem 2.7.1; and

2. Commutation of σ-fields is a rather special property; it typically does not hold.

A seemingly different condition is commutation of filtrations. An N-parameter filtration $\mathcal{F} = (\mathcal{F}_t;\ t \in \mathbb{N}_0^N)$ is **commuting**[5] if for every $s, t \in \mathbb{N}_0^N$ and for all bounded $\mathcal{F}_t$-measurable random variables Y,

$$\mathbb{E}[Y\,|\,\mathcal{F}_s] = \mathbb{E}[Y\,|\,\mathcal{F}_{s \wedge t}], \qquad \text{a.s.}$$

---

[5] This is also known as condition (F4).

Exercise 3.4.1 Construct several (e.g., six or seven) 2-parameter filtrations that do not commute. □

The following is the first sign of the connections between commuting filtrations, orthomartingales, and martingales. It states that if the underlying filtration is commuting, many martingales are also orthomartingales. As such, this serves as a converse to Proposition 3.2.1.

Theorem 3.4.1 *Suppose $\mathcal{F}$ is an N-parameter commuting filtration. Then, for all bounded random variables Z and all $t \in \mathbb{N}_0^N$,*

$$\mathbb{E}\Big[\cdots \mathbb{E}\big\{\mathbb{E}(Z \mid \mathcal{F}^1_{t(1)}) \mid \mathcal{F}^2_{t(2)}\big\} \cdots \Big| \mathcal{F}^N_{t(N)}\Big] = \mathbb{E}[Z \mid \mathcal{F}_t], \quad a.s.,$$

where $\mathcal{F}^1, \ldots, \mathcal{F}^N$ denote the marginal filtrations of $\mathcal{F}$.

Proof To simplify the demonstration, suppose $N = 2$ and write $M_{i,j}$ for $M_{(i,j)}$, etc. By Doob's 1-parameter martingale convergence theorem (Theorem 1.7.1),

$$\mathbb{E}\big[\mathbb{E}\{Z \mid \mathcal{F}^1_i\} \mid \mathcal{F}^2_j\big] = \mathbb{E}\big[\lim_{k \to \infty} \mathbb{E}\{Z \mid \mathcal{F}_{i,k}\} \mid \mathcal{F}^2_j\big]$$

$$= \lim_{k \to \infty} \mathbb{E}\big[\mathbb{E}\{Z \mid \mathcal{F}_{i,k}\} \mid \mathcal{F}^2_j\big]$$

$$= \lim_{k \to \infty} \lim_{\ell \to \infty} \mathbb{E}\big[\mathbb{E}\{Z \mid \mathcal{F}_{i,k}\} \mid \mathcal{F}_{\ell,j}\big],$$

where all of the convergences are taking place in $L^1(\mathbb{P})$. Recall that $\mathcal{F}$ is commuting and $Y = \mathbb{E}[Z \mid \mathcal{F}_{i,k}]$ is $\mathcal{F}_{i,k}$-measurable. This implies that a.s.,

$$\mathbb{E}[Y \mid \mathcal{F}_{\ell,j}] = \mathbb{E}[Y \mid \mathcal{F}_{(i,k) \wedge (\ell,j)}] = \mathbb{E}[Y \mid \mathcal{F}_{i \wedge \ell, k \wedge j}] = \mathbb{E}[Z \mid \mathcal{F}_{i \wedge \ell, k \wedge j}].$$

The ultimate equality follows from the towering property of conditional expectations; cf. equation (1) of Section 1.1, Chapter 1. We have shown that for every $i, j \geq 0$ and for all bounded random variables Z,

$$\mathbb{E}\big[\mathbb{E}\{Z \mid \mathcal{F}^1_i\} \mid \mathcal{F}^2_j\big] = \lim_{k \to \infty} \lim_{\ell \to \infty} \mathbb{E}[Z \mid \mathcal{F}_{i \wedge \ell, k \wedge j}], \quad a.s.,$$

which, almost surely, equals $\mathbb{E}\{Z \mid \mathcal{F}_{i,j}\}$. □

The following result better explains the term *commuting*.

Corollary 3.4.1 *The marginal filtrations $\mathcal{F}^1, \ldots, \mathcal{F}^N$ of an N-parameter commuting filtration $\mathcal{F}$ all commute.*

Proof Consider a bounded random variable Z and a $t \in \mathbb{N}_0^N$. By relabeling the parameter coordinates, we see that for all $\pi \in \Pi_N$ (the collection of all permutations of $\{1, \ldots, N\}$),

$$\mathbb{E}\Big[\cdots \mathbb{E}\big\{\mathbb{E}(Z \mid \mathcal{F}^{\pi(1)}_{t(\pi(1))}) \mid \mathcal{F}^{\pi(2)}_{t(\pi(2))}\big\} \cdots \Big| \mathcal{F}^{\pi(N)}_{t(\pi(N))}\Big] = \mathbb{E}[Z \mid \mathcal{F}_t], \quad a.s.$$

Since the right-hand side is not affected by our choice of π, this implies the result. □

***Exercise* 3.4.2** Prove that the filtration $\mathcal{F}$ of the Dubins–Pitman counterexample of Section 3.3 is not commuting. □

***Exercise* 3.4.3** The intention of this exercise is to show that under commutation, Lemma 3.2.1 is sharp. One way to make this precise is as follows: Suppose $\mathcal{F}^1, \ldots, \mathcal{F}^N$ are N independent 1-parameter filtrations. That is, for all $t \in \mathbb{N}_0^N$, $\mathcal{F}_{t(1)}^1, \ldots, \mathcal{F}_{t(N)}^N$ are independent σ-fields. For all $t \in \mathbb{N}_0^N$, define $\mathcal{F}_t = \cap_{\ell=1}^N \mathcal{F}_{t(\ell)}^{(\ell)}$ and prove that $\mathcal{F} = (\mathcal{F}_t;\ t \in \mathbb{N}_0^N)$ is a commuting N-parameter filtration. Compute its marginal filtrations. This is related to Exercise 2.1.1. □

3.5 Martingales

In Section 3.2 we showed that orthomartingales are martingales. We now show that if the underlying filtration is commuting, the converse also holds. In particular, under commutation, the powerful machinery of orthosmartingales becomes available to martingales; see Corollary 3.5.1 below.

Theorem 3.5.1 *Suppose that $\mathcal{F}$ is an N-parameter commuting filtration and that $M = (M_t;\ t \in \mathbb{N}_0^N)$ is adapted to $\mathcal{F}$. Then, the following are equivalent:*

(i) M is an orthosubmartingale with respect to the marginals of $\mathcal{F}$; and

(ii) M is a submartingale with respect to $\mathcal{F}$.

Proof Proposition 3.2.1 shows that $(i) \implies (ii)$. We now show that under commutation, (ii) implies (i). Let us suppose (ii) holds. To simplify the exposition, we also suppose $N = 2$ and write $M_{i,j}$ for $M_{(i,j)}$, etc. By the one-parameter martingale convergence theorem,

$$\mathbb{E}[M_{i+1,j} \mid \mathcal{F}_i^1] = \lim_{k \to \infty} \mathbb{E}[M_{i+1,j} \mid \mathcal{F}_{i,k}], \quad \text{a.s.}$$

Since $\mathcal{F}$ is commuting, for all $k > j$, $\mathbb{E}[M_{i+1,j} \mid \mathcal{F}_{i,k}] = \mathbb{E}[M_{i+1,j} \mid \mathcal{F}_{i,j}]$. Since M is a submartingale,

$$\mathbb{E}[M_{i+1,j} \mid \mathcal{F}_i^1] = \lim_{k \to \infty} \mathbb{E}[M_{i+1,j} \mid \mathcal{F}_{i,j}] \geq M_{i,j}.$$

That is, for any fixed $j \geq 0$, $(M_{i,j};\ i \geq 0)$ is a one-parameter submartingale with respect to $\mathcal{F}^1$. Relabeling the order of the parameters proves the result. □

As an immediate but important consequence, we obtain the following.

Corollary 3.5.1 *Suppose $M = (M_t;\ t \in \mathbb{N}_0^N)$ is a submartingale with respect to the N-parameter commuting filtration $\mathcal{F}$. Then, Cairoli's maximal inequalities (Theorems 2.3.1 and 2.4.1 and 2.5.1) all hold for M. Moreover, so do Cairoli's convergence theorems (Theorems 2.7.1 and 2.8.1).*

Exercise 3.5.1 Suppose $X = (X_t;\ t \in \mathbb{N}_0^N)$ is an N-parameter submartingale with respect to a commuting filtration $\mathcal{F} = (\mathcal{F}_t;\ t \in \mathbb{N}_0^N)$. Show that there exist a martingale $M = (M_t;\ t \in \mathbb{N}_0^N)$ and an adapted process $A = (A_t;\ t \in \mathbb{N}_0^N)$ such that $X_t = M_t + A_t$. Moreover, A can be chosen so that it is nondecreasing with respect to the partial order $\preccurlyeq$. That is, whenever $s \preccurlyeq t$, $A_s \leq A_t$. This is from Cairoli (1971).
(HINT: Write $M_t = \sum_{s \preccurlyeq t} \xi_s$ and try to imitate the given proof of Doob's decomposition; cf. Exercise 1.2.2 and Supplementary Exercise 1.) □

Exercise 3.5.2 Suppose $\mathcal{F} = (\mathcal{F}_n;\ n \geq 0)$ is a one-parameter filtration and suppose $X = (X_t;\ t \in \mathbb{N}_0^N)$ is a sequence of random variables indexed by N parameters. Assume that

(i) there exists a (nonrandom, finite) constant $K > 0$ such that a.s., $\sup_t |X_t| \leq K$; and

(ii) the sectorial limit of X_t exists, which we denote by X_∞.

Prove that no matter how we let $n, t^{(1)}, \ldots, t^{(N)} \to \infty$, $\mathbb{E}[X_t \mid \mathcal{F}_n] \to \mathbb{E}[X_\infty \mid \mathcal{F}_n]$, a.s. In words, show that $\mathbb{E}[X_\infty \mid \mathcal{F}_n]$ is the sectorial limit of $\mathbb{E}[X_t \mid \mathcal{F}_n]$, as (n, t) goes to infinity in $\mathbb{N}^{N+1}$. This is based on Blackwell and Dubins (1962, Theorem 2). □

3.6 Conditional Independence

Suppose $\mathcal{G}_1, \ldots, \mathcal{G}_{k+1}$ are sub-σ-fields of $\mathcal{G}$. We say that $\mathcal{G}_1, \ldots, \mathcal{G}_k$ are **conditionally independent given** $\mathcal{G}_{k+1}$ if

$$\mathbb{E}\Big[\prod_{\ell=1}^k Y_\ell \,\Big|\, \mathcal{G}_{k+1}\Big] = \prod_{\ell=1}^k \mathbb{E}[Y_\ell \mid \mathcal{G}_{k+1}]$$

whenever $Y_1, \ldots, Y_k$ are bounded random variables that are measurable with respect to $\mathcal{G}_1, \ldots, \mathcal{G}_k$, respectively.

Exercise 3.6.1 Consider an ordinary random walk $S = (S_n;\ n \geq 0)$ described in Exercise 1.1.1. Fix some $n \geq 1$ and define the following σ-fields: (a) Let $\mathcal{F}_n$ be the σ-field generated by $(S_j;\ 0 \leq j \leq n)$; and (b) let $\mathcal{G}_n$ denote the σ-field generated by $(S_{n+j};\ j \geq 0)$. Show that $\mathcal{F}_n$ and $\mathcal{G}_n$ are conditionally independent, given $\sigma(S_n)$, the σ-field generated by S_n. □

The following characterization theorem describes an important connection between conditional independence and commutation.

Theorem 3.6.1 *For a given filtration* $\mathcal{F} = (\mathcal{F}_t;\ t \in \mathbb{N}_0^N)$, *the following are equivalent:*

(i) $\mathcal{F}$ *is commuting; and*

(ii) for all $s, t \in \mathbb{N}_0^N$, $\mathcal{F}_t$ and $\mathcal{F}_s$ are conditionally independent, given $\mathcal{F}_{s \wedge t}$.

Proof Suppose that for all $t \in \mathbb{N}_0^N$, Y_t is a bounded $\mathcal{F}_t$-measurable random variable. By the towering property of conditional expectations (equation (1), Section 1.1),

$$\mathbb{E}[Y_t Y_s \,|\, \mathcal{F}_{t \wedge s}] = \mathbb{E}\big[Y_t \, \mathbb{E}\{Y_s \,|\, \mathcal{F}_t\} \,\big|\, \mathcal{F}_{t \wedge s}\big] = \mathbb{E}\big[Y_t \, \mathbb{E}\{Y_s \,|\, \mathcal{F}_{t \wedge s}\} \,\big|\, \mathcal{F}_{t \wedge s}\big], \qquad \text{a.s.}$$

Thus, $(i) \Rightarrow (ii)$. Conversely, supposing that (ii) holds,

$$\begin{aligned}
\mathbb{E}[Y_t Y_s] &= \mathbb{E}\big[\mathbb{E}\{Y_t Y_s \,|\, \mathcal{F}_{t \wedge s}\}\big] \\
&= \mathbb{E}\big[\mathbb{E}\{Y_t \,|\, \mathcal{F}_{t \wedge s}\} \cdot \mathbb{E}\{Y_s \,|\, \mathcal{F}_{t \wedge s}\}\big] \\
&= \mathbb{E}\big[\mathbb{E}\{Y_t \,|\, \mathcal{F}_{t \wedge s}\} \cdot Y_s\big].
\end{aligned}$$

Since $\mathcal{F}_{t \wedge s} \subset \mathcal{F}_s$ and the above holds for all bounded $\mathcal{F}_s$-measurable random variables Y_s, $\mathbb{E}[Y_t \,|\, \mathcal{F}_{t \wedge s}] = \mathbb{E}[Y_t \,|\, \mathcal{F}_s]$, almost surely. This shows that (ii) implies (i), and hence (ii) is equivalent to (i). □

At this point it would be a good idea to take another look at Figures 1.1, 1.2, and 1.3 and interpret commutation pictorially, at least in the interesting case where the N-parameter filtration in question is the history of a certain stochastic process.

We conclude this section, and in fact this chapter, with a series of indispensable exercises on conditionally independent σ-fields. Henceforth, let $\mathcal{G}_1, \mathcal{G}_2$ denote two σ-fields that are conditionally independent, given another σ-field $\mathcal{G}_3$.

***Exercise* 3.6.2** Prove that $\mathcal{G}_1 \vee \mathcal{G}_3$ and $\mathcal{G}_2 \vee \mathcal{G}_3$ are conditionally independent, given $\mathcal{G}_3$. Use this to prove that the following are equivalent:

(i) $\mathcal{G}_1$ and $\mathcal{G}_2$ are conditionally independent, given $\mathcal{G}_3$; and

(ii) for all bounded $\mathcal{G}_1$-measurable random variables X_1,

$$\mathbb{E}[X_1 \,|\, \mathcal{G}_2 \vee \mathcal{G}_3] = \mathbb{E}[X_1 \,|\, \mathcal{G}_3], \qquad \text{a.s.}$$

In the general theory of random fields, condition (ii) is referred to as the Markov property for the triple $(\mathcal{G}_1, \mathcal{G}_3, \mathcal{G}_2)$; see Rozanov (1982, Section 1, Chapter 2). □

***Exercise* 3.6.3** Show that $\mathcal{G}_1$ and $\mathcal{G}_2$ are conditionally independent, given $\mathcal{G}_4$, for any other σ-field $\mathcal{G}_4 \supset \mathcal{G}_3$ such that $\mathcal{G}_4 \subset \mathcal{G}_1 \vee \mathcal{G}_2$. Construct an example to show that this last condition cannot be dropped.
(HINT: Let $\mathcal{G}_1$ and $\mathcal{G}_2$ denote the σ-fields generated by ξ_1 and ξ_2, respectively, where ξ_1, ξ_2 are i.i.d. ± 1 with probability $\frac{1}{2}$ each. Consider $\mathcal{G}_3 = \{\varnothing, \Omega\}$ and let $\mathcal{G}_4$ denote the σ-field generated by $\xi_1 + \xi_2$.) □

40   1. Discrete-Parameter Martingales

***Exercise* 3.6.4** Suppose $\mathcal{H}_n$ ($n \geq 1$) is a decreasing sequence of σ-fields such that for all $n \geq 1$, $\mathcal{G}_1$ and $\mathcal{G}_2$ are conditionally independent, given $\mathcal{H}_n$. Prove that $\mathcal{G}_1$ and $\mathcal{G}_2$ are conditionally independent, given $\cap_{n \geq 1} \mathcal{H}_n$. □

***Exercise* 3.6.5** Verify that whenever $\mathcal{G}_1$ and $\mathcal{G}_2$ are conditionally independent, given $\mathcal{G}_3$, then $\mathcal{G}_1 \cap \mathcal{G}_2 \subset \mathcal{G}_3$.
(WARNING: You will need to assume that the underlying probability space, together with all of the mentioned σ-fields, is complete.) □

4 Supplementary Exercises

1. Consider a collection of real numbers $a = (a_t; \, t \in \mathbb{N}_0^N)$. Show that corresponding to a, there exists a unique sequence $\delta = (\delta_t; \, t \in \mathbb{N}_0^N)$ such that (i) for all $t \in \mathbb{N}_0^N$, $a_t = \sum_{s \preccurlyeq t} \delta_s$; and (ii) for all $t \in \mathbb{N}_0^N$, δ_t depends only on $(a_s; \, s \preccurlyeq t)$. This is an aspect of the so-called **inclusion–exclusion** formula of J. H. Poincaré, and the δ's are referred to as the **increments** of the sequence a.

2. Suppose $M = (M_n; \, n \geq 0)$ is a nonnegative uniformly integrable 1-parameter submartingale with respect to a filtration $\mathcal{F} = (\mathcal{F}_n; \, n \geq 0)$.

   (i) Use Doob's convergence theorem and Fatou's lemma to prove that for any $\mathcal{F}$-stopping time T, $\mathbb{E}[M_T \mathbf{1}_{(T<\infty)}] < \infty$.

   (ii) Conclude that whenever T is a finite $\mathcal{F}$-stopping time, $(M_{T \wedge n}; \, n \geq 0)$ is uniformly integrable.

   (iii) Conclude that for *all* uniformly integrable submartingales $M = (M_n; \, n \geq 0)$ and for all stopping times T,
   $$\mathbb{E}\{|M_T|\mathbf{1}_{(T<\infty)}\} \leq \sup_n \mathbb{E}\{|M_n|\} < \infty.$$
   Moreover, when M is a uniformly integrable martingale, then for all stopping times T, $\mathbb{E}[M_T \mathbf{1}_{(T<\infty)}] = \mathbb{E}[M_0]$.

3. As observed by S. D. Chatterji, Doob's convergence theorem for martingale is also a consequence of Doob's maximal inequalities; cf. Chatterji (1967, 1968) and Lamb (1973). In this exercise you may not appeal to the martingale convergence theorem.
   Throughout, the underlying probability space is $(\Omega, \mathcal{F}_\infty, \mathbb{P})$ and $\mathcal{F} = (\mathcal{F}_n; \, n \geq 0)$ denotes a filtration of sub-σ-fields of $\mathcal{F}_\infty$ with $\mathcal{F}_\infty = \vee_n \mathcal{F}_n$.

   (i) Suppose $(Y_n; \, n \geq 0)$ is a uniformly integrable martingale with respect to the filtration $\mathcal{F}$. For all $A \in \mathcal{F}_n$, define $\mu(A) = \mathbb{E}[Y_n \mathbf{1}_A]$ and verify that μ is a (signed) measure on $\vee_n \mathcal{F}_n$. Furthermore, μ is absolutely continuous with respect to $\mathbb{P}$. Conclude the existence of a random variable $Y_\infty \in L^1(\mathbb{P})$ such that for all $n \geq 0$, $Y_n = \mathbb{E}[Y_\infty \mid \mathcal{F}_n]$, a.s.

   (ii) Let F denote the collection of all $Z \in L^1(\mathbb{P})$ such that for some $n = n(Z)$, Z is $\mathcal{F}_n$-measurable. Prove that F is dense in $L^1(\mathbb{P})$.

(iii) Prove that whenever Y is the uniformly integrable martingale of (i), then as $n \to \infty$, Y_n converges a.s. and in $L^1(\mathbb{P})$ to Y_∞.

(iv) Prove that whenever $M = (M_n;\ n \geq 0)$ is a uniformly integrable martingale on an arbitrary probability space $(\Omega, \mathcal{G}, \mathbb{P})$ and with respect to an arbitrary filtration $\mathcal{F}_1, \mathcal{F}_2, \ldots$, there exists $M_\infty \in L^1(\mathbb{P})$ such that M_n converges to M_∞ a.s. and in $L^1(\mathbb{P})$.

(v) Let $M = (M_n;\ n \geq 0)$ denote a martingale that is bounded in $L^1(\mathbb{P})$. That is, $\sup_n \mathbb{E}\{|M_n|\} < \infty$. Prove that there exists $M_\infty \in L^1(\mathbb{P})$ such that M_n converges to M_∞, a.s.

(vi) Deduce Doob's martingale convergence theorem from (v) and Doob's decomposition, Exercise 1.2.1.

(HINT: For part (i), check that $\mu(A) = \lim_{k \to \infty} \mathbb{E}[Y_k \mathbf{1}_A]$. For part (ii), consider Z's of the form $Z = \mathbf{1}_E$, where $E \in \mathcal{F}_n$. For part (iii), choose $Y'_\infty \in F$ such that $\mathbb{E}\{|Y_\infty - Y'_\infty|\} \leq \varepsilon$. Define $Y'_n = \mathbb{E}[Y'_\infty \mid \mathcal{F}_n]$ and note that $Y'_n \to Y'_\infty$, a.s. Finally, estimate $\mathbb{P}(\sup_n |Y_n - Y'_n| \geq \varepsilon)$, via Doob's maximal inequality. For part (iv), let $T = \inf(n \geq 0 :\ |M_n| \geq R)$ and first prove that $n \mapsto M_{T \wedge n}$ is uniformly integrable. Finish by picking R so large that $\mathbb{P}(T < \infty) < \varepsilon$ for ε small enough.)

4. Suppose $M = (M_n;\ n \geq 0)$ is a reversed martingale with respect to a reversed filtration $\mathcal{F} = (\mathcal{F}_n;\ n \geq 0)$.

(i) Verify that $\lim_{n \to \infty} \mathbb{E}[M_n]$ exists but may be $-\infty$.

(ii) If $\lim_n \mathbb{E}[M_n] > -\infty$, prove the existence of a positive finite constant A such that $\sup_n \mathbb{P}(|M_n| \geq \lambda) \leq \frac{A}{\lambda}$, for all $\lambda > 0$.

(iii) Show that whenever $\lim_n \mathbb{E}[M_n] > -\infty$, $(M_k^!;\ k \geq 0)$ is uniformly integrable.

(iv) Prove that whenever $n > m$ and $\lambda > 0$,
$$0 \geq \mathbb{E}[M_n \mathbf{1}_{(M_n \leq -\lambda)}] \geq \mathbb{E}[M_n - M_m] + \mathbb{E}[M_m \mathbf{1}_{(M_n < -\lambda)}].$$

(v) Conclude that $\lim_n \mathbb{E}[M_n] > -\infty$ is equivalent to the uniform integrability of M.

(HINT: For part (ii), check that $\mathbb{E}[M_n^+ \mathbf{1}_{(M_n^+ \geq \lambda)}] \leq \mathbb{E}[M_1^+ \mathbf{1}_{(M_n^+ \geq \lambda)}]$.)

5. Consider the hypercube $I_{0,1} = [0,1]^N$. We can subdivide $I_{0,1}$ into 2^N equal-sized subhypercubes of side $\frac{1}{2}$ each and denote them by $J_{1,1}, \ldots, J_{1,2^N}$, in some order. For instance, supposing that $N = 2$, we could take $J_{1,1} = [0, \frac{1}{2}]^2$, $J_{1,2} = [0, \frac{1}{2}] \times [\frac{1}{2}, 1]$, $J_{1,3} = [\frac{1}{2}, 1] \times [0, \frac{1}{2}]$ and $J_{1,4} = [\frac{1}{2}, 1]^2$. Next, we toss 2^N independent p-coins (i.e., probability of heads equals $p \in]0, 1[$)—one coin for every one of the J's—and declare $J_{1,j}$ as *open* if the jth coin landed heads. Let N_1 denote the number of open hypercubes among $J_{1,1}, \ldots, J_{1,2^N}$ and write these (randomly selected) open hypercubes as $I_{1,1}, \ldots, I_{1,N_1}$ in some order. We also define $K_1 = \cup_{j=1}^{N_1} I_{1,j}$, which is a random compact subset of $[0,1]^N$. Next, we subdivide each of the $I_{1,j}$'s ($1 \leq j \leq N_1$) into 2^N equal-sized subhypercubes of side $\frac{1}{4}$. According to newly introduced independent p-coins, we declare each of them open or not as before to obtain a compact set $K_2 \subset K_1$. We can continue this construction inductively: Having constructed K_n as a random union of hypercubes $J_{n,1}, \ldots, J_{n,N_n}$, we subdivide each $J_{n,i}$ into 2^N equal-sized subhypercubes and keep them or not according to the parity of i.i.d. p-coins, all tossed independently of the previous one. Define K_{n+1} to be the union of the hypercubes kept in this $(n+1)$st stage and continue the process indefinitely to obtain a decreasing family of random compact sets $K_1 \supset K_2 \supset \cdots$. Let $K = \cap_n K_n$ denote the limiting compact set and prove that there exists $p_c \in]0,1[$ such that for all $p > p_c$, $K \neq \varnothing$, with positive probability, whereas for all $p < p_c$, $K = \varnothing$, a.s. Can you compute p_c?

This process of creating random Cantor-like sets was coined **fractal percolation** by Mandelbrot (1982).

(HINT: Consider a hidden branching process.)

6. Consider an N-parameter martingale $M = (M_t;\ t \in \mathbb{N}_0^N)$ with respect to an N-parameter filtration $\mathcal{F} = (\mathcal{F}_t;\ t \in \mathbb{N}_0^N)$. We do not require $\mathcal{F}$ to be commuting here.

(i) Verify that whenever $\mathbb{E}[M_t^2] < \infty$ for all $t \in \mathbb{N}_0^N$, then for all $s \preccurlyeq t$, both in $\mathbb{N}_0^N$,
$$\mathbb{E}[(M_t - M_s)^2] = \mathbb{E}[M_t^2] - \mathbb{E}[M_s^2].$$

(ii) In particular, demonstrate that whenever $\sup_t \mathbb{E}[M_t^2] < \infty$, $t \mapsto M_t$ is a Cauchy sequence in $L^2(\mathbb{P})$.

(iii) Deduce the martingale convergence theorem when $\sup_t \mathbb{E}[M_t^2] < \infty$.

Thus, while we essentially need the commutation of $\mathcal{F}$ to prove the existence of a.s. limits, the existence of $L^2(\mathbb{P})$ limits is another matter.

7. Let $\mathcal{F} = (\mathcal{F}_t;\ t \in \mathbb{N}_0^N)$ denote an N-parameter filtration. An $\mathbb{N}_0^N \cup \{+\infty\}$-valued random variable is an $\mathcal{F}$-**stopping point** if for all $t \in \mathbb{N}_0^N$, $(T \preccurlyeq t) \in \mathcal{F}_t$.

(i) Let $X = (X_t;\ t \in \mathbb{N}_0^N)$ be adapted to $\mathcal{F}$ and hold integers $t^{(2)}, \ldots, t^{(N)} \geq 0$ fixed. Prove, then, that for all Borel sets $E \subset \mathbb{R}$, T is an $\mathcal{F}$-stopping point, where $T^{(j)} = t^{(j)}$ for all $N \geq j \geq 2$ and $T^{(1)} = \inf(t^{(1)} \geq 0 : X_t \in E)$. (You need to interpret $\inf \varnothing = +\infty$ and identify $(\infty, t^{(2)}, \ldots, t^{(N)})$ with the point $+\infty$ in the one-point compactification of $\mathbb{N}_0^N$.)

(ii) Prove that whenever $M = (M_t;\ t \in \mathbb{N}_0^N)$ is an N-parameter submartingale with respect to $\mathcal{F}$ and whenever $T \preccurlyeq S$ are bounded stopping points, then $\mathbb{E}[M_T] \leq \mathbb{E}[M_S]$.

(iii) If T is an $\mathcal{F}$-stopping point, let $\mathcal{F}_T$ denote the collection of all $s \in \mathbb{N}_0^N$ such that for all events A, $A \cap (T = s) \in \mathcal{F}_s$.

(a) Prove that $\mathcal{F}_T$ is a σ-field.

(b) Show that T is $\mathcal{F}_T$ measurable and that $(T = s)$ can, equivalently, be replaced by $(T \succ s)$ in the definition of $\mathcal{F}_T$.

(c) If S and T are $\mathcal{F}$-stopping points, $S \vee T$ is also an $\mathcal{F}$-stopping point.

(d) (Hard) Construct an example of a 2-parameter filtration $\mathcal{F}$ and two $\mathcal{F}$-stopping points S and T such that $S \wedge T$ is *not* an $\mathcal{F}$-stopping point.

8. Let $\mathcal{F}$ denote an N-parameter filtration. A collection of $\mathcal{F}$-stopping points $(\gamma(n);\ n \geq 0)$ is an **optional increasing path** if $\gamma(1) \preccurlyeq \gamma(2) \cdots$, almost surely. Extend a part of Exercise 2.2.2 by showing that when M is an N-parameter martingale with respect to $\mathcal{F}$ and when γ is an optional increasing path, $M \circ \gamma$ is a one-parameter martingale with respect to $\mathcal{F} \circ \gamma$, where the $\circ$ notation is taken from Exercise 2.2.2.

9. Let $\mathfrak{R}$ denote the collection of all sets of the form $[0,t]$, where $t \in \mathbb{N}_0^N$; let $(\Omega, \mathcal{G}, \mathbb{P})$ designate a probability space; and let $\mathcal{F} = (\mathcal{F}_t;\ t \in \mathbb{N}_0^N)$ denote an N-parameter filtration of sub–σ-field of $\mathcal{G}$.

A **stopping domain** D is a subset of $\mathbb{N}_0^N$ such that (a) D is a (random) denumerable union of elements of $\mathfrak{R}$; and (b) for all $t \in \mathbb{N}_0^N$, $(t \in D) \in \mathcal{F}_t$. Its **interior** D° is defined as the collection of all $t \in D$ such that there exists $s \prec t$ that is also in D. Demonstrate the truth of the following:

(i) Whenever $N = 1$, $D \setminus D^\circ$ is a stopping time with respect to $\mathcal{F}$.

(ii) Any nonrandom union of elements of $\mathfrak{R}$ is a stopping domain.

(iii) If D_1 and D_2 are stopping domains, so are $D_1 \cap D_2$ and $D_1 \cup D_2$.

(iv) Suppose X is an adapted stochastic process. Let K be a measurable set and define
$$D_K = \bigcup \left\{ [0,t]:\ t \in \mathbb{N}_0^N,\ X_s \in K^\complement \text{ for all } s \prec t \right\}.$$
Then, D_K is a stopping domain.

(v) If D is a stopping domain, the following is a σ-field:
$$\mathcal{F}_D = (A \in \mathcal{G}:\ A \cap (t \notin D^\circ) \notin \mathcal{F}_t \text{ for all } t \in \mathbb{N}_0^N).$$

(vi) If $D_1 \subset D_2$ are stopping domains, $\mathcal{F}_{D_1} \subset \mathcal{F}_{D_2}$.

(vii) If D is a stopping domain, then for all $t \in \mathbb{N}_0^N$, $(t \in D) \in \mathcal{F}_D$.

(This is due to J. B. Walsh.)

10. (Hard. Continued from Supplementary Exercise 9.)
Let $\mathcal{F} = (\mathcal{F}_t;\ t \in \mathbb{N}_0^N)$ denote an N-parameter filtration on some probability space and let $\mathcal{F}^1, \cdots, \mathcal{F}^N$ denote its marginal filtrations.

(i) For all $t \in \mathbb{N}_0^N$, let $\mathcal{F}_t^\star = \bigvee_{j=1}^N \mathcal{F}_{t(j)}^j$ and prove that $\mathcal{F}^\star = (\mathcal{F}_t^\star;\ t \in \mathbb{N}_0^N)$ is an N-parameter filtration.

(ii) If D is a stopping domain, define $\mathcal{F}_D^\star$ to be the collection of all events A such that for all $t \in \mathbb{N}_0^N$, $A \cap (t \notin D^\circ) \notin \mathcal{F}_t^\star$. Prove that (a) $\mathcal{F}_D^\star$ is a σ-field; (b) $(t \in D) \in \mathcal{F}_D^\star$, for all $t \in \mathbb{N}_0^N$; and (c) for all stopping domains $D_1 \subset D_2$, $\mathcal{F}_{D_1}^\star \subset \mathcal{F}_{D_2}^\star$.

(iii) An adapted process $M = (M_t;\ t \in \mathbb{N}_0^N)$ is a **strong martingale** if for all $s \prec t$, both in $\mathbb{N}_0^N$, $M_t \in L^1(\mathbb{P})$ and $\mathbb{E}[\delta_t \mid \mathcal{F}_s^\star] = 0$, where the δ's denote the increments of M in the sense of Supplementary Exercise 1. Prove that whenever $D_1 \subset D_2$ are stopping domains,
$$\mathbb{E}\left[\sum_{t \in D_2} \delta_t \,\bigg|\, \mathcal{F}_{D_1}^\star \right] = \sum_{t \in D_1} \delta_t,$$
almost surely.

44   1. Discrete-Parameter Martingales

This exercise constitutes J. B. Walsh's **optional stopping theorem** for strong martingales. It opens the door to a rich theory of strong martingales that runs parallel to the one-parameter theory.

11. (Hard) Suppose M denotes an N-parameter orthomartingale. For all $i \in \{1, \ldots, N\}$, all $t \in \mathbb{N}_0^N$, and for all $a < b$, define $U_t^i[a, b]$ to be the number of upcrossings of $[a, b]$ made by the numerical sequence $(M_s;\ s^{(j)} = t^{(j)}$ for all $j \neq i$, $s^{(i)} \leq t^{(i)})$. Prove that

$$\sup_{t \in \mathbb{N}_0^N} \mathbb{E}\{\max_{1 \leq i \leq N} U_t^i[a, b]\} < \infty,$$

as long as $\sup_t \mathbb{E}(|M_t|\{\ln_+ |M_t|\}^{N-1}) < \infty$. Using this, describe another proof of Cairoli's second convergence theorem (Theorem 2.8.1).

12. (Hard) Let $\mathcal{F}$ denote a commuting N-parameter filtration and suppose that M is a submartingale with respect to $\mathcal{F}$. In this exercise we prove Cairoli's second convergence theorem for M, using the ideas of the 1-parameter proof of S. D. Chatterji; cf. Supplementary Exercise 3.

  (i) If $(Y_t;\ t \in \mathbb{N}_0^N)$ is adapted and uniformly integrable, there exists $Y_\star \in L^1(\mathbb{P})$ such that for all $t \in \mathbb{N}_0^N$, $Y_t = \mathbb{E}[Y_\star \mid \mathcal{F}_t]$, a.s.

  (ii) Let F denote the collection of all $Z \in L^1(\mathbb{P})$ such that for some $t = t(Z)$, Z is $\mathcal{F}_t$-measurable. Prove that F is dense in $L^1(\mathbb{P})$.

  (iii) Prove that whenever Y is the uniformly integrable martingale of (i), Y_t has an a.s. and $L^1(\mathbb{P})$ limit, as $t \to \infty$.

  (iv) Returning to the submartingale M, assume that

$$\sup_t \mathbb{E}(|M_t|\{\ln_+ |M_t|\}^{N-1}) < \infty,$$

  and show that by (i)–(iii), M converges a.s. as $t \to \infty$.

5 Notes on Chapter 1

Section 1 The basic theory of discrete-parameter martingales is covered in most introductory graduate texts in probability theory such as (Billingsley 1995; Chow and Teicher 1997; Chung 1974; Durrett 1991) and the more advanced books (Dudley 1989; Stroock 1993). Many of the one-parameter exercises of this chapter are standard and borrowed from these texts. For more specialized treatments, together with various applications, see Garsia (1970, 1973), Hall and Heyde (1980), and Neveu (1975). The earlier development of the subject, together with a wealth of information, can be found in Doob (1990).

A comprehensive account of 1-parameter martingale theory, in both discrete and continuous time, is Dellacherie and Meyer (1982).

In the preamble the claim was made that maximal inequalities are critical to the existence of strong convergence theorems. This can be made quite precise; cf. Burkholder (1964).

The inequalities of Sections 1.3–1.5 are all part of one "master inequality" involving Orlicz norms; see Neveu (1975, Appendix). Exercise 1.2.1 is due to J. L. Doob; see Doob (1990) and Krickeberg (1963, 1965). An abstract theory of martingales can be found in Krickeberg and Pauc (1963).

It is interesting that a nonadapted variant of Exercise 1.2.2 arises in the analysis of the game of craps; cf. Ethier (1998) for further details.

Corollary 1.2.1 is what is most commonly called Doob's stopping time theorem; see Dellacherie and Meyer (1982, Sec. 2, Ch. V).

Exercise 1.7.5 is the 0-1 law of Hewitt and Savage (1955). The argument outlined borrows its ideas from Hoeffding (1960) and arose from the fundamental work of W. Hoeffding on the so-called U-statistics; see Serfling (1980) for further details.

Exercise 3.5.2 was originally presented for $N = 1$ in Blackwell and Dubins (1962, Theorem 2) in order to rigorously prove the following intriguing principle: *"given infinite information, any two people who agree on all things possible and all things impossible, assign equal probabilities to all possible events."* The fact that this exercise also holds for $N > 1$ is a remark in Föllmer (1984a).

Exercise 3.5.2 is also sometimes called Hunt's lemma; cf. Dellacherie and Meyer (1982, Theorem 45, Ch. V).

Section 2 The literature on multiparameter martingales is truly massive. This is partly due to the fact that there are entirely different aspects to this theory. Various facets of the theory of multiparameter martingales not treated in this book, in particular those with applications to ergodic theory and optimization, can be found in (Edgar and Sucheston 1992; Cairoli and Dalang 1996).

The astute reader of this section will note that the multiparameter results of Section 2 are proved by several one-parameter projections. See Cairoli and Dalang (1996) and Sucheston (1983).

Section 3 This section, together with Section 3, can be viewed as a systematic exposition of the works of R. Cairoli and J. B. Walsh on this subject. This material, in a less expanded form and together with many other aspects of the theory of multiparameter processes, already appears in the extensive notes of J. B. Walsh; see Walsh (1986a, 1986b).

Section 3.3 is a slightly simplified version of the counterexample of L. E. Dubins and J. W. Pitman. The mentioned counterexample, in its full generality, is presented in Exercise 3.3.3. Exercise 3.3.4 is ascribed to S. Janson in Walsh (1986b).

The multiparameter martingale convergence theorems here are part of a metatheorem that was made precise in Sucheston (1983). In words, it states that we can sometimes work one parameter at a time. An alternative approach to Cairoli's convergence theorem that is closer, in spirit, to the discussion of this chapter can be found in Bakry (1981a). One can also prove such theorems by using the rather general theory of *amarts* or asymptotic martingales; see Edgar and Sucheston (1992) for a pedagogic treatment as well as an extensive bibliography of the relevant literature.

As mentioned within it, the main idea behind Supplementary Exercise 3 is borrowed from Chatterji, (1968, 1967), where the martingale convergence theorem is proved for some Banach-space-valued martingales. In the context of the original finite-dimensional convergence theorem, this proof was rediscovered later in Lamb (1973).

Supplementary Exercises 11 and 12 are new; see Khoshnevisan (2000) for yet another approach.

Exercises 3.6.2–3.6.5 form an integral part in the study of splitting fields and Markov properties of random fields; see Rozanov (1982), for instance.

There is a large body of work that extends and studies martingales indexed by partially order sets. A good general reference for those motivated by classical analysis is Walsh (1986b, Ch. 1). Those motivated by the general theory of random fields are studied in (Hürzeler 1985; Ivanova and Mertsbakh 1992; Song 1988), among many other references; see also the volume Fouque et al. (1996) for some of the recent developments. A comprehensive treatment of set-indexed martingales can be found in the recent book of Ivanoff and Merzbach (2000).

Section 4 Supplementary Exercise 7 is borrowed from Walsh (1986b); see also Mazziotto and Szpirglas (1983, 1982).

A self-contained treatment of multiparameter optional stopping theory and related results is Cairoli and Dalang (1996).

Supplementary Exercise 8 is part of the general theory of Walsh (1981); see also Cairoli and Dalang (1996).

2
Two Applications in Analysis

One of the unifying themes of this book is its systematic applications of martingales and, in particular, maximal inequalities. While most of our intended applications are in probability theory, in this chapter we sidetrack slightly and show probabilistic/martingale proofs of two fundamental analytic theorems.

The first is a theorem that states that Haar functions form an orthonormal basis for $L^p[0,1]$ for any $p > 1$. That is, there exists a prescribed collection of piecewise flat functions $(H_\ell;\ \ell \geq 1)$ such that any function $f \in L^p[0,1]$ can be written as $f = \sum_\ell \langle f, H_\ell \rangle H_\ell$, where $\langle \bullet, \bullet \rangle$ denotes the Hilbertian inner product on $L^2[0,1]$, extended to L^p, and the infinite sum converges in $L^p[0,1]$.

The second theorem of interest to us is the celebrated differentiation theorem of H. Lebesgue; it states that indefinite integrals of functions in $L^1[0,1]$ are almost everywhere differentiable.

After gathering some preliminary information, this chapter carefully states and proves the mentioned results. Moreover, it describes some multidimensional extensions of these works that will use the methods of multiparameter martingales.

1 Haar Systems

Suppose $f : [0,1]^N \to \mathbb{R}$ is Lebesgue measurable and for all $p > 0$, define the L^p norm of f by

$$\|f\|_p = \left(\int_{[0,1]^N} |f(x)|^p\, dx \right)^{\frac{1}{p}},$$

where $dx = \mathrm{Leb}(dx)$ denotes N-dimensional Lebesgue measure. A probabilistic way to think of this is as follows: Given a random variable U uniformly distributed over $[0,1]^N$, $\|f\|_p^p = \mathbb{E}\{|f(U)|^p\}$. Moreover, for any Borel set $A \subset \mathbb{R}$,
$$\mathrm{Leb}[f^{-1}(A)] = \mathbb{P}(f(U) \in A),$$
where $f^{-1}(A) = \{t \in [0,1]^N : f(t) \in A\}$.

Let $L^p[0,1]^N$ denote the collection of all such f's for which $\|f\|_p < \infty$. When $p \geq 1$, the space $L^p[0,1]^N$ is a complete vector space in the metric defined by its norm, if we identify functions that are equal almost everywhere. We can also define $L^\infty[0,1]^N$ to be the collection of all bounded measurable functions $f : [0,1]^N \to \mathbb{R}$, with suitable identification made for a.e. equivalence.

When $f \in L^\infty[0,1]^N$, define $\|f\|_\infty$ to be the **essential supremum**, ess $\sup_{t \in [0,1]^N} |f(t)|$. Recall that for any measurable $g : [0,1]^N \to \mathbb{R}$,

$$\operatorname*{ess\,sup}_{t \in [0,1]^N} g(t) = \inf\left\{r \in \mathbb{R} : \mathrm{Leb}[g^{-1}(r,\infty)] = 0\right\},$$

where $\inf \varnothing = \infty$.

What do the elements of $L^p[0,1]^N$ look like, when $p \geq 1$? To answer this, let us begin with the simplest case, $N = 1$.

1.1 The 1-Dimensional Haar System

Let
$$h_{0,0}(t) = 1, \qquad 0 \leq t \leq 1,$$
and for all $k \geq 1$, define
$$h_{k,0}(t) = \begin{cases} 2^{\frac{1}{2}(k-1)}, & \text{if } 0 \leq t \leq 2^{-k}, \\ -2^{\frac{1}{2}(k-1)}, & \text{if } 2^{-k} < t \leq 2^{-k+1}, \\ 0, & \text{otherwise.} \end{cases}$$

For any $0 \leq j \leq 2^{k-1} - 1$, define
$$h_{k,j}(t) = h_{k,0}(t - j2^{-k+1}), \qquad t \geq 0.$$

That is, for all $0 \leq j \leq 2^{k-1} - 1$, $h_{k,j}$ is $h_{k,0}$ shifted to the right by the amount $j2^{-k+1}$. The collection $\{h_{k,j}; k \geq 1, 0 \leq j \leq 2^{k-1} - 1\} \cup \{h_{0,0}\}$ is the standard **Haar system** on $[0,1]$, and the $h_{k,j}$'s are the **Haar functions** on $[0,1]$. In order to simplify the forthcoming formulæ, let us define the index set
$$\Gamma(k) = \{0, 1, \ldots, 2^{(k \vee 1) - 1} - 1\}.$$

Thus, the Haar system is compactly represented as $\{h_{k,j}; k \geq 0, j \in \Gamma(k)\}$.

Note that the Haar functions are in $L^p[0,1]$ for all $p > 0$. It is also a simple matter to check that we have the following **orthonormality property**: For all $k, m \geq 0$, $j \in \Gamma(k)$, and $n \in \Gamma(m)$,

$$\int_0^1 h_{k,j}(r) h_{m,n}(r)\,dr = \begin{cases} 1, & \text{if } (k,j) = (m,n), \\ 0, & \text{otherwise.} \end{cases}$$

When it makes sense, define the **inner product** $\langle f, g \rangle = \int_0^1 f(r)g(r)\,dr$. As is usual in Lebesgue's theory, this induces a norm $\|f\|_2 = \langle f, f \rangle^{\frac{1}{2}}$ and renders $L^2[0,1]$ a real separable Hilbert space. In particular, $\langle h_{k,j}, h_{n,m} \rangle = 0$ if $(k,j) \neq (n,m)$, and otherwise it is equal to $\|h_{k,j}\|_2^2 = 1$.

In this section we will show that the Haar system spans $L^1[0,1]$, and consequently, it also spans $L^p[0,1]$ for any $p \geq 1$. Thus, in light of the orthonormality property mentioned above, our goal is to show that the Haar system is an orthonormal basis for $L^p[0,1]$ for any $p \geq 1$. Introduce a random variable U that is uniformly distributed on $[0,1]$, and define the **dyadic filtration**

$$\mathcal{F}_k = \bigvee_{j \in \Gamma(k)} \sigma\{h_{k,j}(U)\}, \qquad k \geq 0,$$

where $\sigma\{\cdots\}$ denotes the σ-field generated by the objects in the braces. The following is an important exercise.

Exercise **1.1.1** Prove that the dyadic filtration is indeed a filtration. Also prove that $\{\operatorname{sgn} h_{k,j}(U);\ k \geq 1, j \in \Gamma(k)\}$ is a collection of mean-zero independent ± 1 valued random variables, where $\operatorname{sgn} x$ equals 1 if $x \geq 0$ and it equals -1 if $x < 0$.
(HINT: Consider first the binary representation of U.) □

Let us consider an arbitrary $f \in L^1[0,1]$. Note that for all $k \geq 1$ and all $j \in \Gamma(k)$,

$$|\langle f, h_{k,j} \rangle| \leq \|h_{k,j}\|_\infty \cdot \|f\|_1 = 2^{\frac{1}{2}(k-1)} \|f\|_1 < \infty. \tag{1}$$

Therefore, the following is well-defined:

$$M_n(f) = \sum_{k=0}^n \sum_{j \in \Gamma(k)} h_{k,j}(U) \langle f, h_{k,j} \rangle.$$

(What about the $k = 0$ term?)

Exercise **1.1.2** Verify that whenever $q > k$ are both integers, $M_k(\mathbf{1}_{[0,2^{-q}]}) = 2^{-q+k} \mathbf{1}_{[0,2^{-q}]}(U)$, a.s. On the other hand, when $q \leq k$, show that $M_k(\mathbf{1}_{[0,2^{-q}]}) = \mathbf{1}_{[0,2^{-q}]}(U)$, a.s.
(HINT: When $q > k$, check that $M_k(\mathbf{1}_{[0,2^{-q}]}) = 2^{-q} \beta_k$, where

$$\beta_k = 1 + \sum_{j=1}^k 2^{j-1} \{\mathbf{1}_{[0,2^{-j}]}(U) - \mathbf{1}_{[2^{-j}, 2^{-j+1}]}(U)\},$$

and use induction.) □

The key observation of this subsection is the following:

Proposition 1.1.1 *If $f \in L^1[0,1]$, then $(M_n(f); n \geq 0)$ is a martingale. In fact, for all $n \geq 0$,*

$$M_n(f) = \mathbb{E}[f(U) \,|\, \mathcal{F}_n], \quad \text{a.s.}$$

Proof By Exercise 1.1.1, the $h_{k,j}$'s are i.i.d. mean-zero random variables. The martingale property is a simple consequence of this. What we are after is the stated representation of $M_n(f)$ as a conditional expectation; this also implies the asserted martingale property of M.

Since $\mathcal{F}_0$ is the trivial σ-field, the assertion of the theorem holds for $n = 0$. We will show that it does for $n \geq 1$ as well.

Note that $f \mapsto M_n(f)$ and $f \mapsto \mathbb{E}[f(U) \,|\, \mathcal{F}_n]$ are both linear maps. First, we prove the result when f has the following form:

$$f(t) = \mathbf{1}_{[0, 2^{-q}]}(t), \quad t \in [0,1],$$

where $q \geq 1$ is a fixed integer. By directly computing, we see that for any two integers $k \geq q$ and $j \in \Gamma(k)$, the following holds almost surely:

$$\mathbb{E}[f(U) \,|\, \mathcal{F}_k]\mathbf{1}_{[j2^{-k},(j+1)2^{-k}]}(U)$$
$$= \mathbb{P}(U \leq 2^{-q} \,|\, j2^{-k} \leq U \leq (j+1)2^{-k})\mathbf{1}_{[j2^{-k},(j+1)2^{-k}]}(U)$$
$$= \mathbf{1}_{[j2^{-k},(j+1)2^{-k}]}(U) \times \begin{cases} 0, & \text{if } j \geq 2^{-q+k}, \\ 1, & \text{if } 0 \leq j \leq 2^{-q+k} - 1. \end{cases} \quad (2)$$

On the other hand, if $k < q$, then for all integers $j \in \Gamma(k)$,

$$\mathbb{E}[f(U) \,|\, \mathcal{F}_k]\mathbf{1}_{[j2^{-k},(j+1)2^{-k}]}(U) = \begin{cases} 2^{-q+k}\mathbf{1}_{[0, 2^{-k}]}(U), & \text{if } j = 0, \\ 0 & \text{if } j \geq 1. \end{cases} \quad (3)$$

It is easy to see directly that when $k \geq q$, $M_k(f)\mathbf{1}_{[j2^{-k},(j+1)2^{-k}]}(U)$ equals the right-hand side of (2), and when $k < q$, it equals the right-hand side of (3); see Exercise 1.1.2 above. This verifies the result when f is of the form, $f = \mathbf{1}_{[0,2^{-q}[}$ for an integer $q \geq 0$. The same argument works to establish the result when f is of the form $f = \mathbf{1}_{[j2^{-q},(j+1)2^{-q}[}$, for positive integers j and q. The remainder of our proof is an exercise in measure theory. By the asserted linearity, the result holds when f is simple and dyadic, i.e., when it is of the form $f = \sum_{j=1}^m \alpha_j \mathbf{1}_{[j2^{-q},(j+1)2^{-q}[}$, where m and q are positive integers and $\alpha_1, \ldots, \alpha_m$ are real-valued. Any continuous function can be sandwiched between two simple dyadic functions. Therefore, the result holds for all continuous f. Since continuous functions are dense in $L^1[0,1]$, the proposition follows. □

It is an immediate consequence of Proposition 1.1.1 that if $f \in L^1[0,1]$, then $(M_n(f); n \geq 1)$ is a uniformly integrable martingale. Therefore, by the (one-parameter) martingale convergence theorem (Theorem 1.7.1, Chapter 1), it converges almost surely and in $L^1(\mathbb{P})$. In fact,

$$\lim_{n\to\infty} M_n(f) = \mathbb{E}[f(U) \mid \vee_n \mathcal{F}_n] = f(U),$$

almost surely, and $\lim_n \mathbb{E}\{|M_n(f) - f(U)|\} = 0$. In analytical terms, we have proved the following result:

Theorem 1.1.1 *If $f \in L^1[0,1]$, the following limit exists for Lebesgue almost all $t \in [0,1]$ and in $L^1[0,1]$:*

$$f(t) = \lim_{n\to\infty} \sum_{k=0}^{n} \sum_{j\in\Gamma(k)} h_{k,j}(t) \langle f, h_{k,j} \rangle.$$

That is, any function in $L^1[0,1]$ can be thought of as an infinite linear combination of Haar functions. Thus, we have shown that the Haar system forms a basis for $L^1[0,1]$. An analogous result holds in $L^p[0,1]$ when $p \in (1,\infty)$.

***Exercise* 1.1.3** Prove that whenever $f \in L^p[0,1]$ for $p \in [1,\infty)$, the convergence in Theorem 1.1.1 holds almost everywhere and in $L^p[0,1]$. □

1.2 The N-Dimensional Haar System

The Haar system on $[0,1]$ extends nicely to form a basis for $L^p[0,1]^N$ ($p \geq 1$). Recall that for any integer $k \geq 0$,

$$\Gamma(k) = \begin{cases} \{0, \ldots, 2^{k-1} - 1\}, & \text{if } k \geq 1, \\ \{0\}, & \text{if } k = 0, \end{cases}$$

and $\{h_{k,j}; k \geq 0, j \in \Gamma(k)\}$ is the Haar system on $[0,1]$ described in the previous section. Let us begin with the N-dimensional analogue of $\Gamma(k)$. For all $k \in \mathbb{N}_0^N$, define

$$\Gamma^N(k) = \Gamma(k^{(1)}) \times \cdots \times \Gamma(k^{(N)}).$$

Suppose $k \in \mathbb{N}_0^N$ and $j \in \Gamma^N(k)$. For all $t \in [0,1]^N$, we can define

$$h_{k,j}^N(t) = \prod_{\ell=1}^{N} h_{k^{(\ell)}, j^{(\ell)}}(t^{(\ell)}), \qquad t \in [0,1]^N.$$

The **Haar system on** $[0,1]^N$ is the collection $\{h_{k,j}^N; k \in \mathbb{N}_0^N, j \in \Gamma^N(k)\}$. Note that h^1 is our old Haar system on $[0,1]$ from Section 1.1.

Next, take a random vector V that is uniformly distributed on $[0,1]^N$, and define the **dyadic filtration**

$$\mathcal{F}_t = \bigvee_{0 \preccurlyeq k \preccurlyeq t} \sigma(h_{k,j}^N(V);\ j \in \Gamma^N(k)), \qquad t \in \mathbb{N}_0^N.$$

Exercise **1.2.1** Show that the above dyadic filtration is indeed an N-parameter commuting filtration. $\square$

For all $f \in L^1[0,1]^N$ and $t \in \mathbb{N}_0^N$, we define $M_t(f)$ by

$$M_t(f) = \sum_{0 \preccurlyeq k \preccurlyeq t} \sum_{j \in \Gamma^N(k)} h_{k,j}^N(V)\langle f, h_{k,j}^N \rangle,$$

where $\langle f, g \rangle = \int_{[0,1]^N} f(t)g(t)\,dt$, whenever it makes sense. Note that $M_t(f)$ is well-defined as long as $f \in L^1[0,1]^N$. Indeed, using equation (1) of Section 1.1, for all $k \in \mathbb{N}^N$ and $j \in \Gamma^N(k)$,

$$|\langle f, h_{k,j}^N \rangle| \leq 2^{\frac{1}{2}(\sum_{\ell=1}^N k^{(\ell)} - N)} \|f\|_1.$$

Our multidimensional analogue of Proposition 1.1.1 is the following.

Proposition 1.2.1 $\mathcal{F}$ *is a commuting filtration. Moreover, if* $f \in L^1[0,1]^N$, *then*

$$M_t(f) = \mathbb{E}[f(V) \mid \mathcal{F}_t], \qquad t \in \mathbb{N}_0^N.$$

Proof That $\mathcal{F}$ is a filtration is immediate. Clearly, the coordinates of V are i.i.d. random variables, each uniformly picked from $[0,1]$. Therefore, by Exercise 1.1.1, $\{h_{k,j}^N(V);\ k \in \mathbb{N}_0^N, j \in \Gamma^N(k)\}$ is a collection of mean-zero independent random variables. Exercise 1.2.1 shows that $\mathcal{F}$ is commuting. In fact, we can be more careful and write $\mathcal{F}$ in terms of its marginal filtrations: For all $t \in [0,1]^N$,

$$\mathcal{F}_t = \bigvee_{\ell=1}^N \mathcal{F}_{t^{(\ell)}}^\ell, \qquad t \in \mathbb{N}_0^N, \tag{1}$$

where the marginal filtrations $\mathcal{F}^1, \ldots, \mathcal{F}^N$ are independent one-dimensional dyadic filtrations. It remains to demonstrate the representation of $M_t(f)$ as the mentioned conditional expectation. By appealing to arguments of measure theory, it suffices to prove the result when f is of the form $f(t) = \prod_{\ell=1}^N f_\ell(t^{(\ell)})$, $t \in [0,1]^N$, where $f_1, \ldots, f_N$ are bounded, measurable functions from $[0,1]$ to $\mathbb{R}$. For f of this form, equation (1) implies that almost surely, $\mathbb{E}[f(V) \mid \mathcal{F}_t] = \prod_{\ell=1}^N \mathbb{E}[f_\ell(V^{(\ell)}) \mid \mathcal{F}_{t^{(\ell)}}^\ell]$. By Proposition 1.1.1,

$$\mathbb{E}[f(V) \mid \mathcal{F}_t] = \prod_{\ell=1}^N M_{t^{(\ell)}}(f_\ell), \qquad \text{a.s.,}$$

where $(M_k(f_\ell);\ k \geq 0)$ is the one-parameter process defined in Section 1.1. (Note: The notation is being slightly abused. When $t \in \mathbb{N}_0^N$ and $f : [0,1]^N \to \mathbb{R}$, $M_t(f)$ is the stochastic process of this section. When $t \in \mathbb{N}_0^1$ and $f : [0,1] \to \mathbb{R}$, $M_t(f)$ is the one defined in Section 1.1.) To finish, let us observe the following simple calculation for f of the above form:

$$M_t(f) = \prod_{\ell=1}^{N} M_{t^{(\ell)}}(f_\ell).$$

This follows from the form of the Haar system together with the observation that when $f = \prod_\ell f_\ell$, $\langle f, h_{k,j}^N \rangle = \prod_{\ell=1}^{N} \langle f_\ell, h_{k^{(\ell)}, j^{(\ell)}} \rangle$. A final word of caution: The first inner product above is an integral over $[0,1]^N$, while the second one is over $[0,1]$. This completes our proof. □

***Exercise* 1.2.2** If $f \in L^1[0,1]^N$, prove that $(M_{t,\ldots,t}(f);\ t \in \mathbb{N}_0)$ is a uniformly integrable one-parameter martingale with respect to the one-parameter filtration $(\mathcal{F}_{t,\ldots,t};\ t \in \mathbb{N}_0^1)$. □

We can argue as in the previous section and use Exercise 1.2.2 to deduce the following:

Theorem 1.2.1 *Suppose $f \in L^1[0,1]^N$. Then,*

$$f(t) = \lim_{n \to \infty} \sum_{0 \preccurlyeq k \preccurlyeq (n,\ldots,n)} \sum_{j \in \Gamma^N(k)} h_{k,j}^N(t) \langle f, h_{k,j}^N \rangle,$$

for Lebesgue almost all $t \in [0,1]^N$ and in $L^1[0,1]^N$.

***Exercise* 1.2.3** Complete the above proof of Theorem 1.2.1. □

Note that we have used only the one-parameter theory to obtain the above. Suppose next that f is $L \ln_+^{N-1} L$-bounded. We can use the multi-parameter martingale convergence theorem (Corollary 2.8.1 of Chapter 1) and its $L^p[0,1]^N$ extensions (Exercise 1.1.2) to obtain the following sharper form of Theorem 1.2.1.

Theorem 1.2.2 *Suppose $f : [0,1]^N \to \mathbb{R}$ satisfies*

$$\int_{[0,1]^N} |f(t)| \cdot [\ln_+ |f(t)|]^{N-1}\, dt < \infty. \tag{2}$$

Then,

$$f(t) = \lim_{s \to \infty} \sum_{0 \preccurlyeq k \preccurlyeq s} \sum_{j \in \Gamma^N(k)} h_{k,j}^N(t) \langle f, h_{k,j}^N \rangle,$$

for Lebesgue almost all $t \in [0,1]^N$ and in $L^1[0,1]^N$. Moreover, if $f \in L^p[0,1]^N$ for some $p \in (1,\infty)$, then convergence also holds in $L^p[0,1]^N$.

54   2. Two Applications in Analysis

***Exercise* 1.2.4** Complete the above proof of Theorem 1.2.2. □

Theorem 1.2.1 shows that the Haar system forms a basis for $L^1[0,1]^N$. On the other hand, Theorem 1.2.2 implies that for $p > 1$, $L^p[0,1]^N$ is spanned by the Haar system, in a uniform fashion.

2 Differentiation

We will discuss two differentiation theorems of classical analysis:

1. Lebesgue's differentiation theorem; and
2. the Jessen–Zygmund–Marcinkiewicz theorem.

2.1 Lebesgue's Differentiation Theorem

Given a continuous function $f : [0,1]^N \to \mathbb{R}$, and any $\Delta, t_0 \in \mathbb{R}^N_+$,

$$\left| \frac{1}{\prod_{j=1}^N \Delta^{(j)}} \int_{[t_0, t_0+\Delta]} f(t)\, dt - f(t_0) \right| \leq \sup_{t \in [t_0, t_0+\Delta]} |f(t) - f(t_0)|.$$

Therefore, for any $t_0 \in \,]0,1[^N$,

$$\lim_{\substack{\Delta \in (0,\infty)^N_+ : \\ \Delta \to 0}} \frac{1}{\prod_{j=1}^N \Delta^{(j)}} \int_{[t_0, t_0+\Delta]} f(t)\, dt = f(t_0). \tag{1}$$

There is need for a technical aside here: As we have stated things, $f(t)$ may not be defined for all $t \in [t_0, t_0 + \Delta]$. To be completely careful, we will extend the definition of any $f : [0,1]^N \to \mathbb{R}$ by defining $f(t)$ to be 0 where $t \notin [0,1]^N$. It is instructive to check that (1) holds uniformly over all choices of $t_0 \in \,]0,1[^N$.

Equation (1) says that—at least for continuous functions—derivatives are anti-integrals. Lebesgue's differentiation theorem shows that, as long as the integrals are well-defined, this is always almost the case.

In its simplest form, Lebesgue's differentiation theorem states the following:

Theorem 2.1.1 (Lebesgue's Differentiation Theorem) *For any function $f \in L^1[0,1]^N$, and for almost all $t \in [0,1]^N$,*

$$\lim_{\varepsilon \to 0+} \varepsilon^{-N} \int_{[t, t+(\varepsilon, \ldots, \varepsilon)]} f(s)\, ds = f(t).$$

It is important to note that the approximation in Lebesgue's differentiation theorem is not as strong as the one in (1). The latter allows for

N-dimensional rectangles whose sides collapse possibly at different rates, while the former is a statement about N-dimensional hypercubes; in particular, the sides are all of equal length.

The key to the above theorem is a **maximal inequality**. To describe it, define the **maximal operator** M by

$$Mf(t) = \sup_{\varepsilon > 0} \varepsilon^{-N} \int_{[t, t+(\varepsilon,\ldots,\varepsilon)]} |f(s)|\, ds, \qquad t \in [0,1]^N,$$

where $f(t) = 0$ if $t \notin [0,1]^N$, for simplicity. The required maximal inequality is the following:

Proposition 2.1.1 *Suppose $f \in L^1[0,1]^N$. Then for all $\lambda > 0$,*

$$\mathrm{Leb}\{t \in [0,1]^N : Mf(t) \geq \lambda\} \leq \frac{4^N}{\lambda} \|f\|_1.$$

Proof Applying Proposition 1.2.1 to $|f|$ and noting the absolute values, we see that for U uniformly distributed on $[0,1]^N$ and $\mathcal{F}$ the corresponding dyadic filtration,

$$M_t(f) = \mathbb{E}\bigl\{|f(U)| \,\big|\, \mathcal{F}_t\bigr\}, \qquad t \in \mathbb{N}_0^N.$$

See Section 1.2 for definitions. In particular, $M(f) = (M_t(f); t \in \mathbb{N}_0^N)$ is an N-parameter martingale with respect to $\mathcal{F}$. For all $k \in \mathbb{N}^N$ and $j \in \Gamma^N(k)$, define $I_{k,j}$ to be the interior of the (necessarily closed) support of the Haar function $h_{k,j}^N$. That is,

$$I_{k,j} = \prod_{\ell=1}^N [j^{(\ell)} 2^{-k^{(\ell)}}, (j^{(\ell)}+1) 2^{-k^{(\ell)}}[, \qquad k \in \mathbb{N}^N,\ j \in \Gamma^N(k). \qquad (2)$$

The following is an instructive exercise in conditional expectations: For all $t, k \in \mathbb{N}^N$ with $k \preccurlyeq t$ and for all $j \in \Gamma^N(k)$,

$$\mathbb{E}\bigl[|f(U)| \,\big|\, \mathcal{F}_t\bigr] \mathbf{1}_{(U \in I_{k,j})} = \frac{1}{\mathrm{Leb}(I_{k,j})} \int_{I_{k,j}} |f(s)|\, ds \cdot \mathbf{1}_{(U \in I_{k,j})}, \qquad \text{a.s.}$$

Define $p: \mathbb{R} \to \mathbb{R}^N$ by $p(x) = (x, \ldots, x)$, and let

$$X_k(t) = 2^{Nk} \sum_{j \in \Gamma^N(k)} \int_{I_{p(k),j}} |f(s)|\, ds \cdot \mathbf{1}_{I_{p(k),j}}(t), \qquad k \geq 1,\ t \in \mathbb{R}_+^N.$$

Since $\mathrm{Leb}(I_{p(k),j}) = 2^{-Nk}$ and the $I_{k,j}$'s are disjoint, we have shown that for all integers $k \geq 1$, $M_{p(k)}(f) = X_k(U)$, a.s. In particular, we have shown that $(X_k(U);\ k \geq 1)$ is a nonnegative martingale with respect to the filtration

$\mathcal{F} \circ p = (\mathcal{F}_{p(k)}; k \geq 1)$. Therefore, by the weak (1,1) inequality (Theorem 1.3.1, Chapter 1), for all $\lambda > 0$,

$$\mathbb{P}\Big(\sup_{k \geq 1} X_k(U) \geq \lambda\Big) \leq \frac{1}{\lambda} \sup_{k \geq 1} \mathbb{E}[X_k(U)]$$

$$= \frac{1}{\lambda} \sup_{k \geq 1} \Big[2^{Nk} \sum_{j \in \Gamma^N(k)} \int_{I_{p(k),j}} |f(s)|\, ds \cdot \mathbb{P}(U \in I_{p(k),j}) \Big]$$

$$= \frac{\|f\|_1}{\lambda}, \tag{3}$$

since the closure of the union over the j's of the $I_{p(k),j}$'s is all of $[0,1]^N$. We finish our proof by using an interpolation argument that relates (3) to the maximal function. For all integers $k \geq 2$ and every $t \in]0,1[^N$, we can uniquely find an integer $j \in \Gamma^N(k)$ such that $t \in I_{p(k),j}$. Therefore, there exists a unique j' such that $[t, t+p(2^{-k})]$ is a subset of $I_{p(k-1),j'}$. (This is why we required $k \geq 2$ and not 1.) In particular,

$$2^{Nk} \int_{[t,t+p(2^{-k})]} |f(s)|\, ds \leq 2^N X_{k-1}(t).$$

Now take $\varepsilon \in]0, \tfrac{1}{4}[$. We can find an integer $k \geq 2$ such that $2^{-k-1} \leq \varepsilon \leq 2^{-k}$. Consequently,

$$\varepsilon^{-N} \int_{[t,t+p(\varepsilon)]} |f(s)|\, ds \leq 2^{N(k+1)} \int_{[t,t+p(2^{-k})]} |f(s)|\, ds \leq 4^N X_{k-1}(t),$$

thus establishing the pointwise inequality

$$\sup_{0 < \varepsilon < \tfrac{1}{4}} \varepsilon^{-N} \int_{[t,t+p(\varepsilon)]} |f(s)|\, ds \leq 4^N \sup_{k \geq 1} X_k(t), \qquad t \in]0,1[^N.$$

On the other hand, it is easy to see that

$$\sup_{\tfrac{1}{4} \leq \varepsilon} \varepsilon^{-N} \int_{[t,t+p(\varepsilon)]} |f(s)|\, ds \leq 4^N \|f\|_1.$$

Thus, whenever $Mf(t) \geq \lambda$, $\sup_{k \geq 1} X_k(t) \geq 4^{-N}\lambda$. In other words, for all $\lambda \geq 4^N \|f\|_1$,

$$\mathrm{Leb}\{t \in [0,1]^N : Mf(t) \geq \lambda\} \leq \mathrm{Leb}\Big\{t \in [0,1]^N : \sup_{k \geq 1} X_k(t) \geq 4^{-N}\lambda\Big\}$$

$$= \mathbb{P}\Big(\sup_{k \geq 1} X_k(U) \geq 4^{-N}\lambda\Big)$$

$$\leq \frac{4^N}{\lambda} \|f\|_1,$$

by equation (3). This completes our proof. $\square$

Exercise 2.1.1 Prove that for all $p > 1$, there exists a finite, positive constant C such that whenever $f \in L^p[0,1]^N$, $\|Mf\|_p \leq C\|f\|_p$. This is called the **Hardy–Littlewood maximal inequality.** □

We are ready to verify Theorem 2.1.1.

Proof of Theorem 2.1.1 Throughout this proof, for any $g \in L^1[0,1]^N$, and for all $\varepsilon > 0$, we write

$$g_\varepsilon(t) = \varepsilon^{-N} \int_{[t,t+(\varepsilon,\ldots,\varepsilon)]} g(s)\,ds, \qquad t \in \,]0,1[^N,$$

and extend g_ε to a function on $[0,1]^N$ by continuity. It should be recognized that whenever g is continuous, then $\lim_{\varepsilon \to 0} g_\varepsilon = g$, uniformly on compacts.

If $f, g \in L^1[0,1]^N$, an application of the triangle inequality yields

$$\|f_\varepsilon - g_\varepsilon\|_1 \leq \|f - g\|_1.$$

On the other hand, for any $f \in L^1[0,1]^N$, we can always find continuous g^n such that $\lim_{n \to \infty} g^n = f$ in $L^1[0,1]^N$. Thus, we can use the previous display to establish

$$\|f_\varepsilon - f\|_1 \leq \|f_\varepsilon - g^n_\varepsilon\|_1 + \|g^n_\varepsilon - g^n\|_1 + \|f - g^n\|_1$$
$$\leq 2\|f - g^n\|_1 + \|g^n_\varepsilon - g^n\|_1.$$

Now, first let $\varepsilon \to 0$ to deduce $\limsup_{\varepsilon \to 0} \|f_\varepsilon - f\|_1 \leq 2\|f - g^n\|_1$. Then, let $n \to \infty$ to see that $f_\varepsilon \to f$ in $L^1[0,1]^N$. It remains to prove almost everywhere convergence.

Recall, once more, that if g is continuous, then $\lim_{\varepsilon \to 0} g_\varepsilon = g$, pointwise. In particular, for *any* continuous function $g : [0,1]^N \to \mathbb{R}$,

$$0 \leq \limsup_{\varepsilon \to 0} f_\varepsilon - \liminf_{\varepsilon \to 0} f_\varepsilon \leq 2M(|f - g|),$$

pointwise. Consequently, by Proposition 1.1.1, for any $\lambda > 0$,

$$\text{Leb}\left\{t \in [0,1]^N : \limsup_{\varepsilon \to 0} f_\varepsilon(t) - \liminf_{\varepsilon \to 0} f(t) \geq \lambda\right\} \leq \frac{2 \cdot 4^N}{\lambda} \|f - g\|_1.$$

Since this holds for an arbitrary continuous g, the above is, in fact, equal to 0 by density. Let $\lambda \downarrow 0$ along a rational sequence to conclude that $\lim_{\varepsilon \to 0} f_\varepsilon$ exists almost everywhere. On the other hand, by the already-proved L^1-convergence, there exists a subsequence $\varepsilon' \to 0$ such that $f_{\varepsilon'} \to f$, almost everywhere. Hence, $\lim_{\varepsilon \to 0} f_\varepsilon = f$ almost everywhere, as desired. □

Exercise 2.1.2 Prove an $L^p[0,1]^N$ version of Theorem 2.1.1; cf. Exercise 1.1.2 for the 1-parameter version of this. □

Exercise 2.1.3 Verify that Lebesgue's differentiation theorem (Theorem 2.1.1) remains true if we replace $[0,1]^N$ by $\mathbb{R}^N$.

□

2.2 A Uniform Differentiation Theorem

As we saw from its proof, Theorem 2.1.1 is actually a one-parameter, or one-dimensional, result. Roughly speaking, it says that if $f \in L^1[0,1]^N$, then its "distribution function" is Lebesgue almost everywhere differentiable. There is a much stronger result, due to B. Jessen, J. Marcinkiewicz, and A. Zygmund, that holds under the stronger assumption that f is in $L(\ln_+ L)^{N-1}$. To simplify the notation, for the remainder of this section define
$$\Psi(x) = x(\ln_+ x)^{N-1}, \qquad x > 0.$$

Theorem 2.2.1 *Suppose $f : [0,1]^N \to \mathbb{R}$ is a measurable function that satisfies the integrability condition $\Psi \circ |f| \in L^1[0,1]^N$. Then, for Lebesgue almost all $t \in [0,1]^N$,*
$$\lim_{\substack{\Delta \in]0,1[^N: \\ \Delta \to 0}} \frac{1}{\prod_{j=1}^N \Delta^{(j)}} \int_{[t,t+\Delta]} f(s)\,ds = f(t).$$

The main point of this theorem is that the sides of the rectangles $[t, t+\Delta]$ need not go to zero at the same rate, as long as
$$\int_{[0,1]^N} |f(x)| \cdot \{\ln_+ |f(x)|\}^{N-1}\,dx < \infty.$$

It is also worth mentioning that in the above, the domain of definition of the function f is implicitly extended to all of $\mathbb{R}^N$ by defining $f(t) \equiv 0$ if $t \notin [0,1]^N$.

Just like its one-parameter counterpart (Theorem 2.1.1), Theorem 2.2.1 is a consequence of a maximal inequality. Define the multiparameter **maximal operator** $\widetilde{M}$ by
$$\widetilde{M}f(t) = \sup_{\Delta \in]0,\frac{1}{4}[^N} \frac{1}{\prod_{j=1}^N \Delta^{(j)}} \int_{[t,t+\Delta]} |f(s)|\,ds,$$

where $f(t) = 0$ if $t \notin [0,1]^N$. The N-parameter analogue of Proposition 1.2.1 is the following:

Proposition 2.2.1 *Suppose $f : [0,1]^N \to \mathbb{R}_+$ is a measurable function such that $\Psi \circ f \in L^1[0,1]^N$. Then, for all $\lambda > 0$,*
$$\text{Leb}\{t \in [0,1]^N : \widetilde{M}f(t) \geq \lambda\} \leq \frac{2^N}{\lambda}\left(\frac{e}{e-1}\right)^{N-1}\{(N-1) + \|\Psi \circ f\|_1\}.$$

Proof First, note that $f \geq 0$, by definition. Having made this remark, we now adapt the presented proof of Proposition 2.1.1 to the present multiparameter setting. Recall equation (2), Section 2.1, for the definition of $I_{k,j}$, $k \in \mathbb{N}^N$, $j \in \Gamma^N(k)$ and define

$$\widetilde{X}_k(t) = 2^{\sum_{\ell=1}^N k^{(\ell)}} \cdot \sum_{j \in \Gamma^N(k)} \int_{I_{k,j}} f(s) ds \cdot \mathbf{1}_{I_{k,j}}(t), \qquad t \in \mathbb{R}_+^N, \ k \in \mathbb{N}^N.$$

For any $k \in \mathbb{N}^N$ and $j \in \Gamma^N(k)$, $\mathrm{Leb}(I_{k,j}) = 2^{-\sum_{\ell=1}^N k^{(\ell)}}$. Thus, the argument of the presented proof of Proposition 2.1.1 shows that $\widetilde{X} = (\widetilde{X}_k(U); \ k \in \mathbb{N}^N)$ is a nonnegative N-parameter martingale with respect to the dyadic filtration $\mathcal{F}$. Moreover, by Proposition 1.2.1, $\mathcal{F}$ is commuting. Consequently (Corollary 3.5.1 of Chapter 1), $\widetilde{X}$ is an orthomartingale with respect to the marginal filtrations of $\mathcal{F}$. By Theorem 2.5.1 of Chapter 1, for all $\lambda > 0$,

$$\mathbb{P}\left(\sup_{k \in \mathbb{N}^N} \widetilde{X}_k(U) \geq \lambda\right) \leq \frac{1}{\lambda}\left(\frac{e}{e-1}\right)^{N-1}\left\{(N-1) + \mathbb{E}\left[\Psi(\widetilde{X}_k(U))\right]\right\}.$$

On the other hand, since Ψ is convex nondecreasing, by Jensen's inequality,

$$\mathbb{E}\left[\Psi(\widetilde{X}_k(U))\right] \leq \sum_{j \in \Gamma^N(k)} \Psi\left(\int_{I_{k,j}} f(s)\, ds\right) \leq \|\Psi \circ f\|_1.$$

(Why?) Thus, we have shown that for all $\lambda > 0$,

$$\mathbb{P}\left(\sup_{k \in \mathbb{N}^N} \widetilde{X}_k(U) \geq \lambda\right) \leq \frac{1}{\lambda}\left(\frac{e}{e-1}\right)^{N-1}\{(N-1) + \|\Psi \circ f\|_1\}. \tag{1}$$

Next, we will relate this to the maximal function.

Recall the function $p: \mathbb{R} \to \mathbb{R}^N$ given by $p(t) = (t, \ldots, t)$. For all $k \in \mathbb{N}^N$, define α_k to be the N-dimensional vector whose ℓth coordinate is $2^{-k^{(\ell)}}$. Next, for any $\Delta \in \,]0, \tfrac{1}{4}[^N$, find $k \succcurlyeq p(2)$ such that $\alpha_{k+p(1)} \preccurlyeq \Delta \preccurlyeq \alpha_k$. By monotonicity, for all $t \in \,]0, 1[^N$,

$$\frac{1}{\prod_{j=1}^N \Delta^{(j)}} \int_{[t,t+\Delta]} f(s)\, ds \leq \frac{1}{\prod_{j=1}^N \alpha_{k+p(1)}^{(j)}} \int_{[t,t+\alpha_k]} f(s)\, ds = 2^N \widetilde{X}_{k-p(1)}(t).$$

The result follows from (1). $\square$

Exercise **2.2.1** For all $p \in (1, \infty)$, prove the existence of a positive, finite constant C that depends on p such that for all $f \in L^p[0,1]^N$, $\|\widetilde{M}f\|_p \leq C\|f\|_p$. $\square$

2. Two Applications in Analysis

We are ready for the following.

Proof of Theorem 2.2.1 Throughout, for all integrable functions $g : [0,1]^N \to \mathbb{R}$, and for all $\Delta \in \,]0,\infty[^N$, we write

$$g_\Delta(t) = \frac{1}{\prod_{j=1}^N \Delta^{(j)}} \int_{[t,t+\Delta]} g(s)\,ds,$$

$$\widetilde{\Theta} g(t) = \limsup_{\Delta \to 0} g_\Delta(t) - \liminf_{\Delta \to 0} g_\Delta(t),$$

for all $t \in \,]0,1[^N$. We can extend the domain of g_Δ to all of $[0,1]^N$ by continuity.

Given a function f as in the statement of the theorem, we can find continuous functions $g^n : [0,1]^N \to \mathbb{R}$ such that

$$\lim_{n \to \infty} \|g^n - f\|_1 = 0.$$

On the other hand, $\Psi(|x-y|) \leq |\Psi(x) - \Psi(y)|$; cf. equation (3), Section 2.8, Chapter 1. Thus, by uniform integrability, for any fixed $\delta > 0$,

$$\lim_{n \to \infty} \left\| \Psi \circ \left(\frac{|g^n - f|}{\delta} \right) \right\|_1 = 0.$$

Now, for each fixed n,

$$\widetilde{\Theta} f(t) \leq \widetilde{\Theta} g^n(t) + 2\widetilde{M}(|f - g^n|)(t) = 2\delta \widetilde{M}\left(\frac{|f-g^n|}{\delta}\right)(t), \qquad t \in \,]0,1[^N.$$

The latter equality holds from continuity of g^n. Consequently, we can apply Proposition 2.2.1 to deduce that for any $\lambda > 0$,

$$\mathrm{Leb}\{t \in [0,1]^N : \widetilde{\Theta} f(t) \geq \lambda\}$$
$$\leq \frac{2^{N+1}\delta}{\lambda} \left(\frac{e}{e-1}\right)^{N-1} \left\{ (N-1) + \left\| \Psi \circ \left(\frac{|f-g^n|}{\delta} \right) \right\|_1 \right\}.$$

We can let $n \to \infty$, $\delta \to 0^+$, and $\lambda \to 0^+$, in this order, to deduce that $\widetilde{\Theta} f = 0$, almost everywhere. Equivalently, $\lim_{\Delta \to 0} f_\Delta$ exists almost everywhere. To show that this limit equals f almost everywhere, we can simply apply Lebesgue's differentiation theorem (Theorem 2.1.1, why?). □

Exercise 2.2.2 Prove an $L^p[0,1]^N$ version of Theorem 2.2.1. □

Exercise 2.2.3 Prove that Theorem 2.2.1 remains true if we replace $[0,1]^N$ by $\mathbb{R}^N$ everywhere. □

3 Supplementary Exercises

1. Given a sequence $(c_{k,j};\ k \geq 1, j \in \Gamma(k))$ of real numbers, define the function $F_n(t) = \sum_{k=0}^n \sum_{j \in \Gamma(k)} c_{k,j} h_{k,j}(t)$, $(t \in [0,1])$. Supposing that $\sum_{k=0}^\infty 2^{\frac{1}{2}(k-1)} \max_{j \in \Gamma(k)} |c_{k,j}| < +\infty$, prove that as $n \to \infty$, F_n converges uniformly and in $L^p[0,1]$ for all $p \geq 1$.

2. Show that Theorem 1.1.1 on Haar function expansions implies Lebesgue's differentiation theorem, Theorem 2.1.1.

3. Suppose $(I_t;\ t \geq 0)$ is a one-parameter monotonic family of rectangles in $[0,1]^N$ with sides parallel to the axes. That is, (a) for any $t \geq 0$, I_t is of the form $[a^{(1)}, b^{(1)}] \times \cdots \times [a^{(N)}, b^{(N)}] \subset [0,1]^N$; and (b) if $s \leq t$, $I_s \subset I_t$. For all functions $f : [0,1]^N \to \mathbb{R}$, define the maximal function $\mathcal{M}f$ by

$$\mathcal{M}f(x) = \sup_{t \geq 0} \frac{1}{\mathrm{Leb}(I_t)} \int_{I_t} |f(x-y)|\, dy, \qquad x \in [0,1]^N,$$

where $f(z) = 0$, for all $z \notin [0,1]^N$. Show that whenever $f \in L^p[0,1]^N$ for some $p > 1$, then $\mathcal{M}f \in L^p[0,1]^N$. In fact, show that for all $p > 1$, there exists a finite constant $C_p > 0$ such that for all $f \in L^p[0,1]^N$, $\|\mathcal{M}f\|_p \leq C_p \|f\|_p$. This is from Zygmund (1988, Ch. XVII).

4. For all $\varepsilon > 0$ and $t \in \mathbb{R}^N$, let $\mathbf{R}_\varepsilon(t)$ denote the collection of all cubes of side ε that contain the point t and whose sides are parallel to the axes, i.e., $R \in \mathbf{R}_\varepsilon(t)$ if and only if $R = [a, a+(\varepsilon, \ldots, \varepsilon)]$ for some $a \in \mathbb{R}^N$ such that $a \preccurlyeq t \preccurlyeq a+(\varepsilon, \ldots, \varepsilon)$. We can define the enhanced maximal operator $\mathfrak{M}$ as follows: For all $f \subset L^1(\mathbb{R}^N)$,

$$\mathfrak{M}f(t) = \sup_{\varepsilon > 0}\ \sup_{R \in \mathbf{R}_\varepsilon(t)} \varepsilon^{-N} \int_R |f(s)|\, ds.$$

(a) Verify that for all $t \in \mathbb{R}^N$, $2^{-N} \mathfrak{M}f\!\left(t - (\varepsilon, \ldots, \varepsilon)\right) \leq Mf(t) \leq \mathfrak{M}f(t)$.

(b) Demonstrate the following extension of Proposition 2.1.1: For all $f \in L^1(\mathbb{R}^N)$ and all $\lambda > 0$, $\mathrm{Leb}\{t \in \mathbb{R}^N : \mathfrak{M}f(t) \geq \lambda\} \leq \frac{8^N}{\lambda} \|f\|_1$.

(c) Conclude the following enhancement of Lebesgue's differentiation theorem: Whenever $f \in L^1(\mathbb{R}^N)$, for almost all $t \in \mathbb{R}^N$,

$$\lim_{\varepsilon \to 0+}\ \sup_{R \in \mathbf{R}_\varepsilon(t)} \left|\varepsilon^{-N} \int_R f(s)\, ds - f(t)\right| = 0.$$

(d) Formulate and prove the corresponding enhancement of the uniform differentiation theorem (Theorem 2.2.1).

5. A function $f : \mathbb{R}^d \to \mathbb{R}$ is said to be **harmonic** if it is continuous and if for all $x \in \mathbb{R}^d$ and all $\varepsilon > 0$, $f(x) = (2\varepsilon)^{-d} \int_{\mathcal{B}(x;\varepsilon)} f(y)\, dy$, where $\mathcal{B}(x;r)$ denotes the open ℓ^∞ ball of radius r about x. Show that the only bounded harmonic functions on $\mathbb{R}^d$ are constants. This is **Liouville's theorem** of classical potential theory.

(HINT: For ε of the form $\varepsilon = 2^{-n}$, interpret the right-hand side as a conditional expectation with respect to the dyadic filtration. An alternative proof can be based on Haar function expansions.)

6. Suppose $(\Omega, \mathcal{F}, \mathbb{P})$ is a probability space.

   (i) Show that if $Y \in L^1(\mathbb{P})$, then $A \mapsto \mathbb{Q}(A) = \mathbb{E}[Y \mathbf{1}_A]$ defines a measure on the $\mathcal{F}$-measurable subsets of Ω.

   (ii) Suppose that for every $n \geq 1$, $\mathcal{C}_n = \{C_{1,n}, C_{2,n}, \ldots\}$ is a denumerable collection of subsets of $\mathcal{F}$ that cover Ω. Let $\mathcal{F}_n$ define the σ-field generated by the elements of $\mathcal{C}_n$ and show that
   $$\mathbb{E}[Y \mid \mathcal{F}_n] = \sum_{k \geq 1:\ \mathbb{P}(C_{k,n}) > 0} \frac{\mathbb{Q}(C_{k,n})}{\mathbb{P}(C_{k,n})} \mathbf{1}_{C_{k,n}}, \qquad n \geq 1.$$

   (iii) If $\mathcal{C}_{n+1} \supset \mathcal{C}_n$, conclude that the $\lim_{n \to \infty} \mathbb{E}[Y \mid \mathcal{F}_n]$ exists a.s. and in $L^1(\mathbb{P})$.

   (iv) Suppose Y is $\vee_n \mathcal{F}_n$-measurable. Then, show that for $\mathbb{P}$-almost every $\omega \in \Omega$,
   $$\lim_{n \to \infty} \frac{\mathbb{Q}(C_{k(\omega),n})}{\mathbb{P}(C_{k(\omega),n})} = Y(\omega),$$
   where $k(\omega)$ is the unique $k \geq 1$ such that $\omega \in C_{k(\omega),n}$. (Since $Y = d\mathbb{Q}/d\mathbb{P}$ in the sense of the Radon–Nikodým theorem, the above is a probabilistic proof of the differentiation theorem, which is, in fact, *equivalent* to the Radon–Nikodým theorem.)

   (v) Obtain the following form of Lebesgue's differentiation theorem on $\mathbb{R}^1$: Every function $F : \mathbb{R} \to \mathbb{R}$ that is of bounded variation is a.e. differentiable in the sense that $\lim_{\varepsilon \to 0^+} \varepsilon^{-1}\{F(x+\varepsilon) - F(x)\}$ exists for almost every $x \in \mathbb{R}$. This is due to F. Riesz; see Riesz and Sz.-Nagy (1955, Ch. 1) and also (Stein 1970; Stein and Weiss 1971).

   (vi) Obtain the following form of Lebesgue's differentiation theorem on a separable metric space (S, d): Given a σ-finite measure μ on the Borel field of S, define for all $f \in L^1(\mu)$,
   $$T_r f(a) = \frac{1}{\mu(\mathcal{B}_d(a; r))} \int_{\mathcal{B}_d(a; r)} f(y) \mu(dy) \qquad a \in S,\ r > 0,$$
   where $\mathcal{B}_d(a; r) = \{x \in \mathbb{R}^d : d(a, r) < r\}$ denotes the ball of radius r about $a \in S$. Prove that as $r \to 0^+$, $T_r f \to f$, μ-almost everywhere.

7. In this exercise we will derive a decomposition of Calderón and Zygmund (1952) that is an important tool in the study of singular integrals.

   (i) Suppose $M = (M_n;\ n \geq 0)$ is a nonnegative one-parameter submartingale with respect to a filtration $\mathcal{F} = (\mathcal{F}_n;\ n \geq 0)$. Suppose M has subexponential growth in the sense that there exists a finite, nonrandom constant $C > 0$ such that for all $n \geq 0$, $M_{n+1} \leq CM_n$. For any $\lambda > 0$, define $T_\lambda = \inf(n \geq 0 : M_n > \lambda)$, where $\inf \varnothing = +\infty$. Check that:

   (a) T_λ is a stopping time for each $\lambda > 0$.

(b) on $(T_\lambda < \infty)$, $\lambda < M_{T_\lambda} \leq C\lambda$.

(ii) Suppose $\mathcal{F}$ is the dyadic filtration on $[0,1]^N$ and consider the martingale M given by $M = \mathbb{E}[f(V) \mid \mathcal{F}_n]$, where $f : [0,1]^N \to \mathbb{R}_+$ and V is uniformly chosen from $[0,1]^N$. Use (i) to prove the following: For each $\alpha \geq 0$, there exists a decomposition $\{\Phi, \Gamma\}$, of $[0,1]^N$ such that:

(a) $[0,1]^N = \Phi \cup \Gamma$, $\Phi \cap \Gamma = \varnothing$;

(b) $f(x) \leq \lambda$, for Lebesgue almost every $x \in \Phi$;

(c) there exists a countable collection of cubes $\{\Gamma_k;\ k \geq 1\}$ with disjoint interiors such that $\Gamma = \cup_{k \geq 1} \Gamma_k$;

(d) for all $k \geq 1$,
$$\lambda \leq \frac{1}{\text{Leb}(\Gamma_k)} \int_{\Gamma_k} f(x)\, dx \leq 2^N \lambda.$$

(HINT: Show that the dyadic martingale of (ii) satisfies (i) with $C = 2^N$. You may want also to note that $(T_\lambda < \infty) = \cup_k (T_\lambda = k)$ and $(T_\lambda = k) \in \mathcal{F}_k$.)

4 Notes on Chapter 2

Section 1 It is safe to say that a probabilist's interpretation of Haar functions is Exercise 1.1.1. This exercise can sometimes even be an exercise for the undergraduate student and is a part of the folklore of stochastic processes. The significance of this exercise was showcased by P. Lévy's construction of Brownian motion, where Supplementary Exercise 1 was essentially introduced and used; cf. (Lévy 1965; Ciesielski 1961). See also Ciesielski (1959), Neveu (1975, III-3), and Ciesielski and Musielak (1959), where you will also find many related works, including proofs for Theorem 1.1.1 and Exercise 1.1.2.

Section 2 The differentiation theorem of Lebesgue, as well as that of Jessen, Zygmund, and Marcinkiewicz, is completely classical. See Zygmund (1988, Ch. XVII) for this and related developments. The constants in Proposition 2.2.1 are better than those in the analysis literature, and its probabilistic proof is new. However, the real content of this proof is classical and has already been utilized by Jessen et al.

A proof of Theorem 2.2.1 where $\Delta \in \mathbb{Q}_+^N$ is dyadic can be found in Walsh (1986b), where Cairoli's convergence theorem is directly applied. It is also possible to use the theory of martingales that are indexed by directed sets in order to prove a more general result than Theorem 2.2.1; see Shieh (1982).

An excellent modern source for results relating to this chapter, and much more, is Stein (1993).

Section 3 The probabilistic proof outlined in Supplementary Exercise 7 is from Gundy (1969).

3
Random Walks

Those cannot remember the past are condemned to repeat it.
—Santayanna

Random walks entered mathematics early on through the analysis of gambling and other games of chance. To cite a typical example, let X_0 denote the initial fortune of a certain gambler and let X_n stand for the amount won (if $X_n \geq 0$) or lost (if $X_n \leq 0$) the nth time that the gambler places a bet. In the simplest gambling situations, the X_n's are i.i.d., and the gambler's fortune at time n is described by the partial sum $S_n = \sum_{j=0}^{n} X_j$. The stochastic process $S = (S_n; n \geq 0)$ is called a one-dimensional random walk and lies at the heart of modern, as well as classical, probability theory. This chapter is a study of some properties of systems of such walks.

The main problem addressed here is, *under what conditions does the random walk return to 0 infinitely often?* To see how this may come up, suppose the gambler plays *ad infinitum* and has an unbounded credit line. We then wish to know under what conditions the gambler can break even, infinitely many times, as he or she plays on. In the language of the theory of Markov chains, we wish to know when the state 0 is *recurrent*.

The analogous problem for systems of random walks is more intricate and is the subject of much of this chapter: Suppose the X_j's are i.i.d. random vectors in d-space. Then, the d-dimensional random walk models the movement of a small particle in a homogeneous medium. Suppose we have N particles, each of which paints every point that it visits. If each individual particle uses a distinct color, under what conditions do the N random lines created by the N random particles cross paths infinitely many times? These are some of the main problems that are taken up in this chapter.

1 One-Parameter Random Walks

The stochastic process $S = (S_n;\, n \geq 1)$ is a **random walk** if it has stationary, independent increments. To put it another way, we consider independent, identically distributed random variables $X_1, X_2, \ldots$, all taking values in $\mathbb{R}^d$, and define the corresponding random walk $n \mapsto S_n$ as $S_n = \sum_{i=1}^n X_i$ ($n = 1, 2, \ldots$). Clearly, $X_1 = S_1$, and for all $n \geq 2$, $X_n = S_n - S_{n-1}$, when $n \geq 2$. Thus, we are justified in calling the X_i's the **increments** of S. This is a review section on one-parameter random walks; we develop the theory with an eye toward multiparameter extensions that will be developed in the remainder of this chapter.

1.1 Transition Operators

Suppose $S = (S_n;\, n \geq 1)$ is a d-dimensional random walk with increments $X = (X_n;\, n \geq 1)$. For all $n \geq 1$, define $\mathcal{F}_n$ to be the σ-field generated by $X_1, \ldots, X_n$. It is simple to see that $\mathcal{F}_n$ is precisely the σ-field generated by $S_1, \ldots, S_n$. In the notation of Chapter 1, we have shown that $\mathcal{F} = (\mathcal{F}_n;\, n \geq 1)$ is the history of the stochastic process S.

It is always the case that the study of the stochastic process S is equivalent to the analysis of probabilities of the form

$$\mathbb{P}(S_{n_1} \in E_1, S_{n_2} \in E_2, \ldots, S_{n_k} \in E_k),$$

where $k, n_1, \ldots, n_k \geq 1$ are integers and $E_1, \ldots, E_k$ are measurable subsets of $\mathbb{R}^d$. These probabilities are called the **finite-dimensional distributions** of S. It turns out that the finite-dimensional distributions of the random walk S are completely determined by the collection $\mathbb{P}(X_1 + x \in E)$, where $E \subset \mathbb{R}^d$ is measurable and $x \in \mathbb{R}^d$. A precise form of such a statement is called the Markov property; we shall come to this later. Bearing this discussion in mind, we define for all measurable functions $f : \mathbb{R}^d \to \mathbb{R}$, all $n \geq 1$, and $x \in \mathbb{R}^d$,

$$\mathcal{T}_n f(x) = \mathbb{E}\big[f(S_n + x)\big].$$

In particular, note that for all Borel sets $E \subset \mathbb{R}^d$, $\mathcal{T}_1 \mathbb{1}_E(x) = \mathbb{P}(X_1 + x \in E)$. Thus, once we know the operator $\mathcal{T}_n$, we know how to compute these probabilities. We begin our study of random walks by first analyzing these operators.

Note that $\mathcal{T}_n$ is a **bounded linear operator**: For all bounded measurable $f, g : \mathbb{R}^d \to \mathbb{R}$, $n \geq 1$, $x \in \mathbb{R}^d$ and all $\alpha, \beta \in \mathbb{R}$,

(i) $\sup_{x \in \mathbb{R}^d} |\mathcal{T}_n f(x)| \leq \sup_{x \in \mathbb{R}^d} |f(x)|$;

(ii) $\mathcal{T}_n(\alpha f + \beta g)(x) = \alpha \mathcal{T}_n f(x) + \beta \mathcal{T}_n g(x)$; and

(iii) $x \mapsto \mathcal{T}_n f(x)$ is measurable.

Next, we interpret $\mathcal{T}_n$ in terms of the conditional distributions of S.

Lemma 1.1.1 *For all $n, k \geq 1$ and all bounded measurable $f : \mathbb{R}^d \to \mathbb{R}$,*
$$\mathbb{E}\big[f(S_{k+n}) \,\big|\, \mathcal{F}_k\big] = \mathbb{E}\big[f(S_{k+n}) \,\big|\, S_k\big] = \mathcal{T}_n f(S_k), \qquad a.s.$$
In particular, for all $x \in \mathbb{R}^d$, $n, k \geq 1$, and all bounded measurable $f : \mathbb{R}^d \to \mathbb{R}$, $\mathcal{T}_{n+k} f(x) = \mathcal{T}_n(\mathcal{T}_k f)(x) = \mathcal{T}_k(\mathcal{T}_n f)(x)$.

In functional-analytic language, $(\mathcal{T}_n;\ n \geq 1)$ is a **semigroup** of operators. To see what the above lemma means, take $f = \mathbf{1}_E$ for some Borel set $E \subset \mathbb{R}^d$. The above says that if k denotes the current time,

1. given the present position S_k, any future position S_{k+n} is conditionally independent of the past positions $S_1, \ldots, S_{k-1}$; and

2. $\mathcal{T}_n \mathbf{1}_E(S_k)$ is the conditional probability of making a transition to E in n steps, given $\mathcal{F}_k$.

Motivated by this, we call $\mathcal{T}_n$ the ***n*-step transition operator of** S.

Proof of Lemma 1.1.1 Note that $S_{k+n} - S_k = \sum_{j=k+1}^{k+n} X_j$ is (a) independent of $\mathcal{F}_k$; and (b) has the same distribution as $S_n = \sum_{j=1}^n X_j$. Thus,
$$\mathbb{E}\big[f(S_{k+n}) \,\big|\, \mathcal{F}_k\big] = \mathbb{E}\big[f(S_{k+n} - S_k + S_k) \,\big|\, \mathcal{F}_k\big] = \int f(x + S_k)\,\mathbb{P}(S_n \in dx)$$
$$= \mathcal{T}_n f(S_k),$$
almost surely. From this, we also can conclude the equality regarding the conditional expectation $\mathbb{E}[f(S_{k+n}) \,|\, S_k]$. Applying the preceding to $f(\bullet + x)$, we obtain $\mathbb{E}[f(x + S_{k+n}) \,|\, \mathcal{F}_k] = \mathcal{T}_n f(x + S_k)$, almost surely. Taking expectations, we deduce that $\mathcal{T}_{k+n} f(x) = \mathcal{T}_k(\mathcal{T}_n f)(x)$. The rest follows from reversing the roles of k and n. □

Digression If we define $S_0 = 0$, then for any $x \in \mathbb{R}^d$, we can, and should, think of $x + S$ as our random walk started at x. In particular, S itself should be thought of as the random walk started at the origin. The above lemma suggests the following interpretation: Given the position of the process at time k, the future trajectories of our walk are those of a random walk started at S_k. The following is a more precise formulation of this and is a version of the so-called **Markov property** of S that was alluded to earlier.

Theorem 1.1.1 (The Markov Property) *Fix integers $k \geq 1$, $n \geq 2$ and bounded measurable functions $f_1, \ldots, f_n : \mathbb{R}^d \to \mathbb{R}$. Then, the following holds with probability one:*
$$\mathbb{E}\bigg[\prod_{\ell=1}^n f_\ell(S_{k+\ell} - S_k)\,\bigg|\, \mathcal{F}_k\bigg] = \mathbb{E}\bigg[\prod_{\ell=1}^n f_\ell(S_\ell)\bigg].$$

68   3. Random Walks

In other words, for any $k \geq 1$, the process $n \mapsto S_{n+k} - S_k$ is (i) independent of $\mathcal{F}_k$; and (ii) has the same finite-dimensional distributions as the original process S.

Recalling that we think of $n \mapsto S_n + x$ as a random walk with increments $X_1, X_2, \ldots$ that starts at $x \in \mathbb{R}^d$, we readily obtain the following useful interpretation of the above.

Corollary 1.1.1 *Suppose $k \geq 1$ is a fixed integer. Then, conditionally on $\sigma(S_k)$, $(S_{k+n};\ k \geq 0)$ is a random walk whose increments have the same distribution as X_1. Moreover, the σ-field generated by $(S_{k+n};\ n \geq 0)$ is conditionally independent of $\mathcal{F}_k$, given $\sigma(S_k)$.*

See Section 3.6 of Chapter 1 for information on conditional independence.

***Exercise* 1.1.1** Carefully prove Corollary 1.1.1. □

Proof of Theorem 1.1.1 Since it depends only on $X_{k+1}, \ldots, X_{k+n}$, the random variable $\prod_{\ell=1}^n f_\ell(S_{k+\ell} - S_k)$ is independent of $(X_1, \ldots, X_k)$ and hence of $\mathcal{F}_k$. (Why?) As a result, with probability one,

$$\mathbb{E}\Big[\prod_{\ell=1}^n f_\ell(S_{k+\ell} - S_k) \,\Big|\, \mathcal{F}_k\Big] = \mathbb{E}\Big[\prod_{\ell=1}^n f_\ell(S_{k+\ell} - S_k)\Big].$$

On the other hand, the sequence $(X_{k+1}, \ldots, X_{k+n})$ has the same distribution as the sequence $(X_1, \ldots, X_n)$. After performing a little algebra, we can reinterpret this statement as follows: The distribution of the $\mathbb{R}^{nd}$-valued random vector $(S_{k+1} - S_k, \ldots, S_{k+n} - S_k)$ is the same as that of $(S_1, \ldots, S_n)$. In particular, we have $\mathbb{E}[\prod_{\ell=1}^n f_\ell(S_{k+\ell} - S_k)] = \mathbb{E}[\prod_{\ell=1}^n f_\ell(S_\ell)]$, which proves the result. □

It is clear that Corollary 1.1.1 extends the conditional independence assertion of Lemma 1.1.1. However, the latter lemma also contains information on the transition operators, to which we now return.

Corollary 1.1.2 *The transition operators, in fact $\mathcal{T}_1$, uniquely determine the finite-dimensional distributions and vice versa.*

Proof By the very definition of $\mathcal{T}_n$, if we know all finite-dimensional distributions, we can compute $\mathcal{T}_n f(x)$ for all measurable $f : \mathbb{R}^d \to \mathbb{R}_+$, all $n \geq 1$, and all $x \in \mathbb{R}^d$. The converse requires an honest proof. Consider the following proposition:

(Π_n) For all measurable $f_1, \ldots, f_n : \mathbb{R}^d \to \mathbb{R}_+$, $\mathbb{E}[\prod_{\ell=1}^n f_\ell(S_\ell)]$ can be computed from $\mathcal{T}_1$.

Our goal is to show that (Π_n) holds for all $n \geq 1$. We will prove this by using induction on n: Lemma 1.1.1 shows that (Π_1) is true. Thus, we suppose that $(\Pi_1), \ldots, (\Pi_{n-1})$ hold and venture to prove (Π_n). By Lemma 1.1.1, for all measurable $f_1, \ldots, f_n : \mathbb{R}^d \to \mathbb{R}_+$,

$$\mathbb{E}\Big[\prod_{\ell=1}^n f_\ell(S_\ell) \,\Big|\, \mathcal{F}_{n-1}\Big] = \prod_{\ell=1}^{n-1} f_\ell(S_\ell) \cdot \mathcal{T}_1 f(S_{n-1}) = \prod_{\ell=1}^{n-1} g_\ell(S_\ell),$$

where $g_i = f_i$ for all $1 \leq i \leq n-2$ and $g_{n-1}(x) = f_{n-1}(x) \cdot \mathcal{T}_1 f_{n-1}(x)$. Taking expectations, we see that $\mathbb{E}[\prod_{\ell=1}^n f_\ell(S_\ell)] = \mathbb{E}[\prod_{i=1}^{n-1} g_i(S_i)]$. By Π_{n-1}, this can be written entirely in terms of $\mathcal{T}_1$, thus proving (Π_n). □

Exercise **1.1.2** Find an explicit recursive formula for $\mathbb{E}[\prod_{\ell=1}^n f_\ell(S_\ell)]$ in terms of $\mathcal{T}_1$. □

1.2 The Strong Markov Property

Let $S = (S_k;\, k \geq 1)$ denote a d-dimensional random walk with history $\mathcal{F} = (\mathcal{F}_k;\, k \geq 1)$ and increment process $X = (X_k;\, k \geq 1)$. The **strong Markov property** of S states that for any finite stopping time T (with respect to the filtration $\mathcal{F}$), the stochastic process $(S_{k+T} - S_T;\, k \geq 1)$ is independent of $\mathcal{F}_T$ and has the same finite-dimensional distributions as the process S. Roughly speaking, this means that the process $(S_{k+T};\, k \geq 1)$ is conditionally independent of $\mathcal{F}_T$ given S_T and is, in distribution, the random walk S started at S_T.

Theorem 1.2.1 (The Strong Markov Property) *Suppose T is a stopping time with respect to $\mathcal{F}$. Given integers $n, k \geq 1$ and bounded, measurable $f_1, \ldots, f_n : \mathbb{R}^d \to \mathbb{R}$,*

$$\mathbb{E}\Big[\prod_{\ell=1}^n f_\ell(S_{T+\ell} - S_T) \,\Big|\, \mathcal{F}_T\Big] \mathbf{1}_{(T<\infty)} = \mathbb{E}\Big[\prod_{\ell=1}^n f_\ell(S_\ell)\Big] \mathbf{1}_{(T<\infty)}, \qquad a.s.$$

Remarks (i) Given the transition operators, the above expression can be computed using Corollaries 1.1.1 and 1.1.2; see Exercise 1.1.2.

(ii) It is important to realize that the stopping time condition cannot be removed in general, as the following clearly shows.

Exercise **1.2.1** Consider the simple walk on $\mathbb{Z}^1$. Here, the increments $X_1, X_2, \ldots$ take the values ± 1 with probability $\frac{1}{2}$ each. Consider the $\mathbb{N}_0 \cup \{\infty\}$-valued random variable $L = \sup(k \geq 0 : S_k \leq -\frac{1}{2}k)$, where $\sup \varnothing = 0$. That is, L designates the last time that the random walk goes below the line $y = -\frac{1}{2}x$.

(i) Show that with probability one, $L < \infty$ and that L is *not* a stopping time with respect to the history of the process S.

(ii) Verify that $\mathcal{F}_L$ is a σ-field and that the process $j \mapsto S_{j+L} - S_L$ is independent of $\mathcal{F}_L$, where $\mathcal{F}_L = \big(A \in \vee_n \mathcal{F}_n : A \cap (L \leq j) \in \mathcal{F}_j$, for all $j \geq 0\big)$ is defined as if L were a stopping time.

(iii) Show that the stochastic process $j \mapsto S_{L+j} - S_L$ does *not* have the same finite-dimensional distributions as S.

This is a part of a deep result of Williams (1970, 1974).
(HINT: for part (i), you can use a limit theorem; for part (ii), condition on the value of L.) □

Proof of Theorem 1.2.1 For all $\ell \geq 1$, $S_{T+\ell} - S_T = \sum_{j=T+1}^{T+\ell} X_j$. Since for all $j \geq 1$, the event $(T = j)$ is $\mathcal{F}_T$-measurable,

$$\mathbb{E}\bigg[\prod_{\ell=1}^n f_\ell(S_{T+\ell} - S_T) \,\bigg|\, \mathcal{F}_T\bigg] \mathbf{1}_{(T<\infty)} = \sum_{j=1}^\infty \mathbb{E}\bigg[\prod_{\ell=1}^n f_\ell(S_{T+\ell} - S_T) \,\bigg|\, \mathcal{F}_T\bigg] \mathbf{1}_{(T=j)}$$

$$= \sum_{j=1}^\infty \mathbb{E}\bigg[\prod_{\ell=1}^n f_\ell(S_{j+\ell} - S_j) \,\bigg|\, \mathcal{F}_T\bigg] \mathbf{1}_{(T=j)},$$

almost surely. Regarding $j \geq 1$ as fixed, define $Y = \prod_{\ell=1}^n f_\ell(S_{j+\ell} - S_j)$ and for all $k \geq 1$, let $M_k = \mathbb{E}[Y \mid \mathcal{F}_k]$. By Theorem 1.1.1, Chapter 1, with probability one, $M_j \mathbf{1}_{(T=j)} = M_T \mathbf{1}_{(T=j)} = \mathbb{E}[Y \mid \mathcal{F}_T]\mathbf{1}_{(T=j)}$. Thus,

$$\mathbb{E}\bigg[\prod_{\ell=1}^n f_\ell(S_{T+\ell} - S_T) \,\bigg|\, \mathcal{F}_T\bigg] \mathbf{1}_{(T<\infty)} = \sum_{j=1}^\infty \mathbb{E}\bigg[\prod_{\ell=1}^n f_\ell(S_{j+\ell} - S_j) \,\bigg|\, \mathcal{F}_j\bigg] \mathbf{1}_{(T=j)}.$$

By the stationarity and the independence of the increments of S, the above equals $\mathbb{E}[\prod_{\ell=1}^n f_\ell(S_\ell)]\mathbf{1}_{(T<\infty)}$, as desired. □

1.3 Recurrence

Suppose S is a d-dimensional random walk with increment process X and history $\mathcal{F}$. Throughout this section we assume that the X's are taking values in the d-dimensional integer lattice $\mathbb{Z}^d$.

A point $x \in \mathbb{Z}^d$ is said to be **recurrent** if $\mathbb{P}(S_k = x$ infinitely often$) > 0$. *When is a point $x \in \mathbb{Z}^d$ recurrent?* In this subsection we will resolve this when x is the origin of $\mathbb{Z}^d$. Since it is the starting position of the random walk, the origin is a very special point; see the Digression in Section 1.1. Recurrence properties of a general point $x \in \mathbb{Z}^d$ are discussed in Section 1.6 below.

Recalling that $\inf \varnothing = \infty$, let $\tau_1 = \inf(j \geq 1 : S_j = 0)$; that is, τ_1 is the first time the random walk visits 0. Iteratively define $\tau_{k+1} = \inf(j \geq 1+\tau_k : S_j = 0)$, for $k \geq 1$. It is easy to see that $\tau_1, \tau_2, \ldots$ are stopping times. One should think of τ_1 ($\tau_2, \ldots$) as the first (second, etc.) time the random walk visits the origin. Among other things, this sequence of visitation times has the following property.

Lemma 1.3.1 *Fix $n, j \geq 1$. On $(\tau_n < \infty)$,*

$$\mathbb{P}(\tau_{n+1} - \tau_n = j \mid \mathcal{F}_{\tau_n}) = \mathbb{P}(\tau_1 = j), \qquad a.s.$$

Suppose we knew that with probability one, $\tau_n < \infty$ for all $n \geq 1$. The above lemma asserts that in this case, $\tau_1, \tau_2 - \tau_1, \ldots$ is a sequence of independent, identically distributed random variables (why?). Since $\tau_n = \tau_1 + \sum_{j=2}^{n}(\tau_j - \tau_{j-1})$, $\tau = (\tau_n;\ n \geq 1)$ is then identified as a random walk with nonnegative increments.

Proof This is a consequence of the strong Markov property (see Theorem 1.2.1). In fact, since $S_{\tau_n} = 0$ on $(\tau_n < \infty)$,

$$\mathbb{P}(\tau_{n+1} - \tau_n = j \mid \mathcal{F}_{\tau_n})\mathbf{1}_{(\tau_n < \infty)}$$
$$= \mathbb{P}(S_{\tau_n + \ell} \neq 0 \text{ for all } 1 \leq \ell \leq j-1,\ S_{\tau_n + j} = 0 \mid \mathcal{F}_{\tau_n})\mathbf{1}_{(\tau_n < \infty)}$$
$$= \mathbb{P}(S_{\tau_n + \ell} - S_{\tau_n} \neq 0 \text{ for all } 1 \leq \ell \leq j-1,\ S_{\tau_n + j} - S_{\tau_n} = 0 \mid \mathcal{F}_{\tau_n})\mathbf{1}_{(\tau_n < \infty)}$$
$$= \mathbb{P}(S_\ell \neq 0 \text{ for all } 1 \leq \ell \leq j-1,\ S_j = 0\ )\mathbf{1}_{(\tau_n < \infty)}$$
$$= \mathbb{P}(\tau_1 = j)\mathbf{1}_{(\tau_n < \infty)}.$$

The strong Markov property (Theorem 1.2.1) is used in the penultimate line. This proves the result. □

In particular, upon summing Lemma 1.3.1 over all integers $j \geq 1$, we arrive at the following: For all $n \geq 2$,

$$\mathbb{P}(\tau_n < \infty) = \mathbb{P}(\tau_n - \tau_{n-1} < \infty, \tau_{n-1} < \infty)$$
$$= \mathbb{E}\left[\mathbb{P}(\tau_n - \tau_{n-1} < \infty \mid \mathcal{F}_{\tau_{n-1}})\mathbf{1}_{(\tau_{n-1} < \infty)}\right]$$
$$= \mathbb{P}(\tau_1 < \infty) \cdot \mathbb{P}(\tau_{n-1} < \infty).$$

By induction,

$$\mathbb{P}(\tau_n < \infty) = \{\mathbb{P}(\tau_1 < \infty)\}^n. \qquad (1)$$

With the unambiguous understanding that $\infty \leq \infty$, we can deduce that the τ_n's are nondecreasing. Continuity properties of probability measures then imply that

$$\mathbb{P}(0 \text{ is recurrent}) = \lim_{n \to \infty} \mathbb{P}(\tau_n < \infty) = \lim_{n \to \infty} \{\mathbb{P}(\tau_1 < \infty)\}^n.$$

Taking equation (1) into account, we have proven the following:

Proposition 1.3.1 *The following are equivalent:*

(i) 0 is recurrent;

(ii) $\mathbb{P}(S_k = 0 \text{ infinitely often}) = 1$; and

72   3. Random Walks

(iii) $\mathbb{P}(\tau_1 < \infty) = 1$.

Informally, we are stating that if starting from the origin we are sure of returning to the origin, then we will do so infinitely many times. This is an example of the strong Markov property at its finest.

1.4 Classification of Recurrence

A natural question is, *how do the finite-dimensional distributions of a $\mathbb{Z}^d$-valued random walk influence the recurrence of the point 0?* For all integers $n \geq 1$, define

$$R_n = 1 + \sum_{k=1}^{n} \mathbf{1}_{(S_k = 0)}.$$

Recalling the Digression of Section 1.1, we think of S as starting from the origin, so that at time 0, S is at 0. Viewed as such, R_n denotes the total number of visits to the origin by time n. Note that $R_\infty = \lim_{n \to \infty} R_n$ is a random variable taking values in $\mathbb{N} \cup \{\infty\}$. Proposition 1.3.1 can be restated as follows: $\mathbb{P}(R_\infty = \infty) \in \{0, 1\}$. Moreover, this probability is 1 if and only if 0 is recurrent.

The key to our analysis of recurrence turns out to be $\mathbb{E}[R_\infty] = 1 + \sum_{k=1}^{\infty} \mathbb{P}(S_k = 0)$. In fact, we have the following result, due to G. Pólya, K. L. Chung, and W. H. J. Fuchs, which appeared in Chung and Fuchs (1951) in full generality; see (Pólya 1921; Chung and Ornstein 1962) for some related results. Supplementary Exercise 9 contains a complete statement of the above results: the so-called Chung–Fuchs theorem.

Theorem 1.4.1 (The Pólya Criterion) *The point 0 is recurrent if and only if $\sum_{k=1}^{\infty} \mathbb{P}(S_k = 0) = \infty$.*

Informally, S will hit 0 infinitely often if it is expected to do so. For our proof, we need the the following simple and powerful lemma, first found in Paley and Zygmund (1932).

Lemma 1.4.1 (Paley–Zygmund Lemma) *Suppose Z is an almost surely nonnegative random variable. Then for all $\varepsilon \in\,]0, 1[$,*

$$\mathbb{P}(Z \geq \varepsilon \mathbb{E}[Z]) \geq (1 - \varepsilon)^2 \frac{\{\mathbb{E}[Z]\}^2}{\mathbb{E}[Z^2]},$$

provided that all of the mentioned expectations exist.

Exercise **1.4.1** Prove the Paley–Zygmund lemma.
(HINT: Apply the Cauchy–Schwarz inequality to $\mathbb{E}[Z \mathbf{1}_{(Z \geq \varepsilon \mathbb{E}[Z])}]$.)   □

Exercise **1.4.2** If Z is a nonnegative random variable that is also in $L^2(\mathbb{P})$, show that $\mathbb{P}(Z=0) \leq \mathrm{Var}(Z)/\{\mathbb{E}[Z]\}^2$, where Var denotes the variance. □

Exercise **1.4.3** Suppose $\mathsf{E}_1, \mathsf{E}_2, \ldots$ are measurable events such that $\sum_j \mathbb{P}(\mathsf{E}_j) = +\infty$. Prove that whenever

$$\liminf_{n\to\infty} \frac{\sum_{j=1}^n \sum_{k=1}^n \mathbb{P}(\mathsf{E}_j \cap \mathsf{E}_k)}{\left\{\sum_{j=1}^n \mathbb{P}(\mathsf{E}_j)\right\}^2} < \infty,$$

then $\mathbb{P}(\mathsf{E}_n \text{ infinitely often}) > 0$. This is from (Chung and Erdős 1952; Kochen and Stone 1964).
(HINT: Consider the first two moments of $J_n = \sum_{j=1}^n \mathbf{1}_{\mathsf{E}_j}$.) □

Proof of Theorem 1.4.1 We have already made the observation that $R_\infty \geq 1$ and $\mathbb{E}[R_\infty - 1] = \sum_{k=1}^\infty \mathbb{P}(S_k = 0)$. (Since $R_\infty = \lim_n R_n$, a.s., this is a consequence of the monotone convergence theorem of measure theory.) Thus, $\sum_k \mathbb{P}(S_k = 0) < \infty$ if and only if $\mathbb{E}[R_\infty - 1] < \infty$. Consequently, $\sum_k \mathbb{P}(S_k < \infty) < \infty$ certainly implies that $R_\infty < \infty$, a.s.; that is to say that 0 is not recurrent. Next, we suppose that $\sum_k \mathbb{P}(S_k = 0) = \infty$. It is clear that $\mathbb{E}[R_n - 1] = \sum_{k=1}^n \mathbb{P}(S_k = 0)$ and that this sequence explodes as $n \to \infty$. We now estimate $\mathbb{E}[(R_n - 1)^2]$, viz.,

$$\mathbb{E}\big[(R_n-1)^2\big] = \mathbb{E}\Big[\sum_{k=1}^n \mathbf{1}_{(S_k=0)}\Big] + 2\mathbb{E}\Big[\sum\sum_{1\leq k<\ell\leq n} \mathbf{1}_{(S_k=0)}\mathbf{1}_{(S_\ell=0)}\Big]$$
$$= \mathbb{E}[R_n - 1] + 2\sum\sum_{1\leq k<\ell\leq n}' \mathbb{P}(S_k=0)\mathbb{P}(S_{\ell-k}=0),$$

by the Markov property (Theorem 1.1.1). Relabeling the last summation and possibly adding more nonnegative terms, we arrive at the estimate

$$\mathbb{E}\big[(R_n-1)^2\big] \leq \mathbb{E}[R_n - 1] + 2\big(\mathbb{E}[R_n-1]\big)^2.$$

Since $R_n - 1 \in \mathbb{N}_0$, $(R_n - 1 > 0) = (R_n \geq 2)$. Applying Lemma 1.4.1 first, and then the above estimate, in this order, we arrive at the following:

$$\mathbb{P}(\tau_1 \leq n) = \mathbb{P}(R_n \geq 2) \geq \frac{\big(\mathbb{E}[R_n-1]\big)^2}{\mathbb{E}[R_n-1] + 2\big(\mathbb{E}[R_n-1]\big)^2},$$

where $\tau_1 = \inf\big(j \geq 1 : S_j = 0\big)$. Since $\lim_n \mathbb{E}[R_n] = \infty$, this implies that $\mathbb{P}(\tau_1 < \infty) \geq \frac{1}{2}$. By Proposition 1.3.1, whenever $\mathbb{P}(\tau < \infty)$ is positive, it is, in fact, 1. This completes our proof. □

While it was meant to bring forth a powerful technique, our demonstration of Theorem 1.4.1 is not the fastest method for getting there, as we see next.

Exercise 1.4.4 Let N denote the total number of returns to zero. That is, $N = \sum_{k=0}^{\infty} \mathbb{1}_{(S_k=0)}$. Show that N is a geometric random variable with mean p^{-1}, where $p = \mathbb{P}(\exists k \geq 1 : S_k = 0)$. Use this to verify Pólya's criterion.
(HINT: Show that for all $k \geq 1$, $\mathbb{P}(N \geq k) = p\mathbb{P}(N \geq k-1)$.) □

1.5 Transience

When a point $x \in \mathbb{Z}^d$ is not recurrent for the $\mathbb{Z}^d$-valued random walk S, we say that it is **transient**. It is easy to see that $0 \in \mathbb{Z}^d$ is transient if and only if
$$\lim_{n \to \infty} \mathbb{P}(S_k = 0 \text{ for some } k \geq n) = 0.$$
Thus, a natural measure for the strength of the transience of the origin is the rate at which $\mathbb{P}(S_k = 0 \text{ for some } k \geq n)$ goes to 0 as n goes to infinity. The following sheds much light on this rate.

Theorem 1.5.1 *If the origin is transient for the $\mathbb{Z}^d$-valued random walk S, the following holds for every integer $n \geq 1$:*
$$\frac{1}{2}\mathcal{T} \leq \mathbb{P}(S_k = 0 \text{ for some } k \geq n) \leq 8\mathcal{T},$$
where
$$\mathcal{T} = \frac{\sum_{j=n}^{\infty} \mathbb{P}(S_j = 0)}{1 + \sum_{j=1}^{\infty} \mathbb{P}(S_j = 0)}.$$

This theorem makes the point that as $n \to \infty$, $\mathbb{P}(S_k = 0 \text{ for some } k \geq n)$ goes to zero like a constant multiple of $\sum_{j \geq n} \mathbb{P}(S_j = 0)$.

Remarks

1. This can be sharpened; see Supplementary Exercise 1.

2. Throughout this subsection we implicitly use the notation of Section 1.3 and Section 1.4.

3. It can be shown that $\mathbb{P}(\tau_1 = \infty) = \{1 + \sum_{k=1}^{\infty} \mathbb{P}(S_k = 0)\}^{-1}$; see Supplementary Exercise 1. This is the probability of never hitting 0.

Proof By transience and by Theorem 1.4.1, $\sum_{j=1}^{\infty} \mathbb{P}(S_j = 0) < \infty$. For all $n \geq 1$, let
$$Z = \sum_{j=n}^{\infty} \mathbb{1}_{(S_j=0)} = R_\infty - R_{n-1},$$
where $R_0 = 1$. Clearly, $\mathbb{E}[Z] = \sum_{j=n}^{\infty} \mathbb{P}(S_j = 0)$, which we know is finite. Recall our proof of Theorem 1.4.1; the method used there to estimate

$\mathbb{E}[(R_n-1)^2]$ can be used here to show that

$$\mathbb{E}[Z^2] \leq 2\sum_{\ell=n}^{\infty} \mathbb{P}(S_\ell = 0) \cdot \left\{1 + \sum_{j=1}^{\infty} \mathbb{P}(S_j = 0)\right\}. \tag{1}$$

Since $(Z > 0) \subseteq (S_k = 0$ for some $k \geq n)$, we obtain the lower bound from the Paley–Zygmund lemma (Lemma 1.4.1).

For the upper bound on the probability, define

$$M_k = \mathbb{E}\left[\sum_{j=n}^{\infty} \mathbf{1}_{(S_j=0)} \,\middle|\, \mathcal{F}_k\right], \qquad k \geq n.$$

It is not hard to check that $M = (M_k;\ k \geq n)$ is a martingale. Moreover, for all $k \geq n$,

$$M_k \geq \mathbb{E}\left[\sum_{j=k}^{\infty} \mathbf{1}_{(S_j=0)} \,\middle|\, \mathcal{F}_k\right] \cdot \mathbf{1}_{(S_k=0)} = \left\{1 + \sum_{j=1}^{\infty} \mathbb{P}(S_{j+k}-S_k = 0 \mid \mathcal{F}_k)\right\} \cdot \mathbf{1}_{(S_k=0)}.$$

We have used the monotone convergence theorem to write the conditional expectation and the sum of the conditional probabilities. By the Markov property (Corollary 1.1.1), $M_k \geq \{1 + \sum_{j=1}^{\infty} \mathbb{P}(S_j = 0)\} \cdot \mathbf{1}_{(S_k=0)}$, almost surely. Taking suprema over all $k \geq n$ and squaring, we obtain the following:

$$\mathbf{1}_{(S_k=0 \text{ for some } k\geq n)} \leq \left\{1 + \sum_{j=1}^{\infty} \mathbb{P}(S_j = 0)\right\}^{-2} \cdot \sup_{k\geq n} M_k^2. \tag{2}$$

By Doob's strong $(2,2)$ inequality (Theorem 1.4.1, Chapter 1),

$$\mathbb{E}\left[\sup_{k\geq n} M_k^2\right] \leq 4 \sup_{k\geq n} \mathbb{E}[M_k^2].$$

Therefore, by taking expectations in equation (2), we obtain

$$\mathbb{P}(S_k = 0 \text{ for some } k \geq n) \leq 4\left\{1 + \sum_{k=1}^{\infty} \mathbb{P}(S_j = 0)\right\}^{-2} \sup_{k\geq n} \mathbb{E}[M_k^2].$$

Jensen's inequality shows that for any $k \geq n$, $\mathbb{E}[M_k^2] \leq \mathbb{E}[Z^2]$. Consequently, equation (1) implies the result. □

1.6 Recurrence of Possible Points

We now return to the question of when a general point $x \in \mathbb{Z}^d$ is recurrent. To illustrate the potential complications, consider the following simple example.

Example Suppose $d = 1$ and the $X_1, X_2, \ldots$ are independent, identically distributed random variables taking the values ± 2 with probability $\frac{1}{2}$ each. Let $S = (S_k;\ k \geq 1)$ denote the random walk whose increments are $X_1, X_2, \ldots$; i.e., $S_n = X_1 + \cdots + X_n$, for all $n \geq 1$. It should be absolutely clear that the point $x = 1$ is not recurrent. In fact, odd values can never be visited by S, and even values can. On the other hand, by the central limit theorem, $\limsup_n S_n = -\liminf_n S_n = +\infty$, almost surely. A little thought reveals that for any even number x, there are infinitely many n's such that $S_n = x$. □

***Exercise* 1.6.1** Use the central limit theorem to show that, in the above example, $\limsup_n S_n = -\liminf_n S_n = +\infty$, a.s. □

In the previous example we constructed a random walk for which all of the even numbers are recurrent, while the odd numbers can never be reached. This property turns out to be typical. To explore this phenomenon in greater depth, suppose S is a $\mathbb{Z}^d$-valued random walk. An $x \in \mathbb{Z}^d$ is **possible** if there exists an integer $k \geq 1$ such that $\mathbb{P}(S_k = x) > 0$. If x is not possible, it is deemed **impossible.** Clearly, impossible points are not, and can never be, visited. Therefore, any discussion of recurrence must be reduced to the possible points. What do the possible points of a random walk look like? Below is a prefatory result that will be elaborated upon in the next section.

Lemma 1.6.1 *The collection of all possible points of a $\mathbb{Z}^d$-valued random walk is an additive semigroup of $\mathbb{Z}^d$.*

Proof Suppose the random walk is denoted by S and $x_1, x_2 \in \mathbb{Z}^d$ are possible for S. By definition, there exist $k_1, k_2 \in \mathbb{Z}^d$ such that $p_i = \mathbb{P}(S_{k_i} = x_i) > 0$ for $i = 1, 2$. Since $\mathbb{P}(S_{k_1+k_2} - S_{k_1} = x_2) = \mathbb{P}(S_{k_2} = x_2) = p_2$, by the Markov property (Corollary 1.1.1),

$$\mathbb{P}(S_{k_1+k_2} = x_2 + x_1) \geq \mathbb{P}(S_{k_1} = x_1, S_{k_1+k_2} - S_{k_1} = x_2) = p_1 p_2 > 0.$$

This proves the lemma. □

The following is a very important exercise.

***Exercise* 1.6.2** Let S denote a random walk on $\mathbb{Z}^d$ whose increment process is X. We say that S is **symmetric** if X_1 and $-X_1$ have the same distributions. Prove that whenever S is a symmetric random walk on $\mathbb{Z}^d$, the set of its possible values forms an additive subgroup of $\mathbb{Z}^d$. In particular, argue that the origin is always possible. □

Lemma 1.6.2 *The collection of all recurrent points is an additive subgroup of $\mathbb{Z}^d$. In particular, if there are any recurrent points, 0 is one of them.*

Proof We will show that whenever x and y are recurrent, so is $x-y$. Let τ denote the first hitting time of y. That is, $\tau = \inf(k \geq 1 : S_k = y)$, where $\inf \emptyset = \infty$. Thanks to the recurrence of y, τ is finite and $S_\tau = y$, a.s.; cf. Proposition 1.3.1. Consequently, the strong Markov property (Theorem 1.2.1) implies the following (why?):

$$\mathbb{P}(S_k = x - y \text{ for infinitely many } k \geq 1)$$
$$= \mathbb{P}(S_{k+\tau} - S_\tau = x - y \text{ for infinitely many } k \geq 1)$$
$$= \mathbb{P}(S_{k+\tau} = x \text{ for infinitely many } k \geq 1)$$
$$= \mathbb{P}(S_k = x \text{ for infinitely many } k \geq 1),$$

which is equal to one, thanks to the recurrence of x, together with Proposition 1.3.1. This completes our proof. □

Theorem 1.6.1 *Suppose S is a $\mathbb{Z}^d$-valued random walk. If $x \in \mathbb{Z}^d$ is possible and $y \in \mathbb{Z}^d$ is recurrent, $x-y$ is recurrent. In particular, the following are equivalent:*

(i) 0 is recurrent;

(ii) all possible points x are recurrent with probability one.

Note that the condition (ii) subsumes the assumption that x is possible and that Theorem 1.6.1 extends Lemma 1.6.2.

Proof To begin, let us argue that the first assertion of the theorem implies the equivalence of (i) and (ii). Suppose (ii) holds, first. Then, for any possible point x, $0 = x - x$ is recurrent, by the first assertion of the theorem, thus proving (i). Conversely, if (i) holds, by the first assertion of the theorem and by Lemma 1.6.2, for any possible point x, $x = x - 0$ is recurrent. We have shown that $(i) \Leftrightarrow (ii)$ and are left to verify that for all possible points x and all recurrent points y, $x - y$ is recurrent. Holding such x and y fixed, define $\sigma_1 = \inf(k \geq 1 : S_k = y)$, $\sigma_2 = \inf(k \geq K_0 + \tau_1 : S_k = y)$, ..., where K_0 is a fixed constant that is to be chosen later on in this proof. (For now, you can think of $K_0 = 1$, in which case σ_j denotes the jth time the random walk hits y.) In general, for all $j \geq 1$, we define $\sigma_{j+1} = \inf(k \geq K_0 + \sigma_j : S_k = y)$, where $\inf \emptyset = \infty$, as usual. Since y is recurrent, $\sigma_j < \infty$ for all $j \geq 1$, with probability one. Now we define the events $\mathsf{E}_1, \mathsf{E}_2, \ldots$ as

$$\mathsf{E}_n = (S_k = x \text{ for some } \sigma_n < k < \sigma_{n+1}), \quad n \geq 1.$$

As k varies between σ_n and σ_{n+1}, the process S_k makes a loop, starting from y and ending at y. This loop is called an **excursion** from y, and E_n denotes the event that in the nth excursion from y, the random walk hits

x at some point. Equivalently,

$$\mathsf{E}_n = \big(S_{k+\sigma_n} - S_{\sigma_n} = x - y \text{ for some } 1 \leq k \leq \sigma_{n+1} - \sigma_n\big).$$

You should check that as a consequence of the strong Markov property, $\mathsf{E}_1, \mathsf{E}_2, \ldots$ are independent events and all have the same probability $\mathbb{P}(\mathsf{E}_1)$; cf. Theorem 1.2.1. Now is the time to choose K_0. Since x is possible, by choosing K_0 large enough, we can ensure that $\mathbb{P}(\mathsf{E}_1) > 0$. (Why?) Thus, by the Borel–Cantelli lemma, $\mathbb{P}(\mathsf{E}_n \text{ infinitely often}) = 1$. In particular, x is recurrent and, thanks to Lemma 1.6.2, so is $x - y$, as desired. □

1.7 Recurrence–Transience Dichotomy

Let S denote a $\mathbb{Z}^d$-valued random walk and P denote the collection of its possible values. According to Theorem 1.6.1, either all $x \in P$ are recurrent or they are all transient. This is the **recurrence–transience dichotomy**. The impossible values, of course, are never visited and have no effect on the structure of the random walk. On the other hand, at least in the presence of some recurrent values, all elements of P are recurrent and P is an additive group (Lemma 1.6.2 and Theorem 1.6.1).

Thus, when $P \neq \varnothing$, we can view S as a *Markov chain* on the group P. A little group theory will show that quite a bit more is true. Indeed, recall that $\mathbb{Z}^d$ is a free abelian group.[1] Since all subgroups of free abelian groups are free abelian,[2] Lemma 1.6.1 shows that P is itself a free abelian group. If $k \in \{1, \ldots, d\}$ denotes the rank of P, then P is *isomorphic* to $\mathbb{Z}^k$ (why an isomorphism and not just a homomorphism?). For us, this means that there exists a $k \times k$ invertible matrix $\mathbf{A}$ such that $\mathbf{A}P = \mathbb{Z}^k$. Since $S_n \in P$, a.s. for all $n \geq 1$, $\mathbf{A}S = \big(\mathbf{A}S_n;\ n \geq 1\big)$ is a random walk on $\mathbb{Z}^k$ and all points in $\mathbb{Z}^k$ are possible for this walk. Since $\mathbf{A}^{-1}$ exists, all statements about the P-valued Markov chain S translate to statements for the $\mathbb{Z}^k$-valued random walk $\mathbf{A}S$, and vice versa. Thus, it is no essential loss in generality to assume that S is itself a $\mathbb{Z}^d$-valued random walk for which all points in $\mathbb{Z}^d$ are possible.

---

[1] Let $\mathfrak{G}$ be a class of groups. Consider some $G \in \mathfrak{G}$ whose generator is the set $g = \{x_i;\ i \in I\}$. Recall that G is **freely generated** by g (within the class $\mathfrak{G}$) if for any group $G' \in \mathfrak{G}$ that is generated by $\{y_i;\ i \in I\}$, the map $x_i \mapsto y_i$ extends to a homomorphism (i.e., operation-preserving) $G \to G'$. The cardinality of I is the **rank** of G, and G is **free** within $\mathfrak{G}$. A **free abelian group** is a group that is free within the class of all abelian groups. While general free groups do not have much rank structure in a "dimensional" sense, free abelian groups do.

[2] This is an immediate consequence of the free abelian group theorem: Each subgroup of a free *abelian* group is itself a free abelian group. (Why is it a consequence?) See Kargapolov and Merzljakov (1979, Theorem 7.1.4, Chapter 3),

Proposition 1.7.1 *Suppose S is a $\mathbb{Z}^d$-valued random walk and let φ denote the characteristic function of the increments $\varphi(\xi) = \mathbb{E}[e^{i\xi \cdot X_1}]$, $\xi \in \mathbb{R}^d$. Then,*

$$\sum_{k=1}^{\infty} \mathbb{P}(S_k = 0) = (2\pi)^{-d} \lim_{\lambda \uparrow 1} \int_{[-\pi,\pi]^d} \operatorname{Re}\left(\frac{\varphi(\xi)}{1 - \lambda\varphi(\xi)}\right) d\xi.$$

Combining the above with Theorem 1.6.1, we conclude the following.

Corollary 1.7.1 *Suppose S is a $\mathbb{Z}^d$-valued random walk for which all points are possible. Let φ denote the characteristic function of the increments of S. Then, all $x \in \mathbb{Z}^d$ are transient, a.s., unless $\lim_{\lambda \uparrow 1} \int_{[-\pi,\pi]^d} \operatorname{Re}\{1 - \lambda\varphi(\xi)\}^{-1} d\xi < \infty$, in which case all points are recurrent, a.s.*

Bearing in mind the discussion in the beginning of this subsection, what the above states is that for *any* random walk on $\mathbb{Z}^d$, either all possible points are recurrent, or all possible points are transient (why?). In the latter case, we say that the random walk is **recurrent** and in the former case, **transient**. It is important to point out that if S is a transient walk for which all points are possible, then with probability one, $\lim_{n \to \infty} |S_n| = \infty$. The converse also holds, as the following shows.

Exercise **1.7.1** *S is transient if and only if $|S_n| \to \infty$, a.s.* □

Proof of Proposition 1.7.1 By the inversion theorem of Fourier analysis on the torus (or by the inversion theorem for discrete random variables), for all $k \geq 1$, $\mathbb{P}(S_k = 0) = (2\pi)^{-d} \int_{[-\pi,\pi]^d} \{\varphi(\xi)\}^k d\xi$. Thus, for all $\lambda \in \,]0, 1[$,

$$\sum_{k=1}^{\infty} \lambda^k \mathbb{P}(S_k = 0) = (2\pi)^{-d} \lambda \int_{[-\pi,\pi]^d} \operatorname{Re}\left(\frac{\varphi(\xi)}{1 - \lambda\varphi(\xi)}\right) d\xi,$$

since the left-hand side is real-valued. (Check this calculation!) To finish, simply let $\lambda \uparrow 1$. □

In fact, the following (surprisingly) subtle fact holds:[3]

$$\lim_{\lambda \uparrow 1} \int_{[-\pi,\pi]^d} \operatorname{Re}\left(\frac{\varphi(\xi)}{1 - \lambda\varphi(\xi)}\right) d\xi = \int_{[-\pi,\pi]^d} \operatorname{Re}\left(\frac{\varphi(\xi)}{1 - \varphi(\xi)}\right) d\xi.$$

We will not have need for this.

---

[3] Cf. Ornstein (1969) and Stone (1969). For a more complete result, see Port and Stone (1971b, Theorem 16.2).

2 Intersection Probabilities

A collection of N (≥ 2) independent $\mathbb{Z}^d$-valued random walks $S^1, S^2, \ldots, S^N$ are said to **intersect** if there exists $t \in \mathbb{N}^N$ such that $S^1_{t(1)} = \cdots = S^N_{t(N)}$. If we think of S^i_k as the position of particle i at time k, then $S^1, \ldots, S^N$ intersect if and only if the particle trajectories cross at some point. It should be recognized that such intersections are different from the **collisions** of $S^1, \ldots, S^N$. The latter happens when there exists $k \in \mathbb{N}$ such that $S^1_k = S^2_k = \cdots = S^N_k$. In words, $S^1, \ldots, S^N$ intersect if the trajectories of $S^1, \ldots, S^N$ intersect, while they collide if the particles $S^1_k, \ldots, S^N_k$ collide at some time k.

In light of the development in Section 1, collision problems are simpler to analyze. For instance, two independent random walks S^1 and S^2 collide infinitely often if and only if 0 is recurrent for the random walk $k \mapsto S^1_k - S^2_k$. In this section we study the more intricate problem of intersections of independent random walks.

Define the multiparameter $\mathbb{Z}^{dN}$-valued process $S = (S_t;\ t \in \mathbb{N}^N)$ by

$$S_t = (S^1_{t(1)}, \ldots, S^N_{t(N)}), \qquad t \in \mathbb{N}^N.$$

This means that the first d coordinates of S_t match those of $S^1_{t(1)}$, the second d coordinates of S_t are the coordinates of $S^2_{t(2)}$, and so on. It is apparent that for any $m \geq 1$ (finite or infinite) the ranges of $S^1, \ldots, S^N$ intersect m times if and only if S hits the diagonal of $\mathbb{Z}^{Nd}$ m times. If we write any $x \in \mathbb{Z}^{Nd}$ as $x = (x^1, \ldots, x^N)$ with $x^i \in \mathbb{Z}^d$, then the diagonal of $\mathbb{Z}^{Nd}$ is the set $\mathrm{diag}(\mathbb{Z}^{Nd}) = \{x \in \mathbb{Z}^{Nd} : x^1 = \cdots = x^N\}$. In direct product notation, we can write $x \in \mathbb{Z}^{Nd}$ as $x = x^1 \otimes \cdots \otimes x^N$, where $x^i \in \mathbb{Z}^d$. (For example, $(1,2,3,4) = (1,2) \otimes (3,4) = 1 \otimes 2 \otimes 3 \otimes 4$.) Since $S_t = S^1_{t(1)} \otimes \cdots \otimes S^N_{t(N)}$, we sometimes write the stochastic process S as $S = S^1 \otimes \cdots \otimes S^N$ and refer to $S^1, \ldots, S^N$ as the **coordinate processes** of S. To write things more explicitly, consider $N = 2$. Then, $S = S^1 \otimes S^2$ is a two-parameter process defined by $S_{(i,j)} = (S^1_i, S^2_j)$, $i, j \geq 1$. This means that the first d coordinates of $S_{(i,j)}$ are the d coordinates of S^1_i, and the next d coordinates of $S_{(i,j)}$ are those of S^2_j.

Henceforth, we will assume that all points are possible for $S^1, \ldots, S^N$. See Section 1.7 for a discussion of this assumption and how it can be essentially made without loss of generality.

2.1 Intersections of Two Walks

Let S^1 and S^2 denote two independent random walks on $\mathbb{Z}^d$ and let $S = S^1 \otimes S^2$ denote the associated 2-parameter process. We are interested in knowing when S hits the diagonal of $\mathbb{Z}^{2d}$ finitely often. In other words, we ask, "when is $\sum_{j=n}^\infty \sum_{k=m}^\infty \mathbb{1}_{(S^1_j = S^2_k)}$ finite for all choices of $n, m \geq 1$?" At the time of writing this book, this question seems unanswerable for completely general walks S^1 and S^2. However, we will give a comprehensive answer when S^1 and S^2 are both symmetric, i.e., when S_j^1 (respectively S_j^2) has the same distribution as $-S_j^1$ (respectively $-S_j^2$) for all $j \geq 1$; cf. Exercise 2.1.3 below for a further refinement.

According to the recurrence–transience dichotomy (Corollary 1.7.1 and its proceeding discussion), S^1 is either recurrent or transient, a.s. First, we address the easy case where S^1 (or equivalently, S^2) is recurrent.

Lemma 2.1.1 *If either of S^1 or S^2 is recurrent, then with probability one, there are infinitely many intersections.*

Exercise 2.1.1 Prove Lemma 2.1.1. □

According to Lemma 2.1.1, in our study of the intersections of S^1 and S^2 we can confine ourselves to the transient case.

Henceforth, S^1 and S^2 are symmetric walks, and $S_0^1 = S_0^2 = 0$.

Consider the function

$$G_\lambda(a,b) = \mathbb{E}\Big[\sum_{j=0}^\infty \sum_{k=0}^\infty \lambda^{j+k} \mathbf{1}_{(S_j^1 + a = S_k^2 + b)}\Big], \quad \lambda \in {]0,1[},\ a,b \in \mathbb{Z}^d. \quad (1)$$

Theorem 2.1.1 *Suppose S^1 and S^2 are symmetric, independent, transient random walks in $\mathbb{Z}^d$. Then, the following are equivalent:*

(i) $\lim_{\lambda \uparrow 1} G_\lambda(0,0) = +\infty$;

(ii) $\mathbb{P}(\sum_{j=1}^\infty \sum_{k=1}^\infty \mathbf{1}_{(S_j^1 = S_k^2)} < \infty) > 0$;

(iii) $\mathbb{P}(\sum_{j=1}^\infty \sum_{k=1}^\infty \mathbf{1}_{(S_j^1 = S_k^2)} < \infty) = 1$; *and*

(iv) $\sum_{j=1}^\infty \sum_{k=1}^\infty \mathbb{P}(S_j^1 = S_k^2) < \infty$.

The following technical lemma lies at the heart of Theorem 2.1.1 and seems to require symmetry.

Lemma 2.1.2 *Let φ^1 and φ^2 denote the characteristic functions of the increments of S^1 and S^2, respectively. Then, for all $\lambda \in {]0,1[}$,*

$$\sup_{a,b \in \mathbb{Z}^d} G_\lambda(a,b) = G_\lambda(0,0).$$

Proof By the inversion formula for characteristic functions,

$$\begin{aligned}
\mathbb{P}(S_j^1 + a = S_k^2 + b) &= (2\pi)^{-d} \int_{[-\pi,\pi]^d} e^{-i\xi \cdot (b-a)} \mathbb{E}[e^{i\xi \cdot S_j^1}] \mathbb{E}[e^{-i\xi \cdot S_k^2}]\, d\xi \\
&= (2\pi)^{-d} \int_{[-\pi,\pi]^d} e^{-i\xi \cdot (b-a)} \{\varphi^1(\xi)\}^j \{\varphi^2(-\xi)\}^k\, d\xi \\
&= (2\pi)^{-d} \int_{[-\pi,\pi]^d} e^{-i\xi \cdot (b-a)} \{\varphi^1(\xi)\}^j \{\varphi^2(\xi)\}^k\, d\xi.
\end{aligned}$$

In the last line, symmetry is used. Therefore,

$$G_\lambda(a,b) = (2\pi)^{-d} \int_{[-\pi,\pi]^d} e^{-i\xi\cdot(b-a)} \frac{1}{1-\lambda\varphi^1(\xi)} \frac{1}{1-\lambda\varphi^2(\xi)} d\xi. \quad (2)$$

On the other hand, $-1 \leq \varphi^1(\xi), \varphi^2(\xi) \leq 1$, which implies that $\{1 - \lambda\varphi^1(\xi)\}^{-1} \times \{1 - \lambda\varphi^2(\xi)\}^{-1}$ is nonnegative. Since $|e^{-i\xi\cdot(b-a)}| \leq 1$, the lemma follows. □

It may be helpful to note that the one-parameter version of this lemma *always* holds:

Exercise 2.1.2 Suppose S is a random walk on $\mathbb{Z}^d$ with $S_0 = 0$, and define for all $\lambda \in]0,1[$, $G_\lambda(a) = \mathbb{E}[\sum_{k=0}^\infty \lambda^k \mathbf{1}_{(S_k=a)}]$. Prove that even if S is not symmetric, $G_\lambda(a) \leq G_\lambda(0)$ for all $a \in \mathbb{Z}^d$ and all $\lambda \in]0,1[$. (HINT: Consider the first hitting time of a.) □

Proof of Theorem 2.1.1 It is clear that $(iii) \Rightarrow (ii)$. Conversely, it is not hard to check that $(ii) \Rightarrow (iii)$, thanks to the Hewitt–Savage 0-1 law; cf. Exercise 1.7.5, Chapter 1. Since $(i) \Leftrightarrow (iv) \Rightarrow (iii)$ is clear, it remains to prove that if (iv) fails, then so will (iii).

Define for all $n \geq 1$,

$$J_\lambda = \sum_{j=0}^\infty \sum_{k=0}^\infty \lambda^{j+k} \mathbf{1}_{(S_j^1 = S_k^2)}.$$

Note that $\mathbb{E}[J_\lambda] = G_\lambda(0,0)$; cf. equation (2).

Since (iv) is assumed to fail, $\lim_{\lambda \uparrow 1} \mathbb{E}[J_\lambda] = +\infty$. Our strategy, then, is to show the existence of a nontrivial constant A_1 such that

$$\mathbb{E}[J_\lambda^2] \leq A_1 \left(\mathbb{E}[J_\lambda]\right)^2, \quad \lambda \in]0,1[. \quad (3)$$

Assuming this, we can finish our proof: Apply equation (3) and the Paley–Zygmund lemma (Lemma 1.4.1) to see that

$$\mathbb{P}\left(\sup_{\lambda \in]0,1[} J_\lambda = +\infty\right) \geq \lim_{\lambda \uparrow 1} \mathbb{P}(J_\lambda \geq \tfrac{1}{2}\mathbb{E}[J_\lambda]) \geq A_1^{-1},$$

which is positive. Thus, it remains to verify equation (3).

We can write $\mathbb{E}[J_\lambda^2] \leq 2(T_1 + T_2)$, where

$$T_1 = \sum_{i \leq i'} \sum_{j \leq j'} \lambda^{i+i'+j+j'} \mathbb{P}(S_i^1 = S_j^2, S_{i'}^1 = S_{j'}^2),$$

$$T_2 = \sum_{i \leq i'} \sum_{j' \leq j} \lambda^{i+i'+j+j'} \mathbb{P}(S_i^1 = S_j^2, S_{i'}^1 = S_{j'}^2).$$

(Why?) Next, we write $T_1 = T_{11} + T_{12} + T_{13}$, where

$$T_{11} = \sum_{i<i'}\sum\sum_{j<j'}\sum \lambda^{i+i'+j+j'} \mathbb{P}(S_i^1 = S_j^2, S_{i'}^1 = S_{j'}^2),$$

$$T_{12} = \sum_{i=0}^{\infty}\sum\sum_{j<j'} \lambda^{2i+j+j'} \mathbb{P}(S_i^1 = S_j^2 = S_{j'}^2),$$

$$T_{13} = \sum_{j=0}^{\infty}\sum\sum_{i<i'} \lambda^{i+i'+2j} \mathbb{P}(S_i^1 = S_j^2 = S_{i'}^1).$$

Similarly, we write $T_2 = T_{21} + T_{12} + T_{13}$, where

$$T_{21} = \sum_{i<i'}\sum\sum_{j'<j}\sum \lambda^{i+i'+j+j'} \mathbb{P}(S_i^1 = S_j^2, S_{i'}^1 = S_{j'}^2).$$

We now estimate the T_{ij}'s in turn.

By the Markov property,

$$T_{11} = \sum_{i<i'}\sum\sum_{j<j'}\sum \lambda^{i+i'+j+j'} \mathbb{P}(S_i^1 = S_j^2)\mathbb{P}(S_{i'-i}^1 = S_{j'-j}^2)$$

$$= \sum_{i<i'}\sum\sum_{j<j'}\sum \lambda^{2i+2j+(i'-i)+(j'-j)} \mathbb{P}(S_i^1 = S_j^2)\mathbb{P}(S_{i'-i}^1 = S_{j'-j}^2)$$

$$\leq \bigl(\mathbb{E}[J_\lambda]\bigr)^2.$$

On the other hand,

$$T_{12} \leq \sum_{i=0}^{\infty}\sum\sum_{j<j'} \lambda^{i+j+j'} \mathbb{P}(S_i^1 = S_j^2)\mathbb{P}(S_{j'-j}^2 = 0) \leq A_2 \mathbb{E}[J_\lambda],$$

where $A_2 = \sum_{i=0}^{\infty} \mathbb{P}(S_i^2 = 0)$. Of course, since S^2 is transient, $A_2 < +\infty$; cf. Theorem 1.4.1. In similar fashion we obtain $T_{13} \leq A_3 \mathbb{E}[J_\lambda]$, where $A_3 = \sum_{i=0}^{\infty} \mathbb{P}(S_i^1 = 0)$ is finite as well. Since we have assumed (iv) of the theorem, our job is complete, once we show that there exists a nontrivial constant A_4 such that for all $n \geq 1$,

$$T_{21} \leq A_4 \bigl(\mathbb{E}[J_\lambda]\bigr)^2. \tag{4}$$

Indeed, from this, equation (3), and hence the theorem, follows.

We observe that T_{21} equals

$$\sum_{i<i'}\sum\sum_{j'<j}\sum \lambda^{i+i'+j+j'} \mathbb{P}\bigl(S_i^1 = S_{j'}^2 + [S_j^2 - S_{j'}^2], S_i^1 + [S_{i'}^1 - S_i^1] = S_{j'}^2\bigr)$$

$$= \sum_{i<i'}\sum\sum_{j'<j}\sum \lambda^{i+i'+j+j'} \mathbb{P}\bigl(S_i^1 = S_{j'}^2 + \bar{S}_{j-j'}^2, S_i^1 + \bar{S}_{i'-i}^1 = S_{j'}^2\bigr),$$

where $(\bar{S}_u^1, \bar{S}_v^2)$ is an independent copy of (S_u^1, S_v^2) for any two integers $u, v \geq 1$. This is a consequence of the Markov property; cf. Corollary 1.1.1. Consequently,

$$T_{21} = \sum_{i<i'} \sum_{j'<j} \sum \sum \lambda^{i+i'+j+j'} \mathbb{P}\big(S_i^1 = S_{j'}^2 + \bar{S}_{j-j'}^2 \,,\, \bar{S}_{i'-i}^1 = \bar{S}_{j-j'}^2\big)$$

$$\leq \sum_{i=0}^{\infty} \sum_{j=0}^{\infty} \sum_{u=0}^{\infty} \sum_{v=0}^{\infty} \lambda^{i+j+u+v} \mathbb{P}\big(S_i^1 = S_j^2 + \bar{S}_v^2 \,,\, \bar{S}_u^1 = -\bar{S}_v^2\big)$$

$$= \sum_{u=0}^{\infty} \sum_{v=0}^{\infty} \lambda^{u+v} \mathbb{E}\big[G_\lambda(0, \bar{S}_v^2) \mathbb{1}_{(\bar{S}_u^1 = -\bar{S}_v^2)}\big],$$

by independence and by equation (1). Thanks to Lemma 2.1.2,

$$T_{21} \leq G_\lambda(0,0) \sum_{u=0}^{\infty} \sum_{v=0}^{\infty} \lambda^{u+v} \mathbb{P}(S_u^1 = -S_v^2)$$

$$= \mathbb{E}[J_\lambda] \sum_{u=0}^{\infty} \sum_{v=0}^{\infty} \lambda^{u+v} \mathbb{P}(S_u^1 = -S_v^2)$$

$$= \big(\mathbb{E}[J_\lambda]\big)^2.$$

To follow up, the first line follows from the fact that $(\bar{S}_u^1, \bar{S}_v^2)$ has the same distribution as (S_u^1, S_v^2). The second line is from the definition of J_λ, and the third line follows from the symmetry hypothesis of the theorem. This verifies equation (4) and completes our task. □

Exercise **2.1.3** A characteristic function φ, on $\mathbb{R}^d$, is said to satisfy the **sector condition** if there exists a constant $A > 0$ such that

$$|\mathrm{Im}\,\varphi(\xi)| \leq A\{1 + |\mathrm{Re}\,\varphi(\xi)|\}, \qquad \xi \in \mathbb{R}^d.$$

Suppose S^1 and S^2 are independent random walks on $\mathbb{Z}^d$, whose increments have characteristic functions that satisfy the sector condition. Prove that Theorem 2.1.1 remains valid in this setting. □

Theorem 2.1.1 states that, under the given conditions, the trajectories[4] of S^1 and S^2 intersect infinitely many times if and only if $\sum_{j,k \geq 1} \mathbb{P}(S_j^1 = S_k^2) = \infty$. By a summability argument (see the described proof of Proposition 1.7.1), the latter can be written as follows.

Proposition 2.1.1 *We have*

$$\sum_{j,k=1}^{\infty} \mathbb{P}(S_j^1 = S_k^2) = (2\pi)^{-d} \lim_{\lambda \uparrow 1} \int_{[-\pi,\pi]^d} \frac{\varphi^1(\xi)}{1 - \lambda \varphi^1(\xi)} \frac{\varphi^2(\xi)}{1 - \lambda \varphi^2(\xi)} \, d\xi.$$

---

[4]Throughout this book, the *trajectories* of a stochastic process $(X_t;\ t \in T)$ are the realizations of the (random) function $t \mapsto X_t$, for any index set T.

***Exercise* 2.1.4** Verify Proposition 2.1.1. □

2.2 An Estimate for Two Walks

Let S^1 and S^2 be two independent $\mathbb{Z}^d$-valued random walks. According to Theorem 2.1.1, we can conclude that $\sum_{j,k\geq 1} \mathbb{P}(S_j^1 = S_k^2) < \infty$ is a necessary and sufficient condition for

$$\lim_{n,m\to\infty} \mathbb{P}(S_j^1 = S_k^2 \text{ for some } (j,k) \succcurlyeq (n,m)) = 0,$$

provided that the random walks are symmetric. We now explore the rate at which the above probability tends to 0, under the extra condition that there exists C_0 such that whenever $\mathbb{P}(S_i^1 = S_j^2) > 0$,

$$\mathbb{P}(S_i^1 = S_j^2 + a) \leq C_0 \mathbb{P}(S_i^1 = S_j^2). \tag{1}$$

This is a unimodality-type condition and is verified, for instance, when S^1 and S^2 are so-called simple random walks; cf. Section 3.

Theorem 2.2.1 *Suppose S^1 and S^2 are two symmetric and independent $\mathbb{Z}^d$-valued random walks that satisfy condition (1). If $\sum_{j,k\geq 1} \mathbb{P}(S_j^1 = S_k^2) < \infty$, there exist nontrivial constants C_1 and C_2 such that for all $n, m \geq 1$,*

$$C_1 \sum_{j=n}^{\infty} \sum_{k=m}^{\infty} \mathbb{P}(S_j^1 = S_k^2) \leq \mathbb{P}(S_j^1 = S_k^2 \text{ for some } (j,k) \succcurlyeq (n,m))$$

$$\leq C_2 \sum_{j=n}^{\infty} \sum_{k=m}^{\infty} \mathbb{P}(S_j^1 = S_k^2).$$

Proof Define for all $n, m \geq 1$,

$$J_{n,m} = \sum_{j=n}^{\infty} \sum_{k=m}^{\infty} \mathbb{1}_{(S_j^1 = S_k^2)}.$$

Arguing as we did in Theorem 2.1.1, we can show that there exist nontrivial constants C_3 and C_4 such that for all $n, m \geq 1$, $\mathbb{E}[J_{n,m}^2] \leq C_3(\mathbb{E}[J_{n,m}])^2 + C_4 \mathbb{E}[J_{n,m}]$; this uses (1), as well as symmetry. Since $\mathbb{E}[J_{n,m}]$ goes to zero as $n, m \to \infty$, we can deduce the existence of a finite constant C_1 such that

$$\mathbb{E}[J_{n,m}^2] \leq \frac{\mathbb{E}[J_{n,m}]}{C_1}. \tag{2}$$

The details are delegated to Supplementary Exercise 6. By the Paley–Zygmund lemma (Lemma 1.4.1),

$$\mathbb{P}(J_{n,m} > 0) \geq C_1 \mathbb{E}[J_{n,m}],$$

which is the desired probability lower bound.

To demonstrate the corresponding upper bound, for all $n, m \geq 1$, let $\mathcal{F}_{n,m}$ define the σ-field generated by $((S_i^1, S_j^2); 1 \leq i \leq n, 1 \leq j \leq m)$. By Exercise 3.4.2 of Chapter 1, $\mathcal{F} = (\mathcal{F}_{n,m}; n, m \geq 1)$ is a commuting filtration in the sense of Chapter 1. Fix $(n, m) \in \mathbb{N}^2$ and define

$$M_{p,q} = \mathbb{E}(J_{n,m} \,|\, \mathcal{F}_{p,q}), \qquad (p, q) \succcurlyeq (n, m).$$

By the Markov property (Corollary 1.1.1),

$$M_{p,q} \geq \sum_{i=p}^{\infty} \sum_{j=q}^{\infty} \mathbb{P}(S_i^1 = S_j^2 \,|\, \mathcal{F}_{p,q}) \mathbf{1}_{(S_p^1 = S_q^2)}$$

$$= \left\{ \sum_{i=1}^{\infty} \sum_{j=1}^{\infty} \mathbb{P}(S_{i+p}^1 - S_p^1 = S_{j+q}^2 - S_q^2 \,|\, \mathcal{F}_{p,q}) + 1 \right\} \mathbf{1}_{(S_p^1 = S_q^2)}$$

$$= \left\{ \sum_{i=1}^{\infty} \sum_{j=1}^{\infty} \mathbb{P}(S_i^1 = S_j^2) + 1 \right\} \mathbf{1}_{(S_p^1 = S_q^2)}$$

$$= \sqrt{\frac{16}{C_2 C_1}} \, \mathbf{1}_{(S_p^1 = S_q^2)}.$$

(This defines C_2.) It is clear that $M = (M_t; t \in \mathbb{N}^2)$ is a two-parameter martingale with respect to the (commuting) filtration $\mathcal{F}$. Thus, by Cairoli's strong $(2, 2)$ inequality (Theorem 2.3.1 and Corollary 3.5.1 of Chapter 1),

$$\mathbb{P}(S_q^1 = S_q^2 \text{ for some } (p, q) \succcurlyeq (n, m)) \leq \frac{C_2 C_1}{16} \mathbb{E}\!\left[\sup_{(p,q) \succcurlyeq (n,m)} M_{p,q}^2\right]$$

$$\leq C_2 C_1 \mathbb{E}[J_{n,m}^2].$$

The probability upper bound follows from this and equation (1). □

2.3 Intersections of Several Walks

We are ready to consider the general problem of when and how often N independent random walks in $\mathbb{Z}^d$ intersect, when $N \geq 2$ is an arbitrary integer. This will be achieved by extending the two-parameter methods of Section 2.1 to N parameters.

Let $S^1, \ldots, S^N$ denote N independent $\mathbb{Z}^d$-valued random walks. The following can be proved in complete analogy to Lemma 2.1.1.

Lemma 2.3.1 *If any one of the coordinate processes is recurrent and if the trajectories of the remaining $N - 1$ coordinate processes intersect infinitely many times, then for all $t \in \mathbb{N}^N$, $\sum_{s \succcurlyeq t} \mathbf{1}_{(S_{s(1)}^1 = \cdots = S_{s(N)}^N)} = \infty$, almost surely.*

***Exercise* 2.3.1** Prove Lemma 2.3.1. □

In particular, we need to consider only the case where all of the coordinate processes are transient. By Theorem 1.4.1, this happens precisely when $\sum_{k=1}^{\infty} \mathbb{P}(S_k^i = 0) < \infty$, for all $i = 1, \ldots, N$, a condition that we will assume tacitly from now on.

Let $S_0^1 = \cdots = S_0^N = 0$ and define the N-variable version of equation (1) of Section 2.2 as

$$G_\lambda(a_1, \ldots, a_N) = \mathbb{E}\left[\sum_{0 \le i_1, \ldots, i_N} \lambda^{i_1 + \cdots + i_N} \mathbf{1}_{(S_{i_1}^1 + a_1 = S_{i_2}^2 + a_2 = \cdots = S_{i_N}^N + a_N)} \right], \quad (1)$$

where $a_1, \ldots, a_N \in \mathbb{Z}^d$ and $\lambda \in]0, 1[$. One can prove the following.

Proposition 2.3.1 *Suppose $S^1, \ldots, S^N$ are N symmetric and independent $\mathbb{Z}^d$-valued random walks whose increments have characteristic functions $\varphi^1, \ldots, \varphi^N$, respectively. Then, $G_\lambda(a_1, \ldots, a_N) \le G_\lambda(0, \ldots, 0)$ for all $a_1, \ldots, a_N \in \mathbb{Z}^d$. Moreover,*

$$G_\lambda(0, \ldots, 0) = (2\pi)^{-d(N-1)} \int_{[-\pi, \pi]^{d(N-1)}} F(\xi; \lambda) \, d\xi,$$

where for all $\xi \in [-\pi, \pi]^{d(N-1)}$ and all $\lambda \in]0, 1[$,

$$F(\xi; \lambda) = \frac{1}{1 - \lambda \varphi^1\left(-\sum_{\ell=1}^{N-1} \xi^{(\ell)}\right)} \cdot \prod_{j=2}^{N-1} \frac{1}{1 - \lambda \varphi^j(\xi)}.$$

***Exercise* 2.3.2** Prove Proposition 2.3.1. □

Theorem 2.3.1 *Suppose $S^1, \ldots, S^N$ are symmetric, independent, $\mathbb{Z}^d$-valued, transient random walks. Then, the following are equivalent:*

(i) With positive probability, $\sum_{t \in \mathbb{N}^N} \mathbf{1}_{(S_{t(1)}^1 = \cdots = S_{t(N)}^N)} < \infty$;

(ii) With probability one, $\sum_{t \in \mathbb{N}^N} \mathbf{1}_{(S_{t(1)}^1 = \cdots = S_{t(N)}^N)} < \infty$; and

(iii) $\sum_{t \in \mathbb{N}^N} \mathbb{P}(S_{t(1)}^1 = \cdots = S_{t(N)}^N) < \infty$.

We provide only a sketch of the proof.

Sketch of Proof In light of the presented proof of Theorem 2.1.1, $(iii) \Rightarrow (ii) \Leftrightarrow (i)$ follows readily; it remains to show that if (iii) fails, then so does (ii).

For all $n \ge 1$, define

$$J_\lambda = \sum_{s \in \mathbb{N}_0^N} \lambda^{s^{(1)} + \cdots + s^{(N)}} \mathbf{1}_{(S_{s(1)}^1 = \cdots = S_{s(N)}^N)}, \qquad \lambda \in]0, 1[.$$

Our goal is to show that if (iii) fails, $\sup_{\lambda \in]0,1[} J_\lambda = \infty$, with positive probability. It is this argument that we merely sketch. Since $\lim_{\lambda \uparrow 1} \mathbb{E}[J_\lambda] = +\infty$, it suffices to exhibit a finite C_1 such that $\mathbb{E}[J_\lambda^2] \leq C_1 (\mathbb{E}[J_\lambda])^2$ for all $\lambda \in]0,1[$. Once this is accomplished, the remainder of our argument follows our proof of Theorem 2.1.1 quite closely.

Clearly,

$$\mathbb{E}[J_\lambda^2] = \sum\sum_{i_1, i'_1 \geq 0} \cdots \sum\sum_{i_N, i'_N \geq 0} \lambda^{\sum_{\ell=1}^N (i_\ell + i'_\ell)} \mathbb{P}\big(S_{i_1}^1 = \cdots = S_{i_N}^N, S_{i'_1}^1 = \cdots = S_{i'_N}^N\big).$$

Let us consider the contribution to the above when $i_1 = i'_1$:

$$\sum_{i_1=0}^\infty \sum_{i_2, i'_2 \geq 0} \cdots \sum_{i_N, i'_N \geq 0} \lambda^{2i_1 + \sum_{\ell=2}^N (i_\ell + i'_\ell)}$$
$$\times \mathbb{P}\big(S_{i_1}^1 = S_{i_2}^2 = \cdots = S_{i_N}^N, S_{i_1}^1 = S_{i'_2}^2 = \cdots = S_{i'_N}^N\big)$$
$$\leq (N-1)! \sum_{i_1=0}^\infty \sum_{0 \leq i_2 \leq i'_2} \cdots \sum_{0 \leq i_N \leq i'_N} \lambda^{\sum_{\ell=1}^N i_\ell + \sum_{\ell=2}^N (i'_\ell - i_\ell)}$$
$$\times \mathbb{P}\big(S_{i_1}^1 = \cdots = S_{i_N}^N\big) \prod_{\ell=2}^N \mathbb{P}(S_{i'_\ell - i_\ell}^\ell = 0)$$
$$\leq C_2 \mathbb{E}[J_\lambda],$$

for some finite constant C_2 that is independent of $\lambda \in [0,1]$. By symmetry,

$$\mathbb{E}[J_\lambda^2] \leq C_3 \mathbb{E}[J_\lambda] + \sum\sum_{\substack{0 \leq i_1, i'_1 \\ i_1 \neq i'_1}} \cdots \sum\sum_{\substack{0 \leq i_N, i'_N \\ i_N \neq i'_N}} \lambda^{\sum_{\ell=1}^N i_\ell + \sum_{\ell=2}^N (i'_\ell - i_\ell)} \mathcal{Q},$$

where $\mathcal{Q} = \mathbb{P}(S_{i_1}^1 = \cdots = S_{i_N}^N, S_{i'_1}^1 = \cdots = S_{i'_N}^N)$. A little thought shows that, over the range in question,

$$\mathcal{Q} = \mathbb{P}\begin{pmatrix} S_{i_1 \wedge i'_1}^1 + \bar{S}_{i_1 - (i_1 \wedge i'_1)}^1 = \cdots = S_{i_N \wedge i'_N}^N + \bar{S}_{i_N - (i_N \wedge i'_N)}^N \\ \text{and} \\ S_{i_1 \wedge i'_1}^1 + \bar{S}_{i'_1 - (i'_1 \wedge i_1)}^1 = \cdots = S_{i_N \wedge i'_N}^N + \bar{S}_{i'_N - (i'_N \wedge i_N)}^N \end{pmatrix},$$

where $(S_{u(1)}^1, \ldots, S_{u(N)}^N)$ and $(\bar{S}_{u(1)}^1, \ldots, \bar{S}_{u(N)}^N)$ are independent copies of one another for each $u \in \mathbb{N}^N$. Solving, we get

$$\mathcal{Q} \leq \mathbb{P}\begin{pmatrix} S_{i_1 \wedge i'_1}^1 + \bar{S}_{i_1 - (i_1 \wedge i'_1)}^1 = \cdots = S_{i_N \wedge i'_N}^N + \bar{S}_{i_N - (i_N \wedge i'_N)}^N \\ \text{and} \\ |\bar{S}_{|i'_1 - i_1|}^1| = \cdots = |\bar{S}_{|i'_N - i_N|}^N| \end{pmatrix}.$$

The rest of the proof follows from changing variables ($j_\ell = |i'_\ell - i_\ell|$) and follows the $N = 2$ argument very closely, except that we now use Proposition 2.3.1 in place of Lemma 2.1.2. $\square$

2.4 An Estimate for N Walks

We consider the problem of the previous subsection in the case where the number of intersections is finite, almost surely. Under the hypotheses of Theorem 2.3.1, this is to say that $\sum_{t \in \mathbb{N}^N} \mathbb{P}(S^1_{t(1)} = \cdots = S^N_{t(N)}) < \infty$. The question that we address now is, *how large is* $\mathbb{P}(S^1_{s(1)} = \cdots = S^N_{s(N)}$ *for some* $s \succcurlyeq t$) *when* $t \in \mathbb{N}^N$ *is large, coordinatewise?* When $N = 2$, this was achieved in Theorem 2.2.1; the general case follows under the following unimodality analogue of equation (1) of Section 2.2: There exists a finite constant C_0 such that

$$\sup_{a_1,\ldots,a_N \in \mathbb{Z}^d} \mathbb{P}(S^1_{i_1} + a_1 = \cdots = S^N_{i_N} + a_N) \leq C_0 \mathbb{P}(S^1_{i_1} = \cdots = S^N_{i_N}), \quad (1)$$

as long as the right-hand side is positive.

Theorem 2.4.1 *Suppose* $S^1, \ldots, S^N$ *are independent* $\mathbb{Z}^d$*-valued random walks and for all* $t \in \mathbb{N}^N$, *let* $\psi(t) = \mathbb{P}(S^1_{t(1)} = \cdots = S^N_{t(N)})$, *and assume that these walks satisfy condition (1) above. If* $\sum_{t \in \mathbb{N}^N} \psi(t) < \infty$, *there exist finite constants* C_1 *and* C_2 *such that for all* $t \in \mathbb{N}^N$,

$$C_1 \sum_{s \succcurlyeq t} \psi(s) \leq \mathbb{P}(S^1_{s(1)} = \cdots = S^N_{s(N)} \text{ for some } s \succcurlyeq t) \leq C_2 \sum_{s \succcurlyeq t} \psi(s).$$

One can prove this by finding a suitable N-parameter modification of the two-parameter argument used to prove Theorem 2.2.1.

Exercise 2.4.1 (Hard) Prove Theorem 2.4.1. □

3 The Simple Random Walk

Nearest-neighborhood random walks on $\mathbb{Z}^d$ are random walks that can move only to the nearest point in $\mathbb{Z}^d$. Indeed, let $(e_1, \ldots, e_d)$ denote the usual basis for $\mathbb{R}^d$. That is, for all $i, j \in \{1, \ldots, d\}$, $e^{(i)}_j$ equals 1 if $i = j$, and it equals 0 otherwise. Consider a $\mathbb{Z}^d$-valued random walk $S = (S_k; k \geq 1)$ with increments $X_1, X_2, \ldots$. We say that S is a **nearest-neighborhood random walk** if with probability one, $X_1 \in \{\pm e_1, \ldots, \pm e_d\}$. Nearest-neighborhood random walks form some of the most common models for the motion of a randomly moving particle. An important member of this family of random walks is the simple random walk. A random walk S is said to be simple if it is truly unbiased in its motion. More precisely, S is a **simple random walk** if $\mathbb{P}(X_1 = e_1) = \mathbb{P}(X_1 = -e_1) = \cdots = \mathbb{P}(X_1 = e_d) = \mathbb{P}(X_1 = -e_d) = (2d)^{-1}$. In this section we put the general theory of Section 2 to test by way of explicit calculations.

Let us recall that for all $x \in \mathbb{R}^k$, $|x| = \max_{1 \leq \ell \leq k} |x^{(\ell)}|$ and $\|x\| = \{\sum_{\ell=1}^k |x^{(\ell)}|^2\}^{\frac{1}{2}}$ denote the ℓ^∞ and ℓ^2 norms of x, respectively.

3.1 Recurrence

We now wish to study the recurrence properties of a simple random walk S in $\mathbb{Z}^d$ with increments $X_1, X_2, \ldots$. The following elementary result is a first step in this direction.

Lemma 3.1.1 *All points are possible for a simple random walk.*

Exercise 3.1.1 Prove Lemma 3.1.1. □

Thus, according to Corollary 1.7.1, S is either recurrent or transient. In order to decide which is the case, we first need a technical lemma.

Lemma 3.1.2 *The integral $\int_{[0,1]^d} \|\xi\|^{-\beta} d\xi$ is finite if and only if $\beta < d$.*

Proof Recall that $\|\xi\| = \{\sum_{j=1}^{N}(\xi^{(j)})^2\}^{\frac{1}{2}}$, while $|\xi| = \max_{1 \leq j \leq N} |\xi^{(j)}|$. The traditional approach to this sort of problem is to estimate the integral in polar coordinates; we will do this in probabilistic language. First, note that $|\xi| \leq \|\xi\| \leq d^{\frac{1}{2}}|\xi|$. Therefore, $\int_{[0,1]^d} \|\xi\|^{-\beta} d\xi < \infty$ if and only if $\int_{[0,1]^d} |\xi|^{-\beta} d\xi < \infty$. Let U be a random variable that is uniformly picked on $[0,1]^d$. The problem is to decide when $\mathbb{E}\{|U|^{-\beta}\}$ is finite. On the other hand, a direct calculation shows that $\sum_{n \geq 1} \mathbb{P}(|U|^{-\beta} \geq n) = \sum_{n \geq 1} n^{-d/\beta}$, which is finite iff $d > \beta$. □

Theorem 3.1.1 *Let S denote the simple random walk in $\mathbb{Z}^d$. Then S is recurrent if $d \leq 2$; otherwise, S is transient.*

Proof Let φ denote the characteristic function of X_1. It is easy to check that

$$\varphi(\xi) = \frac{1}{d} \sum_{\ell=1}^{d} \cos(\xi^{(\ell)}), \qquad \xi \in \mathbb{R}^d. \tag{1}$$

Since $\varphi(\xi) \geq 0$ (and is, of course, real) for all $\xi \in [-1,1]^d$, we can apply the bounded and monotone convergence theorems to Proposition 1.7.1 to see that

$$\sum_{n=1}^{\infty} \mathbb{P}(S_n = 0) = (2\pi)^{-d} \int_{[-\pi,\pi]^d} \frac{\varphi(\xi)}{1 - \varphi(\xi)} d\xi.$$

(Why?) Equivalently, we apply symmetry to deduce

$$1 + \sum_{n=1}^{\infty} \mathbb{P}(S_n = 0) = \pi^{-d} \int_{[0,\pi]^d} \left(1 - \frac{1}{d} \sum_{\ell=1}^{d} \cos(\xi^{(\ell)})\right)^{-1} d\xi.$$

By Theorem 1.4.1, it suffices to show that the above integral is finite if and only if $d \geq 3$. Owing to Taylor's theorem with remainder, for all y there exists a λ between 0 and y such that

$$\cos(y) = 1 - \frac{y^2}{2} + \frac{\lambda^4}{12}.$$

Hence, for all $y \in [0,1]$,

$$1 - \frac{y^2}{2} \leq \cos(y) \leq 1 - y^2\left(\frac{1}{2} - \frac{1}{12}\right) = 1 - \frac{5}{12}y^2. \tag{2}$$

This, in turn, implies the inequality

$$2d \int_{[0,1]^d} \|\xi\|^{-2} \, d\xi \leq \int_{[0,1]^d} \left(1 - \frac{1}{d}\sum_{\ell=1}^{d} \cos(\xi^{(\ell)})\right)^{-1} d\xi \leq \frac{12d}{5} \int_{[0,1]^d} \|\xi\|^{-2} \, d\xi.$$

Since d is an integer, by Lemma 3.1.2, $\int_{[0,1]^d}(1 - d^{-1}\sum_{\ell=1}^{d}\cos(\xi^{(\ell)}))^{-1} d\xi$ is finite if and only if $d \geq 3$. Our proof is concluded once we show that $\int_K (1 - d^{-1}\sum_{\ell=1}^{d}\cos(\xi^{(\ell)}))^{-1} d\xi < \infty$, where $K = [0,\pi]^d \setminus [0,1]^d$. To observe this, note that whenever $\xi \in K$, there is at least one $\ell \in \{1,\ldots,N\}$ such that $\cos(\xi^{(\ell)}) \leq \cos(1)$. For such ξ's, we can conclude that

$$1 - d^{-1} \sum_{\ell=1}^{d} \cos(\xi^{(\ell)}) \geq d^{-1}[1 - \cos(1)].$$

Since $\cos(1) < 1$,

$$\int_K \left(1 - \frac{1}{d}\sum_{\ell=1}^{d} \cos(\xi^{(\ell)})\right)^{-1} d\xi \leq \frac{d}{1 - \cos(1)} \operatorname{Leb}(K). \tag{3}$$

Clearly, $\operatorname{Leb}(K) \leq \pi^d < \infty$, which proves the result. $\square$

Theorem 3.1.1 is deeply related to the following:

Exercise 3.1.2 (Hard) If S denotes the simple walk in $\mathbb{Z}^d$, then there exists a finite constant $C > 1$ such that for all $n \geq 1$,

$$C^{-1} n^{-\frac{d}{2}} \leq \mathbb{P}(S_{2n} = 0) \leq \sup_{a \in \mathbb{Z}^d} \mathbb{P}(S_{2n} = a) \leq C n^{-\frac{d}{2}}.$$

(HINT: Use the inversion theorem for characteristic functions and write $\mathbb{P}(S_{2n} = 0)$ as $(2\pi)^{-\frac{d}{2}} \int_{[-\pi,\pi]^d} \mathbb{E}[e^{i\xi \cdot S_{2n}}] \, d\xi$. Use the fact that S has i.i.d. increments and expand this integral near $\xi = 0$. Alternatively, look at Durrett (1991) under "local central limit theorem.") $\square$

Exercise 3.1.3 Use Exercise 3.1.2, together with Theorem 1.4.1, to construct an alternative proof of Theorem 3.1.1. $\square$

3.2 Intersections of Two Simple Walks

Given two independent $\mathbb{Z}^d$-valued simple random walks, when do their trajectories intersect infinitely often? In other words, if the random walks are denoted by S^1 and S^2, when can we conclude that $\sum_{j,k\geq 1} \mathbf{1}_{(S_j^1 = S_k^2)} = +\infty$?

Theorem 3.2.1 *Suppose S^1 and S^2 are independent simple random walks in $\mathbb{Z}^d$. With probability one, the trajectories of S^1 and S^2 intersect infinitely often if and only if $d \leq 4$.*

Proof When $d \leq 2$, S^1 and S^2 are recurrent; cf. Theorem 3.1.1. By Lemma 2.1.1, we can assume with no loss of generality that $d \geq 3$, i.e., that S^1 and S^2 are transient. Let φ denote the characteristic function of the increments of S^1 and/or S^2, since they have the same distribution. By Corollary 1.7.1, $\lim_{\lambda \uparrow 1} \int_{[-\pi,\pi]^d} \{1 - \lambda \varphi(\xi)\}^{-1} d\xi < \infty$. Since $\varphi(\xi) \geq 0$ for all $\xi \in [-1,1]^d$, the bounded and monotone convergence theorems together show us that

$$\int_{[-\pi,\pi]^d} \frac{1}{1 - \varphi(\xi)} d\xi < \infty.$$

Once again applying the bounded and monotone convergence theorems, this time via Proposition 2.1.1, we obtain the following:

$$(2\pi)^d \sum_{j,k=1}^{\infty} \mathbb{P}(S_j^1 = S_k^2) = \int_{[-\pi,\pi]^d} \frac{[\varphi(\xi)]^2}{\{1 - \varphi(\xi)\}^2} d\xi$$

$$= \int_{[-\pi,\pi]^d} \{1 - \varphi(\xi)\}^{-2} d\xi$$

$$- \int_{[-\pi,\pi]^d} \{1 + \varphi(\xi)\}\{1 - \varphi(\xi)\}^{-1} d\xi.$$

The second integral is finite. In fact, it is positive and bounded above by $2 \int_{[-\pi,\pi]^d} \{1 - \varphi(\xi)\}^{-1} d\xi < +\infty$. Thanks to symmetry and by Theorem 2.1.1, it suffices to show that $\int_{[0,\pi]^d} \{1 - \varphi(\xi)\}^{-2} d\xi < \infty$ if and only if $d \geq 5$.

Following the demonstration of Theorem 3.1.1, we split the integral in two parts: where $\xi \in [0,1]^d$ and where $\xi \in K = [0,\pi]^d \setminus [0,1]^d$. As in the derivation of equation (3) of Section 3.1, $\int_K \{1 - \varphi(\xi)\}^{-2} d\xi \leq d^2 \pi^d \{1 - \cos(1)\}^{-2}$, which is always finite. It remains to show that $\int_{[0,1]^d} \{1 - \varphi(\xi)\}^{-2} d\xi$ is finite if and only if $d \geq 5$.

Using equation (2) of Section 3.1,

$$(2d)^2 \int_{[0,1]^d} \|\xi\|^{-4} d\xi \leq \int_{[0,1]^d} \{1 - \varphi(\xi)\}^{-2} d\xi \leq \left(\frac{12d}{5}\right)^2 \int_{[0,1]^d} \|\xi\|^{-4} d\xi.$$

We obtain the result from Lemma 3.1.2. □

The next question that we address is, *when do three or more independent simple random walks intersect infinitely many times?* When $d \geq 5$, the above theorem states that the answer is *never*, a.s. On the other hand, when $d \leq 2$, Theorem 3.1.1 implies that the random walks in question are recurrent; Lemmas 2.1.1 and 2.3.1 together show that any number of such random walks will intersect infinitely many times, a.s. Thus, the only dimensions of interest are $d = 3$ and $d = 4$. In the next two subsections we will study these in detail.

Exercise **3.2.1** Use Exercise 3.1.3 and Theorem 1.4.1 together to find an alternative proof of Theorem 3.2.1. □

Exercise **3.2.2** Show that if S^1 and S^2 denote two independent simple walks in $\mathbb{Z}^d$ where $d \geq 5$, there exists a finite constant $C > 1$ such that for all $n \geq 1$,
$$C^{-1} n^{-\frac{1}{2}(d-4)} \leq \mathbb{P}(S_i^1 = S_j^2 \text{ for some } i, j \geq n) \leq C n^{-\frac{1}{2}(d-4)}.$$
(HINT: Use Exercise 3.1.3 and Theorem 2.2.1.) □

3.3 Three Simple Walks

By Theorem 3.2.1 of the previous subsection, two independent $\mathbb{Z}^4$-valued simple random walks will intersect infinitely many times. We now address the problem for three such walks.

Theorem 3.3.1 *Suppose S^1, S^2, and S^3 are independent $\mathbb{Z}^d$-valued simple random walks. The trajectories of S^1, S^2, and S^3 will a.s. intersect infinitely often if and only if $d \leq 3$.*

Our proof relies on two technical lemmas regarding the function $E_\beta^d : \mathbb{R}^d \to \mathbb{R}_+$ that is defined as follows:
$$E_\beta^d(y) = \int_{\substack{\xi \in \mathbb{R}^d: \\ \|\xi\| \leq 1}} \|y - \xi\|^{-\beta} \|\xi\|^{-\beta} \, d\xi, \qquad y \in \mathbb{R}^d. \tag{1}$$

Lemma 3.3.1 *Suppose $\beta < d < 2\beta$. Then, there are two finite and positive constants C_1 and C_2 that depend only on β and d such that for all $y \in \mathbb{R}^d$ with $\|y\| \leq 1$,*
$$C_1 \|y\|^{d-2\beta} \leq E_\beta^d(y) \leq C_2 \|y\|^{d-2\beta}.$$

Proof Fix some $y \in \mathbb{R}^d$ with $\|y\| \leq 1$. Evidently,
$$E_\beta^d(y) \geq \int_{\|\xi\| \leq \|y\|} \|\xi - y\|^{-\beta} \cdot \|\xi\|^{-\beta} \, d\xi.$$

Over the region of integration, $\|\xi - y\| \leq \|\xi\| + \|y\| \leq 2\|y\|$. Hence,
$$E_\beta^d(y) \geq 2^{-\beta} \|y\|^{-\beta} \int_{\|\xi\| \leq \|y\|} \|\xi\|^{-\beta} \, d\xi = 2^{d-\beta} \int_{\|\zeta\| \leq 1} \|\zeta\|^{-\beta} d\zeta \cdot \|y\|^{d-2\beta},$$

which gives the desired lower bound with $C_1 = 2^{d-\beta} \int_{\|\zeta\| \leq 1} \|\zeta\|^{-\beta} d\zeta$. (By Lemma 3.1.2, C_1 is finite and positive.) Next, we proceed with the upper bound. Write
$$E_\beta^d(y) = T_1 + T_2 + T_3,$$
where
$$T_1 = \int_{\substack{\|\xi - y\| \leq \frac{1}{2}\|y\| \\ \|\xi\| \leq 1}} \|\xi - y\|^{-\beta} \cdot \|\xi\|^{-\beta} \, d\xi,$$

$$T_2 = \int_{\substack{\|\xi - y\| > \frac{1}{2}\|y\| \\ \|\xi\| \leq 2\|y\| \wedge 1}} \|\xi - y\|^{-\beta} \|\xi\|^{-\beta} \, d\xi,$$

$$T_3 = \int_{\substack{\|\xi - y\| > \frac{1}{2}\|y\| \\ 2\|y\| \leq \|\xi\| \leq 1}} \|\xi - y\|^{-\beta} \|\xi\|^{-\beta} \, d\xi.$$

We estimate the above in order. When $\|\xi - y\| \leq \frac{1}{2}\|y\|$, by the triangle inequality, $\|\xi\| \geq \frac{1}{2}\|y\|$. Thus,
$$T_1 \leq 2^\beta \|y\|^{-\beta} \int_{\|\zeta\| \leq \frac{1}{2}\|y\|} \|\zeta\|^{-\beta} \, d\zeta = 2^{2\beta - d} \int_{\|\zeta\| \leq 1} \|\zeta\|^{-\beta} \, d\zeta \, \|y\|^{d - 2\beta}.$$

By Supplementary Exercise 7,
$$T_1 \leq \frac{2^{2\beta - d} d \omega_d}{d - \beta} \|y\|^{d - 2\beta}, \tag{2}$$

where ω_d denotes the d-dimensional Lebesgue measure of the ball $\{z \in \mathbb{R}^d : \|z\| \leq 1\}$. Similarly,
$$T_2 \leq 2^\beta \|y\|^{-\beta} \int_{\|\zeta\| \leq 2\|y\|} \|\zeta\|^{-\beta} d\zeta = 2^d \int_{\|\zeta\| \leq 1} \|\zeta\|^{-\beta} d\zeta \, \|y\|^{d - 2\beta}.$$

Another application of Supplementary Exercise 7 leads us to the bound
$$T_2 \leq \frac{2^d d \omega_d}{d - \beta} \|y\|^{d - 2\beta}. \tag{3}$$

It remains to estimate T_3. First, we note that if $\|\xi - y\| \geq \frac{1}{2}\|y\|$ and $\|\xi\| \geq 2\|y\|$, then certainly $\|\xi - y\| \leq \|\xi\| + \|y\| \leq \frac{3}{2}\|\xi\|$. Thus,
$$T_3 \leq \left(\frac{3}{2}\right)^\beta \int_{\|\xi - y\| \geq \frac{1}{2}\|y\|} \|\xi - y\|^{-2\beta} d\xi$$
$$\leq \left(\frac{3}{2}\right)^\beta \int_{\|\zeta\| > \frac{1}{2}\|y\|} \|\zeta\|^{-2\beta} d\zeta$$
$$\leq 3^\beta 2^{-d} \|y\|^{d - 2\beta} \int_{\|\zeta\| > 1} \|\zeta\|^{-2\beta} \, d\zeta.$$

To finish, we are left to show that $\int_{\|\zeta\|>1}\|\zeta\|^{-2\beta}d\zeta < \infty$. This is easy to do: Since $d < 2\beta$, by Supplementary Exercise 7,

$$\int_{\|\zeta\|>1}\|\zeta\|^{-2\beta}d\zeta = d\omega_d \int_1^\infty r^{d-1-2\beta}dr = \frac{d\omega_d}{2\beta - d}.$$

To summarize, we have shown that $T_3 \le 3^\beta 2^{-d}\omega_d d(2\beta - d)^{-1}\|y\|^{d-2\beta}$. Combining this with (3) and (4), we obtain $E_\beta^d \le C_2\|y\|^{d-2\beta}$ with

$$C_2 = d\omega_d\left\{\frac{2^d + 2^{2\beta-d}}{d-\beta} + \frac{3^\beta 2^{-d}}{2\beta - d}\right\}.$$

Since C_2 is clearly finite and positive, this concludes our proof. □

Going over the above argument with some care, we can also decide what happens when $d = 2\beta$.

Lemma 3.3.2 *There exists a finite and positive constant C that depends only on d such that for all $y \in \mathbb{R}^d$ with $\|y\| \le 1$, $E_{\frac{d}{2}}^d(y) \le C \ln(4/\|y\|)$.*

Proof In the notation of our proof of Lemma 3.3.1, write $E_{d/2}^d(y) = T_1 + T_2 + T_3$. Since they still hold for $d = 2\beta$, equations (2) and (3) together show that $T_1 + T_2 \le C_1 \le C_1 \ln(4/\|y\|)$, with $C_1 = (2^{d+1} + 2)\omega_d$. Still proceeding with our proof of Lemma 3.3.1 and using $\beta = \frac{d}{2}$, we obtain,

$$T_3 \le \left(\frac{3}{2}\right)^{\frac{d}{2}} \int_{2 \ge \|\xi\| \ge \frac{1}{2}\|y\|} \|\xi\|^{-d}d\xi$$

$$= d\omega_d\left(\frac{3}{2}\right)^{\frac{d}{2}} \int_{\frac{1}{2}\|y\|}^2 r^{-1}dr$$

$$= d\omega_d\left(\frac{3}{2}\right)^{\frac{d}{2}} \ln\left(\frac{4}{\|y\|}\right).$$

We have used Supplementary Exercise 7 once more and obtained the desired result with $C = C_1 + (\frac{3}{2})^{\frac{d}{2}}d\omega_d$. □

Exercise 3.3.1 Prove that Lemma 3.3.2 is sharp, up to a constant. That is, prove that $\liminf_{\|y\|\to 0^+}\{\ln(1/\|y\|)\}^{-1}E_{\frac{d}{2}}^d(y) > 0$. □

We are ready for the following.

Proof of Theorem 3.3.1 When $d \le 2$, the simple random walk is recurrent (Theorem 3.1.1). Thus, Lemmas 2.1.1 and 2.3.1 tell us that the trajectories of S^1, S^2, and S^3 intersect infinitely many times. (Why?) On the other hand, if $d \ge 5$, then by Theorem 3.2.1, the trajectories of S^1 and

S^2 intersect only finitely many times. In particular, so do the trajectories of S^1, S^2, and S^3. Thus, it remains to focus our attention on $d \in \{3, 4\}$.

Let $\mathfrak{S} = (2\pi)^{2d} \sum_{i,j,k=1}^{\infty} \mathbb{P}(S_i^1 = S_j^2 = S_k^3)$. Thanks to Theorem 2.3.1, we need to show that $\mathfrak{S} < \infty$ when $d = 4$ while $\mathfrak{S} = \infty$ when $d = 3$. In order to do this, we begin with the identity

$$\mathfrak{S} = \lim_{\lambda \uparrow 1} \int_{[-\pi,\pi]^{2d}} \frac{\varphi(\xi_1 + \xi_2)}{1 - \lambda\varphi(\xi_1 + \xi_2)} \frac{\varphi(\xi_1)}{1 - \lambda\varphi(\xi_1)} \frac{\varphi(\xi_2)}{1 - \lambda\varphi(\xi_2)} d\xi_1 d\xi_2.$$

(We have implicitly used the fact that φ is real-valued. Why?) While for every $\xi_1, \xi_2 \in [-1, 1]^d$, $\varphi(\xi_1), \varphi(\xi_2) \geq 0$, it is not always true that $\varphi(\xi_1 + \xi_2) \geq 0$. To regain positivity, we split the above integral into two parts: Let I_1 denote the above integral taken over $[-\frac{1}{2}, \frac{1}{2}]^{2d}$, and I_2 the integral over $K = [-\pi, \pi]^{2d} \setminus [-\frac{1}{2}, \frac{1}{2}]^{2d}$. We estimate I_2 first. Since cosines are bounded above by 1,

$$|I_2| \leq \lim_{\lambda \uparrow 1} \int_K \frac{1}{1 - \lambda\varphi(\xi_1 + \xi_2)} \frac{1}{1 - \lambda\varphi(\xi_1)} \frac{1}{1 - \lambda\varphi(\xi_2)} d\xi_1 d\xi_2.$$

Note that whenever $\xi_1, \xi_2 \in K$, then for all $1 \leq \ell \leq d$,

(a) $\cos(\xi_1^{(\ell)} + \xi_2^{(\ell)}) \leq \cos(\frac{1}{2}) < 1$;

(b) $\cos(\xi_1^{(\ell)}) \leq \cos(\frac{1}{2}) < 1$; and

(c) $\cos(\xi_2^{(\ell)}) \leq \cos(\frac{1}{2}) < 1$.

Hence,

$$|I_2| \leq (2\pi)^{2d} \{1 - \cos(\tfrac{1}{2})\}^{-3} < \infty.$$

Thus, we need to show that $|I_1|$ is finite when $d = 4$ and is infinite when $d = 3$. This is where positivity comes into play: If $\xi_1, \xi_2 \in [-\frac{1}{2}, \frac{1}{2}]^{2d}$, then $\varphi(\xi_1), \varphi(\xi_2)$, and $\varphi(\xi_1 + \xi_2)$ are all nonnegative. By the monotone convergence theorem,

$$I_1 = \int_{[-\frac{1}{2},\frac{1}{2}]^{2d}} \frac{\varphi(\xi_1 + \xi_2)}{1 - \varphi(\xi_1 + \xi_2)} \frac{\varphi(\xi_1)}{1 - \varphi(\xi_1)} \frac{\varphi(\xi_2)}{1 - \varphi(\xi_2)} d\xi_1 d\xi_2.$$

Moreover, if $\xi \in [-\frac{1}{2}, \frac{1}{2}]$, then $0 < \cos(\frac{1}{2}) \leq \cos(\xi) \leq 1$. We have arrived at the bound $\{\cos(\frac{1}{2})\}^3 I_1' \leq I_1 \leq I_1'$, where

$$I_1' = \int_{[-\frac{1}{2},\frac{1}{2}]^{2d}} (1 - \varphi(\xi_1 + \xi_2))^{-1} (1 - \varphi(\xi_1))^{-1} (1 - \varphi(\xi_2))^{-2} d\xi_1 d\xi_2.$$

Since $I_1' \geq 0$, we want to show that I_1' is finite if $d = 4$ but is infinite if $d = 3$. By equations (1) and (2) of Section 3.1, $(2d)^3 I_1'' \leq I_1' \leq (\frac{12d}{5})^3 I_1''$, where

$$I_1'' = \int_{[-\frac{1}{2},\frac{1}{2}]^{2d}} \|\xi_1 + \xi_2\|^{-2} \|\xi_1\|^{-2} \|\xi_2\|^{-2} d\xi_1 d\xi_2.$$

Our goal now is to show that I_1'' is finite if $d = 4$ and is infinite if $d = 3$. By Fubini's theorem and symmetry,

$$I_1'' = \int_{[-\frac{1}{2},\frac{1}{2}]^{2d}} \|\xi_1 - \xi_2\|^{-2} \|\xi_1\|^{-2} \|\xi_2\|^{-2} \, d\xi_1 \, d\xi_2$$

$$\leq \int_{[-\frac{1}{2},\frac{1}{2}]^{2d}} E_2^d(\xi_1) \|\xi_1\|^{-2} \, d\xi_1.$$

If $d = 4$, by Lemma 3.3.2 there exists a finite and positive constant C_1 such that

$$I_1'' \leq C_1 \int_{[-1,1]^4} \ln\left(\frac{4}{\|\xi\|}\right) \|\xi\|^{-2} \, d\xi.$$

Since $\ln(4/\|\xi\|) \leq 4/\|\xi\|$ for all $\xi \in \mathbb{R}^4$ with $\|\xi\| \leq 1$,

$$I_1'' \leq C_1 \int_{[-1,1]^d} \|\xi\|^{-3} \, d\xi,$$

which is finite, thanks to Lemma 3.1.2. If $d = 3$, by Lemma 3.3.2 there exists a finite positive constant C_2 such that

$$I_1'' \geq C_2 \int_{[-\frac{1}{2},\frac{1}{2}]^3} \|\xi\|^{-3} \, d\xi.$$

Since $d = 3$, Lemma 3.1.2 shows us that $I_1'' = \infty$. This concludes our proof. □

3.4 Several Simple Walks

Throughout, let us fix an integer $N \geq 4$ and consider N independent simple walks, $S^1, \ldots, S^N$, all taking values in $\mathbb{Z}^d$. If $d \leq 2$, such random walks are recurrent (Theorem 3.1.1). By Lemma 2.1.1, when $d \leq 2$, the trajectories of $S^1, \ldots, S^N$ intersect infinitely often, a.s. Next, suppose $d \geq 4$. In this case, the trajectories of S^1, S^2, and S^3 intersect finitely often, a.s. (Theorem 3.3.1). Therefore, the same holds for $S^1, \ldots, S^N$. The only case that remains to be analyzed is $d = 3$.

Theorem 3.4.1 *The trajectories of four or more independent simple walks in $\mathbb{Z}^3$ will almost surely intersect at most finitely many times.*

Our proof is an imitation of those in the previous sections but requires one more technical lemma.

Lemma 3.4.1 *For all $y \in \mathbb{R}^3$ define*

$$F(y) = \int_{\substack{\xi \in \mathbb{R}^3: \\ \|\xi\| \leq 1}} \|\xi - y\|^{-1} \|\xi\|^{-2} \, d\xi.$$

Then, for all $y \in \mathbb{R}^3$ with $\|y\| \leq 1$, $F(y) \leq 20\pi \ln(4/\|y\|)$.

Proof We follow closely the arguments used in the given proofs of Lemmas 3.3.1 and 3.3.2. Write $F(y) = T_1 + T_2 + T_3$, where

$$T_1 = \int_{\substack{\|\xi-y\| \leq \frac{1}{2}\|y\| \\ \|\xi\| \leq 1}} \|\xi - y\|^{-1} \|\xi\|^{-2} \, d\xi,$$

$$T_2 = \int_{\substack{\|\xi-y\| > \frac{1}{2}\|y\| \\ \|\xi\| \leq 2\|y\| \wedge 1}} \|\xi - y\|^{-1} \|\xi\|^{-2} \, d\xi,$$

$$T_3 = \int_{\substack{\|\xi-y\| > \frac{1}{2}\|y\| \\ 2\|y\| \leq \|\xi\| \leq 1}} \|\xi - y\|^{-1} \|\xi\|^{-2} \, d\xi.$$

We estimate each as in the demonstrations of Lemmas 3.3.1 and 3.3.2. To estimate T_1, use $\|\xi - y\| \geq \|y\|/2$ to obtain

$$T_1 \leq 4\|y\|^{-2} \int_{\|\xi-y\| \leq \frac{1}{2}\|y\|} \|\xi - y\|^{-1} \, d\xi\| = \int_{\|r\| \leq 1} \|r\|^{-1} dr.$$

By Supplementary Exercise 7, $T_1 \leq 2\pi \leq 2\pi \ln(4/\|y\|)$. We have used the elementary fact that $\omega_3 = \frac{4\pi}{3}$. Likewise,

$$T_2 = 2\|y\|^{-1} \int_{\|\xi\| \leq 2\|y\|} \|\xi\|^{-2} \, d\xi = 4 \int_{\|\xi\| \leq 1} \|\xi\|^{-2} \, d\xi = 8\pi.$$

Since $8\pi \leq 8\pi \ln(4/\|y\|)$, it remains to show that $T_3 \leq 9\pi \ln(4/\|y\|)$. Use $\|\xi - y\| \leq 3\|\xi\|$ to obtain

$$T_3 \leq \frac{9}{4} \int_{1 \geq \|\xi-y\| \geq \frac{1}{2}\|y\|} \|\xi - y\|^{-3} \, d\xi \leq \frac{9}{4} \int_{2 \geq \|\zeta\| \geq \frac{1}{2}\|y\|} \|\zeta\|^{-3} d\zeta.$$

By Exercise 3.4.1 below this equals $9\pi \ln(4/\|y\|)$, as desired. □

Exercise 3.4.1 For any $\varepsilon \in]0, 2[$, compute $\int_{2 \geq \|\zeta\| \geq \varepsilon} \|\zeta\|^{-3} \, d\zeta$. □

Exercise 3.4.2 Show that Lemma 3.4.1 is sharp, up to a constant. That is, $\liminf_{\|y\| \to 0^+} F(y)/\ln(1/\|y\|) > 0$. □

We are ready to prove the theorem.

Proof of Theorem 3.4.1 It suffices to consider only $N = 4$ and to show that

$$\sum_{i,j,k,\ell=0}^{\infty} \mathbb{P}(S_i^1 = S_j^2 = S_k^3 = S_\ell^4) < \infty,$$

where $S_0^1 = S_0^2 = S_0^3 = S_0^4 = 0$. However, symmetry and Proposition 2.3.1 together show that this is the same as showing that

$$\lim_{\lambda \uparrow 1} \int_{[-\pi,\pi]^9} \frac{\varphi(\xi_1 + \xi_2 + \xi_3)}{1 - \lambda\varphi(\xi_1 + \xi_2 + \xi_3)} \prod_{j=1}^{3} \frac{\varphi(\xi_j)}{1 - \lambda\varphi(\xi_j)} \, d\xi_1 \, d\xi_2 \, d\xi_3 < \infty.$$

We split the above integral into two parts. Let I_1 be the integral over $[-\frac{1}{3}, \frac{1}{3}]^9$ and I_2 the integral over $K = [-\pi, \pi]^9 \setminus [-\frac{1}{3}, \frac{1}{3}]^9$. The same argument used to prove Theorem 3.3.1 goes through unhindered to show that

$$|I_2| \leq (2\pi)^9 \left[1 - \cos(\tfrac{1}{3})\right]^{-4} < \infty.$$

It suffices to show that I_1 is finite. When $\xi_i \in [-\frac{1}{3}, \frac{1}{3}]^3$ ($i = 1, 2, 3$), $\varphi(\xi_i)$ is positive ($i = 1, 2, 3$). Moreover, so is $\varphi(\xi_1 + \xi_2 + \xi_3)$. By the monotone convergence theorem,

$$I_1 = \int_{[-\frac{1}{3}, \frac{1}{3}]^9} \frac{\varphi(\xi_1 + \xi_2 + \xi_3)}{1 - \varphi(\xi_1 + \xi_2 + \xi_3)} \prod_{j=1}^{3} \frac{\varphi(\xi_j)}{1 - \varphi(\xi_j)} \, d\xi_1 \, d\xi_2 \, d\xi_3$$

$$\leq \iiint_{\|\xi_1\|, \|\xi_2\|, \|\xi_3\| \leq 1} \left(1 - \varphi(\xi_1 + \xi_2 + \xi_3)\right)^{-1} \prod_{\ell=1}^{3} \left(1 - \varphi(\xi_\ell)\right)^{-1} d\xi_1 \, d\xi_2 \, d\xi_3.$$

Employing equations (1) and (2) of Section 3.1, we deduce that $I_1 \leq (\frac{36}{5})^4 J$, where

$$J = \iiint_{\|\xi_1\|, \|\xi_2\|, \|\xi_3\| \leq 1} \|\xi_1 + \xi_2 + \xi_3\|^{-2} \|\xi_1\|^{-2} \|\xi_2\|^{-2} \|\xi_3\|^{-2} \, d\xi_1 \, d\xi_2 \, d\xi_3.$$

We propose to show that $J < \infty$. Using symmetry and the definition of E_β^d (equation (1) of Section 3.3),

$$J \leq \iint_{\|\xi_1\|, \|\xi_2\| \leq 1} E_2^3(\xi_1 + \xi_2) \|\xi_1\|^{-2} \|\xi_2\|^{-2} \, d\xi_1 \, d\xi_2.$$

Lemma 3.3.1 can be applied with $d = 3$ and $\beta = 2$ to show us the existence of a positive and finite constant C such that $J \leq C \int_{\|\xi\| \leq 1} F(\xi) \|\xi\|^{-2} \, d\xi$. By Supplementary Exercise 7, and by Lemma 3.4.1 above,

$$J \leq 20\pi C \int_{\|\xi\| \leq 1} \ln\left(\frac{4}{\|\xi\|}\right) \|\xi\|^{-2} \, d\xi,$$

which is finite, by Supplementary Exercise 7. $\square$

4 Supplementary Exercises

1. Show that the inequalities of Theorem 1.5.1 can be sharpened to the following: $\mathbb{P}(S_k = 0 \text{ for some } k \geq n) = Q_n \{1 + Q_1\}^{-1}$, where $Q_n = \sum_{n=1}^{\infty} \mathbb{P}(S_j = 0)$.

2. Refine an aspect of Exercise 3.1.2 by showing that when S denotes the simple walk on $\mathbb{Z}^d$, $\lim_{n\to\infty}(2n)^{\frac{d}{2}}\mathbb{P}(S_{2n}=0) = (2\pi)^{-\frac{d}{2}}$.
This is a part of the local central limit theorem. You should compare this to the classical central limit theorem of A. de Moivre and P.-S. Laplace by looking at the density function of a mean-zero Gaussian random variable with the same variance as S_{2n}.

3. Let S denote a transient random walk on $\mathbb{Z}^d$ with $S_0 = 0$ and define T_x to be the first time S hits x. That is, $T_x = \inf(k \geq 0 : S_k = x)$. In the notation of Section 1.4, show that $\mathbb{E}[R_{T_x}] = \sum_{k=0}^{\infty}\{\mathbb{P}(S_k = 0) - \mathbb{P}(S_k = -x)\}$.
(HINT: By transience, $R_\infty < \infty$, a.s. Now we can write $R_\infty = \sum_{k=0}^{T_x-1}\mathbf{1}_{(S_k=0)} + \sum_{k=T_x}^{\infty}\mathbf{1}_{(S_k=0)}$ and use the strong Markov property.)

4. Show that for any random walk S on $\mathbb{Z}^d$ and for all integers $n, k \geq 1$, $\mathbb{E}[R_n^k] \leq k!\{\mathbb{E}[R_n]\}^k$. In particular, obtain the large deviation bound

$$\mathbb{P}\left(\frac{R_n}{\mathbb{E}[R_n]} \geq \lambda\right) \leq \frac{1}{1-\delta}e^{-\delta\lambda}, \qquad \lambda > 0,$$

where δ is an arbitrary number strictly between 0 and 1.

5. (*Mixing*) Much of the theory for independent random variables goes through with fewer hypotheses than independence. We explore one such possibility in this exercise.

A sequence of random variables $\xi_1, \xi_2, \ldots$ is said to be φ-**mixing** if

$$\sup_{i\geq 1}\sup_{\substack{E\in\mathcal{F}_{[i+n,\infty[}\\F\in\mathcal{F}_{[1,i]}}}\left|\mathbb{P}(E\mid F) - \mathbb{P}(E)\right| \leq \varphi(n),$$

where $\mathcal{F}_A$ is the σ-field generated by $\{\xi_i; i \in A\}$, and $\lim_{n\to\infty}\varphi(n) = 0$. Note that if the ξ_i's are independent, then they are φ-mixing for any φ that vanishes at infinity.

(i) Prove that the tail σ-field $\mathcal{T} = \cap_n \mathcal{F}_{[n,\infty[}$ is trivial.

(ii) Show that whenever $\sum_n \varphi(n) < +\infty$,

$$\mathbb{P}(\xi_n = 0 \text{ infinitely often}) = \begin{cases} 0, & \text{if } \sum_{n=1}^{\infty}\mathbb{P}(\xi_n = 0) < +\infty,\\ 1, & \text{if } \sum_{n=1}^{\infty}\mathbb{P}(\xi_n = 0) = +\infty. \end{cases}$$

6. Verify equation (1) of Section 2.3.

7. Suppose U is chosen uniformly at random from $\mathbb{D}_m = \{\xi \in \mathbb{R}^m : \|\xi\| \leq 1\}$.

(i) Show that the density function of $\|U\|$ at $x \in [0,1]$ is mx^{m-1}.

(ii) Use the previous part to prove the following integration-by-parts formula: For all integrable functions $f : [0,1] \to [0,1]$,

$$\int_{\mathbb{D}_m} f(\|u\|)\, du = m\omega_m \cdot \int_0^1 s^{m-1}f(s)\, ds,$$

where ω_m denotes Lebesgue's (m-dimensional) measure of $\mathbb{D}_m$.

(iii) Show that

$$\omega_m = \begin{cases} \dfrac{m\pi^{\frac{m}{2}}}{(m/2)!} & \text{if } m \text{ is even,} \\ \dfrac{2^{\frac{1}{2}(m+1)}\pi^{\frac{1}{2}(m-1)}}{1\cdot 3\cdot 5\cdots(m-2)}, & \text{if } m \text{ is odd and } m>1. \end{cases}$$

8. **(Hard)** Let S denote the simple walk on $\mathbb{Z}^2$ and let $S_0 = 0$.
   (i) When $d=1$, use Supplementary Exercise 2 to deduce that with probability one,
   $$\lim_{n\to\infty} \frac{1}{\ln n} \sum_{k=1}^{n} \frac{\mathbf{1}_{(S_k=0)}}{k^{\frac{1}{2}}} = \frac{1}{2\sqrt{2\pi}}.$$
   (ii) Prove that when $d=2$,
   $$\lim_{n\to\infty} \frac{1}{\ln\ln n} \sum_{k=2}^{n} \frac{\mathbf{1}_{(S_k=0)}}{\ln k} = \frac{1}{4\pi}.$$

This is due to Erdős and Taylor (1960a, 1960b).
(HINT: For part (i), start by proving that the expected value of the limit theorem holds. Then, prove that the variance of the given sum is bounded by $C\ln n$, for some finite constant $C>0$. Use the Borel–Cantelli lemma to obtain the a.s. convergence along the subsequence $n_k = \exp(k^2)$. To conclude part (i), estimate the sum for $n_k \le n \le n_{k+1}$ by the end values of n. Part (ii) is proved similarly, but the variance estimate is now given by a bound of $C\ln\ln n$, and the subsequence should be changed to $n_k = \exp(e^{k^2})$.)

9. **(Hard)** Suppose $X_1, X_2, \ldots$ denote i.i.d. random variables that take their values in $\mathbb{R}^d$ and define the corresponding random walk $S_n = \sum_{j=1}^{n} X_j$ ($n \ge 1$). We say that 0 is recurrent if for all $\varepsilon > 0$, $\mathbb{P}(|S_n| < \varepsilon \text{ infinitely often}) > 0$.
   (i) Verify that when $\mathbb{P}(X_1 \in \mathbb{Z}^d) = 1$, our two notions of recurrence are one and the same.
   (ii) Show that 0 is recurrent if and only if for all $\varepsilon > 0$, $\mathbb{P}(|S_n| < \varepsilon \text{ infinitely often}) = 1$.
   (iii) Define $S_0 = 0$ and prove that for all $n \ge 1$ and all $\varepsilon > 0$,
   $$\sum_{j=0}^{n} \mathbb{P}(|S_j| \le 2\varepsilon) \le 16^d \sum_{j=0}^{n} \mathbb{P}(|S_j| \le \varepsilon).$$
   (iv) Show that the following are all equivalent:
   (a) 0 is recurrent;
   (b) for some $\varepsilon > 0$, $\sum_{j=1}^{\infty} \mathbb{P}(|S_j| \le \varepsilon) = +\infty$;
   (c) for all $\varepsilon > 0$, $\sum_{j=1}^{\infty} \mathbb{P}(|S_j| \le \varepsilon) = +\infty$.

(HINT: For part (iii), cover $[-2\varepsilon, 2\varepsilon]^d$ with 16^d cubes of side $\frac{1}{2}\varepsilon$ and apply the Markov property.)

10. Given a transient random walk S on $\mathbb{Z}^d$ with $S_0 = 0$, define for each $a \in \mathbb{Z}^d$, $u(a) = \mathbb{E}[\sum_{k=0}^{\infty} \mathbf{1}_{(S_k+a=0)}]$.
  (i) Check that $u(0) = \mathbb{E}[R_\infty]$ and show that $u(a)$ is finite for all $a \in \mathbb{Z}^d$.
  (ii) Show that $m \mapsto u(S_m)$ is a supermartingale.
  (HINT: Apply Lemma 1.1.1 to $f(x) = \mathbf{1}_{\{0\}}(x)$.)

11. (Hard) Let S denote the simple walk on $\mathbb{Z}^d$.
  (i) In the case $d \geq 3$, prove that there are finite positive constants $C_1 < C_2$ such that for all $n \geq 1$,
  $$C_1 n^{-\frac{1}{2}(d-2)} \leq \mathbb{P}(S_i = 0 \text{ for some } i \geq n) \leq C_2 n^{-\frac{1}{2}(d-2)}.$$
  (ii) Let $d = 2$ and suppose $c_1, c_2, \ldots$ is a nondecreasing sequence such that $\lim_{n \to \infty} c_n = +\infty$ and $\limsup_{n \to \infty} c_n/n < \infty$. Show that when $d = 2$, there exist finite positive constants $C_1 < C_2$ such that for all $n \geq 1$,
  $$C_1 \frac{c_n}{n \ln c_n} \leq \mathbb{P}(S_i = 0 \text{ for some } n \leq i \leq n + c_n) \leq C_2 \frac{c_n}{n \ln c_n}.$$
  (HINT: For the lower bound, consider the first two moments of $\sum_{j=n}^{n+c_n} \mathbf{1}_{(S_j=0)}$. For the upper bound, estimate the conditional expectation of $\sum_{j=n}^{n+2c_n} \mathbf{1}_{(S_j=0)}$, given $\mathcal{F}_m$, where m is between n and $n+c_n$.)

In different forms and to various extents, this can be found in Benjamini et al. (1995), Erdős and Taylor (1960a, 1960b), Lawler (1991), and Révész (1990).

12. Let $\tau_1, \tau_2, \ldots$ denote the first, second, $\ldots$ hitting times of 0 by a $\mathbb{Z}^d$-valued random walk S. The goal of this exercise is an exact computation of the distribution of τ_1.
  (i) Show that for all $\lambda > 0$ and for all integers $n \geq 1$, $\mathbb{E}[e^{-\lambda \tau_n}] = (\mathbb{E}[e^{-\lambda \tau_1}])^n$.
  (ii) Let $S_0 = \tau_0 = 0$ and for all $\lambda > 0$, define $V_\lambda = \sum_{k=0}^{\infty} e^{-\lambda k} \mathbf{1}_{(S_k=0)}$. Show that $V_\lambda = \sum_{n=0}^{\infty} e^{-\lambda \tau_n}$ and conclude the following identity for the Laplace transform of τ_1: $\mathbb{E}[e^{-\lambda \tau_1}] = 1 - \{\sum_{k=0}^{\infty} e^{-\lambda k} \mathbb{P}(S_k = 0)\}^{-1}$.
  (iii) Show that when S is the simple walk on $\mathbb{Z}^d$,
  $$\lim_{\lambda \to 0^+} \lambda^{\frac{1}{2}} \sum_{k=0}^{\infty} e^{-\lambda k} \mathbb{P}(S_k = 0) = \sqrt{2}$$
  when $d = 1$, and when $d = 2$,
  $$\lim_{\lambda \to 0^+} \frac{1}{\ln(\frac{1}{\lambda})} \sum_{k=0}^{\infty} e^{-\lambda k} \mathbb{P}(S_k = 0) = \frac{1}{2\pi}.$$
  (HINT: Consider the distribution function $F(k) = \sum_{j \leq k} \mathbb{P}(S_j = 0)$. Apply the Tauberian theorem Theorem 2.1.1, Appendix B, together with Supplementary Exercise 2.)

Such results are a part of the folklore of random walks; for instance, read Chung and Hunt (1949) with care. In the above forms, they can be found in Khoshnevisan (1994), where you can also find further applications to measure the zero set of random walks.

13. (Continued from Supplementary Exercise 12)

(i) Let S denote the simple walk on $\mathbb{Z}^d$. In the notation of Supplementary Exercise 12, show that when $d = 1$, τ_n/n^2 converges in distribution to a nonnegative random variable τ_∞ whose Laplace transform is $\mathbb{E}[e^{-\zeta \tau_\infty}] = \exp(-\sqrt{\zeta})$.

(HINT: Use the convergence theorem for Laplace transforms (cf. Theorem 1.2.1, Appendix B). The random variable τ_∞ is the so-called **stable** random variable of index $\frac{1}{2}$ and will reappear later in Section 3.2, Chapter 10.)

(ii) Conclude that when $d = 1$, $R_n/\sqrt{n}$ converges in distribution to the absolute value of a standard Gaussian random variable.

(HINT: Since τ is the inverse function to R, roughly speaking, $\mathbb{P}(R_n \geq \lambda \sqrt{n}) = \mathbb{P}(\tau_{\lambda \sqrt{n}} \leq n)$. You need to make this work by a series of inequalities.)

14. (Hard) Suppose S^1 and S^2 are two independent simple walks on $\mathbb{Z}^4$. Consider a nondecreasing sequence $c_1, c_2, \ldots$ such that $\lim_{n \to \infty} c_n = +\infty$ and $\limsup_{n \to \infty} c_n/n < \infty$. Show the existence of two positive finite constants $C_1 < C_2$ such that for all $n \geq 1$,

$$C_1 \left(\frac{c_n}{n}\right)^2 \cdot \frac{1}{\ln c_n} \leq \mathbb{P}(S_i^1 = S_j^2 \text{ for some } n \leq i, j \leq n + c_n) \leq C_2 \left(\frac{c_n}{n}\right)^2 \cdot \frac{1}{\ln c_n}.$$

(You should first study Supplementary Exercise 11.)

5 Notes on Chapter 3

Section 1 The references (Ornstein 1969; Spitzer 1964; Révész 1990; Revuz 1984) are excellent resources for the fine and general structure of one-parameter random walks, Markov chains, and their connections to ergodic theory and potential theory.

The argument of Section 1.7 that reduces attention to the set of possible points is quite old, but often goes unmentioned when $d > 1$, perhaps to avoid discussions relating to free abelian groups.

Much of the material of this section, and, in fact, chapter, can be extended to random walks on locally compact abelian groups. A comprehensive account of the potential-theoretic aspects of this can be found in Port and Stone (1971a, 1971b).

The basic message of the investigations of recurrence for random walks is that a point is recurrent for the walk if and only if the walk is expected to hit that point infinitely often. The number of times the random walk hits a given point is the so-called local time at that point. There are limit theorems associated with such local times; they can be viewed as refinements of the notion of recurrence, among other things; see Bass and Khoshnevisan (1993b, 1993c, 1995), Borodin (1986, 1988), Csáki and Révész (1983), Csörgő and Révész (1984, 1985, 1986), Kesten and Spitzer (1979), Jacod (1998), Khoshnevisan (1992, 1993), Knight (1981), Perkins (1982), and Révész (1981).

3. Random Walks

Section 2 In the probability literature, the study of the intersections of random walk trajectories goes back at least to Dvoretzky and Erdős (1951), as well as Erdős and Taylor (1960b, 1960a), and Dvoretzky et al. (1950, 1954, 1958, 1957). Related results, together with references to the physics literature, can be found in (Madras and Slade 1993; Lawler 1991).

In this section we essentially showed that the intersections are recurrent if and only if the walks are expected to intersect infinitely many times, at least as long as all of the intervening walks are symmetric. At this time it is not known whether Theorem 2.3.1 holds without any symmetry, or sector-type, hypotheses.

Further analysis of the number of intersections of random walks leads to a so-called intersection local time that is the main subject of Le Gall et al. (1989), Le Gall and Rosen (1991), Lawler (1991), Rosen (1993), and Stoll (1987, 1989). Some very general results can be found in (Bass and Khoshnevisan 1992a; Dynkin 1988).

Many of the quantitative results of this section are new.

Section 3 The results of this section are all classical and can be found in the pre-60's references cited under Section 2 above. For further refinements, see Lawler (1991). Many of the presented proofs in this section are new. Further related works, but in a genuine multiparameter context, can be found in Etemadi (1977).

A variant of Exercise 3.2.1 can be found in Lawler (1991, Theorem 3.3.2).

Section 4 A variant of Supplementary Exercise 14 can be found in Lawler (1991, Theorem 3.3.2).

Supplementary Exercise 5 seems to be new. However, much is known about sums of mixing random variables. A good starting place for this is Billingsley (1995).

4
Multiparameter Walks

The discussions of Chapter 3 revolved around multiparameter processes that are formed by considering systems of independent one-parameter random walks. In this chapter we consider properties of genuinely multiparameter random walks. For an example of such a process, suppose each "site" $t \in \mathbb{N}^N$ corresponds to an independent particle that is negatively charged with probability p and positively charged with probability $1-p$. Let $X_t = 1$ if the particle at site $t \in \mathbb{N}^N$ is negatively charged; otherwise, set $X_t = 0$. Then, the total number of negatively charged particles in the rectangle $[0,t]$ is precisely $\sum_{s \preccurlyeq t} X_s$. When the X's are general i.i.d. random variables, this defines a general N-parameter random walk. To summarize the main results of this chapter, let us first suppose that the increments of the multiparameter random walk have the same distribution as some random variable ξ. Then, a rough summary of the main results of this chapter is as follows:

The strong law of large numbers holds iff $\mathbb{E}[|\xi|\{\ln_+ |\xi|\}^{N-1}] < \infty$;
the law of the iterated logarithm holds iff $\mathbb{E}[|\xi|^2\{\ln_+ |\xi|\}^{N-1}] < \infty$,

with the condition on the law of the iterated logarithm being nearly optimal; cf. Exercise 2.2.1 and Supplementary Exercise 2, as well as the Notes at the end of the chapter. Both of these results are natural extensions of 1-parameter results, with which the reader is expected to be familiar. We prove these results by establishing connections to martingales, maximal inequalities, etc.

1 The Strong Law of Large Numbers

In this section we discuss the strong law of large numbers for mean-zero, finite-variance multiparameter random walks. We first recall the requisite 1-parameter background.

If $X_1, X_2, \ldots$ are i.i.d. mean-zero random variables, the random walk $S = (S_n;\ n \geq 1)$ is defined by $S_n = X_1 + \cdots + X_n$ $(n \geq 1)$. Supposing that the X's have mean zero, the weak law of large numbers states that $\frac{1}{n}S_n$ converges to 0 in probability, as $n \to \infty$. A much more interesting fact—the strong law of large numbers—states that this convergence holds almost surely.

In like manner, in the multiparameter setting we will define the increments X_s ($s \in \mathbb{N}^N$), together with a walk $S_t = \sum_{s \preccurlyeq t} X_s$ ($t \in \mathbb{N}^N$). Note that for any fixed $t \in \mathbb{N}^N$, S_t is a sum of $\langle t \rangle$ i.i.d. random variables, where

$$\langle t \rangle = \prod_{j=1}^{N} t^{(j)}, \qquad t \in \mathbb{N}^N. \tag{1}$$

Thus, as soon as the X's have mean zero, by the weak law of large numbers,

$$\lim_{t \to \infty} \frac{S_t}{\langle t \rangle} = 0, \qquad \text{in probability.}$$

A more interesting mode of convergence is convergence with probability one, where the analogous law of large numbers is much more delicate. For instance, we have already seen, in the preamble to this chapter, that the necessary and sufficient condition for this strong law is the integrability of $X_1\{\ln_+ |X_1|\}^{N-1}$ and not X_1. Informally, this difference between the 1-parameter and and multiparameter theories arises because in the latter case there are many ways in which the actual time parameter $\langle t \rangle$ can go to infinity.

1.1 Definitions

Fix an integer $N \geq 1$ and consider independent, identically distributed $\mathbb{R}^d$-valued random variables indexed by $\mathbb{N}^N$: $X = (X_t;\ t \in \mathbb{N}^N)$. The corresponding **multiparameter random walk** $S = (S_t;\ t \in \mathbb{N}^N)$ is defined by

$$S_t = \sum_{s \preccurlyeq t} X_s, \qquad t \in \mathbb{N}^N.$$

In analogy to the one-parameter theory, the X's are referred to as the **increments** of the underlying random walk S.

Corresponding to the multiparameter walk S, we define the N-parameter filtration $\mathcal{F} = (\mathcal{F}_t;\ t \in \mathbb{N}^N)$ to be the **history** of S. That is, for all $t \in \mathbb{N}^N$, $\mathcal{F}_t$ designates the σ-field generated by $(X_r;\ r \preccurlyeq t)$. If $\mathcal{F}^1, \ldots, \mathcal{F}^N$ stand for the marginal filtrations of $\mathcal{F}$, after recalling Section 2.6 of Chapter 1, it quickly follows that $\mathcal{F}^1,\ldots,\mathcal{F}^N$ are the orthohistories of S.

1.2 Commutation

An important property of the filtration $\mathcal{F}$ is that it is commuting. This will be a consequence of the following.

Lemma 1.2.1 (Inclusion–Exclusion Formula) *Given a sequence of real numbers* $X = (X_t;\ t \in \mathbb{N}^N)$, *let* $S_t = \sum_{s \preccurlyeq t} X_s$, $t \in \mathbb{N}^N$. *Then, for all* $t \in \mathbb{N}^N$,

$$X_t = \sum_{r \in \{0,1\}^N} (-1)^{\sum_{\ell=1}^N r^{(\ell)}} S_{t-r},$$

where $S_t = 0$ *if* $t \notin \mathbb{N}^N$.

Proof We proceed by induction on N. First, suppose $N = 1$. Clearly, for all $t \in \mathbb{N}^N$,

$$X_t = S_t - S_{t-1} = \sum_{r \in \{0,1\}} (-1)^r S_{t-r}.$$

Thus, the result holds when $N = 1$. Next, we argue by induction on N (≥ 2). Assuming that the result holds for $N - 1$, we prove it for N. Write any $t \in \mathbb{N}^N$ as $t = (t^{(1)}, t')$, where $t' \in \mathbb{N}^{N-1}$ is given by $t' = (t^{(2)}, \ldots, t^{(N)})$. We will write $X_{t^{(1)},t'}$ for $X_t = X_{(t^{(1)},t')}$. The same convention applies to S_t. Clearly, $S_t = \sum_{s' \preccurlyeq t'} Y_{s'}(t^{(1)})$ ($t \in \mathbb{N}^N$), where $Y_r(k) = \sum_{j=1}^k X_{j,r}$ ($r \in \mathbb{N}^{N-1}$). By the induction hypothesis,

$$Y_{t'}(t^{(1)}) = \sum_{r \in \{0,1\}^{N-1}} (-1)^{\sum_{\ell=1}^{N-1} r^{(\ell)}} S_{t^{(1)}, t'-r}, \qquad t \in \mathbb{N}^N.$$

On the other hand, $X_t = Y_{t'}(t^{(1)}) - Y_{t'}(t^{(1)} - 1)$, where for any $t' \in \mathbb{N}^{N-1}$, $Y_{t'}(k) = 0$ whenever $k \notin \mathbb{N}^N$. In summary,

$$X_t = \sum_{r \in \{0,1\}^{N-1}} (-1)^{\sum_{\ell=1}^{N-1} r^{(\ell)}} S_{t^{(1)}, t'-r} - \sum_{r \in \{0,1\}^{N-1}} (-1)^{\sum_{\ell=1}^{N-1} r^{(\ell)}} S_{t^{(1)}-1, t'-r},$$

which is a restatement of the lemma. □

A corollary of the above is the commutation property of $\mathcal{F}$.

Proposition 1.2.1 *If* $\mathcal{F} = (\mathcal{F}_t;\ t \in \mathbb{N}^N)$ *denotes the history of a multiparameter random walk* $S = (S_t;\ t \in \mathbb{N}^N)$ *with increments* $X = (X_t;\ t \in \mathbb{N}^N)$, *then:*

(i) for all $t \in \mathbb{N}^N$, $\mathcal{F}_t$ *is the σ-field generated by* $(X_r;\ r \preccurlyeq t)$; *and*

(ii) $\mathcal{F}$ is a commuting filtration.

Proof Temporarily, define $\mathcal{X}_t$ to be the σ-field generated by $(X_r;\ r \preccurlyeq t)$. Since $S_t = \sum_{r \preccurlyeq t} X_r$, $\mathcal{X}_t \subset \mathcal{F}_t$ for all $t \in \mathbb{N}^N$. The converse inclusion also holds and follows from Lemma 1.2.1. Next, we prove *(ii)*.

Applying Theorem 3.6.1 of Chapter 1, we set out to show that for all $t, s \in \mathbb{N}^N$, $\mathcal{F}_t$ and $\mathcal{F}_s$ are conditionally independent, given $\mathcal{F}_{s \wedge t}$. (The mentioned theorem is about stochastic processes indexed by $\mathbb{N}_0^N$, while we have processes indexed by $\mathbb{N}^N$. To overcome this trivial difficulty, simply "shift back" time by $(1, \ldots, 1)$.)

We need to show the following: For all $s, t \in \mathbb{N}^N$ and all bounded random variables Y_s and Y_t that are $\mathcal{F}_s$ and $\mathcal{F}_t$ measurable, respectively,

$$\mathbb{E}[Y_t Y_s \,|\, \mathcal{F}_{s \wedge t}] = \mathbb{E}[Y_t \,|\, \mathcal{F}_{s \wedge t}] \cdot \mathbb{E}[Y_s \,|\, \mathcal{F}_{s \wedge t}], \qquad \text{a.s.}$$

It suffices to do so for $s \neq t$. Henceforth, we fix two different $s, t \in \mathbb{N}^N$. By Lemma 1.2.1, for all $r \in \mathbb{N}^N$, $\mathcal{F}_r$ is the σ-field generated by $(X_s;\ s \preccurlyeq r)$. By a standard argument of measure theory, we need only consider Y_t and Y_s of the form $Y_t = \prod_{n=0}^{k} f_n(X_{u_n})$ and $Y_s = \prod_{m=0}^{j} g_m(X_{v_m})$, where $f_n, g_m : \mathbb{R}^d \to [0, 1]$ ($0 \leq n \leq k$, $0 \leq m \leq j$) are bounded and measurable functions, $v_m \preccurlyeq s$, and $u_n \preccurlyeq t$ ($0 \leq n \leq k$, $0 \leq m \leq j$). By relabeling indices and possibly introducing more variables, we may assume the existence of $0 \leq M \leq j$ and $0 \leq N \leq k$ such that for all $n \leq N$ and $m \leq M$, $u_n \preccurlyeq s \wedge t$ and $v_m \preccurlyeq s \wedge t$. On the other hand, for values of $n > N$ (respectively $m > M$) $u_n \not\preccurlyeq s \wedge t$ (respectively $v_m \not\preccurlyeq s \wedge t$). Recall that $t \neq s$, $u_n \preccurlyeq t$ and $v_m \preccurlyeq s$ ($0 \leq n \leq k$, $0 \leq m \leq j$). Thus, $(u_n;\ N < n \leq k)$ and $(v_m;\ M < m \leq j)$ are necessarily disjoint sequences. Using the independence of the X's, we can conclude the following:

$$\mathbb{E}[Y_s Y_t \,|\, \mathcal{F}_{s \wedge t}] = \prod_{n=0}^{N} f_n(X_{u_n}) \prod_{m=0}^{M} g_m(X_{v_m})$$

$$\times \mathbb{E}\bigg[\prod_{n=N+1}^{k} f_n(X_{u_n})\bigg] \mathbb{E}\bigg[\prod_{m=M+1}^{j} g_m(X_{v_m})\bigg]$$

$$= \mathbb{E}[Y_t \,|\, \mathcal{F}_{s \wedge t}]\, \mathbb{E}[Y_s \,|\, \mathcal{F}_{s \wedge t}].$$

This proves the result. □

It follows from the above that—when they are in $\mathbb{R}$ and integrable— mean-zero, multiparameter random walks are martingales with respect to a commuting filtration. In particular, they enjoy the general properties of martingales with respect to commuting filtrations. Put in more precise terms, we may state the following lemma:

Lemma 1.2.2 *Suppose S is an $\mathbb{R}$-valued, N-parameter random walk with increments $X = (X_t;\ t \in \mathbb{N}^N)$. If, in addition, $X_{(1,\ldots,1)}$ has mean 0, then S is an N-parameter martingale with respect to a commuting filtration.*

***Exercise* 1.2.1** Complete the proof of Lemma 1.2.2. □

***Exercise* 1.2.2** In the context of Lemma 1.2.2, suppose further that the X_t's have a finite variance σ^2. Show that $t \mapsto S_t^2 - \sigma^2 \langle t \rangle$ is a mean-zero martingale with respect to the same commuting filtration as S. □

***Exercise* 1.2.3** Suppose S is an N-parameter random walk with mean-zero increments, each of which has a finite moment generating function. Given $\alpha \in \mathbb{R}$, find a nonrandom function ψ such that $t \mapsto \psi(t) \cdot e^{\alpha S_t}$ is an N-parameter martingale with mean 1. □

1.3 A Reversed Orthomartingale

Let $S = (S_t;\ t \in \mathbb{N}^N)$ denote an N-parameter random walk that takes values in $\mathbb{R}^d$. Recalling equation (1) in the introductory portion of this section, let X denote the increments process and define the "*sample average*"

$$A_t = \frac{S_t}{\langle t \rangle}, \qquad t \in \mathbb{N}^N. \tag{1}$$

The main goal of this subsection is to show that A is a reversed N-parameter orthomartingale. To do so, we shall need N (one-parameter) reversed filtrations. For $1 \le i \le N$, define $\mathcal{R}^i = (\mathcal{R}_k^i;\ k \ge 1)$ as follows: For all $k \ge 1$, let $\mathcal{R}_k^i$ denote the σ-field generated by $(S_t;\ t \in \mathbb{N}^N$, such that $t^{(i)} \ge k)$. To illustrate this better, consider $N = 2$. Then, $\mathcal{R}_k^1$ is the σ-field generated by $(S_{n,m};\ n \ge k,\ m \ge 1)$, and $\mathcal{R}_k^2$ is the σ-field generated by $(S_{n,m};\ n \ge 1,\ m \ge k)$. Continuing with the discussion for $N = 2$, the inclusion–exclusion formula (Lemma 1.2.1) implies that for any $k \ge 1$, $\mathcal{R}_k^1$ is the σ-field generated by the following two collections of random variables:

(a) $(S_{k,m};\ m \ge 1)$; and

(b) $(X_{n,m};\ n \ge k+1, m \ge 1)$.

Likewise, $\mathcal{R}_k^2$ is the σ-field generated by

(c) $(S_{k,n};\ n \ge 1)$; and

(d) $(X_{n,m};\ n \ge 1, m \ge k+1)$.

Now let us suppose $d = 1$, so that X, S, and A are all real-valued stochastic processes. By the stationarity and the independence of the increments, for any $k \ge 1$,

$$\mathbb{E}[X_{1,1} \mid \mathcal{R}_k^1] = \cdots = \mathbb{E}[X_{k,1} \mid \mathcal{R}_k^1].$$

Since $X_{1,1} + \cdots + X_{k,1} = S_{k,1}$, this implies that

$$\mathbb{E}[X_{1,1} \mid \mathcal{R}_k^1] = \frac{S_{k,1}}{k} = A_{k,1}.$$

More generally, this argument shows that for all $k, n \geq 1$,

$$A_{k,n} = \frac{1}{n}\mathbb{E}[X_{1,1} + \cdots + X_{1,n} \mid \mathcal{R}^1_k].$$

Similarly,

$$A_{k,n} = \frac{1}{k}\mathbb{E}[X_{1,1} + \cdots + X_{k,1} \mid \mathcal{R}^2_n].$$

Hence, when $N = 2$, A is a 2-parameter reversed orthomartingale with respect to the reversed filtrations $\mathcal{R}^1$ and $\mathcal{R}^2$. The same proof works in any number of temporal dimensions ($N > 1$). In fact, we have the following result.

Lemma 1.3.1 *Suppose $S = (S_t; t \in \mathbb{N}^N)$ is a multiparameter walk with $\mathbb{R}$-valued increments $X = (X_t; t \in \mathbb{N}^N)$. Let $A = (A_t; t \in \mathbb{N}^N)$ be the sample average process given by equation (1). If $X_{(1,\ldots,1)}$ is integrable, then A is a reversed orthomartingale with respect to the reversed filtrations $\mathcal{R}^1, \ldots, \mathcal{R}^N$ defined above. Moreover, for all nondecreasing convex $\Phi : \mathbb{R} \to \mathbb{R}_+$,*

$$\sup_{t \in \mathbb{N}^N} \mathbb{E}[\Phi(A_t)] \leq \mathbb{E}[\Phi(X_{(1,\ldots,1)})],$$

as long as $\mathbb{E}[\Phi(X_{(1,\ldots,1)})] < +\infty$.

Exercise 1.3.1 Prove Lemma 1.3.1 when $N \geq 3$. □

1.4 Smythe's Law of Large Numbers

We now state and prove the strong law of large numbers of Smythe (1973) for multiparameter random walks.

Theorem 1.4.1 (Smythe's Law of Large Numbers) *Suppose S is an N-parameter random walk with real-valued increments $X = (X_t; t \in \mathbb{N}^N)$. Let X_0 be a random variable with the same distribution as X_t ($t \in \mathbb{N}^N$) and recall A from equation (1) of Section 1.3. If $\mathbb{E}\{|X_0|(\ln_+ |X_0|)^{N-1}\} < \infty$, then almost surely,*

$$\lim_{t \to \infty} A_t = \mathbb{E}[X_0].$$

Conversely, suppose $\mathbb{P}(L = \infty) > 0$, where $L = \limsup_t |A_t|$. Then,

$$\mathbb{E}\{|X_0|(\ln_+ |X_0|)^{N-1}\} = \infty.$$

To demonstrate this, we develop two preliminary real-variable facts concerning the following function:

$$I_N(x) = \int_{[1,\infty[^N} \mathbb{1}_{[1,x]}(\langle t \rangle) \frac{dt}{\langle t \rangle}, \qquad x \geq 1. \tag{1}$$

Lemma 1.4.1 *For any random variable Z,*

$$\frac{1}{2}\mathbb{E}\{|Z|I_{N-1}(\tfrac{1}{2}|Z|)\} \leq \int_{[1,\infty[^N} \mathbb{P}(|Z| \geq \langle t \rangle)\, dt \leq \mathbb{E}\{|Z|I_{N-1}(|Z|)\},$$

where $N \geq 1$ is an arbitrary integer.

Proof It is evident that

$$\int_{[1,\infty[^N} \mathbb{P}(|Z| \geq \langle t \rangle)\, dt = \int_{[1,\infty[^{N-1}} \int_1^\infty \mathbb{P}(|Z| \geq s\langle v \rangle)\, ds\, dv$$

$$= \int_{[1,\infty[^{N-1}} \int_{\langle v \rangle}^\infty \mathbb{P}(|Z| \geq u)\, du\, \frac{dv}{\langle v \rangle}$$

$$= \mathbb{E}\Big[\int_{[1,\infty[^{N-1}} \mathbb{1}_{[1,|Z|]}(\langle v \rangle)\{|Z| - \langle v \rangle\}\, \frac{dv}{\langle v \rangle}\Big].$$

The inequality $|Z| - \langle v \rangle \leq |Z|$ gives the upper bound of the lemma. On the other hand, if $\langle v \rangle \leq \tfrac{1}{2}|Z|$, we have $|Z| - \langle v \rangle \geq \tfrac{1}{2}|Z|$, thus deriving the converse bound of the lemma. □

Our second preliminary result is an estimation of the function I_N.

Lemma 1.4.2 *For all $x > e$ and all integers $N \geq 1$,*

$$N^{-N}(\ln x)^N \leq I_N(x) \leq (\ln x)^N.$$

This is proved by means similar to those used to verify Lemma 1.4.1. As such, we leave the argument as an exercise.

Exercise **1.4.1** Prove Lemma 1.4.2. □

Proof of Theorem 1.4.1 To prove half of the theorem, suppose $|X_0|(\ln_+ |X_0|)^{N-1}$ is integrable. Certainly, X_0 is also integrable. Thus, Lemma 1.3.1 shows that A is a reversed orthomartingale. Moreover, applying the latter lemma to $\Phi(t) = t(\ln_+ t)^{N-1}$, we can deduce that $\sup_{t\in\mathbb{N}^N} \mathbb{E}[|A_t|(\ln_+ |A_t|)^{N-1}]$ is finite. By Theorem 2.9.1 of Chapter 1, A_t has a limit, a.s. To finish our proof of the sufficiency, we need to identify this limit as $\mathbb{E}[X_0]$. However, by Kolmogorov's (one-parameter) strong law of large numbers, $\lim_{t\to\infty} A_t = \mathbb{E}[X_0]$, almost surely. This proves the sufficiency.

To prove the converse, let us suppose that with positive probability, $L = \infty$. Since $(L = \infty)$ is a tail event for independent random variables, we must have $L = \infty$, a.s. (Why?) For all $t \in \mathbb{Z}^N \setminus \mathbb{N}^N$, define $A_t = 0$. With this convention in mind, the inclusion–exclusion formula (Lemma 1.2.1) shows us that for all $t \in \mathbb{N}^N$,

$$\frac{X_t}{\langle t \rangle} = \sum_{r \in \{0,1\}^N} (-1)^{\sum_{j=1}^N r^{(j)}} A_{t-r}.$$

Since $\mathbb{P}(L = \infty) = 1$, $\limsup_{t\to\infty} \langle t\rangle^{-1}|X_t| = +\infty$, almost surely. By the Borel–Cantelli lemma for independence events this is equivalent to the following: For all $\lambda > 0$,

$$\sum_{t\in\mathbb{N}^N} \mathbb{P}\bigl(|X_t| \geq \lambda\langle t\rangle\bigr) = +\infty.$$

Since X_t has the same distribution as X_0, the integral test of calculus can be employed to see that

$$\int_{[1,\infty[^N} \mathbb{P}\bigl(|X_0| \geq \langle t\rangle\bigr)\, dt = \infty.$$

The result follows, after we show that for all $N \geq 1$,

$$\int_{[1,\infty[^N} \mathbb{P}\bigl(|X_0| \geq \langle t\rangle\bigr)\, dt = +\infty \iff \mathbb{E}\{|X_0|(\ln_+ |X_0|)^{N-1}\} = +\infty.$$

But this follows from Lemmas 1.4.1 and 1.4.2. $\square$

Exercise **1.4.2** In the context of Smythe's law of large numbers, suppose further that $X_0 \in L^p(\mathbb{P})$ for some $p > 1$ and show that as $t \to \infty$, A_t converges to $\mathbb{E}[X_0]$ in $L^p(\mathbb{P})$ as well. $\square$

Exercise **1.4.3** Complete our proof of Smythe's law of large numbers by showing that $L = \infty$ implies that a.s., $\limsup_{t\to\infty} \langle t\rangle^{-1}|X_t| = +\infty$. $\square$

Exercise **1.4.4** Refine Lemma 1.4.1 by showing that

$$\int_{[1,\infty[^N} \mathbb{P}(Z \geq \langle t\rangle)\, dt$$
$$= \sum_{j=1}^{N-1} (-1)^{j+1}\mathbb{E}\{ZI_{N-j}(Z)\} + (-1)^{N-1}\mathbb{E}\{Z\mathbf{1}_{[1,\infty[}(Z)\},$$

where Z is any almost surely nonnegative random variable, and $N \geq 1$ is an integer. $\square$

2 The Law of the Iterated Logarithm

Suppose S is an N-parameter random walk in $\mathbb{R}$ with increments $X = (X_t;\ t \in \mathbb{N}_0^N)$. Given that $\mathbb{E}\{|X_0|(\ln_+ |X_0|)^{N-1}\} < \infty$, the strong law of large numbers (Theorem 1.4.1) shows that as $\langle t\rangle \to \infty$, $S_t \approx \langle t\rangle \cdot \mathbb{E}[X_0]$, almost surely.

One can refine this by seeking to estimate the size of the difference $S_t - \langle t \rangle \mathbb{E}[X_0]$. When the increments have sufficient integrability, this refinement is the so-called law of the iterated logarithm (written LIL).

As a warm-up, we first work with the special case where the random walk is one-parameter Gaussian. This shows some of the salient features of what makes the LIL happen in a setting that is free of many of the technical difficulties inherent to such an undertaking.

The general law of the iterated logarithm will be stated and proved following the one-parameter Gaussian case, and will take up the remainder of this chapter.

2.1 The One-Parameter Gaussian Case

This subsection proves the following law of the iterated logarithm in the one-parameter Gaussian setting: the simplest of its kind.

Theorem 2.1.1 (LIL–Gaussian Case) *Let $S = (S_k;\ k \geq 1)$ denote an $\mathbb{R}$-valued random walk whose increments $X = (X_k;\ k \geq 1)$ are standard Gaussian random variables. Then,*

$$\limsup_{k \to \infty} \frac{S_k}{\sqrt{2k \ln \ln k}} = 1, \quad \text{a.s.}$$

The usual proof of the above LIL relies on three key steps:

(i) a maximal inequality;

(ii) tail estimates for deviation probabilities; and

(iii) a blocking argument.

We will address each in order. The first step relies on Lévy's maximal inequality, which holds for all symmetric random walks.[1]

Lemma 2.1.1 (Lévy's Maximal Inequality) *Suppose S denotes a real-valued, one-parameter symmetric random walk. Then, for all $k \geq 1$ and $\lambda > 0$,*

$$\mathbb{P}\Big(\max_{1 \leq j \leq k} |S_j| \geq \lambda\Big) \leq 2\mathbb{P}(|S_k| \geq \lambda).$$

Proof Fix $k \geq 1$ and $\lambda > 0$ and define $\tau = \inf(j \geq 1 :\ |S_j| \geq \lambda)$, with the usual convention that $\inf \varnothing = \infty$. Note that $\mathbb{P}(\max_{1 \leq j \leq k} |S_j| \geq \lambda) = \mathbb{P}(\tau \leq k)$, so that

$$\mathbb{P}\Big(\max_{1 \leq j \leq k} |S_j| \geq \lambda\Big) = \mathbb{P}(|S_k| \geq \lambda) + \mathbb{P}(|S_k| < \lambda\,,\ \tau \leq n)$$
$$= \mathbb{P}(|S_k| \geq \lambda) + \mathbb{P}(|S_k - S_\tau + S_\tau| < \lambda\,,\ \tau \leq n).$$

---

[1] That is, the increments are symmetric. Recall that a random variable Z is **symmetric** if Z and $-Z$ have the same distribution.

Since τ is a stopping time for the history of S, by the strong Markov property (Theorem 1.2.1 of Chapter 3), $S_k - S_\tau$ is independent of S_τ and $(\tau \leq n)$. This and the symmetry of the increments together imply that we can replace $S_k - S_\tau$ by $S_\tau - S_k$ and not change the probability. That is,

$$\mathbb{P}\left(\max_{1 \leq j \leq k} |S_j| \geq \lambda\right) = \mathbb{P}(|S_k| \geq \lambda) + \mathbb{P}(|2S_\tau - S_k| < \lambda, \, \tau \leq n).$$

On the other hand, on $(\tau < \infty)$, $|S_\tau| \geq \lambda$. Thus, thanks to the triangle inequality, on $(\tau < \infty)$,

$$(|2S_\tau - S_k| < \lambda) \subset (|S_k| \geq \lambda).$$

This has the desired result. □

Exercise 2.1.1 Suppose S is an N-parameter random walk whose increments X are symmetric random variables. Prove that for all $t \in \mathbb{N}^N$ and for all $\lambda > 0$,

$$\mathbb{P}\left(\max_{s \preccurlyeq t} |S_s| \geq \lambda\right) \leq 2^N \mathbb{P}(|S_t| \geq \lambda).$$

This is due to P. Lévy when $N = 1$, and Wichura (1973) for $N > 1$. (HINT: Use the method used in our proof of Lemma 2.1.1, one parameter at a time.) □

The second step in our proof of the theorem is a probability tail estimate.

Lemma 2.1.2 *Let S denote a real-valued, one-parameter Gaussian random walk. Then, if x_n is a sequence that tends to infinity,*

$$\lim_{k \to \infty} \frac{1}{x_k} \ln \mathbb{P}\left(S_k > \sqrt{k x_k}\right) = -\frac{1}{2}.$$

Proof For any $k \geq 1$, S_k is a Gaussian random variable with mean 0 and variance k. Thus, for any $x > 0$,

$$\mathbb{P}\left(S_k > x\sqrt{k}\right) = (2\pi)^{-\frac{1}{2}} \int_x^\infty e^{-\frac{1}{2}u^2} du.$$

On the other hand, it is easy to check from this (via L'Hôpital's rule of calculus) that

$$\lim_{x \to \infty} x e^{\frac{1}{2}x^2} \mathbb{P}\left(S_k > x\sqrt{k}\right) = (2\pi)^{-\frac{1}{2}},$$

and that the convergence rate must be independent of the choice of $k \geq 1$. □

***Exercise* 2.1.2** (Hard) Show that Lemma 2.1.2 continues to hold if we replace the Gaussian walk by the simple walk, at least as long as $x_k = \alpha\sqrt{2k \ln \ln k}$ for $\alpha > 0$ fixed.
(HINT: One way to proceed is by proving an upper and a lower bound for $\mathbb{P}(S_n \geq \alpha\sqrt{2n \ln \ln n})$. For a method of getting the lower bound, peek ahead to Lemma 2.10.1 below. To obtain an upper bound, you can:

(i) compute the moment generating function of S_n; and

(ii) prove H. Chernoff's inequality: For all $\alpha > 0$,

$$\mathbb{P}(S_n > x) \leq 2e^{-\alpha n}\mathbb{E}[e^{\alpha S_n}].$$

Optimize over the α of part (ii) to get a working upper bound for the probability in question.) □

We are ready to derive our Gaussian LIL, using the third step: the blocking argument.

Proof of Theorem 2.1.1 We will first prove that for any $\varepsilon > 0$,

$$\limsup_{k \to \infty} \frac{|S_k|}{\sqrt{2k \ln \ln k}} \leq 1 + \varepsilon, \qquad \text{a.s.} \tag{1}$$

Fix any $\varepsilon > 0$, and for any other fixed $1 < b < (1+\varepsilon)^2$, define $b_k = \lfloor b^k \rfloor$ ($k \geq 1$). By Lemma 2.1.1,

$$\mathbb{P}\left(\max_{1 \leq j \leq b_{k+1}} |S_j| \geq (1+\varepsilon)\sqrt{2b_k \ln \ln b_k}\right) \leq 2\mathbb{P}\left(|S_{b_{k+1}}| \geq \sqrt{b_{k+1} x_{k+1}}\right),$$

where $x_{k+1} = 2(1+\varepsilon)^2 \frac{b_k}{b_{k+1}} \ln \ln b_k$. To this we can apply Lemma 2.1.2 and deduce

$$\mathbb{P}\left(\max_{1 \leq j \leq b_{k+1}} |S_j| \geq (1+\varepsilon)\sqrt{2b_k \ln \ln b_k}\right) \leq \exp\left\{-(1+\delta_k)\frac{(1+\varepsilon)^2}{b} \ln k\right\},$$

where δ_k is some term that goes to 0 as $k \to \infty$. Since $b > (1+\varepsilon)^2$, the above can be summed over k, and we obtain, from the Borel–Cantelli lemma, that with probability one,

$$\max_{1 \leq j \leq b_{k+1}} |S_j| \leq \sqrt{2(1+\varepsilon)b_k \ln \ln b_k},$$

for all but finitely many k. Therefore, for all $m \in [b_k, b_{k+1}]$ sufficiently large,

$$|S_m| \leq \max_{1 \leq j \leq b_{k+1}} |S_j| \leq \sqrt{2(1+\varepsilon)b_k \ln \ln b_k} \leq \sqrt{2(1+\varepsilon)m \ln \ln m}.$$

This verifies equation (1).

We now work toward the converse. Fixing $b > 1$, $\varepsilon > 0$, define

$$\mathsf{E}_k = \left(S_{b_{k+1}} - S_{b_k} \geq (1-\varepsilon)\sqrt{2(b_{k+1} - b_k)\ln\ln b_k} \right).$$

The E_k's are independent events. On the other hand, $S_{b_{k+1}} - S_{b_k}$ and $S_{b_{k+1}-b_k}$ have the same distribution. Thus, by Lemma 2.1.2, applied to $x_n = 2(1-\varepsilon)^2 \ln\ln b_n$,

$$\mathbb{P}(\mathsf{E}_k) \geq \exp\left\{ -(1+\eta_k)(1-\varepsilon)\ln k \right\},$$

where η_k is some term that goes to 0 as $k \to \infty$. The upshot of this is that $\sum_k \mathbb{P}(\mathsf{E}_k) = +\infty$. Therefore, owing to the independence of the E_k's and by the Borel–Cantelli lemma for independent events, $\mathbb{P}(\mathsf{E}_k$ infinitely often$) = 1$. Equivalently,

$$\limsup_{k\to\infty} \frac{S_{b_{k+1}} - S_{b_k}}{\sqrt{2b_{k+1}\ln\ln b_{k+1}}} \geq (1-\varepsilon)\sqrt{1 - \frac{1}{b}}, \quad \text{a.s.}$$

Since $b > 1$, equation (1) shows that $\limsup_{k\to\infty}(2b_{k+1}\ln\ln b_{k+1})^{-\frac{1}{2}}|S_{b_k}| \leq b^{-\frac{1}{2}}$, a.s. Therefore, with probability one,

$$\limsup_{n\to\infty} \frac{S_n}{\sqrt{2n\ln\ln n}} \geq \limsup_{k\to\infty} \frac{S_{b_{k+1}}}{\sqrt{2b_{k+1}\ln\ln b_{k+1}}}$$
$$\geq (1-\varepsilon)\sqrt{1 - \frac{1}{b}} - \sqrt{\frac{1}{b}}.$$

Since this holds for all $b > 1$, and since equation (1) holds for all $\varepsilon > 0$, the result follows. □

We conclude this subsection with the following *very important exercise*.

Exercise 2.1.3 (Khintchine's LIL) Prove that the 1-parameter LIL (Theorem 2.1.1) still holds when Gaussian walks are replaced by simple ones, that is, when the increments are ± 1 with probability $\frac{1}{2}$ each. □

2.2 The General LIL

We define the function Λ by

$$\Lambda(x) = \begin{cases} \sqrt{2x\ln\ln x}, & \text{if } x \geq 4, \\ 1, & \text{otherwise.} \end{cases} \quad (1)$$

There is nothing special about the 4 in this definition. All that we need is a function that equals $\sqrt{2x\ln\ln x}$ for x large, and Λ is one such function.

Theorem 2.2.1 (The Law of the Iterated Logarithm) *Let S denote a real-valued, N-parameter random walk with increments $X = (X_t; \; t \in \mathbb{N}_0^N)$. If $\mathbb{E}[X_0] = 0$, $\mathbb{E}[X_0^2] = 1$ and $\mathbb{E}[X_0^2 (\ln_+ |X_0|)^{N-1}] < \infty$, then with probability one,*

$$\limsup_{t \to \infty} \frac{S_t}{\Lambda(\langle t \rangle)} = \sqrt{N}, \tag{2}$$

where Λ is given by (1) above, and $\langle t \rangle$ is defined in (1) in the introduction of Section 1.

We will prove this result in Sections 2.3 through 2.10. Our proof follows loosely along the original 1-parameter proof of A. I. Khintchine for simple walks. However, there are a number of technical difficulties that are overcome by applying the elegant method of de Acosta (1983). Before proving the direct half of the LIL, it should be pointed out that there is an essential converse.

***Exercise* 2.2.1** Show that in the setting of the LIL (Theorem 2.2.1),

$$\mathbb{E}\left[\frac{X_0^2 (\ln_+ |X_0|)^{N-1}}{\ln_+ \ln_+ |X_0|}\right] = +\infty \implies \limsup_{t \to \infty} \frac{|S_t|}{\Lambda(\langle t \rangle)} = +\infty.$$

This is due to Wichura (1973).
(HINT: Adapt our proof of the "converse" in Smythe's law of large numbers; cf. Theorem 1.4.1.) □

To conclude this subsection we point out a restatement of the above that is obtained from a real-variable argument.

Lemma 2.2.1 *Under the conditions of Theorem 2.2.1, equation (2) holds if and only if*

$$\limsup_{\langle t \rangle \to \infty} \frac{S_t}{\Lambda(\langle t \rangle)} = \sqrt{N}, \quad \text{a.s.}$$

***Exercise* 2.2.2** Prove Lemma 2.2.1. □

2.3 Summability

In this subsection we present two summability lemmas. The first is the multiparameter extension of **Kronecker's lemma** of classical probability theory. When $N = 1$, a more general version of this is attributed to L. Kronecker; cf. Supplementary Exercise 6.

Lemma 2.3.1 (Part of Kronecker's Lemma) *Suppose $(x_t; \; t \in \mathbb{N}^N)$ and $a = (a_t; \; t \in \mathbb{N}^N)$ denote two collections of nonnegative real numbers. Suppose that a is nondecreasing with respect to the partial order $\preccurlyeq$,*

$\lim_{t \to \infty} a_t = \infty$, and $\sum_{t \in \mathbb{N}^N} x_t/a_t < \infty$. Then,

$$\lim_{t \to \infty} \frac{1}{a_t} \sum_{s \preccurlyeq t} x_s = 0.$$

Our second result is a quantitative version of **summation by parts**.

Lemma 2.3.2 (Summation by Parts) *Suppose $\{a_i\}$, $\{b_i\}$, $\{c_i\}$, and $\{d_i\}$ are sequences of positive real numbers such that:*

(i) $b_1, b_2, \ldots$ is a nondecreasing sequence; and

(ii) $c_k \le \sum_{j=1}^{k} a_j \le d_k$, for all $k \ge 1$.

Then, for all $k \ge 1$,

$$\sum_{j=1}^{k} c_j \left[\frac{b_{j+1} - b_j}{b_j b_{j+1}} \right] + \frac{c_k}{b_k} \le \sum_{j=1}^{k} \frac{a_j}{b_j} \le \sum_{j=1}^{k} d_j \left[\frac{b_{j+1} - b_j}{b_j b_{j+1}} \right] + \frac{d_k}{b_k}.$$

Exercise **2.3.1** Prove the above form of the Kronecker's Lemma 2.3.1, as well as the summation-by-parts Lemma 2.3.2. □

2.4 Dirichlet's Divisor Lemma

For any positive integer k, let $D_k^{(2)}$ denote the number of its divisors. The following result of J. P. G. L. Dirichlet is an elementary fact from analytic number theory:

$$\lim_{n \to \infty} \frac{1}{n \ln n} \sum_{k=1}^{n} D_k^{(2)} = 1.$$

That is, a number that is uniformly picked from $\{1, \ldots, n\}$ has, on average, about $\ln n$ divisors. We will need a quantitative version of this, which makes precise the fact that $\sum_{k=1}^{n} D_k^{(2)}$ is of the order $n \ln n$. The key to this is the observation that $D_k^{(2)} = \sum_{t \in \mathbb{N}^2} \mathbf{1}_{\{k\}}(t^{(1)} t^{(2)})$. More generally, define

$$D_k^{(N)} = \sum_{t \in \mathbb{N}^N} \mathbf{1}_{\{k\}}(\langle t \rangle), \qquad k \ge 1. \tag{1}$$

Lemma 2.4.1 *For any two integers $n, N \ge 1$,*

$$\sum_{k=1}^{n} D_k^{(N)} \le 2^N n (N \ln 2 + \ln n)^{N-1}.$$

Remark While we need only the above upper bound on $\sum_{k=1}^{n} D_k^{(N)}$, much more can be done; cf. Supplementary Exercise 1 for a sample, or see Karatsuba (1993, Theorems 3 and 7, Ch. 1) for very detailed estimates when $N = 2$.

Proof For each $t \in \mathbb{N}^N$ define $Q(t) = \{s \in \mathbb{N}^N : s \not\geq t, |s - t| < 1\}$, and recall that for all $x \in \mathbb{R}^N$, $|x| = \max_{1 \leq j \leq N} |x^{(j)}|$ denotes the maximum modulus norm. Thus, $Q(t)$ denotes the cube of side 1 in $\mathbb{R}^N$ whose "lower end point" is t. Since Lebesgue's measure of $Q(t)$ is 1,

$$\sum_{k=1}^{n} D_k^{(N)} = \sum_{t \in \mathbb{N}^N} \int_{Q(t)} \mathbb{1}_{\{1,\ldots,n\}}(\langle t \rangle) \, ds$$

$$\leq \sum_{t \in \mathbb{N}^N} \int_{Q(t)} \mathbb{1}_{[1, 2^N n]}(\langle s \rangle) \, ds$$

$$= \int_{[1,\infty[^N} \mathbb{1}_{[1, 2^N n]}(\langle s \rangle) \, ds.$$

To obtain the inequality above we used the fact that for $t \not\geq (1, \ldots, 1)$ and $s \in Q(t)$,

$$\langle s \rangle \leq \prod_{j=1}^{N} [t^{(j)} + 1] \leq 2^N \langle t \rangle.$$

In summary, $\sum_{k=1}^{n} D_k^{(N)} \leq J_N(2^N n)$, where for all $x > 1$,

$$J_N(x) = \int_{[1,\infty[^N} \mathbb{1}_{[1,x]}(\langle s \rangle) \, ds. \tag{2}$$

Since $J_N(x) \leq x(\ln x)^{N-1}$, this proves the lemma. □

***Exercise* 2.4.1** Complete the above by showing that $J_N(x) \leq x(\ln x)^{N-1}$. □

2.5 Truncation

Define

$$\beta(x) = \begin{cases} \sqrt{\dfrac{x}{\ln \ln x}}, & \text{if } x \geq 4, \\ 1, & \text{otherwise.} \end{cases} \tag{1}$$

Now, for any $\eta > 0$, we can truncate the increment X_t as

$$X_t(\eta) = X_t \mathbb{1}_{(|X_t| \leq \eta \beta(\langle t \rangle))}, \qquad t \in \mathbb{N}^N. \tag{2}$$

4. Multiparameter Walks

These are independent, but not identically distributed, random variables, which have the advantage of being bounded. Based on them we can build the mean-zero independent increments process $S(\eta)$ as

$$S_t(\eta) = \sum_{s \preccurlyeq t} \{X_s(\eta) - \mathbb{E}[X_s(\eta)]\}, \qquad t \in \mathbb{N}^N. \tag{3}$$

In this section we make our first move by reducing the LIL to a problem about sums of *bounded* independent random variables that are not necessarily identically distributed. Namely, we have the following:

Proposition 2.5.1 *For any $\eta > 0$,*

$$\lim_{t \to \infty} \frac{|S_t - S_t(\eta)|}{\Lambda(\langle t \rangle)} = 0, \qquad a.s.$$

To prove it we need a supporting lemma.

Lemma 2.5.1 *The following are equivalent:*

(i) $\mathbb{E}[X_0^2 (\ln_+ |X_0|)^{N-1}] < \infty$;

(ii) for all $\eta > 0$,

$$\sum_{t \in \mathbb{N}^N} \frac{\mathbb{E}[|X_t - X_t(\eta)|]}{\Lambda(\langle t \rangle)} < \infty.$$

Proof of Lemma 2.5.1 We will prove that $(i) \Rightarrow (ii)$; the reverse implication is proved similarly.

Define the right inverse to β as

$$\beta^{-1}(a) = \sup\{x > 0 : \beta(x) < a\}.$$

By equation (1) above, there exists a constant C_1 such that for all $x \geq 0$,

$$\beta^{-1}(a) \leq C_1 a^2 \ln_+ \ln_+ a, \qquad \forall a \geq 0. \tag{4}$$

(Why?) On the other hand,

$$\sum_{t \in \mathbb{N}^N} \frac{\mathbb{E}[|X_t - X_t(\eta)|]}{\Lambda(\langle t \rangle)} = \mathbb{E}\bigg[|X_0| \cdot \sum_{t \in \mathbb{N}^N} \frac{\mathbb{1}_{(|X_0| \geq \eta \beta(\langle t \rangle))}}{\Lambda(\langle t \rangle)}\bigg].$$

To this we apply equation (4) with $a = |X_0|\eta^{-1}$, and deduce

$$\sum_{t \in \mathbb{N}^N} \frac{\mathbb{E}[|X_t - X_t(\eta)|]}{\Lambda(\langle t \rangle)}$$

$$\leq \mathbb{E}\bigg[|X_0| \cdot \sum_{t \in \mathbb{N}^N} \frac{\mathbb{1}_{(\langle t \rangle \leq C_1 X_0^2 \eta^{-2} \ln_+ \ln_+(X_0^2 \eta^{-2}))}}{\Lambda(\langle t \rangle)}\bigg]$$

$$= \mathbb{E}\bigg[|X_0| \cdot \sum_{k=1}^{\infty} \frac{D_k^{(N)}}{\Lambda(k)} \mathbb{1}_{(k \leq C_1 X_0^2 \eta^{-2} \ln_+ \ln_+(X_0^2 \eta^{-2}))}\bigg], \tag{5}$$

where $D_k^{(N)}$ was defined in equation (1) of Section 2.4. By Dirichlet's divisor lemma (Lemma 2.4.1), there exists a constant C_2 such that for all $j \geq 1$, $\sum_{k=1}^{j} D_k^{(N)} \leq C_2 j(\ln_+ j)^{N-1}$. Thus, summation-by-parts (Lemma 2.3.2) yields a constant C_3 such that for any $a \geq 1$,

$$\sum_{1 \leq k \leq a} \frac{D_k^{(N)}}{\Lambda(k)} \leq C_3 \sum_{1 \leq k \leq a} k(\ln_+ k)^{N-1} \left[\frac{\Lambda(k+1) - \Lambda(k)}{(\Lambda(k))^2}\right] + C_3 \frac{(\ln_+ a)^{N-1}}{\Lambda(a)}.$$

By Taylor's expansion we can find a constant C_4 such that for all integers $k \geq 1$, $\Lambda(k+1) - \Lambda(k) \leq C_4 k^{-\frac{1}{2}}(\ln \ln k)^{\frac{1}{2}}$. Using this, and a few more lines of calculations, we deduce the existence of a constant C_5 such that for all $a > 0$,

$$\sum_{0 \leq k \leq a} \frac{D_k^{(N)}}{\Lambda(k)} \leq C_5 \frac{\sqrt{a}(\ln_+ a)^{N-1}}{\sqrt{\ln_+ \ln_+ a}}.$$

Hence, equation (5) implies that

$$\sum_{t \in \mathbb{N}^N} \frac{\mathbb{E}\{|X_t - X_t(\eta)|\}}{\Lambda(\langle t \rangle)} \leq C_5 \mathbb{E}\left\{|X_0| \cdot \frac{\sqrt{Y}(\ln_+ Y)^{N-1}}{\sqrt{\ln_+ \ln_+ Y}}\right\},$$

where $Y = C_1 X_0^2 \eta^{-2} \ln_+ \ln_+(X_0^2 \eta^{-2})$. Since $\mathbb{E}[X_0^2(\ln_+ |X_0|)^{N-1}]$ is finite, so is the display above. □

Using the previous lemma we can finish our proof of Proposition 2.5.1.

Proof of Proposition 2.5.1 Since $\mathbb{E}[X_t] = 0$,

$$\sum_{t \in \mathbb{N}^N} \left|\frac{\mathbb{E}[X_t(\eta)]}{\Lambda(\langle t \rangle)}\right| = \sum_{t \in \mathbb{N}^N} \left|\frac{\mathbb{E}[X_t \mathbf{1}_{(|X_t| > \eta \beta(\langle t \rangle))}]}{\Lambda(\langle t \rangle)}\right| \leq \sum_{t \in \mathbb{N}^N} \frac{\mathbb{E}\{|X_t - X_t(\eta)|\}}{\Lambda(\langle t \rangle)},$$

which is finite, thanks to Lemma 2.5.1. Another application of the latter lemma yields

$$\sum_{t \in \mathbb{N}^N} \frac{|X_t - \{X_t(\eta) - \mathbb{E}[X_t(\eta)]\}|}{\Lambda(\langle t \rangle)} < +\infty, \qquad \text{a.s.}$$

Kronecker's lemma (Lemma 2.3.1) now shows that $(i) \Rightarrow (ii)$. □

Exercise 2.5.1 Verify the converse half of Lemma 2.5.1, i.e., show that $(ii) \Rightarrow (i)$. □

2.6 Bernstein's Inequality

According to Proposition 2.5.1, we can reduce the LIL to proving a law of the iterated logarithm for bounded, independent random variables. In this

4. Multiparameter Walks

section we present a deep inequality of F. Bernstein that states that sums of mean-zero independent random variables have Gaussian tails.

Proposition 2.6.1 (Bernstein's Inequality) *Let $V_1, V_2, \ldots, V_n$ be independent random variables with zero means and with $\max_{1 \le i \le n} |V_i| \le \mu$, for some deterministic constant $\mu > 0$. Then, for all choices of $\lambda > 0$,*

$$\mathbb{P}\left(\left|\sum_{i=1}^n V_i\right| \ge \lambda\right) \le 2\exp\left\{-\frac{1}{2}\frac{\lambda^2}{\sum_{i=1}^n \mathrm{Var}(V_i) + \mu\lambda}\right\}.$$

Proof By Taylor's expansion, for any $\zeta > 0$,

$$e^{\zeta V_i} = 1 + \zeta V_i + \frac{\zeta^2 V_i^2}{2} + \frac{\zeta^3 (V_i^\star)^3}{6} e^{\zeta V_i^\star},$$

where $|V_i^\star| \le |V_i|$, and $V_i^\star V_i \ge 0$. Consequently,

$$e^{\zeta V_i} \le 1 + \zeta V_i + \frac{\zeta^2 V_i^2}{2}\left[1 + \frac{\zeta\mu e^{\zeta\mu}}{3}\right], \quad \text{a.s.}$$

We take expectations to see that

$$\mathbb{E}\{e^{\zeta V_i}\} \le 1 + \frac{1}{2}\zeta^2 \mathrm{Var}(V_i)\left[1 + \frac{1}{3}\zeta\mu e^{\zeta\mu}\right] \le \exp\left\{\frac{1}{2}\zeta^2 \mathrm{Var}(V_i)\left[1 + \frac{1}{3}\zeta\mu e^{\zeta\mu}\right]\right\},$$

since $1 + x \le e^x$, for $x \ge 0$. In particular, the independence of the V_i's leads to

$$\mathbb{E}\{e^{\zeta \sum_{i=1}^n V_i}\} \le \exp\left\{\frac{1}{2}\zeta^2 \sigma_n^2\left[1 + \frac{1}{3}\zeta\mu e^{\zeta\mu}\right]\right\},$$

where we write $\sigma_n^2 = \sum_{i=1}^n \mathrm{Var}(V_i)$, for brevity. Thus, by Chebyshev's inequality, for all $\zeta, \lambda > 0$,

$$\mathbb{P}\left\{\sum_{i=1}^n V_i \ge \lambda\right\} \le \exp\left\{\frac{\zeta^2 \sigma_n^2}{2}\left[1 + \frac{1}{3}\zeta\mu e^{\zeta\mu}\right] - \zeta\lambda\right\}. \tag{1}$$

Now we choose

$$\zeta = \frac{\lambda}{\sigma_n^2 + \lambda\mu}.$$

Since $\frac{1}{3}e^{\zeta\mu} \le \frac{e}{3} \le 1$,

$$\mathbb{P}\left\{\sum_{i=1}^n V_i \ge \lambda\right\} \le \exp\left\{\frac{\zeta^2 \sigma_n^2}{2}[1 + \zeta\mu] - \zeta\lambda\right\}$$

$$= \exp\left\{\frac{\lambda^2}{2(\sigma_n^2 + \lambda\mu)} \cdot \frac{\sigma_n^2}{\sigma_n^2 + \lambda\mu}\left[1 + \frac{\lambda\mu}{\sigma_n^2 + \lambda\mu}\right] - \frac{\lambda^2}{\sigma_n^2 + \lambda\mu}\right\}$$

$$= \exp\left\{\frac{\lambda^2}{2(\sigma_n^2 + \lambda\mu)} \cdot \frac{1}{1+\eta}\left[\frac{1 + 2\eta}{1+\eta}\right] - \frac{\lambda^2}{\sigma_n^2 + \lambda\mu}\right\},$$

where $\eta = \lambda\mu/\sigma_n^2$. Consequently,

$$\mathbb{P}\Big\{\sum_{i=1}^n V_i \geq \lambda\Big\} \leq \exp\Big\{-\frac{\lambda^2}{2(\sigma_n^2+\lambda\mu)}\cdot\Big(2-\frac{1}{1+\eta}\Big[\frac{1+2\eta}{1+\eta}\Big]\Big)\Big\}.$$

It remains to prove that

$$2-\frac{1}{1+\eta}\Big[\frac{1+2\eta}{1+\eta}\Big] \geq 1.$$

But this is an elementary fact. □

One can use a different choice of ζ in (1), to get other inequalities. For instance, see the following exercise.

Exercise 2.6.1 Prove that in Proposition 2.6.1 there exists a constant C, independent of the distribution of the V_i's, such that for all $\lambda > 0$,

$$\mathbb{P}\Big\{\Big|\sum_{i=1}^n V_i\Big| \geq \lambda\Big\} \leq C\exp\Big(-\frac{\lambda}{\max\big[\sum_{j=1}^n \mathrm{Var}\,V_i\,,\,\mu^2\big]^{\frac{1}{2}}}\Big).$$

In its essence, this is due to A. N. Kolmogorov. □

Exercise 2.6.2 Improve Proposition 2.6.1 by finding a constant $\kappa \in\,]0,1[$, such that

$$\mathbb{P}\Big\{\Big|\sum_{i=1}^n V_i\Big| \geq \lambda\Big\} \leq 2\exp\Big(-\frac{\lambda^2}{2\big[\sum_{j=1}^n \mathrm{Var}\,V_i + \kappa\mu\lambda\big]}\Big).$$

Usually, one refers to Bernstein's inequality as this with $\kappa = \frac{1}{3}$. □

2.7 Maximal Inequalities

In this subsection we prove two maximal inequalities. When used together with Bernstein's inequality (Proposition 2.6.1), their implications are analogous to those of Lemma 2.1.1.

Lemma 2.7.1 *Suppose $Z = (Z_t;\, t \in \mathbb{N}^N)$ is a collection of $\mathbb{R}$-valued independent random variables with mean zero and $\sigma_\star^2 = \sup_{t\in\mathbb{N}^N} \mathbb{E}[Z_t^2] < \infty$. Then, whenever $\lambda \geq 2^{N+1}\sigma_\star$,*

$$\sup_{t\in\mathbb{N}^N} \mathbb{P}\Big(\max_{s \preccurlyeq t}\Big|\sum_{r \preccurlyeq s} Z_r\Big| \geq \lambda\sqrt{\langle t\rangle}\Big) \leq \frac{1}{2}.$$

Our next maximal inequality is a variant of classical inequalities of P. Lévy and G. Ottaviani; cf. (Pyke 1973; Shorack and Smythe 1976; Wichura 1973) for other variants. When the random walk in question is symmetric, an even better inequality is found in Exercise 2.1.1 above.

Lemma 2.7.2 *In the notation of Lemma 2.7.1 above, for any $x, y > 0$, and for all $t \in \mathbb{N}^N$,*

$$\mathbb{P}\Big(\max_{s \preccurlyeq t}\Big|\sum_{r \preccurlyeq s} Z_r\Big| \geq x+Ny\Big)\Big\{\mathbb{P}\Big(\max_{s \preccurlyeq t}\Big|\sum_{r \preccurlyeq s} Z_r\Big| \leq x\Big)\Big\}^N \leq \mathbb{P}\Big(\Big|\sum_{s \preccurlyeq t} Z_s\Big| \geq y\Big).$$

Proof of Lemma 2.7.1 Throughout, we write $T_t = \sum_{s \preccurlyeq t} Z_s$.

For any $t \in \mathbb{N}^N$, define $\mathcal{F}_t$ to be the σ-field generated by $(T_r;\ r \preccurlyeq t)$. Our proof of Proposition 1.2.1 can be mimicked to show that $\mathcal{F} = (\mathcal{F}_t;\ t \in \mathbb{N}^N)$ is a commuting filtration, since in the latter proposition the identical distribution of the increments was not needed. Moreover, the arguments used in Lemma 1.2.2 go through (verbatim) to show that $t \mapsto T_t$ is a martingale with respect to $\mathcal{F}$. By Corollary 3.5.1 and Theorem 2.3.1—both of Chapter 1—for all $t \in \mathbb{N}^N$,

$$\mathbb{E}\Big[\max_{s \preccurlyeq t} T_s^2\Big] \leq 4^N \mathbb{E}[T_t^2] = 4^N \sum_{s \preccurlyeq t} \mathbb{E}[Z_s^2] \leq 4^N \sigma_\star^2 \cdot \langle t \rangle.$$

In the last equality we used the fact that the increments of T_t are mean-zero independent random variables. The result follows from the above and Chebyshev's inequality. $\square$

Proof of Lemma 2.7.2 We will prove this when $N = 1$ and $N = 2$. The general case is deferred to Exercise 2.7.1 below.

First, suppose $N = 1$, and consider the stopping time

$$\tau = \inf(k \geq 1 : |Z_k| \geq x + y),$$

where, as usual, $\inf \varnothing = +\infty$. Clearly,

$$\mathbb{P}\Big(\max_{k \leq n}\Big|\sum_{i=1}^k Z_i\Big| \geq x+y\Big) = \mathbb{P}(\tau \leq n).$$

We decompose the above as follows, all the time writing $T_n = \sum_{i=1}^n Z_i$ for brevity:

$$\mathbb{P}(\tau \leq n) \leq \mathbb{P}(|T_n| \geq x) + \mathbb{P}(\tau \leq n,\ |T_n| \leq x).$$

On the other hand, on $(\tau \leq n)$, we can write

$$|T_n| = |T_n - T_\tau + T_\tau| \geq |T_\tau| - \max_{i \leq n}|T_{i+\tau} - T_\tau| \geq x+y - \max_{i \leq n}|T_{i+\tau} - T_\tau|.$$

Thus,

$$\mathbb{P}(\tau \leq n) \leq \mathbb{P}(|T_n| \geq x) + \mathbb{P}\Big(\tau \leq n,\ \max_{i \leq n}|T_{i+\tau} - T_\tau| \geq y\Big).$$

The above, together with the strong Markov property (Theorem 1.2.1 of Chapter 3), yields

$$\mathbb{P}(\tau \leq n) \leq \mathbb{P}(|T_n| \geq x) + \mathbb{P}(\tau \leq n)\mathbb{P}\left(\max_{i \leq n}|T_i| \geq y\right).$$

Solving, we obtain

$$\mathbb{P}\left(\max_{i \leq n}|T_i| \geq x+y\right) \cdot \mathbb{P}\left(\max_{i \leq n}|T_i| \leq y\right) \leq \mathbb{P}(|T_n| \geq x), \quad (1)$$

which is the desired result when $N = 1$. Note that equation (1) holds even if the increments of T are d-dimensional; in that case, $|\cdots|$ denotes, as usual, the ℓ^∞-norm on $\mathbb{R}^d$.

We now proceed with our proof for $N = 2$. Here, we write $T_{n,m} = \sum_{i \leq n} \sum_{j \leq m} Z_{i,j}$, where $Z_{i,j}$ is written in place of $Z_{(i,j)}$, etc.

For any fixed integer $m \geq 1$, $i \mapsto T'_i = (T_{i,1}, \ldots, T_{i,m})$ is an m-dimensional, one-parameter random walk. Since equation (1) applies in this case, we have

$$\mathbb{P}\left(\max_{i \leq n}|T'_i| \geq x+2y\right) \cdot \mathbb{P}\left(\max_{i \leq n}|T'_i| \leq 2y\right) \leq \mathbb{P}(|T'_n| \geq x).$$

Equivalently, in terms of our two-parameter walk, this states that

$$\mathbb{P}\left(\max_{(i,j) \preccurlyeq (n,m)}|T_{i,j}| \geq x+2y\right) \cdot \mathbb{P}\left(\max_{(i,j) \preccurlyeq (n,m)}|T_{i,j}| \leq y\right)$$
$$\leq \mathbb{P}\left(\max_{j \leq m}|T_{n,j}| \geq x+y\right).$$

The last term is a maximal inequality involving the one-parameter, one-dimensional walk $j \mapsto T_{n,j}$. Equation (1) implies

$$\mathbb{P}\left(\max_{j \leq m}|T_{n,j}| \geq x+y\right) \cdot \mathbb{P}\left(\max_{(i,j) \preccurlyeq (n,m)}|T_{i,j}| \leq y\right)$$
$$\leq \mathbb{P}\left(\max_{j \leq m}|T_{n,j}| \geq x+y\right) \cdot \mathbb{P}\left(\max_{j \leq m}|T_{n,j}| \leq y\right)$$
$$\leq \mathbb{P}(|T_{n,m}| \geq x).$$

This and the preceding display together demonstrate the result for the case $N = 2$. □

Exercise 2.7.1 Prove the above for general $N > 1$. □

2.8 A Number-Theoretic Estimate

In Section 2.4 we saw Dirichlet's divisor numbers $D_k^{(N)}$ ($k \geq 1$). Now we introduce a variant of these.

For any $k \geq 1$, define

$$R_k^{(N)} = \sum_{t \in \mathbb{N}^N} \mathbb{1}_{\{k\}}\left(\sum_{j=1}^N t^{(j)}\right). \tag{1}$$

Note that $R_k^{(N)} = 0$, unless $k \geq N$. We proceed to find a quantitative estimate for the cumulative sums of the function $k \mapsto R_k^{(N)}$.

Lemma 2.8.1 *For any $k \geq 1$, $\sum_{\ell=1}^k R_\ell^{(N)} \leq k^N/N!$.*

Proof Define $R_\ell^{(N)} = 0$, for all negative integers ℓ, and note that for all $k \geq 1$,

$$R_k^{(N)} = \sum_{t \in \mathbb{N}^{N-1}} \sum_{\ell=1}^\infty \mathbb{1}_{\{k-\ell\}}\left(\sum_{j=1}^{N-1} t^{(j)}\right) = \sum_{\ell=1}^\infty R_{k-\ell}^{(N-1)} = \sum_{\ell=1}^{k-1} R_\ell^{(N-1)}. \tag{2}$$

Therefore, for all $k \geq 1$,

$$\sum_{j=1}^k R_j^{(N)} = \sum_{j=1}^k \sum_{\ell=1}^{j-1} R_\ell^{(N-1)}. \tag{3}$$

We can now perform mathematical induction on N. Clearly, both bounds hold for $N = 1$ and, in fact, $\sum_{\ell=1}^k R_\ell^{(1)} = k$. Assuming that the result holds for $N-1$, we proceed to show that it does for N.

By (2) and by the induction hypothesis,

$$\sum_{j=1}^k R_j^{(N)} \leq \sum_{j=1}^k \frac{(j-1)^{N-1}}{(N-1)!} \leq \frac{1}{(N-1)!} \int_0^k x^{N-1} dx,$$

which is the desired upper bound. □

Lemma 2.8.1 is sharp, asymptotically, as $k \to \infty$:

Exercise 2.8.1 *If $x_+ = \max(x, 0)$, show that*

$$\sum_{\ell=1}^k R_\ell^{(N)} \geq \frac{(k-N+1)_+^N}{N!}.$$

□

Historically, Dirichlet's divisor lemma (Lemma 2.4.1) and Lemma 2.8.1 arise in analytic number theory through the theory of integer points. Indeed, consider Lemma 2.4.1 with $N = 2$ and observe that $R_k^{(2)}$ denote the number of points of integral coordinates in a large triangle.

***Exercise* 2.8.2** Verify that as $k \to \infty$, $k^{-2} R_k^{(2)} \to \frac{1}{2}$. □

***Exercise* 2.8.3** Continuing with Exercise 2.8.2, let G_k denote the number of points of integral coordinates in the centered disk of radius k on the plane. Show that as $k \to \infty$, $k^{-2} G_k \to \pi$. This calculation, together with some of its refinements, are due to C.-F. Gauss. See Karatsuba (1993, Theorems 2 and 6, Ch. 1) for this and more. □

For our applications to the LIL, integer points come in via summability criteria such as the following important exercise.

***Exercise* 2.8.4** Show that $\sum_{k=1}^{\infty} k^{-\gamma} R_k^{(N)} < +\infty \iff \gamma > N$. □

2.9 Proof of the LIL: The Upper Bound

In light of Proposition 2.5.1, it suffices to show that for any $\eta > 0$, almost surely,

$$\limsup_{t \to \infty} \frac{S_t(\eta)}{\Lambda(\langle t \rangle)} \leq \sqrt{\tfrac{1}{4} N^2 \eta^2 + N} + \tfrac{1}{2} N \eta, \tag{1}$$

where $S(\eta)$ is the truncation of the walk S; cf. equations (1)–(3) of Section 2.5. The upper bound of the LIL follows upon appealing to Proposition 2.5.1, and letting $\eta \downarrow 0$ along a rational sequence, in the above display.

We work toward establishing tail estimates for the walk $S(\eta)$.

Since $\mathbb{E}\{|X_t(\eta)|^2\} \leq \mathbb{E}\{|X_t|^2\} = 1$, Lemma 2.7.1 shows us that

$$\inf_{t \in \mathbb{N}^N} \mathbb{P}\Big(\max_{s \preccurlyeq t} |S_s(\eta)| \leq 2^{N+2} \sqrt{\langle t \rangle}\Big) \geq \frac{1}{2}.$$

Therefore, by Lemma 2.7.2, for all $t \in \mathbb{N}^N$, and all $\alpha > 0$,

$$\mathbb{P}\Big\{\max_{s \preccurlyeq t} |S_s(\eta)| \geq \alpha \Lambda(\langle t \rangle) + 2^{N+2} \sqrt{\langle t \rangle}\Big\} \leq 2^N \mathbb{P}\Big(|S_t(\eta)| \geq \alpha \Lambda(\langle t \rangle)\Big).$$

The right-hand side can be estimated by Bernstein's inequality (Proposition 2.6.1). Indeed, since $\text{Var} S_t(\eta) \leq \text{Var} S_t = \langle t \rangle$, Bernstein's inequality and the previous display together show us that for all $t \in \mathbb{N}^N$, and all $\alpha > 0$,

$$\mathbb{P}\Big\{\max_{s \preccurlyeq t} |S_s(\eta)| \geq \alpha \Lambda(\langle t \rangle) + 2^{N+2} \sqrt{\langle t \rangle}\Big\}$$
$$\leq 2^{N+1} \exp\Big(-\frac{\alpha^2 \langle t \rangle \ln_+ \ln_+ \langle t \rangle}{\langle t \rangle + \alpha \eta \Lambda(\langle t \rangle) \beta(\langle t \rangle)}\Big),$$

where β is defined in equations (1) of Section 2.5. In particular,

$$\mathbb{P}\Big\{\max_{s \preccurlyeq t} |S_s(\eta)| \geq \alpha \Lambda(\langle t \rangle) + 2^{N+2} \sqrt{\langle t \rangle}\Big\}$$
$$\leq 2^{N+1} \exp\Big(-\frac{\alpha^2}{1+\alpha\eta} \ln_+ \ln_+ \langle t \rangle\Big) \leq 2^{N+1} \big(\ln_+ \langle t \rangle\big)^{-\frac{\alpha^2}{1+\alpha\eta}}, \tag{2}$$

and the last inequality is an equality if $\ln_+ \ln_+ \langle t \rangle = \ln \ln \langle t \rangle$. This is the desired probability estimate; we will use it via a blocking argument. To define the blocks, let us fix some $\delta > 0$, and for all $t \in \mathbb{N}^N$, let $r_t \in \mathbb{N}^N$ be the point

$$r_t^{(j)} = \lfloor \exp(\delta t^{(j)}) \rfloor, \qquad j = 1, \ldots, N. \tag{3}$$

Note that $\ln \langle r_t \rangle \geq \delta \sum_{j=1}^N t^{(j)} - \delta$. In light of (2), we deduce the existence of a constant C such that

$$\sum_{t \in \mathbb{N}^N} \mathbb{P}\left\{ \max_{s \preccurlyeq r_t} |S_s(\eta)| \geq \alpha \Lambda(\langle r_t \rangle) + 2^{N+2}\sqrt{\langle r_t \rangle} \right\} \leq C \sum_{t \in \mathbb{N}^N} \left(\sum_{j=1}^N t^{(j)} \right)^{-\frac{\alpha^2}{1+\alpha\eta}}$$

$$= C \sum_{k=1}^\infty k^{-\frac{\alpha^2}{1+\alpha\eta}} R_k^{(N)},$$

where $R_k^{(N)}$ was defined in equation (1), Section 2.8. Thanks to Exercise 2.8.4, these sums converge if

$$\frac{\alpha^2}{1+\alpha\eta} > N. \tag{4}$$

We choose α, η such that they satisfy equation (4) and deduce, from the Borel–Cantelli lemma, that almost surely,

$$\max_{s \preccurlyeq r_t} |S_s(\eta)| \leq \alpha \Lambda(\langle r_t \rangle) + 2^{N+2}\sqrt{\langle r_t \rangle}, \quad \text{for all but finitely many } t \in \mathbb{N}^N.$$

Now, to blocking: For any $\nu \in \mathbb{N}^N$, we can find $t \in \mathbb{N}^N$ such that $r_t \leq \nu \leq r_{t+1}$, where $(t+1)^{(j)} = t^{(j)} + 1$. Thus, with probability one, for all but a finite number of ν's in $\mathbb{N}^N$,

$$|S_\nu(\eta)| \leq \max_{s \preccurlyeq r_{t+1}} |S_s(\eta)| \leq \alpha \Lambda(\langle r_{t+1} \rangle) + 2^{N+2}\sqrt{\langle r_{t+1} \rangle}$$

$$\leq \alpha e^{\delta N/2}(1 + o_\nu)\Lambda(\langle \nu \rangle),$$

where $o_\nu \to 0$ as $\langle \nu \rangle \to \infty$. In other words, assuming that condition (4) holds, $\limsup_{\nu \to \infty} |\Lambda(\langle \nu \rangle)|^{-1} S_\nu(\eta) \leq \alpha e^{\delta N/2}$, a.s. Since δ is arbitrary, we can let $\delta \to 0$ along a rational sequence and see that this lim sup is bounded above by α. We can even let α converge to the positive solution of $\alpha^2 = N(1+\alpha\eta)$ and not change the validity of this bound. This verifies equation (1) and completes our proof of the upper bound in the LIL. □

2.10 A Moderate Deviations Estimate

To prove the lower half of the LIL, we need the following one-parameter result, which states that at least half of Lemma 2.1.2 holds only under the conditions of the central limit theorem. This is, in general, sharp; see Supplementary Exercise 5 below.

Lemma 2.10.1 *Let $S = (S_k; \ k \geq 1)$ denote an $\mathbb{R}$-valued random walk with increments $X = (X_k; \ k \geq 1)$. If $\mathbb{E}[X_1] = 0$ and $\mathbb{E}[X_1^2] = 1$, then for all $\alpha > 0$,*

$$\liminf_{k \to \infty} \frac{1}{\ln \ln k} \ln \mathbb{P}(S_k \geq \alpha\sqrt{2k \ln \ln k}) \geq -\alpha^2.$$

Proof To clarify the essential ideas for this proof, we first present a rough outline. This discussion will be made rigorous subsequently.

Fix some $t > 0$ and working informally, we can write $S_k = Y_1 + \cdots + Y_{t \ln \ln k}$, where $Y_1 = S_{k/(t \ln \ln k)}$, and for all $j \in \{2, \ldots, t \ln \ln k\}$, $Y_j = S_{jk/(t \ln \ln k)} - Y_{j-1} - \cdots - Y_1$. Note that $Y_1, Y_2, \ldots, Y_{t \ln \ln k}$ are independent and all have the same distribution as $S_{k/(t \ln \ln k)}$. (This is where our proof is rough; it is rigorous only if $t \ln \ln k$ and $k/(t \ln \ln k)$ are both integers. Nonetheless, this idea of "blocking" the sum into independent blocks is correct in spirit.)

Clearly, if all the Y_j's are larger than $\frac{\alpha}{t}\sqrt{2k/\ln \ln k}$, then $S_k \geq \alpha\sqrt{2k \ln \ln k}$. Thus, using the i.i.d. structure of the Y's, we can estimate probabilities as follows:

$$\mathbb{P}(S_k \geq \alpha\sqrt{2k \ln \ln k}) \geq \left\{\mathbb{P}\left(Y_1 \geq \frac{\alpha}{t}\sqrt{\frac{2k}{\ln \ln k}}\right)\right\}^{t \ln \ln k}$$

$$= \left\{\mathbb{P}\left(S_{k/(t \ln \ln k)} \geq \frac{\alpha}{t}\sqrt{\frac{2k}{\ln \ln k}}\right)\right\}^{t \ln \ln k}.$$

Let N denote a standard Gaussian random variable. By the central limit theorem, for all $\varepsilon > 0$, there exists k_0 such that for all $k \geq k_0$,

$$\mathbb{P}\left(S_{k/(t \ln \ln k)} \geq \frac{\alpha}{t}\sqrt{\frac{2k}{\ln \ln k}}\right) \geq \mathbb{P}\left(N \geq \alpha\sqrt{\frac{2}{t}}\right) - \varepsilon.$$

Consequently, for all $\varepsilon > 0$,

$$\liminf_{k \to \infty} \frac{1}{\ln \ln t} \ln \mathbb{P}(S_k \geq \alpha\sqrt{2k \ln \ln k}) \geq t \ln \mathbb{P}\left(N \geq \alpha\sqrt{\frac{2}{t}}\right) - \varepsilon.$$

We can let $\varepsilon \to 0$, and appeal to standard tail estimates, e.g., Supplementary Exercise 11, to finish.

As we have already seen, the above is a complete proof, except that $t \ln \ln k$ and/or $k/(t \ln \ln k)$ need not be integer-valued for some choices of t and k. To get around this, fix $t > 0$ and define for all $k \geq 4$,

$$a_k = \left\lfloor \frac{k}{t \ln \ln k} \right\rfloor, \qquad b_k = \lfloor t \ln \ln k \rfloor.$$

Since $a_k b_k \leq k$ for all k large enough, we can then write $S_k = Y_1 + \cdots + Y_{b_k} + (S_k - Y_{b_k})$, where $Y_1 = S_{a_k}$, and for all $2 \leq j \leq b_k$, $Y_j = S_{ja_k} - Y_{j-1}$.

For all $\delta > 0$,

$$\mathbb{P}\left(S_k \geq \left(\frac{\alpha - \delta}{t}\right) b_k \sqrt{\frac{2k}{\ln \ln k}}\right)$$

$$\geq \left\{\mathbb{P}\left(Y_1 \geq \left(\frac{\alpha}{t}\right)\sqrt{\frac{2k}{\ln \ln k}}\right)\right\}^{b_k} \mathbb{P}\left(|S_k - Y_{b_k}| \leq \delta\sqrt{2k \ln \ln k}\right).$$

On the other hand, $S_k - Y_{b_k}$ has the same distribution as $S_{k-a_k b_k}$ which has mean zero and variance $(k - a_k b_k)$. By Chebyshev's inequality,

$$\mathbb{P}\left(|S_k - Y_{b_k}| \leq \delta\sqrt{2k \ln \ln k}\right) \geq 1 - \frac{k - a_k b_k}{2\delta^2 k \ln \ln k}.$$

Since $a_k \geq k(t \ln \ln k)^{-1} - 1$ and $b_k \geq t \ln \ln k - 1$,

$$\frac{k - a_k b_k}{k \ln \ln k} \leq \frac{1}{t(\ln \ln k)^2} + \frac{t}{+k \ln \ln k}.$$

That is, $\mathbb{P}(|S_k - Y_{b_k}| \leq \delta\sqrt{2k \ln \ln k})$ goes to 1 as $k \to \infty$. By changing α to $\alpha + \delta$, we obtain

$$\liminf_{k \to \infty} \frac{1}{b_k} \mathbb{P}\left(S_k \geq \alpha t^{-1} b_k \sqrt{\frac{2k}{\ln \ln k}}\right)$$

$$\geq \liminf_{k \to \infty} \ln \mathbb{P}\left(S_{a_k} \geq (\alpha + \delta) t^{-1} \sqrt{\frac{2k}{\ln \ln k}}\right)$$

$$= \ln \mathbb{P}\left(\mathcal{N} \geq (\alpha + \delta)\sqrt{\frac{2}{t}}\right),$$

thanks to the central limit theorem and the fact that $\lim_{k \to \infty} (a_k \ln \ln k / k) = t^{-1}$. Since $\lim_{k \to \infty} (b_k / \ln \ln k) = t$, for all $\alpha' < \alpha$ and all $t > 0$,

$$\liminf_{k \to \infty} \frac{1}{\ln \ln k} \mathbb{P}\left(S_k \geq \alpha' \sqrt{2k \ln \ln k}\right) \geq \frac{1}{t} \ln \mathbb{P}\left(\mathcal{N} \geq (\alpha + \beta)\sqrt{\frac{2}{t}}\right).$$

As $t \to 0$ we arrive at the following: For all $0 < \alpha' < \alpha$ and all $\beta > 0$,

$$\liminf_{k \to \infty} \frac{1}{\ln \ln k} \mathbb{P}\left(S_k \geq \alpha' \sqrt{2k \ln \ln k}\right) \geq -(\alpha + \beta)^2.$$

Letting $\beta \downarrow 0$ and $\alpha \downarrow \alpha'$, this proves the result with α' replacing α. □

2.11 Proof of the LIL: The Lower Bound

To this end, let us fix some $\delta > 2$ and for all $r \in \mathbb{N}^N$, define

$$\mathsf{B}(r) = \left\{t \in \mathbb{N}^N : \text{ for all } 1 \leq j \leq N, \ e^{\delta r^{(j)}} \leq t^{(j)} < e^{\delta(r^{(j)}+1)}\right\}. \tag{1}$$

Note the appearance of r_t in the above, where r_t were defined in equation (3) in our proof of the upper bound of the LIL (Section 2.9).

Whenever $r, t \in \mathbb{N}^N$ satisfy $t \in \mathsf{B}(r)$, then two things must happen:

1. $\exp(\delta \sum_{j=1}^N r^{(j)}) \leq \langle t \rangle \leq \exp(\delta + \delta \sum_{j=1}^N r^{(j)})$; and

2. $t \preccurlyeq \mathsf{L}(r)$, where $\mathsf{L}(r)$ is defined to be the point whose jth coordinate is $\mathsf{L}^{(j)}(r) = \lfloor \exp(\delta + \delta r^{(j)}) \rfloor$.

In words, $\mathsf{L}(r)$ is the unique largest element of $\mathsf{B}(r)$ with respect to the partial order $\preccurlyeq$. It is important also to recognize that

$$\#\mathsf{B}(r) = \{1 - e^{-\delta}\}^{-N} \langle \mathsf{L}(r) \rangle, \tag{2}$$

where $\#\mathsf{B}(r)$ denotes the cardinality of $\mathsf{B}(r)$. Fix $\eta \in\,]0, 1[$ and define $\alpha > 0$ by

$$\alpha = \sqrt{\frac{N}{1 - \eta}}. \tag{3}$$

By Lemma 2.10.1, there exists an integer k_0 (which may depend on η and δ) such that whenever $\#\mathsf{B}(r) \geq k_0$,

$$\mathbb{P}\bigg\{ \sum_{t \in \mathsf{B}(r)} X_t \geq \alpha \Lambda\big(\#\mathsf{B}(r)\big) \bigg\} \geq \big[\ln \#\mathsf{B}(r)\big]^{-\alpha^2(1-\varepsilon)}.$$

Also, by equation (2), $\ln \#\mathsf{B}(r) = (1 + o_r)\delta \sum_{j=1}^N r^{(j)}$, where $o_r \to 0$ as $\sum_{j=1}^N r^{(j)} \to \infty$. Therefore, there exists a constant C_0 such that for all but finitely many r,

$$\mathbb{P}\bigg\{ \sum_{t \in \mathsf{B}(r)} X_t \geq \alpha \Lambda\big(\#\mathsf{B}(r)\big) \bigg\} \geq C_0 \bigg[\sum_{j=1}^N r^{(j)}\bigg]^{-(1-\varepsilon)\alpha^2} = C_0 \bigg[\sum_{j=1}^N r^{(j)}\bigg]^{-N},$$

thanks to equations (2) and (3). Consequently, there exist constants C_1 and C_2 such that

$$\sum_{r \in \mathbb{N}^N} \mathbb{P}\bigg\{ \sum_{t \in \mathsf{B}(r)} X_t \geq \alpha \Lambda\big(\#\mathsf{B}(r)\big) \bigg\} \geq C_1 \sum_{k \geq C_2} k^{-N} R_k^{(N)},$$

where $R_k^{(N)}$ is defined by equation (1) of Section 2.8. By Exercise 2.8.4, $\sum_{r \in \mathbb{N}^N} \mathbb{P}(\mathsf{E}_r) = +\infty$, where E_r is the event that $\sum_{t \in \mathsf{B}(r)} X_t \geq \alpha \Lambda(\#\mathsf{B}(r))$. Since the E_r's are independent events, the Borel–Cantelli lemma tells us that infinitely many of the E_r's occur with probability one.

To recapitulate, for all $\delta > 2$ and all α and $\eta > 0$ satisfying equation (3),

$$\limsup_{\#\mathsf{B}(s) \to \infty} \frac{\sum_{t \in \mathsf{B}(s)} X_t}{\Lambda(\#\mathsf{B}(s))} \geq \alpha, \quad \text{a.s.}$$

By equation (2), almost surely,

$$\limsup_{\langle L(s)\rangle \to \infty} \frac{\sum_{t\in B(s)} X_t}{\Lambda(\langle L(s)\rangle)} \geq \alpha\{1-e^{-\delta}\}^{-\frac{N}{2}}. \tag{4}$$

Pick any $s \in \mathbb{N}^N$ such that $\langle s\rangle$ is very large, and note that

$$\left|S_{L(s)} - \sum_{t\in B(s)} X_t\right| \leq \sum_r^{(s)} |S_{L(r)}|, \tag{5}$$

where the sum $\sum_r^{(s)}$ is taken over all $r \in \mathbb{N}^N$ such that for all but one $1 \leq j \leq N$, $r^{(j)} = s^{(j)}$; for the one exceptional j, $r^{(j)} = s^{(j)} - 1$. (There are N summands in this sum.) By the already-proven upper bound to the LIL (Section 2.9),

$$\left|S_{L(s)} - \sum_{t\in B(s)} X_t\right| \leq (1+\theta_s)N^{\frac{1}{2}}\sum_r^{(s)} \Lambda(\langle r\rangle) \leq (1+\kappa_s)N^{\frac{3}{2}}e^{-\frac{1}{2}N\delta}\Lambda(\langle s\rangle),$$

where $\theta_s, \kappa_s \to 0$ as $\langle s\rangle \to \infty$. By equations (3)–(5), the following holds, a.s.:

$$\limsup_{t\to\infty} \frac{S_t}{\Lambda(\langle t\rangle)} \geq \limsup_{\langle s\rangle \to \infty} \frac{S_{L(s)}}{\Lambda(\langle s\rangle)}$$

$$\geq \alpha\{1-e^{-\delta}\}^{-\frac{N}{2}} - N^{\frac{3}{2}}e^{-\frac{1}{2}N\delta}$$

$$= \sqrt{\frac{N}{1-\eta}}\{1-e^{-\delta}\}^{-\frac{N}{2}} - N^{\frac{3}{2}}e^{-\frac{1}{2}N\delta}.$$

Moreover, this holds a.s., simultaneously for all rational numbers $\delta > 2$ and $\eta > 0$. The desired lower bound to the LIL follows upon letting $\delta \to \infty$ and $\eta \to 0$, in this order, and along rational sequences. □

3 Supplementary Exercises

1. Prove that as $n \to \infty$, $\sum_{k=1}^n D_k^{(N)}/n(\ln n)^{N-1}$ converges to 1.

2. Suppose $X_1, X_2, \ldots$ are i.i.d. random variables such that with probability one, $\limsup_{n\to\infty}(n \ln\ln n)^{-\frac{1}{2}}|\sum_{j=1}^n X_j| < \infty$. We wish to show that this implies $\mathbb{E}[X_1^2] < \infty$. That is, in the one-parameter setting, LIL is equivalent to the existence of two finite moments.

   (i) Prove there exists a nonrandom $M > 0$ such that the above $\limsup$ is bounded above by M.

(ii) Prove that it suffices to show this when the X's are symmetric.

(iii) Suppose the X's are symmetric, fix $c > 0$, and define $\overline{X}_i = X_i \mathbf{1}_{(|X_i| \le c)} - X_i \mathbf{1}_{(|X_i| > c)}$. Show that $(\overline{X}_i;\ i \ge 1)$ is a copy of $(X_i;\ i \ge 1)$, i.e., that X and $\overline{X}$ have the same finite-dimensional distributions.

(iv) Prove that almost surely, $\limsup_{n \to \infty} (n \ln \ln n)^{-\frac{1}{2}} |\sum_{j=1}^n X_i \mathbf{1}_{(|X_j| \le c)}| \le 2M$.

(v) Use the LIL to show that $\mathbb{E}[X_1^2 \mathbf{1}_{(|X_1| \le c)}] \le 4M^2$ and conclude that $\mathbb{E}[X_1^2] < \infty$.

This result is due to Strassen (1966), while the above-sketched proof is from Feller (1969).
(HINT: For part (ii), introduce independent copies X_i' of X_i and consider $X_i - X_i'$ instead. For part (iv), note that $2X\mathbf{1}_{(|X| \le c)} = X + \overline{X}$.)

3. Let $S = (S_k;\ k \ge 0)$ denote the simple walk on $\mathbb{Z}$, starting from $0 \in \mathbb{Z}$. That is, the increments of S are ± 1 with probability $\frac{1}{2}$ each. Prove that for any $n \ge 1$, $\max_{1 \le k \le n} S_k$ has the same distribution as $|S_n|$. This is the **reflection principle** of D. André; see Feller (1968).
(HINT: Study the described proof of Lemma 2.7.2 carefully.)

4. Suppose $(X_j;\ j \ge 0)$ are i.i.d. random variables with mean 0 and variance 1. If $S_n = \sum_{j=1}^n X_j$, prove the following moderate small-deviations bound: For each $\lambda > 0$,

$$\limsup_{n \to \infty} \frac{1}{\ln \ln n} \ln \mathbb{P}\Big(\max_{1 \le k \le n} |S_k| \le \lambda \sqrt{\frac{n}{\ln \ln n}} \Big) \le \ln \mathbb{P}(|\mathcal{N}| \le 2\lambda),$$

where $\mathcal{N}$ is a standard Gaussian random variable. Conclude that with probability one, $\liminf_{n \to \infty} (\ln \ln n / n)^{-\frac{1}{2}} \max_{1 \le k \le n} |S_k| > 0$. This is a part of the statement of Chung (1948).

5. (Hard) Our goal, here, is to show the following **moderate deviations principle**: If $V, V_1, V_2, \ldots$ are i.i.d. random variables with mean zero and variance one, and if they have a finite moment generating function in a neighborhood of the origin, then for any sequence x_n tending to infinity, such that $\lim_{n \to \infty} \frac{1}{n} x_n = 0$,

$$\lim_{n \to \infty} \frac{1}{x_n} \ln \mathbb{P}\big(V_1 + \cdots + V_n \ge \sqrt{n x_n}\big) = -\frac{1}{2}.$$

(i) First, show that even without the assumption on the local existence of a moment generating function, $\liminf_{n \to \infty} x_n^{-1} \ln \mathbb{P}(V_1 + \cdots + V_n \ge \sqrt{n x_n}) \ge -\frac{1}{2}$.

(ii) Use a blocking argument to show that under the stated conditions of the exercise, the corresponding $\limsup \le -\frac{1}{2}$.

(HINT: For (i), mimic the derivation of Lemma 2.10.1. For part (ii), start, as in the first part, by blocking $\sum_{j \le n} V_j = \sum_{j \le a_n} Y_{j,n}$, where $Y_{1,n}, \ldots, Y_{a_n,n}$ are i.i.d. random variables, and a_n is chosen carefully. Then, write

$$\mathbb{P}\Big(\sum_{j \le n} V_j \ge \sqrt{n x_n} \Big) = \mathbb{P}\Big(\sum_{j \le a_n} Z_{j,n} \ge a_n \Big),$$

where $Z_{j,n} = a_n(nx_n)^{-\frac{1}{2}} Y_{j,n}$. Use the exponential use of Chebyshev's inequality as was done for Bernstein's inequality (Proposition 2.6.1), together with the central limit theorem.)

6. (Hard) Prove that Kronecker's lemma (Lemma 2.3.1) remains true even if the x_s's are no longer assumed to be nonnegative. When $N = 1$, this is the classical and standard form of Kronecker's lemma. When $N \geq 2$, this is due to Cairoli and Dalang (1996, Lemma 2, Ch. 1).
(HINT: First, do this for $N = 1$. Then, proceed by induction on N, using the inclusion–exclusion formula, Lemma 1.2.1.)

7. Suppose $\varepsilon_1, \varepsilon_2, \ldots$ are i.i.d. Rademacher random variables. That is, $\mathbb{P}(\varepsilon_1 = 1) = \mathbb{P}(\varepsilon_1 = -1) = \frac{1}{2}$.

   (i) If $a_1, a_2, \ldots$ is an arbitrary sequence of numbers, prove that with probability one, $\limsup_{n\to\infty} (2v_n \ln \ln v_n)^{-\frac{1}{2}} \sum_{j=1}^n \varepsilon_j a_j = 1$, where $v_n = \sum_{j=1}^n a_j^2$.

   (ii) Suppose $X_1, X_2, \ldots$ are i.i.d. and are totally independent of the sequence of ε's. Conclude that $\limsup_{n\to\infty} (2V_n \ln \ln V_n)^{-\frac{1}{2}} \sum_{j=1}^n \varepsilon_j X_j = 1$, where $V_n = \sum_{j=1}^n X_j^2$.

   (iii) Prove that if the X_i's are symmetric, $\limsup_{n\to\infty} (2V_n \ln \ln V_n)^{-\frac{1}{2}} \sum_{j=1}^n X_j = 1$, almost surely.

   (iv) Show that even without symmetry, $\limsup_{n\to\infty} (V_n \ln \ln V_n)^{-\frac{1}{2}} \sum_{j=1}^n X_j > 0$, almost surely.
   (HINT: Symmetrize.)

   (v) Verify that when the X_i's are symmetric and have variance 1, the LIL follows.

Derive an N-parameter version of this result.
This is an example of a "self-normalized LIL" and is essentially due to Griffin and Kuelbs (1991). Roughly speaking, it states that in the symmetric case, the LIL is a consequence of the strong law.

8. In the context of Smythe's law of large numbers, show that whenever $\mathbb{E}[|X_0|^p] < \infty$ for some $p > 1$, A_t converges to $\mathbb{E}[X_0]$ in $L^p(\mathbb{P})$.

9. In the context of Smythe's law of large numbers, show that for any $p > 1$, $\mathbb{E}[\sup_t |A_t|^p] < \infty$ if and only if $\mathbb{E}[|X_0|^p] < \infty$, whereas $\mathbb{E}[\sup_t |A_t|] < \infty$ if and only if $\mathbb{E}[|X_0|\{\ln_+ |X_0|\}^N] < \infty$. (Note the power of N and not $N-1$.)
This is, in various forms, due to (Burkholder 1962; Gabriel 1977; Gut 1979); see also Cairoli and Dalang (1996, Theorem 2.4.2, Chapter 2).

10. The moment conditions of the strong law and the LIL are crucial, as the following shows. Recall that a random variable Y has a standard **Cauchy** distribution if it has the probability density function (with respect to Lebesgue's measure) $f(x) = \frac{1}{\pi}(1+x^2)^{-1}$, $(x \in \mathbb{R})$. Given i.i.d. (standard) Cauchy random variables $X_1, X_2, \ldots$, define $S_n = \sum_{j=1}^n X_j$ to be the corresponding random walk.

   (i) Verify that for each $n \geq 1$, $\frac{1}{n} S_n$ has a standard Cauchy random distribution.
   (HINT: Consider characteristic functions.)

(ii) (Hard) Suppose $a_1, a_2, \ldots$ is a nondecreasing sequence of positive reals such that $\lim_n a_n = +\infty$. Prove that the following dichotomy holds:

$$\mathbb{P}(S_n > na_n \text{ infinitely often }) = \begin{cases} 0, & \text{if } \sum_n a_{2^n}^{-1} < \infty, \\ 1, & \text{if } \sum_n a_{2^n}^{-1} = +\infty. \end{cases}$$

(iii) Conclude that for any nondecreasing function h, $\limsup_{n\to\infty} S_n/\{nh(n)\}$ is 0 or infinity (a.s.).

(iv) Derive an N-parameter version of this "summability test."

11. If $\mathcal{N}$ is standard Gaussian, prove that for all $\lambda > 0$, $\mathbb{P}(|\mathcal{N}| \geq \lambda) \leq 2e^{-\lambda^2/2}$. Moreover, prove that $\lim_{\lambda\to\infty} \lambda e^{\lambda^2/2} \mathbb{P}(|\mathcal{N}| \geq \lambda) = \sqrt{2/\pi}$.

4 Notes on Chapter 4

Section 1 The proof given here of Smythe's law of large numbers is close to those in (Edgar and Sucheston 1992; Smythe 1973; Walsh 1986b). For a survey of some of these results (five of them, to be exact), see Smythe (1974a). To see further related results, refer to (Bass and Pyke 1984c; Kwon 1994; Pyke 1973; Shorack and Smythe 1976; Smythe 1974b). The method of reduction to the one-parameter setting is, in fact, a metatheorem; cf. Sucheston (1983). Alternative derivations of Smythe's law can be found in (Cairoli and Dalang 1996; Khoshnevisan 2000).

There is a rich multiparameter ergodic theory that essentially runs parallel to the limit theory of this chapter. See, for example, (Krengel and Pyke 1987; Frangos and Sucheston 1986; Sucheston 1983; Stein 1961; Ricci and Stein 1992).

Section 2 The one-parameter law of the iterated logarithm has a long and distinguished history. The literature is so large that it would be impossible to give an extensive bibliography here. Ledoux and Talagrand (1991) and Stout (1974) are good sources for further reading in different directions. For two particularly beautiful works about this general topic, see (Erdős 1942; Rogers and Taylor 1962).

There are many LILs that are related to the material of this chapter and, in fact, the entire book. As a sampler, we mention (Bass 1985; Bass and Pyke 1984b; Li and Wu 1989; Paranjape and Park 1973; Park 1975; Wichura 1973).

We have already seen that when $N = 1$, the LIL holds if and only if the increments of the random walk have a finite second moment. When $N \geq 2$, the $L^2\{\ln L\}^{N-1}$ can be improved to the following *necessary and sufficient* condition:

$$\mathbb{E}\left[|X|^2 \frac{(\ln_+ |X|)^{N-1}}{\ln_+ \ln_+ |X|}\right] < +\infty.$$

See Wichura (1973); for an infinite-dimensional extension, see Li and Wu (1989). It is intriguing that while the strong law always holds iff the increments are in $L(\ln_+ L)^{N-1}$, the necessary and sufficient condition for the existence of the LIL depends on whether or not the number of parameters, N, is greater than or equal to 2!

The proof of the LIL given in this section is the most elegant one that I know, and is a multiparameter adaptation of de Acosta (1983). Lemma 2.10.1 is borrowed from de Acosta (1983) and de Acosta and Kuelbs (1981). Moreover, there are related inequalities to those that appear here that you can learn from (Etemadi 1991; Pyke 1973; Shorack and Smythe 1976).

Section 3 Supplementary Exercise 5 is the starting point for a number of interesting directions of work on random walks. You can explore related themes under the general headings of the law of the iterated logarithm and moderate (and sometimes large) deviations in (Dembo and Zeitouni 1998; Petrov 1995), for instance.

5
Gaussian Random Variables

Classical probability theory has shown us that $\mathbb{R}^d$-valued Gaussian random variables appear naturally as limits of random walks that take their values in $\mathbb{R}^d$. Later on, in Chapter 6, we shall see that such limit theorems are part of an elegant abstract theory. At the heart of such a theory is an understanding of Gaussian variables that take values in a rather general state space. This chapter is concerned with the development of a theory of Gaussian variables that is sufficiently general for our later needs.

1 The Basic Construction

In this section we will construct a Gaussian random variable that takes values in the space of continuous linear functionals on a certain separable Hilbert space. This construction is sufficiently general to yield many interesting objects such as "*Brownian motion*," "*Brownian sheet*," and "*white noise*." We begin our discussion with preliminaries on finite-dimensional Gaussian random variables.

1.1 Gaussian Random Vectors

An $\mathbb{R}^d$-valued random variable g is said to be a **Gaussian random variable** (or equivalently a **Gaussian random vector**) if there are a vector $\mu \in \mathbb{R}^d$ and a matrix $\Sigma \in \mathbb{R}^d \times \mathbb{R}^d$ such that for all $t \in \mathbb{R}^d$, $g \cdot t = \sum_{j=1}^d t^{(j)} g^{(j)}$ is an $\mathbb{R}$-valued Gaussian random variable with mean

$\mu \cdot t$ and variance $t'\Sigma t$; here and throughout, we view t as a *column* vector, and t' denotes the transpose of t. Equivalently, g is a Gaussian random variable if for all $t \in \mathbb{R}^d$ and $\xi \in \mathbb{R}$,

$$\mathbb{E}\Big[\exp\big(i\xi(g\cdot t)\big)\Big] = \exp\Big(i\xi \sum_{j=1}^{d} t^{(j)}\mu^{(j)} - \frac{\xi^2}{2}\sum_{j=1}^{d}\sum_{k=1}^{d} t^{(j)}t^{(k)}\Sigma^{(j,k)}\Big).$$

The vector μ is called the **mean vector**, and Σ is called the **covariance matrix** of g. Since for all $t \in \mathbb{R}^d$, $0 \le \text{Var}(g \cdot t) = t'\Sigma t$, Σ is always a nonnegative definite matrix.

Throughout this book we will consistently write $g \sim \mathcal{N}_d(\mu, \Sigma)$ to mean that g is an $\mathbb{R}^d$-valued Gaussian random vector with mean vector μ and covariance matrix Σ. In particular, $\mathcal{N}_1(0,1)$ corresponds to the standard Gaussian distribution on the real line.

***Exercise* 1.1.1** If Σ is nonsingular, show that the distribution of g is absolutely continuous with respect to Lebesgue's measure and its density at $x \in \mathbb{R}^d$ is

$$\frac{1}{(2\pi)^{d/2}\sqrt{\det \Sigma}} \exp\Big\{-\frac{1}{2}(x-\mu)\Sigma^{-1}(x-\mu)'\Big\},$$

where Σ^{-1} (respectively $\det \Sigma$) denotes the inverse (respectively determinant) of Σ. □

***Exercise* 1.1.2 (Mill's Ratios)** Suppose that $d = 1$ and that the mean and variance of g are 0 and 1, respectively. Use the explicit form of the density of g (Exercise 1.1.1) and L'Hôpital's rule to deduce that as $\lambda \to \infty$,

$$\lambda e^{\frac{1}{2}\lambda^2} \mathbb{P}(g > \lambda) \to (2\pi)^{-\frac{1}{2}}.$$

Conclude the existence of a finite constant $C > 1$ such that for all $\lambda > 1$, $C^{-1}\lambda^{-1}\exp(-\frac{1}{2}\lambda^2) \le \mathbb{P}(g > \lambda) \le C\lambda^{-1}\exp(-\frac{1}{2}\lambda^2)$. In this regard, see also Supplementary Exercise 11, Chapter 4. □

In order to better understand the roles of μ and Σ, fix some $i \in \{1, \ldots, d\}$ and choose $t \in \mathbb{R}^d$ according to

$$t^{(j)} = \begin{cases} 1, & \text{if } j = i, \\ 0, & \text{otherwise,} \end{cases} \quad j = 1, \ldots, d.$$

For this choice of t we have $g \cdot t = g^{(i)} \sim \mathcal{N}_1(\mu^{(i)}, \Sigma^{(i,i)})$. In other words, μ is the vector of the means of g, viewed coordinatewise, and the diagonal of Σ is the vector of the variances of g, also viewed coordinatewise.

To identify the off-diagonal elements of Σ, fix distinct $i, j \in \{1, \ldots, d\}$, and choose $t \in \mathbb{R}^d$ as

$$t^{(\ell)} = \begin{cases} 1, & \text{if } \ell = i, \\ -1, & \text{if } \ell = j, \\ 0, & \text{otherwise,} \end{cases} \quad \ell = 1, \ldots, d.$$

We see that $g \cdot t = g^{(i)} - g^{(j)}$ is an $\mathbb{R}$-valued Gaussian random variable with variance $\Sigma^{(i,i)} + \Sigma^{(j,j)} - \Sigma^{(i,j)} - \Sigma^{(j,i)}$. Furthermore, thanks to the preceding paragraph, $\Sigma^{(\ell,\ell)} = \text{Var}[g^{(\ell)}]$. Thus,

$$\text{Var}[g^{(i)} - g^{(j)}] = \text{Var}[g^{(i)}] + \text{Var}[g^{(j)}] - \{\Sigma^{(i,j)} + \Sigma^{(j,i)}\}.$$

Equivalently, $\text{Cov}[g^{(i)}, g^{(j)}] = \Sigma_\star^{(i,j)}$, where $\Sigma_\star = \frac{1}{2}\{\Sigma^{(i,j)} + \Sigma^{(j,i)}\}$ is the **symmetrization** of Σ. An important property of matrix symmetrization is that it preserves quadratic forms; i.e.,

$$t'\Sigma_\star t = t'\Sigma t, \qquad t \in \mathbb{R}^d.$$

(Check!) In other words, whenever $g \sim \mathcal{N}_d(\mu, \Sigma)$, then $g \sim \mathcal{N}_d(\mu, \Sigma_\star)$. Moreover, one can determine $\Sigma_\star$ uniquely as the symmetric matrix of the covariances of g. From now on, whenever we write $g \sim \mathcal{N}_d(\mu, \Sigma)$, μ and Σ designate the mean vector and matrix of the covariances of g, respectively. This justifies the use of the terms "*mean vector*" and "*covariance matrix*."

According to the above discussion, we can assume, without loss of generality, that Σ is a symmetric matrix, and we have also seen that Σ is positive definite. It turns out that these properties of Σ together completely characterize $\mathbb{R}^d$-valued Gaussian random variables in the following distributional sense.

Theorem 1.1.1 *Given any vector $\mu \in \mathbb{R}^d$ and any symmetric, positive definite matrix $\Sigma \in \mathbb{R}^d \times \mathbb{R}^d$, one can construct (on some probability space) an $\mathbb{R}^d$-valued random variable $g \sim \mathcal{N}_d(\mu, \Sigma)$. Conversely, if $g \sim \mathcal{N}_d(\mu, \Sigma)$, then Σ can be taken to be symmetric and is always real and positive definite.*

Proof Since the converse part was proved in the previous paragraph, we proceed by deriving the direct half.

Suppose $\mu \in \mathbb{R}^d$ and Σ is a symmetric, positive definite $(d \times d)$ matrix. By elementary matrix algebra, $\Sigma = P'\Lambda P$, where P is a real, orthogonal $(d \times d)$ matrix and Λ is a diagonal $(d \times d)$ matrix of eigenvalues with $\Lambda^{(i,i)} \geq 0$ for all $1 \leq i \leq d$. Let $A = \Lambda^{\frac{1}{2}} P$, where $\Lambda^{\frac{1}{2}}$ denotes the $(d \times d)$ diagonal matrix whose (i,i)th coordinate is $\sqrt{\Lambda^{(i,i)}}$, and note that A is a real matrix that satisfies $\Sigma = A'A$. Let Z be an $\mathbb{R}^d$-valued vector of independent standard Gaussian random variables constructed on some probability space and define $g = \mu + A'Z$. By directly computing the characteristic function of $t \cdot g$ for $t \in \mathbb{R}^d$, we can see that $g \sim \mathcal{N}_d(\mu, \Sigma)$, as desired. $\square$

The following is an important consequence of the explicit form of the characteristic function of $g \cdot t$:

Corollary 1.1.1 *Suppose $g \sim \mathcal{N}_d(\mu, \Sigma)$, where $\Sigma^{(i,j)} = 0$ for $i \neq j$. Then $g^{(1)}, \ldots, g^{(d)}$ are independent, $\mathbb{R}$-valued, Gaussian random variables.*

***Exercise* 1.1.3** Prove Corollary 1.1.1. □

***Exercise* 1.1.4** The family of Gaussian distributions is closed under convergence in distribution. More precisely, suppose $g_1, g_2, \ldots$ are mean-zero Gaussian random variables in $\mathbb{R}$ that converge in distribution to some g. Prove that g is a Gaussian random variable. □

1.2 Gaussian Processes

Given an abstract index set T, a process $X = (X_t;\, t \in T)$ is a real-valued **Gaussian process** if for all finite choices of $t_1, \ldots, t_k \in T$, $(X_{t_1}, \ldots, X_{t_k})$ is an $\mathbb{R}^k$-valued Gaussian random variable. To each such Gaussian process we can associate a **mean function** $\mu(t) = \mathbb{E}[X_t]$ $(t \in T)$ and a **covariance function** $\Sigma(s,t) = \mathbb{E}[X_s X_t]$ $(s, t \in T)$. When T is a measurable subset of $\mathbb{R}^N$, we may also refer to X as a **Gaussian random field** to conform to the historical use of this term.

The intention of this subsection is to decide when a Gaussian random process exists. Since Gaussian random vectors are, in a trivial sense, random processes, this description should be viewed as a natural extension of Theorem 1.1.1.

We say that a function $f : T \times T \to \mathbb{R}$ is **symmetric** if for all $s, t \in T$, $f(s,t) = f(t,s)$. It is **positive definite** if for all $s_1, \ldots, s_n \in T$ and all $\xi_1, \ldots, \xi_n \in \mathbb{R}$,

$$\sum_{i=1}^n \sum_{j=1}^n \xi_i f(s_i, s_j) \xi_j \geq 0. \tag{1}$$

We first state two elementary results.

Lemma 1.2.1 *Any symmetric, positive definite function f is nonnegative on the diagonal in the sense that $f(x,x) \geq 0$ for all x.*

***Exercise* 1.2.1** Prove Lemma 1.2.1. □

***Exercise* 1.2.2** Suppose T is a measurable space, and $f : T \to \mathbb{R}$ is measurable. If μ is a measure on the measurable subsets of T, and if f is positive definite, show that

$$\iint f(s,t)\, \mu(ds)\mu(dt) \geq 0,$$

provided that the integral is well-defined. □

Lemma 1.2.2 *The covariance function of a Gaussian process is symmetric and positive definite.*

Proof Let $X = (X_t;\ t \in T)$, and let Σ denote our Gaussian process and its covariance function, respectively. To show symmetry, we need only note that for all $s, t \in T$, $\mathbb{E}[X_s X_t] = \mathbb{E}[X_t X_s]$. To show that Σ is positive definite, fix $s_1, \ldots, s_n \in T$ and $\xi_1, \ldots, \xi_n \in \mathbb{R}$ and observe that

$$\mathbb{E}\Big[\Big\{\sum_{j=1}^n \xi_j X(s_j)\Big\}^2\Big] = \sum_{i=1}^n \sum_{j=1}^n \xi_i \Sigma(s_i, s_j) \xi_j.$$

Since the left-hand side is positive, so is the right-hand side. □

In fact, the properties of Lemma 1.2.2 characterize Gaussian processes in the following sense.

Theorem 1.2.1 *Given an abstract set T, an arbitrary function $\mu : T \to \mathbb{R}$, and a symmetric, positive definite function $\Sigma : T \times T \to \mathbb{R}$, there exists a Gaussian process $X = (X_t;\ t \in T)$ with mean and covariance functions μ and Σ, respectively.*

This theorem asserts that such a process exists on *some* probability space, as the following exercise shows.

Exercise **1.2.3** Let Ω be a denumerable set and let $\mathcal{F}$ denote a σ-field on the subsets of Ω. Show that one cannot define a real-valued Gaussian random variable on the measure space $(\Omega, \mathcal{F})$. □

Proof of Theorem 1.2.1 Suppose we could prove the result for the mean-zero case; i.e., suppose we can construct a Gaussian process $Y = (Y_t;\ t \in T)$ with mean and covariance functions 0 and Σ, respectively. Then, $X_t = Y_t + \mu(t)$ is the Gaussian process mentioned in the statement of the theorem. Thus, we can assume, without loss of generality, that $\mu(t) = 0$, for all $t \in T$.

Whenever $F = \{t_1, \ldots, t_k\}$ is a finite subset of T, define the $(k \times k)$ matrix Σ_F by

$$\Sigma_F^{(i,j)} = \Sigma(t_i, t_j), \qquad 1 \le i, j \le k.$$

Clearly, Σ_F is a symmetric, positive definite $(k \times k)$ matrix. By Theorem 1.1.1, on some probability space we can construct an $\mathbb{R}^k$-valued random variable $Z_F \sim \mathcal{N}_k(0, \Sigma_F)$ whose distribution is, for all measurable $A \subset \mathbb{R}^k$, defined as

$$\mu_F(A) = \mathbb{P}(Z_F \in A).$$

Suppose $k > 1$ and let $F_1 = \{t_1, \ldots, t_{k-1}\}$. In particular, $F_1 \subset F$ and Z_{F_1} has the same distribution as the first $k - 1$ coordinates of Z_F. Thus, for all measurable $A_1, \ldots, A_{k-1} \subset \mathbb{R}$,

$$\mu_{F_1}(A_1 \times \cdots \times A_{k-1}) = \mu_F(A_1 \times \cdots \times A_{k-1} \times \mathbb{R}).$$

142   5. Gaussian Random Variables

In fact, if F_0 is any subset of F of cardinality m ($\leq k$),
$$\mu_{F_0}(A_1 \times \cdots \times A_m) = \mu_F(B_1 \times \cdots \times B_k),$$
where $B_j = A_j$ for all j such that $t_j \in F_0$, and $B_j = \mathbb{R}$, otherwise.

In this way we have created probability measures μ_F, on $\mathbb{R}^F$ such that the family $(\mu_F;\ F \subset T,\ \text{finite})$ is a consistent family. By Kolmogorov's existence theorem (Theorem 2, Appendix A), on an appropriate probability space we can construct a real-valued process $X = (X_t;\ t \in T)$ such that for all finite $F \subset T$, the distribution of $(X_t;\ t \in F)$ is μ_F. This is our desired stochastic process. □

1.3 White Noise

Our first nontrivial example of a Gaussian process will be one-dimensional white noise on $\mathbb{R}^N$. This will be a Gaussian process indexed by the Borel field on $\mathbb{R}^N$, which we write as T. (We could also let T denote the collection of all Lebesgue measurable sets in $\mathbb{R}^N$.) Let us begin with the form of the covariance function.

Lemma 1.3.1 *For all $A, B \in T$, define $\Sigma(A, B) = \text{Leb}(A \cap B)$. Then, $\Sigma : T \times T \to \mathbb{R}$ is symmetric and positive definite.*

Proof It suffices to show that Σ is positive definite. Since $\Sigma(A, B) = \int_{\mathbb{R}^N} \mathbf{1}_A(u)\mathbf{1}_B(u)du$, Fubini's theorem shows that for all $A_i \in T$ and all $\xi_i \in \mathbb{R}$ ($1 \leq i \leq n$),
$$\sum_{i,j=1}^n \xi_i \Sigma(A_i, A_j)\xi_j = \int_{\mathbb{R}^N} \left(\sum_{i=1}^n \mathbf{1}_{A_i}(u)\xi_i\right)^2 du \geq 0.$$
This proves the lemma. □

(One-dimensional) **white noise** on $\mathbb{R}^N$ is defined to be a mean-zero Gaussian process $\mathbb{W} = (\mathbb{W}(A);\ A \in T)$ whose covariance function is $\Sigma(A, B) = \text{Leb}(A \cap B)$ ($A, B \in T$). By Theorem 1.2.1 and by Lemma 1.3.1 above, white noise exists as a well-defined process. The following captures some of its salient features.

Theorem 1.3.1 *Let $\mathbb{W} = (\mathbb{W}(A);\ A \in T)$ be white noise on $\mathbb{R}^N$.*

(a) For all disjoint $A, B \in T$, $\mathbb{W}(A)$ and $\mathbb{W}(B)$ are independent.

(b) For all $A, B \in T$, $\mathbb{W}(A \cup B) = \mathbb{W}(A) + \mathbb{W}(B) - \mathbb{W}(A \cap B)$, a.s.

(c) If $A_1, A_2, \ldots \in T$ are disjoint and $\sum_{i=1}^\infty \text{Leb}(A_i) < \infty$, then a.s.,
$$\mathbb{W}\left(\bigcup_{i=1}^\infty A_i\right) = \sum_{i=1}^\infty \mathbb{W}(A_i).$$

1 The Basic Construction   143

Remark It is tempting to think that $\mathbb{W}$ is almost surely a σ-finite, signed measure on the Borel field of $\mathbb{R}^N$. This is not the case (Supplementary Exercise 2), since the null sets in *(a)*, *(b)*, and *(c)* above depend on the choice of the sets $A, B, A_1, A_2, \ldots$, and there are uncountably many such null sets. However, $\mathbb{W}$ *is* a σ-finite, signed, $L^2(\mathbb{P})$-valued measure; cf. Exercise 1.3.1 below.

Proof If $A, B \in \mathcal{T}$ are disjoint, $\mathbb{E}[\mathbb{W}(A)\mathbb{W}(B)] = \text{Leb}(A \cap B) = 0$. Corollary 1.1.1 (i.e., "uncorrelated $\Rightarrow$ independent") implies part *(a)*.

Suppose $A, B \in \mathcal{T}$ are disjoint and define $D = \mathbb{W}(A \cup B) - \mathbb{W}(A) - \mathbb{W}(B)$. By squaring D and taking expectations, and since $A \cap B = \varnothing$,

$\mathbb{E}[D^2]$
$= \mathbb{E}\{|\mathbb{W}(A \cup B)|^2\} - 2\mathbb{E}[\mathbb{W}(B)\mathbb{W}(A \cup B)] - 2\mathbb{E}[\mathbb{W}(A)\mathbb{W}(A \cup B)]$
$\quad + \mathbb{E}\{|\mathbb{W}(A)|^2\} + 2\mathbb{E}[\mathbb{W}(A)\mathbb{W}(B)] + \mathbb{E}\{|\mathbb{W}(B)|^2\}$
$= \text{Leb}(A \cup B) - 2\text{Leb}(B) - 2\text{Leb}(A)$
$\quad + \text{Leb}(A) + 2\text{Leb}(A \cap B) + \text{Leb}(B) = 0.$

This verifies *(b)* when A and B are disjoint. In fact, we can use this, together with induction, to deduce that whenever $A_1, \ldots, A_k \in \mathcal{T}$ are disjoint, $\mathbb{W}(\cup_{i=1}^k A_i) = \sum_{i=1}^k \mathbb{W}(A_i)$. Now we prove *(b)* in general.

Write $A \cup B = A_1 \cup A_2 \cup A_3$, where $A_1 = A \cap B^{\complement}$, $A_2 = B \cap A^{\complement}$, and $A_3 = A \cap B$. Since A_1, A_2, and A_3 are disjoint,

$$\mathbb{W}(A \cup B) = \mathbb{W}(A_1) + \mathbb{W}(A_2) + \mathbb{W}(A_3), \quad \text{a.s.} \tag{1}$$

Similarly,
$$\begin{aligned} \mathbb{W}(A) &= \mathbb{W}(A_1) + \mathbb{W}(A_3), \\ \mathbb{W}(B) &= \mathbb{W}(A_2) + \mathbb{W}(A_3). \end{aligned} \tag{2}$$

Part *(b)* follows from (1) and (2), and it remains to prove *(c)*.

For all $n \geq 1$, define $M_n = \sum_{i=1}^n \mathbb{W}(A_i)$ and $\varepsilon_n = \mathbb{W}(\cup_{i=1}^\infty A_i) - M_n$. By *(b)*, $M_n = \mathbb{W}(\cup_{i=1}^n A_i)$ and $\varepsilon_n = \mathbb{W}(\cup_{i=n+1}^\infty A_i)$, a.s. On the other hand,

$$\mathbb{E}[\varepsilon_n^2] = \text{Leb}\left(\bigcup_{i=n+1}^\infty A_i\right) = \sum_{i=n+1}^\infty \text{Leb}(A_i),$$

since the A_i's are disjoint. By the summability assumption on Lebesgue's measure of the A_i's, ε_n goes to 0 in $L^2(\mathbb{P})$ as $n \to \infty$. Equivalently, $M_n \to \mathbb{W}(\cup_{i=1}^\infty A_i)$ in $L^2(\mathbb{P})$.

To show a.s. convergence, note that $M = (M_n; n \geq 1)$ is a martingale, since M_n is a sum of n independent mean-zero random variables. Moreover,

$$\mathbb{E}[M_n^2] = \sum_{i=1}^n \text{Leb}(A_i),$$

which is bounded. By the martingale convergence theorem (Theorem 1.7.1, Chapter 1), M_n converges in $L^2(\mathbb{P})$ and almost surely. Since we have already identified the $L^2(\mathbb{P})$ limit, the result follows. □

Exercise **1.3.1** Show that for all Borel sets $A_1 \supset A_2 \supset \cdots$ that satisfy $\cap_n A_n = \emptyset$, $\lim_{n\to\infty} \mathbb{E}\{|\mathbb{W}(A_n)|^2\} = 0$. Conclude that $\mathbb{W}$ can be viewed as a σ-finite, $L^2(\mathbb{P})$-valued, signed measure. □

Exercise **1.3.2** Improve Theorem 1.3.1(a) by checking that whenever $\text{Leb}(A \cap B) = 0$, $\mathbb{W}(A)$ is independent of $\mathbb{W}(B)$. □

1.4 The Isonormal Process

Let T designate the Borel field on $\mathbb{R}^N$, and $\mathbb{W} = (\mathbb{W}(A); A \in T)$ be white noise on $\mathbb{R}^N$. While it is not a measure almost surely (Remark of Section 1.3), Exercise 1.3.1 shows that $\mathbb{W}$ *is* a random $L^2(\mathbb{P})$-valued, signed measure. This alone makes it possible to define a reasonable integral $W(h) = \int h(a) \mathbb{W}(ds)$, which is sometimes called the **Wiener integral** of h. The map $h \mapsto W(h)$ is the isonormal process whose construction is taken up in this subsection.

A function $f : \mathbb{R}^N \to \mathbb{R}$ is an **elementary function** if for some $E \in T$ with $\text{Leb}(E) < \infty$, $f = \mathbf{1}_E$. Finite linear combinations of elementary functions are **simple functions**.

Recall that $L^2(\mathbb{R}^N)$ denotes the collection of all measurable functions $f : \mathbb{R}^N \to \mathbb{R}$ such that $\int_{\mathbb{R}^N} |f(s)|^2 \, ds < \infty$. For all $f, g \in L^2(\mathbb{R}^N)$, let $\langle f, g \rangle = \int_{\mathbb{R}^N} f(s)g(s) \, ds$ and $\|f\| = \langle f, f \rangle^{1/2}$. If we identify the elements of $L^2(\mathbb{R}^N)$ that are almost everywhere equal, then $L^2(\mathbb{R}^N)$ is a complete, separable metric space in the metric $d(f, g) = \|f - g\|$.

Lemma 1.4.1 *Simple functions are dense in $L^2(\mathbb{R}^N)$.*

Proof Fix an arbitrary $\varepsilon > 0$ and $f \in L^2(\mathbb{R}^N)$. We intend to show that there exists an elementary function S such that $\|S - f\| \leq \varepsilon$. By the dominated convergence theorem, there exists an $R > 0$ such that

$$\|f \mathbf{1}_{\mathbb{R}^N \setminus [-R,R]^N}\| = \left(\int_{|s| \geq R} |f(s)|^2 \, ds\right)^{\frac{1}{2}} \leq \frac{\varepsilon}{2}. \tag{1}$$

Fix such an R and define

$$S(s) = \begin{cases} \sum_{j=-\lfloor R/\eta \rfloor - 1}^{\lfloor R/\eta \rfloor + 1} j\eta \, \mathbf{1}_{[j\eta, (j+1)\eta]}(f(s)), & \text{if } |s| \leq R, \\ 0, & \text{otherwise,} \end{cases}$$

where $\eta = \frac{1}{2}(2R)^{-N}\varepsilon$, and $\lfloor \bullet \rfloor$ denotes the greatest integer function. (What does S look like?) We claim that S is indeed a simple function. Let $E_j = f^{-1}([j\eta, (j+1)\eta])$ and note that $E_j \in T$, since f is measurable. Once we demonstrate that $\text{Leb}(E_j) < \infty$, this verifies that S is a simple function. On the other hand, by Chebyshev's inequality,

$$\text{Leb}(E_j) \leq \text{Leb}(s \in \mathbb{R}^N : |f(s)| \geq j\eta) \leq \frac{\|f\|^2}{j^2 \eta^2} < \infty.$$

Consequently, S is a simple function. Moreover, for all $s \in [-R, R]^N$, $f(s) - \eta \leq S(s) \leq f(s)$. In particular,

$$\|S - f \mathbf{1}_{[-R,R]^N}\| \leq (2R)^N \eta = \frac{\varepsilon}{2}.$$

Combining this with equation (1) and using the triangle inequality, we see that $\|S - f\| \leq \varepsilon$, which is the desired result. $\square$

Suppose $h : \mathbb{R}^N \to \mathbb{R}$ is an elementary function. If we write $h = \mathbf{1}_E$ for $E \in \mathcal{T}$, we can define

$$W(h) = \int h(s)\, \mathbb{W}(ds) = \mathbb{W}(E).$$

For any two elementary functions f and g, let $W(f + g) = W(f) + W(g)$. This is well-defined, as the following shows.

Lemma 1.4.2 *Suppose $f + g = p + q$, where f, g, p, q are elementary functions. Then, $W(f) + W(g) = W(p) + W(q)$, almost surely.*

Proof By symmetry, it suffices to consider the case where $g \equiv 0$. Suppose $f = p + q$ and $f = \mathbf{1}_F$, $p = \mathbf{1}_P$ and $q = \mathbf{1}_Q$, where $F, P, Q \in \mathcal{T}$. We seek to show that a.s., $D = 0$, where,

$$D = \mathbb{W}(F) - \mathbb{W}(P) - \mathbb{W}(Q).$$

Since $\mathbf{1}_F = \mathbf{1}_P + \mathbf{1}_Q$, P and Q must be disjoint. Thus, writing $F = (F \cap P) \cup (F \cap Q)$ leads to

$$\mathbb{E}[D^2] = \mathbb{E}\big[\{\mathbb{W}(F \cap P) - \mathbb{W}(P) + \mathbb{W}(F \cap Q) - \mathbb{W}(Q)\}^2\big]$$
$$= \mathbb{E}\big[\{\mathbb{W}(F \cap P) - \mathbb{W}(P)\}^2\big] + \mathbb{E}\big[\{\mathbb{W}(F \cap Q) - \mathbb{W}(Q)\}^2\big].$$

We have used Theorem 1.3.1 in this calculation. The same theorem can be applied again to show that

- $\mathbb{E}[\{\mathbb{W}(F \cap P) - \mathbb{W}(P)\}^2] = \text{Leb}(F^\complement \cap P)$;
- $\mathbb{E}[\{\mathbb{W}(F \cap Q) - \mathbb{W}(Q)\}^2] = \text{Leb}(F^\complement \cap Q)$.

Thus, $\mathbb{E}[D^2] = \text{Leb}\{F^\complement \cap (P \cup Q)\}$, which is zero, since $f = p + q$. $\square$

Finally, if $\alpha, \beta \in \mathbb{R}$ and f, g are elementary functions, define $W(\alpha f + \beta g) = \alpha W(f) + \beta W(g)$. By Lemma 1.4.2, this definition is well-defined.

Moreover, we have now defined $W(h)$ for all simple functions h. We want to use Lemma 1.4.1 to define $W(h)$ for all $h \in L^2(\mathbb{R}^N)$. The key to this is the following; it will show that $W : L^2(\mathbb{P}) \to L^2(\mathbb{R}^N)$ is an isometry.

Lemma 1.4.3 *If $h : \mathbb{R}^N \to \mathbb{R}$ is a simple function, then $\mathbb{E}\{|W(h)|^2\} = \|h\|^2$.*

Proof We can write h as $h = \sum_{j=1}^n c_j \mathbf{1}_{E_j}$, where the c_j's are real numbers and $E_j \in \mathcal{T}$ are disjoint and have finite Lebesgue measure. By Theorem 1.3.1,

$$\mathbb{E}\{|W(h)|^2\} = \sum_{j=1}^n c_j^2 \,\mathrm{Leb}(E_j).$$

Since the E_j's are disjoint, this equals $\|h\|^2$. □

By Lemma 1.4.1, simple functions are dense in $L^2(\mathbb{R}^N)$. Thus, for any $f \in L^2(\mathbb{R}^N)$, there are simple functions s_n such that $\lim_{n\to\infty} \|s_n - f\| = 0$. In particular, $(s_n; \, n \geq 1)$ is a Cauchy sequence in $L^2(\mathbb{R}^N)$. By Lemma 1.4.3, $(W(s_n); \, n \geq 1)$ is a Cauchy sequence in $L^2(\mathbb{P})$. Since the latter is complete, $\lim_{n\to\infty} W(s_n)$ exists in $L^2(\mathbb{P})$. We denote this limit by $W(f)$. The stochastic process $(W(h); \, h \in L^2(\mathbb{R}^N))$ is called the **isonormal process**.

Theorem 1.4.1 *The isonormal process $W = (W(h); \, h \in L^2(\mathbb{R}^N))$ is a mean-zero Gaussian process indexed by $L^2(\mathbb{R}^N)$ such that for all $h_1, h_2 \in L^2(\mathbb{R}^N)$, $\mathbb{E}[W(h_1)W(h_2)] = \langle h_1, h_2 \rangle$. Moreover, for all $\alpha, \beta \in \mathbb{R}$ and for every $f, g \in L^2(\mathbb{R}^N)$,*

$$W(\alpha f + \beta g) = \alpha W(f) + \beta W(g), \qquad \text{a.s.}$$

Proof The asserted linearity follows from the construction of W. We need to show that for all $t \in \mathbb{R}^k$ and all all $h_1, \ldots, h_k \in L^2(\mathbb{R}^N)$, $\sum_{i=1}^k t^{(i)} W(h_i) \sim \mathcal{N}_1(0, \sigma^2)$, where $\sigma^2 = \sum_{i=1}^k \sum_{j=1}^k t^{(i)} t^{(j)} \langle h_i, h_j \rangle$. However, by the asserted linearity of this result, $\sum_{i=1}^k t^{(i)} W(h_i) = W(h_\star)$, a.s., where $h_\star = \sum_{i=1}^k t^{(i)} h_i \in L^2(\mathbb{R}^N)$. Therefore, it suffices to show that for all $h \in L^2(\mathbb{R}^N)$, $W(h)$ is a Gaussian random variable with mean 0 and variance $\|h\|^2$. If h is elementary, this follows from Lemma 1.4.3 and the fact that white noise $\mathbb{W}$ is a mean-zero Gaussian process. Since $L^2(\mathbb{R}^N)$-limits of Gaussian random variables are Gaussian, the result follows; cf. Exercise 1.1.4. □

An alternative approach to the above construction is to directly *define* the isonormal process as a Gaussian process indexed by $L^2(\mathbb{R}^N)$ whose mean and covariance functions are 0 and $\Sigma(f, g) = \langle f, g \rangle$, respectively. Then, one can define white noise by $\mathbb{W}(A) = W(\mathbf{1}_A)$; see Supplementary

Exercise 3 for details.

We conclude this subsection by introducing more notation. Let W and $\mathbb{W}$ denote the isonormal process and the white noise of this subsection, respectively. Given a Borel set $A \subseteq \mathbb{R}^N$ and a measurable function $f : \mathbb{R}^N \to \mathbb{R}$, we write $\int_A f(u)\,\mathbb{W}(du)$ for $W(\mathbf{1}_A f) = \int \mathbf{1}_A(u) f(u)\,\mathbb{W}(du)$. We say that $f \in L^2_{loc}(\mathbb{R}^N)$ if for all compact sets $K \subset \mathbb{R}^N$, $\mathbf{1}_K f \in L^2(\mathbb{R}^N)$. If $f \in L^2_{loc}(\mathbb{R}^N)$, we can define the process $W(f) = (W_t(f);\ t \in \mathbb{R}^N_+)$ by $W_t(f) = \int_{[0,t]} f(u)\,\mathbb{W}(du)$ ($t \in \mathbb{R}^N_+$). The following is an immediate corollary of Theorem 1.4.1.

Corollary 1.4.1 *If $f \in L^2_{loc}(\mathbb{R}^N)$ is fixed, $W(f) = (W_t(f);\ t \in \mathbb{R}^N_+)$ is a Gaussian process with mean function 0 and covariance function $\Sigma(s,t) = \int_{[0,s\wedge t]} |f(u)|^2\,du$, where $W_t(f) = \int_{[0,t]} f(u)\,\mathbb{W}(du)$.*

The following is well worth trying at this point. The connection to martingales will be further elaborated upon later on.

***Exercise* 1.4.1** Prove that in Corollary 1.4.1, $t \mapsto W_t(f)$ is a mean-zero, N-parameter martingale. Use this to define $W_t(f)$, and hence $W(f)$, for any $f \in L^p(\mathbb{R}^N)$ ($p > 1$). □

1.5 The Brownian Sheet

The **one-dimensional Brownian sheet** indexed by $\mathbb{R}^N_+$ is a Gaussian process $B = (B_t;\ t \in \mathbb{R}^N_+)$ with mean 0 and covariance function

$$\Sigma(s,t) = \prod_{\ell=1}^{N} (s^{(\ell)} \wedge t^{(\ell)}), \qquad s,t \in \mathbb{R}^N_+.$$

For any integer $d \geq 1$, the **d-dimensional Brownian sheet** indexed by $\mathbb{R}^N_+$ is the process $B = (B_t;\ t \in \mathbb{R}^N_+)$, where $B^{(i)} = (B^{(i)}_r;\ r \geq 0)$ ($1 \leq i \leq d$) are independent, 1-dimensional, N-parameter Brownian sheets.[1]

Recall that when $N = 1$, the one-dimensional Brownian sheet is more commonly called Brownian motion. Likewise, the d-dimensional, 1-parameter Brownian sheet is more commonly known as the **d-dimensional Brownian motion.** For our current purposes, it suffices to study the 1-dimensional, N-parameter Brownian sheet, which is what we now concentrate on.

Since $\Sigma(s,t) = \text{Leb}\big([0,s] \cap [0,t]\big)$, Lemma 1.3.1 shows that Σ is indeed positive definite and such a process is well-defined and exists on some probability space. In fact, by checking covariances, we can immediately deduce the following useful representation of Čentsov (1956).

---

[1] The processes $X^1, X^2, \ldots$, indexed by some set T, are **independent** if for all finite sets $F_1, F_2 \ldots \subset T$, $(X^j_t;\ t \in F_j)$ are independent random vectors ($j = 1, 2, \ldots$).

Theorem 1.5.1 (Čentsov's Representation) *Consider a white noise on $\mathbb{R}^N$ denoted by $\mathbb{W} = (\mathbb{W}(A);\ A \subset \mathbb{R}^N,\ Borel)$. Then, $B = (B_t;\ t \in \mathbb{R}_+^N)$ is a Brownian sheet, where $B_t = \mathbb{W}([0,t])$.*

***Exercise* 1.5.1** Verify the details of the proof of Theorem 1.5.1. □

If W denotes the isonormal process obtained from $\mathbb{W}$, then $\mathbb{W}(A) = W(\mathbf{1}_A)$ for all Borel sets $A \subset \mathbb{R}^N$ that have finite Lebesgue's measure. In particular, the Brownian sheet of the theorem has the representation $B_t = W(\mathbf{1}_{[0,t]})$. Recall from Section 1.4 that we may also write this as $B_t = \int_{[0,t]} \mathbb{W}(ds)$. Therefore, *very* loosely speaking,

$$\mathbb{W}([0,t]) = \frac{\partial^N}{\partial t^{(1)} \cdots \partial t^{(N)}} B_t.$$

While one can make sense of this derivative as a so-called generalized function, it does not exist in the usual way: B turns out to be a.s. nowhere differentiable. However, this observation motivates the following convenient notation: For all $f \in L^2(\mathbb{R}_+^N)$, define

$$\int f(s)\, dB_s = \int f(s)\, \mathbb{W}(ds) = W(f).$$

Similarly, for all $f \in L^2_{loc}(\mathbb{R}^N)$ and all $t \in \mathbb{R}_+^N$,

$$\int_{[0,t]} f(s)\, dB_s = \int_{[0,t]} f(s)\, \mathbb{W}(ds) = W(\mathbf{1}_{[0,t]} f).$$

We will use the above notations interchangeably.

Let us conclude with a few exercises on the basic structure of the Brownian sheet.

***Exercise* 1.5.2** If B denotes a 2-parameter Brownian sheet, show that for any fixed $t > 0$, $s \mapsto t^{-1/2} B_{s,t}$ and $s \mapsto t^{-1/2} B_{t,s}$ are each Brownian motions. Find an N-parameter analogue when $N > 2$. □

***Exercise* 1.5.3** If B is a 2-parameter Brownian sheet, so are $(s,t) \mapsto sB_{(1/s),t}$, $(s,t) \mapsto tB_{s,(1/t)}$, and $(s,t) \mapsto stB_{(1/s),(1/t)}$. These are examples of time inversion. Formulate and verify the general N-parameter results. Note that, in general, there are $2^N - 1$ different ways of inverting time. □

2 Regularity Theory

Suppose $X = (X_t;\ t \in T)$ is a stochastic process that is indexed by some metric space T. In this section we derive general conditions under which the random function $t \mapsto X_t$ is continuous, a.s.

A useful way of establishing continuity is to verify that on every compact subset of T, $t \mapsto X_t$ is, with probability one, uniformly continuous. Since T is a general index set, we begin by reexamining the compactness, or, more generally, total boundedness, of T.

2.1 Totally Bounded Pseudometric Spaces

Given an abstract set T, a function $d : T \times T \to \mathbb{R}_+$ is said to be a **pseudometric** if it satisfies the following conditions:

(i) (symmetry) for all $s, t \in T$, $d(s,t) = d(t,s)$;

(ii) (triangle inequality) for all $r, s, t \in T$, $d(s,t) \leq d(s,r) + d(r,t)$; and

(iii) for all $s \in T$, $d(s,s) = 0$.

In particular, the difference between a pseudometric and a proper metric is that the former need not separate points, i.e., $d(s,t) = 0$ need not imply $s = t$. For example, if we do not identify functions up to almost everywhere equality, its usual metric is, in fact, a pseudometric for any given L^p space. As a less trivial example, consider the space $C[0,1]$ of real-valued, continuous functions on $[0,1]$, and define

$$d(f,g) = \inf_{x,y \in [0,1]} |f(x) - g(y)|, \qquad f, g \in C[0,1].$$

It is an easy matter to check that this is a pseudometric on $C[0,1]$. The typical probabilistic example of a pseudometric is given by the following.

***Exercise* 2.1.1** Given a stochastic process $(X_t; t \in T)$, and given $p \geq 1$, $d(s,t) = [\mathbb{E}\{|X_s - X_t|^p\}]^{1/p}$ defines a pseudometric on T, provided that $X_t \in L^p(\mathbb{P})$ for all $t \in T$. Also prove that, quite generally, $d(s,t) = \mathbb{E}\{|X_s - X_t| \wedge 1\}$ defines a pseudometric on T. □

We say that (T, d) is a **pseudometric space** in order to emphasize the fact that d is the pseudometric on T. The (open) **ball** of radius $r > 0$ about $t \in T$ is denoted by $\mathcal{B}_d(t; r)$ and defined by

$$\mathcal{B}_d(t; r) = \{s \in T : d(s,t) < r\}.$$

We say that (T, d) is **totally bounded** if for all $\varepsilon > 0$, there exists an integer m and there are distinct $t_1, \ldots, t_m \in T$ such that $\mathcal{B}_d(t_1; \varepsilon), \ldots, \mathcal{B}_d(t_m; \varepsilon)$ cover T. Moreover, any such collection $(t_1, \ldots, t_m)$ is an **ε-net** for T.

Any ε-net $(t_1, \ldots, t_m)$ has the important property that for all $t \in T$, there exists $1 \leq i \leq m$ such that $d(t, t_i) \leq \varepsilon$. We will appeal to this interpretation later on.

For complete metric spaces, totally boundedness is easily characterized, as the following standard fact from general topology shows. We will not need it very much in the sequel, and refer to Munkres (1975, Theorem 3.1, Chapter 7) for a proof.

Theorem 2.1.1 *A metric space (T, d) is compact if and only if it is complete and totally bounded.*

Example 1 Suppose $T = [-1, 1]^N$ is endowed with the metric $d(s, t) = |s - t|$ $(s, t \in \mathbb{R}^N)$; recall that this is the ℓ^∞-metric. In this metric, the "ball" $\mathcal{B}_d(t; r)$ is an open hypercube of side $\frac{1}{2}r$ centered at $t \in \mathbb{R}^N$. We now seek to find a good ε-net in T.

For any $\varepsilon > 0$, let T_ε denote the collection of all points $s \in T$ such that $\varepsilon s \in \mathbb{Z}^N$. This is a natural ε-net for T, and a little thought shows that the cardinality of T_ε satisfies

$$\lim_{\varepsilon \to 0} \varepsilon^{-N} \# T_\varepsilon = 1;$$

see Supplementary Exercise 4 for details. It should be intuitively clear that the T_ε in the above example leads to the best possible ε-cover for $T = [-1, 1]^N$. As such, $\# T_\varepsilon \approx \varepsilon^{-N}$ ought to be a good measure of the size of T. □

We define the **metric entropy** $D(\bullet; T, d)$ of (T, d) as follows:[2] For each $\varepsilon > 0$, $D(\varepsilon; T, d)$ denotes the minimum number of balls of radius $\varepsilon > 0$ required to cover all of T. It is clear that $\varepsilon \mapsto D(\varepsilon; T, d)$ is nonincreasing, and unless T is a finite set, $\lim_{\varepsilon \to 0} D(\varepsilon; T, d) = +\infty$. Moreover, the rate at which the metric entropy of (T, d) blows up near $\varepsilon = 0$ is, in fact, a gauge for the size of T, and in many cases is estimable. For instance, consider $T = [-1, 1]^N$ with $d(s, t) = |s - t|$. Then, according to Example 1, $D(\varepsilon; [-1, 1]^N, d) \approx \varepsilon^{-N}$ when $\varepsilon \approx 0$.

Another measure of the size of (T, d) is the **Kolmogorov capacitance** $K(\bullet; T, d)$:[3] For all $\varepsilon > 0$, $K(\varepsilon; T, d)$ denotes the maximum number m of points $t_1, \ldots, t_m \in T$ such that for all $i \neq j$, $d(t_i, t_j) \geq \varepsilon$. The functions D and K are related by the following inequalities; see Dudley (1984, Theorem 6.0.1).

Lemma 2.1.1 *Given a totally bounded metric space (T, d), for each $\varepsilon > 0$,*

$$D(\varepsilon; T, d) \leq K(\varepsilon; T, d) \leq D(\tfrac{1}{2}\varepsilon; T, d).$$

Proof Fix $\varepsilon > 0$ and let $D = D(\varepsilon; T, d)$, $D' = D(\frac{1}{2}\varepsilon; T, d)$, and $K = K(\varepsilon; T, d)$. We can find a maximal collection $t_1, \ldots, t_K \in T$ such that for all $i \neq j$, $d(t_i, t_j) \geq \varepsilon$. The maximality of this collection, together with

---

[2] The full power of metric entropy was brought to light in the work of R. M. Dudley; hence, the letter D. See Dudley (1973, 1984), where D is replaced by N throughout.

[3] Capacitance was motivated by the groundbreaking work of Shannon (1948) in information theory; for a discussion of Kolmogorov's contribution, see Tihomirov (1963).

an application of the triangle inequality, shows that for any $t \in T$ there exists $1 \leq i \leq K$ such that $d(t, t_i) < \varepsilon$. In other words, the balls $\mathcal{B}_d(t_i; \varepsilon)$ $(1 \leq i \leq K)$ cover T, and since D is the cardinality of the minimal such covering, $D \leq K$. Conversely, we can find $\sigma_1, \ldots, \sigma_{D'} \in T$ such that $\mathcal{B}_d(\sigma_i; \frac{1}{2}\varepsilon)$ cover T for $1 \leq i \leq D'$. Let $B_i = \mathcal{B}_d(\sigma_i; \frac{1}{2}\varepsilon)$ and note that whenever $s_1, s_2 \in T$ satisfy $d(s_1, s_2) \geq \varepsilon$, then s_1 and s_2 cannot be in the same B_j for any $j \leq D'$. Thus, we have that $K \leq D'$, as desired. □

As the following examples show, Lemma 2.1.1 provides a useful way to estimate the metric entropy and/or the Kolmogorov capacitance of a totally bounded set T.

Example 2 Recall that a function $f : [0,1] \to \mathbb{R}$ is a **contraction** if it is measurable and

$$|f(s) - f(t)| \leq |s - t|, \qquad s, t \in [0, 1].$$

Let T denote the collection of all contractions $f : [0,1] \to \mathbb{R}$ such that $f(0) = 0$, and endow T with the maximum modulus metric

$$d(f, g) = \sup_{0 \leq t \leq 1} |f(t) - g(t)|, \qquad f, g \in T.$$

Next, we introduce A. N. Kolmogorov's estimates for $D(\varepsilon; T, d)$ and $K(\varepsilon; T, d)$. We shall see from these estimates that $D(\varepsilon; T, d)$ and $K(\varepsilon; T, d)$ are finite, and the total boundedness of (T, d) follows readily.

To begin with, note that f maps $[0, 1]$ into $[0, 1]$. Indeed, if $f \in T$, then for all $s \in [0, 1]$,

$$|f(s)| = |f(s) - f(0)| \leq s \leq 1.$$

Now for all $k > 0$, let T_k denote the collection of all piecewise linear functions $f \in T$ such that for all integers $0 \leq j \leq k$, $kf(j/k) \in \mathbb{N}_0$. Since the elements of T_k are contractions, $\#(T_k) = 3^{k+1}$.[4] Now if $f, g \in T_k$ are distinct, $d(f, g) \geq k^{-1}$; this leads to the inequality $K(k^{-1}; T, d) \geq 3^k$. Next, we use a monotonicity argument to estimate $K(\varepsilon; T, d)$ for *any* ε: Whenever $\varepsilon \in]0, 1[$, then, $\varepsilon \in [(k+1)^{-1}, k^{-1}]$, for some integer $k \geq 1$. Since $\varepsilon \mapsto K(\varepsilon; T, d)$ and $k \mapsto 3^{-k}$ are nonincreasing,

$$K(\varepsilon; T, d) \geq 3^{1/\varepsilon}, \qquad 0 < \varepsilon < 1. \tag{1}$$

On the other hand, for any $f \in T$ we can define $\pi_k f$ to be the piecewise linear function in T_k such that $\pi_k f(jk^{-1}) = \lfloor kf(jk^{-1}) \rfloor$, for all $0 \leq j \leq k$.

---

[4]To verify this, think of $\#(T_k)$ as the total number of ways to construct piecewise linear contractions $f : [0,1] \to [0,1]$ such that $f(0) = 0$ and $kf(jk^{-1}) \in \mathbb{N}_0$, for all $j = 0, \ldots, k$. Once $f(k^{-1}), \ldots, f(jk^{-1})$ have been constructed, there are only three possibilities for constructing $f((j+1)k^{-1})$: $f(jk^{-1})$ or $f(jk^{-1}) \pm k^{-1}$.

The fact that f is a contraction implies that $d(f, \pi_k f) \leq k^{-1}$. Since $\pi_k f \in T_k$ for all contractions f, we can conclude that the balls $(\mathcal{B}_d(f; k^{-1}); f \in T_k)$ have radius at most k^{-1} and cover T. Consequently, $D(k^{-1}; T, d) \leq \#(T_k) = 3^{k+1}$. By another monotonicity argument,

$$D(\varepsilon; T, d) \leq 3^{1+(1/\varepsilon)}, \qquad 0 < \varepsilon < 1. \tag{2}$$

Upon combining (1) and (2) with Lemma 2.1.1, we obtain the following:

$$\frac{1}{2}\ln 3 \leq \liminf_{\varepsilon \to 0} \varepsilon \ln D(\varepsilon; T, d) \leq \limsup_{\varepsilon \to 0} \varepsilon \ln D(\varepsilon; T, d) \leq \ln 3,$$

$$\ln 3 \leq \liminf_{\varepsilon \to 0} \varepsilon \ln K(\varepsilon; T, d) \leq \limsup_{\varepsilon \to 0} \varepsilon \ln K(\varepsilon; T, d) \leq 2\ln 3,$$

the point being that, very roughly speaking, both K and D behave like $\exp(c\varepsilon^{-1})$ in the present infinite-dimensional setting. This should be compared to the polynomial growth of the entropy as outlined in Example 1 above. □

Example 3 For any fixed $c > 0$, let T_c denote the class of all differentiable functions $f : [0, 1] \to \mathbb{R}$ such that $f(0) = 0$ and $\sup_{0 \leq s \leq 1} |f'(s)| \leq c$, and endow T_c with the metric $d(f, g) = \sup_{0 \leq t \leq 1} |f(t) - g(t)|$. It is easy to see that (T_c, d) is a totally bounded metric space. Indeed, $(c^{-1}f; f \in T_c)$ is a subset of the collection of all contractions on $[0, 1]$, which, by Example 2 above, is totally bounded in the same metric d. □

We conclude this subsection with the following corollary of Lemma 2.1.1.

Corollary 2.1.1 *Suppose $T \subset \mathbb{R}^N_+$ is measurable and d is a pseudometric on T. If there exist $C, \alpha > 0$ such that for all $s, t \in T$, $d(s, t) \leq C|s - t|^\alpha$, then there exists $r_0 > 0$ such that for all $r \in [0, r_0]$, $D(r; T, d) \leq C^{N/\alpha} \operatorname{Leb}(T) r^{-N/\alpha}$, where $\operatorname{Leb}(T)$ is Lebesgue's measure of T.*

Proof Let ∂ denote the metric $\partial(s, t) = |s - t|$, and define

$$\varepsilon_0 = \tfrac{1}{2} \sup\{\partial(s, t) : s, t \in T\}.$$

Whenever $0 < \varepsilon < \varepsilon_0$ and $t_1, \ldots, t_m \in T$ are such that $\partial(t_i, t_j) > \varepsilon$ ($i \neq j$), then $\mathcal{B}_\partial(t_i; \tfrac{1}{2}\varepsilon)$ are disjoint and $\bigcup_{i=1}^m \mathcal{B}_\partial(t_i; \tfrac{1}{2}\varepsilon) \subset T$. Hence,

$$m\varepsilon^N = \operatorname{Leb}\left[\bigcup_{i=1}^m \mathcal{B}_\partial(t_i; \tfrac{1}{2}\varepsilon)\right] \leq \operatorname{Leb}(T).$$

Equivalently, $m \leq \operatorname{Leb}(T)\varepsilon^{-N}$. Since m is arbitrary, we can maximize over all such possible m's to see that for all $0 < \varepsilon < \varepsilon_0$, $K(\varepsilon; T, \partial) \leq \operatorname{Leb}(T)\varepsilon^{-N}$. By Lemma 2.1.1, for all $0 < \varepsilon < \varepsilon_0$, $D(\varepsilon; T, \partial) \leq \operatorname{Leb}(T)\varepsilon^{-N}$. On the other hand, $d(s, t) \leq C\{\partial(s, t)\}^\alpha$. Thus, any ε-net with respect to ∂ is a $C\varepsilon^\alpha$-net with respect to d. That is, for all $0 < \varepsilon < \varepsilon_0$, $D(C\varepsilon^\alpha; T, d) \leq D(\varepsilon; T, \partial) \leq \operatorname{Leb}(T)\varepsilon^{-N}$. The result follows with $r_0 = C\varepsilon_0^\alpha$. □

Exercise 2.1.2 Consider the collection of continuous functions $f : [0,1] \to \mathbb{R}$ such that for all $s, t \in [0,1]$, $|f(s) - f(t)| \leq C|t-s|^\alpha$ for some $\alpha > 0$ and some finite $C > 0$.

(a) Show that α is necessarily in $[0,1]$.

(b) (Hard) Compute the best possible upper bound for the metric entropy of this collection.

(HINT: To show that your upper bound is best possible, consider global helices: A function f is a **global helix** of order α on $[0,1]$ if there exists a finite constant $C > 1$ such that for all open intervals $I \subset [0,1]$ of length ℓ, $\sup_{s,t \in I} |f(s) - f(t)| \geq C\ell^\alpha$. Compute a lower bound for the metric entropy of global helices on $[0,1]$. You may assume, without proof, the existence of global helices for any $\alpha \in \,]0,1[$.) □

2.2 Modifications and Separability

The transition from discrete-parameter to continuous-parameter processes could potentially pose a great number of technical problems. However, a good number of such issues would vanish if the process $t \mapsto X_t$ ($t \in T$) were continuous. Of course, this makes sense only if the parameter set T is topological.

We begin this subsection with the following simple but instructive example, which implies that one cannot be too ambitious in setting out such goals.

Example Let $B = (B_t; \ t \geq 0)$ denote a Brownian motion. Recall from Section 1.5 that B is a Gaussian process indexed by $T = [0, \infty[$ with mean and covariance functions $\mu(t) = 0$ and $\Sigma(s,t) = s \wedge t$, respectively ($s, t \in T$). Suppose U is picked at random, uniformly in $[0,1]$, and that U is independent of B. Of course, we may need to enlarge the underlying probability space to construct such a U. On this possibly enlarged space we can define a stochastic process $Y = (Y_t; \ t \geq 0)$ as follows:

$$Y_t = \begin{cases} B_t, & \text{if } U \neq t, \\ 12, & \text{if } U = t, \end{cases} \quad t \geq 0.$$

Note that for all $t_1, \ldots, t_k \geq 0$,

$$\mathbb{P}(Y_{t_i} \neq B_{t_i}, \text{ for some } 1 \leq i \leq k) = \mathbb{P}(U \in \{t_1, \ldots, t_k\}) = 0.$$

This shows that Y has the same finite-dimensional distributions as Brownian motion; hence, Y is a Brownian motion.

Supposing that B is continuous, we now show that Y is a.s. *not* continuous. Indeed, by the independence of U and Y, and using the uniformity of U in $[0,1]$, we obtain

$$\mathbb{P}(B_U = 12) = \int_0^1 \mathbb{P}(B_t = 12)\, dt = 0.$$

In summary, we have shown that any continuous Brownian motion B can be *modified* to produce a Brownian motion Y that is *not* continuous. □

The above example suggests that it is virtually impossible to show that the process X itself is continuous, since a priori we know only its finite-dimensional distributions. However, it may still be possible to modify X, without altering its finite-dimensional distributions, so as to produce a continuous process. For instance, in Example 2, if B were continuous, then it could be viewed as such a modification of Y.

In this subsection we explore some of the fundamental properties of modifications of a given stochastic process.

Given an arbitrary index set T, two real-valued stochastic processes $X = (X_t;\, t \in T)$ and $Y = (Y_t;\, t \in T)$ are said to be **modifications** of each other if for all $t \in T$, $\mathbb{P}(X_t = Y_t) = 1$.

Following the special definition of Section 1.1, Chapter 3, we will define the **finite-dimensional distributions** of a real-valued stochastic process X as the collection of all the distributions of the $\mathbb{R}^k$-dimensional random variables $(X_{t_1}, \ldots, X_{t_k})$, as we vary $k \geq 1$ and $t_1, \ldots, t_k \in T$, and begin with the remark that modifying a stochastic process does not alter its finite-dimensional distributions.

Lemma 2.2.1 *If $X = (X_t;\, t \in T)$ is a modification of $Y = (Y_t;\, t \in T)$, then X and Y have the same finite-dimensional distributions.*

Exercise **2.2.1** Verify Lemma 2.2.1. □

But which modification of X, if any, is useful? To answer this, we first observe that unless T is at most denumerable, some very natural objects such as $\sup_{t \in T} X_t$ can be nonmeasurable. For instance, if $\lambda \in \mathbb{R}$,

$$\left(\omega:\, \sup_{t \in T} X_t(\omega) \geq \lambda\right) = \bigcup_{t \in T} \left(\omega:\, X_t(\omega) \geq \lambda\right)$$

need not be measurable when T is uncountable.

In the remainder of this section we will see that, under some technical conditions on T, one can construct a *continuous* modification $Y = (Y_t;\, t \in T)$ of X. By Lemma 2.2.1, Y and X have the same finite-dimensional distributions, and under mild conditions such as separability of T, $\sup_{t \in T} Y_t$ is indeed measurable in such cases. However, even when continuous modifications do not exist, one can often obtain a modification with nice measurability properties, e.g., such that $\sup_{t \in T} Y_t$ is measurable.

We will close this subsection with the more general, but weaker, property of separability for real-valued multiparameter processes; it is good enough to handle many such measurability issues, even when continuous modifications do not exist.

A stochastic process $X = (X_t;\ t \in \mathbb{R}_+^N)$ is said to be **separable** if there exists an at most countable collection $T \subset \mathbb{R}_+^N$ and a null set Λ such that for all closed sets $A \subset \mathbb{R}$ and all open sets $I \subset \mathbb{R}_+^N$ of the form $I = \prod_{\ell=1}^N]\alpha^{(\ell)}, \beta^{(\ell)}[$, where $\alpha^{(\ell)} \leq \beta^{(\ell)}$ ($1 \leq \ell \leq N$) are rational or infinite,

$$\big(\omega : X_s(\omega) \in A \text{ for all } s \in I \cap T\big) \setminus \big(\omega : X_s(\omega) \in A \text{ for all } s \in I\big) \subset \Lambda.$$

It is important to note the order of the quantifiers: The choice of Λ and T can be made independently of that of A and I.

Recall that a probability space is **complete** if all subsets of null sets are themselves null sets. Also, recall that one can always complete the underlying probability space at no cost. Consequently, we have the following.

Lemma 2.2.2 *Suppose $X = (X_t;\ t \in \mathbb{R}_+^N)$ is a separable stochastic process and suppose the underlying probability space is complete. Then for any open rectangle $I \subset \mathbb{R}_+^N$ and any $\tau \in \mathbb{R}_+^N$, the following are random variables: $\sup_{t \in I} X_t$, $\inf_{t \in I} X_t$, $\limsup_{t \to \tau} X_t$, and $\liminf_{t \to \tau} X_t$.*

Exercise **2.2.2** Verify Lemma 2.2.2. □

Thus far, we have seen that it is advantageous to study a separable modification Y of a process X, if and when it exists, since Y has the same finite-dimensional distributions as X but has fewer measure-theoretic problems. When can we find such modifications? The following remarkable fact, due to J. L. Doob, says that the answer is *always*; see Doob (1990, Chapter II) for extensions and other variants.

Theorem 2.2.1 (Doob's Separability Theorem) *Any stochastic process $X = (X_t;\ t \in \mathbb{R}_+^N)$ has a separable modification.*

The proof of Doob's separability theorem is long and is divided into two steps, which are stated below as technical lemmas.

Throughout, $\mathcal{R}$ denotes the collection of all open rectangles in $\mathbb{R}_+^N$ with rational or infinite endpoints, $\mathcal{J}$ denotes the collection of all closed intervals in $\mathbb{R}$ with rational or infinite endpoints, and $\mathcal{C}$ denotes the collection of all closed subsets of $\mathbb{R}$.

Lemma 2.2.3 *There exists a countable set $S \subset \mathbb{R}_+^N$ such that for any fixed $t \in \mathbb{R}_+^N$, the following is a null set:*

$$N_t = \bigcup_{A \in \mathcal{J}} \big(\omega :\ X_s(\omega) \in A \text{ for all } s \in S,\ X_t(\omega) \notin A\big).$$

Proof. Temporarily, fix some Borel set $A \subset \mathbb{R}$ and let $t_1 = 0 \in \mathbb{R}_+^N$ and

$$\varepsilon_1 = \sup_{t \in \mathbb{R}_+^N} \mathbb{P}(X_0 \in A, \ X_t \notin A).$$

Having constructed distinct $t_1, \ldots, t_k \in \mathbb{R}_+^N$ and $\varepsilon_1 \geq \varepsilon_2 \geq \cdots \geq \varepsilon_k$ define

$$\varepsilon_{k+1} = \sup_{t \in \mathbb{R}_+^N} \mathbb{P}(X_{t_j} \in A \text{ for all } 1 \leq j \leq k \ X_t \notin A).$$

Clearly, $\varepsilon_1 \geq \varepsilon_2 \geq \cdots \geq \varepsilon_{k+1}$. Moreover, we can always choose some $t_{k+1} \in \mathbb{R}_+^N \setminus \{t_1, \ldots, t_k\}$ such that

$$\mathbb{P}(X_{t_j} \in A \text{ for all } 1 \leq j \leq k, \ X_{t_{k+1}} \notin A) \geq \frac{\varepsilon_{k+1}}{2}.$$

Thus,

$$\sum_{k=2}^{\infty} \varepsilon_k \leq 2 \sum_{k=2}^{\infty} \mathbb{P}(X_{t_j} \in A \text{ for all } 1 \leq j \leq k-1, \ X_{t_k} \notin A)$$
$$= 2\mathbb{P}(X_{t_k} \notin A \text{ for some } k \geq 2) < +\infty.$$

In particular, $\lim_{k \to \infty} \varepsilon_k = 0$. In other words, we have shown that for any Borel set $A \in \mathbb{R}_+^N$, there exists a countable set T_A such that

$$\sup_{t \in \mathbb{R}_+^N} \mathbb{P}(X_s \in A \text{ for all } s \in T_A, \ X_t \notin A) = 0.$$

To finish, define $S = \bigcup_{A \in \mathfrak{I}} T_A$, which is clearly countable, since $\mathfrak{I}$ is. □

Lemma 2.2.4 *For each $t \in \mathbb{R}_+^N$,*

$$\bigcup_{A \in \mathfrak{C}} (\omega : X_s(\omega) \in A \text{ for all } s \in S, \ X_t(\omega) \notin A) \subset N_t.$$

Proof. Note that any $A \in \mathfrak{C}$ can be written as $A = \cap_{n=1}^{\infty} A_n$, where $A_n \in \mathfrak{I}$. By Lemma 2.2.3, for any such $A \in \mathfrak{C}$ and $A_1, A_2, \ldots \in \mathfrak{I}$,

$$(\omega : X_s(\omega) \in A \text{ for all } s \in S, \ X_t(\omega) \notin A_n)$$
$$\subset (\omega : X_s(\omega) \in A_n \text{ for all } s \in S, \ X_t(\omega) \notin A_n)$$
$$\subset \bigcup_{E \in \mathfrak{I}} (\omega : X_s(\omega) \in E \text{ for all } s \in S, \ X_t(\omega) \notin E)$$
$$\subset N_t.$$

Since $A = \cap_{n \geq 1} A_n$, the results follows. □

2 Regularity Theory

Proof of Theorem 2.2.1[5] Let $I \in \mathcal{R}$ and apply Lemma 2.2.4 to the stochastic process $(X_t;\ t \in I)$ to conclude the existence of null sets $N_t(I)$ ($t \in \mathbb{R}_+^N$) and a countable set $S_I \subset I$ such that

$$\bigcup_{A \in \mathcal{C}} \big(\omega:\ X_s(\omega) \in A \text{ for all } s \in S_I\ ,\ X_t(\omega) \in A\big) \subset N_t(I).$$

Since $\mathcal{R}$ is countable, $S_\star = \bigcup_{I \in \mathcal{R}} S_I$ is countable. Similarly, for all $t \in \mathbb{R}_+^N$, $\Lambda_t = \bigcup_{I \in \mathcal{R}} N_t(I)$ is a null set. Define

$$R_I(\omega) = \overline{\{X_s(\omega);\ s \in I \cap S_\star\}}.$$

In words, R_I is the closure of the image of I under the random function X. Note that R_I may include the values of $\pm\infty$. Moreover, R_I is closed and nonempty in $\mathbb{R} \cup \{\pm\infty\}$. For any $t \in \mathbb{R}_+^N$, define the random set

$$R_t = \bigcap_{I \in \mathcal{R}:\ t \in I} R_I.$$

Clearly, $R_t \subset \mathbb{R} \cup \{\pm\infty\}$ is closed and nonempty, for all $\omega \in \Omega$. Moreover,

$$\text{if } t \in \mathbb{R}_+^N \text{ and } \omega \notin \Lambda_t,\ \text{ then } X_t(\omega) \in R_t(\omega). \tag{1}$$

We are ready to construct the desired modification of X. For all $\omega \in \Omega$ and all $t \in S_\star$, define $\widetilde{X}_t(\omega) = X_t(\omega)$. If $t \notin S_\star$ and $\omega \notin \Lambda_t$, also define $\widetilde{X}_t(\omega) = X_t(\omega)$. Finally, whenever $t \notin S_\star$ and $\omega \in \Lambda_t$, define $\widetilde{X}_t(\omega)$ to be some designated element of $R_t(\omega)$. Since $\mathbb{P}(\Lambda_t) = 0$ for each $t \in \mathbb{R}_+^N$, $\mathbb{P}(X_t = \widetilde{X}_t) = 1$, which is to say that $\widetilde{X} = (\widetilde{X}_t:\ t \in \mathbb{R}_+^N)$ is a modification of X. It remains to show that $\widetilde{X}$ is separable. Fix $A \in \mathcal{C}$ and $I \in \mathcal{R}$, and suppose ω satisfies the following:

$$\widetilde{X}_s(\omega) \in A \text{ for all } s \in I \cap S_\star.$$

If $s \in I \cap S_\star$ but $\omega \notin \Lambda_s$, $\widetilde{X}_s(\omega) = X_s(\omega) \in R_s(\omega) \subset R_I(\omega) \subset A$, since A is closed. Similarly, if $s \in I$, but $s \notin S_\star$ and $\omega \notin \Lambda_s$, then $\widetilde{X}_s(\omega) = X_s(\omega) \in R_I(\omega) \subset A$, by equation (1). Define

$$\Lambda = \bigcup_{s \in S_\star} \Lambda_s.$$

Since $S_\star$ is countable, Λ is a null set; it is also chosen independently of all $A \in \mathcal{C}$ and $I \in \mathcal{R}$. Finally, we have shown that

$$\big(\omega:\ \widetilde{X}_s(\omega) \in A \text{ for all } s \in I \cap S_\star\big) \cap \Lambda \subset \big(\omega:\ \widetilde{X}_s(\omega) \in A \text{ for all } s \in I\big).$$

---

[5] This can be skipped at first reading.

Thus,

$$\left(\omega : \widetilde{X}_s(\omega) \in A \text{ for all } s \in I \cap S_\star\right) \supset \left(\omega : \widetilde{X}_s(\omega) \in A \text{ for all } s \in I\right)$$

implies that

$$\left(\omega : \widetilde{X}_s(\omega) \in A \text{ for all } I \cap S_\star\right) \setminus \left(\widetilde{X}_s(\omega) \text{ for all } s \in I\right) \subset \Lambda. \quad (2)$$

Finally, any open rectangle $J \subset \mathbb{R}_+^N$ can be written as $J = \cup_{n=1}^{\infty} I_n$ where $I_n \in \mathcal{R}$. Since Λ is chosen independently of all $I \in \mathcal{R}$ and all closed sets A, we conclude that for all open rectangles $J \subset \mathbb{R}_+^N$ and all closed sets $A \subset \mathbb{R}$,

$$\left(\omega : \widetilde{X}_s(\omega) \in A \text{ for all } J \cap S_\star\right) \setminus \left(\widetilde{X}_s(\omega) \text{ for all } s \in J\right) \subset \Lambda.$$

This shows the separability of $\widetilde{X}$. □

2.3 Kolmogorov's Continuity Theorem

Given a totally bounded pseudometric space (T, d), we seek to find conditions under which a stochastic process $X = (X_t; \, t \in T)$ has a continuous modification.[6]

If X has a continuous modification, then it is clearly **continuous in probability** in the sense that for any $\varepsilon > 0$ and any $t \in T$,

$$\lim_{s \to t} \mathbb{P}(|X_s - X_t| \geq \varepsilon) = 0.$$

By Chebyshev's inequality, this automatically holds if there exists $p > 0$ such that for all $t \in T$,

$$\lim_{s \to t} \mathbb{E}\{|X_s - X_t|^p\} = 0.$$

Our next theorem—known as **Kolmogorov's continuity theorem**—states that if $\mathbb{E}\{|X_s - X_t|^p\} \to 0$ quickly enough, then X has a continuous modification.

Theorem 2.3.1 (Kolmogorov's Continuity Theorem) *Consider a process $X = (X_t; \, t \in T)$ where (T, d) is a totally bounded pseudometric space, and suppose there exists $p > 0$ and a nondecreasing, continuous function $\Psi : \mathbb{R}_+ \to \mathbb{R}_+$ such that $\Psi(0) = 0$ and for all $s, t \in T$,*

$$\mathbb{E}\{|X_s - X_t|^p\} \leq \Psi(d(s,t)).$$

Then, X has a continuous modification, provided that there exists a nondecreasing, measurable function $f : \mathbb{R}_+ \to \mathbb{R}_+$ such that:

---

[6]Detailed historical background and related results, together with extensions from metric entropy to majorizing measures, can be found in Adler (1990), Dudley (1973, 1984, 1989), Ledoux (1996), Ledoux and Talagrand (1991), and their combined references.

(i) $\int_0^1 r^{-1} f(r)\, dr < \infty$; and

(ii) $\int_0^1 D(r; T, d) \Psi(2r) \{f(r)\}^{-p}\, dr < +\infty$.

Remark If $Y = (Y_t;\ t \in T)$ denotes any modification of X, it is not a priori clear that the event $(t \mapsto Y_t$ is continuous) is measurable. This is a part of the assertion of Theorem 2.3.1.

We will prove Theorem 2.3.1 in the next subsection. However, for now, let us state and prove a useful consequence.

Corollary 2.3.1 *Let $X = (X_t;\ t \in \mathbb{R}_+^N)$ be a stochastic process that satisfies the following for some $C, p > 0$ and $\gamma > 1 + p$:*

$$\mathbb{E}\{|X_s - X_t|^p\} \leq C|t - s|^N \left[\ln_+ \frac{1}{|t - s|}\right]^{-\gamma}, \qquad t, s \in \mathbb{R}_+^N.$$

Then, X has a continuous modification.

Proof We first show that for all $\tau \in \mathbb{N}^N$, $(X_t;\ t \in [0, \tau])$ has a continuous modification. Fix such a τ, let $T = [0, \tau]$, and define the pseudometric d by $d(s, t) = |s - t|$, $s, t \in T$. In fact, d is a metric on T and (T, d) is totally bounded.

To show total boundedness, we bound the metric entropy of (T, d). For any integer $n \geq 1$, let T_n denote the collection of all points $t \in T \cap n^{-1} \mathbb{Z}^N$. Since $\tau \in \mathbb{N}^N$, the cardinality of T_n is precisely $C_\tau n^N$, where $C_\tau = \prod_{\ell=1}^N (\tau^{(\ell)} + 1)$. On the other hand, T_n is easily seen to be the maximal $\frac{1}{n}$-net for T. Recalling Kolmogorov's capacitance K from Section 2.1, it follows that $K(\frac{1}{n}; T, d) = C_\tau n^N$. By a monotonicity argument, for all $n \geq 2$ and all $n^{-1} \leq r \leq (n-1)^{-1}$,

$$D(r; T, d) \leq K(r; T, d) \leq K(\tfrac{1}{n}; T, d) = C_\tau n^N \leq C_\tau \left(\frac{1}{r} + 1\right)^N.$$

We have used Lemma 2.1.1 to compare entropy to capacitance. The above discussion implies that for all $0 < r < 1$,

$$D(r; T, d) \leq 2^N C_\tau r^{-N}. \tag{1}$$

By the assumption of Corollary 2.3.1, for all $s, t \in T$, $\mathbb{E}\{|X_s - X_t|^p\} \leq \Psi(d(s, t))$, where $\Psi(x) = C x^N \{\ln_+(1/x)\}^{-\gamma}$. Let $f(r) = \{\ln_+(1/r)\}^{-\theta}$, where $1 < \theta < p^{-1}(\gamma - 1)$ is a fixed number. By equation (1),

$$\int_0^1 \frac{D(r; T, d) \Psi(2r)}{r\{f(r)\}^p}\, dr \leq 4^N C\, C_\tau \int_0^1 \frac{1}{r} \left[\ln_+ \left(\frac{1}{2r}\right)\right]^{\theta p - \gamma}\, dr,$$

which is finite, since $\theta p - \gamma < -1$. Since $\int_0^1 r^{-1} f(r)\, dr < \infty$, Kolmogorov's continuity theorem implies that for all $\tau \in \mathbb{N}^N$, the process $(X_t;\ t \in [0, \tau])$ has a continuous modification.

160   5. Gaussian Random Variables

We conclude this demonstration by proving our result for $(X_t;\ t \in \mathbb{R}_+^N)$; this is done by appealing to a **patching argument**.

Let $Y^\tau = (Y_t^\tau;\ t \in [0,\tau])$ denote the mentioned modification of $(X_t;\ t \in [0,\tau])$. For any $\tau \preccurlyeq \sigma$, both in $\mathbb{N}^N$, Y^τ and Y^σ agree on $[0,\tau]$; i.e., $\mathbb{P}(Y_t^\tau = Y_t^\sigma) = 1$. Therefore, by continuity, $\mathbb{P}(Y_s^\tau = Y_s^\sigma,\ \text{for all } s \in [0,\tau]) = 1$. Using the ortholimit notation of Chapter 1, Section 2.7, define $Y_t = \lim_{\sigma \rightsquigarrow \infty} Y_t^\sigma$, $t \in \mathbb{R}_+^N$. It is easily seen that this limit exists, is continuous, and agrees with each Y^τ on $[0,\tau]$. As such, Y is the desired modification of X. (This is called a patching argument, since Y is constructed in *patches* $(Y^\sigma;\ \sigma \in \mathbb{N}^N)$.) □

2.4 Chaining

In this subsection we prove Kolmogorov's continuity theorem (Theorem 2.3.1). Throughout, the assumptions of Theorem 2.3.1 are in force.

If $D_n = D(2^{-n}; T, d)$ denotes the metric entropy of T evaluated at 2^{-n}, we can find D_n balls of radius 2^{-n} that cover T; let $\mathcal{B}_n$ denote the collection of these balls. For every $t \in T$, there exists a well-defined element of $\mathcal{B}_n$ that contains t; we denote this ball by $B_n(t)$. Note that we can always choose this consistently in the sense that whenever $s \in B_n(t)$, then $B_n(t) = B_n(s)$. Since $B_n(t) \cap B_{n+1}(t)$ is open and contains t, in an arbitrary but fixed way

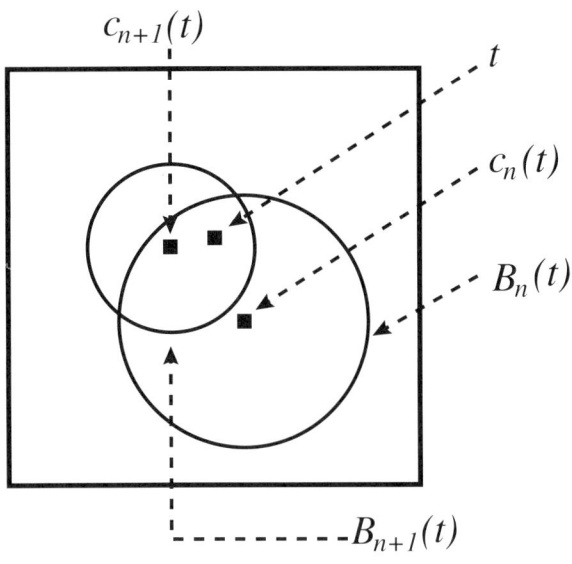

Figure 5.1: Covering by balls

we choose some point $c_n(t) \in B_n(t) \cap B_{n+1}(t)$ and let $\mathcal{C}_n = (c_n(t);\ t \in T)$ designate the totality of these points. (We can think of $c_n(t)$ as the "center" of $B_n(t)$ and $\mathcal{C}_n$ as the collection of these centers. However, strictly speaking, this interpretation need not be correct.) It is important to note that the cardinality of $\mathcal{C}_n$ is at most that of $\mathcal{B}_n$, which is D_n. In particular, $\mathcal{C}_n$ is a finite set.

We mention three more properties of the elements of $\mathcal{C}_n$, the last one of which we just stated explicitly:

(P1) since $c_n(t) \in B_n(t)$, $d(c_n(t), t) \leq 2^{-n}$, for all $t \in T$;

(P2) for all $t \in T$ and all $n \geq 1$, $B_{n+1}(c_n(t)) = B_{n+1}(t)$, which implies that
$$c_{n+1}(t) = c_{n+1}(c_n(t));$$

(P3) $\#\mathcal{C}_n \leq D_n$, where $\#K$ denotes the cardinality of any set K.

For each integer $n \geq 1$, define the stochastic processes $X^n = (X_t^n;\ t \in T)$ and $Y = (Y_t;\ t \in T)$ as follows:
$$X_t^n = X_{c_n(t)} \quad \text{and} \quad Y_t = \limsup_{n \to \infty} X_{c_n(t)}, \quad t \in T.$$

The process Y will turn out to be an almost surely continuous modification of X, while X^n is a *discretization* of it. We will remodify Y later, in order to obtain a continuous modification of X. First, we show that we can even discuss the continuity of Y in a measurable way.

Lemma 2.4.1 *The event $(t \mapsto Y_t$ is continuous$)$ is measurable.*

Proof The event in question can be written as follows:
$$\bigcap_{\varepsilon \in \mathbb{Q}_+} \bigcup_{\delta \in \mathbb{Q}_+} \bigcap_{\substack{s,t \in \cup_{n=1}^\infty \mathcal{C}_n: \\ d(s,t) \leq \delta}} (|Y_s - Y_t| \leq \varepsilon),$$

which is measurable, since $\cup_{n=1}^\infty \mathcal{C}_n$ is at most countable. $\square$

The key estimate is the following; it shows that for large m, Y is very close to X^m, uniformly over all $t \in T$.

Lemma 2.4.2 *There exists an integer $m_0 \geq 1$ such that for all $\lambda > 0$ and all integers $m \geq m_0$,*
$$\mathbb{P}\left(\sup_{k \geq m} \sup_{t \in T} |Y_t - X_t^k| \geq \lambda\right) \leq 2\lambda^{-p} \int_0^{2^{-m-1}} \frac{D(r; T, d)\Psi(2r)}{r\{f(r)\}^p}\, dr.$$

In particular, with probability one, $\lim_{k \to \infty} \sup_{t \in T} |Y_t - X_t^k| = 0$.

Proof By the first assertion, for all $\lambda > 0$,
$$\lim_{m \to \infty} \mathbb{P}\left(\sup_{k \geq m} \sup_{t \in T} |Y_t - X_t^k| \geq \lambda\right) = 0.$$

Since the event inside the probability is decreasing in m, the second assertion follows. Thus, it suffices to demonstrate the asserted probability estimate.

For any two integers $n > m \geq 1$, $\sup_{t \in T} |X_t^n - X_t^m|$ is a maximum over a finite set. As such, it is a random variable. Having checked measurability, we use what is called a chaining argument. Note that

$$\max_{m \leq k \leq n} |X_t^k - X_t^m| \leq \sum_{j=m+1}^{n} |X_{c_j(t)} - X_{c_{j-1}(t)}|.$$

In words, we have *chained* X_t^n to X_t^m by observing X over appropriate elements of $\mathcal{C}_m, \ldots, \mathcal{C}_n$. For all integers $j \geq 1$, define $w_j = f(2^{-j})$. Since f is nondecreasing,

$$\sum_{j=1}^{\infty} w_j = \sum_{j=1}^{\infty} \int_{2^{-j}}^{2^{-j+1}} 2^j f(2^{-j}) \, dr \leq 2 \int_0^1 \frac{f(r)}{r} \, dr < \infty.$$

Thus, $m_0 = \inf(k \geq 1 : \sum_{j=k}^{\infty} w_j \leq 1)$ is well-defined and is a finite integer. For all $\lambda > 0$ and all $m \geq m_0$

$$\mathbb{P}\Big(\sup_{t \in T} \max_{m \leq k \leq n} |X_t^k - X_t^m| \geq \lambda\Big)$$

$$\leq \mathbb{P}\Big(\sum_{j=m+1}^{n} |X_{c_j(t)} - X_{c_{j-1}(t)}| \geq \lambda \text{ for some } t \in T \Big)$$

$$\leq \sum_{j=m+1}^{n} \mathbb{P}\big(|X_{c_j(\alpha)} - X_\alpha| \geq w_j \lambda \text{ for some } \alpha \in \mathcal{C}_j\big) \quad (1)$$

$$\leq \sum_{j=m+1}^{n} \sum_{\alpha \in \mathcal{C}_j} \mathbb{P}\big(|X_{c_j(\alpha)} - X_\alpha| \geq w_j \lambda\big)$$

$$\leq \sum_{j=m+1}^{n} D_j \max_{\alpha \in \mathcal{C}_j} \mathbb{P}\big(|X_{c_j(\alpha)} - X_\alpha| \geq w_j \lambda\big). \quad (2)$$

Equations (1) and (2) follow from (P2) and (P3), respectively. By Chebyshev's inequality and the assumptions of Kolmogorov's continuity theorem (Theorem 2.3.1), for all $\lambda > 0$ and all $m \geq m_0$,

$$\mathbb{P}\Big(\sup_{t \in T} \max_{m \leq k \leq n} |X_t^k - X_t^m| \geq \lambda\Big)$$

$$\leq \lambda^{-p} \sum_{j=m+1}^{n} D_j w_j^{-p} \max_{\alpha \in \mathcal{C}_j} \mathbb{E}\{|X_{c_j(\alpha)} - X_\alpha|^p\}$$

$$\leq \lambda^{-p} \sum_{j=m+1}^{n} D_j \{f(2^{-j})\}^{-p} \max_{\alpha \in \mathcal{C}_j} \Psi(d(c_j(\alpha), \alpha)).$$

By the monotonicity of Ψ and by (P1), $\Psi(d(c_j(\alpha), \alpha)) \leq \Psi(2^{-j})$. Hence, for all $\lambda > 0$ and all $m \geq m_0$,

$$\mathbb{P}\left(\sup_{t \in T} \max_{m \leq k \leq n} |X_t^k - X_t^m| \geq \lambda\right) \leq \lambda^{-p} \sum_{j=m+1}^{n} D_j \{f(2^{-j})\}^{-p} \Psi(2^{-j}).$$

Thus, for all $m \geq m_0$,

$$\mathbb{P}\left(\sup_{t \in T} \sup_{k \geq m} |X_t^k - X_t^m| \geq \lambda\right) \leq \lambda^{-p} \sum_{j=m+1}^{\infty} D_j \{f(2^{-j})\}^{-p} \Psi(2^{-j})$$

$$= \lambda^{-p} \sum_{j=m+1}^{\infty} \int_{2^{-j-1}}^{2^{-j}} \frac{D(2^{-j}; T, d) \Psi(2^{-j}) 2^j}{\{f(2^{-j})\}^p} \, dr$$

$$\leq 2 \sum_{j=m+1}^{\infty} \int_{2^{-j-1}}^{2^{-j}} \frac{D(r; T, d) \Psi(2r)}{r \{f(r)\}^p} \, dr$$

$$= 2 \int_{0}^{2^{-m-1}} \frac{D(r; T, d) \Psi(2r)}{r \{f(r)\}^p} \, dr. \quad (3)$$

Since $Y_t = \limsup_{k \to \infty} X_t^k$, for any $k \geq m \geq 1$ and all $t \in T$,

$$|Y_t - X_t^k| \leq \sup_{j \geq m} |X_t^j - X_t^m|.$$

The lemma is a consequence of this fact, together with equation (3). □

Lemma 2.4.3 Y is a modification of X.

Proof To start with, X is continuous in probability. In fact, by Chebyshev's inequality, for all $\varepsilon > 0$, $\mathbb{P}(|X_s - X_t| \geq \varepsilon) \leq \varepsilon^{-p} \Psi(d(s,t))$. Since Ψ is continuous and $\Psi(0) = 0$, it follows that as $s \to t$, X_s converges to X_t in probability. In particular, for each $t \in T$, as $n \to \infty$, $X_t^n = X_{c_n(t)} \to X_t$, in probability. By equation (3) there exists a finite m_0 such that for all $m \geq m_0$ and all $\lambda > 0$,

$$\mathbb{P}(|X_t^m - X_t| \geq \lambda) = \lim_{n \to \infty} \mathbb{P}(|X_t^m - X_t^n| \geq \lambda)$$

$$\leq \mathbb{P}\left(\sup_{t \in T} \sup_{k \geq m} |X_t^k - X_t^m| \geq \lambda\right)$$

$$\leq \lambda^{-p} \int_{0}^{2^{-m-1}} \frac{D(r; T, d) \Psi(2r)}{r \{f(r)\}^p} \, dr,$$

which goes to zero as $m \to \infty$. This and Lemma 2.4.2 together show that for each $t \in T$, $\mathbb{P}(Y_t = X_t) = 1$, proving the result. □

Now that we know that Y is a modification of X, we use Lemma 2.4.2 to provide an estimate for the **modulus of continuity** of Y.

Lemma 2.4.4 *Recall the integer m_0 of Lemma 2.4.2. For all $0 < \delta < 2^{-m_0}$ and all $\lambda > 0$,*

$$\mathbb{P}\left(\sup_{\substack{s,t \in T: \\ d(s,t) \leq \delta}} |Y_s - Y_t| \geq \lambda\right) \leq 6^p \lambda^{-p} \int_0^\delta \frac{D(r;T,d)\Psi(2r)}{r\{f(r)\}^p} dr.$$

Proof For all $s, t \in T$, whenever $s \in B_m(t)$, then $c_m(s) = c_m(t)$. In particular, if $s \in B_m(t)$, then $X_s^m = X_t^m$. Hence, for any $m \geq 1$,

$$\sup_{t \in T} \sup_{s \in B_m(t)} |Y_s - Y_t| \leq \sup_{t \in T} |Y_t - X_t^m| + \sup_{s \in T} |Y_s - X_s^m|$$

$$= 2 \sup_{t \in T} |Y_t - X_t^m|.$$

Applying this, we can deduce that for every integer $m \geq 1$ and for all $s, t \in T$ with $d(s,t) \leq 2^{-m}$,

$$|Y_s - Y_t| \leq |Y_t - Y_{c_m(t)}| + |Y_{c_m(t)} - Y_{c_m(s)}| + |Y_s - Y_{c_m(s)}|$$

$$\leq 3 \sup_{t \in T} \sup_{s \in B_m(t)} |Y_t - Y_s|$$

$$\leq 6 \sup_{t \in T} |Y_t - X_t^m|.$$

We have used the fact that whenever $d(s,t) \leq 2^{-m}$, then $c_m(t) \in B_n(c_m(s))$; see (P2). Lemma 2.4.2 implies that for all $m \geq m_0$,

$$\mathbb{P}\left(\sup_{\substack{s,t \in T: \\ d(s,t) \leq 2^{-m}}} |Y_s - Y_t| \geq \lambda\right) \leq 6^p \lambda^{-p} \int_0^{2^{-m-1}} \frac{D(r;T,d)\Psi(2r)}{r\{f(r)\}^p} dr. \quad (4)$$

Now, if $\delta < 2^{-m_0}$, we can find an integer $m \geq m_0$ such that $2^{-m-1} \leq \delta \leq 2^{-m}$. Hence,

$$\sup\{|Y_s - Y_t| : s, t \in T, d(s,t) \leq \delta\} \leq \sup\{|Y_s - Y_t| : s, t \in T, d(s,t) \leq 2^{-m}\}.$$

Similarly, $\int_0^{2^{-m-1}} \{\cdots\} dr \leq \int_0^\delta \{\cdots\} dr$, where $\{\cdots\}$ represents the integrand in equation (4). Equation (4) itself now implies the lemma. $\square$

We are now ready to prove Kolmogorov's continuity theorem.

Proof of Theorem 2.3.1 By Lemma 2.4.4 and by the continuity properties of $\mathbb{P}$, for all $\lambda > 0$,

$$\mathbb{P}\left(\lim_{\delta \to 0^+} \sup_{\substack{s,t \in T: \\ d(s,t) \leq \delta}} |Y_s - Y_t| \geq \lambda\right) = \lim_{\delta \to 0^+} \mathbb{P}\left(\sup_{\substack{s,t \in T: \\ d(s,t) \leq \delta}} |Y_s - Y_t| \geq \lambda\right) = 0.$$

Since $\lambda > 0$ is arbitrary, we see that Y is an a.s. continuous modification of X. To finish the proof, we remodify Y as follows. Consider the event Ω_0 defined to be the collection of all $\omega \in \Omega$ such that $t \mapsto Y_t(\omega)$ is continuous. By Lemma 2.4.1, Ω_0 is measurable, and we have just shown that $\mathbb{P}(\Omega_0) = 0$. For all $\omega \notin \Omega_0$, define the process $\widetilde{Y}_t(\omega) = Y_t(\omega)$ for all $t \in T$, and for all $\omega \in \Omega_0$, let $\widetilde{Y}_t(\omega) \equiv 0$ for all $t \in T$. Clearly, $\widetilde{Y} = (\widetilde{Y}_t;\ t \in T)$ is continuous. Moreover, it is not hard to see that $\widetilde{Y}$ is a modification of X, since it is a modification of Y. Theorem 2.3.1 follows for this modification. □

2.5 Hölder-Continuous Modifications

Roughly speaking, Kolmogorov's continuity theorem (Theorem 2.3.1) states that whenever X_s and X_t are sufficiently close uniformly in $L^p(\mathbb{P})$, then X has a continuous modification. In this subsection we further investigate this principle by studying an important special case.

Given a totally bounded pseudometric space (T, d) and a continuous function $y : T \to \mathbb{R}$, define the **modulus of continuity**[7] $\omega_{y,T} : \mathbb{R}_+ \to \mathbb{R}_+$ of y by

$$\omega_{y,T}(\delta) = \sup_{\substack{s,t \in T:\\ d(s,t) \leq \delta}} |y(s) - y(t)|, \qquad \delta \geq 0.$$

We say that y is **Hölder continuous of order** $q > 0$ if

$$\limsup_{\delta \to 0^+} \delta^{-q} \omega_{y,T}(\delta) < +\infty.$$

Stated in terms of the function y, y is Hölder continuous of order $q > 0$ if there are constants $C, \delta_0 > 0$ such that whenever $0 < \delta < \delta_0$, for all $s, t \in T$ satisfying $d(s,t) \leq \delta$, $|y(s) - y(t)| \leq C|s-t|^q$; see Supplementary Exercise 5 for details.

When T is not totally bounded, we say that y is Hölder continuous of order $q > 0$, if for all totally bounded $S \subset T$, $\limsup_{\delta \to 0^+} \delta^{-q} \omega_{y,S}(\delta) < +\infty$.

In Section 2.4 we have shown that under some conditions on the process X, there exists a modification Y such that $\lim_{\delta \to 0^+} \omega_{Y,T}(\delta) = 0$. That is, we have shown that Y is continuous. Our next theorem is an N-parameter refinement of Corollary 2.3.1. It shows that under enough smoothness in $L^p(\mathbb{P})$, X has a Hölder continuous modification.

Theorem 2.5.1 Let $X = (X_t;\ t \in \mathbb{R}_+^N)$ denote a stochastic process that satisfies the following for some $C, p > 0$ and $\gamma > N$:

$$\mathbb{E}\{|X_s - X_t|^p\} \leq C|s-t|^\gamma, \qquad s, t \in \mathbb{R}_+^N.$$

---

[7] This notion is due to H. Lebesgue.

Then, there exists a modification $Y = (Y_t; t \in \mathbb{R}_+^N)$ of X that is Hölder continuous of any order $q \in [0, p^{-1}(\gamma - N)[$.

Proof For all $s, t \in \mathbb{R}_+^N$, define $d(s,t) = |s - t|$. While $(\mathbb{R}_+^N, d)$ is not totally bounded, $([a,b], d)$ is, as long as $a \prec b$ are both in $\mathbb{R}_+^N$. Fixing such $a, b \in \mathbb{R}_+^N$, we aim to show that $(X_t; t \in [a,b])$ has a modification $Y = (Y_t; t \in [a,b])$ that is Hölder continuous of any order $q < p^{-1}(\gamma - N)$. The patching argument used to complete the proof of Corollary 2.3.1 can be invoked to finish our proof.

The process $Y = (Y_t; t \in [a,b])$ is the one provided to us by Section 2.4. By the proof of Corollary 2.3.1, there exists a constant $C_{a,b}$ such that for all $0 < r < 1$, $D(r; [a,b], d) \leq 2^N C_{a,b} r^{-N}$.

Fix $\theta > 0$ and define $f(r) = \{\ln_+(1/r)\}^{-\theta}$. Since $\int_0^1 r^{-1} f(r)\, dr < +\infty$, we can apply Lemma 2.4.4 with $\Psi(x) = Cx^\gamma$ to see that there exists a constant $0 < \delta_0 < 1$ such that for all $\lambda > 0$ and all $0 < \delta < \delta_0$,

$$\mathbb{P}(\omega_{Y,[a,b]}(\delta) \geq \lambda) \leq 2^{N+\gamma} C_{a,b} C 6^p \lambda^{-p} \int_0^\delta r^{-N+\gamma-1} \{\ln(1/r)\}^{p\theta}\, dr.$$

Fix two arbitrary constants q and Q that satisfy $0 < q < Q < p^{-1}(\gamma - N)$. Since $\delta_0 < 1$, then for all $\lambda > 0$ and $0 < \delta < \delta_0$,

$$\mathbb{P}(\omega_{Y,[a,b]}(\delta) \geq \lambda) \leq \Gamma \lambda^{-p} \delta^{Qp}, \tag{1}$$

where $\Gamma = 2^{N+\gamma} C_{a,b} C 6^p \int_0^1 r^{\gamma-N-Qp-1} \left(\ln(1/r)\right)^{p\theta} dr < \infty$. We can apply equation (1) with $\delta = e^{-n}$ and $\lambda = e^{-(n+1)q}$ to deduce that for all integers $n > \ln(1/\delta_0)$,

$$\mathbb{P}\left(\omega_{Y,[a,b]}(e^{-n}) \geq e^{-(n+1)q}\right) \leq e^{pq} \Gamma e^{-np(Q-p)}.$$

Since $Q > p$, the summation of the above over n is finite. By the Borel–Cantelli lemma, there exists a finite random variable n_0 such that a.s., for all $n \geq n_0$,

$$\omega_{Y,[a,b]}(e^{-n}) \leq e^{-(n+1)q}.$$

We now use a **monotonicity argument:** For all $\delta < e^{-n_0}$, there exists some integer $n_\delta \geq n_0$ with $e^{-(n_\delta+1)} \leq \delta \leq e^{-n_\delta}$. Thus, outside one set of $\mathbb{P}$-measure 0, for all $0 < \delta < e^{-n_0}$,

$$\omega_{Y,[a,b]}(\delta) \leq \omega_{Y,[a,b]}(e^{-n_\delta}) \leq e^{-(n_\delta+1)q} \leq \delta^q.$$

In particular, $\limsup_{\delta \to 0^+} \delta^{-q} \omega_{Y,[a,b]}(\delta) \leq 1$, almost surely. We have shown the existence of a modification of X that is a.s. Hölder continuous. We can remodify this, as we did in the proof of Kolmogorov's continuity theorem, to obtain a Hölder continuous modification and finish our derivation. □

***Exercise* 2.5.1** (Hard) In the setting of Theorem 2.5.1, show that for all $\tau \in \mathbb{R}_+^N$, $0 < q < p$, and all $Q \in]0, p^{-1}(\gamma - N)[$, there exists a finite constant C (which depends on p, q, Q, γ and τ) such that for all $\delta \in [0, 1[$,

$$\mathbb{E}\left\{\sup_{s,t \in [0,\tau]:\ |s-t| \leq \delta} |X_s - X_t|^q\right\} \leq C\delta^{Qp}.$$

(HINT: Use integration by parts, in conjunction with equation (1).) $\square$

***Exercise* 2.5.2** In the setting of Theorem 2.5.1, show that for all $0 < q < p$ and all $\tau \in \mathbb{R}_+^N$, $\mathbb{E}\{\sup_{t \in [0,\tau]} |X_t|^q\} < \infty$. $\square$

***Exercise* 2.5.3** There are other, more relaxed, notions of modulus of continuity. For instance, consider the **integral modulus of continuity** $\Omega_f(\varepsilon)$ of an integrable function f, defined by

$$\Omega_f(\varepsilon) = \sup_{0 \leq y \leq \varepsilon} \int |f(x+y) - f(x)|\,dx.$$

Prove that for every integrable function f, $\lim_{\varepsilon \to 0} \Omega_f(\varepsilon) = 0$.
(HINT: First do this for simple functions f, and then appeal to density.) $\square$

2.6 The Entropy Integral

If (T, d) denotes a totally bounded pseudometric space, Theorem 2.5.1 shows one instance of the following general principle: If the stochastic process $X = (X_t;\ t \in T)$ is very smooth in distribution, then it has a modification with some nice continuity properties. In this subsection we illustrate another class of processes for which the above principle holds true.

We say that the pseudometric d is a **modulus of continuity for X in probability** if there exists a strictly decreasing function $\Phi : \mathbb{R}_+ \to [0, 1]$ such that $\lim_{\lambda \to \infty} \Phi(\lambda) = 0$ and for all $\lambda > 0$,

$$\sup_{s,t \in T} \mathbb{P}\{|X_s - X_t| > d(s,t)\lambda\} \leq \Phi(\lambda).$$

In order to emphasize the dependence on the function Φ, we will say that (d, Φ) is a modulus of continuity for X in probability. This definition is motivated by the following simple result.

Lemma 2.6.1 *If (d, Φ) is a modulus of continuity for X in probability, then X is continuous in probability. Conversely, if X is continuous in probability and if (T, d) is a compact topological space, X has a modulus of continuity in probability.*

***Exercise* 2.6.1** Prove Lemma 2.6.1. $\square$

Throughout the remainder of this subsection we will suppose that (d, Φ) is a modulus of continuity for X in probability with a corresponding function Φ. Let Φ^{-1} denote the inverse function to Φ; this exists, since Φ is strictly decreasing. For any measurable function $f : \mathbb{R}_+ \to \mathbb{R}_+$, we define

$$\Theta_f(\delta) = \int_0^\delta \Phi^{-1}\left(\frac{f(r)}{D(\frac{1}{2}r; T, d)}\right) dr, \qquad \delta > 0.$$

Next, we show that whenever $\Theta_f(1) < \infty$ for a suitable function f, then X has a continuous modification whose modulus of continuity is Θ_f; see Section 2.5 for the latter.

Theorem 2.6.1 *Consider a process $X = (X_t;\ t \in T)$ indexed by a totally bounded pseudometric space (T, d). Suppose (d, Φ) is a modulus of continuity for X in probability, and that there exists a nondecreasing function $f : \mathbb{R}_+ \to \mathbb{R}_+$ such that:*

(a) $\int_0^{1/2} r^{-1} f(r)\, dr < \infty$; and

(b) $\Theta_f(\frac{1}{2}) < \infty$.

Then, X has a continuous modification $Y = (Y_t;\ t \in T)$. Finally, suppose that condition (a) is replaced by the stronger condition

(c) $\int_0^{1/2} r^{-1} \ln(\frac{1}{r}) f(r)\, dr < \infty$.

Then,

$$\limsup_{\delta \to 0^+} \frac{\omega_{Y,T}(\delta)}{\Theta_f(\delta)} \leq 12.$$

Proof[8] We follow the arguments of Section 2.4 very closely, but now we pick the weights $w_1, w_2, \ldots$ with more care. We will delineate the important steps of the proof, all the while using much of the notation of Section 2.4. Define

$$w_j = 2^{-j} \Phi^{-1}\left(\frac{f(2^{-j})}{D(2^{-j}; T, d)}\right), \qquad j \geq 1.$$

An important property of the w_j's is that for all integers $n > m \geq 1$,

$$\sum_{j=m+1}^n w_j = \sum_{j=m+1}^n \int_{2^{-j}}^{2^{-j+1}} \Phi^{-1}\left(\frac{f(2^{-j})}{D(2^{-j}; T, d)}\right) dr \leq 2\Theta_f(2^{-m-1}). \qquad (1)$$

We have used the monotonicity properties of metric entropy, Φ^{-1}, and f. In analogy to Section 2.4, define

$$m_0 = \inf\left(k \geq 1 : \sum_{j=k}^\infty w_j \leq 1\right).$$

---

[8] This can be skipped at first reading.

The above implies that m_0 is a well-defined finite integer. Arguing as in equations (1) and (2) of Section 2.4, we see that for all integers $n > m \geq m_0$,

$$\mathbb{P}\Big\{\sup_{t\in T}\max_{m\leq k\leq n}|X_t^k - X_t^m| \geq 2\Theta_f(2^{-m-1})\Big\}$$
$$\leq \sum_{j=m+1}^{n}\sum_{\alpha\in\mathcal{C}_j}\mathbb{P}(|X_{c_j(\alpha)} - X_\alpha| \geq w_j) \leq \sum_{j=m+1}^{n}\sum_{\alpha\in\mathcal{C}_j}\Phi\Big(\frac{w_j}{d(c_j(\alpha),\alpha)}\Big).$$

We have implicitly used Lemma 2.6.1, together with the convention that $\Phi(t/0) = 0$, for all $t \geq 0$. By (P3) of Section 2.4, $d(c_j(\alpha),\alpha) \leq 2^{-j}$. Since Φ is decreasing, for all $n > m \geq m_0$,

$$\mathbb{P}\Big\{\sup_{t\in T}\max_{m\leq k\leq n}|X_t^k - X_t^m| \geq 2\Theta_f(2^{-m-1})\Big\}$$
$$\leq \sum_{j=m+1}^{n}\sum_{\alpha\in\mathcal{C}_j}\Phi(2^j w_j) = \sum_{j=m+1}^{n} f(2^{-j})\sum_{\alpha\in\mathcal{C}_j(\alpha)}\frac{1}{D(2^{-j};T,d)}$$
$$\leq \sum_{j=m+1}^{n} f(2^{-j}).$$

The last line follows from (P3) of Section 2.4. Thus, for all $n > m \geq m_0$,

$$\mathbb{P}\{\sup_{t\in T}\max_{m\leq k\leq n}|X_t^k - X_t^m| \geq 2\Theta_f(2^{-m-1})\} \leq \sum_{j=m+1}^{n}\int_{2^{-j}}^{2^{-j+1}}\frac{1}{2^{-j}}f(2^{-j})\,dr$$
$$\leq 2\int_0^{2^{-m}}\frac{f(r)}{r}\,dr.$$

Now we use the same modification provided us by Lemma 2.4.3. The proof of Lemma 2.4.4 shows that for all $m \geq m_0$,

$$\mathbb{P}\Big\{\sup_{\substack{s,t\in T:\\d(s,t)\leq 2^{-m}}}|Y_s - Y_t| \geq 12\Theta_f(2^{-m-1})\Big\} \leq 2\int_0^{2^{-m}}\frac{f(r)}{r}\,dr. \qquad (2)$$

Since $\int_0^{\frac{1}{2}} r^{-1}f(r)\,dr < +\infty$, the right-hand side of equation (2) goes to 0 as $m \to \infty$; similarly, $\lim_{m\to\infty}\Theta_f(2^{-m-1}) = 0$. Consequently, for all $1 > \varepsilon > 0$, there exists some $m_1(\varepsilon) \geq m_0$ such that $m \geq m_1(\varepsilon)$, $\Theta_f(2^{-m-1}) \leq \frac{1}{12}\varepsilon$. By equation (2), for all $\varepsilon > 0$ and all $m \geq m_1(\varepsilon)$,

$$\mathbb{P}\Big(\sup_{\substack{s,t\in T:\\d(s,t)\leq 2^{-m}}}|Y_s - Y_t| \geq \varepsilon\Big) \leq 2\int_0^{2^{-m}}\frac{f(r)}{r}\,dr.$$

By the continuity properties of $\mathbb{P}$, for all $\varepsilon > 0$,

$$\lim_{m\to\infty}\mathbb{P}\Big(\sup_{\substack{s,t\in T:\\d(s,t)\leq 2^{-m}}}|Y_s - Y_t| \geq \varepsilon\Big) = \mathbb{P}\Big(\limsup_{m\to\infty}\sup_{\substack{s,t\in T:\\d(s,t)\leq 2^{-m}}}|Y_s - Y_t| \geq \varepsilon\Big) = 0.$$

In particular, $\lim_{m\to\infty} \omega_{Y,T}(2^{-m}) = 0$, a.s. . We can now apply a monotonicity argument to complete the proof of a.s. continuity: For all $2^{-m} \le \delta \le 2^{-m+1}$,
$$\omega_{Y,T}(2^{-m}) \le \omega_{Y,T}(\delta) \le \omega_{Y,T}(2^{-m+1}).$$
This shows that $\lim_{\delta\to 0+} \omega_{Y,T}(\delta) = 0$. In fact, a little more holds true, provided that $\int_0^1 r^{-1} \ln(\frac{1}{r}) f(r)\, dr < \infty$. To see this, we use equation (2) to see that

$$\sum_{m=m_0}^{\infty} \mathbb{P}\{\omega_{Y,T}(2^{-m}) \ge 12\Theta_f(2^{-m-1})\} \le 2 \sum_{m=1}^{\infty} \int_0^{2^{-m}} \frac{f(r)}{r}\, dr$$

$$= 2 \int_0^{\frac{1}{2}} \sum_{m=1}^{\infty} \mathbf{1}_{[0,2^{-m}]}(r) \frac{f(r)}{r}\, dr$$

$$\le \frac{2}{\ln 2} \int_0^{\frac{1}{2}} \frac{f(r)\ln(\frac{1}{r})}{r}\, dr,$$

which is finite. By the Borel–Cantelli lemma, there exists a random variable n_0 such that a.s., for all $m \ge n_0$, $\omega_{Y,T}(2^{-m}) \le 12\Theta_f(2^{-m-1})$. By a monotonicity argument we have established the existence of an a.s. continuous modification of X with an a.s. modulus of continuity of Θ_f. To remove the a.s., we remodify the process further as in the proof of Kolmogorov's continuity theorem. □

2.7 Dudley's Theorem

Let $X = (X_t;\ t \in T)$ be a mean-zero Gaussian process, indexed by some set T. That is, for all $t \in T$, $\mathbb{E}[X_t] = 0$. Let Σ denote the covariance function of X:
$$\Sigma(s,t) = \mathbb{E}[X_s X_t], \qquad s,t \in T.$$

We wish to find conditions under which X has a continuous modification.[9] This question is always well posed when T is a pseudometric space. On the other hand, the process X itself induces a natural pseudometric on T as follows:
$$d(s,t) = \sqrt{\mathbb{E}\{|X_s - X_t|^2\}}, \qquad s,t \in T. \tag{1}$$

According to Supplementary Exercise 1, (T,d) is a pseudometric space, and it turns out that as far as the Gaussian process X is concerned, d is the correct pseudometric on T.[10]

---

[9] The general nonzero mean-μ case can be obtained from this by considering $X_t + \mu(t)$.
[10] This is a part of the general theory of Gaussian processes. For example, see (Adler 1990; Ledoux 1996; Ledoux and Talagrand 1991).

2 Regularity Theory

Exercise 2.7.1 The pseudometric d can be computed from the covariance function as follows:
$$d(s,t) = \sqrt{\Sigma(s,s) + \Sigma(t,t) - 2\Sigma(s,t)}, \qquad s,t \in T.$$
□

Define the function $\Phi : \mathbb{R} \to [0,1]$ by
$$\Phi(\lambda) = e^{-\frac{1}{2}\lambda^2}, \qquad \lambda > 0.$$
Clearly, Φ is strictly decreasing, and its inverse function is
$$\Phi^{-1}(t) = \sqrt{2\ln(1/t)}, \qquad t \in [0,1].$$

Lemma 2.7.1 *(d, Φ) is a modulus of continuity for X in probability.*

Exercise 2.7.2 Prove Lemma 2.7.1. □

Dudley's theorem is the following sufficient condition for the continuity of a Gaussian process.

Theorem 2.7.1 (Dudley's Theorem) *Suppose $X = (X_t; \, t \in T)$ is a mean-zero Gaussian process indexed by the pseudometric space (T, d). If (T, d) is totally bounded and if $\int_0^1 \sqrt{\ln D(r; T, d)}\, dr < \infty$, then X has a continuous modification $Y = (Y_t; \, t \in T)$. Moreover, there exists a universal constant $C > 0$ such that*
$$\limsup_{\delta \to 0^+} \frac{\omega_{Y,T}(\delta)}{\int_0^\delta \sqrt{\ln D(\frac{1}{2}r; T, d)}\, dr + C\delta\sqrt{\ln\ln(1/\delta)}} \leq 24.$$

This theorem relies on the following real-variable lemma.

Lemma 2.7.2 *There exists a constant $c > 0$ such that for all $0 < \delta < e^{-2}$,*
$$\int_0^\delta \sqrt{\ln\ln(\tfrac{1}{r})}\, dr \leq c\delta\sqrt{\ln\ln(1/\delta)}.$$

Exercise 2.7.3 Prove Lemma 2.7.2.
(HINT: You can try L'Hôpital's rule, for instance.) □

Proof of Theorem 2.7.1 Fix some $\theta > 2$ and let
$$f(r) = \left|\ln_+\left(\frac{1}{r}\right)\right|^{-\theta}, \qquad r > 0.$$
It is easy to see that condition *(a)* of Theorem 2.6.1 holds and, using the notation of Section 2.6,
$$\Theta_f(\delta) = \int_0^\delta \sqrt{2\ln D(\tfrac{1}{2}r; T, d) + 2\theta\ln\ln(1/r)}\, dr, \qquad \delta > 0.$$
The result follows from the inequality $|a+b| \leq 2(|a|+|b|)$, and from Lemma 2.7.2. □

Exercise 2.7.4 Prove that, in Dudley's theorem, $C = \sqrt{2}$ works. □

3 The Standard Brownian Sheet

This section is concerned with the study of continuity properties of the Brownian sheet of Section 1.5. Recall that $B = (B_t;\ t \in \mathbb{R}_+^N)$ is a Brownian sheet if it is a mean-zero Gaussian process indexed by $\mathbb{R}_+^N$ that has the following covariance function:

$$\Sigma(s,t) = \mathbb{E}[B_s B_t] = \prod_{\ell=1}^{N} \left(s^{(\ell)} \wedge t^{(\ell)}\right), \qquad s,t \in \mathbb{R}_+^N.$$

3.1 Entropy Estimate

Let $B = (B_t;\ t \in \mathbb{R}_+^N)$ denote the N-parameter Brownian sheet. Since it is a Gaussian process, by the discussion of Section 2.7 there is a natural metric that B defines on its parameter space. This is given by

$$d(s,t) = \sqrt{\mathbb{E}\{|B_s - B_t|^2\}} = \sqrt{\mathbb{E}\{(B_s)^2 + (B_t)^2 - 2 B_s B_t\}}$$
$$= \left\{ \prod_{i=1}^{N} s^{(i)} + \prod_{j=1}^{N} t^{(j)} - 2 \prod_{k=1}^{N} \left(s^{(k)} \wedge t^{(k)}\right) \right\}^{\frac{1}{2}}.$$

Recall that when $N = 1$, B is called Brownian motion. In this case,

$$d(s,t) = \sqrt{s + t - 2st} = \sqrt{|s - t|}.$$

However, such nice expressions do not exist when $N > 1$. On the other hand, as far as the continuity properties of B are concerned, all that matters is the behavior of $d(s,t)$ when s and t are close. In other words, we need to know how fast $d(s,t)$ goes to 0 as $s \to t$. In order to do this, define the **symmetric difference** $A \triangle B$ of any two sets $A, B \in \mathbb{R}^N$ by

$$A \triangle B = \left(A \cap B^\complement\right) \cup \left(A^\complement \cap B\right).$$

Lemma 3.1.1 *For all $s, t \in \mathbb{R}_+^N$, $d(s,t) = \text{Leb}([0,t] \triangle [0,s])$. Moreover,*

$$A\sqrt{|s - t|} \leq d(s,t) \leq A'\sqrt{|s - t|},$$

where

$$A = \max_{1 \leq k \leq N} \prod_{\substack{j=1 \\ j \neq k}}^{N-1} \sqrt{t^{(j)} \wedge s^{(j)}} \quad \text{and} \quad A' = \max_{1 \leq k \leq N} \prod_{\substack{j=1 \\ j \neq k}}^{N-1} \sqrt{t^{(j)} \vee s^{(j)}}.$$

Proof One can use the explicit expression for $d(s,t)$ and messy calculations to arrive at this estimate. However, there is a conceptual method that avoids many of the algebraic pitfalls.

Recall from Section 1.4 the isonormal process $W = (W(h); \, h \in L^2(\mathbb{R}^N))$. By Čentsov's representation (Theorem 1.5.1), B can be written as $B_t = W(\mathbf{1}_{[0,t]})$; thus, $B_t - B_s = W(\mathbf{1}_{[0,t]}) - W(\mathbf{1}_{[0,s]})$. On the other hand, W enjoys linearity properties (Theorem 1.4.1). Hence, $B_t - B_s = W(\mathbf{1}_{[0,t]} - \mathbf{1}_{[0,s]})$, a.s. A little thought shows that

$$(\mathbf{1}_{[0,t]} - \mathbf{1}_{[0,s]})^2 = \mathbf{1}_{[0,t] \triangle [0,s]}.$$

Therefore, $B_t - B_s = W(\mathbf{1}_{[0,t] \triangle [0,s]})$, a.s. (why?), and Theorem 1.4.1 implies

$$\mathbb{E}\{|B_t - B_s|^2\} = \mathrm{Leb}([0,t] \triangle [0,s]),$$

as desired. To estimate the above, note that $r \in [0,t] \triangle [0,s]$ if and only if there exists some integer $1 \le k \le N$ such that $s^{(k)} \wedge t^{(k)} < r^{(k)} < s^{(k)} \vee t^{(k)}$. In particular,

$$[0,t] \triangle [0,s] \subset \bigcup_{k=1}^{N} \left\{ r \in [0, s \curlyvee t] : \, s^{(k)} \wedge t^{(k)} < r^{(k)} < s^{(k)} \vee t^{(k)} \right\},$$

where $s \curlyvee t \in \mathbb{R}^N$ is defined by $(s \curlyvee t)^{(j)} = s^{(j)} \vee t^{(j)}$, $1 \le j \le N$. The upper bound on $d(s,t)$ readily follows from this. Similarly, for any $1 \le k \le N$,

$$[0,t] \triangle [0,s] \supset \left\{ r \in [0, s \curlywedge t] : \, s^{(k)} \wedge t^{(k)} < r^{(k)} < s^{(k)} \vee t^{(k)} \right\},$$

which proves the corresponding lower bound. $\square$

Combining the above with Corollary 2.1.1, we obtain the following estimate on metric entropy after some algebraic manipulations.

Corollary 3.1.1 *For any* $a \prec b$ *both in* $\mathbb{R}_+^N$, *there exist finite positive constants* $C, r_0 > 0$ *such that for all* $0 < r < r_0$,

$$D(r; [a,b], d) \le C r^{-2N}.$$

Exercise 3.1.1 Prove Corollary 3.1.1. $\square$

3.2 Modulus of Continuity

Armed with the entropy estimate of Section 3.1 (see Corollary 3.1.1), we now estimate the modulus of continuity of the Brownian sheet.

Theorem 3.2.1 *An N-parameter Brownian sheet $B = (B_t; t \in \mathbb{R}_+^N)$ has a continuous modification $\beta = (\beta_t; t \in \mathbb{R}_+^N)$. Moreover, β can be constructed so that for any $a \prec b$ both in $\mathbb{R}_+^N$, the following holds:*

$$\limsup_{\delta \to 0^+} \sup_{\substack{s,t \in [a,b]:\\ |s-t| \le \delta}} \frac{|\beta_s - \beta_t|}{\sqrt{\delta \ln(1/\delta)}} \le 24N|b|^{\frac{N}{2}}.$$

In particular, β is Hölder continuous of any order $\gamma < \frac{1}{2}$.

From now on, *any* continuous Brownian sheet is referred to as the **standard Brownian sheet**, for clarity. Since a 1-parameter Brownian sheet is called Brownian motion, by a **standard Brownian motion** we mean a continuous Brownian motion.

Proof The asserted Hölder continuity is an immediate consequence of the bound on the modulus of continuity of β that we prove next.

By a patching argument, it suffices to show that for all $a \prec b$ both in $\mathbb{R}_+^N$, $(B_t; t \in [a,b])$ has a continuous modification $\beta = (\beta : t \in [a,b])$; see the proof of Corollary 2.3.1 for the details of a patching argument. Now apply Corollary 3.1.1 to deduce the existence of constants $c > 0$ and $r_0 > 0$ such that for all $0 < r < r_0$, $D(r; [a,b], d) \le cr^{-2N}$. Without loss of any generality, we can assume that $r_0 < 1$. Since metric entropy is nonincreasing,

$$\int_0^1 \sqrt{\ln D(r; [a,b], d)}\, dr$$
$$\le \int_0^{r_0} \sqrt{\ln c + 2N \ln(1/r)}\, dr + \int_{r_0}^1 \sqrt{\ln D(r_0; T, d)}\, dr,$$

which is finite. Dudley's theorem (Theorem 2.7.1) establishes the existence of a continuous modification β. To obtain the stated lim sup bound, we use Corollary 3.1.1 together with L'Hôpital's rule of elementary calculus to see that

$$\limsup_{\delta \to 0^+} \frac{\int_0^\delta \sqrt{\ln D(r; T, d)}\, dr + C\delta \sqrt{\ln \ln(1/\delta)}}{\delta \sqrt{2N \ln(1/\delta)}} \le 1,$$

where C is the constant in Dudley's theorem. Thus, the latter result implies that

$$\limsup_{\delta \to 0^+} \frac{\sup_{s,t \in [a,b]:\, d(s,t) \le \delta} |\beta_s - \beta_t|}{\delta \sqrt{\ln(1/\delta)}} \le 24\sqrt{2N}.$$

By Lemma 3.1.1, whenever $|s-t| \le \varepsilon = N^{-1}|b|^{-N}\delta^2$, then $d(s,t) \le \delta$. Thus,

$$\sup_{s,t \in [a,b]:|s-t| \le \varepsilon} |\beta_s - \beta_t| \le \sup_{s,t \in [a,b]: d(s,t) \le \delta} |\beta_s - \beta_t|.$$

The upper bound on the lim sup follows. □

The modulus of continuity of the Brownian sheet (Theorem 3.2.1) is sharp up to a constant:

Exercise 3.2.1 Suppose $N = 1$ and consider a standard Brownian motion B. Show that for any $0 \leq a \leq b$, a.s.,

$$\liminf_{\delta \to 0^+} \sup_{\substack{s,t \in [a,b]: \\ |s-t| \leq \delta}} \frac{|B_s - B_t|}{\sqrt{2\delta \ln(1/\delta)}} \geq 1.$$

This is due to P. Lévy.
(HINT: Fix $\varepsilon > 0$, define

$$\mathsf{E}_j = \{|B_{\frac{1}{n}(j+1)} - B_{\frac{1}{n}j}|^2 \leq (2+\varepsilon)\tfrac{1}{n}\ln(\tfrac{1}{n})\},$$

and show that $\lim_{n \to \infty} \mathbb{P}(\cap_{j=0}^n \mathsf{E}_j) = 0$.) □

Exercise 3.2.2 Let B denote the standard N-parameter Brownian sheet. Prove the following converse to Theorem 3.2.1:

$$\liminf_{\delta \to 0^+} \sup_{\substack{s,t \in [a,b]: \\ |s-t| \leq \delta}} \frac{|B_s - B_t|}{\sqrt{2\delta \ln(1/\delta)}} \geq 1,$$

almost surely. □

4 Supplementary Exercises

1. In the context of equation (1) of Section 2.7, verify that (T, d) is a pseudometric space.

2. Let $\mathcal{D}_n$ denote the collection of all dyadic cubes of side 2^{-n} in $[0,1]^N$. That is, $I \in \mathcal{D}_n$ if and only if I is of the form $I = \prod_{\ell=1}^N [j^{(\ell)}2^{-n}, (j^{(\ell)}+1)2^{-n}]$, where $j \in \mathbb{N}_0^N$ satisfies $0 \leq j^{(\ell)} \leq 2^n$.

   (i) If μ is any finite signed measure on $[0,1]^N$, prove that $\lim_{n \to \infty} \sum_{I \in \mathcal{D}_n} \{\mu(I)\}^2 = 0$.

   (ii) Let $\mathbb{W}$ denote a white noise on $\mathbb{R}_+^N$. Prove that a.s.,

   $$\lim_{n \to \infty} \sum_{I \in \mathcal{D}_n} \{\mathbb{W}(I)\}^2 = 1.$$

   Use the above to prove that white noise is a.s. not a σ-finite signed measure. This is essentially due to P. Lévy.
   (HINT: Compute the mean and variance.)

3. For this exercise you need to know some elementary functional analysis.

Let H denote a Hilbert space with inner product $\langle \bullet, \bullet \rangle$. We can then define the **isonormal process** $W = (W(h); h \in H)$ as a Gaussian process with mean function 0 and covariance function given as follows: For all $h, g \in H$, $\mathbb{E}[W(h)W(g)] = \langle h, g \rangle$.

(i) Use Kolmogorov's existence theorem (Theorem 2, Appendix A) to prove directly that W exists.

(ii) Given two linear subspaces H_1 and H_2 of H, check that $(W(h); h \in H_1)$ and $(W(h); h \in H_2)$ are independent if and only if H_1 and H_2 are orthogonal; i.e., for all $h_1 \in H_1$ and $h_2 \in H_2$, $\langle h_1, h_2 \rangle = 0$.

(iii) Suppose H has a countable orthonormal basis; i.e., there exist $\psi_1, \psi_2, \ldots \in H$ such that every $f \in H$ has the representation $f = \sum_{i=1}^{\infty} \psi_i \langle f, \psi_i \rangle$, where the convergence takes place in H. Then, there exist independent standard Gaussian variables $g_1, g_2, \ldots$ such that for any $h \in H$, $W(h) = \sum_{i=1}^{\infty} g_i \psi_i$, where the convergence holds in $L^2(\mathbb{P})$.

4. If $d(s,t) = |s-t|$ for $s, t \in [-1,1]^N$, verify that $\lim_{\varepsilon \to 0} \varepsilon^N D(\varepsilon; [-1,1]^N, d) = 1$.

5. Consider a totally bounded metric space (T, d). Prove that a function $f : T \to \mathbb{R}$ is Hölder continuous of order $q > 0$ if and only if there exist two finite positive constants C and ε_0 such that for all $s, t \in T$ with $d(s,t) \leq \varepsilon_0$, $|f(s) - f(t)| \leq C|s-t|^q$. Prove that unless f is a constant, $q \leq 1$.

6. Suppose $X = (X_t; t \in T)$ is a Gaussian process. Prove that for all $p > 0$ and all $s \in T$, $\mathbb{E}\{|X_s|^p\} = k_p \{\mathbb{E}[|X_s|^2]\}^{p/2}$, where $k_p = 2^{p/2} \pi^{-1/2} \Gamma(\frac{1}{2}(p+1))$. Conclude that for all $s, t \in T$,

$$\mathbb{E}\{|X_s - X_t|^p\} = k_p [\mathbb{E}\{|X_s - X_t|^2\}]^{p/2}.$$

This is an $L^p(\mathbb{P})$ variant of Exercise 2.7.1.

7. Demonstrate the following variant of Exercise 2.5.1: Under the conditions of Exercise 2.5.1, for all $\tau \succ 0$ in $\mathbb{R}_+^N$, all $0 < q < p$, and for every $Q \in {]}0, p^{-1}(\gamma - N)[$,

$$\mathbb{E}\left(\sup_{s,t \in [0,\tau]} \frac{|X_s - X_t|^q}{|s-t|^{Qp}} \right) < +\infty.$$

8. (Hard) Let X and Y be two d-dimensional Gaussian random vectors such that for all $1 \leq i, j \leq d$, $\mathbb{E}[X^{(i)}] = \mathbb{E}[Y^{(j)}] = 0$, $\mathbb{E}[(X^{(i)})^2] = \mathbb{E}[(Y^{(i)})^2]$, and $\mathbb{E}[X^{(i)} X^{(j)}] \leq \mathbb{E}[Y^{(i)} Y^{(j)}]$. We intend to demonstrate the following comparison inequality of Slepian (1962) for all $\lambda > 0$:

$$\mathbb{P}\left(\max_{1 \leq i \leq d} X^{(i)} \geq \lambda \right) \geq \mathbb{P}\left(\max_{1 \leq i \leq d} Y^{(i)} \geq \lambda \right). \tag{1}$$

This will be done in a number of steps.

(i) Let $F[f]$ denote the Fourier transform of f. That is, for all $f : \mathbb{R}^d \to \mathbb{R}$,
$$F[f](\xi) = \int e^{i\xi \cdot x} f(x)\,dx, \qquad \xi \in \mathbb{R}^d,$$
if, say, $f \in L^1(\mathbb{R}^d)$. Show that whenever f is continuously differentiable and vanishes outside a compact set,
$$F\left[\frac{\partial f}{\partial x^{(i)}}\right](\xi) = -i\xi^{(i)} F[f](\xi),$$
where $\xi \in \mathbb{R}^d$ and $1 \le i \le d$. Conclude that if f is twice continuously differentiable and vanishes outside a compact set,
$$F\left[\frac{\partial^2 f}{\partial x^{(i)} \partial x^{(j)}}\right](\xi) = -\xi^{(i)} \xi^{(j)} F[f](\xi).$$

(ii) For all $t \in [0,1]$, define $Z_t = \sqrt{1-t}\, X + \sqrt{t}\, Y$. Verify that $Z = (Z_t;\ t \in [0,1])$ is an $\mathbb{R}^d$-valued Gaussian process indexed by $[0,1]$. That is, for all finite $T \subset [0,1]$, $(Z_t;\ t \in T)$ is a Gaussian random vector. Moreover, $\alpha_{i,j}(t) = \mathbb{E}[Z_t^{(i)} Z_t^{(j)}]$ satisfies the differential inequality $\alpha'_{i,j}(t) \ge 0$.

(iii) Let φ_t denote the probability density of Z_t (with respect to Lebesgue's measure). Show that
$$F[\varphi_t](\xi) = \exp\left\{-\frac{1}{2} \sum_{i=1}^d \sum_{j=1}^d \xi^{(i)} \xi^{(j)} \alpha_{i,j}(t)\right\}.$$

(iv) Use the inversion theorem for Fourier transforms, combined with (iii), to deduce that
$$\frac{d\varphi_t}{dt}(x) = \frac{1}{2} \sum_{i=1}^d \sum_{j=1}^d \alpha'_{i,j}(t) \frac{\partial^2 \varphi_t}{\partial x^{(i)} \partial x^{(j)}}(x).$$

(v) Prove that
$$\frac{\partial^2 \varphi_t}{\{\partial x^{(i)}\}^2}(x) = |x^{(i)}|^2 \varphi_t(x) \ge 0.$$
Conclude that
$$\frac{d\varphi_t}{dt}(x) \ge \frac{1}{2} \sum_{\substack{i,j=1 \\ i \ne j}}^d \alpha'_{i,j}(t) \frac{\partial^2 \varphi_t}{\partial x^{(i)} \partial x^{(j)}}(x).$$

(vi) Prove that
$$\frac{d}{dt} \mathbb{P}(\max_{1 \le i \le d} Z_t^{(i)} \le \lambda) = \int_{-\infty}^{\lambda} \cdots \int_{-\infty}^{\lambda} \frac{d\varphi_t}{dt}(x)\,dx$$
$$\ge \frac{1}{2} \sum_{\substack{i,j=1 \\ i \ne j}}^d \alpha'_{i,j}(t) \int_{-\infty}^{\lambda} \cdots \int_{-\infty}^{\lambda} \frac{\partial^2 \varphi_t}{\partial x^{(i)} \partial x^{(j)}}(x)\,dx.$$

Moreover, show that the above is nonnegative and derive equation (1) above.

(HINT: In (vi), to show the final positivity, show that the multidimensional integral is a probability.)

9. Apply Slepian's inequality (Supplementary Exercise 8) to derive the following improvement of Exercise 3.2.2: If B denotes the N-parameter standard Brownian sheet, then with probability one,

$$\liminf_{\delta \to 0^+} \sup_{\substack{s,t \in [a,b]:\\ |s-t| \leq \delta}} \frac{|B_s - B_t|}{\sqrt{2N\delta \ln(1/\delta)}} \geq 1.$$

This is a part of Orey and Pruitt (1973, Theorem 2.4); see also Esquível (1996). (HINT: The increments of B have positive correlations. Compare them to a Gaussian process whose increments are independent.)

10. Let $(\xi_{i,j};\ i, j \geq 0)$ denote i.i.d. standard Gaussian random variables and recall from Chapter 2 the N-dimensional Haar functions $(h_{k,j}^N;\ k \geq 0, j \in \Gamma^N(k))$.

(i) If $g_1, g_2, \ldots$ are i.i.d. standard Gaussian random variables, there exists a finite $C > 0$ such that for all $n \geq 1$, $\mathbb{E}\{\max_{j \leq n} |g_j|\} \leq C\sqrt{\ln_+ n}$.

(ii) Show that with probability one, $\sum_{k=0}^\infty \sum_{j \in \Gamma^N(k)} \xi_{k,j} h_{k,j}^N(t)$ converges uniformly in $t \in [0,1]^N$.

(iii) Check that $t \mapsto B_t$ is a Brownian sheet with t restricted to $[0,1]^N$, where $B_t = \sum_{k=0}^\infty \sum_{j \in \Gamma^N(k)} \xi_{k,j} h_{k,j}^N(t)$.

When $N = 1$, this series expansion of the Brownian sheet is due to P. Lévy and simplifies an older one, due to N. Wiener, who used it to prove the existence of Brownian motion; this and related results can be found in Adler (1990, Section 3.3, Chapter 3) and Itô and McKean (1974, Section 1.5, Chapter 1).
(HINT: For part (i), integrate the inequality $\mathbb{P}(\max_{j \leq n} |g_j| \geq \lambda) \leq n\mathbb{P}(|g_1| \geq \lambda)$. For part (ii), have a look at Supplementary Exercise 1, Chapter 2.)

5 Notes on Chapter 5

Section 1 This section's construction of Gaussian processes and variables only scratches the surface of this theory. See Janson (1997) for a very rich, as well as modern, account.

The isonormal process makes rigorous objects (such as white noise) that have been lurking in the physics and engineering literature for a long time, and was developed in Segal (1954); an excellent resource for this and its relation to abstract Gaussian measures is Dudley (1973).

The Brownian sheet and related multiparameter processes were first discovered in the theory of mathematical statistics; see Kitagawa (1951). Modern accounts can now be found in (Adler 1990; Dudley 1984; Ledoux and Talagrand 1991; Gänssler 1983; Pollard 1984); see also Adler and Pyke (1997) for a related set of results. An interesting recent appearance of the Brownian sheet in statistical mechanics is presented in Kuroda and Manaka (1987, 1988, 1998).

Section 2 In its present form, the notion of modifications and separability is due to J. L. Doob. The classic text Doob (1990, Chapter II) contains a very thorough account as well as a wealth of references to older literature.

As mentioned within the text, metric entropy is due to R. M. Dudley; what we call Kolmogorov entropy is sometimes also called ε-entropy. This was motivated by a related notion of a λ-capacity from information theory. The latter was discovered in Shannon (1948); see also Shannon and Weaver (1949).

Example 2 of this section is a special case of more general estimates of Kolmogorov; see Dudley (1984, Theorem 7.1.1). Kolmogorov's infinite-dimensional calculations were preceded by the finite-dimensional computations of Muroga (1949).

The continuity theorems of this chapter are essentially borrowed from Dudley (1973); they are the sharp form of classical results of A. N. Kolmogorov, as well as those of Garsia et al. (1971). The metric entropy integral condition has since been replaced by a "majoring measure" condition due to X. Fernique and, later, to M. Talagrand. This level of refinement is not used in this book, although it is well within reach of the methods here; see (Adler 1990; Ledoux 1996; Ledoux and Talagrand 1991) for more detailed information.

Section 3 Theorem 3.2.1 can be sharpened. In fact, the lim sup there equals the lim inf, and the lower bound in Supplementary Exercise 9 is sharp. See Orey and Pruitt (1973, Theorem 2.4) and Csörgő and Révész (1978, 1981) for related results, as well as further refinements.

6
Limit Theorems

Suppose $X_1, X_2, \ldots$ are independent, identically distributed $\mathbb{R}^d$-valued random variables and consider, as usual, the random walk $k \mapsto S_k = X_1 + \cdots + X_k$. If $\mathbb{E}[X_1] = 0$ and $\mathbb{E}[X_1^2] < +\infty$, the classical central limit theory on $\mathbb{R}^d$ implies that as $n \to \infty$, the random vector $n^{-1/2} S_n$ converges in distribution to an $\mathbb{R}^d$-valued Gaussian random vector. It turns out that much more is true; namely, as $n \to \infty$, the distribution of the process $t \mapsto n^{-1/2} S_{\lfloor nt \rfloor}$ starts to approximate that of a suitable Gaussian process. This approximation is good enough to show that various functionals of the random walk path converge in distribution to those of the limiting Gaussian process.

To do any of this, we first view the process S as a random element of a certain space of functions. Thus, we begin with general facts about random variables that take their values in a topological space.

1 Random Variables

This chapter starts with some general results on random variables that take values in a topological space T. Throughout, $\mathcal{B}(T)$ will denote the Borel σ-field on T, which is the σ-field generated by all open subsets of T. Oftentimes, we assume further that T is a metric space; i.e., we are really assuming the existence of a metric d on T that is compatible with the topology of T. The astute reader will notice that much of the material of this section is in parallel with the theory of $\mathbb{R}^d$-valued random variables.

1.1 Definitions

Let T be a topological space, endowed with its Borel field $\mathcal{B}(T)$. A T-valued **random variable** X on the probability space $(\Omega, \mathcal{G}, \mathbb{P})$ is just a measurable map $X : \Omega \to T$. In other words, X is a T-valued random variable if for all $E \in \mathcal{B}(T)$, $(\omega \in \Omega : X(\omega) \in E) \in \mathcal{G}$. The latter is an **event** and will be written as $(X \in E)$ for the sake of brevity. For example, using this notation, we need only write $\mathbb{P}(X \in E)$ instead of the more cumbersome $\mathbb{P}(\omega \in \Omega : X(\omega) \in E)$.

The calculus of T-valued random variables follows the same rules as that of $\mathbb{R}^d$-valued random variables, and is derived by similar means. Therefore, many of the results of this section are left as exercises that the first-time reader is encouraged to attempt.

Theorem 1.1.1 *Let T_1 and T_2 be topological spaces with Borel fields $\mathcal{B}(T_1)$ and $\mathcal{B}(T_2)$, respectively. If X is a T_1-valued random variable and $f : T_1 \to T_2$ is measurable, then $f(X)$ is a T_2-valued random variable.*

***Exercise* 1.1.1** Prove Theorem 1.1.1. $\square$

The above has many important consequences, an example of which is the following.

Corollary 1.1.1 *Suppose T_1, T_2, and T_3 are topological spaces. If X_i is a T_i-valued random variable ($i = 1, 2$), and if $f : T_1 \times T_2 \to T_3$ is product measurable, then $f(X_1, X_2)$ is a T_3-valued random variable.*

***Exercise* 1.1.2** Prove Corollary 1.1.1. $\square$

In particular, when $X_1, \ldots, X_n$ are T-valued random variables, and given any function $f : T^n \to T$, $Y = f(X_1, \ldots, X_n)$ is a random variable. As an example, consider a linear topological space T over $\mathbb{R}$. If $X_1, \ldots, X_n$ are T-valued random variables, then so is $S_n = \frac{1}{n} \sum_{j=1}^n X_j$. This is a familiar object and motivates the need for the following theorem.

Theorem 1.1.2 *Suppose T is a complete metric space and $X_1, X_2, \ldots$ is a.s. a Cauchy sequence of T-valued random variables. Then, there exists a T-valued random variable X such that $X = \lim_{n \to \infty} X_n$, a.s.*

Proof Since T is complete and $(X_n; n \geq 1)$ is a.s. a Cauchy sequence, then for $\mathbb{P}$-almost all $\omega \in \Omega$, $\lim_{n \to \infty} X_n(\omega)$ exists. Thus, there exists an event $\Omega_0 \in \mathcal{G}$ such that $\mathbb{P}(\Omega_0) = 1$ and for all $\omega \in \Omega_0$, $X(\omega) = \lim_{n \to \infty} X_n(\omega)$ exists. Fix some arbitrary $t \in T$ and define $X(\omega) = t$ for all $\omega \in \Omega \setminus \Omega_0$, so that $X(\omega)$ is now defined for all $\omega \in \Omega$. Suppose that $G \subset T$ is open and that for some $\omega \in \Omega_0$, $X(\omega) \in G$. Then, there exists $n_0 \geq 1$ such that for all $n \geq n_0$, $X_n(\omega) \in G$. Similarly, the converse also holds. In summary,

$$(X \in G) \cap \Omega_0 = \bigcup_{n_0=1}^{\infty} \bigcap_{n=n_0}^{\infty} (X_n \in G) \cap \Omega_0. \tag{1}$$

If $t \in G$, $(X \in G) = \Omega_0^{\complement} \cup (X \in G) \cap \Omega_0$. Otherwise, $(X \in G) = (X \in G) \cap \Omega_0$. In any case, equation (1) shows that $(X \in G) \in \mathcal{G}$, for all open sets $G \subset T$. Since $\mathcal{B}(T)$ is generated by open sets, X is a T-valued random variable. □

1.2 Distributions

Suppose T is a topological space with its Borel field $\mathcal{B}(T)$. To every T-valued random variable X we can associate a **distribution** $\mathbb{P} \circ X^{-1}$ defined by

$$\mathbb{P} \circ X^{-1}(E) = \mathbb{P}(X \in E), \qquad E \in \mathcal{B}(T).$$

The following characterizes an important property of the distribution of X.

Lemma 1.2.1 *If X is a T-valued random variable, then $\mathbb{P} \circ X^{-1}$ is a probability measure on $(T, \mathcal{B}(T))$.*

Exercise 1.2.1 Prove Lemma 1.2.1. □

Thus, for every T-valued random variable X, there exists a probability measure $\mathbb{P} \circ X^{-1}$; the converse is also true.

Theorem 1.2.1 *Let μ be a probability measure on a topological space T. There exists a probability space $(\Omega, \mathcal{G}, \mathbb{P})$ on which there is a T-valued random variable X whose distribution, $\mathbb{P} \circ X^{-1}$, is μ.*

Proof Let $(\Omega, \mathcal{G}, \mathbb{P}) = (T, \mathcal{B}(T), \mu)$, and define X to be the coordinate function on T; that is, $X(\omega) = \omega$, for all $\omega \in T$. It follows that for all $E \in \mathcal{B}(T)$, $\mathbb{P} \circ X^{-1}(E) = \mathbb{P}(X \in E) = \mu(E)$, as desired. □

Let us conclude this subsection with a change of variables formula.

Theorem 1.2.2 *Suppose X is a T-valued random variable where T is a topological space. Then, for all bounded, continuous functions $f : T \to \mathbb{R}$,*

$$\mathbb{E}[f(X)] = \int_T f(\omega) \, \mathbb{P} \circ X^{-1}(d\omega).$$

Exercise 1.2.2 Prove Theorem 1.2.2.
(HINT: Consider, first, functions of the form $f(x) = \mathbb{1}_A(x)$.) □

1.3 Uniqueness

Suppose $T = \mathbb{R}$ and let X be an $\mathbb{R}$-valued random variable. In order to "know" the measure $\mathbb{P} \circ X^{-1}$, all we need are the probabilities $\mathbb{P} \circ X^{-1}(E)$ for an appropriately large class of Borel sets E. For example, it is sufficient to know $\mathbb{P} \circ X^{-1}(E)$ for all E of the form $]-\infty, x]$ where $x \in \mathbb{R}$. In this case, $\mathbb{P} \circ X^{-1}(]-\infty, x]) = \mathbb{P}(X \leq x)$ is the familiar cumulative distribution function of X. More generally, when $T = \mathbb{R}^d$, we only need to know the "distribution function" $x \mapsto \mathbb{P}(X \preccurlyeq x)$ (why?).

When T is a general topological space, such notions do not easily extend themselves. In order to generalize to a topological T, first notice that when $x \in T = \mathbb{R}^d$, $\mathbb{P}(X \preccurlyeq x) = \mathbb{E}[\mathbf{1}_{]-\infty,x]}(X)]$, where $]-\infty, x] = \prod_{\ell=1}^{d}]-\infty, x^{(\ell)}]$. Since one can approximate $\mathbf{1}_{]-\infty,x]}$ by a bounded, continuous function arbitrarily well, it follows that when $T = \mathbb{R}^d$, knowing the collection

$$\{\mathbb{E}[f(X)]; \ f : \mathbb{R}^d \to \mathbb{R} \text{ bounded, continuous}\}$$

amounts to knowing the entire distribution of X. One attractive feature of this formulation is that since continuity is a topological phenomenon, one can just as easily work on more general topological spaces. One way to state this is as follows: Let $C_b(T)$ denote the collection of all bounded, continuous functions $f : T \to \mathbb{R}$.

Theorem 1.3.1 *Consider probability measures $\mathbb{P}_1$ and $\mathbb{P}_2$ on $(T, \mathcal{B}(T))$, where T is a metric space. If $\int_T f(\omega) \, \mathbb{P}_1(d\omega) = \int_T f(\omega) \, \mathbb{P}_2(d\omega)$ for all $f \in C_b(T)$, then $\mathbb{P}_1 = \mathbb{P}_2$.*

Before proving Theorem 1.3.1, we note the following important result.

Corollary 1.3.1 *Let T be a topological space, and consider any T-valued random variable X. Then, the collection $\{\mathbb{E}[f(X)]; f \in C_b(T)\}$ uniquely determines the distribution of X.*

In other words, as f varies over $C_b(T)$, $f \mapsto \mathbb{E}[f(X)]$ plays the role of the cumulative distribution function of X.

The key technical step in the proof of Theorem 1.3.1 is Urysohn's lemma of general topology, which we state here in the context of metric spaces; see Munkres (1975, Chapter 4, Section 4-3) for this, and for further extensions.

Lemma 1.3.1 (Urysohn's Lemma) *If A and B are two disjoint closed subsets of a metric space T, there exists a continuous function $f : T \to [0, 1]$ such that for all $\omega \in A$, $f(\omega) = 1$, and for all $\omega \in B$, $f(\omega) = 0$.*

In words, disjoint closed subsets of a metric space can be **separated** by continuous functions.

Proof of Theorem 1.3.1 It suffices to show that for all closed sets $F \subset T$,

$$\int_T \mathbf{1}_F(\omega)\, \mathbb{P}_1(d\omega) = \int_T \mathbf{1}_F(\omega)\, \mathbb{P}_2(d\omega).$$

Had $\mathbf{1}_F$ been continuous, we would be done. Since this is clearly not the case, we approximate $\mathbf{1}_F$ by a bounded continuous function.

For any $\varepsilon > 0$, define $F_\varepsilon = \{x \in T : d(\{x\}, F) < \varepsilon\}$, where the distance $d(A, B)$ between any two sets A and B is defined as

$$d(A, B) = \inf\{d(x, y) : x \in A,\ y \in B\}. \tag{1}$$

Since F is closed, $x \mapsto d(\{x\}, F)$ is continuous; thus, $F_\varepsilon^\complement$ is closed. By Lemma 1.3.1, we can find a continuous function $f : T \to [0, 1]$ such that $f = 1$ on F and $f = 0$ on $F_\varepsilon^\complement$. Since $0 \leq f \leq 1$, we have shown that $\mathbf{1}_F \leq f \leq \mathbf{1}_{F_\varepsilon}$. Thus,

$$\mathbb{P}_1(F_\varepsilon) = \int_T \mathbf{1}_{F_\varepsilon}(\omega)\, \mathbb{P}_1(d\omega) \geq \int_T f(\omega)\, \mathbb{P}_1(d\omega) \geq \mathbb{P}_2(F).$$

On the other hand, if $0 < \varepsilon_1 \leq \varepsilon_2$, then $F_{\varepsilon_1} \subset F_{\varepsilon_2}$ and $\cap_{\varepsilon > 0} F_\varepsilon = F$. Thus,

$$\mathbb{P}_1(F) = \lim_{\varepsilon \to 0} \mathbb{P}_1(F_\varepsilon) \geq \mathbb{P}_2(F).$$

Reversing the roles of $\mathbb{P}_1$ and $\mathbb{P}_2$, we obtain the result. $\square$

2 Weak Convergence

Let $(\mu_n;\ 1 \leq n \leq \infty)$ denote a sequence of probability measures on a topological space T. We say that μ_n **converges weakly** to μ_∞ if for all bounded, continuous functions $f : T \to \mathbb{R}$,

$$\lim_{n \to \infty} \int_T f(\omega)\, \mu_n(d\omega) = \int_T f(\omega)\, \mu_\infty(d\omega),$$

and write this as $\mu_n \Longrightarrow \mu_\infty$. Sometimes, when μ_n is the distribution of a T-valued random variable X_n ($1 \leq n \leq \infty$), we may say, instead, that X_n converges weakly to X_∞, and write this as $X_n \Longrightarrow X_\infty$. Note that we do not require the X_n's to be defined on the same probability space.

Suppose that for all $1 \leq n \leq \infty$, X_n is defined on a probability space $(\Omega_n, \mathcal{F}_n, \mathbb{P}_n)$. By Theorem 1.2.2, $X_n \Longrightarrow X_\infty$ if and only if for all continuous, bounded functions $f : T \to \mathbb{R}$, $\lim_{n \to \infty} \mathbb{E}_n[f(X_n)] = \mathbb{E}_\infty[f(X_\infty)]$, where $\mathbb{E}_n$ denotes the expectation operator corresponding to $\mathbb{P}_n$ ($1 \leq n \leq \infty$). Since this is distracting, we often abuse the notation by writing $\lim_{n \to \infty} \mathbb{E}[f(X_n)] = \mathbb{E}[f(X)]$; it should be clear which expectation applies to what random variable.

2.1 The Portmanteau Theorem

The following result characterizes weak convergence of probability measures.

Theorem 2.1.1 (The Portmanteau Theorem) *Suppose $(\mu_n;\ 1 \leq n \leq \infty)$ denotes a collection of probability measures on a topological space T. The following are equivalent:*

(i) $\mu_n \Longrightarrow \mu_\infty$;

(ii) for all closed sets $F \subset T$, $\limsup_{n \to \infty} \mu_n(F) \leq \mu_\infty(F)$;

(iii) for all open sets $G \subset T$, $\liminf_{n \to \infty} \mu_n(G) \geq \mu_\infty(G)$; and

(iv) for all measurable $A \subset T$ such that $\mu_\infty(\partial A) = 0$, $\lim_{n \to \infty} \mu_n(A) = \mu_\infty(A)$.

Remarks (a) ∂A is the topological boundary of A; it is defined as $\partial A = \bar{A} \setminus A^\circ$, where $\bar{A}$ and A° are the closure and interior of A, respectively.

(b) At least when $T = \mathbb{R}^d$, Theorem 2.1.1 should be a familiar result.

Proof If F is closed, $F^\complement$ is open and vice versa. Since $\mu(U) = 1 - \mu(U^\complement)$, for all $U \in \mathcal{B}(T)$, the equivalence of (ii) and (iii) follows. Now we show that (ii) and (iii) together imply (iv).

Using the notation of the above remark (a), $\mu_n(A^\circ) \leq \mu_n(A) \leq \mu_n(\bar{A})$. Since A° is open and $\bar{A}$ is closed, (ii) and (iii) together show that

$$\mu_\infty(A^\circ) \leq \liminf_{n \to \infty} \mu_n(A) \leq \limsup_{n \to \infty} \mu_n(A) \leq \mu_\infty(\bar{A}).$$

Since $\mu_\infty(\bar{A}) = \mu_\infty(A^\circ) + \mu_\infty(\partial A) = \mu_\infty(A^\circ)$, property (iv) follows. Next, we show that $(iv) \Rightarrow (iii)$.

In the notation of equation (1) of Section 1.3, for any $F \subset T$ and for all $\varepsilon > 0$, define

$$F_\varepsilon = \{y \in T : d(\{y\}, F) < \varepsilon\}. \tag{1}$$

This is an open set for any $\varepsilon > 0$; cf. the proof of Theorem 1.3.1. Since $\mu_1, \mu_2, \ldots, \mu_\infty$ can have no more than a denumerable number of atoms, we can choose a sequence $\varepsilon_k \to 0$ such that for all $1 \leq n \leq \infty$ and all $k \geq 1$, $\mu_n(\partial F_{\varepsilon_k}) = 0$. By (iv), $\lim_{n \to \infty} \mu_n(F_{\varepsilon_k}) = \lim_{n \to \infty} \mu_n(\overline{F_{\varepsilon_k}}) = \mu_\infty(\overline{F_{\varepsilon_k}})$. Thus, (iv) implies that for any measurable $F \subset T$,

$$\limsup_{n \to \infty} \mu_n(F) \leq \lim_{n \to \infty} \mu_n(\overline{F_{\varepsilon_k}}) = \mu_\infty(\overline{F_{\varepsilon_k}}).$$

If, in addition, F is closed, then $\cap_k \overline{F_{\varepsilon_k}} = F$, and we obtain $(iv) \Rightarrow (ii)$ from the above. Finally, we show that $(i) \Rightarrow (ii)$.

If F is a closed subset of T, by Urysohn's lemma (Lemma 1.3.1), we can find a continuous function $f: T \to [0,1]$ such that $f = 1$ on F and $f = 0$ on $F_\varepsilon^\complement$, where F_ε is defined in equation (1) above. Since $f \geq \mathbf{1}_F$, $\mu_n(F) \leq \int_T f(\omega)\, \mu_n(d\omega)$. By (i), as $n \to \infty$, this converges to $\int_T f(\omega)\mu_\infty(d\omega) \leq \mu_\infty(F_\varepsilon)$, thanks to the inequality $f \leq \mathbf{1}_{F_\varepsilon}$. This proves $(i) \Rightarrow (ii)$, and it remains to verify the converse.

Fix any continuous function $f: T \to \mathbb{R}$ such that for some integer $k > 0$, $-k \leq f(\omega) \leq k$, for all $\omega \in T$. For all real numbers $m > 1$,

$$\int_T f(\omega)\, \mu_n(d\omega) \leq \sum_{j \in \mathbb{Z}: |j| \leq km} \frac{j}{m} \mu_n\left(\omega \in T: \frac{j-1}{m} \leq f(\omega) \leq \frac{j}{m}\right).$$

On the other hand, the collection $\mu_1^{-1} \circ f, \mu_2^{-1} \circ f, \ldots, \mu_\infty^{-1} \circ f$ has at most a denumerable number of atoms. Thus, we can find $m_k \to \infty$ such that

$$\mu_n^{-1} \circ f\left(\left\{\frac{j}{m_k}\right\}\right) = \mu_n\left(\omega \in T: f(\omega) = \frac{j}{m_k}\right) = 0,$$

for all $j, k \geq 1$ and all $n = 1, 2, \ldots, \infty$. Since $(\omega \in T: \frac{j-1}{m} \leq f(\omega) \leq \frac{j}{m})$ is closed, then by (ii), and after applying the above display,

$$\limsup_{n \to \infty} \int_T f(\omega)\, \mu_n(d\omega) \leq \sum_{j \in \mathbb{Z}: |j| \leq km} \frac{j}{m_k} \mu_\infty\left(\omega \in T: \frac{j-1}{m_k} \leq f(\omega) \leq \frac{j}{m_k}\right)$$

$$\leq \int_T f(\omega)\, \mu_\infty(d\omega) + \frac{1}{m_k}.$$

Sending $m_k \to \infty$ demonstrates $(ii) \Rightarrow (i)$ and completes this proof. $\square$

Alternatively, one can write the portmanteau theorem in terms of random variables.

Theorem 2.1.2 *Suppose $(X_n; 1 \leq n \leq \infty)$ is a sequence of random variables that take their values in a topological space T. The following are equivalent:*

(i) $X_n \Longrightarrow X_\infty$;

(ii) for all closed sets $F \subset T$, $\limsup_{n \to \infty} \mathbb{P}(X_n \in F) \leq \mathbb{P}(X_\infty \in F)$;

(iii) for all open sets $G \subset T$, $\liminf_{n \to \infty} \mathbb{P}(X_n \in G) \geq \mathbb{P}(X_\infty \in G)$; and

(iv) for any measurable $A \subset T$ such that $\mathbb{P}(X_\infty \in \partial A) = 0$,

$$\lim_{n \to \infty} \mathbb{P}(X_n \in A) = \mathbb{P}(X_\infty \in A).$$

Henceforth, any reference to the portmanteau theorem will be to either Theorem 2.1.1 or 2.1.2, whichever naturally applies.[1]

***Exercise* 2.1.1** Check Theorem 2.1.2. □

***Exercise* 2.1.2** Given a collection of probability measures $(\mu_n; 1 \leq n \leq \infty)$ on a topological space T, extend the portmanteau theorem by showing that for all bounded upper semicontinuous functions $f : T \to \mathbb{R}$,

$$\liminf_{n \to \infty} \int_T f(\omega)\, \mu_n(d\omega) \geq \int_T f(\omega)\, \mu_\infty(d\omega).$$

You may recall that f is upper semicontinuous if for all real α, $(\omega : f(\omega) < \alpha)$ is an open set in T. □

2.2 The Continuous Mapping Theorem

If T_1 and T_2 are two topological spaces, it is not difficult to see that whenever $(x_n; 1 \leq n \leq \infty)$ are elements of T_1 such that $\lim_{n \to \infty} x_n = x_\infty$, then for any continuous function $f : T_1 \to T_2$, $\lim_{n \to \infty} f(x_n) = f(x_\infty)$. The following is the weak convergence analogue of such a fact.

Theorem 2.2.1 (The Continuous Mapping Theorem) *If $X_n \Longrightarrow X_\infty$ and $f : T_1 \to T_2$ is continuous, then $f(X_n) \Longrightarrow f(X_\infty)$.*

***Exercise* 2.2.1** Prove the continuous mapping theorem.
(HINT: Consider Corollary 1.1.1.) □

2.3 Weak Convergence in Euclidean Space

We now specialize to the case $T = \mathbb{R}^d$, with which the reader is assumed to be familiar. In concordance with the rest of the book, $\mathbb{R}^d$ is metrized by $d(x, y) = |x - y|$, $(x, y \in \mathbb{R}^d)$, which provides us with the usual Euclidean topology on $\mathbb{R}^d$ and the corresponding Borel field. Any probability measure μ on $\mathbb{R}^d$ gives rise to a **cumulative distribution function** F defined by

$$F(t) = \mu(r \in \mathbb{R}^d : r \preccurlyeq t), \qquad t \in \mathbb{R}^d.$$

The following two results are elementary.

Theorem 2.3.1 *Two probability measures on $\mathbb{R}^d$ are equal if and only if their cumulative distribution functions are.*

---

[1] We reiterate that since the X_n's need not be defined on the same probability space, the notational remarks of Section 2 preceding Theorem 2.1.1 apply also to Theorem 2.1.2.

Theorem 2.3.2 *Suppose $\mu_1, \ldots, \mu_\infty$ is a possibly infinite sequence of probability measures on $\mathbb{R}^d$ whose cumulative distribution functions are $F_1, \ldots, F_\infty$, respectively. Then, $\mu_n \Longrightarrow \mu_\infty$ if and only if whenever F_∞ is continuous at a point $t \in \mathbb{R}^d$, we have $\lim_{n \to \infty} F_n(t) = F_\infty(t)$.*

***Exercise* 2.3.1** Prove Theorems 2.3.1 and 2.3.2. □

In other words, we may state the following corollary:

Corollary 2.3.1 *Weak convergence in $\mathbb{R}^d$ is the same as convergence in distribution in the classical sense.*

2.4 Tightness

When does a family of probability measures converge weakly? To answer this, we first need the notion of tightness.

Suppose T is a topological space and $(\mathbb{P}_\alpha;\ \alpha \in A)$ is a collection of probability measures defined on $\mathcal{B}(T)$, where A is some indexing set. We say that $(\mathbb{P}_\alpha;\ \alpha \in A)$ is **tight** if for all $\varepsilon \in\ ]0,1[$, there exists a compact set $\Gamma_\varepsilon \subset T$ such that
$$\sup_{\alpha \in A} \mathbb{P}_\alpha(\Gamma_\varepsilon^\complement) \leq \varepsilon.$$

Sometimes when $\mathbb{P}_\alpha$ denotes the distribution of a T-valued random variable X_α ($\alpha \in A$) we may refer to the family $(X_\alpha;\ \alpha \in A)$ as tight when $(\mathbb{P}_\alpha;\ \alpha \in A)$ is tight. Finally, we say that $\mathbb{P}$ is tight if the singleton $\{\mathbb{P}\}$ is tight.

***Exercise* 2.4.1** Suppose $(\mathbb{P}_\alpha;\ \alpha \in A)$ is a collection of probability measures on $\mathbb{R}$. Let F_α denote the cumulative distribution function of $\mathbb{P}_\alpha$ ($\alpha \in A$), and prove that $(\mathbb{P}_\alpha;\ \alpha \in A)$ is tight if and only if $\lim_{x \to \infty} F_\alpha(x) = 1$ and $\lim_{x \to -\infty} F_\alpha(x) = 0$, *uniformly* for all $\alpha \in A$. □

***Exercise* 2.4.2** Show that if $\mathbb{P}_\alpha$ is tight for each $\alpha \in A$, and if A is a finite set, then $(\mathbb{P}_\alpha;\ \alpha \in A)$ is tight. □

***Exercise* 2.4.3** Show that if T is σ-compact, i.e., a countable union of compact sets, any finite number of probability measures on T are tight. □

Theorem 2.4.1 *If T is a complete separable metric space, any finite number of probability measures on T are tight.*

Proof By Exercise 2.4.2, we need only show that a single probability measure $\mathbb{P}$ on T is tight. Since T is separable, for any integer $n \geq 1$, we can find a countable sequence of points $x_1, x_2, \ldots \in T$ such that $\cup_{i=1}^\infty B_{i,n}$ covers T, where $B_{i,n}$ denotes the ball (in T) of radius $\frac{1}{n}$ about x_i. Thus, for any probability measure $\mathbb{P}$ on T, $\lim_{m \to \infty} \mathbb{P}(\cup_{i=1}^m B_{i,n}) = 1$. In particular, for any $\varepsilon \in \,]0,1[$, there exists $m(n)$ so large that $\mathbb{P}(\cup_{i=1}^{m(n)} B_{i,n}) \geq 1 - \varepsilon 2^{-n}$, for all $n \geq 1$. An immediate consequence of this is that

$$\mathbb{P}\Big(\bigcup_{n=1}^{\infty} \bigcap_{i=1}^{m(n)} B_{i,n}^{\complement}\Big) \leq \sum_{n=1}^{\infty} \mathbb{P}\Big(\bigcap_{i=1}^{m(n)} B_{i,n}^{\complement}\Big) \leq \varepsilon \sum_{n=1}^{\infty} 2^{-n} \leq \varepsilon.$$

Note that $\Gamma_\varepsilon = \cap_{n=1}^{\infty} \cup_{i=1}^{m(n)} B_{i,n}$ is totally bounded. (Why? See Section 2.1 of Chapter 5.) By Theorem 2.1.1 of Chapter 5, Γ_ε is compact, and our proof is complete. □

Exercise 2.4.4 Let $(X_n; n \geq 1)$ denote a collection of $\mathbb{R}^k$-valued random vectors that are bounded in $L^p(\mathbb{P})$ for some $p > 0$; i.e., $\sup_n \mathbb{E}\{|X_n|^p\} < +\infty$. Prove that $(X_n; n \geq 1)$ is a tight family. □

The following gives the first indication of deep relationships between tightness and weak convergence.

Proposition 2.4.1 *Suppose $(\mathbb{P}_n; 1 \leq n \leq \infty)$ is a collection of probability measures on a complete, separable metric space T such that $\mathbb{P}_n \Longrightarrow \mathbb{P}_\infty$. Then, $(\mathbb{P}_n; n \geq 1)$ is a tight family.*

Proof We will show the slightly stronger fact that $(\mathbb{P}_n; 1 \leq n \leq \infty)$ is a tight family. For any $\varepsilon \in [0,1]$, we can choose a compact set Γ_ε such that $\mathbb{P}_\infty(\Gamma_\varepsilon^{\complement}) \leq \varepsilon$; cf. Theorem 2.4.1. Since $\Gamma_\varepsilon^{\complement}$ is closed, by the portmanteau theorem (Theorem 2.1.2) there exists $n_0(\varepsilon)$ such that for all $n \geq n_0(\varepsilon)$, $\mathbb{P}_n(\Gamma_\varepsilon^{\complement}) \leq 2\varepsilon$. On the other hand, the finite collection $\{\mathbb{P}_n; 1 \leq n \leq n_0(\varepsilon)\}$ is tight; cf. Exercise 2.4.2. Therefore, we can find another compact set K_ε such that for all $1 \leq n \leq n_0(\varepsilon)$, $\mathbb{P}_n(K_\varepsilon^{\complement}) \leq 2\varepsilon$. Concerning the compact set $\Lambda_\varepsilon = K_\varepsilon \cup \Gamma_\varepsilon$, we have shown that for all $1 \leq n \leq \infty$, $\mathbb{P}_n(\Lambda_\varepsilon^{\complement}) \leq 2\varepsilon$; this proves tightness. □

2.5 Prohorov's Theorem

Proposition 2.4.1 shows that on a complete metric space, weak convergence implies tightness. The converse also holds, but in general only along subsequences. This deep fact, discovered by Yu. V. Prohorov, is the subject of this subsection.

Theorem 2.5.1 (Prohorov's Theorem) *Suppose $(\mathbb{P}_n; n \geq 1)$ is a tight collection of probability measures on a complete, separable metric space (T, d). Then, there exists a subsequence n' and a probability measure $\mathbb{P}$ on T such that $\mathbb{P}_{n'} \Longrightarrow \mathbb{P}$.*

Remarks The following remarks are elaborated upon in standard references on weak convergence; cf. Billingsley (1968), for example.

(a) According to Supplementary Exercise 2, weak convergence can be topologized. As such, the above can be restated as follows: *On a complete, separable metric space, a tight collection of probability measures is sequentially compact.*

(b) It can be shown that separability is not needed in the above.

(c) As it is stated, this is only half of Prohorov's theorem. The other half asserts that on a complete, separable metric space (T, d), any sequentially compact family $(\mathbb{P}_n;\ n \geq 1)$ of probability measures on T is tight.

***Exercise* 2.5.1** Verify the following extension of Prohorov's theorem: *For any subsequence $(\mathbb{P}_{n'})$ of the tight family $(\mathbb{P}_n)$, there is a further subsequence $(\mathbb{P}_{n''})$ that converges weakly to a probability measure.* □

Proof of Theorem 2.5.1 in the Compact Case We merely prove Prohorov's theorem when T is compact. Supplementary Exercise 4 provides guidelines for extending this to the general case.

Note that when T is compact, $(\mathbb{P}_n)$ is *always* a tight family. In this case, Theorem 2.5.1 is a reformulation of the Banach–Alaoglu theorem of functional analysis; cf. Rudin (1973, Theorem 3.15), for instance.

Our proof is divided into four easy steps, all the time assuming that T is compact.

Step 1. (*Reduction to $T \subset \mathbb{R}^\infty$*)
In this first step we argue that without loss of generality, T is a compact subset of $\mathbb{R}^\infty$, where the latter is, as usual, endowed with the product topology. A key role is played by the following variant of Urysohn's metrization theorem; cf. Munkres (1975, Theorem 4.1, Chapter 4), for instance.

> **Urysohn's Metrization Theorem** *Any separable metric space T is homeomorphic to a subset of $\mathbb{R}^\infty$.*

That is, there exists a one-to-one continuous function $h : T \to \mathbb{R}^\infty$ whose inverse function $h^{-1} : h(T) \to T$ is also continuous. Let X_n (respectively X) be a T-valued random variable with distribution $\mathbb{P}_n$ (respectively $\mathbb{P}$) We wish to show the existence of a subsequence n' such that for all bounded continuous functions $f : T \mapsto \mathbb{R}$, $\lim_{n' \to \infty} \mathbb{E}[f(X_{n'})] = \mathbb{E}[f(X)]$. By the continuous mapping theorem (Theorem 2.2.1), this is equivalent to showing that for all homeomorphisms $h : T \to \mathbb{R}^\infty$ and all bounded continuous functions $\psi : \mathbb{R}^\infty \to \mathbb{R}$, $\lim_{n' \to \infty} \mathbb{E}[\psi(h(X_{n'}))] = \mathbb{E}[\psi(h(X))]$. Since $h(X_n)$ and $h(X)$ are $\mathbb{R}^\infty$-valued random variables, this reduces our proof of Prohorov's theorem to the case $T \subset \mathbb{R}^\infty$. □

Step 2. (*Separability of the Space of Continuous Functions*)
Henceforth, T is a compact subset of $\mathbb{R}^\infty$, and $C(T)$ denotes the collection

of all continuous functions $f : T \to \mathbb{R}$, metrized by the supremum norm. In this second step of our proof we propose to show that $C(T)$ is separable. To do so, we will need to recall the Stone–Weierstrass theorem.

Recall that $\mathfrak{A} \subset C(T)$ is an **algebra** if $f, g \in \mathfrak{A}$ implies that the product fg is in $\mathfrak{A}$. Also recall that $\mathfrak{A} \subset C(T)$ is said to **separate points** if whenever $x, y \in T$ are distinct, we can find $f \in \mathfrak{A}$ such that $f(x) \neq f(y)$.

> **The Stone–Weierstrass Theorem** *Suppose T is compact and $\mathfrak{A} \subset C(T)$ is an algebra that (i) separates points and (ii) contains all constants. Then, $\mathfrak{A}$ is dense in $C(T)$.*

For a proof, see Royden (1968, Theorem 28, Chapter 9, Section 7), for instance.

Since $T \subset \mathbb{R}^\infty$, we can unambiguously define $\mathfrak{A}_0 \subset C(T)$ as the collection of all polynomials with rational coefficients, i.e., all functions of type $f(x) = \alpha + \beta \prod_{j=1}^n [x^{(j)}]^{\gamma_j}$, where $x \in T$, α, β range over the rationals, and $n, \gamma_1, \ldots, \gamma_n$ range over nonnegative integers. Let $\mathfrak{A}$ denote the smallest algebra that contains $\mathfrak{A}_0$ and apply the Stone–Weierstrass theorem to conclude that $\mathfrak{A}$ is a countable dense subset of $C(T)$, as desired. □

***Exercise* 2.5.2** Prove the following refinement: *If Γ is a compact metric space, $C(\Gamma)$ is separable.* □

Step 3. (*Existence of Subsequential Limits*)
We now show the existence of a subsequence n' such that for every $f \in C(T)$, $\Lambda(f) = \lim_{n' \to \infty} \int f(\omega) \, \mathbb{P}_{n'}(d\omega)$ exists.

Since T is compact, any $f \in C(T)$ is bounded. Thus, for each $f \in C(T)$, $\int f d\mathbb{P}_1, \int f d\mathbb{P}_2, \ldots$ is a bounded sequence in $\mathbb{R}$. In particular, for each $f \in C(T)$, there exists a subsequence n_m such that $\lim_{n_m \to \infty} \int f d\mathbb{P}_{n_m}$ exists. On the other hand, by Step 2, $C(T)$ has a countable dense subset $\mathfrak{A}$. Applying Cantor's diagonalization argument, we can extract *one* subsequence n' such that for all $f \in \mathfrak{A}$,

$$\Lambda(f) = \lim_{n' \to \infty} \int f(\omega) \, \mathbb{P}_{n'}(d\omega) \tag{1}$$

exists. We complete Step 3 by showing that the above holds for all $f \in C(T)$, along the *same* subsequence n'.

We do this by first continuously extending the domain of Λ from $\mathfrak{A}$ to all of $C(T)$. Indeed, note that for all $f, g \in \mathfrak{A}$, $|\Lambda(f) - \Lambda(g)| \leq \sup_{x \in T} |f(x) - g(x)|$. By density, for any $f \in C(T)$, we can find $f_m \in \mathfrak{A}$ such that $\lim_{m \to \infty} f_m = f$, in $C(T)$, i.e., in the supremum norm. Since $f_1, f_2, \ldots$ is a Cauchy sequence, this shows that $\Lambda(f) = \lim_m \Lambda(f_m)$ exists, and $|\Lambda(f) - \Lambda(g)| \leq \sup_{x \in T} |f(x) - g(x)|$ for all $f, g \in C(T)$. Once more, for all $f \in C(T)$, find $f_m \in \mathfrak{A}$ such that $f_m \to f$ in $C(T)$. This leads to the following sequence of bounds:

$$\limsup_{n'\to\infty}\left|\int f(\omega)\,\mathbb{P}_{n'}(d\omega)-\Lambda(f)\right|$$
$$\le |\Lambda(f)-\Lambda(f_m)|+\limsup_{n'\to\infty}\left|\int f(\omega)\,\mathbb{P}_{n'}(d\omega)-\int f_m(\omega)\,\mathbb{P}_{n'}(d\omega)\right|$$
$$\le 2\sup_{x\in T}|f(x)-f_m(x)|.$$

Let $m\to\infty$ to see that equation (1) holds for all $f\in C(T)$, as desired. □

Step 4. (*The Conclusion*)
Prohorov's theorem readily follows upon combining Step 3 with the representation theorem of F. Riesz; see Rudin (1974, Theorem 2.14), for a proof.

> **The Riesz Representation Theorem** *Let T be a compact metric space and let Λ denote a positive and continuous linear functional on $C(T)$. Then, there exists a measure $\mathbb{P}$ on the Borel subsets of T such that for all $f\in C(T)$, $\Lambda(f)=\int f(\omega)\,\mathbb{P}(d\omega)$.*

Exercise **2.5.3** Complete the details of the proof of Prohorov's theorem.
□

3 The Space C

In Chapter 5 we encountered processes that are continuous or have a continuous modification. Among other things, in this chapter we adopted the viewpoint that such processes are, in fact, random variables that take values in some space of continuous functions. We now turn to weak convergence on such spaces.

3.1 Uniform Continuity

Define $C=C([0,1]^N,\mathbb{R}^d)$ to be the collection of all continuous functions $f:[0,1]^N\to\mathbb{R}^d$. It is metrized by

$$d_C(f,g)=\sup_{t\in[0,1]^N}|f(t)-g(t)|,\qquad f,g\in C. \tag{1}$$

Thus, the space $C([0,1]^N)$ of Section 2.5 is none other than C with $d=1$.

According to Prohorov's theorem (Theorem 2.5.1), in order to study weak convergence on C, we need first to understand the structure of its compact subsets. This is described by the following result, whose proof can be found, for example, in Munkres (1975, Theorem 6.1, Chapter 7).

The Arzelá–Ascoli Theorem *A subset C' of C has compact closure if and only if (a) it is equicontinuous and (b) for each $x \in [0,1]^N$, the closure of $C'_x = \{f(x) : f \in C'\}$ is compact in $\mathbb{R}^d$.*

The **modulus of continuity** of $f \in C$ is the function

$$\omega_f(\varepsilon) = \sup_{\substack{s,t \in [0,1]^N : \\ |s-t| \leq \varepsilon}} |f(s) - f(t)|, \qquad \varepsilon > 0. \tag{2}$$

(Compare to the analogous notion defined in Section 2.5 of Chapter 5.) Clearly, a function $f \in C$ is uniformly continuous if $\lim_{\varepsilon \to 0^+} \omega_f(\varepsilon) = 0$.

***Exercise* 3.1.1** Prove that a subset C' of C is equicontinuous if and only if $\lim_{\varepsilon \to 0^+} \sup_{f \in C'} \omega_f(\varepsilon) = 0$. □

This exercise leads to the following reformulation of the Arzelá–Ascoli theorem.

Theorem 3.1.1 (The Arzelá–Ascoli Theorem) *A subset C' of C has compact closure if and only if for all $x \in [0,1]^N$, the closure of $C'_x = \{f(x) : f \in C'\}$ is compact in $\mathbb{R}^d$ and $\lim_{\varepsilon \to 0^+} \sup_{f \in C'} \omega_f(\varepsilon) = 0$.*

Theorem 3.1.1 will be used to characterize tightness on C. But what do C-valued random variables look like? In fact, are there any interesting ones? The following example shows that there are indeed many natural C-valued random variables.

Example Let $g = (g_t; t \in \mathbb{R}_+^N)$ be a real-valued Gaussian process with a continuous mean function μ and some covariance function Σ. According to Chapter 5, when μ is continuous, under some technical conditions on Σ, g has a continuous modification $X = (X_t; t \in \mathbb{R}_+^N)$; see Sections 2.3, 2.5, and 2.6 of Chapter 5. According to Lemma 2.2.1 of Chapter 5, X and g have the same finite-dimensional distributions. Since Gaussian processes are solely described by their finite-dimensional distributions, X is itself a Gaussian process with mean function μ and covariance function Σ. For all $\omega \in \Omega$ and all $t \in [0,1]^N$, let $X(t)(\omega) = X_t(\omega)$. That is, we think of X as the random function $t \mapsto X_t$ restricted to $[0,1]^N$. Continuity of X implies that $\mathbb{P}(X \in C) = 1$, where $C = C([0,1]^N, \mathbb{R})$. Equivalently, X is a C-valued random variable. A concrete example of a C-valued random variable is the standard Brownian sheet; cf. Section 3.2 of Chapter 5 for the requisite continuity results. □

We conclude this subsection with the following technical result, which will tacitly be used from now on.

Lemma 3.1.1 *Suppose X is a C-valued random variable. Then, for any $\varepsilon > 0$, $\omega_X(\varepsilon)$ is an $\mathbb{R}_+$-valued random variable.*

Exercise **3.1.2** Prove Lemma 3.1.1. □

3.2 Finite-Dimensional Distributions

Given that $C = C([0,1]^N, \mathbb{R}^d)$, let d_C be defined by equation (1) of Section 3.1, and suppose that μ_1 and μ_2 are two probability measures on Borel subsets of $\mathbb{R}^d$. According to Theorem 3.1.1, $\mu_1 = \mu_2$ if and only if $\int h\, d\mu_1 = \int h\, d\mu_2$ for all bounded, continuous functions $h : \mathbb{R}^d \to \mathbb{R}$. Let X_1 and X_2 denote random variables whose distributions are μ_1 and μ_2, respectively; cf. Section 1.2. The preceding discussion says that X_1 and X_2 have the same distribution if and only if for all bounded continuous functions $h : C \to \mathbb{R}$, $\mathbb{E}[h(X_1)] = \mathbb{E}[h(X_2)]$. The following is often more useful.

Theorem 3.2.1 *Let X_1 and X_2 be two C-valued random variables. Then, X_1 and X_2 have the same distributions if and only if for all $t_1, \ldots, t_k \in [0,1]^N$, $(X_1(t_1), \ldots, X_1(t_k))$ and $(X_2(t_1), \ldots, X_2(t_k))$ have the same distribution.*

Remarks

(a) Since $X_i \in C$, it is a continuous $\mathbb{R}^d$-valued random function. In particular, we point out that $(X_i(t_1), \ldots, X_i(t_k))$ can be viewed as an $\mathbb{R}^{dk}$-valued random vector; see Theorem 1.1.1.

(b) In the notation of Chapter 2, the above theorem says that X_1 and X_2 have the same distribution if and only if they have the same finite-dimensional distributions.

Proof We have already mentioned that X_1 and X_2 have the same distribution if and only if for all bounded, continuous $h : C \to \mathbb{R}$, $\mathbb{E}[h(X_1)] = \mathbb{E}[h(X_2)]$.

Suppose, first, that X_1 and X_2 have the same distribution. Fix $t_1, \ldots, t_k \in [0,1]^N$ and let $h(x) = f(x(t_1), \ldots, x(t_k))$ ($x \in C$), where $f : \mathbb{R}^{dk} \to \mathbb{R}$ is bounded and continuous. Since $h : C \to \mathbb{R}$ is continuous, this implies that $\mathbb{E}[h(X_1)] = \mathbb{E}[h(X_2)]$, which implies the equality of the finite-dimensional distributions.

Conversely, suppose X_1 and X_2 have the same finite-dimensional distributions. It suffices to show that for all closed sets $F \subset C$,

$$\mathbb{P}(X_1 \in F) \leq \mathbb{P}(X_2 \in F). \tag{1}$$

(Why?) Fix $\varepsilon > 0$ and let $t_1, \ldots, t_k$ be a partition of $[0,1]^N$ of mesh ε. Define the projection operator $\pi_{t_1,\ldots,t_k}$ by $\pi_{t_1,\ldots,t_k} f = (f(t_1), \ldots, f(t_k))$, for all $f \in C$. Since X_1 and X_2 have the same finite-dimensional distributions, in the notation of Sections 1.1 and 1.2,

$$\mathbb{P}_1 \circ \pi^{-1}_{t_1,\ldots,t_k} = \mathbb{P}_2 \circ \pi^{-1}_{t_1,\ldots,t_k}, \tag{2}$$

where $\mathbb{P}_i$ denotes the distribution of X_i ($i = 1, 2$). For any closed $F \subset C$, let

$$\pi_{t_1,\ldots,t_k} F = \{\pi_{t_1,\ldots,t_k} f : f \in F\}.$$

Of course, if $f \in F$, then $\pi_{t_1,\ldots,t_k} f \in \pi_{t_1,\ldots,t_k} F$. On the other hand, if $f \in C$ satisfies $\omega_f(\varepsilon) \leq \eta$ for some $\eta > 0$ and if $\pi_{t_1,\ldots,t_k} f \in \pi_{t_1,\ldots,t_k} F$, then

$$f \in F^\eta = \{g \in C : d_C(\{g\}, F) \leq \eta\},$$

where $d_C(\{g\}, F) = \inf_{h \in F} d_C(h, g)$. We combine these observations as follows: For all $\varepsilon, \eta > 0$,

$$\begin{aligned}
\mathbb{P}(X_1 \in F) &\leq \mathbb{P}(\pi_{t_1,\ldots,t_k} X_1 \in \pi_{t_1,\ldots,t_k} F) \\
&= \mathbb{P}_1 \circ \pi_{t_1,\ldots,t_k}^{-1}(\pi_{t_1,\ldots,t_k} F) \\
&= \mathbb{P}_2 \circ \pi_{t_1,\ldots,t_k}^{-1}(\pi_{t_1,\ldots,t_k} F), \qquad \text{(by (2))} \\
&= \mathbb{P}(\pi_{t_1,\ldots,t_k} X_2 \in \pi_{t_1,\ldots,t_k} F) \\
&\leq \mathbb{P}(X_2 \in F^\eta) + \mathbb{P}(\omega_{X_2}(\varepsilon) > \eta).
\end{aligned}$$

Let $\varepsilon \to 0^+$ and use the Arzelá–Ascoli theorem (Theorem 3.1.1) to see that $\mathbb{P}(X_1 \in F) \leq \mathbb{P}(X_2 \in F^\eta)$. On the other hand, $\eta > 0$ is arbitrary and $\eta \mapsto F^\eta$ is set-theoretically decreasing with $\cap_{\eta \in \mathbb{Q}_+} F^\eta = F$. Equation (1) follows. $\square$

3.3 Weak Convergence in C

Continuing with our discussion of Section 3.2, we let $(\mu_n; 1 \leq n \leq \infty)$ denote a collection of probability measures on C and seek conditions that ensure that $\mu_n \Longrightarrow \mu_\infty$. Bearing the discussion of Section 1.2 in mind, we need to ask, *if $(X_n; 1 \leq n \leq \infty)$ is a sequence of C-valued random variables, when does $X_n \Longrightarrow X_\infty$?* Given the development of Sections 3.1 and 3.2, we may be tempted to conjecture that $X_n \Longrightarrow X_\infty$ if and only if all finite-dimensional distributions of X_n converge to those of X_∞. This is not so, as the following example shows.

Example Let $N = 1$ and $S = \mathbb{R}$. Thus, C is the collection of all continuous functions $f : [0, 1] \to \mathbb{R}$, and d_C is the usual supremum norm on C. We wish to construct a sequence $(X_n; n \geq 1)$ of C-valued random variables such that the finite-dimensional distributions of X_n converge to those of the function 0, and yet X_n does not converge weakly to 0. Here is one such construction: On an appropriate probability space, construct independent random variables $U_1, U_2, \ldots$ all of which are uniformly distributed on the interval $[\frac{1}{2}, \frac{3}{4}]$ (say). Define

$$X_n(t) = \begin{cases} n^2 t - n^2 U_n, & \text{if } 0 \leq U_n \leq t \leq U_n + \frac{1}{n}, \\ -n^2 t + n^2 U_n + 2n, & \text{if } U_n + \frac{1}{n} \leq t \leq U_n + \frac{2}{n} \leq 1, \\ 0, & \text{otherwise.} \end{cases}$$

Recall that almost surely, $U_n \in [\frac{1}{2}, \frac{3}{4}]$. Therefore, after directly plotting the function X_n, we see that for all $n \geq 8$, X_n is a continuous, piecewise linear function on $[0,1]$ such that $X_n(0) = X_n(U_n) = 0$, $X_n(U_n + \frac{1}{n}) = n$, $X_n(U_n + \frac{2}{n}) = X_n(1) = 0$, and between these values, X_n is obtained by linear interpolation. In particular, $\mathbb{P}(X_n \in C) = 1$. Fix any $t_1, \ldots, t_k \in [0,1]$ and consider the event that $(X_n(t_i) \neq 0$, for some $1 \leq i \leq k)$, which is the same as $(t_i \notin [U_n, U_n + \frac{2}{n}]$, for some $1 \leq i \leq k)$. Since the latter has probability at most $\frac{4k}{n}$,

$$\mathbb{P}(X_n(t_1) = \cdots = X_n(t_k) = 0) \geq 1 - \frac{4k}{n}$$

converges to 1 as $n \to \infty$. That is, the finite-dimensional distributions of X_n converge to those of the function 0. Next, we will verify that $X_n \not\Rightarrow 0$. To show this, for all $x \in C$ define $f_1(x) = \sup_{0 \leq t \leq 1} |x(t)|$. The function f_1 has the following two properties:

(a) $f_1 : C \to \mathbb{R}_+$; and

(b) for all $x, y \in C$, $|f(x) - f(y)| \leq \sup_{0 \leq t \leq 1} |x(t) - y(t)| = d_C(x,y)$.

In particular, f_1 is a continuous function from C to $\mathbb{R}_+$. Define $f_2 : \mathbb{R}_+ \to [0,1]$ by $f_2(t) = t \wedge 1$ and let $f = f_2 \circ f_1$. Apparently, $f : C \to [0,1]$ is bounded and continuous, although for all $n \geq 1$, $f(X_n) \equiv 1$ and $f(0) = 0$. Hence, $\mathbb{E}[f(X_n)] \not\to \mathbb{E}[f(0)]$. □

The above example shows that convergence of the finite-dimensional distributions is not sufficiently strong to guarantee weak convergence in C. Indeed, the missing ingredient is tightness, as the following shows.

Proposition 3.3.1 *Suppose $(X_n; 1 \leq n \leq \infty)$ are C-valued random variables. Then, $X_n \Longrightarrow X_\infty$, provided that:*

(i) *the finite-dimensional distributions of X_n converge to those of X_∞; and*

(ii) *(X_n) is a tight sequence.*

Proof Let $\mathbb{P}_n$ denote the distribution of X_n ($1 \leq n \leq \infty$). By Exercise 2.5.1, for any subsequence n' there exists a further subsequence n''

and a probability measure $\mathbb{Q}_\infty$ such that for all bounded continuous functions $f : C \to \mathbb{R}$, $\int f d\mathbb{P}_{n''} \to \int f d\mathbb{Q}_\infty$. It suffices to show that no matter which sequence n' we choose, $\mathbb{Q}_\infty = \mathbb{P}_\infty$ (why?). The convergence of finite-dimensional distributions implies that the finite-dimensional distributions of $\mathbb{Q}_\infty$ and $\mathbb{P}_\infty$ agree. Consequently, the proposition follows from Theorem 3.2.1. □

In the example above, we may see that the trouble comes in via the wild oscillations of the functions X_n. In other words, the family $(X_n; n \geq 1)$ is not equicontinuous, i.e., not tight. The following theorem shows that this lack of tightness is indeed the source of the difficulty.

Theorem 3.3.1 *Let $(X_n; 1 \leq n \leq \infty)$ denote a collection of C-valued random variables. Then $X_n \Longrightarrow X_\infty$ provided that:*

(i) the finite-dimensional distributions of X_n converge to those of X_∞; and

(ii) for all $\varepsilon > 0$, $\lim_{\delta \to 0} \limsup_{n \to \infty} \mathbb{P}(\omega_{X_n}(\delta) \geq \varepsilon) = 0$.

Proof In light of Proposition 3.3.1, it suffices to prove that (X_n) is tight. That is, given any $\varepsilon \in]0, 1[$, we need to produce a compact set Γ_ε such that $\sup_n \mathbb{P}(X_n \notin \Gamma_\varepsilon) \leq \varepsilon$.

Owing to Theorem 2.4.1, condition (ii) of the theorem is equivalent to the following: For all $\varepsilon > 0$, $\lim_{\delta \to 0} \sup_n \mathbb{P}(\omega_{X_n}(\delta) \geq \varepsilon) = 0$. (Why?) Thus, for any $\varepsilon \in]0, 1[$, we can find a sequence $\delta_1 > \delta_2 > \cdots$ such that $\lim_k \delta_k = 0$ and $\sup_n \mathbb{P}(\omega_{X_n}(\delta_k) \geq \frac{1}{k}) \leq \varepsilon 2^{-k-1}$. That is, if we define

$$A_k = \left\{ f \in C : \omega_f(\delta_k) \leq \frac{1}{k} \right\}, \quad k \geq 1,$$

we have $\sup_n \mathbb{P}(X_n \notin \cap_{k \geq 1} A_k) \leq \sum_{k \geq 1} \varepsilon 2^{-k-1} = \frac{1}{2}\varepsilon$. Next, define for all $\lambda > 0$,

$$A_0(\lambda) = \{ f \in C : |f(0)| \leq \lambda \}.$$

By the convergence of finite-dimensional distributions and by the portmanteau theorem (Theorem 2.1.1), for all $\lambda > 0$,

$$\limsup_{n \to \infty} \mathbb{P}\{X_n \notin A_0(\lambda)\} \leq \mathbb{P}\{X_\infty \notin A_0(\lambda)\}.$$

Thus, there exists λ large such that $\sup_n \mathbb{P}\{X_n \notin A_0(\lambda)\} \leq \frac{1}{2}\varepsilon$. If we let Γ_ε be the closure of $A_0(\lambda) \cap \bigcap_{k \geq 1} A_k$, we have $\sup_n \mathbb{P}(X_n \notin \Gamma_\varepsilon) \leq \varepsilon$, and Γ_ε is compact, thanks to the Arzelá–Ascoli theorem (Theorem 3.1.1; why?) This completes our proof. □

Exercise 3.3.1 Suppose (μ_n) is a collection of probability measures on C. Prove that (μ_n) is tight if and only if:

(i) $\lim_{\lambda \to \infty} \limsup_{n \to \infty} \mu_n(f \in C : |f(0)| \geq \lambda) = 0$; and

(ii) for all $\varepsilon > 0$, $\limsup_{\delta \to 0} \limsup_{n \to \infty} \mu_n(f \in C : \omega_f(\delta) \geq \varepsilon) = 0$.

Is there anything special about the condition $|f(0)| \geq \lambda$ in (i)? For instance, can it be replaced by $|f(a)| \geq \lambda$, where $a \in [0,1]^N$ is fixed? $\square$

3.4 Continuous Functionals

Suppose $C = C([0,1]^N, \mathbb{R}^d)$ is defined as before. If $(X_n;\ 1 \leq n \leq \infty)$ are C-valued random variables, we now explore some of the consequence of the statement "$X_n \Longrightarrow X_\infty$."

Recall that a functional is a real-valued function on C, and a functional Λ is continuous if for all $f, f_n \in C$, $\lim_{n \to \infty} d_C(f_n, f) = 0$ implies $\lim_{n \to \infty} \Lambda(f_n) = \Lambda(f)$. (Recall that d_C is the distance that metrizes C; see (1) of Section 3.1.) As the following two examples show, interesting continuous functionals on C abound.

Example 1 Define the functional Λ by $\Lambda(f) = \sup_{t \in [0,1]^N} |f(t)|$. Then, Λ is a continuous functional. In fact, $\Lambda(f) = d_C(f, 0)$, where 0 denotes the 0 function. Moreover, for all $p > 0$, Λ_p defines a continuous functional, where

$$\Lambda_p(f) = \left(\int_{[0,1]^N} |f(t)|^p dt \right)^{1/p}.$$

This follows from the trivial inequality $|\Lambda_p(f) - \Lambda_p(g)| \leq d_C(f, g)$. As another example, consider $d = 1$ and let $\Lambda_+(f) = \sup_{t \in [0,1]^N} f(t)$. Then Λ_+ is a continuous functional. $\square$

Example 2 Fix an integer $k \geq 1$ and a continuous function $\theta : S^k \to \mathbb{R}$. For any fixed $t_1, \ldots, t_k \in [0,1]^N$ and all $f \in C$, define $\Lambda(f) = \theta(f(t_1), \ldots, f(t_k))$. It is not hard to see that Λ is a continuous functional. For instance, when $S = \mathbb{R}$, $\Lambda(f) = \sum_{j=1}^k \xi^{(j)} f(t_j)$ is a continuous functional, where $\xi \in \mathbb{R}^k$ is fixed. $\square$

The following is an immediate consequence of the continuous mapping theorem (Theorem 2.2.1).

Theorem 3.4.1 Suppose that $X_n \Longrightarrow X_\infty$, as elements of C. Then for any continuous functional Λ on C, $\Lambda(X_n)$ converges in distribution to $\Lambda(X_\infty)$.

To better understand Theorem 3.4.1, try the following almost sure version.

200   6. Limit Theorems

***Exercise* 3.4.1** Suppose $X_1, \ldots, X_\infty$ are all $C = C([0,1]^N, \mathbb{R}^d)$-valued random variables such that for each continuous functional Λ on C, $\lim_{n\to\infty} \Lambda(X_n) = \Lambda(X_\infty)$, a.s. Then, show that with probability one, $\lim_{n\to\infty} X_n = X_\infty$, where the convergence takes place in C. □

Thus, Theorem 3.4.1 is one way to state that $X_n \Longrightarrow X_\infty$ in C if and only if the "distribution" of the entire function $t \mapsto X_n(t)$ converges to that of $t \mapsto X_\infty(t)$, all viewed as functions from $[0,1]^N$ to $\mathbb{R}^d$.

3.5 A Sufficient Condition for Pretightness

If $X_1, X_2, \ldots$ is a sequence of $C = C([0,1]^N, \mathbb{R}^d)$-valued random variables, we can inspect the contents of Proposition 3.3.1, Theorem 3.3.1, and Exercise 3.3.1 to see that verifying the tightness of (X_n) often boils down to finding a sufficient condition for the following: For all $\varepsilon > 0$,

$$\lim_{\delta \to 0} \limsup_{n \to \infty} \mathbb{P}(\omega_{X_n}(\delta) \geq \varepsilon) = 0.$$

When this holds, we say that (X_n) is **pretight**.

In this subsection we find a technical condition for the pretightness of (X_n). This condition will be used in Section 5 in its consideration of multi-parameter random walks. The main result is the following; see Billingsley (1968, Theorem 8.3); see Lachout (1988) for related results.

Theorem 3.5.1 *A collection $X_1, X_2, \ldots$ of $C([0,1]^N, \mathbb{R}^d)$-valued random variables is pretight if for every $\ell = 1, \ldots, N$,*

$$\lim_{\delta \to 0} \limsup_{n \to \infty} \sup_{0 \leq t^{(\ell)} \leq 1} \frac{1}{\delta} \mathbb{P}\Big(\sup |X_n(s) - X_n(t)| \geq \varepsilon\Big) = 0,$$

where the supremum inside the probability is taken over all $s \in [0,1]^N$ and all $t^{(k)} \in [0,1]$ ($k \neq \ell$) such that $t^{(\ell)} \leq s^{(\ell)} \leq t^{(\ell)} + \delta$ and for all $k \neq \ell$, $s^{(k)} = t^{(k)}$.

It is important to note that the supremum inside the probability is taken only over the values of $s^{(1)}, \ldots, s^{(N)}$ and $t^{(k)}$, where $k \neq \ell$; the supremum over the $t^{(\ell)}$ is outside the probability, and this alone makes such a result useful.

Proof We will prove this for $N = 1$ only; when $N > 1$, the proof is similar but should be attempted by the reader for better understanding. Thus, throughout this proof we let $N = 1$, and set out to prove that the following implies the pretightness of (X_n): For all $\varepsilon > 0$,

$$\lim_{\delta \to 0} \limsup_{n \to \infty} \sup_{0 \leq t \leq 1} \frac{1}{\delta} \mathbb{P}\Big(\sup_{s \in [0,1]:\, t \leq s \leq t+\delta} |X_n(s) - X_n(t)| \geq \varepsilon\Big) = 0. \quad (1)$$

To prove this, we will use a simplified version of the chaining argument of Chapter 5.

Fix some $\delta > 0$ and define $\Gamma_\delta = \{0, \delta, 2\delta, \ldots, \lfloor \frac{1}{\delta} \rfloor \delta\}$. In particular, note that for all $s \in [0,1]$ there exists a unique $\gamma_s \in \Gamma_\delta$ such that $\gamma_s \leq s \leq \gamma_s + \delta$. By the triangle inequality, for all $s, t \in [0,1]$ with $|s-t| \leq \delta$,

$$|X_n(s) - X_n(t)|$$
$$\leq |X_n(s) - X_n(\gamma_s)| + |X_n(t) - X_n(\gamma_t)| + |X_n(\gamma_s) - X_n(\gamma_t)|$$
$$\leq 3 \sup_{r \in [0,1]} |X_n(r) - X_n(\gamma_r)|.$$

(The last inequality uses the continuity of X. Why?) Since the cardinality of Γ_δ is no more than $1 + \frac{1}{\delta}$,

$$\mathbb{P}(\omega_{X_n}(\delta) \geq \varepsilon) \leq \left(1 + \frac{1}{\delta}\right) \max_{\gamma \in \Gamma_\delta} \mathbb{P}\left(\sup_{s \in [0,1]:\, \gamma \leq s \leq \gamma + \delta} |X_n(s) - X_n(\gamma)| \geq \frac{\varepsilon}{3}\right).$$

The theorem follows readily from this and from equation (1). □

Exercise 3.5.1 Prove Theorem 3.5.1 for *all* $N \geq 1$. □

4 Invariance Principles

Let $\xi_1, \xi_2, \ldots$ denote independent, identically distributed random variables that have mean μ and variance σ^2 and let $S = (S_n;\, n \geq 1)$ denote the associated random walk, i.e., $S_n = \sum_{j=1}^n \xi_j$ ($n \geq 1$). The classical central limit theorem states that as $n \to \infty$, $n^{-1/2}\sigma^{-1}\{S_n - n\mu\}$ converges in distribution to a standard Gaussian random variable. Combining the notation of Section 2.3 with Section 1.1 of Chapter 5, we can write this as

$$\frac{S_n - n\mu}{\sqrt{n}} \Longrightarrow \mathcal{N}_1(0, \sigma^2),$$

where $\Longrightarrow$ denotes weak convergence in $\mathbb{R}$; see Chapter 5 for the notation on Gaussian distributions.

Exercise 4.0.2 Show that for any choice of $0 \leq t_1 < \cdots < t_k \leq 1$,

$$\left(\frac{S_{\lfloor nt_1 \rfloor} - nt_1\mu}{\sqrt{n}\,\sigma}, \ldots, \frac{S_{\lfloor nt_k \rfloor} - nt_k\mu}{\sqrt{n}\,\sigma}\right) \Longrightarrow (Z_{t_1}, \ldots, Z_{t_k}),$$

where $\Longrightarrow$ denotes weak convergence in $\mathbb{R}^k$ and $(Z_{t_1}, \ldots, Z_{t_k})$ is an $\mathbb{R}^k$-valued Gaussian random vector with mean vector 0 and covariance matrix Σ, where $\Sigma^{(i,j)} = t_i \wedge t_j$ ($1 \leq i, j \leq k$). □

In the above exercise, the limiting probability measure (or the random vector Z) is reminiscent of a Brownian motion sampled at fixed times $t_1, \ldots, t_k$. That is, if Z is a Brownian motion, $(Z_{t_1}, \ldots, Z_{t_k})$ has the same finite-dimensional distributions as the limiting vector in Exercise 4.0.2 above, for any choice of $0 \leq t_1 < \cdots < t_k \leq 1$. It is natural to ask whether the stochastic process $t \mapsto n^{-1/2}\sigma^{-1}(S_{\lfloor nt \rfloor} - nt\mu)$ ($t \in [0,1]$) converges weakly to Brownian motion in $C([0,1], \mathbb{R})$, as $n \to \infty$. Unfortunately, the random function $f_n(t) = S_{\lfloor nt \rfloor}$ ($0 \leq t \leq 1$) does not belong to $C([0,1], \mathbb{R})$. However, we are interested only in the random walk values $f_n(n/k)$, where $0 \leq k \leq n$. Thus, we can alternatively study the stochastic process $\mathbf{S}_n = (\mathbf{S}_n(t); \, t \in [0,1])$ defined by

$$\mathbf{S}_n(t) = \frac{S_{\lfloor nt \rfloor} + (nt - \lfloor nt \rfloor)\xi_{\lfloor nt \rfloor + 1} - nt\mu}{\sqrt{n}\,\sigma}, \qquad 0 \leq t \leq 1. \tag{1}$$

That is, $\mathbf{S}_n$ is a random continuous function on $[0,1]$ such that for all $0 \leq k \leq n$,

$$\mathbf{S}_n\left(\frac{k}{n}\right) = \frac{S_k - k\mu}{\sqrt{n}\,\sigma},$$

and between these values, $\mathbf{S}_n$ is defined by linear interpolation. (To see things clearly, set $\mu = 0$ and $\sigma = 1$.) In particular,

$$\mathbb{P}\{\mathbf{S}_n \in C([0,1], \mathbb{R})\} = 1.$$

Donsker's theorem states that the C-valued process $\mathbf{S}_n$ converges weakly to standard Brownian motion on $[0,1]$. Assuming this result for the moment, we can then combine the first example of Section 3.4 with Theorem 3.4.1 to prove results such as the following: For all $\lambda \geq 0$,

$$\lim_{n \to \infty} \mathbb{P}\left\{ \max_{1 \leq k \leq n} (S_k - k\mu) \geq \sqrt{n}\,\sigma\lambda \right\} = \mathbb{P}\left(\sup_{0 \leq t \leq 1} B_t \geq \lambda \right), \tag{2}$$

where $B = (B_t; \, t \geq 0)$ denotes a standard Brownian motion. This is an example of an **invariance principle**: The limiting distribution is independent of the distributions of the ξ's, as long as the latter are i.i.d., with mean μ and variance σ^2. Sometimes, such invariance principles can themselves be used to compute the limiting distribution; for examples of this technique see Supplementary Exercises 5, 6, and 7.

In this section we will prove the aforementioned theorem of M. Donsker and a multiparameter extension due to (Bickel and Wichura 1971; Pyke 1973). Donsker's theorem, in a slightly different setting, can be found in Donsker (1952). For general theory, and for detailed historical accounts, see (Billingsley 1968; Dudley 1989; Ethier and Kurtz 1986).

4.1 Preliminaries

Let $\xi = (\xi_t; \, t \in \mathbb{N}^N)$ denote a collection of independent, identically distributed random variables with mean 0 and variance 1. As in Chapter 5, we associate to ξ a random walk $S = (S_t;\ t \in \mathbb{N}^N)$ defined by

$$S_t = \sum_{r \preccurlyeq t} \xi_r, \qquad t \in \mathbb{N}^N.$$

In the case $N = 1$, there is a clear way to linearly interpolate and define a continuous function $\mathbf{S}_n(t)$ that agrees with $n^{-1/2} S_{nt}$ for values of $t = 0, \frac{1}{n}, \frac{2}{n}, \ldots, 1$; cf. the discussion preceding Section 4.1. When $N > 1$, "linear interpolation" is more arduous but still possible, as we shall see next.

For all $t \in [0,1]^N$, let

$$\Xi(t) = \mathrm{Leb}\left([[t],t]\right) \xi_{[t]+(1,\ldots,1)},$$

where $[t] = (\lfloor t^{(1)} \rfloor, \ldots, \lfloor t^{(N)} \rfloor)$, and define the process $\mathbf{S}_n = (\mathbf{S}_n(t);\ t \in [0,1]^N)$ by the following Stieltjes integral:

$$\mathbf{S}_n(t) = n^{-\frac{N}{2}} \int_{[0,nt]} \Xi(ds), \qquad n \geq 0,\ t \in [0,1]^N.$$

Proposition 4.1.1 *Whenever $n \geq 1$:*

(i) for all $t \in [0,1]^N$, $\mathbf{S}_n(\frac{1}{n}[nt]) = n^{-\frac{N}{2}} S_{[nt]}$, a.s.;

(ii) $\mathbb{P}\{\mathbf{S}_n \in C([0,1]^N, \mathbb{R})\} = 1$; and

(iii) with probability one, $\mathbf{S}_n$ is a linear function, coordinatewise.

In particular, when $N = 1$, $\mathbf{S}_n$ is the same as equation (2) of the beginning of this section, with $\mu = 0$ and $\sigma = 1$.

Proof We will perform the explicit computation (i). Assertions (ii) and (iii) are proved analogously. For all $s \in [0,1]^N$, define

$$Q(s) = \{t \in [0,1]^N : s \preccurlyeq t \prec s + (1,\ldots,1)\}.$$

Then,

$$\mathbf{S}_n\left(\frac{[nt]}{n}\right) = n^{-\frac{N}{2}} \sum_{0 \preccurlyeq s \prec [nt]} \int_{Q(s)} \Xi(dr).$$

If $r \in Q(s)$, then $\Xi(r) = \mathrm{Leb}([s,r])\xi_{s+(1,\ldots,1)}$. Therefore, on $Q(s)$, the (random) signed measure Ξ is absolutely continuous with respect to Lebesgue's measure, and $\Xi(dr)/dr = \xi_{s+(1,\ldots,1)}$. Thus,

$$\mathbf{S}_n\left(\frac{[nt]}{n}\right) = n^{-\frac{N}{2}} \sum_{0 \preccurlyeq s \prec [nt]} \xi_{s+(1,\ldots,1)},$$

which proves (i). □

Exercise 4.1.1 Complete the proof of Proposition 4.1.1. □

The goal of this section is to prove the following result.

Theorem 4.1.1 *As $n \to \infty$, $\mathbf{S}_n$ converges weakly in $C([0,1]^N, \mathbb{R})$ to a standard Brownian sheet.*

The above is an immediate consequence of Theorem 3.3.1, once we show that (a) the finite-dimensional distributions of $\mathbf{S}_n$ converge to those of a standard Brownian sheet; and (b) $(\mathbf{S}_n)$ is pretight; cf. also Section 3.5 above. We shall verify (a) and (b) in Sections 4.2 and 4.3, respectively.

4.2 Finite-Dimensional Distributions

In this subsection we continue with the discussion of Section 4.1 and prove the following:

Proposition 4.2.1 *The finite-dimensional distributions of $\mathbf{S}_n$ converge to those of a standard Brownian sheet.*

Our proof requires three technical lemmas. Throughout, $B = (B_t;\, t \in \mathbb{R}_+^N)$ denotes a standard Brownian sheet.

Lemma 4.2.1 *Let X be a random variable with mean 0 and variance ν^2. Then,*

$$\left| \mathbb{E}[e^{iX}] - 1 + \frac{\nu^2}{2} \right| \le \frac{7}{2} \mathbb{E}[X^2(1 \wedge |X|)].$$

Proof By Taylor's theorem, for any real number a, $e^{ia} = 1 + ia - \tfrac{1}{2}a^2 - \tfrac{i}{6}a_\star^3$, where $|a_\star| \le |a|$. Thus,

$$\left| e^{ia} - 1 - ia + \frac{1}{2}a^2 \right| \le \frac{1}{6}|a|^3 \le \frac{7}{2}|a|^3. \tag{1}$$

The above is useful only if $|a| \le 1$. If $|a| > 1$, we can get a better estimate by simply using the triangle inequality, viz.,

$$\left| e^{ia} - 1 - ia + \frac{1}{2}a^2 \right| \le 2 + |a| + \frac{1}{2}a^2 \le \frac{7}{2}a^2.$$

We obtain the result by combining this with equation (1), plugging in $a = X$ and taking expectations. □

Exercise 4.2.1 Verify Lemma 4.2.1, with $\tfrac{7}{2}$ reduced to 1. This can be improved further still. □

Lemma 4.2.2 *Given $n \ge 1$, consider functions $f_n : \mathbb{N}_+^N \to \mathbb{R}$ such that:*

(i) $\sup_{n \ge 1} \sup_{s \in \mathbb{N}_+^N} |f_n(s)| < \infty$; *and*

(ii) for some finite $p > 0$, $\lim_{n \to \infty} n^{-N} \sum_{s \preccurlyeq (n,\ldots,n)} f_n^2(s) = p$.

Then, $Y_n = n^{-\frac{N}{2}} \sum_{s \preccurlyeq (n,\ldots,n)} f_n(s) \xi_s$ converges in distribution to a Gaussian random variable with mean 0 and variance p.

Proof By the independence of the ξ's and by Lemma 4.2.1, for any $\theta \in \mathbb{R}$,

$$\mathbb{E}\left[e^{i\theta Y_n}\right] = \prod_{s \in \mathbb{N}_+^N : |s| \leq n} \mathbb{E}\left[e^{i\theta n^{-N/2} \xi_s f_n(s)}\right]$$

$$= \prod_{s \in \mathbb{N}_+^N : |s| \leq n} \left\{1 - \frac{\theta^2 n^{-N} f_n^2(s)}{2}(1 + \varepsilon_n(s))\right\},$$

where $\varepsilon_n : \mathbb{N}_+^N \to [-1, 1]$, $\lim_{n \to \infty} \varepsilon_n(s) = 0$, for all $s \in \mathbb{N}_+^N$, and $|s| = \max_{1 \leq j \leq N} |s^{(j)}|$, as always. Thus, by Taylor's expansion,

$$\mathbb{E}\left[e^{i\theta Y_n}\right] = \exp\left(\sum_{s \in \mathbb{N}_+^N : |s| \leq n} \ln\left\{1 - \frac{\theta^2 n^{-N} f_n^2(s)(1 + \varepsilon_n(s))}{2}\right\}\right)$$

$$= \exp\left\{-\frac{\theta^2}{2n^N} \sum_{s \in \mathbb{N}_+^N : |s| \leq n} f_n^2(s)(1 + \delta_n(s))\right\},$$

where $\delta_1, \delta_2, \ldots : \mathbb{N}_+^N \to \mathbb{R}$ is a bounded sequence of functions that satisfies $\lim_{n \to \infty} \delta_n(s) = 0$, for all $s \in \mathbb{N}_+^N$. By the dominated convergence theorem, $\lim_{n \to \infty} \mathbb{E}[e^{i\theta Y_n}] = e^{-\theta^2 p/2}$, and the result follows from the convergence theorem for characteristic functions. $\square$

Lemma 4.2.3 *For all $t_1, \ldots, t_k \in [0, 1]^N$,*

$$\left(\mathbf{S}_n\left(\frac{[nt_1]}{n}\right), \ldots, \mathbf{S}_n\left(\frac{[nt_k]}{n}\right)\right) \Longrightarrow (B_{t_1}, \ldots, B_{t_k}),$$

where $\Longrightarrow$ denotes weak convergence in $\mathbb{R}^k$ and B is the standard Brownian sheet.

You may recall that for $s \in \mathbb{R}_+^N$, $[s] \in \mathbb{R}_+^N$ denotes the point whose ith coordinate is $\lfloor s^{(i)} \rfloor$.

Proof Owing to Proposition 4.1.1, we need to show that for all $t_1, \ldots, t_k \in [0, 1]^N$,

$$n^{-\frac{N}{2}}(S_{[nt_1]}, \ldots, S_{[nt_k]}) \Longrightarrow (B_{t_1}, \ldots, B_{t_k}),$$

where $\Longrightarrow$ denotes weak convergence in $\mathbb{R}^k$. By Exercise 4.2.2 below, it suffices to show that for all $\alpha_1, \ldots, \alpha_k \in \mathbb{R}$,

$$n^{-\frac{N}{2}} \sum_{i=1}^{k} \alpha_i S_{[nt_i]} \Longrightarrow \sum_{i=1}^{k} \alpha_i B_{t_i},$$

where $\Longrightarrow$ now denotes weak convergence in $\mathbb{R}$. Recall from Chapter 5, Section 1.1, that $Z = \sum_{i=1}^{k} \alpha_i B_{t_i}$ is an $\mathbb{R}$-valued Gaussian random variable. Moreover, it has mean zero, and a direct computation reveals that the variance of Z is

$$\sigma^2 = \sum_{i=1}^{k}\sum_{j=1}^{k} \alpha_i \alpha_j \prod_{\ell=1}^{n} \left(t_i^{(\ell)} \wedge t_j^{(\ell)} \right).$$

Thus, our goal is to show that $Y_n = n^{-\frac{N}{2}} \sum_{i=1}^{k} \alpha_i S_{[nt_i]}$ converges in distribution to $\mathcal{N}_1(0, \sigma^2)$. On the other hand,

$$Y_n = n^{-\frac{N}{2}} \sum_{i=1}^{k} \alpha_i \sum_{s \preccurlyeq [nt_i]} \xi_s = n^{-\frac{N}{2}} \sum_{s \in \mathbb{N}^N : |s| \le n} f_n(s) \xi_s,$$

where $f_n(s) = \sum_{i=1}^{k} \alpha_i \mathbf{1}_{[0,[nt_i]]}(s)$. Since

$$n^{-N} \sum_{s \preccurlyeq (n,\ldots,n)} f_n^2(s) = n^{-N} \sum_{i=1}^{k}\sum_{j=1}^{k} \alpha_i \alpha_j \sum_{s \preccurlyeq (n,\ldots,n)} \mathbf{1}_{[0,[nt_i]\wedge [nt_j]]}(s),$$

Riemann sum approximations show that $n^{-N} \sum_{s \preccurlyeq (n,\ldots,n)} f_n^2(s)$ converges to σ^2, and the result follows from Lemma 4.2.2. □

Exercise 4.2.2 If $(X_n;\, n \le \infty)$ are $\mathbb{R}^d$-valued random variables, then X_n converges weakly to X_∞ if and only if for all $\alpha \in \mathbb{R}^d$, the random variable $\alpha \cdot X_n$ converges in distribution to $\alpha \cdot X_\infty$. This is called the Cramér–Wald device. □

We are ready for our proof of Proposition 4.2.1.

Proof of Proposition 4.2.1 In light of Lemma 4.2.3, we need only show that for each $t \in [0,1]^N$, as $n \to \infty$, $\mathbf{S}_n(t) - \mathbf{S}_n([nt]/n) \to 0$, in $L^2(\mathbb{P})$. (Why?)

If $\Delta_n = \partial \mathbb{N}^N \cap [0, [nt] + (1,\ldots,1)]$, one can check directly that

$$\mathbf{S}_n(t) - \mathbf{S}_n\left(\frac{[nt]}{n}\right) = n^{-\frac{N}{2}} \sum_{s \in \Delta_n} L_{s,t}\, \xi_s,$$

where $L_{s,t}$ is the Lebesgue measure of a certain subrectangle of the cube $\{r \in \mathbb{R}_+^N : s - (1,\ldots 1) \preccurlyeq r \preccurlyeq s\}$ (plot a picture!). All that we need to know of the L's is the simple fact that $|L_{s,t}| \le 1$. In particular, since the ξ's are i.i.d. with mean 0 and variance 1,

$$\mathbb{E}\left\{ \left| \mathbf{S}_n(t) - \mathbf{S}_n(\tfrac{[nt]}{n}) \right|^2 \right\} = n^{-N} \sum_{s \in \Delta_n} L_{s,t}^2 \le n^{-N} \#\Delta_n,$$

where $\#A$ denotes the cardinality of the set A. One can check that $\Delta_n \subset [0, [nt] + (1, \ldots, 1)] \setminus [0, [nt]]$. Therefore, for all $t \in \mathbb{N}^N$,

$$\mathbb{E}\left\{\left|\mathbf{S}_n(t) - \mathbf{S}_n(\tfrac{[nt]}{n})\right|^2\right\} \leq n^{-N}\left\{\prod_{\ell=1}^N [nt^{(\ell)} + 1] - \prod_{\ell=1}^N \lfloor nt^{(\ell)} \rfloor\right\}$$

$$= \prod_{\ell=1}^n \frac{\lfloor nt^{(\ell)} \rfloor}{n} \cdot \left\{\prod_{\ell=1}^N \left(\frac{\lfloor nt^{(\ell)} + 1 \rfloor}{\lfloor nt^{(\ell)} \rfloor}\right) - 1\right\}$$

$$\leq \prod_{\ell=1}^N \left(\frac{nt^{(\ell)} + 1}{nt^{(\ell)} - 1}\right) - 1$$

$$\leq \left(\frac{n+1}{n-1}\right)^N - 1,$$

since the expression in the penultimate line is increasing in $t^{(\ell)} \geq 1$ for all ℓ. As $n \to \infty$, this bound goes to 0, and our result follows. $\square$

4.3 Pretightness

Having verified convergence of the finite-dimensional distribution, we now prove the following:

Proposition 4.3.1 $\mathbf{S}_1, \mathbf{S}_2, \ldots$ *is pretight.*

Proof We will prove this only for $N = 2$. The case $N = 1$ is much easier, and $N \geq 3$ is proved by similar means as $N = 2$, but the notation is more cumbersome.

To prove pretightness, we will verify the condition of Theorem 3.5.1. A little thought shows that it suffices to verify that condition with $\ell = 1$; the $\ell = 2$ case is similar. Thus, we need only show that for all $\varepsilon > 0$,

$$\lim_{\delta \to 0} \limsup_{n \to \infty} \sup_{0 \leq t \leq 1} \mathbb{P}\left(\sup_{\substack{s \in [0,1]:\\ t \leq s \leq t+\delta}} \sup_{u \in [0,1]} |\mathbf{S}_n(s,u) - \mathbf{S}_n(t,u)| \geq \varepsilon\right) = 0. \quad (1)$$

Of course, in the above, $\mathbf{S}_n(s,u) = \mathbf{S}_n(p)$, where $p = (s,u) \in [0,1]^2$.

For notational simplicity, define for all $t \in [0,1]$ and $\delta > 0$,

$$\mathcal{Q}_n(t,\delta) = \sup_{\substack{s \in [0,1]:\\ t \leq s \leq t+\delta}} \sup_{u \in [0,1]} |\mathbf{S}_n(s,u) - \mathbf{S}_n(t,u)|. \quad (2)$$

By Proposition 4.1.1, $u \mapsto \mathbf{S}_n(s,u)$ is piecewise linear on each of the intervals $u \in [\tfrac{k}{n}, \tfrac{k+1}{n}]$ ($0 \leq k \leq n-1$). This leads to the following bound:

$$\mathcal{Q}_n(t,\delta) = \sup_{\substack{s \in [0,1]:\\ t \leq s \leq t+\delta}} \sup_{u \in [0,1]} \left|\mathbf{S}_n\left(s, \frac{\lfloor nu \rfloor}{n}\right) - \mathbf{S}_n\left(t, \frac{\lfloor nu \rfloor}{n}\right)\right|.$$

208   6. Limit Theorems

Hence,

$$\mathcal{Q}_n(t,\delta) \leq \sup_{\substack{s\in[0,1]:\\ t\leq s\leq t+\delta}} \sup_{u\in[0,1]} \left| \mathbf{S}_n\left(\frac{\lfloor ns\rfloor}{n}, \frac{\lfloor nu\rfloor}{n}\right) - \mathbf{S}_n\left(\frac{\lfloor nt\rfloor}{n}, \frac{\lfloor nu\rfloor}{n}\right) \right|$$

$$+ \sup_{\substack{s\in[0,1]:\\ t\leq s\leq t+\delta}} \sup_{u\in[0,1]} \left| \mathbf{S}_n\left(\frac{\lfloor ns\rfloor}{n}, \frac{\lfloor nu\rfloor}{n}\right) - \mathbf{S}_n\left(s, \frac{\lfloor nu\rfloor}{n}\right) \right|$$

$$+ \sup_{\substack{s\in[0,1]:\\ t\leq s\leq t+\delta}} \sup_{u\in[0,1]} \left| \mathbf{S}_n\left(\frac{\lfloor nt\rfloor}{n}, \frac{\lfloor nu\rfloor}{n}\right) - \mathbf{S}_n\left(t, \frac{\lfloor nu\rfloor}{n}\right) \right|.$$

Now we can use linearity in t (and s) to deduce that

$$\mathcal{Q}_n(t,\delta) \leq \sup_{\substack{s\in[0,1]:\\ t\leq s\leq t+\delta}} \sup_{u\in[0,1]} \left| \mathbf{S}_n\left(\frac{\lfloor ns\rfloor}{n}, \frac{\lfloor nu\rfloor}{n}\right) - \mathbf{S}_n\left(\frac{\lfloor nt\rfloor}{n}, \frac{\lfloor nu\rfloor}{n}\right) \right|$$

$$+ \sup_{\substack{s\in[0,1]:\\ t\leq s\leq t+\delta}} \sup_{u\in[0,1]} \left| \mathbf{S}_n\left(\frac{\lfloor ns\rfloor}{n}, \frac{\lfloor nu\rfloor}{n}\right) - \mathbf{S}_n\left(\frac{\lfloor ns\rfloor+1}{n}, \frac{\lfloor nu\rfloor}{n}\right) \right|$$

$$+ \sup_{\substack{s\in[0,1]:\\ t\leq s\leq t+\delta}} \sup_{u\in[0,1]} \left| \mathbf{S}_n\left(\frac{\lfloor nt\rfloor}{n}, \frac{\lfloor nu\rfloor}{n}\right) - \mathbf{S}_n\left(\frac{\lfloor nt\rfloor+1}{n}, \frac{\lfloor nu\rfloor}{n}\right) \right|.$$

Using Proposition 4.1.1 once more, we can relate this to the multiparameter random walk, all the time remembering that $N=2$:

$$\sup_{u\in[0,1]} |\mathbf{S}_n(s,u) - \mathbf{S}_n(t,u)| \leq \frac{1}{n} \sup_{u\in[0,1]} |S_{\lfloor ns\rfloor,\lfloor nu\rfloor} - S_{\lfloor nt\rfloor,\lfloor nu\rfloor}|$$

$$+ \frac{1}{n} \sup_{u\in[0,1]} |S_{\lfloor ns\rfloor,\lfloor nu\rfloor} - S_{\lfloor ns\rfloor+1,\lfloor nu\rfloor}|$$

$$+ \frac{1}{n} \sup_{u\in[0,1]} |S_{\lfloor nt\rfloor,\lfloor nu\rfloor} - S_{\lfloor nt\rfloor+1,\lfloor nu\rfloor}|.$$

Thus, for any $\delta > 0$ and any $t \in [0,1]$,

$$\mathcal{Q}_n(t,\delta) \leq \frac{1}{n} \max_{1\leq k\leq \lfloor n\delta\rfloor+1} \max_{\substack{u\in\mathbb{N}:\\ u\leq n}} |S_{\lfloor nt\rfloor+k,u} - S_{\lfloor nt\rfloor,u}|$$

$$+ \frac{2}{n} \sup_{t\leq s\leq t+\delta} \max_{\substack{u\in\mathbb{N}:\\ u\leq n}} |S_{\lfloor ns\rfloor,u} - S_{\lfloor ns\rfloor+1,u}|.$$

By the triangle inequality, for the above values of s and u we have

$$|S_{\lfloor ns\rfloor,u} - S_{\lfloor ns\rfloor+1,u}| \leq |S_{\lfloor ns\rfloor,u} - S_{\lfloor nt\rfloor,u}| + |S_{\lfloor nt\rfloor,u} - S_{\lfloor ns\rfloor+1,u}|$$

$$\leq 2 \max_{1\leq k\leq \lfloor n\delta\rfloor+2} |S_{k+\lfloor nt\rfloor,u} - S_{\lfloor nt\rfloor,u}|.$$

In particular, for any $t \in [0,1]$, $n \geq 1$, and $\delta > 0$,

$$\mathcal{Q}_n(t,\delta) \leq 5n^{-1} \max_{\substack{1 \leq k \leq \lfloor n\delta \rfloor + 2 \\ 1 \leq j \leq n}} |S_{k+\lfloor nt \rfloor, j} - S_{\lfloor nt \rfloor, j}|.$$

By the stationarity of the increments of S, for all $t \in [0,1]$, $\delta > 0$, $n \geq 1$, and $\lambda > 0$,

$$\mathbb{P}(\mathcal{Q}_n(t,\delta) \geq \lambda) \leq \mathbb{P}\left(\max_{\substack{1 \leq k \leq \lfloor n\delta \rfloor + 2 \\ 1 \leq j \leq n}} |S_{k,j}| \geq \frac{1}{5}\lambda n \right),$$

which is independent of $t \in [0,1]$! Thus, we can apply Lemma 2.7.2 of Chapter 4 to see that for all $\alpha, \varepsilon > 0$,

$$\mathbb{P}(\mathcal{Q}_n(t,\delta) \geq \alpha + 2\varepsilon)$$
$$\leq \mathbb{P}\left(|S_{\lfloor n\delta \rfloor + 2, n}| \geq \frac{\alpha n}{5}\right) \times \left[\mathbb{P}\left(\max_{\substack{1 \leq k \leq \lfloor n\delta \rfloor + 2 \\ 1 \leq j \leq n}} |S_{k,j}| \leq \frac{\varepsilon n}{5} \right) \right]^{-2}. \quad (3)$$

Note that $\mathbb{E}[S^2_{\lfloor n\delta \rfloor + 2, n}] = n(\lfloor n\delta \rfloor + 2) \leq 2\delta n^2$, as long as $n \geq \delta^{-1}$. Thus, we can apply Lemma 2.7.1 of Chapter 4 with $N = 2$, $\sigma_\star = 1$, and $Z_t = X_t$ to see that for all $\lambda > 8$ and for all $n \geq \delta^{-1}$,

$$\mathbb{P}\left(\max_{\substack{1 \leq k \leq \lfloor n\delta \rfloor + 2 \\ 1 \leq j \leq n}} |S_{k,j}| \leq \sqrt{2\delta} \lambda n \right) \geq \frac{1}{2}.$$

Plugging this into equation (3) with $\lambda = 10$, we see that for all $\varepsilon, \alpha > 0$, with $0 < \delta < e^{-200}\varepsilon^2$, and for all $n \geq \delta^{-1}$,

$$\mathbb{P}(\mathcal{Q}_n(t,\delta) \geq \alpha + 2\varepsilon) \leq 4\mathbb{P}\left(|S_{\lfloor n\delta \rfloor + 2, n}| \geq \frac{\alpha n}{5}\right).$$

By Proposition 4.1.1, $n^{-1} S_{\lfloor n\delta \rfloor + 2, n} = \mathbf{S}_n\left(\frac{\lfloor n\delta \rfloor}{n} + \frac{2}{n}, 1\right)$. In addition, Proposition 4.2.1 and its proof together show that as $n \to \infty$, this converges in distribution to $B_{\delta,1}$, where B is the standard Brownian sheet. Therefore, for all $\varepsilon, \alpha > 0$ such that $0 < \delta < e^{-200}\varepsilon^2$,

$$\limsup_{n \to \infty} \sup_{t \in [0,1]} \mathbb{P}(\mathcal{Q}_n(t,\delta) \geq \alpha + \varepsilon 2) \leq 4\mathbb{P}\left(|B_{\delta,1}| \geq \frac{\alpha}{5}\right).$$

Since $\delta^{-1} B_{\delta,1}$ is Gaussian with mean 0 and variance 1, Supplementary Exercise 11 of Chapter 4 implies that if $0 < \delta < e^{-200}\varepsilon^2$,

$$\limsup_{n \to \infty} \sup_{t \in [0,1]} \mathbb{P}(\mathcal{Q}_n(t,\delta) \geq \alpha + 2\varepsilon) \leq 8 \exp\left(-\frac{\alpha^2}{50\delta^2}\right).$$

In particular, for all $\alpha, \varepsilon > 0$,

$$\lim_{\delta \to 0} \limsup_{n \to \infty} \frac{1}{\delta} \sup_{t \in [0,1]} \mathbb{P}(\mathcal{Q}_n(t, \delta) \geq \alpha + 2\varepsilon) = 0.$$

In light of equation (2), we obtain equation (1), and hence the result when $N = 2$. □

Exercise 4.3.1 Verify Proposition 4.3.1 when $N \neq 2$. □

5 Supplementary Exercises

1. Suppose $X_1, X_2, \ldots, X_\infty$ are S-valued random variables, where (S, d) is a separable metric space, and that for all $\varepsilon > 0$, $\lim_{n \to \infty} \mathbb{P}\{d(X_n, X_\infty) \geq \varepsilon\} = 0$. Prove that $X_n \Longrightarrow X_\infty$. That is, show that convergence in probability implies weak convergence.

2. For any probability measure $\mathbb{P}$ on a separable metric space T, consider sets of the form

$$U_{f,\varepsilon}(\mathbb{P}) = \left\{ \mathbb{Q} \in \mathcal{P}(T) : \left| \int_T f(\omega)\, \mathbb{Q}(d\omega) - \int_T f(\omega)\, \mathbb{P}(d\omega) \right| < \varepsilon \right\},$$

where $f : T \to \mathbb{R}$ is bounded and continuous, $\mathcal{P}(T)$ denotes the collection of all probability measures on T, and $\varepsilon > 0$. Show that the topology generated by such sets $U_{f,\varepsilon}(\mathbb{P})$ topologizes weak convergence.

3. In the notation of Section 3.3, suppose $(X_n; 1 \leq n \leq \infty)$ is a sequence of C-valued random variables such that as $n \to \infty$, X_n converges weakly in C to X_∞. Prove that:
   (i) for all $t_1, \ldots, t_k \in [0,1]^N$, $(X_n(t_1), \ldots, X_n(t_k))$ converges weakly in $\mathbb{R}^k$ to $(X_\infty(t_1), \ldots, X_\infty(t_k))$; and
   (ii) for all $\varepsilon > 0$, $\lim_{\delta \to 0+} \limsup_{n \to \infty} \mathbb{P}(\omega_{X_n}(\delta) \geq \varepsilon) = 0$.

4. The intention of this exercise is to complete the proof of Prohorov's theorem in the noncompact case; cf. Theorem 2.5.1.
   (i) Show that the tightness of $(\mathbb{P}_n)$ implies the existence of compact sets $K_1 \subset K_2 \subset \cdots$, all subsets of T, such that for all $m \geq 1$, $\sup_n \mathbb{P}_n(K_m^\complement) \leq (2m)^{-1}$.
   (ii) For all $n, m \geq 1$, define $\mathbb{Q}_n^m(\bullet) = \mathbb{P}_n(\bullet \cap K_m)/\mathbb{P}_n(K_m)$. Prove that there exists a subsequence n' and for each m, there exists a probability measure $\mathbb{Q}_\infty^m$ on K_m such that as $n \to \infty$, $\mathbb{Q}_n^m \Longrightarrow \mathbb{Q}_\infty^m$, for each $m \geq 1$.
   (iii) Prove that for all $m \geq 1$ and all Borel sets $E \subset K_m$, $\mathbb{Q}_\infty^m(E) = \mathbb{Q}_\infty^{m+1}(E)$. That is, the probability measures $\mathbb{Q}_\infty^m$ are nested.
   (iv) Use (iii) to complete the proof of Theorem 2.5.1.

5. Suppose $S = (S_n; n \geq 1)$ denotes the simple walk on $\mathbb{Z}^1$. Prove that as $n \to \infty$, $n^{-1/2} \max_{k \leq n} S_k$ converges weakly in $\mathbb{R}$ to $|\mathcal{N}|$, where $\mathcal{N}$ denotes a standard Gaussian random variable. Use this to conclude the following: Given arbitrary i.i.d. random variables $X_1, X_2, \ldots$, with mean 0 and variance 1, let $S_n = \sum_{j=1}^n X_j$ to see that for all $\lambda > 0$,

$$\lim_{n \to \infty} \mathbb{P}\left(\max_{1 \leq k \leq n} S_k \leq \lambda \sqrt{n}\right) = \sqrt{\frac{2}{\pi}} \int_0^\lambda e^{-\frac{1}{2}x^2} \, dx.$$

Relate this to equation (2) of the preamble to Section 4.
(HINT: Start with Supplementary Exercise 3, Chapter 4.)

6. Let $(S_n; n \geq 1)$ be a mean-zero variance-one random walk (1-parameter) and let $\mathbf{S}_n$ be the associated process that was defined in Section 4.1 (once more, $N = 1$).
   (i) Apply Theorem 4.1.1 to verify that as $n \to \infty$, the random variables $\int_0^1 \mathbf{S}_n(t) \, dt$ converge in distribution to $\int_0^t B_s \, ds$.
   (ii) Use the above to prove that as $n \to \infty$, $n^{-3/2} \sum_{k=1}^n S_k$ converges in distribution (in $\mathbb{R}$) to $\int_0^1 B_s \, ds$.
   (iii) Find the latter distribution in terms of its probability density function. Refine this by proving that if S is an N-parameter random walk with mean 0 and variance 1, there exists $\alpha_n \to 0$ such that $\alpha_n \sum_{t \preccurlyeq (n,\ldots,n)} S_t$ has a distributional limit. Find α_n and identify this limit.

(HINT: For part (iii), start by proving that $\int_0^t B_s \, ds$ is Gaussian random variable. In fact, any bounded linear functional of the random function $t \mapsto B_t$ is Gaussian.)

7. Suppose that S is an N-parameter random walk with mean 0 and variance 1, and that $f : \mathbb{R} \to \mathbb{R}$ is a continuous function. Find $\alpha_n \to \infty$ such that as $n \to \infty$, $\alpha_n \sum_{t \preccurlyeq (n,\ldots,n)} f(S_t)$ converges in distribution. Identify the limiting distribution.

8. Let $U_1, U_2, \ldots$ denote i.i.d. random variables, all chosen uniformly from $[0,1]$. Define $\alpha_n(t) = \frac{1}{n} \sum_{j=1}^n \mathbf{1}_{[0,t]}(U_j)$, $(n \geq 1, t \in [0,1])$. The random distribution function α_n is called the **empirical distribution function** for the "data" $\{U_1, \ldots, U_n\}$.
   (i) Verify that the finite-dimensional distributions of the process $t \mapsto \sqrt{n}(\alpha_n(t) - t)$ $(t \in [0,1])$ converge, as $n \to \infty$, to those of $t \mapsto B_t - tB_1$ $(t \in [0,1])$, where B denotes standard Brownian motion.
   (HINT: Use a multidimensional central limit theorem for multinomials.)
   (ii) We would like to state that as $n \to \infty$, $t \mapsto \sqrt{n}(\alpha_n(t) - t)$ $(t \in [0,1])$ converges in C to $t \mapsto B_t - tB_t$ $(t \in [0,1])$. Unfortunately, $\alpha_n \notin C([0,1])$. To get around this, show that there are random functions $A_n, B_n \in C([0,1])$ such that (a) $A_n(t) \leq \alpha_n(t) \leq B_n(t)$ for all $n \geq 1$ and all $t \in [0,1]$; (b) $\sup_{t \in [0,1]} |A_n(t) - B_n(t)| \to 0$; and (c) A_n and B_n both converge weakly in $C([0,1])$, as $n \to \infty$, to $t \mapsto B_t - tB_1$ $(t \in [0,1])$.
   (HINT: Approximate the random step function $t \mapsto \alpha_n(t)$ by random piecewise linear functions.)

(iii) Prove that as $n \to \infty$, $\sqrt{n} \sup_{t \in [0,1]} |\alpha_n(t) - t|$ converges weakly. Identify this limit.

(iv) Prove that as $n \to \infty$, $\alpha_n(t) \to t$, uniformly in $t \in [0,1]$, a.s.
(HINT: Convergence in probability follows from (iii). For the a.s. convergence, work from first principles.)

This, the Glivenko–Cantelli theorem, is one of the fundamental results of empirical processes. See (Billingsley 1968; Dudley 1984; Gänssler 1983; Pollard 1984) for various refinements, and for related discussions. The Gaussian process $(B_t - tB_1;\ t \in [0,1])$ is called the **Brownian bridge** on $[0,1]$.

9. Let $B = (B_t;\ t \in \mathbb{R}_+^N)$ denote the standard N-parameter Brownian sheet and define a stochastic process $B^\circ = (B_t^\circ;\ t \in [0,1]^N)$, indexed by $[0,1]^N$, as follows: $B_t^\circ = B_t - \prod_{j=1}^N t^{(j)} B_{(1,\ldots,1)}$, $(t \in [0,1]^N)$. This is the "Brownian sheet pinned to zero at time $(1,\ldots,1)$" (or the N-parameter Brownian bridge.)

(i) Check that B° is a Gaussian process, and compute its mean and covariance functions.

(ii) Check that the *entire* process B° is independent of $B_{(1,\ldots,1)}$.

(iii) Define measures $(\mathbb{P}_\varepsilon;\ \varepsilon \in\ ]0,1[)$ on Borel subsets of $C([0,1]^N)$ by

$$\mathbb{P}_\varepsilon(\bullet) = \mathbb{P}(B \in \bullet \mid 0 \le B_{(1,\ldots,1)} \le \varepsilon).$$

Show that $\mathbb{P}_\varepsilon$ is a probability measure on $C([0,1]^N)$ and prove that for all closed sets $F \subset C([0,1]^N)$, $\lim_{\varepsilon \to 0+} \mathbb{P}_\varepsilon(F) \le \mathbb{P}(B^\circ \in F^\eta)$, where F^η is the closed η-enlargement of F. Conclude that $\mathbb{P}_\varepsilon$ converges in $C([0,1]^N)$ to the distribution of B°. Intuitively speaking, this states that the pinned process B° is the process B, conditioned to be 0 at time $t = (1,\ldots,1)$.

(iv) What can you say about the asymptotic behavior of the measures $\mathbb{Q}_\varepsilon$ defined by $\mathbb{Q}_\varepsilon(A) = \mathbb{P}(B \in A \mid |B_{(1,\ldots,1)}| \le \varepsilon)$?

(v) Let $B = (B_{(s,t)};\ s,t \ge 0)$ denote the 2-parameter Brownian sheet and consider the 2-parameter process $B^| = (B^|_{(s,t)};\ s,t \in [0,1])$ defined as

$$B^|_{(s,t)} = B_{(s,t)} - sB_{(1,t)}, \qquad 0 \le s,t \le 1.$$

Provide a weak convergence justification for the statement that $B^|$ is B conditioned to be 0 on the line $\{(1,t) : t \in [0,1]\}$.
(HINT: First show that the 2-parameter process $B^|$ is independent of the process $(B_{(1,t)};\ t \in [0,1])$.)

10. Let $B = (B_t;\ t \ge 0)$ denote the standard 1-dimensional Brownian motion.

(i) For all $n \ge 1$, define $V_n(t) = \sum_{0 \le j \le 2^n t}(B_{(j+1)2^{-n}} - B_{j2^{-n}})^2$, whenever $t \ge 0$ is of the form $t = j2^{-n}$, and for other $t \ge 0$, define the random function V_n by linearly interpolating these values between t's of the form $j2^{-n}$, $j = 0, 1, \ldots$. Prove that with probability one, as $n \to \infty$, $V_n(t)$ converges to t, uniformly on t-compacta.

(ii) Find constants μ, α, and σ such that as $n \to \infty$, the process $t \mapsto \sigma^{-1} n^\alpha \{V_n(t) - \mu t\}$ converges weakly to Brownian motion in $C([0,1])$.

Does this have an N-parameter extension? You should also consult Supplementary Exercise 2, Chapter 5.
(HINT: For part (i), first compute the mean and the variance. Next, show that for any finite set $F \subset [0,1]$ and all $\varepsilon > 0$, $\mathbb{P}(\max_{t \in F} |V_n(t) - t| \geq \varepsilon)$ is small. Interpolate between the points in a sufficiently well chosen $F = F_n$.)

6 Notes on Chapter 6

Sections 1–3 Our construction of abstract random variables is completely standard. The material on weak convergence is modeled after Billingsley (1968) and is a subset of the rich development there. As a warning to the functional analysis aficionado, we should mention that in the probability literature, "weak convergence" translates to "*weak-⋆ convergence*" of analysis (and *not* "*weak convergence*"!) Theorem 2.4.1 is from Oxtoby and Ulam (1939) The term "*tightness*" is due to Lucien LeCam; cf. LeCam (1957). Our proof of Prohorov's theorem in the compact case is philosophically different from the usual ones that can be found, for instance, in (Billingsley 1968; Rudin 1973), although there are some similarities.

Section 5 Supplementary Exercise 9 is modeled after Billingsley (1968, equation (11.31)), whereas Supplementary Exercise 10 is classical, and a part of the folklore of the subject. For a variant on random sceneries, see Khoshnevisan and Lewis (1998).

Part II
Continuous-Parameter Random Fields

7
Continuous-Parameter Martingales

The second part of this book starts with a continuous-parameter extension of the discrete-parameter theory of Chapter 1. Our use of the term "extension" is quite misleading. Indeed, we will quickly find that in order to carry out these "extensions," one needs a good understanding of the regularity of the sample functions of multiparameter stochastic processes; this will require a great effort. However, we will be rewarded for our hard work, since it will lead to a successful continuous-parameter theory that, in many ways, probes much more deeply than its discrete-parameter counterpart. Moreover, this theory lies at the foundations of nearly all of the random fields that arise throughout the rest of this book and a great deal more. Viewed as such, this chapter is simply indispensable for those who wish to read on.

We will also discuss elements of hyperbolic stochastic partial differential equations as an interesting area with ready applications.

1 One-Parameter Martingales

In continuous-time, and in informal terms, we can introduce an $\mathbb{R}_+^N$-indexed martingale as we defined an $\mathbb{N}_0^N$-indexed martingale, but replacing $\mathbb{N}_0^N$-valued parameters by $\mathbb{R}_+^N$-valued ones. While this is simple enough to understand, attempts at developing a viable continuous-time theory will quickly encounter a number of technical and conceptual problems. These difficulties are easier to isolate in the simpler one-parameter setting, which will be our starting point. This presentation will be extended to the multi-parameter setting afterwards.

1.1 Filtrations and Stopping Times

Motivated by Chapter 1, we say that a collection $\mathcal{F} = (\mathcal{F}_t;\ t \geq 0)$ of sub–σ-fields of the underlying σ-field $\mathcal{G}$ is a (one-parameter) **filtration** if for all $0 \leq s \leq t$, $\mathcal{F}_s \subset \mathcal{F}_t$. We emphasize that, here, the variables s and t are $\mathbb{R}_+$-valued.

Suppose $X = (X_t;\ t \geq 0)$ is an S-valued stochastic process, where (S, d) is a metric space. We say that X is **adapted** to the filtration $\mathcal{F}$ if for all $t \geq 0$, X_t is $\mathcal{F}_t$-measurable. We also say that a $[0, \infty]$-valued random variable T is a **stopping time** if

$$(T \leq t) \in \mathcal{F}_t, \qquad t \geq 0.$$

In order to highlight the dependence of T on the filtration $\mathcal{F}$, we may sometimes refer to T as an $\mathcal{F}$-stopping time. The following exercise shows that some of the important properties of stopping times are preserved in the transition from discrete to continuous-time.

***Exercise* 1.1.1** Prove that whenever $T_1, T_2, \ldots$ are stopping times, so are $\inf_{n \geq 1} T_n$, $\sup_{n \geq 1} T_n$, and $\sum_{i=1}^{\infty} T_i$. Moreover, show that all nonrandom times are stopping times. □

For any stopping time T, we define

$$\mathcal{F}_T = \left(A \in \bigvee_{t \geq 0} \mathcal{F}_t :\ A \cap (T \leq t) \in \mathcal{F}_t,\ \text{for all } t \geq 0 \right).$$

The collection $\mathcal{F}_T$ is, in some sense, similar to its discrete-time counterpart. For example:

(i) $\mathcal{F}_T$ is a σ-field;

(ii) if T is nonrandom, say $T = k$, then $\mathcal{F}_T = \mathcal{F}_k$;

(iii) T is an $\mathcal{F}_T$-measurable random variable; and

(v) if $T \leq S$ are both $\mathcal{F}$-stopping times, then $\mathcal{F}_T \subset \mathcal{F}_S$.

***Exercise* 1.1.2** Verify that the above properties hold true. □

Letting $X_T(\omega) = X_{T(\omega)}(\omega)$, we would also like to know that whenever X is adapted to $\mathcal{F}$, and if T is an $\mathcal{F}$-stopping time, then X_T is $\mathcal{F}_T$-measurable. Here, we meet our first stumbling block, since X_T is, in general, not $\mathcal{F}_T$-measurable.

Example Let T be a strictly positive random variable with an absolutely continuous distribution, and define

$$X_t = \begin{cases} 0, & \text{if } 0 \leq t < T, \\ \xi, & \text{if } t = T, \\ 1, & \text{if } t > T, \end{cases}$$

where $\xi > 1$ and is independent of T. (We can arrange it so that $T(\omega) > 0$ and $\xi(\omega) > 1$ for all $\omega \in \Omega$. Thus, there is no need for a.s. statements here.)

Define the filtration $\mathcal{F} = (\mathcal{F}_t;\, t \geq 0)$ as follows: For all $t \geq 0$, $\mathcal{F}_t$ denotes the σ-field generated by the collection of random variables $(X_r;\, 0 \leq r \leq t)$. By its very definition, X is adapted to $\mathcal{F}$, and we note also that

$$T = \inf(s > 0 :\, X_s > 0).$$

Therefore:

(P1) $T(\omega) > t$ if and only if for all $0 \leq s \leq t$, $X_s(\omega) = 0$.

(P2) Conversely, $T(\omega) < t$ if and only if for all $s \geq t$, $X_s(\omega) = 1$.

This shows that T is a stopping time, since for any $t \geq 0$,

$$(T > t) = (X_t = 0) \in \mathcal{F}_t.$$

On the other hand, not only is X_T not $\mathcal{F}_T$-measurable, but in fact, X_T is independent of $\mathcal{F}_T$. To see this, fix some $t > 0$ and consider bounded, measurable functions $g, f_1, \ldots, f_n : \mathbb{R}_+ \to \mathbb{R}$ and a collection of points $s_1, \ldots, s_n$ such that for some $k \geq 1$, $0 \leq s_1 < \cdots < s_k \leq t < s_{k+1} < \cdots < s_n$. Since T has an absolutely continuous distribution,

$$\mathbb{E}\Big[\prod_{i=1}^n f_i(X_{s_i}) g(\xi) \mathbf{1}_{(T>t)}\Big] = \sum_{\ell=k+1}^n \mathbb{E}\Big[\prod_{i=1}^n f_i(X_{s_i}) g(\xi) \mathbf{1}_{(s_\ell < T < s_{\ell+1})}\Big]$$
$$+ \mathbb{E}\Big[\prod_{i=1}^n f_i(X_{s_i}) g(\xi) \mathbf{1}_{(t < T < s_{k+1})}\Big],$$

where $s_{n+1} = \infty$, for notational convenience. Using (P1), (P2), and the independence of ξ from T, we can deduce that

$$\mathbb{E}\Big[\prod_{i=1}^n f_i(X_{s_i}) g(\xi) \mathbf{1}_{(T>t)}\Big] = \sum_{\ell=k+1}^n \prod_{i=1}^\ell f_i(0) \cdot \prod_{j=\ell+1}^n f_j(1)\, \mathbb{E}\big[g(\xi) \mathbf{1}_{(s_\ell < T < s_{\ell+1})}\big]$$
$$+ \prod_{i=1}^k f_i(0) \cdot \prod_{j=k+1}^n f_j(1)\, \mathbb{E}\big[g(\xi) \mathbf{1}_{(t < T < s_{k+1})}\big]$$
$$= \mathbb{E}[g(\xi)] \cdot \mathbb{E}\Big[\prod_{i=1}^n f_i(X_{s_i}) \mathbf{1}_{(T>t)}\Big].$$

By a monotone class argument, for any bounded, $\vee_{t \geq 0} \mathcal{F}_t$-measurable random variable Z,

$$\mathbb{E}[Z g(\xi) \mathbf{1}_{(T>t)}] = \mathbb{E}[g(\xi)] \cdot \mathbb{E}[Z \mathbf{1}_{(T>t)}].$$

Take $Z = \mathbf{1}_A$, where $A \in \vee_{t \geq 0} \mathcal{F}_t$, to see that $X_T = \xi$ is independent of $\mathcal{F}_T$ (why?). In fact, we have shown the surprising fact that X_T is independent of $\vee_{t \geq 0} \mathcal{F}_t$. That is, X_T is independent of the entire process X. □

This example shows one of the peculiarities of the theory of continuous-time stochastic processes: If $t \mapsto X_t$ is ill-behaved, X_T need not be $\mathcal{F}_T$-measurable. (In the example, $t \mapsto X_t$ has a discontinuity of the second kind at the random point $t = T$.) The following result demonstrates a kind of converse to this.

Theorem 1.1.1 *Suppose X is a right-continuous, S-valued stochastic process that is adapted to a filtration $\mathcal{F}$. Then, for all measurable functions $f : S \to \mathbb{R}$ and for all $\mathcal{F}$-stopping times T, $f(X_T)\mathbf{1}_{(T<\infty)}$ is an $\mathcal{F}_T$-measurable, real-valued random variable.*

A word of caution: $f(X_T)\mathbf{1}_{(T<\infty)}$ is *defined* to be 0 on the event $(T = \infty)$. Our proof of this theorem relies on the following approximation result.

Lemma 1.1.1 *Let $\mathcal{F} = (\mathcal{F}_t;\ t \geq 0)$ be a filtration and T an $\mathcal{F}$-stopping time. For all $n \geq 1$, define*

$$T^n = \begin{cases} \sum_{k=0}^{\infty} (k+1) 2^{-n} \mathbf{1}_{(k2^{-n} \leq T < (k+1)2^{-n})}, & \text{if } T < \infty, \\ +\infty, & \text{otherwise.} \end{cases}$$

Then:

(i) for each $n \geq 1$, T^n is an $\mathcal{F}$-stopping time; and

(ii) $T^n \downarrow T$, as $n \to \infty$.

Exercise **1.1.3** Prove Lemma 1.1.1. □

Proof of Theorem 1.1.1 Note that X_T is defined only on $(T < \infty)$, while $(T = \infty)$ need not be empty. Therefore, we need to consider two classes of ω's: those for which $T(\omega) < \infty$ and those for which $T(\omega) = \infty$. To get around this, introduce a point $\delta \notin S$, called the cemetery state. Let $S' = S \cup \{\delta\}$ and topologize S' by declaring $E \cup \{\delta\}$ open whenever $E \subset S$ is open. Endow S' with the induced Borel field and extend any $f : S \to \mathbb{R}$ to a function on S' by $f(\delta) = 0$.

It is easy to see that if $f : S \to \mathbb{R}$ is measurable, its extension to S'—still denoted by f—is a measurable function from S' into $\mathbb{R}$. We also extend the process X by defining $X_\infty(\omega) = \delta$ for all $\omega \in \Omega$. With this extension in mind, we need to show that for all Borel sets $E \subset \mathbb{R}$ and all $t \geq 0$, $(f(X_T) \in E) \cap (T \leq t) \in \mathcal{F}_t$. In fact, upon writing $(T \leq t)$ as $(T = t) \cup (T < t)$, we need only show that for all Borel sets $E \subset \mathbb{R}$ and all $t > 0$,

$$(f(X_T) \in E) \cap (T < t) \in \mathcal{F}_t.$$

On the other hand, since $f^{-1}(E) = (s \in S' : f(s) \in E)$ is a measurable subset of S', it suffices to show that for all Borel sets $F \subset S'$,

$$(X_T \in F) \cap (T < t) \in \mathcal{F}_t, \qquad t \geq 0. \tag{1}$$

The collection of all sets F that satisfy the above is a monotone class. Therefore, we need only verify (1) for all open sets F.

Let T^n be as in Lemma 1.1.1 and note that for any open $F \subset S'$,

$$(X_T \in F) \cap (T < t) = \bigcup_{m=1}^{\infty} \bigcap_{n=m}^{\infty} (X_{T^n} \in F) \cap (T^n < t).$$

(Why?) This equals

$$\bigcup_{m=1}^{\infty} \bigcap_{n=m}^{\infty} \bigcup_{\substack{k \in \mathbb{N}^N : \\ k2^{-n} < t}} (X_{k2^{-n}} \in F) \cap (T^n = k2^{-n}),$$

which is clearly in $\mathcal{F}_t$. □

1.2 Entrance Times

Suppose $X = (X_t;\, t \geq 0)$ is a random process that takes values in a metric space (S, d), and is adapted to a filtration $\mathcal{F} = (\mathcal{F}_t;\, t \geq 0)$. The **entrance (or hitting) time** of any set $E \subset S$ is defined as

$$T_E = \inf(s \geq 0 :\ X_s \in E).$$

We now show that in many cases of interest, T_E is a stopping time.

Theorem 1.2.1 *Suppose X is a right-continuous, S-valued stochastic process that is adapted to a filtration $\mathcal{F}$. Then, for all open sets $E \subset S$, T_E is a stopping time. If $t \mapsto X_t$ is continuous, then for all closed sets $E \subset S$, T_E is a stopping time.*

Proof Since E is open and X is right-continuous, whenever $\omega \in (T_E \leq t)$, either $X_t(\omega) \in E$ or there exists a rational $r < t$ such that $X_r(\omega) \in E$. That is,

$$(T_E \leq t) = \bigcup_{r \in \mathbb{Q}_+ \cap [0, t[} (X_r \in E) \cup (X_t \in E),$$

which is in $\mathcal{F}_t$ and shows that T_E is a stopping time.

For the second assertion, we suppose that $t \mapsto X_t$ is continuous and note that for all $t \geq 0$,

$$(T_E \leq t) = \bigcap_{\varepsilon \in \mathbb{Q}_+} \bigcup_{s \in [0,t] \cap \mathbb{Q}_+} (d(\{X_s\}, E) \leq \varepsilon),$$

where $d(F, E) = \inf\{d(x, y);\ x \in F, y \in E\}$. Since $\{x \in S :\ d(\{x\}, E) \leq \varepsilon\}$ is closed, it is a measurable subset of S. Thus, $(T \leq t) \in \mathcal{F}_t$, as desired. □

1.3 Smartingales and Inequalities

A stochastic process M is said to be a **submartingale** with respect to a filtration $\mathcal{F}$ if:

(i) M is adapted to $\mathcal{F}$;

(ii) for all $t \geq 0$, $\mathbb{E}\{|M_t|\} < +\infty$; and

(iii) for all $t, s \geq 0$, $\mathbb{E}\{M_{t+s} \mid \mathcal{F}_s\} \geq M_s$, a.s.

It is said to be **supermartingale** if $-M$ is a submartingale, and it is a **martingale** if it is both a sub- and a supermartingale. If M is either a sub- or a supermartingale, we refer to it as a **smartingale**.

At first glance, it may seem that the theory of continuous-time smartingales is the same as its discrete-time relative. However, the example of Section 1.1 shows that unless $t \mapsto M_t$ is well-behaved, one cannot possibly hope for an extensive theory. On the other hand, if M is right-continuous (say), then it has many nice properties. We list some of them below.

Theorem 1.3.1 *If M is a nonnegative, right-continuous submartingale:*

(i) *(Weak (1,1) inequality) for all $t, \lambda > 0$,*

$$\mathbb{P}\left(\sup_{0 \leq s \leq t} M_s \geq \lambda\right) \leq \frac{1}{\lambda} \mathbb{E}[M_t \mathbf{1}_{(\sup_{0 \leq s \leq t} M_s \geq \lambda)}];$$

(ii) *(Strong (p,p) inequality) for all $p > 1$ and all $t \geq 0$,*

$$\mathbb{E}\left[\sup_{0 \leq s \leq t} M_s^p\right] \leq \left(\frac{p}{p-1}\right)^p \mathbb{E}[M_t^p]; \text{ and}$$

(iii) *(The $L \ln L$ inequality) for all $t \geq 0$,*

$$\mathbb{E}\left[\sup_{0 \leq s \leq t} M_s\right] \leq \left(\frac{e}{e-1}\right)\{1 + \mathbb{E}[M_t \ln_+ M_t]\},$$

where $\ln_+ x = \ln(x \wedge 1)$.

Proof Consider finite sets $F_k \subset [0,t] \cap \mathbb{Q}_+$ ($k \geq 1$) such that as $k \to \infty$, F_k increases to $[0,t] \cap \mathbb{Q}_+$. By right continuity, as $k \to \infty$, $\sup_{s \in F_k} M_s$ converges upwards to $\sup_{0 \leq s \leq t} M_s$. To prove (i), apply Theorem 1.3.1 of Chapter 1 to $\sup_{s \in F_k} M_s$ and use the monotone convergence theorem to take limits. Parts (ii) and (iii) follow from applying a similar argument, and using Theorems 1.4.1 and 1.5.1 of Chapter 1, respectively. $\square$

1.4 Regularity

Section 1.3 shows that right-continuous smartingales in continuous-time have many of the desirable properties of discrete-parameter martingales. In this subsection we seek conditions that guarantee that our smartingale has a right-continuous modification that is itself a smartingale. (Recall Section 2.2 of Chapter 5 for the definition of modifications.)

Throughout, we assume that the underlying probability space is complete. If not, we can always complete it without changing anything critical.

Given a filtration $\mathcal{F}$, we can define a filtration $\mathcal{F}^\star$, in two stages, as follows: First, define $\overline{\mathcal{F}}_t$ as the $\mathbb{P}$-completion of $\mathcal{F}_t$ for all $t \geq 0$. That is, $\overline{\mathcal{F}}_t$ is the σ-field generated by the collection of all sets of the form $A \cup B$, where $A \in \mathcal{F}_t$ and B is a subset of a $\mathbb{P}$-null set. Clearly, $\overline{\mathcal{F}}$ is a filtration. Now we can define

$$\mathcal{F}_t^\star = \bigcap_{s>t} \overline{\mathcal{F}}_s, \qquad t \geq 0.$$

In words, $\mathcal{F}_t^\star$ is defined as the "right-continuous extension" of $\mathcal{F}_t$ for each $t \geq 0$.

It should not be hard to check that $\mathcal{F}^\star$ is also a filtration. The main result of this section is the following:

Theorem 1.4.1 *Suppose M is a submartingale with respect to the filtration $\mathcal{F}^\star$. If $t \mapsto \mathbb{E}[M_t]$ is right-continuous, then M has a right-continuous modification.*

Remarks (i) The said modification is necessarily a right-continuous submartingale with respect to $\mathcal{F}^\star$; cf. Exercise 1.4.1 below.

(ii) If $Y \in L^1(\mathbb{P})$, then $M_t = \mathbb{E}[Y \mid \mathcal{F}_t^\star]$ is a martingale for any version of this conditional expectation. Theorem 1.4.1 implies that there exists a version of this conditional expectation such that $t \mapsto \mathbb{E}[Y \mid \mathcal{F}_t^\star]$ is right-continuous; see Exercise 1.4.1.

(iii) $\mathcal{F}^\star$ is complete and right-continuous; i.e., for all $t \geq 0$, $\mathcal{F}_t^\star = \cap_{s>t} \mathcal{F}_s^\star$.

Exercise **1.4.1** Verify the claims of the above remarks. □

Theorem 1.4.1 is proved in two stages that are stated as the following lemmas.

Lemma 1.4.1 *With probability one, $\lim_{r \downarrow t : r \in \mathbb{Q}_+} M_r$ exists, simultaneously for all $t \geq 0$.*

Note the order of the quantifiers, namely, that there exists one null set Θ such that for all $\omega \notin \Theta$, $\lim_{r \downarrow t : r \in \mathbb{Q}_+} M_r(\omega)$ exists for all $t \geq 0$.

Proof The measurability of the event in question is part of the statement of the lemma, and will be proved shortly. Note that this assertion is not obvious, since the event in question involves an uncountable number of t's.

For any two real numbers $a < b$, and for any finite set $F \subset [0, \infty[$, define $U_F[a,b]$ to be the number of upcrossings of the interval $[a,b]$ made by the sequence $(M_r;\ r \in F)$. Whenever $G \subset [0, \infty[$, let $U_G[a,b] = \sup_{F \subset G,\ F \text{ finite}} U_F[a,b]$. Recall the upcrossing inequality (Theorem 1.6.1, Chapter 1), and observe that by taking finite sets $F_k \uparrow \mathbb{Q}_+ \cap [0, n]$, we deduce for all integers $n \geq 1$,

$$\mathbb{E}\{U_{[0,n]}[a,b]\} \leq \frac{|a| + \mathbb{E}\{|M_n|\}}{b - a}.$$

Let Ω_0 denote the collection of all ω's such that for some pair of rational numbers $a < b$, and for some integer $n \geq 1$, $U_{[0,n]}[a,b] = \infty$. The above shows that $\mathbb{P}(\Omega_0) = 0$. For all $a < b$, define

$$N_{a,b} = \bigcup_{t \geq 0} \left(\liminf_{r \downarrow t:\ r \in \mathbb{Q}_+} M_r \leq a < b \leq \limsup_{r \downarrow t:\ r \in \mathbb{Q}_+} M_r \right).$$

Since the above is an uncountable union, $N_{a,b}$ need not be measurable. However, inspecting the event ω by ω, we can deduce that $\cup_{a < b:\, a,b \in \mathbb{Q}_+} N_{a,b} \subset \Omega_0$. Since Ω_0 is $\mathbb{P}$-null and our probability space is complete, $\mathbb{P}(\cup_{a < b:\, a,b \in \mathbb{Q}_+} N_{a,b}) = 0$, which is the desired result. $\square$

Define the process $M^+ = (M_{t+};\ t \geq 0)$ by $M_{t+} = \lim_{r \downarrow t:\ r \in \mathbb{Q}_+} M_r$; the existence of this limit is guaranteed by Lemma 1.4.1. Clearly, M^+ is adapted to $\mathcal{F}^\star$. Moreover, by Lemma 1.4.1, M^+ is a.s. right-continuous.

CAVEAT The random variable $M_t^+ = M_{t+}$ is *not* the same as $(M_t)^+$, the latter being the positive part of M_t in most accounts.

The following contains some of its other properties.

Lemma 1.4.2 *Let M be a submartingale with respect to the filtration $\mathcal{F}^\star$, and let M^+ be as above. Then:*

(i) for all $t \geq 0$, $M_{t+} \in L^1(\mathbb{P})$;

(ii) for each $t \geq 0$, $M_t \leq M_{t+}$, a.s.;

(iii) if $t \mapsto \mathbb{E}[M_t]$ is right-continuous, M^+ is a modification of M;

(iv) M^+ is a submartingale with respect to $\mathcal{F}^\star$; and

(v) if M is a martingale, so is M^+.

Proof Fix a $t \geq 0$ and let $\rho_1, \rho_2, \ldots$ be an enumeration of positive rationals in $[t, t+1]$ such that as $n \to \infty$, $\rho_n \downarrow t$. Clearly, $(M_{\rho_n}; n \geq 1)$ is a reversed submartingale, and $\sup_n \mathbb{E}\{|M_{\rho_n}|\} \leq \mathbb{E}\{|M_{t+1}|\} < +\infty$. By Supplementary Exercise 4 of Chapter 1, $\lim_{n \to \infty} M_{\rho_n}$ exists, almost surely and in $L^1(\mathbb{P})$; this implies *(i)*.

To prove *(ii)*, we merely note that there exists one null set, outside of which for all $n \geq 1$, $M_t \leq \mathbb{E}[M_{\rho_n} | \mathcal{F}_t^\star]$, a.s. Assertion *(ii)* follows readily from this.

To prove *(iii)*, first note that by convergence in $L^1(\mathbb{P})$, $\mathbb{E}[M_{t+}] = \lim_{n \to \infty} \mathbb{E}[M_{\rho_n}]$. The assumed right continuity of $s \mapsto \mathbb{E}[M_s]$ implies that $\mathbb{E}[M_{t+} - M_t] = 0$. Since $M_{t+} - M_t \geq 0$, a.s. (part *ii*), and has mean 0, by Supplementary Exercise 3, $\mathbb{P}(M_{t+} = M_t) = 1$.

To prove *(iv)*, we begin by fixing an s and a t such that $0 \leq s < t$. Note that almost surely, $M_p \leq \mathbb{E}[M_q | \mathcal{F}_p^\star]$, simultaneously for all rationals $p < q$. By Exercise 1.4.1, $\cap_{p > s: p \in \mathbb{Q}_+} \mathcal{F}_p^\star = \mathcal{F}_s^\star$. Consequently, we can take $p \downarrow s$ ($p \in \mathbb{Q}_+$) and use the discrete-parameter convergence theorem for reversed martingales (Supplementary Exercise 4, Chapter 1) to see that with probability one, $M_{s+} \leq \mathbb{E}[M_q | \mathcal{F}_s^\star]$, simultaneously over all rational $q > s$. Let $q \downarrow t$ ($q \in \mathbb{Q}_+$) and use $L^1(\mathbb{P})$ convergence to see that M^+ is a submartingale.

Part *(v)* follows from *(iv)*, by considering both M and $-M$. □

We are ready for our proof of Theorem 1.4.1.

Proof of Theorem 1.4.1 Let Ω_1 be the collection of all ω's such that for some $t \geq 0$, $\lim_{r \downarrow t: r \in \mathbb{Q}_+} M_r(\omega)$ does *not* exist. By Lemma 1.4.1, Ω_1 is a $\mathbb{P}$-null set. For all $t \geq 0$ and all ω, define

$$N_t(\omega) = \begin{cases} M_{t+}(\omega), & \text{if } \omega \notin \Omega_1, \\ 0, & \text{if } \omega \in \Omega_1. \end{cases}$$

By Lemmas 1.4.1 and 1.4.2, the process $N = (N_t; t \geq 0)$ is a right-continuous modification of M that is also a submartingale. □

Thus far, we have shown that if M is a smartingale with respect to the larger filtration $\mathcal{F}^\star$, then it has a right-continuous modification. Since $\mathcal{F}^\star$ is larger than $\mathcal{F}$, it may appear that the class of right-continuous $\mathcal{F}^\star$-smartingales is substantially more restrictive than the collection of right-continuous $\mathcal{F}$-smartingales. The following shows that this is not so.

Lemma 1.4.3 *If M is a right-continuous submartingale with respect to a filtration $\mathcal{F}$, then M is a right-continuous submartingale with respect to $\mathcal{F}^\star$.*

Exercise **1.4.2** Prove Lemma 1.4.3. □

From now on, we say that the filtration $\mathcal{F}$ satisfies **the usual conditions** if for all $t \geq 0$, $\mathcal{F}_t = \mathcal{F}_t^\star$. Theorem 1.4.1 states that if M is a smartingale with respect to a filtration that satisfies the usual conditions, and if $t \mapsto \mathbb{E}[M_t]$ is right-continuous, then M has a right-continuous modification that is itself a smartingale. Note that when M is a martingale, $t \mapsto \mathbb{E}[M_t]$ is automatically continuous, and right-continuous modifications exist automatically.

1.5 Measurability of Entrance Times

In this subsection we state the following deep theorem, due to G. A. Hunt and C. Dellacherie, to various degrees of generality. Its proof relies on several results from measure theory that would take too long to develop. Since we have only one use for this theory (namely the following result), we omit a proof. However, a self-contained derivation can be found in Bass (1995, Theorem 2.8, Ch. II), and in Dellacherie and Meyer (1978, Theorem 50, Ch. IV).

Theorem 1.5.1 *Suppose $\mathcal{F}$ is a filtration that satisfies the usual conditions, and X is a right-continuous, adapted, S-valued stochastic process, where S is a complete, separable metric space. Then, for all Borel sets $E \subset S$, $T_E = \inf(s \geq 0 : X_s \in E)$ is a stopping time.*

When X is continuous and E is closed, we do not need $\mathcal{F}$ to satisfy the usual conditions, and S need not be complete and separable; cf. Theorem 1.2.1.

1.6 The Optional Stopping Theorem

We are in a position to state and prove our first main result for continuous-time martingales. It is the natural continuous-time extension of the discrete-time result of Section 1.2, Chapter 1, and will be called the **optional stopping theorem**.

Theorem 1.6.1 (The Optional Stopping Theorem) *Suppose M is a right-continuous submartingale with respect to a given filtration $\mathcal{F}$. Whenever T_1 and T_2 are a.s. bounded $\mathcal{F}$-stopping times with $T_1 \leq T_2$, a.s., then*

$$M_{T_1} \leq \mathbb{E}[M_{T_2} \,|\, \mathcal{F}_{T_1}], \text{ a.s.}$$

Proof Since T_1 and T_2 are almost surely bounded, there exists a nonrandom integer $K > 0$ such that with probability one, $T_1, T_2 \leq K$. Let T_1^n and T_2^n be the stopping times of Lemma 1.1.1, so that with probability one, for all $n \geq 1$, $T_1^n, T_2^n \in \{k2^{-n}; 1 \leq k \leq K2^n\}$, and as $n \to \infty$, $T_1^n \downarrow T_1$ and $T_2^n \downarrow T_2$. From their construction, it is clear that for all $n \geq 1$, $T_1^n \leq T_2^n$; cf. Lemma 1.1.1. Note that (1) $(M_{k2^{-n}}; 1 \leq k \leq K2^n)$ is a discrete-parameter submartingale with respect to the filtration $\mathcal{F}^n = (\mathcal{F}_{k2^{-n}}; 1 \leq k \leq K2^n)$; and (2) thanks to their particular construction, T_1^n and T_2^n are both $\mathcal{F}^n$-stopping times. The optional stopping theorem (Theorem 1.6.1, Chapter 1) implies that with probability one, $M_{T_1^n} \leq \mathbb{E}[M_{T_2^n} \mid \mathcal{F}_{T_1^n}]$. Since T_2^n decreases as n increases, for all integers $n \geq m \geq 1$,

$$M_{T_1^n} \leq \mathbb{E}[M_{T_2^m} \mid \mathcal{F}_{T_1^n}], \qquad \text{a.s.}$$

Let $n \to \infty$ and use Lemma 1.1.1, together with right continuity of M, to see that $M_{T_1} \leq \limsup_{n \to \infty} \mathbb{E}[M_{T_2^m} \mid \mathcal{F}_{T_1^n}]$, a.s. On the other hand, since T_1^n is decreasing in n and since T_2^m is bounded above by $K + 2^{-m}$, $(\mathbb{E}[M_{T_2^m} \mid \mathcal{F}_{T_1^n}]; n \geq 1)$ is a uniformly integrable, reversed submartingale. In particular, with probability one, it converges a.s. and in $L^1(\mathbb{P})$; see Supplementary Exercise 4, Chapter 1. That is,

$$M_{T_1} \leq \mathbb{E}\left[M_{T_2^m} \,\Big|\, \bigcap_{n=1}^{\infty} \mathcal{F}_{T_1^n}\right].$$

By Theorem 1.1.1, M_{T_1} is $\mathcal{F}_{T_1}$-measurable. Since $\mathcal{F}_{T_1} \subset \cap_{n=1}^{\infty} \mathcal{F}_{T_1^n}$, we can take conditional expectations of the above inequality, given $\mathcal{F}_{T_1}$, to obtain

$$M_{T_1} \leq \mathbb{E}[M_{T_2^m} \mid \mathcal{F}_{T_1}], \qquad \text{a.s.} \tag{1}$$

On the other hand, $(M_{T_2^m}; m \geq 1)$ is a uniformly integrable, reversed submartingale. Therefore, as $m \to \infty$, $M_{T_2^m}$ converges a.s. and in $L^1(\mathbb{P})$. Since $T_2^m \downarrow T_2$ and M is right-continuous, $\lim_{m \to \infty} M_{T_2^m} = M_{T_2}$, a.s. and in $L^1(\mathbb{P})$. In particular, equation (1) implies the theorem. $\square$

Part of the assertion of the previous theorem is that $\mathbb{E}\{|M_T|\} < +\infty$ for all bounded stopping times T. This is implicit in the proof of Theorem 1.6.1.

Next is our first application of the optional stopping theorem.

Corollary 1.6.1 *Suppose T_1 and T_2 are a.s. finite stopping times. Then,*

$$\mathbb{E}\Big(\mathbb{E}[Y \mid \mathcal{F}_{T_1}] \,\Big|\, \mathcal{F}_{T_2}\Big) = \mathbb{E}[Y \mid \mathcal{F}_{T_1 \wedge T_2}], \qquad \text{a.s.}$$

***Exercise* 1.6.1** Verify Corollary 1.6.1. $\square$

This is an example of **commutation** of filtrations in continuous-time. We will return to this concept in greater depth later on. For the time being, we will be satisfied with the following reformulation.

Corollary 1.6.2 *If M is a right-continuous submartingale with respect to a filtration $\mathcal{F}$, then for any stopping time T, $(M_{T \wedge t};\ t \geq 0)$ is a right-continuous submartingale with respect to $\mathcal{F}$.*

Proof We need the following two facts, which are proved in Exercise 1.6.2 below.

(a) for any integrable random variable Y and for all $s \geq 0$, a.s.,
$$\mathbb{E}[Y \mid \mathcal{F}_s]\mathbf{1}_{(T>s)} = \mathbb{E}[Y \mid \mathcal{F}_{T \wedge s}]\mathbf{1}_{(T>s)}, \quad \text{a.s.; and}$$

(b) $M_T \mathbf{1}_{(T \leq s)}$ is $\mathcal{F}_s$-measurable.

Since $(T > s), (T \leq s) \in \mathcal{F}_s$, for all $s \geq 0$, we conclude that for all $t \geq s \geq 0$,

$$\begin{aligned}
\mathbb{E}[M_{T \wedge t} \mid \mathcal{F}_s] &= \mathbb{E}[M_{T \wedge t}\mathbf{1}_{(T \leq s)} \mid \mathcal{F}_s] + \mathbb{E}[M_{T \wedge t}\mathbf{1}_{(T>s)} \mid \mathcal{F}_s] \\
&= \mathbb{E}[M_T \mathbf{1}_{(T \leq s)} \mid \mathcal{F}_s] + \mathbb{E}[M_{T \wedge t} \mid \mathcal{F}_s]\mathbf{1}_{(T>s)} \\
&= M_T \mathbf{1}_{(T \leq s)} + \mathbf{1}_{(T>s)}\mathbb{E}[M_{T \wedge t} \mid \mathcal{F}_{T \wedge s}] \quad \text{(by (a) and (b))} \\
&= M_{T \wedge s}\mathbf{1}_{(T \leq s)} + \mathbf{1}_{(T>s)}\mathbb{E}[M_{T \wedge t} \mid \mathcal{F}_{T \wedge s}] \\
&\geq M_{T \wedge s},
\end{aligned}$$

almost surely. We have used the optional stopping theorem in the last line; cf. Theorem 1.6.1. □

***Exercise* 1.6.2** Prove claims (a) and (b) of the proof of Corollary 1.6.2. □

1.7 Brownian Motion

Recall from Chapter 5 that an $\mathbb{R}$-valued process $B = (B_t;\ t \geq 0)$ is (standard) **Brownian motion** if it is a continuous Gaussian process with mean 0 and covariance function given by $\mathbb{E}[B_s B_t] = s \wedge t$ $(s,t \geq 0)$.

Note that for all $t \geq s \geq r \geq 0$, $\mathbb{E}[(B_t - B_s)B_r] = 0$. By Corollary 1.1.1, Chapter 5, for all $0 \leq r_1, \ldots, r_k \leq s$, $B_t - B_s$ is independent of $(B_{r_1}, \ldots, B_{r_k})$. Hence, $B_t - B_s$ is independent of $\mathcal{H}_s$, which is defined as the σ-field generated by $(B_r;\ 0 \leq r \leq t)$.

In fact, one can do a little better. By relabeling the indices, we see that for all $t \geq s$ and all $\varepsilon > \eta > 0$, $B_{t+\varepsilon} - B_{s+\varepsilon}$ is independent of $\mathcal{H}_{s+\eta}$. Since the choice of $\eta < \varepsilon$ is arbitrary, $B_{t+\varepsilon} - B_{s+\varepsilon}$ is independent of $\mathcal{H}_{s+}$ (recall the notation from Section 1.4). Equivalently, for all $A \in \mathcal{H}_{s+}$, and for all bounded, continuous functions $f : \mathbb{R} \to \mathbb{R}$,

$$\mathbb{E}[f(B_{t+\varepsilon} - B_{s+\varepsilon})\mathbf{1}_A] = \mathbb{P}(A) \cdot \mathbb{E}[f(B_{t+\varepsilon} - B_{s+\varepsilon})]. \tag{1}$$

We can replace A by $A \cup \Lambda$ with no change, where $A \in \mathcal{H}_{s+}$ and Λ is a $\mathbb{P}$-null set. Hence, we see that $B_{t+\varepsilon} - B_{s+\varepsilon}$ is independent of $\mathcal{F}_s$, where $\mathcal{F} = (\mathcal{F}_t; t \geq 0)$ denotes the smallest filtration that (1) contains $\mathcal{H}$; and (2) satisfies the usual conditions. According to the notation of Section 1.4, $\mathcal{F}_t = \mathcal{H}_t^\star$, for all $t \geq 0$, and we refer to $\mathcal{F}$ as the **history of** B.

By the a.s. (right) continuity of B, we can let $\varepsilon \to 0$ to see that (1) holds for all $t > s \geq 0$, all $A \in \mathcal{F}_s$, and with $\varepsilon = 0$. This can be stated in other words as follows.

1 One-Parameter Martingales

Theorem 1.7.1 (Stationary, Independent Increments Property) *If B denotes a Brownian motion with history $\mathcal{F}$, then for all $t \geq s \geq 0$, $B_t - B_s$ is independent of $\mathcal{F}_s$, and $B_t - B_s \sim \mathcal{N}_1(0, t-s)$.*

As a corollary, we obtain the following.

Corollary 1.7.1 *If B denotes a Brownian motion, the following are martingales with respect to the history $\mathcal{F}$ of B:*

(i) $t \mapsto B_t$;

(ii) $t \mapsto B_t^2 - t$; and

(iii) $t \mapsto \exp(\alpha B_t - \frac{1}{2} t \alpha^2)$, *where $\alpha \in \mathbb{R}$ is fixed.*

Exercise **1.7.1** Prove Corollary 1.7.1. □

Next, we mention a second corollary of Theorem 1.7.1.

Corollary 1.7.2 *With probability one,*

$$\limsup_{n \to \infty} B_n = +\infty, \quad \text{and} \quad \liminf_{n \to \infty} B_n = -\infty.$$

Exercise **1.7.2** Prove Corollary 1.7.2. □

Once we have identified enough martingales, we can use the optional stopping theorem to make certain computations possible. To produce a class of interesting examples, let $T_{a,b} = T_{\{a,b\}} = \inf(s \geq 0 : B_s \in \{a, b\})$ be the entrance time of $\{a, b\}$, where $a < 0 < b$. Since $t \mapsto B_t$ is continuous, Corollary 1.7.2 shows that for all $a < 0 < b$, $\mathbb{P}(T_{a,b} < \infty) = 1$. Furthermore, $T_{a,b}$ is an $\mathcal{F}$-stopping time (Theorem 1.2.1), and $(B_{T_{a,b} \wedge t}; t \geq 0)$ is a continuous martingale with respect to $\mathcal{F}$ (Corollary 1.6.2). In fact, since $\sup_{t \geq 0} |B_{T_{a,b} \wedge t}| \leq |a| \vee |b|$, $t \mapsto B_{T_{a,b} \wedge t}$ is a bounded continuous martingale. By the bounded convergence theorem, $\mathbb{E}[B_{T_{a,b}}] = 0$. Since $B_{T_{a,b}} \in \{a, b\}$, almost surely, the latter expectation can be written as

$$0 = a\mathbb{P}(B_{T_{a,b}} = a) + b\Big\{1 - \mathbb{P}(B_{T_{a,b}} = a)\Big\}.$$

Upon solving the above algebraic equation, we obtain the following:

Corollary 1.7.3 *For all $a < 0 < b$,*

$$\mathbb{P}(B_{T_{a,b}} = a) = \frac{b}{b-a}, \quad \mathbb{P}(B_{T_{a,b}} = b) = \frac{-a}{b-a}.$$

Corollary 1.7.3 is the solution to the **gambler's ruin problem**: Suppose the gambler's fortune at time t is B_t, where negative fortune means loss. If b is the house limit and $-a$ is all that the gambler owns, $\mathbb{P}(B_{T_{a,b}} = a)$ is the probability of ruin for the gambler and $\mathbb{P}(B_{T_{a,b}} = b)$ is the probability of ruin for the house. Compare to Supplementary Exercise 2.

We conclude this subsection with a final computation involving Brownian motion. Let $T_a = \inf(s \geq 0 : B_s = a)$, and apply Theorem 1.1.1 and Corollary 1.7.2 to deduce that T_a is an almost surely finite stopping time for all $a \in \mathbb{R}$. In this language, Corollary 1.7.3 shows that for all $a < 0 < b$, $\mathbb{P}(T_a < T_b) = b/(b-a)$. Arguing as in Corollary 1.7.3, we can apply Corollary 1.7.1 *(iii)* with $\alpha = \sqrt{2\lambda}$ to compute the Laplace transform of T_a.

Corollary 1.7.4 *For all $a \in \mathbb{R}$ and all $\lambda > 0$, $\mathbb{E}[e^{-\lambda T_a}] = e^{-a\sqrt{2\lambda}}$.*

Exercise **1.7.3** Prove Corollary 1.7.4. In fact, it is possible to check directly that the probability density of T_a is

$$\mathbb{P}(T_a \in dt) = \frac{a e^{-a^2/2t}}{\sqrt{2\pi}\, t^{\frac{3}{2}}} dt,$$

where $t \geq 0$, and $a \in \mathbb{R}$. □

1.8 Poisson Processes

A real-valued stochastic process $(X_t;\ t \geq 0)$ is a (time-homogeneous) **Poisson process with rate** $\lambda > 0$ if:

(i) for each $s, t \geq 0$, $X_{t+s} - X_t$ is independent of $\mathcal{F}_t = \mathcal{H}_t^\star$, where $\mathcal{H}_t$ denotes the σ-field generated by $(X_r;\ 0 \leq r \leq t)$ (cf. Section 1.4); and

(ii) for all $s, t \geq 0$, $X_{t+s} - X_t$ has a Poisson distribution with mean λs; i.e.,

$$\mathbb{P}(X_{t+s} - X_t = k) = \frac{1}{k!} e^{-\lambda s}(\lambda s)^k, \qquad \forall k = 0, 1, \ldots.$$

We will always, and implicitly, use a separable modification of X, whose existence is guaranteed by Theorem 2.2.1 of Chapter 5.

In words, a Poisson process with rate λ is a stochastic process with stationary, independent increments that are themselves Poisson random variables. In light of Theorem 1.7.1, Poisson processes are closely related to Brownian motion. However, there are also obvious differences. For example, by (ii) above, $t \mapsto X_t$ is almost surely increasing.

Let $\tau_0 = 0$ and for all $k \geq 1$, define

$$\tau_k = \inf(s > \tau_{k-1} : X_s - X_{\tau_{k-1}} \geq 1), \qquad k \geq 1.$$

1 One-Parameter Martingales

***Exercise* 1.8.1** The τ_i's are $\mathcal{F}$-stopping times. □

Since X is almost surely increasing, the τ_i's are the times at which X increases. The following is an important first step in the analysis of Poisson processes.

Lemma 1.8.1 *The random variables* $(\tau_k - \tau_{k-1}; \; k \geq 1)$ *are i.i.d. exponential random variables with mean* λ^{-1} *each.*

Proof Since $t \mapsto X_t$ is a.s. increasing, for all $t \geq 0$,
$$\mathbb{P}(\tau_1 > t) = \mathbb{P}(X_t = 0) = e^{-\lambda t}.$$
That is, τ_1 has an exponential distribution with mean λ^{-1}. We now proceed with induction. Supposing the result is true for some $k \geq 1$, it suffices to show that it is true for $k+1$.

Since τ_k has an absolutely continuous distribution, for all Borel sets $A \subset \mathbb{R}$,
$$\mathbb{P}(\tau_{k+1} - \tau_k > t, \tau_k \in A) = \lambda \int_A \mathbb{P}(\tau_{k+1} - \tau_k > t \mid \tau_k = s) e^{-\lambda s} \, ds. \quad (1)$$
On the other hand, on $(\tau_k = s)$, $\tau_{k+1} - \tau_k > t$ if and only if $X_{t+s} - X_s = 0$. By (ii) of the definition of Poisson processes, this latter event is independent of $\mathcal{F}_s$ and hence of $(\tau_k = s)$, since τ_k is a stopping time. Thus, by (i),
$$\mathbb{P}(\tau_{k+1} - \tau_k > t \mid \tau_k \in A) = \mathbb{P}(X_{t+s} - X_s = 0) = e^{-\lambda t}.$$
The result follows from (1). □

Now consider a right-continuous modification of X that we continue to write as X. This process is itself a Poisson process with rate λ. The following exercise shows that, once it is suitably 'compensated', X becomes a martingale.

***Exercise* 1.8.2** Prove that the stochastic process $(X_t - \lambda t; \; t \geq 0)$ is a right-continuous martingale. □

Since $X_0 = 0$, by the optional stopping theorem (Theorem 1.6.1), for all $n \geq 1$,
$$\mathbb{E}[X_{\tau_1 \wedge n} - \lambda(\tau_1 \wedge n)] = 0.$$
On the other hand, $|X_{\tau_1 \wedge n}| \leq 1$, almost surely. Thus, by the monotone and the dominated convergence theorems, $\mathbb{E}[X_{\tau_1}] = \lambda \mathbb{E}[\tau_1]$, which equals 1, thanks to Lemma 1.8.1 above. Since $X_{\tau_1} \geq 1$ (right continuity), this shows that $X_{\tau_1} = 1$, almost surely; cf. Supplementary Exercise 3 for details. In summary, we have shown that with probability one, at the first jump time, the process always jumps to 1. This can be generalized as follows.

Lemma 1.8.2 *With probability one, for all* $n \geq 1$, $X_{\tau_n} = n$.

***Exercise* 1.8.3** Check that the above extension is valid. □

Now we can combine Lemmas 1.8.1 and 1.8.2 to see that Poisson processes of rate $\lambda > 0$ exist. Moreover, they have a simple construction:

(a) Since $X_0 \geq 0$ and $\mathbb{E}[X_0] = 0$, $X_0 = 0$, i.e., the process starts at 0.

(b) For all $0 \leq s < \tau_1$, $X_s = 0$, where τ_1 has an exponential distribution with mean λ^{-1}.

(c) For all $\tau_1 \leq s < \tau_2$, $X_s = 1$, where $\tau_2 - \tau_1$ is independent of τ_1 and has an exponential distribution with mean λ^{-1}. More generally, for all $\tau_k \leq s < \tau_{k+1}$, $X_s = k$ and $\tau_{k+1} - \tau_k$ is an exponentially distributed random variable with mean λ^{-1} that is independent of $\tau_1, \ldots, \tau_k$.

In other words, we can always construct a right-continuous Poisson process with rate $\lambda > 0$ as follows.

Proposition 1.8.1 *Let $\xi_1, \xi_2, \ldots$ be independent exponential random variables with mean λ^{-1}. If $\gamma_n = \sum_{i=1}^{n} \xi_i$ ($n \geq 1$), the process $Y = (Y_t; \, t \geq 0)$ is a Poisson process with rate λ, where*

$$Y_t = \sum_{n=1}^{\infty} \mathbf{1}_{(\gamma_n \leq t)}, \qquad t \geq 0.$$

***Exercise* 1.8.4** Suppose $U_1, U_2, \ldots$ are i.i.d. random variables, all uniformly picked from the interval $[0, 1]$. Consider the **empirical distribution function** F_n, described by

$$F_n(t) = \sum_{j=1}^{n} \mathbf{1}_{(U_j \leq t)}, \qquad t \in [0, 1], \, n \geq 1.$$

This is a random distribution function on $[0, 1]$ for every $n \geq 1$. Let N_n be an independent Poisson random variable with mean n, and define

$$Y_n(t) = F_{N_n}(t), \qquad t \in [0, 1], \, n \geq 1.$$

(i) Check that conditional on $(N_n = n)$, Y_n has the same finite-dimensional distributions as F_n.

(ii) Prove that Y_n is a Poisson process (indexed by $[0, 1]$) of rate n.

This is from Kac (1949). □

2 Multiparameter Martingales

In the previous section we discussed the general construction and regularity theory of one-parameter martingales in continuous-time. We are now in position to construct multiparameter martingales in continuous-time. We do this by first introducing some concepts whose discrete-time counterparts appeared in Chapter 1.

2.1 Filtrations and Commutation

Recall that $(\Omega, \mathcal{G}, \mathbb{P})$ is our underlying probability space. A collection $\mathcal{F} = (\mathcal{F}_t;\, t \in \mathbb{R}_+^N)$ is said to be an (N-parameter) **filtration** if $\mathcal{F}$ is a collection of sub-σ-fields of $\mathcal{G}$ with the property that whenever $s \preccurlyeq t$, both in $\mathbb{R}_+^N$, then $\mathcal{F}_s \subset \mathcal{F}_t$. To each such N-parameter filtration $\mathcal{F}$ we ascribe N *one-parameter* filtrations (in the sense of Section 1.1) $\mathcal{F}^1, \ldots, \mathcal{F}^N$, defined by the following: For every $i \in \{1, \ldots, N\}$,

$$\mathcal{F}_r^i = \bigvee_{t \in \mathbb{R}_+^N:\, t^{(i)} = r} \mathcal{F}_t, \qquad r \geq 0.$$

Exercise 2.1.1 Prove that

$$\mathcal{F}_r^i = \bigvee_{t \in \mathbb{R}_+^N:\, t^{(i)} \leq r} \mathcal{F}_t, \qquad r \geq 0,$$

for every $1 \leq i \leq N$. □

The filtrations $\mathcal{F}^1, \ldots, \mathcal{F}^N$ are called the **marginal filtrations** of $\mathcal{F}$.

We say that the N-parameter filtration $\mathcal{F}$ is **commuting** if for all $s, t \in \mathbb{R}_+^N$ and all bounded $\mathcal{F}_t$-measurable random variables Y,

$$\mathbb{E}[Y \mid \mathcal{F}_s] = \mathbb{E}[Y \mid \mathcal{F}_{s \wedge t}], \quad \text{a.s.}$$

The following characterizes two of the fundamental properties of commuting filtrations.

Theorem 2.1.1 *Suppose* $\mathcal{F} = (\mathcal{F}_t;\, t \in \mathbb{R}_+^N)$ *is a commuting filtration. Then:*

(i) for all bounded random variables Z, and for all $t \in \mathbb{R}_+^N$,

$$\mathbb{E}[Z \mid \mathcal{F}_t] = \mathbb{E}\Big(\cdots \mathbb{E}\big\{\, \mathbb{E}[Z \mid \mathcal{F}_{t^{(1)}}^1] \,\big|\, \mathcal{F}_{t^{(2)}}^2 \big\} \cdots \,\Big|\, \mathcal{F}_{t^{(N)}}^N\Big), \quad \text{a.s.;\ and}$$

(ii) the commutation property of $\mathcal{F}$ is equivalent to the following: For all $s, t \in \mathbb{R}_+^N$, $\mathcal{F}_s$ and $\mathcal{F}_t$ are conditionally independent, given $\mathcal{F}_{s \wedge t}$.

Exercise 2.1.2 Prove Theorem 2.1.1.
(HINT: Consult Theorems 3.4.1 and 3.6.1 of Chapter 1.) □

2.2 Martingales and Histories

Henceforth, we will *always* make the following assumption.

Assumption The underlying probability space $(\Omega, \mathcal{G}, \mathbb{P})$ is complete.

Recall that one can always complete the probability space at no cost. The advantage is a gain in regularity for stochastic processes; cf. Theorem 2.2.1 of Chapter 5.

Suppose $\mathcal{F} = (\mathcal{F}_t; \, t \in \mathbb{R}_+^N)$ is a filtration of sub–σ-fields of $\mathcal{G}$. By completing them, we can, and will, assume that $\mathcal{F}$ is complete in the sense that $\mathcal{F}_t$ is a complete σ-field for each $t \in \mathbb{R}_+^N$.

A real-valued stochastic process $M = (M_t; \, t \in \mathbb{R}_+^N)$ is a **submartingale** (with respect to $\mathcal{F}$) if:

(i) M is **adapted** to $\mathcal{F}$. That is, for all $t \in \mathbb{R}_+^N$, M_t is $\mathcal{F}_t$-measurable.

(ii) M is **integrable**. That is, for all $t \in \mathbb{R}_+^N$, $M_t \in L^1(\mathbb{P})$.

(iii) For all $s \preccurlyeq t$ both in $\mathbb{R}_+^N$, $\mathbb{E}[M_t \,|\, \mathcal{F}_s] \geq M_s$, a.s.

A stochastic process M is a **supermartingale** if $-M$ is a submartingale. It is a **martingale** if it is both a sub- and a supermartingale.

Many of the properties of discrete-time multiparameter super- or submartingales carry through with few changes. For example, if M is a nonnegative submartingale and $\Psi : \mathbb{R}_+ \to \mathbb{R}_+$ is convex and nondecreasing, then $(\Psi(M_t); \, t \in \mathbb{R}_+^N)$ is a nonnegative submartingale, as long as $\Psi(M_t) \in L^1(\mathbb{P})$ for all $t \in \mathbb{R}_+^N$.

What about maximal inequalities? As we have already seen in Chapter 5, $\sup_{s \leq t} M_s$ need not even be a random variable. In order to circumvent this difficulty, we can use Theorem 2.2.1 of Chapter 5 to construct a separable modification $\widetilde{M} = (\widetilde{M}_t; \, t \in \mathbb{R}_+^N)$ of M. The following shows that there is no harm in doing this.

Lemma 2.2.1 *If M is a submartingale (supermartingale), so is $\widetilde{M}$.*

Exercise **2.2.1** Prove Lemma 2.2.1. □

From now on, we will always choose and work with such a separable modification.

Let us conclude this subsection with a brief discussion of histories. In complete analogy to the discrete-time theory, we say that $\mathcal{H} = (\mathcal{H}_t; \, t \in \mathbb{R}_+^N)$ is the **history** of $M = (M_t; \, t \in \mathbb{R}_+^N)$ if for all $t \in \mathbb{R}_+^N$, $\mathcal{H}_t$ is the σ-field generated by $(M_r; \, r \preccurlyeq t)$. It is possible to show that whenever M is a supermartingale (respectively submartingale) with respect to a filtration $\mathcal{F}$, it is also a supermartingale (respectively submartingale) with respect to its history. As a result, when we say that M is a super- or a submartingale with no reference to the corresponding filtration, we are safe in assuming that the underlying filtration is the history of M.

***Exercise* 2.2.2** Verify the above claims. That is, suppose M is a supermartingale with respect to a filtration $\mathcal{F}$. Show that M is a supermartingale with respect to its history. Construct an example where this history is not commuting. □

2.3 Cairoli's Maximal Inequalities

We are ready to state and prove Cairoli's maximal inequalities for super- and submartingales that are indexed by $\mathbb{R}_+^N$.

Theorem 2.3.1 (Cairoli's Weak $L\ln^{N-1}L$-Inequality) *If M is a separable, nonnegative N-parameter submartingale with respect to a commuting filtration, then for any $t \in \mathbb{R}_+^N$ and for all $\lambda > 0$,*

$$\mathbb{P}\Big(\sup_{s \prec t} M_s \geq \lambda\Big) \leq \frac{1}{\lambda}\Big(\frac{e}{e-1}\Big)^{N-1}\Big\{(N-1) + \mathbb{E}\big[M_t(\ln_+ M_t)^{N-1}\big]\Big\}.$$

Theorem 2.3.2 (Cairoli's Strong (p,p) Inequality) *If $M = (M_t; t \in \mathbb{R}_+^N)$ is a separable, nonnegative submartingale with respect to a commuting filtration, then for any $t \in \mathbb{R}_+^N$ and all $p > 1$,*

$$\mathbb{E}\Big[\sup_{s \prec t} M_s^p\Big] \leq \Big(\frac{p}{p-1}\Big)^{Np} \mathbb{E}[M_t^p].$$

We will prove Theorem 2.3.1; Theorem 2.3.2 is proved similarly.

Proof of Theorem 2.3.1 To begin, note that by Lemma 2.2.1 of Chapter 4, $\sup_{s \prec t} M_s$ is a random variable. By separability and countable additivity,

$$\mathbb{P}\Big(\sup_{s \prec t} M_s \geq \lambda\Big) = \sup_{F(t)} \mathbb{P}\Big(\max_{s \in F(t)} M_s \geq \lambda\Big),$$

where $\sup_{F(t)}$ denotes the supremum over all finite sets $F(t) \subset [0,t[$. On the other hand, since $(M_s; s \in F(t))$ is a discrete-parameter submartingale, by Cairoli's weak maximal inequality (Theorem 2.5.1, Chapter 1),

$$\mathbb{P}\Big(\sup_{s \prec t} M_s \geq \lambda\Big)$$
$$\leq \sup_{F(t)} \max_{s \in F(t)} \frac{1}{\lambda}\Big(\frac{e}{e-1}\Big)^{N-1}\Big\{(N-1) + \mathbb{E}\big[M_s(\ln_+ M_s)^{N-1}\big]\Big\}$$
$$= \frac{1}{\lambda}\Big(\frac{e}{e-1}\Big)^{N-1}\Big\{(N-1) + \mathbb{E}\big[M_t(\ln_+ M_t)^{N-1}\big]\Big\}.$$

We have applied Jensen's inequality to obtain the last line. This completes our proof. □

Exercise 2.3.1 Prove Theorem 2.3.2. □

It is not hard to extend the above to include estimates for $\sup_{s \preccurlyeq t} M_s$ (as opposed to $\sup_{s \prec t} M_s$). We will supply the following analogue, which is reminiscent of some of the results of Section 1.

Corollary 2.3.1 *Suppose M is a separable, nonnegative submartingale with respect to a commuting filtration.*

1. *If $t \mapsto \mathbb{E}[M_t(\ln_+ M_t)^{N-1}]$ is right-continuous, then for any $t \in \mathbb{R}_+^N$ and for all $\lambda > 0$,*

$$\mathbb{P}\left(\sup_{s \preccurlyeq t} M_s \geq \lambda\right) \leq \frac{1}{\lambda}\left(\frac{e}{e-1}\right)^{N-1}\left\{(N-1) + \mathbb{E}\left[M_t(\ln_+ M_t)^{N-1}\right]\right\}.$$

2. *If $p > 1$ and $t \mapsto \mathbb{E}[M_t^p]$ is right-continuous, then for all $t \in \mathbb{R}_+^N$,*

$$\mathbb{E}\left[\sup_{s \preccurlyeq t} M_s^p\right] \leq \left(\frac{p}{p-1}\right)^{Np}\mathbb{E}[M_t^p].$$

We have used the following definition implicitly: A function $f : \mathbb{R}_+^N \to \mathbb{R}$ is **right-continuous** (with respect to the partial order $\preccurlyeq$) if for all $t \in \mathbb{R}_+^N$,

$$\lim_{s \succcurlyeq t:\, s \to t} f(s) = f(t).$$

Exercise 2.3.2 Prove Corollary 2.3.1. □

2.4 Another Look at the Brownian Sheet

Continuous N-parameter martingales abound. Indeed, consider a real-valued N-parameter Brownian sheet B. It is possible, then, to produce a multiparameter analogue of Corollary 1.7.1.

Lemma 2.4.1 *The following are N-parameter martingales:*

(a) $t \mapsto B_t$;

(b) $t \mapsto B_t^2 - \prod_{\ell=1}^N t^{(\ell)}$; and

(c) $t \mapsto \exp(\alpha B_t - \frac{1}{2}\alpha^2 \prod_{\ell=1}^N t^{(\ell)})$, where $\alpha \in \mathbb{R}$ is fixed.

Exercise 2.4.1 Verify Lemma 2.4.1. □

In order to use the above (and other) martingales effectively, we need to have access to maximal inequalities. In light of Cairoli's inequalities (Theorems 2.3.1 and 2.3.2 and Corollary 2.3.1), the following important result of R. Cairoli and J. B. Walsh shows that all continuous smartingales with respect to the Brownian sheet filtration satisfy maximal inequalities; see Walsh (1986b).

Theorem 2.4.1 (The Cairoli–Walsh Commutation Theorem) *The Brownian sheet's history is a commuting filtration.*

Proof Let $\mathcal{F}$ denote the history of the Brownian sheet. We are to show that for all $s, t \in \mathbb{R}_+^N$, and for all bounded, $\mathcal{F}_t$-measurable random variables Y, $\mathbb{E}[Y \mid \mathcal{F}_s] = \mathbb{E}[Y \mid \mathcal{F}_{s \wedge t}]$, almost surely. Equivalently, it suffices to show that for all bounded, continuous functions $f_1, \ldots, f_m : \mathbb{R} \to \mathbb{R}$ and all $r_1, \ldots, r_m \preccurlyeq t$ (all in $\mathbb{R}_+^N$),

$$\mathbb{E}\Big[\prod_{j=1}^m f_j(B_{r_j}) \,\Big|\, \mathcal{F}_s\Big] = \mathbb{E}\Big[\prod_{j=1}^m f_j(B_{r_j}) \,\Big|\, \mathcal{F}_{s \wedge t}\Big], \qquad \text{a.s.} \tag{1}$$

(Why?)

For each integer $n \geq 1$, we define a filtration $\mathcal{F}^n = (\mathcal{F}_t^n; t \in \mathbb{R}_+^N)$ as follows: Given $t \in \mathbb{R}_+^N$, define $\mathcal{F}_t^n$ as the σ-field generated by the finite collection $(B_s;\; 2^n s \in \mathbb{Z}_+^N,\; s \preccurlyeq t)$. Note that $\mathcal{F}_t^n \subset \mathcal{F}_t^{n+1}$ for all $t \in \mathbb{R}_+^N$.

Suppose that $n > k \geq 1$ are integers, and that $r_1, \ldots, r_m \preccurlyeq t$ satisfy $2^k r_j \in \mathbb{Z}_+^N$ for all $1 \leq j \leq m$. It is sufficient to prove that

$$\mathbb{E}\Big[\prod_{j=1}^m f_j(B_{r_j}) \,\Big|\, \mathcal{F}_s^n\Big] = \mathbb{E}\Big[\prod_{j=1}^m f_j(B_{r_j}) \,\Big|\, \mathcal{F}_{s \wedge t}^n\Big], \qquad \text{a.s.} \tag{2}$$

Once this is established, we can let $n \to \infty$ and use Doob's martingale convergence theorem (Theorem 1.7.1, Chapter 1) to see that

$$\mathbb{E}\Big[\prod_{j=1}^m f_j(B_{r_j}) \,\Big|\, \bigvee_{n=1}^\infty \mathcal{F}_s^n\Big] = \mathbb{E}\Big[\prod_{j=1}^m f_j(B_{r_j}) \,\Big|\, \bigvee_{n=1}^\infty \mathcal{F}_{s \wedge t}^n\Big], \qquad \text{a.s.}$$

By the a.s. continuity of $t \mapsto B_t$, $\mathcal{F}_t$ and $\mathcal{F}_{s \wedge t}$ are the completions of $\vee_n \mathcal{F}_t^n$ and $\vee_n \mathcal{F}_{s \wedge t}^n$, respectively. Thus, equation (2) would imply that for all bounded, continuous $f_1, \ldots, f_m : \mathbb{R} \to \mathbb{R}$ and all $r_1, \ldots, r_m \preccurlyeq t$ such that $2^k r_j \in \mathbb{Z}_+^N$,

$$\mathbb{E}\Big[\prod_{j=1}^m f_j(B_{r_j}) \,\Big|\, \mathcal{F}_t\Big] = \mathbb{E}\Big[\prod_{j=1}^m f_j(B_{r_j}) \,\Big|\, \mathcal{F}_{s \wedge t}\Big], \qquad \text{a.s.}$$

By the continuity of $t \mapsto B_t$ and the dominated convergence theorem, the above holds simultaneously for every $r_1, \ldots, r_m \preccurlyeq t$ all in $\mathbb{R}_+^N$. That is, we have argued that equation (2) implies equation (1). Thus, in order to conclude our proof, we are left to verify equation (2). On the other hand, (2) is equivalent to the following: For all s,t such that $2^n t, 2^n s \in \mathbb{Z}_+^N$ and for all bounded, $\mathcal{F}_t^n$-measurable random variables Y,

$$\mathbb{E}[Y \mid \mathcal{F}_s^n] = \mathbb{E}[Y \mid \mathcal{F}_{s \wedge t}^n], \quad \text{a.s.} \tag{3}$$

Thus, we are to show that $(\mathcal{F}_t^n;\ 2^n t \in \mathbb{Z}_+^N)$ is a commuting filtration for each $n \geq 1$.

Let $\mathcal{D}_n$ denote the collection of all dyadic cubes of side 2^{-n}. That is, $A \in \mathcal{D}_n$ if and only if there exists $r \in 2^n \mathbb{Z}_+^N$ such that

$$A = \prod_{j=1}^N [r^{(j)} 2^{-n}, (r^{(j)} + 1) 2^{-n}].$$

Note that for all $t \in 2^n \mathbb{Z}_+^N$, $\mathbf{1}_{[0,t]} = \sum_{A \in \mathcal{D}_n:\ A \subset [0,t]} \mathbf{1}_A$. By its very definition, $B_t = W(\mathbf{1}_{[0,t]})$, a.s., where W denotes the isonormal process on $\mathbb{R}_+^N$; cf. Section 1.4 of Chapter 5. Therefore, for all $t \in 2^n \mathbb{Z}_+^N$,

$$B_t = \sum_{\substack{A \in \mathcal{D}_n: \\ A \subset [0,t]}} W(\mathbf{1}_A), \tag{4}$$

almost surely. This follows form Theorem 1.3.1 and the fact that whenever $A \in \mathcal{D}_n$, $W(\mathbf{1}_{\partial A}) = 0$, almost surely. On the other hand, $\{A \in \mathcal{D}_n : A \subset [0,t]\}$ is a collection of cubes whose interiors are disjoint. This disjointness implies that for any two distinct $A_1, A_2 \in \mathcal{D}_n$, we always have $\mathbb{E}[W(\mathbf{1}_{A_1}) W(\mathbf{1}_{A_2})] = 0$. Consequently, Corollary 1.1.1 of Chapter 5 shows us that $(W(\mathbf{1}_A) : A \in \mathcal{D}_n,\ A \subset [0,t])$ is an independent collection of random variables. Let $\mathfrak{m}(A)$ denote the unique largest element of A (with respect to the partial order $\preccurlyeq$). We can rewrite equation (4) as follows: For all $A \in \mathcal{D}_n$,

$$B_{\mathfrak{m}(A)} = \sum_{\mathfrak{m}(A') \in \mathcal{D}_n:\ \mathfrak{m}(A') \preccurlyeq \mathfrak{m}(A)} W(\mathbf{1}_{A'}),$$

almost surely, which means that $B_{\mathfrak{m}} = (B_{\mathfrak{m}(A)};\ A \in \mathcal{D}_n)$ is a multiparameter random walk. By the inclusion–exclusion lemma (Lemma 1.2.1, Chapter 4), $(\mathcal{F}_{\mathfrak{m}(A)}^n;\ A \in \mathcal{D}_n)$ is the history of the multiparameter random walk $B_{\mathfrak{m}}$, and Proposition 1.2.1 of Chapter 4 implies the commutation of $\mathcal{F}^n$. This, in turn, proves equation (3) and completes our proof of Theorem 2.4.1. □

The following variant is an extremely important exercise.

Exercise 2.4.2 Let $\mathcal{F}$ denote the history of the Brownian sheet B and define

$$\mathcal{F}_t^+ = \bigcup_{s \not\succcurlyeq t} \mathcal{F}_s, \quad t \in \mathbb{R}_+^N.$$

Prove that $\mathcal{F}^+ = (\mathcal{F}_t^+;\ t \in \mathbb{R}_+^N)$ is an N-parameter filtration. Is it commuting? This filtration arises in J. B. Walsh's theory of strong martingales; cf. Dozzi (1989, 1991), Imkeller (1988), and Walsh (1979, 1986b). □

The filtration $\mathcal{F}$ that appears in Theorem 2.4.1 is the **complete augmented history** of B.

3 One-Parameter Stochastic Integration

Stochastic integral processes are at the heart of the theory of (local) martingales, as well as its multiparameter extensions. Roughly speaking, a stochastic integral process is a process of the form $t \mapsto \int_0^t X_s\, dM_s$, where M—the integrator—is a continuous, one-parameter local martingale and X—the integrand—is a stochastic process.

We shall soon see that M typically has unbounded variation. Thus, any reasonable definition of $\int X\, dM$ will not be a classical integral and is, rather, a genuine "stochastic integral." Moreover, when X is nonrandom and M is Brownian motion (a very nice continuous martingale, indeed), we would expect the definition of $\int X\, dM$ to agree with that of the stochastic integral of X against white noise, as defined in Chapter 5. That is, when X is a nonrandom function and M is Brownian motion, we should expect $\int X\, dM$ to equal $M(X)$, where M is the isonormal process on $\mathbb{R}$; cf. Section 1.4 of Chapter 5.

Our construction of stochastic integrals tries to mimic that of the isonormal process. While doing this, we are forced to address a number of fundamental issues. First and foremost, we need to understand the class of processes X for which $\int X\, dM$ can be defined. Once this class is identified, our construction is not too different from that of classical integrals in spirit, and can be made in a few steps. This section goes through just such a program.[1] We begin by showing that, viewed as random functions, continuous martingales have rougher paths than many "nice" functions.

3.1 Unbounded Variation

The following shows that aside from trivial cases, continuous martingales cannot be too smooth, where here, smoothness is gauged by having bounded variation.

---

[1] This is only the tip of an iceberg; see (Bass 1995; Chung and Williams 1990; Dellacherie and Meyer 1982; Karatzas and Shreve 1991; Revuz and Yor 1994; Rogers and Williams 1987) for aspects of the general theory of stochastic integration. You should also study (Bass 1998; Dellacherie and Meyer 1988; Fukushima, Ōshima, and Takeda 1994; Hunt 1966; Sharpe 1988) for applications and connections to Markov processes.

240   7. Continuous-Parameter Martingales

Theorem 3.1.1 *If $M = (M_t; t \geq 0)$ is an almost surely continuous martingale of bounded variation, then*

$$\mathbb{P}(M_t = M_0 \text{ for all } t \geq 0) = 1.$$

Continuity is an *indispensable* condition for this unbounded variation property to hold. For instance, see the martingale of Exercise 1.8.2.

Proof. There exists a measurable set $\Lambda \subset \Omega$ such that $\mathbb{P}(\Lambda) = 0$ and for all $\omega \notin \Lambda$, $t \mapsto M_t(\omega)$ is continuous. We will show that for each $t \geq 0$,

$$\mathbb{P}(M_t = M_0) = 1. \tag{1}$$

Let Λ' denote the measurable collection of all $\omega \in \Omega$ such that for some $t \in \mathbb{Q}_+$, $M_t(\omega) \neq M_0(\omega)$. By countable additivity and by equation (1), $\mathbb{P}(\Lambda') = 0$. Consequently, $\mathbb{P}(\Lambda \cup \Lambda') = 0$. On the other hand, if two continuous functions from $\mathbb{R}_+$ into $\mathbb{R}$ agree on all positive rationals, then they are equal everywhere. Thus, $\mathbb{P}(M_t \neq M_0, \text{ for some } t \geq 0) = 0$. Thus, it suffices to derive equation (1). In fact, it is enough to do this when M is a bounded martingale. To show this, we use a **localization argument**: According to Theorem 1.2.1, τ_k is a stopping time, where

$$\tau_k = \inf(s \geq 0 : |M_s| \geq k),$$

and $\inf \varnothing = \infty$. By Corollary 1.6.2, $(M_{\tau_k \wedge t}; t \geq 0)$ is a martingale that is bounded in magnitude by k. If we derive equation (1) for all bounded martingales, we can then deduce that $\mathbb{P}(M_{t \wedge \tau_k} = M_0) = 1$. By path continuity, with probability one, $\lim_{k \to \infty} M_{t \wedge \tau_k} = M_t$, for all $t \leq \sup_k \tau_k$, which implies (1) in general. We are left to prove equation (1) for any almost surely continuous martingale M such that $\sup_{t \geq 0} |M_t| \leq k$ for some nonrandom $k \geq 0$.

Note that for any $t \geq s \geq 0$,

$$\mathbb{E}[(M_t - M_s)^2 \mid \mathcal{F}_s] = \mathbb{E}[M_t^2 - M_s^2 \mid \mathcal{F}_s] - 2\mathbb{E}[M_s(M_t - M_s) \mid \mathcal{F}_s]$$
$$= \mathbb{E}[M_t^2 - M_s^2 \mid \mathcal{F}_s], \tag{2}$$

thanks to the martingale property. Thus,

$$\mathbb{E}[(M_t - M_0)^2] = \mathbb{E}[M_t^2 - M_0^2] = \sum_{j=0}^{n-1} \mathbb{E}\left[M_{\frac{j+1}{n}t}^2 - M_{\frac{j}{n}t}^2\right]$$
$$= \mathbb{E}\left[\sum_{j=0}^{n-1} \left(M_{\frac{j+1}{n}t} - M_{\frac{j}{n}t}\right)^2\right]. \tag{3}$$

Let $\omega_n = \sup_{0 \leq u, v \leq t: |u-v| \leq 1/n} |M_u - M_v|$, and $V_n = \sum_{j=0}^{n-1} |M_{t(j+1)/n} - M_{tj/n}|$. Since M is a.s. continuous, it is uniformly continuous on $[0, t]$, almost surely. In particular, with probability one, $\lim_{n\to\infty} w_n = 0$. Furthermore, since M a.s. has bounded variation, with probability one, $\sup_{n\geq 1} V_n < \infty$. Consequently,

$$\lim_{n\to\infty} \sum_{j=0}^{n-1} \left(M_{\frac{j+1}{n}t} - M_{\frac{j}{n}t}\right)^2 = 0, \text{ a.s.} \tag{4}$$

In light of equation (3), we wish to show that the above a.s. convergence also holds in $L^1(\mathbb{P})$. By equation (4) and uniform integrability, it suffices to show that the sequence under study in equation (4) is bounded in $L^2(\mathbb{P})$. Now,

$$\mathbb{E}\left[\left\{\sum_{j=0}^{n-1}(M_{\frac{j+1}{n}t} - M_{\frac{j}{n}t})^2\right\}^2\right] = T_n^1 + 2T_n^2,$$

where

$$T_n^1 = \sum_{j=0}^{n-1} \mathbb{E}\left[\left(M_{\frac{j+1}{n}t} - M_{\frac{j}{n}t}\right)^4\right], \tag{5}$$

$$T_n^2 = \sum_{i=0}^{n-2} \sum_{j=i+1}^{n-1} \mathbb{E}\left[\left(M_{\frac{j+1}{n}t} - M_{\frac{j}{n}t}\right)^2 \left(M_{\frac{i+1}{n}t} - M_{\frac{i}{n}t}\right)^2\right]. \tag{6}$$

We will show that T_n^1 and T_n^2 are both bounded in n. This ensures the requisite uniform integrability, and implies that equation (4) also holds in $L^1(\mathbb{P})$. Thanks to (3), equation (1) follows, and so does the theorem.

Since M is bounded by k, and keeping equation (5) in mind,

$$T_n^1 \leq 4k^2 \mathbb{E}\left[\sum_{j=0}^{n-1}(M_{\frac{j+1}{n}t} - M_{\frac{j}{n}t})^2\right] = 4k^2 \mathbb{E}\left[\sum_{j=0}^{n-1}(M_{\frac{j+1}{n}t}^2 - M_{\frac{j}{n}t}^2)\right]$$
$$= 4k^2 \mathbb{E}[M_t^2 - M_0^2] \leq 4k^4.$$

The second line uses equation (2), and this shows that the term in equation (5) is bounded in n. Next, we show that the term in equation (6) is bounded, and conclude our argument. Utilizing equation (2) once more,

$$\sum_{j=i+1}^{n-1} \mathbb{E}\left[\left(M_{\frac{j+1}{n}t} - M_{\frac{j}{n}t}\right)^2 \Big| \mathcal{F}_{\frac{i+1}{n}t}\right] = \sum_{j=i+1}^{n-1} \mathbb{E}\left[\left(M_{\frac{j+1}{n}t}^2 - M_{\frac{j}{n}t}^2\right) \Big| \mathcal{F}_{\frac{i+1}{n}t}\right]$$
$$= \mathbb{E}\left[M_t^2 - M_{\frac{i+1}{n}t}^2 \Big| \mathcal{F}_{\frac{i+1}{n}t}\right] \leq k^2.$$

Once more using equation (2), we conclude that for any n, T_n^2 is bounded above by $k^2 \mathbb{E}[M_t^2 - M_0^2] \leq k^4$, which is the desired result. $\square$

3.2 Quadratic Variation

Consider, for the time being, a function $g : \mathbb{R}_+ \to \mathbb{R}$ that is of bounded variation. Very roughly speaking, this bounded variation property states that for "typical values" of $s,t \geq 0$ with $s \approx t$, $|g(s) - g(t)| \approx |s - t|$. In fact, if this property holds for all s,t sufficiently close, then g is also differentiable; cf. Chapter 2 for ways of making this rigorous.

Theorem 3.1.1 asserts that when M is a continuous martingale, e.g., M is a Brownian motion (Corollary 1.7.1), then $|M_t - M_s|$ is much larger than $|s-t|$ for typical values of s and t. We now refine this by showing a precise formulation of the statement that for most $s,t \geq 0$, $|M_t - M_s| \approx \sqrt{|s-t|}$. That is, while they have unbounded variation, continuous martingales have finite **quadratic variation**. More precisely, we state the following theorem.

Theorem 3.2.1 *Suppose M is a continuous martingale with respect to a filtration $\mathcal{F}$ that satisfies the usual conditions. If, in addition, $M_t \in L^2(\mathbb{P})$ for all $t \geq 0$, then there exists a unique nondecreasing, continuous adapted process $[M] = ([M]_t;\ t \geq 0)$ such that:*

(i) $[M]_0 = 0$, a.s.; and

(ii) $t \mapsto M_t^2 - [M]_t$ is a martingale.

Moreover, if $r_{j,n} = j2^{-n}$ $(j, n \geq 0)$, then

$$[M]_t = \lim_{n \to \infty} \sum_{1 \leq j \leq 2^n t} (M_{r_{j,n}} - M_{r_{j-1,n}})^2,$$

where the convergence holds in probability.

The process $[M]$ is the **quadratic variation** of M and uniquely determines M among all continuous $L^2(\mathbb{P})$-martingales. For instance, in light of Corollary 1.7.1, if B denotes standard Brownian motion, then $[B]_t = t$, which has the added—and very special—property of being nonrandom.

We will prove Theorem 3.2.1 in three steps. In the first step we verify uniqueness, while the following two steps are concerned with existence.

Step 1. (Uniqueness) Suppose there exists another continuous adapted process $I = (I_t;\ t \geq 0)$ such that $(M_t^2 - I_t;\ t \geq 0)$ is a martingale and $I_0 = 0$. Since the difference between two martingales is itself a martingale, $t \mapsto [M]_t - I_t$ is a continuous martingale of bounded variation that is 0 at $t = 0$. By Theorem 3.1.1, with probability one, $[M]_t = I_t$ for all $t \geq 0$, which proves the uniqueness of $[M]$. □

Step 2. (Localization) At this stage we show that it suffices to prove Theorem 3.2.1 for continuous bounded martingales. This is done by appealing to localization, an argument that has appeared earlier in our proof of Theorem 3.1.1.

Suppose we have verified Theorem 3.2.1 for all bounded martingales, and let M denote a general continuous $L^2(\mathbb{P})$-martingale. For all $k \geq 1$, define $\tau(k) = \inf(s \geq 0 : |M_s| > k)$; this is a stopping time for each $k \geq 1$, and since M is continuous, $M^k = (M_{t \wedge \tau(k)}; \ t \geq 0)$ is a bounded martingale (Corollary 1.6.2).

The bounded portion of Theorem 3.2.1 assures us of the existence of a nondecreasing, continuous adapted process $I^k = (I^k_t; \ t \geq 0)$ that is the quadratic variation of M^k. By uniqueness, for all $t \in [0, \tau(k)]$, $I^k_t = I^{k+1}_t$. Thus, there exists a nondecreasing, continuous adapted process $I = (I_t; \ t \geq 0)$ such that for each integer $k \geq 1$, $I_t = I^k_t$ for all $t \in [0, \tau(k)]$. The process I is the quadratic variation of M, and the remaining details are checked directly. □

Exercise 3.2.1 Complete the proof of Step 1 by verifying that the process I constructed therein has the asserted properties. □

Step 3. (Proof in the Bounded Case) By Step 2, we can assume that there exists some nonrandom constant $\kappa > 0$ such that $\sup_{t \geq 0} |M_t| \leq \kappa$. Throughout this proof we shall choose an arbitrary constant $T > 0$ that is held fixed. Define

$$Q_n(t) = \sum_{1 \leq j \leq 2^n t} (M_{r_{j,n}} - M_{r_{j-1,n}})^2, \qquad n \geq 1, \ t \geq 0.$$

Note that $t \mapsto Q_n(t)$ is bounded uniformly for all $t \in [0, T]$. We will show that it is Cauchy in $L^2(\mathbb{P})$, also uniformly in $t \in [0, T]$. If $m > n \geq 1$, then $Q_m(t) - Q_n(t)$ equals

$$\sum_{1 \leq j \leq 2^m t} (M_{j 2^{-m}} - M_{(j-1) 2^{-m}})^2 - \sum_{1 \leq \ell \leq 2^n t} (M_{\ell 2^{-n}} - M_{(\ell-1) 2^{-n}})^2.$$

Since dyadic partitions are nested inside one another, the summands (without the squares) in the second sum can be written as

$$M_{\ell 2^{-n}} - M_{(\ell-1) 2^{-n}} = \sum_{2^{m-n}(\ell-1) < j \leq 2^{m-n} \ell} (M_{j 2^{-m}} - M_{(j-1) 2^{-m}}).$$

That is, when $m > n \geq 1$,

$$Q_m(t) - Q_n(t)$$
$$= \sum_{1 \leq \ell \leq 2^n t} \sum_{2^{m-n}(\ell-1) < j \leq 2^{m-n} \ell} (M_{j 2^{-m}} - M_{(j-1) 2^{-m}})^2$$
$$- \sum_{1 \leq \ell \leq 2^n t} \left[\sum_{2^{m-n}(\ell-1) < j \leq 2^{m-n} \ell} (M_{j 2^{-m}} - M_{(j-1) 2^{-m}}) \right]^2$$
$$= -2 \sum_{1 \leq \ell \leq 2^n t} \sum_{2^{m-n}(\ell-1) < j < k \leq 2^{m-n} \ell} (M_{j 2^{-m}} - M_{(j-1) 2^{-m}})$$
$$\times (M_{k 2^{-m}} - M_{(k-1) 2^{-m}}). \qquad (1)$$

Next, we wish to square the above and take expectations. By the martingale property, the off-diagonal terms vanish and we obtain the following (why?):

$$\mathbb{E}\big[\{Q_m(t) - Q_n(t)\}^2\big] = 4 \sum_{1 \le \ell \le 2^n t} \sum_{2^{m-n}(\ell-1) < j < k \le 2^{m-n}\ell}$$
$$\mathbb{E}\Big[(M_{j2^{-m}} - M_{(j-1)2^{-m}})^2 (M_{k2^{-m}} - M_{(k-1)2^{-m}})^2\Big].$$

By equation (2) of Section 3.1, whenever $k > j$,

$$\mathbb{E}[(M_{k2^{-m}} - M_{(k-1)2^{-m}})^2 \,|\, \mathcal{F}_{(j-1)2^{-m}}] = \mathbb{E}[M_{k2^{-m}}^2 - M_{(k-1)2^{-m}}^2 \,|\, \mathcal{F}_{(j-1)2^{-m}}].$$

Thus, we can use this, and telescope the sum over all k's to see that

$$\mathbb{E}[\{Q_m(t) - Q_n(t)\}^2] = 4 \sum_{1 \le \ell \le 2^n t} \sum_{2^{m-n}(\ell-1) < j \le 2^{m-n}\ell}$$
$$\mathbb{E}\big[(M_{j2^{-m}} - M_{(j-1)2^{-m}})^2 (M_{\ell 2^{-n}} - M_{(j-1)2^{-m}})^2\big].$$

For the ℓ and j in the above range and for all $t \in [0, T]$,

$$(M_{\ell 2^{-n}} - M_{(j-1)2^{-m}})^2 \le \sup_{0 \le u, v \le T: |u-v| \le 2^{-n+1}} |M_u - M_v|^2 = \gamma_n.$$

Note that $\gamma_n \le 2\kappa^2$ is a bounded sequence of random variables. Moreover, by the continuity of $t \mapsto M_t$, $\lim_{n \to \infty} \gamma_n = 0$. Finally,

$$\mathbb{E}[\{Q_m(t) - Q_n(t)\}^2]$$
$$\le 4 \sum_{1 \le \ell \le 2^n t} \sum_{2^{m-n}(\ell-1) < j \le 2^{m-n}\ell} \mathbb{E}\big[\gamma_n \cdot (M_{j2^{-m}} - M_{(j-1)2^{-m}})^2\big].$$

As $m > n \to \infty$, the above goes to 0, thanks to Lebesgue's dominated convergence theorem. This uses the mentioned properties of γ_n, together with the following consequence of equation (1) of the previous subsection (cf. Section 3.1):

$$\sum_{1 \le \ell \le 2^n t} \sum_{2^{m-n}(\ell-1) < j \le 2^{m-n}\ell} \mathbb{E}\big[(M_{j2^{-m}} - M_{(j-1)2^{-m}})^2\big]$$
$$= \sum_{1 \le \ell \le 2^n t} \sum_{2^{m-n}(\ell-1) < j \le 2^{m-n}\ell} \mathbb{E}[M_{j2^{-m}}^2 - M_{(j-1)2^{-m}}^2]$$
$$= \mathbb{E}[M_{2^{-n}\lfloor 2^m t \rfloor}^2 - M_0^2]$$
$$\le \mathbb{E}[M_t^2 - M_0^2],$$

by Jensen's inequality, where $\lfloor \bullet \rfloor$ denotes the greatest integer function. We have shown that for each $t \in [0, T]$,

$$\lim_{n, m \to \infty} \mathbb{E}[\{Q_m(t) - Q_n(t)\}^2] = 0.$$

On the other hand, equation (1), and the martingale property, together show that $Q_n - Q_m$ is a martingale (why?). By Doob's strong $(2,2)$ inequality,

$$\lim_{n,m\to\infty} \mathbb{E}\left[\sup_{0\le t\le T} \{Q_m(t) - Q_n(t)\}^2\right] = 0;$$

see Theorem 1.3.1. However, L^2 spaces are complete, $Q_n(0) = 0$, and Q_n is adapted and nondecreasing. Thus, there exists an adapted, nondecreasing process $[M]$ such that $[M]_0 = 0$ and

$$\lim_{n\to\infty} \mathbb{E}\left(\sup_{0\le t\le T} \{Q_n(t) - [M]_t\}^2\right) = 0.$$

Next, we show that $t \mapsto [M]_t$ is continuous. While Q_n is not continuous, the following is (check!):

$$Q_n(t) + \{M_t - M_{2^n \lfloor 2^{-n}t \rfloor}\}^2, \qquad t \ge 0.$$

By the boundedness and continuity of $t \mapsto M_t$,

$$\sup_{0\le t\le T} |M_t - M_{2^n\lfloor 2^{-n}t\rfloor}| = 0,$$

almost surely and in $L^2(\mathbb{P})$. Hence, $t \mapsto [M]_t$ is also continuous. To finish, note that whenever $t > s$,

$$\mathbb{E}[Q_n(t) \mid \mathcal{F}_s] = Q_n(s) + \sum_{2^n s < j \le 2^n t} \mathbb{E}\left[M^2_{r_{j,n}} - M^2_{r_{j-1,n}} \Big| \mathcal{F}_s\right]$$

$$= Q_n(s) + \mathbb{E}\left[M^2_{2^n\lfloor 2^{-n}t\rfloor} - M^2_{2^n\lfloor 2^{-n}s\rfloor} \Big| \mathcal{F}_s\right],$$

almost surely. We have used equation (2) of Section 3.1. Using boundedness and continuity, we can deduce that for $t > s$,

$$\mathbb{E}\{[M]_t \mid \mathcal{F}_s\} = [M]_s + \mathbb{E}\{M_t^2 - M_s^2 \mid \mathcal{F}_s\},$$

almost surely. This demonstrates the martingale assertion and completes our proof of Theorem 3.2.1. □

3.3 Local Martingales

As far as quadratic variation is concerned, local martingales are the most natural extension of the class of $L^2(\mathbb{P})$-martingales. Here, $M = (M_t; t \ge 0)$ is a **local martingale** if there exist stopping times $\tau_1, \tau_2, \ldots$ such that for each $k \ge 1$, $M^k = (M_{t\wedge \tau_k}; t \ge 0)$ is a bounded martingale.[2] Any

---

[2] Thus, very loosely speaking, local martingales are martingales minus the integrability hypothesis. However, this naively understates the role of local martingales, since there are some very important local martingales that are not martingales; cf. Supplementary Exercise 7, Chapter 9.

such collection of τ's is said to be a **localizing sequence**, but when M is continuous, there is always a natural localizing sequence given by the following lemma.

Lemma 3.3.1 (Localization Lemma) *If M is a continuous local martingale, then $\tau_1, \tau_2, \ldots$ is a localizing sequence for M, where*

$$\tau_k = \inf(s \geq 0 : |M_s| > k), \qquad k \geq 1.$$

***Exercise* 3.3.1** Prove the localization lemma. □

Continuous local martingales are uniquely described by their quadratic variation, as the following theorem asserts.

Theorem 3.3.1 *If M is a continuous local martingale, there exists a unique continuous, adapted, nondecreasing process $[M]$ such that $[M]_0 = 0$ and $t \mapsto M_t^2 - [M]_t$ is a local martingale. Moreover, for each $T > 0$, as $n \to \infty$,*

$$\sup_{0 \leq t \leq T} \left| \sum_{1 \leq j \leq 2^n t} (M_{j2^{-n}} - M_{(j-1)2^{-n}})^2 - [M]_t \right| \to 0,$$

in probability.

***Exercise* 3.3.2** Prove Theorem 3.3.1. □

3.4 Stochastic Integration of Elementary Processes

Given an $L^2(\mathbb{P})$ process M that is a continuous martingale with respect to a filtration $\mathcal{F}$, we now wish to define $\int X \, dM$ for a class of nice processes X. As usual, we tacitly assume that $\mathcal{F}$ satisfies the usual conditions.

We say that a stochastic process X is an **elementary process** if there exist nonrandom constants $\beta \geq \alpha > 0$ and a bounded $\mathcal{F}_\alpha$-measurable random variable Θ such that for all $s \geq 0$,

$$X_s = \Theta \mathbf{1}_{]\alpha, \beta]}(s).$$

Note that elementary processes are necessarily adapted.

For such an "elementary integrand," we define the stochastic integral process $M(X) = (M(X)_t; \, t \geq 0)$ as

$$M(X)_t = \Theta \cdot (M_{\beta \wedge t} - M_{\alpha \wedge t}), \qquad t \geq 0.$$

A more suggestive notation for this is

$$M(X)_t = \int_0^t X_s \, dM_s, \qquad t \geq 0,$$

which we also adopt. The following can be checked by direct means.

Lemma 3.4.1 *Suppose M is a continuous $L^2(\mathbb{P})$-martingale. If X is an elementary process, then $M(X)$ is a continuous $L^2(\mathbb{P})$-martingale with $M(X)_0 = 0$ and with quadratic variation*

$$[M(X)]_t = \int_0^t X_s^2 \, d[M]_s, \qquad t \geq 0.$$

***Exercise* 3.4.1** Prove Lemma 3.4.1. □

Since $t \mapsto [M]_t$ is nondecreasing and $t \mapsto M_t$ is continuous, the integral $\int_0^t X_s \, d[M]_s$ is a (random) Stieltjes integral in the classical sense.

We close this subsection with another important exercise.

Lemma 3.4.2 (Polarization) *Consider a continuous $L^2(\mathbb{P})$-martingale M, and suppose X and Y are elementary processes. Then, the following is a continuous martingale:*

$$t \mapsto M(X)_t \, M(Y)_t - \int_0^t X_s Y_s \, d[M]_s.$$

***Exercise* 3.4.2** Verify the above. □

3.5 Stochastic Integration of Simple Processes

Suppose M is a continuous $L^2(\mathbb{P})$-martingale with respect to a filtration $\mathcal{F}$ that satisfies the usual conditions. We now extend our definition of $\int X \, dM$ to include a larger class of integrands than elementary ones.

We say that a process X is a **simple process** if there are a finite number of elementary processes $X^1, \ldots, X^m$ such that for all $s \geq 0$, $X_s = X_s^1 + \cdots + X_s^m$. Simple processes are bounded and adapted, and for such integrands, we define

$$M(X)_t = \int_0^t X_s \, dM_s = \sum_{j=1}^m M(X^j)_t = \sum_{j=1}^m \int_0^t X_s^j \, dM_s, \qquad t \geq 0.$$

The following can be checked directly.

Lemma 3.5.1 *Suppose M is a continuous $L^2(\mathbb{P})$-martingale.*

(a) *If X is a simple process, $M(X)$ is properly defined. That is, the definition of $M(X)$ does not depend on the particular elementary process representation of X.*

(b) *Lemmas 3.4.1 and 3.4.2 continue to hold for simple processes in place of elementary ones.*

Exercise 3.5.1 Prove Lemma 3.5.1. Moreover, show that if X and Y are simple processes, so is $X+Y$, and that a.s., $M(X+Y)_t = M(X)_t + M(Y)_t$, for all $t \geq 0$.
(HINT: You may need Lemmas 3.4.1 and 3.4.2.) □

3.6 Integrating Continuous Adapted Processes

Not surprisingly, when we say that a stochastic process X is **continuous adapted** (with respect to a filtration $\mathcal{F}$ that satisfies the usual conditions), we mean:

1. $s \mapsto X_s$ is continuous; and
2. X is adapted to $\mathcal{F}$.

Our next lemma states that continuous adapted processes can be well approximated by simple ones.

Lemma 3.6.1 *If X is a bounded, continuous adapted process, there exists a sequence of simple processes $X^1, X^2, \ldots$ such that for all $t \geq 0$,*

$$\lim_{n \to \infty} \sup_{0 \leq s \leq t} |X_s^n - X_s| = 0,$$

almost surely and in $L^p(\mathbb{P})$ for all $p > 0$.

Proof Here is one such candidate: For all $n \geq 1$, define

$$X_s^n = \sum_{1 \leq j \leq \lfloor nt \rfloor} X_{\frac{j}{n}} \cdot \mathbf{1}_{]\frac{j}{n}, \frac{j+1}{n}]}(s), \qquad s \geq 0.$$

Since X is adapted and bounded, X^n is a simple process for each n. The other assertion of the lemma follows from the boundedness and continuity of $t \mapsto X_t$, as well as from Lebesgue's dominated convergence theorem. □

We are in position to define $M(X)$, where X is a bounded, continuous and adapted process.

Let $X^1, X^2, \ldots$ denote the approximating simple processes of Lemma 3.6.1 above. Since the difference of two simple processes is itself a simple process, Lemma 3.5.1 shows us that for all $n, m \geq 0$ and all $t \geq 0$,

$$\mathbb{E}\left[\int_0^t (X_s^n - X_s^m)^2 \, d[M]_s\right] = \mathbb{E}[(M(X^n)_t - M(X^m)_t)^2].$$

Of course, $M(X^n) - M(X^m)$ is a martingale. Therefore, we can use Doob's strong $(2,2)$ inequality (Theorem 1.3.1) to see that for all $n, m \geq 1$ and all $t \geq 0$,

$$\mathbb{E}\left[\sup_{0 \leq s \leq t} \{M(X^n)_s - M(X^m)_s\}^2\right] \leq 4\mathbb{E}\left[\int_0^t (X_s^n - X_s^m)^2 \, d[M]_s\right].$$

By Lemma 3.6.1, as $n, m \to \infty$, the above goes to zero. Consequently, there exists a process $M(X)$ such that for all $t \geq 0$,

$$\lim_{n \to \infty} \mathbb{E}\Big[\sup_{0 \leq s \leq t} \{M(X)_s - M(X^n)_s\}^2 \Big] = 0.$$

(Why?) We shall write this process as $M(X)_t = \int_0^t X_s \, dM_s$, and readily deduce that $t \mapsto M(X)_t$ is a continuous $L^2(\mathbb{P})$-martingale.

Exercise **3.6.1** Demonstrate Lemmas 3.4.1 and 3.4.2 for all bounded continuous adapted processes in place of simple ones. Moreover, show that whenever X and Y are bounded continuous adapted processes, so is $X + Y$, and with probability one, $M(X + Y)_t = M(X)_t + M(Y)_t$ for all $t \geq 0$. □

Our next theorem follows from the above and the "localization" methods used in Step 2 of Theorem 3.2.1.

Theorem 3.6.1 *Given an $L^2(\mathbb{P})$-martingale M and a continuous adapted process X, there exists a continuous local martingale $M(X)$ with quadratic variation*

$$[M(X)]_t = \int_0^t X_s^2 \, d[M]_s, \qquad t \geq 0,$$

as long as the above is almost surely finite for each $t \geq 0$.

Exercise **3.6.2** Prove Theorem 3.6.1. □

Remarks (i) We shall refer to $M(X)_t$ as a stochastic integral and also write it as $\int_0^t X_s \, dM_s$. The term "integral" is justified, since $M(X)$ has many of the usual properties of integrals whose integrand is X; cf. Exercise 3.6.1 above, for example.

(ii) Since $M(X)_0 = 0$, the definition of quadratic variation used in conjunction with the above theorem implies that

$$\mathbb{E}\Big[\Big\{ \int_0^t X_s \, dM_s \Big\}^2 \Big] = \mathbb{E}\Big\{ \int_0^t X_s^2 \, d[M]_s \Big\}, \qquad t \geq 0.$$

Recall that the condition that a.s., $\int_0^t X_s^2 \, d[M]_s$ is finite for all $t \geq 0$ is also written as, X is a.s. locally square integrable with respect to $[M]$.

Theorem 3.6.2 *Consider a continuous $L^2(\mathbb{P})$-martingale M. If X and Y are continuous adapted processes that are a.s. locally square integrable with respect to $[M]$, the following is a continuous martingale:*

$$t \mapsto M(X)_t \, M(Y)_t - \int_0^t X_s Y_s \, d[M]_s.$$

Exercise 3.6.3 Prove Theorem 3.6.2. □

Remark If M is a local martingale, $[M]$ is still well-defined, and as long as $\int_0^t X_s^2 \, d[M]_s$ is a.s. finite, one can still define $t \mapsto M(X)_t$ as a local martingale. See Supplementary Exercises 4 and 5 for details.

3.7 Two Approximation Theorems

We continue our discussion of this section by showing that stochastic integrals are approximable by a kind of left-point rule, in a similar way that Riemann sums approximate ordinary Riemann integrals. We shall also verify an analogous fact for the quadratic variations of the stochastic integrals under study.[3] Throughout this subsection M denotes a continuous $L^2(\mathbb{P})$-martingale and $r_{j,n} = j2^{-n}$ ($n = 1, 2\ldots, j = 0, 1, \ldots$).

Theorem 3.7.1 *Suppose X is a bounded and continuous adapted process that is a.s. locally square integrable with respect to $[M]$. For all $t \geq 0$,*

$$\lim_{n \to \infty} \sup_{0 \leq s \leq t} \left| \sum_{1 \leq j \leq 2^n s} X_{r_{j-1,n}}(M_{r_{j,n}} - M_{r_{j-1,n}}) - \int_0^s X_r \, dM_r \right| = 0,$$

where the convergence holds in $L^2(\mathbb{P})$.

The above is *not* true if the left-point rule is replaced by the midpoint rule; cf. Supplementary Exercise 7 for details. We recall that the midpoint rule is obtained by replacing $X_{r_{j-1,n}}$ (in the summand) by $\frac{1}{2}(X_{r_{j,n}} + X_{r_{j-1,n}})$.

Proof Since X is bounded, it is not hard to check directly that

$$\sum_{1 \leq j \leq 2^n s} X_{r_{j-1,n}}(M_{r_{j,n}} - M_{r_{j-1,n}}) = \int_0^s X_r^n \, dM_r,$$

where X^n is the *simple* process defined by

$$X_s^n = \sum_{1 \leq j \leq 2^n s} X_{r_{j-1,n}} \mathbb{1}_{](j-1)2^{-n}, j2^{-n}]}(s), \qquad s \geq 0.$$

Let $\gamma_n = \sup_{0 \leq r \leq t} |X_r^n - X_r|^2$, so that $\gamma_n^{\frac{1}{2}}$ is the greatest error in the approximation X by X^n. By Doob's strong $(2,2)$ inequality (Theorem 1.3.1),

---

[3] To various degrees of generality, these were discovered by K. Itô, and in subsequent works by E. Wong and M. Zakai, as well as by D. L. Fisk.

and Theorem 3.6.1, we can deduce the following.

$$\mathbb{E}\Big[\sup_{0\leq s\leq t}\Big\{\int_0^s X_r^n\,dM_r - \int_0^s X_r\,dM_r\Big\}^2\Big] \leq 4\mathbb{E}\Big[\Big\{\int_0^t (X_r^n - X_r)\,dM_r\Big\}^2\Big]$$

$$= 4\mathbb{E}\Big[\int_0^t (X_r^n - X_r)^2\,d[M]_r\Big]$$

$$\leq 4\mathbb{E}\{\gamma_n \cdot [M]_t\},$$

By Lebesgue's dominated convergence theorem, used together with the boundedness and continuity of $t \mapsto X_t$, as $n \to \infty$, the above goes to 0. □

The second result of this subsection is the analogue of Theorem 3.7.1 for quadratic variations of stochastic integrals.

Theorem 3.7.2 *Suppose X is a bounded and continuous adapted process that is a.s. locally square integrable with respect to $[M]$. For all $t \geq 0$,*

$$\lim_{n\to\infty} \sup_{0\leq s\leq t}\Big|\sum_{1\leq j\leq 2^n s} X_{r_{j-1,n}}(M_{r_{j,n}} - M_{r_{j-1,n}})^2 - \int_0^s X_r\,d[M]_r\Big| = 0,$$

where the convergence holds in probability.

Proof When X is an elementary process, this follows readily from Theorem 3.2.1. Subsequently, it also holds when X is a simple process. The general result follows by approximating bounded and continuous adapted processes by simple ones; cf. Lemma 3.6.1 above. □

3.8 Itô's Formula: Stochastic Integration by Parts

A key result of classical integration theory states that a function $x : \mathbb{R}_+ \to \mathbb{R}$ that has bounded variation satisfies the following integration by parts formula: For all continuously differentiable functions $f : \mathbb{R} \to \mathbb{R}$,

$$f(x(t)) = f(x(0)) + \int_0^t f'(x(s))\,dx(s), \qquad t \geq 0.$$

A remarkable result of K. Itô states that when x is a continuous $L^2(\mathbb{P})$-martingale, the above continues to hold a.s., but with an extra term.[4]

Theorem 3.8.1 (Itô's Formula) *Suppose that M is a continuous $L^2(\mathbb{P})$-martingale. Then, for all twice continuously differentiable functions $f : \mathbb{R} \to \mathbb{R}$,*

$$f(M_t) = f(M_0) + \int_0^t f'(M_s)\,dM_s + \frac{1}{2}\int_0^t f''(M_s)\,d[M]_s, \qquad t \geq 0,$$

---

[4]See (Itô 1944; Kunita and Watanabe 1967).

almost surely.

Proof Clearly, $t \mapsto f'(M_t)$ and $t \mapsto f''(M_t)$ are both continuous adapted processes. Therefore, both integrals are well-defined (why?).

It suffices to show that for each $t \geq 0$, Itô's formula holds almost surely (why?). This is what we will show. By a localization argument, we can assume that there exists a constant $k \geq 0$ such that
$$\|f\|_\infty + \|f'\|_\infty + \|f''\|_\infty \leq k.$$
(Why?) Now recall $r_{j,n} = j2^{-n}$, and write
$$f(M_t) - f(M_0) = \sum_{1 \leq j \leq 2^n t} \{f(M_{r_{j,n}}) - f(M_{r_{j-1,n}})\}.$$
By Taylor's expansion, $f(x) - f(y) = f'(y) \cdot (x-y) + \int_y^x f''(v) \cdot (x-v)\, dv$. Thus,
$$f(M_t) - f(M_0) = \sum_{1 \leq j \leq 2^n t} f'(M_{r_{j-1,n}}) \cdot (M_{r_{j,n}} - M_{r_{j-1,n}})$$
$$+ \sum_{1 \leq j \leq 2^n t} \int_{M_{r_{j-1,n}}}^{M_{r_{j,n}}} f''(v) \cdot (M_{r_{j,n}} - v)\, dv$$
$$= T_1 + T_2.$$
By Theorem 3.7.1, as $n \to \infty$, T_1 converges to $\int_0^t f'(M_s)\, dM_s$ in probability. Therefore, it suffices to show that T_2 converges in probability to $\frac{1}{2} \int_0^t f''(M_s)\, d[M]_s$.

Let $I_{j,n}$ denote the interval whose two endpoints are the maximum and the minimum of $M_{r_{j,n}}$ and $M_{r_{j-1,n}}$. Since $t \mapsto f''(M_t)$ is continuous,
$$\lim_{n \to \infty} \max_{1 \leq j \leq 2^n t} \sup_{v \in I_{j,n}} |f''(v) - f''(M_{r_{j-1,n}})| = 0,$$
almost surely. Moreover, the above is bounded, since f'' is. Hence,
$$\lim_{n \to \infty} \left| T_2 - \frac{1}{2} \sum_{1 \leq j \leq 2^n t} f''(M_{r_{j-1,n}}) \cdot (M_{r_{j,n}} - M_{r_{j-1,n}})^2 \right| = 0,$$
almost surely. Itô's formula now follows from Theorem 3.7.2. □

To see the power of this formula, we apply it to Brownian motion, which is a continuous $L^2(\mathbb{P})$-martingale (Corollary 1.7.1).

Example Let B denote Brownian motion and recall that $[B]_t = t$; cf. Corollary 1.7.1(ii) and the uniqueness part of Theorem 3.2.1. Itô's formula states that for all twice continuously differentiable functions $f : \mathbb{R} \to \mathbb{R}$,
$$f(B_t) = f(0) + \int_0^t f'(B_s)\, dB_s + \frac{1}{2} \int_0^t f''(B_s)\, ds.$$

We apply this to the function $f(x) = x^2$ to obtain the almost sure identity

$$B_t^2 - t = 2 \int_0^t B_s \, dB_s.$$

This gives the explicit form for the martingale term in Corollary 1.7.1 *(ii)*. One can also obtain an integral representation for the martingale of part *(iii)* of the mentioned corollary. Indeed, it turns out that for any $\alpha \in \mathbb{R}$,

$$e^{\alpha B_t - \frac{1}{2} t \alpha^2} = 1 + \alpha \int_0^t \exp\left\{\alpha B_s - \frac{\alpha^2}{2} s\right\} dB_s, \qquad t \geq 0.$$

However, in order to prove this, one needs an "extended form" of Itô's formula that can be found in Supplementary Exercises 8 and 10. □

3.9 The Burkholder–Davis–Gundy Inequality

We have just seen that for any continuous $L^2(\mathbb{P})$-martingale M, $\mathbb{E}[M_t^2] = \mathbb{E}\{[M]_t\}$, $(t \geq 0)$. In particular, we can apply Doob's strong $(2,2)$ inequality (Theorem 1.3.1) to conclude that

$$\mathbb{E}\left[\sup_{t \geq 0} M_t^2\right] \leq 4 \sup_{t \geq 0} \mathbb{E}\{[M]_t\}.$$

In rough terms, we have here an L^2 maximal inequality that relates the "maximal L^2-norm" of M to the $L^1(\mathbb{P})$-norm of the quadratic variation $[M]$.

The main result of this section is more or less half of an inequality of D. L. Burkholder, B. Davis, and R. Gundy that states an L^p analogue of the above estimate. See Burkholder, Davis, and Gundy (1972), as well as the related works of Burkholder and Gundy (1972) and Burkholder (1973, 1975).

Theorem 3.9.1 (The Burkholder–Davis–Gundy Inequality) *If M denotes a continuous $L^2(\mathbb{P})$-martingale with $M_0 = 0$, then for all $p \geq 1$,*

$$\mathbb{E}\left[\sup_{t \geq 0} M_t^{2p}\right] \leq c(p) \cdot \sup_{t \geq 0} \mathbb{E}\{[M]_t^p\},$$

where

$$c(p) = \left(\frac{2p}{2p-1}\right)^{2p^2} p^p (2p-1)^p.$$

Used in conjunction with the optional stopping theorem (Theorem 1.6.1), this has the following important corollary.

Corollary 3.9.1 *If M is a continuous $L^2(\mathbb{P})$-martingale with $M_0 = 0$, then for all $p \geq 1$, and for all stopping times T,*

$$\mathbb{E}\Big[\sup_{t \leq T} M_t^{2p} \mathbf{1}_{(T<\infty)}\Big] \leq c(p) \cdot \mathbb{E}\{[M]_T^p \mathbf{1}_{(T<\infty)}\}.$$

In particular, we have the following.

Corollary 3.9.2 *Let B denote standard Brownian motion. For all finite stopping times T and for all $p \geq 1$,*

$$\mathbb{E}\Big[\sup_{0 \leq t \leq T} B_t^{2p}\Big] \leq c(p) \cdot \mathbb{E}[T^p].$$

Remarks (i) If we consider only nonrandom times T, we can dramatically improve the constant $c(p)$ of Corollary 3.9.2; see Supplementary Exercise 12.

(ii) Theorem 3.9.1 is sharp up to a constant; see Exercise 3.9.2 below.

We defer the proofs of Corollaries 3.9.1 and 3.9.2 to Exercise 3.9.1, and demonstrate Theorem 3.9.1 next.

Proof of Theorem 3.9.1 We can apply localization to reduce the result to the case where M_∞ and $\sup_{t \geq 0} M_t^2$ are both in $L^p(\mathbb{P})$ (why?). Henceforth, we will assume this L^p condition without loss of generality.

By the monotone convergence theorem, $\sup_{t \geq 0} \mathbb{E}\{[M]_t^p\} = \mathbb{E}\{[M]_\infty^p\}$, where $[M]_\infty = \lim_{t \to \infty} [M]_t$ a.s. exists, thanks to monotonicity of quadratic variation. Apply Itô's formula (Theorem 3.8.1) to the function $f(x) = x^{2p}$ to see that for all $t \geq 0$,

$$M_t^{2p} = 2p \int_0^t M_s^{2p-1} dM_s + p(2p-1) \int_0^t M_s^{2p-2} d[M]_s, \quad \text{a.s.}$$

We can take expectations of both sides to obtain

$$\mathbb{E}[M_t^{2p}] = p(2p-1) \mathbb{E}\Big\{\int_0^t M_s^{2p-2} d[M]_s\Big\}$$

$$\leq p(2p-1) \mathbb{E}\Big\{\sup_{0 \leq s \leq t} M_s^{2p-2} \cdot [M]_t\Big\}$$

$$\leq p(2p-1) \Big(\mathbb{E}\Big[\sup_{0 \leq s \leq t} M_s^{2p}\Big]\Big)^{1 - \frac{1}{p}} \cdot (\mathbb{E}\{[M]_\infty^p\})^{\frac{1}{p}},$$

thanks to Hölder's inequality. By Doob's strong $(2, 2)$ inequality (Theorem 1.3.1),

$$\mathbb{E}\Big[\sup_{t \geq 0} M_t^{2p}\Big] \leq \Big(\frac{2p}{2p-1}\Big)^{2p} \sup_{t \geq 0} \mathbb{E}[M_t^{2p}],$$

which implies the result, after a few lines of direct calculations. $\square$

Exercise 3.9.1 Prove Corollaries 3.9.1 and 3.9.2. □

Exercise 3.9.2 (Hard) If M is a continuous $L^2(\mathbb{P})$-martingale, prove that for all $p \geq 2$, there exists a positive and finite constant $\gamma(p)$ such that

$$\sup_{t \geq 0} \mathbb{E}\{[M]_t^p\} \leq \gamma(p) \mathbb{E}\Big[\sup_{t \geq 0} M_t^{2p}\Big].$$

(HINT: Use $M_t^2 = 2\int_0^t M_s\, dM_s + [M]_t$ together with localization and the inequality $|x+y|^p \leq 2^p\{|x|^p + |y|^p\}$. This approach can be found in Revuz and Yor (1994, Proposition 4.4, Ch. VI).) □

4 An Introduction to Stochastic PDEs

Having defined one-parameter stochastic integrals, we can now define multiparameter stochastic integrals with respect to the Brownian sheet. This will allow us to study elements of a class of so-called hyperbolic stochastic partial differential equations (henceforth hyperbolic SPDEs).

As in nonstochastic settings, in a one-parameter setting, SPDEs are called stochastic differential equations, and we have seen at least one of them so far. To wit, let B denote standard Brownian motion and define $E_t = \exp(B_t - \frac{t}{2})$, $t \geq 0$. According to the example of Section 3.8,

$$E_t = 1 + \int_0^t E_s\, dB_s, \qquad t \geq 0.$$

See also Supplementary Exercise 10. Thus, we can think of the process E as the solution to the stochastic differential equation $dE = E\, dB$, subject to $E_0 = 1$.[5]

This section is concerned with the definition, as well as the existence and uniqueness, of solutions to a class of hyperbolic SPDEs. Before embarking on this journey, we need to construct and study the basic properties of multiparameter stochastic integrals. We will do this in Section 4.1 below. Sections 4.2 and 4.3 form a very brief introduction to hyperbolic SPDEs.

Throughout this section $B = (B_t;\ t \in \mathbb{R}_+^N)$ will unwaveringly denote a real-valued, N-parameter Brownian sheet, and $\mathcal{F} = (\mathcal{F}_t;\ t \in \mathbb{R}_+^N)$ is the complete natural history of the process B.

---

[5] The process E plays a critical role in the detailed analysis of continuous martingales. Motivated by this, and the fact that it solves the stochastic differential equation $dE = E\, dB$, H. P. McKean, has dubbed it the **exponential martingale**.

4.1 Stochastic Integration Against the Brownian Sheet

Our construction of stochastic integrals against the Brownian sheet follows that of one-parameter stochastic integrals.

We say that a process $[M]$ is the **quadratic variation** of an N-parameter martingale M if $t \mapsto M_t^2 - [M]_t$ is an N-parameter martingale. With this in mind, we have the following result.

Theorem 4.1.1 *Suppose X is a continuous process such that (i) it is adapted to $\mathcal{F}$; and (ii) for all $t \in \mathbb{R}_+^N$, $\mathbb{E}[\int_{[0,t]} X_s^2 \, ds] < +\infty$. Then, there exists a continuous $L^2(\mathbb{P})$-martingale $B(X) = (B(X)_t; \, t \in \mathbb{R}_+^N)$ such that:*

(a) $B(X)_0 = 0$;

(b) $[B(X)]_t = \int_{[0,t]} X_s^2 \, ds$ *for all $t \in \mathbb{R}_+^N$; and*

(c) $[B(X)]$ *is the a.s. unique adapted process of bounded variation that is zero when $t = 0$.*

Furthermore, whenever X and Y are continuous adapted processes such that $\mathbb{E}[\int_{[0,t]} (X_s^2 + Y_s^2) \, ds] < \infty$:

(i) *with probability one, for all $t \in \mathbb{R}_+^N$, $B(X+Y)_t = B(X)_t + B(Y)_t$; and*

(ii) *as a process indexed by $t \in \mathbb{R}_+^N$, the following is a continuous martingale:*

$$B(X)_t \, B(Y)_t - \int_{[0,t]} X_s Y_s \, ds, \qquad t \in \mathbb{R}_+^N.$$

We shall interchangeably write $\int_{[0,t]} X_s \, dB_s$ for $B(X)_t$.

An attractive feature of this construction is that when X is nonrandom, $B(X)_t$ agrees with $\int_{[0,t]} X_s \, dB_s$ of Chapter 5, Section 1.5 (Supplementary Exercise 13).

Theorem 4.1.1 is proved along the same lines as Theorems 3.6.1 and 3.6.2; see Supplementary Exercise 11 for details.

In order to obtain a multiparameter extension of the Burkholder–Davis–Gundy inequality, let us first note that if M is an N-parameter martingale with respect to $\mathcal{F}$, then for every fixed $t^{(2)}, \ldots, t^{(N)} \geq 0$, $t^{(1)} \mapsto M_t$ is a 1-parameter martingale with respect to the first marginal filtration of $\mathcal{F}$, $\mathcal{F}^1$. In a discrete setting, this follows from Proposition 2.2.1 of Chapter 1; its continuous extension is proved similarly (check!).

Now let us consider the N-parameter martingale $B(X)$ where X is any continuous adapted process such that for all $t \in \mathbb{R}_+^N$, $\mathbb{E}[\int_{[0,t]} X_s^2 \, ds] < +\infty$. Let us fix $t^{(2)}, \ldots, t^{(N)} \geq 0$ and note that $t^{(1)} \mapsto B(X)_t$ and $t^{(1)} \mapsto \{B(X)_t\}^2 - [B(X)]_t$ are 1-parameter martingales; see the previous paragraph. Since Theorem 4.1.1 shows us the explicit form of the quadratic variation $[B(X)]$, we can apply the 1-parameter Burkholder–Davis–Gundy inequality (Theorem 3.9.1) to see that for any $t \in \mathbb{R}_+^N$ and for all $p \geq 1$,

$$\mathbb{E}\Big[\Big\{\int_{[0,t]} X_s \, dB_s\Big\}^{2p}\Big] \leq c(p) \mathbb{E}\Big[\Big\{\int_{[0,t]} X_s^2 \, ds\Big\}^p\Big].$$

Subsequently, we can combine the Cairoli-Walsh commutation theorem (Theorem 2.4.1) together with Cairoli's strong $(2p, 2p)$ inequality (Theorem 2.3.2) and obtain a multiparameter extension of the Burkholder–Davis–Gundy inequality.

Theorem 4.1.2 (The Burkholder–Davis–Gundy Inequality) *Suppose X is continuous and adapted, and for all $t \in \mathbb{R}_+^N$, $\mathbb{E}[\int_{[0,t]} X_s^2 \, ds] < +\infty$. Then, for all $t \in \mathbb{R}_+^N$ and all $p \geq 1$,*

$$\mathbb{E}\Big[\sup_{s \in [0,t]} \Big\{\int_{[0,s]} X_r \, dB_r\Big\}^{2p}\Big] \leq c(p) 4^{pN} \mathbb{E}\Big[\Big\{\int_{[0,t]} X_r^2 \, dr\Big\}^p\Big].$$

That is, whenever the right-hand side is finite, so is the left-hand side, and the above bound holds.

Exercise **4.1.1** Fill in the gaps to prove Theorem 4.1.2. □

4.2 Hyperbolic SPDEs: Some Physical Motivation

We begin our discussion with a heuristic, though compelling, description of a physical model that naturally leads to a large class of hyperbolic stochastic partial differential equations (written as SPDEs).

Consider three smooth functions $\alpha, \beta, b : \mathbb{R}^N \to \mathbb{R}$. The following nonlinear hyperbolic partial differential equation seeks a "solution" $f : \mathbb{R}^N \to \mathbb{R}$ that, in some reasonable sense, satisfies

$$\frac{\partial^N f}{\partial t^{(1)} \cdots \partial t^{(N)}}(t) = \alpha(f(t)) \frac{\partial^N b}{\partial t^{(1)} \cdots \partial t^{(N)}}(t) + \beta(f(t)), \qquad t \in \mathbb{R}^N. \tag{1}$$

To see what this equation means in its simplest nontrivial setting, suppose $N = 2$, $b(t) \equiv 0$, and $\beta(u) \equiv 0$, the zero function. Let us relabel the variables $s = s^{(1)}$ and $y = t^{(2)}$ to get

$$\frac{\partial^2 f}{\partial s \partial y}(s, y) = 0. \tag{2}$$

At this stage we can use D'Alembert's method of characteristics to relate the above to the well-known wave equation of mathematical physics. We change variables as $t = s + y$ and $x = s - y$, so that the (t, x)-plane is a rotation of the (s, y)-plane by $45°$. Apparently,

$$\frac{\partial f}{\partial s} = \frac{\partial f}{\partial t} + \frac{\partial f}{\partial x}.$$

Another round of differentiation, this time with respect to y, yields

$$\frac{\partial^2 f}{\partial s \partial y} = \frac{\partial^2 f}{\partial t^2} - \frac{\partial^2 f}{\partial x^2},$$

and we have transformed equation (2) into $\partial^2 f/\partial t^2 = \partial^2 f/\partial x^2$. Consequently, it can easily be checked that for any fixed $c > 0$, the function $\psi(t,x) = f(t, x/c)$ solves the one-dimensional wave equation of mathematical physics, viz.,

$$\frac{\partial^2 \psi}{\partial t^2} = \frac{1}{c^2} \frac{\partial \psi}{\partial x^2}.$$

In a few words, the solution ψ represents the displacement, or position, of a flexible vibrating string at "time" t in "position" x. The above equation is in terms of some constant c that has a physical interpretation as well. Indeed, one can write $c^2 = T/\rho$, where T is the tension in the string and ρ is the linear string density.

If, in addition, we apply a possibly time-dependent external force of amount $F(t,x)$ per unit length to this string, the equation of the vibrating string changes to

$$\frac{\partial^2 \psi}{\partial t^2} = \frac{\rho}{T} \frac{\partial^2 \psi}{\partial x^2} - \frac{1}{T} F.$$

Conversely, if b satisfies $\partial^2 b/\partial t^2 = \partial^2 b/\partial x^2 + F$, we would expect that the vibrating string problem with external force can be transformed into equation (1) with $\alpha(u) \equiv 1/T$, $\beta(u) = 0$, and $N = 2$.

Still working with the vibrating string problem in the $N = 2$ case, suppose the external force $F(t,x)$ is "truly random." Say, we have a string, and F describes the quantum effect of the surrounding particles.

Since F is obtained by "averaging out" the effect of many i.i.d. particles, it stands to reason that F is a centered Gaussian process. On the other hand, in the "truly random" case, we would expect $(F(t,x); x \in \mathbb{R}, t \geq 0)$ to be an i.i.d. collection of random variables. We also might as well scale things so that $\mathbb{E}[\{F(t,x)\}^2] = 1$ for all $x \in \mathbb{R}$ and all $t \geq 0$. Thus, our model for the external force formally is described by $F(t,x) = -W(\delta_{t,x})$, where W is the isonormal process on $\mathbb{R}^2$ and δ_p denotes Dirac's delta function at the point p (why?).

Since delta functions are not really functions, one needs to work to justify the above model properly. Nonetheless, arguing formally still, the random process b is modeled to satisfy $\partial^2 b/\partial t^2 - \partial^2 b/\partial x^2 = W(\delta_{t,x})$. Reverting to the (s,y)-plane, b must satisfy $\partial^2 b/\partial s \, \partial y = W(\delta_{s,y})$. Integrating (still purely formally) and recalling that W is a random linear operator, we would want $b(s,y) = W\big(\int_0^s \int_0^y \delta_{u,v} \, du \, dv\big)$. On the other hand, as "functions" of

(p,q),[6]

$$\int_0^s \int_0^y \delta_{u,v}(p,q)\,du\,dv = \mathbb{1}_{[0,s]\times[0,y]}(p,q).$$

Let $t = (s,y)$ to see that we want $b(t) = W(\mathbb{1}_{[0,t]})$, whose modification is a Brownian sheet. Thus, another equally formal interpretation of equation (1), when the external force is "truly random" is given by the following equation for $N = 2$:

$$df(t) = \alpha(f(t))\,dB_t + \beta(f(t))\,dt, \qquad t \in \mathbb{R}_+^N,$$

where B denotes the N-parameter Brownian sheet.

We are now in a position to write our model in a sensible way: Find a continuous adapted process $X = (X_t;\ t \in \mathbb{R}_+^N)$ such that

$$dX_t = \alpha(X_t)\,dB_t + \beta(X_t)\,dt.$$

We can interpret this further as the following stochastic integral equation:

$$X_t - X_0 = \int_{[0,t]} \alpha(X_s)\,dB_s + \int_{[0,t]} \beta(X_s)\,ds, \qquad t \in \mathbb{R}_+^N. \qquad (3)$$

This formulation of our hyperbolic SPDE is at least in terms of objects each of which is, in principle, perfectly well defined. Moreover, when $\alpha(u) \equiv 1$ and $\beta(u) \equiv 0$, the solution with $X_0 \equiv 0$ is $X_t = B_t$. That is, when $N = 2$, the Brownian sheet has the physical interpretation of being the solution to the stochastic vibrating string problem in the transformed space, when $\alpha \equiv 1$ and $\beta \equiv 0$.

In the next subsection we show that if α and β have nice regularity features, the general hyperbolic SPDE of equation (3) has a solution that is unique, given the "boundary" value X_0. A notable example is the following special case.

Suppose $X_0 = 0$, $N = 2$, $\alpha(u) \equiv T^{-1}$, and $\beta(u) \equiv 0$. Then this equation, once transformed, solves the following vibrating string problem: "A flat string is subject to a 'truly random' external force that forces it to vibrate. Then, displacement of the string at time t at position x is $X_{t,x/c}$, where $c = T/\rho$ and X solves equation (3)."

By allowing $\beta(u)$ to be a constant other than 0, we can allow for damped strings, and by allowing a general β, we get the full stochastic vibrating string equation with nonlinear damping.

---

[6]Strictly speaking, this is nonsense and needs to be interpreted in the sense of distributions. However, we will not have need for a rigorous interpretation of the above, since we plan to transform things into a rigorously definable equation.

4.3 Hyperbolic SPDEs: Existence and Uniqueness Issues

Let B denote a real-valued, N-parameter Brownian sheet, and consider the following hyperbolic SPDE:

$$X_t = x_0 + \int_{[0,t]} \alpha(X_s)\, dB_s + \int_{[0,t]} \beta(X_s)\, ds, \qquad t \in \mathbb{R}_+^N, \qquad (1)$$

where α and β are suitably chosen functions. We now prove the existence and uniqueness of a solution to the hyperbolic SPDE (1), under some regularity conditions on α and β.

We say that a function $f: \mathbb{R} \to \mathbb{R}$ is **globally Lipschitz** if there exists a constant Γ such that for all $x, y \in \mathbb{R}$, $|f(x) - f(y)| \leq \Gamma |x - y|$. The smallest such Γ is called the **Lipschitz norm** of f and is denoted by $\|f\|_L$, i.e.,

$$\|f\|_L = \sup_{\substack{x,y \in \mathbb{R} \\ x \neq y}} \frac{|f(x) - f(y)|}{|x - y|}.$$

Note that $\|f\|_L \leq \|f'\|_\infty$ if f happens to be differentiable. Moreover, this inequality is not improvable; see Supplementary Exercise 6 in this connection.

Theorem 4.3.1 *Suppose α and β are bounded and globally Lipschitz functions. Then, for every $x_0 \in \mathbb{R}$, the hyperbolic SPDE (1) has a continuous adapted solution that is a.s. unique.*

Almost sure uniqueness of a continuous adapted solution X means that any other continuous adapted solution $\widetilde{X}$ to (1) is necessarily a modification of X.

Before proving this result, we shall state a useful analytical estimate.

Lemma 4.3.1 (Gronwall's Lemma) *Suppose $\varphi_1, \varphi_2, \ldots : \mathbb{R}_+^N \to \mathbb{R}_+^N$ are measurable and nondecreasing in each of their N coordinate variables. Suppose, further, that there exist finite constants $C, T > 0$ such that for all $t \in [0, T]$ and all $n \geq 1$,*

$$\varphi_{n+1}(t) \leq C \int_{[0,t]} \varphi_n(s)\, ds.$$

Then, for all $n \geq 1$ and all $t \in [0, T]$,

$$\varphi_{n+1}(t) \leq \frac{\left(C \prod_{\ell=1}^N t^{(\ell)}\right)^n}{(n!)^N} \varphi_1(t).$$

Exercise 4.3.1 Prove Gronwall's lemma. □

4 Stochastic Partial Differential Equations

Proof of Theorem 4.3.1 We use Picard's iteration of classical ODEs. That is, we show that there is always a fixed-point solution to equation (1), and that it is unique.

Define the process $X_t^0 = x_0$, and iteratively let

$$X_t^{n+1} = x_0 + \int_{[0,t]} \alpha(X_s^n)\, dB_s + \int_{[0,t]} \beta(X_s^n)\, ds.$$

By induction, X^n is continuous, and both integrals are always well-defined, since α and β are bounded and continuous. We proceed by showing that, in a suitable sense, $n \mapsto X_t^n$ is Cauchy, uniformly over t-compacts.

To this end, let us fix a nonrandom $T \in \mathbb{R}_+^N$ and consider for $m, n \geq 0$ and all $t \in [0, T]$,

$$X_t^{m+1} - X_t^{n+1} = \int_{[0,t]} \{\alpha(X_s^m) - \alpha(X_s^n)\}\, dB_s + \int_{[0,t]} \{\beta(X_s^m) - \beta(X_s^n)\}\, ds.$$

Clearly,

$$\sup_{t \in [0,T]} \left| \int_{[0,t]} \{\beta(X_s^m) - \beta(X_s^n)\}\, ds \right| \leq \|\beta\|_L \cdot \int_{[0,T]} |X_s^n - X_s^m|\, ds.$$

Thus, by the Cauchy–Schwarz inequality, for all $p \geq 1$,

$$\mathbb{E}\left[\sup_{t \in [0,T]} \left| \int_{[0,t]} \{\beta(X_s^m) - \beta(X_s^n)\}\, ds \right|^{2p} \right] \leq A_1 \int_{[0,T]} \mathbb{E}\{|X_s^n - X_s^m|^{2p}\}\, ds,$$

where $A_1 = \|\beta\|_L^{2p} \cdot \{\prod_{\ell=1}^N T^{(\ell)}\}^{2p-1}$. Likewise, by the Burkholder–Davis–Gundy inequality (Theorem 4.1.2),

$$\mathbb{E}\left[\sup_{t \in [0,T]} \left| \int_{[0,t]} \{\alpha(X_s^m) - \alpha(X_s^n)\}\, dB_s \right|^{2p} \right]$$
$$\leq c(p) 4^{pN} \mathbb{E}\left[\left(\int_{[0,T]} \{\alpha(X_s^m) - \alpha(X_s^n)\}^2\, ds \right)^p \right]$$
$$\leq c(p) 4^{pN} \|\alpha\|_L^{2p} \cdot \mathbb{E}\left[\left(\int_{[0,T]} |X_s^m - X_s^n|^2\, ds \right)^p \right]$$
$$\leq A_2 \int_{[0,T]} \mathbb{E}\{|X_s^n - X_s^m|^{2p}\}\, ds,$$

where $A_2 = c(p) 4^{pN} \|\alpha\|_L^{2p} \cdot \{\prod_{\ell=1}^{2p-1} T^{(\ell)}\}^{2p-1}$. Combining terms, we have verified the following bound:

$$\mathbb{E}\left\{ \sup_{t \in [0,T]} |X_s^{m+1} - X_s^{n+1}|^{2p} \right\} \leq 4^p (A_1 + A_2) \int_{[0,T]} \mathbb{E}\left\{ \sup_{s \in [0,t]} |X_s^m - X_s^n|^{2p} \right\} dt.$$

We have invoked the elementary inequality $(a+b)^{2p} \le 4^p(a^{2p}+b^{2p})$. Use the above with $m = n+1$ and apply Gronwall's lemma (Lemma 4.3.1) to $\varphi_{n+1}(t) = \mathbb{E}\{\sup_{t\in[0,T]}|X_t^{n+1} - X_t^n|^{2p}\}$ and $C = 4^p(A_1+A_2)$ to see that

$$\mathbb{E}\left\{\sup_{t\in[0,T]} |X_t^{n+1} - X_t^n|^{2p}\right\} \le \frac{(C \prod_{\ell=1}^N T^{(\ell)})^n}{(n!)^N} \mathbb{E}\left\{\sup_{t\in[0,T]} |X_t^1 - X_t^0|^{2p}\right\}.$$

On the other hand, $X_t^1 - X_t^0 = \alpha(x_0)B_t + \beta(x_0)\prod_{\ell=1}^N t^{(\ell)}$. Thus,

$$\mathbb{E}\left\{\sup_{t\in[0,T]} |X_t^1 - X_t^0|^{2p}\right\}$$

$$\le 4^p\left\{|\alpha(x_0)|^{2p} \cdot \mathbb{E}\left[\sup_{t\in[0,T]} B_t^{2p}\right] + |\beta(x_0)|^{2p} \cdot \prod_{\ell=1}^N |T^{(\ell)}|^{2p}\right\}$$

$$= C(p),$$

which is finite. (Why?) This shows that

$$\sum_{n=1}^\infty \mathbb{E}\left\{\sup_{t\in[0,T]} |X_t^{n+1} - X_t^n|^{2p}\right\} < +\infty.$$

In particular, by the completeness of L^p spaces and by the Borel–Cantelli lemma, there exists a process X such that for all $T \in \mathbb{R}_+^N$,

$$\lim_{n\to\infty} \sup_{t\in[0,T]} |X_t^n - X_t| = 0,$$

almost surely and in $L^p(\mathbb{P})$ for all $p \ge 1$. The process X is clearly continuous and adapted, since the X^n's are. Moreover, it is our solution to the hyperbolic SPDE (1); this fact is relegated to Exercise 4.3.2 below. We now show uniqueness.

Suppose Y is another continuous adapted solution. Then, the same argument as above shows that for all $T \in \mathbb{R}_+^N$ and $p \ge 1$, there exists a finite constant A_p such that for all $t \in [0,T]$,

$$\varphi(t) \le A_p \int_{[0,t]} \varphi(s)\, ds,$$

where $\varphi(t) = \mathbb{E}\{\sup_{s\in[0,t]}|X_s - Y_s|^{2p}\}$. Since α and β are bounded, Exercise 4.3.3 shows that $\varphi(t)$ is finite for all $t \in \mathbb{R}_+^N$. In particular, by Gronwall's lemma applied to $C = A_p$ and $\varphi_n = \varphi$ for all n and all $t \in [0,T]$,

$$\varphi(t) \le \frac{D^n}{(n!)^N}\varphi(t),$$

where D is a finite constant. Let $n \to \infty$ to see that $\varphi(t) \equiv 0$. In particular, for all $T \in \mathbb{R}_+^N$, $\mathbb{P}(X_t = Y_t \text{ for all } t \in [0,T]) = 1$. Let $T \rightsquigarrow \infty$, i.e., coordinatewise, to prove the asserted uniqueness. $\square$

Exercise 4.3.2 Show that in the proof of Theorem 4.3.1, $X = \lim_{n\to\infty} X^n$ solves equation (1). □

Exercise 4.3.3 Prove that in the proof of Theorem 4.3.1, the boundedness of α and β guarantees that $\varphi(t) < +\infty$ for all $t \in \mathbb{R}_+^N$.
(HINT: Gronwall's lemma.) □

Exercise 4.3.4 Prove that the solution to the hyperbolic SPDE (1) is, in fact, Hölder continuous.
(HINT: Theorem 2.3.1, Chapter 5.) □

5 Supplementary Exercises

1. Suppose that for each $n \geq 1$, M^n is an N-parameter martingale with respect to one fixed N-parameter filtration $\mathcal{F}$. Suppose further that for every $t \in \mathbb{R}_+^N$, M_t^n converges in $L^1(\mathbb{P})$ to some M_t, as $n \to \infty$. Prove that M is an N-parameter martingale.

2. Suppose $S = (S_n;\ n \geq 0)$ denotes the simple symmetric random walk on $\mathbb{Z}$; cf. Section 3 of Chapter 3. For every $a \in \mathbb{Z}$, define $T_a = \inf(k \geq 0: S_k = a)$ to be the entrance time to $\{a\}$, and show that for all $a, b \geq 0$, $\mathbb{P}(T_{-a} < T_b) = b/(b-a)$. Here is a gambling interpretation: Suppose you gamble on a fair game independently one time after another. Suppose further that on every trial you either lose or win a dollar. Then, S_n denotes your net profit by the nth trial, and the above computes the probability that you reach the house limit before going bankrupt.

3. If $\mathbb{P}(Z \geq 0) = 1$ and $\mathbb{E}[Z] = 0$, show that $\mathbb{P}(Z = 0) = 1$.

4. If M denotes a continuous one-parameter local martingale, show that there exists a unique nondecreasing, continuous adapted process $[M]$ such that with probability one, $[M]_0 = 0$, and $t \mapsto M_t^2 - [M]_t$ is a continuous local martingale.

5. Suppose M is a continuous one-parameter local martingale, and X is a one-parameter continuous adapted process such that almost surely, $\int_0^t X_s^2\, d[M]_s < \infty$ for all $t \geq 0$. Construct a stochastic integral process $t \mapsto M(X)_t$ such that $M(X)$ is a continuous local martingale with quadratic variation $t \mapsto \int_0^t X_s^2\, d[M]_s$. Moreover, for all continuous, adapted processes X and Y such that $\int_0^t (X_s^2 + Y_s^2)\, d[M]_s < +\infty$, a.s. for all $t \geq 0$, with probability one, $M(X+Y)_t = M(X)_t + M(Y)_t$, for all $t \geq 0$.

6. Prove that whenever $f : \mathbb{R} \to \mathbb{R}$ is globally Lipschitz, it has a derivative at almost every point. However, find an example to show that such an f need not have a derivative at every point.
(HINT: For the interval $I = (x, y)$, define $\mu(I) = f(y) - f(x)$, and show that μ extends to an absolutely continuous measure on the Borel subsets of $\mathbb{R}$.)

7. Suppose X (M) is a one-parameter continuous adapted (martingale) process. Let $r_{j,n} = j2^{-n}$, and prove the existence of

$$\int_0^t X_s \circ dM_s = \lim_{n \to \infty} \sum_{1 \leq j \leq 2^n t} \left(\frac{X_{r_{j-1,n}} + X_{r_{j,n}}}{2} \right) \cdot (M_{r_{j,n}} - M_{r_{j-1,n}}).$$

This stochastic integral is called the **Stratonovich integral**, as found in Stratonovich (1966). Identify $\int_0^t X_s \circ dM_s$ in terms of $\int_0^t X_s \, dM_s$ explicitly, and show that for all *twice continuously differentiable* functions $f : \mathbb{R} \to \mathbb{R}$, $f(M_t) = f(M_0) + \int_0^t f'(M_s) \circ dM_s$, for all $t \geq 0$, a.s. That is, the Stratonovich calculus follows the rules of the ordinary calculus of functions while the Itô calculus does not. Finally, show that whenever X has zero quadratic variation, the two integrals agree. That is, for all $t \geq 0$, $\int_0^t X_s \, dM_s = \int_0^t X_s \circ dM_s$, a.s. (HINT: Imitate the proof of Itô's lemma, Theorem 3.8.1.)

8. A d-dimensional Brownian motion B is defined by $B_t = (B_t^{(1)}, \ldots, B_t^{(d)})$, where $B^{(1)}, \ldots, B^{(d)}$ are d independent standard Brownian motions. Prove that for all twice continuously differentiable functions $f : \mathbb{R}^d \to \mathbb{R}$, the following holds a.s. for all $t \geq 0$:

$$f(B_t) = f(0) + \sum_{i=1}^d \int_0^t \frac{\partial^i f}{\partial x^{(i)}}(B_s) \, dB_s^{(i)} + \frac{1}{2} \int_0^t \Delta f(B_s) \, ds,$$

where $\Delta f(x) = \sum_{j=1}^d \frac{\partial^2 f}{\partial (x^{(j)})^2}(x)$ is the **Laplacian** of f.

9. Let M denote a one-parameter continuous $L^2(\mathbb{P})$-martingale with respect to a filtration $\mathcal{F}$ that satisfies the usual conditions. Given an adapted process V of bounded variation, show that for all $f : \mathbb{R} \times \mathbb{R} \ni (x, v) \mapsto f(x, v) \in \mathbb{R}$ that are twice continuously differentiable in x and continuously differentiable in v,

$$f(M_t, V_t) = f(M_0, V_0) + \int_0^t \frac{\partial f}{\partial x}(M_s, V_s) \, dM_s$$
$$+ \int_0^t \frac{\partial f}{\partial v}(M_s, V_s) \, dV_s + \frac{1}{2} \int_0^t \frac{\partial^2 f}{\partial x^2}(M_s, V_s) \, d[M]_s,$$

where the dV integral is defined (ω by ω) as a Stieltjes integral.

10. Given a d-dimensional Brownian motion B, an adapted process V of bounded variation, and a sufficiently smooth function $f : \mathbb{R} \times \mathbb{R} \to \mathbb{R}$,

$$f(B_t, V_t) = f(0, V_0) + \sum_{j=1}^d \int_0^t \frac{\partial f}{\partial x^{(i)}}(B_s, V_s) \, dB_s^{(j)}$$
$$+ \int_0^t \frac{\partial f}{\partial v}(B_s, V_s) \, dV_s + \frac{1}{2} \int_0^t \Delta f(B_s, V_s) \, ds,$$

where Δ is the Laplacian applied to the x variable. Apply this to show that when $d = 1$, $E_t = \exp(B_t - \frac{1}{2}t)$ is a continuous martingale that solves the stochastic differential equation $E_t = 1 + \int_0^t E_s \, dB_s$. (HINT: See Exercise 8.)

5 Supplementary Exercises

11. We wish to prove Theorem 4.1.1. Throughout, $\mathcal{F} = (\mathcal{F}_t;\ t \in \mathbb{R}_+^N)$ denotes the complete augmented history of the Brownian sheet.

(i) We say that an N-parameter process X is **elementary** if it is adapted to $\mathcal{F}$ and if there are $\alpha \preccurlyeq \beta$ in $\mathbb{R}_+^N$ and a bounded $\mathcal{F}_\beta$-measurable random variable Θ such that $X_s = \Theta \mathbf{1}_{]\alpha,\beta]}(s)$ ($s \in \mathbb{R}_+^N$). Prove that Theorem 4.1.1 holds for elementary processes in place of continuous, adapted processes.

(ii) A **simple process** is defined to be a finite linear combination of elementary processes. Extend the previous argument to include integrands that are simple processes.

(iii) Conclude the proof of Theorem 4.1.1 by proving a multiparameter version of Lemma 3.6.1.

12. If B denotes the standard Brownian motion:

(i) Show that for all $\lambda > 0$, $\mathbb{P}(\sup_{0 \le s \le 1} B_s \ge \lambda) = \mathbb{P}(|B_1| \ge \lambda)$.

(ii) Deduce that for all $p, t > 0$, $\mathbb{E}[\sup_{0 \le s \le t} B_s^{2p}] = \mathbb{E}[B_t^{2p}]$. Compute this.

(iii) Improve the constant $c(p)$ of Corollary 3.9.2 to 1 when only nonrandom times T are considered.

(HINT: Part (i) is the reflection principle of D. André; see Supplementary Exercise 3 of Chapter 4 for the discrete case. For part (ii), the expression $\mathbb{E}[B_1^{2p}]$) can be computed directly using the properties of Gaussian densities. However, it can also be computed using Itô's formula. You may wish try this neat approach!)

13. Let B denote the N-parameter Brownian sheet. Show that when X is a continuous nonrandom function, the stochastic integral $B(X)$ of this chapter is a.s. the same as the stochastic integral of Chapter 5.

14. (Hard) Recall that a function $f: \mathbb{R}^d \to \mathbb{R}$ is **locally Lipschitz** if for all $M > 0$, there exists a finite constant $\Gamma_M > 0$ such that for all $x, y \in [-M, M]^d$, $|f(x) - f(y)| \le \Gamma_M |x - y|$. Suppose (i) α and β are locally Lipschitz; and (ii) there exists a finite constant $C > 0$ such that for all $x \in \mathbb{R}$, $|\alpha(x)| + |\beta(x)| \le C\{1 + |x|\}$. Show that the conclusion of Theorem 4.3.1 still holds true.
(HINT: Using the notation of our proof of Theorem 4.3.1, start by showing that for any $t \in \mathbb{R}_+^N$, $\mathbb{E}\{\sup_{r \in [0,t]} |X_r^n|^{2p}\}$ is bounded in n. You may need to derive the following variant of Gronwall's lemma: If $\varphi_{n+1}(t) \le C\{1 + \int_{[0,t]} \varphi_n(s)\,ds\}$, then $\sup_n \varphi_n$ is locally bounded.)

15. Suppose α and β are globally Lipschitz and bounded functions, and consider the SPDE of equation (1) of Section 4.3. Let $X = (X_{x_0, t};\ t \in \mathbb{R}_+^N,\ x_0 \in \mathbb{R}^d)$ denote the solution, viewed as a random function of t as well as the "starting point" x_0. Show that there exists a Hölder continuous modification of $(x_0, t) \mapsto X_{x_0, t}$. This process is a **stochastic flow**, viewed as a function-valued function of x_0.

6 Notes on Chapter 7

Sections 1,3 The material of these two sections is standard fare in the theory of continuous martingales. See (Chung and Williams 1990; Karatzas and Shreve 1991; Revuz and Yor 1994) for three different pedagogic accounts. For an encyclopedic treatment, see Dellacherie and Meyer (1982). Our proof of Lemma 1.8.1 is motivated by ideas from Kunita and Watanabe (1967), which is excellent reading to this day.

Our proof of the Burkholder–Davis–Gundy inequality (Theorem 3.9.1) does not give sharp constants, but has the advantage of being brief.

Theorem 3.9.1 holds for all $p > 0$; see Bass (1987) for a proof, and much more. The fact that Theorem 3.9.1 holds for all $p > 0$ also follows from the fact that it holds for all $p \geq 1$, used in conjunction with the multiplier theorem of J. Marcinkiewicz; see Stein (1993).

Section 2 Walsh (1986b, 1986a) are excellent places to start learning more about the general theory of multiparameter martingales. Since the general literature on this subject is massive, we merely refer the reader to the following references, which are more or less directly related to this chapter: Bakry (1979, 1981b, 1982), Imkeller (1985, 1988), Mazziotto and Merzbach (1985), Mazziotto and Szpirglas (1981, 1982), and Nualart (1985). See Merzbach and Nualart (1985) for a survey of various aspects of the general theory, and Körezlioğlu et al. (1981) for some of the more recent activity.

There is a powerful regularity theorem, due to D. Bakry, that proves the multi-parameter analogue of Theorem 1.4.1. The basic message is that all $L\{\ln_+ L\}^{N-1}$-bounded, N-parameter martingales (with respect to commuting filtrations) have a right-continuous modification. Note that for this modification, Corollary 2.3.1 always holds; cf. Theorem 2.3.2. Bakry's regularity theorem is worked out in detail in Bakry (1979) in the 2-parameter case using Cairoli's inequalities and the general theory of 2-parameter processes, in particular, a section theorem found by E. Merzbach. This general theory is well described in Dozzi (1989, 1991) and Imkeller (1988).

Section 4 The heuristic discussion of Section 4.2 is modeled after aspects of the presentations of Cabaña (1991) and Walsh (1986a, 1986b).

This section presents stochastic integration against the Brownian sheet by a multiparameter extension of Itô integrals. Since such integrals are all that we develop in this book, this approach is sufficient for our needs. However, stochastic integration against more general multiparameter martingales is truly a more complicated story. Much of this is developed in (Imkeller 1988; Nualart 1985; Walsh 1986a) and their combined references. Further extensions of these notions can be found in Nualart (1995).

8
Constructing Markov Processes

Markov processes will provide us with a large class of processes that can be used as building blocks for useful and interesting random fields. This and the following chapter are a brief introduction to Markov processes.

Although in some specialized settings we have already encountered the Markov property, e.g., Chapters 3 and 7, the general theory of Markov processes is substantially more complicated; thus, we restrict attention to a nice class of Markov processes that are known as Feller processes. In order to gel the basic ideas, we begin with the simplest case, which is that of discrete Markov chains. Our treatment of the continuous-time theory will follow, starting with Section 2.

1 Discrete Markov Chains

We start our development of the theory of Markov processes in its simplest setting: discrete Markov chains. These are discrete-time, discrete-space processes that possess the Markov property. Once some of the key concepts and methods are isolated, we proceed with our development of the more complicated continuous-time theory.

1.1 Preliminaries

If S denotes a denumerable set, we say that a stochastic process $X = (X_n;\ n \geq 0)$ is a (discrete) **Markov chain with state space** S if there exists a filtration $\mathcal{F} = (\mathcal{F}_n;\ n \geq 0)$ such that:

268   8. Constructing Markov Processes

1. X is an S-valued process that is adapted to $\mathcal{F}$;
2. for all $n \geq 0$ and all $a \in S$,
$$\mathbb{P}(X_{n+1} = a \mid \mathcal{F}_n) = \mathbb{P}(X_{n+1} = a \mid X_n), \text{ a.s.}[1]$$

Property 2 above is called the **Markov property** of X.

The following is an important exercise.

Exercise 1.1.1 If $\mathcal{H} = (\mathcal{H}_n;\ n \geq 0)$ denotes the history of a Markov chain X, then for all $a \in S$ and all $n \geq 0$, $\mathbb{P}(X_{n+1} = a \mid \mathcal{H}_n) = \mathbb{P}(X_{n+1} = a \mid X_n)$, a.s. □

Thus, as a consequence of Exercise 1.1.1, unless a specific filtration is mentioned, the underlying filtration is tacitly assumed to be the history of X.

For all $n, k \geq 0$ and $a, b \in S$, define,
$$p_{n,n+k}(a,b) = \begin{cases} \mathbb{P}(X_{n+k} = b \mid X_n = a), & \text{if } \mathbb{P}(X_n = a) > 0, \\ 0, & \text{otherwise.} \end{cases}$$

In words, $p_{n,n+k}(a, b)$ denotes the probability that our chain goes from a to b, starting at time n, and ending at time $n + k$.

Recall that on $(X_n = a)$, $p_{n,n+1}(a, b) = \mathbb{P}(X_{n+1} = b \mid X_n)$. Thus, the Markov property can be restated as follows: For all $n \geq 0$ and all $a \in S$,
$$p_{n,n+1}(a, b) = \mathbb{P}(X_{n+1} = b \mid \mathcal{F}_n),$$
on $(X_n = a)$. Equivalently, we have the Markov property if and only if
$$p_{n,n+1}(X_n, b) = \mathbb{P}(X_{n+1} = b \mid \mathcal{F}_n), \text{ a.s.}$$

We say that X is a **time-homogeneous Markov chain** if for all $n, k \geq 0$, $p_{n,n+k} = p_{0,k}$. One can usually reduce attention to time-homogeneous Markov chains, as the following shows.

Lemma 1.1.1 *Let X be an S-valued Markov chain. Define $Y = (Y_n;\ n \geq 0)$ by*
$$Y_n = (n, X_n), \qquad n \geq 0.$$
Then Y is an $\mathbb{N}_0 \times S$-valued, time-homogeneous Markov chain.

Proof Let $\mathcal{F}$ denote the history of X and note that $\mathcal{F}$ is also the history of Y. Clearly,
$$\mathbb{P}(Y_{n+k} = y \mid \mathcal{F}_n) = \mathbb{P}(Y_{n+k} = y \mid Y_n), \qquad \text{a.s.}$$

---
[1] Recall that for any random variable Y and event E, $\mathbb{P}(E \mid Y)$ is shorthand for $\mathbb{P}\{E \mid \sigma(Y)\}$, where $\sigma(Y)$ is the σ-field generated by Y. A similar remark holds for conditional expectations.

That is, we have the Markov property. It suffices to prove time-homogeneity. We can write any $x \in \mathbb{N}_0 \times S$ as $x = (x^{(1)}, x^{(2)})$, where $x^{(1)}$ is a nonnegative integer and $x^{(2)} \in S$. With the above in mind, note that

$$\mathbb{P}(Y_{n+k} = y \mid Y_n = x)$$
$$= \begin{cases} \mathbb{P}(X_{x^{(1)}+k} = y^{(2)} \mid X_{x^{(1)}} = x^{(2)}), & \text{if } y^{(1)} = x^{(1)} + k, \\ 0, & \text{otherwise.} \end{cases}$$

Since this is independent of n, our proof is complete. □

Since the map $X \mapsto Y$ is a tractable one, it is often sufficient to study time-homogeneous Markov chains. This motivates the following simplification.

Convention Unless we state otherwise, Markov chains are to be assumed time-homogeneous.

The **transition probabilities** of a Markov chain X are the collection of the probabilities $p_{0,n}(x,y)$, $n \geq 0$, $x,y \in S$. We will write $p_n(x,y)$ for $p_{0,n}(x,y)$ for brevity. Thus, in this time-homogeneous setting, the Markov property can be recast as follows: For all $n \geq 0$ and all $y \in S$,

$$\mathbb{P}(X_{n+1} = y \mid \mathcal{F}_n) = p_1(X_n, y), \quad \text{a.s.} \tag{1}$$

The function $(x,y) \mapsto p_k(x,y)$ is the **k-step transition function** of X. In words, $p_k(x,y)$ denotes the probability that in k time steps, the process moves from state $x \in S$ to state $y \in S$. The 1-step transition function p_1 is called simply the **transition function** of X, and thanks to (1), it determines all of the finite-dimensional distributions of X. Indeed, for any $a_0, \ldots, a_m \in S$,

$$\mathbb{P}(X_0 = a_0, \ldots, X_m = a_m) = p_1(a_{m-1}, a_m)\mathbb{P}(X_0 = a_0, \ldots, X_{m-1} = a_{m-1}).$$

By iterating this, we see that

$$\mathbb{P}(X_0 = a_0, \ldots, X_m = a_m) = \mathbb{P}(X_0 = a_0) \prod_{\ell=0}^{m-1} p_1(a_\ell, a_{\ell+1}). \tag{2}$$

Consequently, if we know the transition function, as well as the distribution of X_0, we can, in principle, compute the probability of any interesting event pertaining to the process X. From now on we refer to the distribution of X_0 as the **initial distribution** of the Markov chain X.

Let ν denote the initial distribution of X, and for any event E, define

$$\mathbb{P}_\nu(E) = \sum_{a \in S} \nu(\{a\})\, \mathbb{P}(E \mid X_0 = a).$$

270   8. Constructing Markov Processes

Clearly, $\mathbb{P}_\nu$ is a probability measure on S endowed with its Borel sets (in the counting topology). Moreover, suppose we define probability measures $(\mathbb{P}_a;\ a \in S)$ by
$$\mathbb{P}_a(\bullet) = \mathbb{P}(\bullet \mid X_0 = a).$$
Then, the above shows that $\mathbb{P}_\nu(\bullet) = \int \mathbb{P}_x(\bullet)\,\nu(da)$. Thus, we should think of $\mathbb{P}_\nu$ as the natural probability measure for the Markov chain X, given that X_0 has distribution ν. Moreover, it is easy to see that the above definitions are consistent. That is, if we denote the point mass at $x \in S$ by δ_x, then $\mathbb{P}_x = \mathbb{P}_{\delta_x}$. Henceforth, we will write $\mathbb{E}_\nu$ and $\mathbb{E}_x$ for the expectation operators corresponding to $\mathbb{P}_\nu$ and $\mathbb{P}_x$, respectively. Equation (2) can now be recast in the following way.

Theorem 1.1.1 (The Chapman–Kolmogorov Equation) *For any integer $m \geq 0$ and all bounded functions $f : S^{m+1} \to \mathbb{R}$,*

$$\mathbb{E}_\nu\big[f(X_0, \ldots, X_m)\big] = \sum_{a_0, \ldots, a_m \in S} f(a_0, \ldots, a_m)\nu(\{a_0\}) \prod_{\ell=0}^{m-1} p_1(a_\ell, a_{\ell+1}).$$

Example (Random Walks) Suppose $\xi_1, \xi_2, \ldots$ are i.i.d. $\mathbb{Z}^d$-valued random vectors, and recall from Chapter 3 that the corresponding random walk $X = (X_n;\ n \geq 1)$ is defined by $X_n = \sum_{j=1}^n \xi_j$.

Now we introduce an extra $\mathbb{Z}^d$-valued random vector ξ_0, independently of $\xi_1, \xi_2, \ldots$, and with distribution ν. Let us then *redefine*

$$X_n = \sum_{j=0}^n \xi_j, \qquad n \geq 0.$$

In this way we see that the process $X = (X_n;\ n \geq 0)$ is a Markov chain on $S = \mathbb{Z}^d$ with initial measure ν; cf. Theorem 1.1.1 of Chapter 3. Moreover, the transition function of X is given by

$$p_1(x, y) = \mathbb{P}(\xi_1 = y - x), \qquad x, y \in S.$$

As such, Theorem 1.1.1 should be viewed as an extension of Corollary 1.1.2, Chapter 3. □

Exercise **1.1.2** Suppose $X = (X_n;\ n \geq 0)$ denotes a symmetric random walk on $\mathbb{Z}^d$. Recall that this means that $-X$ and X have the same finite-dimensional distributions. If γ denotes the counting (or uniform) measure on $\mathbb{Z}^d$, show that for all $m \geq 0$ and all $f_1, \ldots, f_m : \mathbb{Z}^d \to \mathbb{R}_+$,

$$\mathbb{E}_\gamma\Big[\prod_{j=1}^m f_j(X_j)\Big] = \mathbb{E}_\gamma\Big[\prod_{j=1}^m f_j(X_{m-j})\Big],$$

where $\mathbb{E}_\gamma[\bullet] = \int \mathbb{E}_x[\bullet]\,\gamma(dx)$ is well-defined, although γ is not a probability measure. In brief, conclude that if the initial (nonprobability) measure is γ, the process $(X_n;\ 0 \leq n \leq m)$ has the same finite-dimensional distributions as $(X_{m-n};\ 0 \leq n \leq m)$. This is an example of **time-reversal**. If $\mathbb{P}_\gamma(\bullet) = \mathbb{E}_\gamma[\mathbf{1}_\bullet]$ is the corresponding "probability," what is $\mathbb{P}_\gamma(X_n \in A)$? □

We conclude this subsection with the following variant of Theorem 1.1.1.

Theorem 1.1.2 *Suppose m is a positive integer and $f : S^m \to \mathbb{R}$ is a bounded function. Then, whenever ν denotes the initial measure of the Markov chain X, for any integer $n \geq 0$,*

$$\mathbb{E}_\nu\left[f(X_{n+1}, \ldots, X_{n+m}) \,|\, \mathcal{F}_n\right] = \mathbb{E}_{X_n}[f(X_1, \ldots, X_m)], \qquad \mathbb{P}_\nu\text{-a.s.}$$

Proof We first check this for $m = 1$. Indeed, by equation (1),

$$\mathbb{E}_\nu[f(X_{n+1}) \,|\, \mathcal{F}_n] = \sum_{y \in S} f(y)\, p_1(X_n, y) = \mathbb{E}_{X_n}[f(X_1)],$$

$\mathbb{P}_\nu$-a.s., since for all $x \in S$, $\mathbb{E}_x[f(X_1)] = \sum_{y \in S} f(y)\, p_1(x, y)$. To conclude, let us assume that Theorem 1.1.2 holds for $m = j$. We will show that it also holds for $m = j+1$. In fact, it suffices to prove that for all $y_1, \ldots, y_{j+1} \in S$, the following holds $\mathbb{P}_\nu$-a.s.:

$$\begin{aligned}
\mathbb{P}_\nu(X_{n+1} = y_1, \ldots, X_{n+j+1} = y_{j+1} \,|\, \mathcal{F}_n) \\
= \mathbb{P}_{X_n}(X_1 = y_1, \ldots, X_m = y_m).
\end{aligned} \qquad (3)$$

By equation (1) (or the induction hypothesis), we have the following $\mathbb{P}_\nu$-a.s.:

$$\mathbb{P}_\nu(X_{n+1} = y_1, \ldots, X_{n+j+1} = y_{j+1} \,|\, \mathcal{F}_n)$$

$$= \mathbb{E}_\nu\left[\mathbb{P}_\nu(X_{n+j+1} = y_{j+1} \,|\, \mathcal{F}_{n+j}) \prod_{\ell=1}^{j} \mathbf{1}_{(X_{n+\ell} = y_\ell)} \,\bigg|\, \mathcal{F}_n\right]$$

$$= \mathbb{E}_\nu\left[p_1(X_{n+j}, y_{j+1}) \prod_{\ell=1}^{j} \mathbf{1}_{(X_{n+\ell} = y_\ell)} \,\bigg|\, \mathcal{F}_n\right].$$

Using the induction hypothesis (possibly one more time), we see that the following holds $\mathbb{P}_\nu$-a.s.:

$$\mathbb{P}_\nu(X_{n+1} = y_1, \ldots, X_{n+j+1} = y_{j+1} \,|\, \mathcal{F}_n) = \mathbb{E}_{X_n}\left[p_1(X_j, y_{j+1}) \prod_{\ell=1}^{j} \mathbf{1}_{(X_\ell = y_\ell)}\right].$$

Another application of equation (1) proves equation (3) and hence the result. □

1.2 The Strong Markov Property

Comparing the example of Section 1.1 to the discussion of Section 1.1, Chapter 3, we should expect that *all* Markov chains on the state space S should satisfy a stronger form of the Markov property. To this end, suppose $X = (X_n;\ n \geq 0)$ is a Markov chain on a denumerable state space S with respect to a filtration $\mathcal{F} = (\mathcal{F}_n;\ n \geq 0)$, and let ν denote the initial measure of X. The following is an extension of the Markov property for random walks (Theorem 1.2.1, Chapter 3) and, in a much more general context, is due to Blumenthal (1957). Precursors to this result can be found in (Hunt 1956b; Kinney 1953).

Theorem 1.2.1 (The Strong Markov Property) *For any stopping time T, for all integers $m \geq 1$, and for all bounded functions $f : S^m \to \mathbb{R}$, the following holds $\mathbb{P}_\nu$-almost surely:*

$$\mathbb{E}_\nu\big[f(X_{T+1}, X_{T+2}, \ldots, X_{T+m})\,\big|\,\mathcal{F}_T\big]\mathbf{1}_{(T<\infty)} = \mathbb{E}_{X_T}\big[f(X_1, \ldots, X_m)\big]\mathbf{1}_{(T<\infty)}.$$

Proof If j denotes any positive integer, we have, $\mathbb{P}_\nu$-a.s. on $(T=j)$,

$$\mathbb{E}_\nu\big[f(X_{T+1}, \ldots, X_{T+m})\,\big|\,\mathcal{F}_T\big] = \mathbb{E}_\nu\big[f(X_{j+1}, \ldots, X_{j+m})\,\big|\,\mathcal{F}_T\big]$$
$$= \mathbb{E}_\nu\big[f(X_{j+1}, \ldots, X_{j+m})\,\big|\,\mathcal{F}_j\big].$$

The last step uses Theorem 1.1.1 of Chapter 1. By Theorem 1.1.2,

$$\mathbb{E}_\nu\big[f(X_{T+1}, \ldots, X_{T+m})\,\big|\,\mathcal{F}_T\big]\mathbf{1}_{(T=j)} = \mathbb{E}_{X_j}\big[f(X_1, \ldots, X_m)\big]\mathbf{1}_{(T=j)},$$

$\mathbb{P}_\nu$-a.s. Summing this over all $j \geq 0$, we can conclude the proof. $\square$

1.3 Killing and Absorbing

Let $X = (X_n;\ n \geq 0)$ denote a Markov chain with a denumerable state space S and transition function $p_1 : S \times S \to \mathbb{R}_+$. In this subsection we show how one can perform certain path surgeries on X to obtain new Markov chains.

The idea behind *killing* is as follows: Consider a given set $A \subset S$. We can then construct a new Markov process $Y = (Y_k;\ k \geq 0)$ by letting Y equal X until the first time the latter enters the set A. After that, we send Y to a so-called cemetery state. We now proceed with more caution.

Let $\Delta \notin S$ be a fixed point, called the **cemetery** or the **coffin state**, and define the enlarged state space $S_\Delta = S \cup \{\Delta\}$. Recall the entrance times

$$T_A = \inf(j \geq 0 :\ X_j \in A),$$

where $\inf \varnothing = +\infty$, as usual. Finally, we can define the killed process Y as

$$Y_n = \begin{cases} X_n, & \text{if } n < T_A, \\ \Delta, & \text{if } n \geq T_A. \end{cases}$$

The process Y is said to be X **killed upon entering** A. Intuitively speaking, Y equals X until the latter process first enters A. At that time, we "*kill it and send it to the cemetery* Δ."

Theorem 1.3.1 *The process Y is a Markov chain with state space $S_\Delta \setminus A$, whose transition function q_1 is given by the following formula:*

$$q_1(x,y) = \begin{cases} p_1(x,y), & \text{if } x,y \in S \setminus A, \\ \sum_{a \in A} p_1(x,a), & \text{if } y = \Delta, \\ \mathbb{1}_{\{\Delta\}}(y), & \text{if } x = \Delta. \end{cases}$$

Remark Once the transition function is found, we can then compute any or all of the k-step transition functions by Theorem 1.1.1; cf. Supplementary Exercise 2.

Proof Let ν denote the initial measure of X. Since $(T_A > n)$ is $\mathcal{F}_n$-measurable, for any $y \in S \setminus A$, $\mathbb{P}_\nu$-a.s.,

$$\mathbb{P}_\nu(Y_{n+1} = y \mid \mathcal{F}_n)\mathbb{1}_{(T_A > n)} = \mathbb{P}_\nu(X_{n+1} = y \mid \mathcal{F}_n)\mathbb{1}_{(T_A > n)} = p_1(X_n, y)\mathbb{1}_{(T_A > n)},$$

by (1). Thus, for any $y \in S \setminus A$, $\mathbb{P}_\nu$-a.s.,

$$\mathbb{P}(Y_{n+1} = y \mid \mathcal{F}_n)\mathbb{1}_{(T_A > n)} = p_1(Y_n, y)\mathbb{1}_{(T_A > n)},$$

$\mathbb{P}_\nu$-almost surely. Similarly, we can show that $\mathbb{P}_\nu$-a.s.,

$$\mathbb{P}_\nu(Y_{n+1} = \Delta \mid \mathcal{F}_n)\mathbb{1}_{(T_A > n)} = \sum_{a \in A} p_1(Y_n, a)\mathbb{1}_{(T_A > n)},$$

$$\mathbb{P}_\nu(Y_{n+1} = y \mid \mathcal{F}_n)\mathbb{1}_{(T_A \leq n)} = \mathbb{1}_{\{\Delta\}}(y)\mathbb{1}_{(T_A \leq n)}.$$

The result follows readily from the above together with the fact that $(T_A > n) = (Y_n \in S \setminus A)$. $\square$

There is another way to kill the chain X. For any given $\lambda \in]0,1[$, let $\mathbf{g}_\lambda$ denote a geometric random variable with parameter λ that is totally independent of the process X. Recall that the distribution of $\mathbf{g}_\lambda$ is given by

$$\mathbb{P}(\mathbf{g}_\lambda = k) = (1-\lambda)\lambda^{k-1}, \qquad k = 1, 2, \ldots.$$

We can define the process **X killed at rate** λ to be given by $X^\lambda = (X_n^\lambda; n \geq 0)$, where

$$X_n^\lambda = \begin{cases} X_n, & \text{if } \mathbf{g}_\lambda > n, \\ \Delta, & \text{if } \mathbf{g}_\lambda \leq n. \end{cases}$$

Theorem 1.3.2 *The process X^λ is a Markov chain on S_Δ whose transition function p_1^λ is given by*

$$p_1^\lambda(x,y) = \begin{cases} \lambda p_1(x,y), & \text{if } x,y \in S, \\ (1-\lambda)\mathbb{1}_{\{\Delta\}}(y), & \text{if } x = \Delta. \end{cases}$$

It is worthwhile to provide an intuitive interpretation for the above procedure. Given that X is at x at time n, in the next step it goes to (some place) $y \in S$ with probability $p_1(x,y)$. On the other hand, given that $X_n^\lambda = x$, we toss an independent λ-coin;[2] if the coin lands heads, we then send X^λ to (some place) $y \in S$ with probability $p_1(x,y)$. If the coin lands tails, we kill X^λ once and for all.

Proof Clearly, the history of X^λ is $\mathcal{F}^\lambda = (\mathcal{F}_n^\lambda;\ n \geq 0)$, where $\mathcal{F}_n^\lambda = \mathcal{F}_{n \wedge g_\lambda}$. We can extend the domain of definition of any function $f : S \to \mathbb{R}$ to S_Δ by defining $f(\Delta) = 0$. Now consider arbitrary bounded functions $f_1, \ldots, f_{n+1} : S \to \mathbb{R}$. Since $X_0^\lambda = X_0$, for any initial measure ν (for X_0),

$$\mathbb{E}_\nu\left[\prod_{j=1}^{n+1} f_j(X_j^\lambda)\right] = \mathbb{E}_\nu\left[\prod_{j=1}^{n+1} f_j(X_j) \mathbb{1}_{(g_\lambda > n+1)}\right] = \lambda^{n+1}\mathbb{E}_\nu\left[\prod_{j=1}^{n+1} f_j(X_j)\right].$$

We have used the simple fact that $\mathbb{P}_\nu(g_\lambda > n+1) = \lambda^{n+1}$. Let $f_{n+1}(x) = \mathbb{1}_{\{y\}}(x)$, where $y \in S$ is fixed. By a monotone class argument, for any initial measure ν, the following holds $\mathbb{P}_\nu$-a.s.:

$$\mathbb{P}_\nu(X_{n+1}^\lambda = y \mid \mathcal{F}_n^\lambda) = \lambda \mathbb{P}_\nu(X_{n+1} = y \mid \mathcal{F}_n) = \lambda p_1(X_n, y).$$

(Why?) The last step uses equation (1) and completes the proof. □

A related concept is that of *absorption* at some fixed state $a \in S$. The idea here is to construct a process Z that equals X until it reaches the set $\{a\}$. After that, Z remains in $\{a\}$. Recalling that we write T_a for $T_{\{a\}}$, we can define this more precisely as

$$Z_n = \begin{cases} X_n, & \text{if } n < T_a, \\ a, & \text{if } n \geq T_a. \end{cases}$$

The process $Z = (Z_k;\ k \geq 0)$ is X **absorbed at state** a. We then have the following:

Theorem 1.3.3 *Let Z be X absorbed at some state $a \in S$. Then, Z is a Markov chain on S whose transition function γ_1 is given by the following:*

$$\gamma_1(x,y) = \begin{cases} p_1(x,y), & \text{if } x \neq a, \\ \mathbb{1}_{\{a\}}(y), & \text{if } x = a. \end{cases}$$

Exercise **1.3.1** Prove Theorem 1.3.3. □

---

[2] A λ-coin is one that tosses heads with probability λ.

1.4 Transition Operators

Recalling from the example of Section 1.1 that random walks are Markov chains, we now wish to extend the notion of transition operators developed in Chapter 3, Section 1.1.

Let $X = (X_n;\ n \geq 0)$ denote a Markov chain on a denumerable state space S. Throughout, we define $\mathcal{F} = (\mathcal{F}_n;\ n \geq 0)$ to be any filtration with respect to which X is a Markov chain. (It may help to think of $\mathcal{F}$ as the history of X.) We define the **transition operators** $(\mathcal{T}_n;\ n \geq 0)$ as follows: For any bounded function $f : S \to \mathbb{R}$,

$$\mathcal{T}_n f(x) = \mathbb{E}_x[f(X_n)], \qquad n \geq 0.$$

In particular, $\mathcal{T}_n \mathbf{1}_A(x)$ denotes the conditional probability that X_n is in A, given that $X_0 = x$. It is clear that $\mathcal{T}_n$ is a linear operator for any $n \geq 0$; cf. Section 1.1 of Chapter 3. Moreover, if $f : S \to \mathbb{R}$ is bounded,

$$\mathcal{T}_n \mathcal{T}_m f(x) = \mathbb{E}_x[\mathcal{T}_m f(X_n)] = \mathbb{E}_x\big\{\mathbb{E}_{X_n}[f(X_m)]\big\} = \mathbb{E}_x\big[f(X_{n+m})\big],$$

by Theorem 1.1.2. We have shown that $\mathcal{T}_{n+m} = \mathcal{T}_n \mathcal{T}_m = \mathcal{T}_m \mathcal{T}_n$. This is the **semigroup property** of the transition operators and should be familiar from Lemma 1.1.1 of Chapter 3. There are three further properties of $(\mathcal{T}_n;\ n \geq 0)$ that are elementary and yet deserve to be mentioned at this stage. Namely:

(i) For every bounded function, $f : S \to \mathbb{R}$ and for all $x \in S$, $\mathcal{T}_0 f(x) = f(x)$.

(ii) For all $n \geq 0$ and all $x \in S$, $\mathcal{T}_n \mathbf{1}(x) = 1$, where $\mathbf{1}(y) = 1$ is the function 1, identically.

(iii) For all $n \geq 0$, $\mathcal{T}_n$ is a **nonnegative operator**. That is, whenever $f(x) \geq 0$ for all $x \in S$, then $\mathcal{T}_n f(x) \geq 0$, for all $x \in S$.

Exercise **1.4.1** Verify the above conditions (i)–(iii). □

If $\mathcal{T} = (\mathcal{T}_n;\ n \geq 0)$ is any semigroup of linear operators on the space of bounded functions $f : S \to \mathbb{R}$ for which (i)–(iii) all hold, we say that $\mathcal{T}$ is a **Markov semigroup**. The following is the main result of this subsection.

Theorem 1.4.1 *Suppose $\mathcal{T} = (\mathcal{T}_n;\ n \geq 0)$ is a collection of linear operators on the space of bounded functions $f : S \to \mathbb{R}$. Then, $\mathcal{T}$ is the transition operator of a Markov chain on S if and only if $\mathcal{T}$ is a Markov semigroup.*

Proof If $\mathcal{T}$ is the transition operator of a Markov chain on S, then we have already shown that it is a Markov semigroup. It suffices to prove the converse. Throughout, we will assume that $\mathcal{T}$ is a Markov semigroup.

For every $x, y \in S$ and for every integer $n \geq 0$, define

$$p_n(x, y) = \mathcal{T}_n \mathbf{1}_{\{y\}}(x).$$

It is easy to see that $y \mapsto p_n(x,y)$ is a probability function on S. In fact, p_n will end up being the n-step transition probability of the Markov chain that we are establishing. With this in mind, we first define probability measures $(\mathbb{Q}_x^n;\ x \in S)$ on S^n as follows: For all $A \subset S^n$ and all $a_0 \in S$,

$$\mathbb{Q}_{a_0}^n(A) = \sum_{a \in S^n} \mathbb{1}_A(a) \prod_{\ell=0}^{n-1} p_1\left(a^{(\ell)}, a^{(\ell+1)}\right).$$

This should be compared to the Chapman–Kolmogorov equation (Theorem 1.1.1). It is not hard to see that for each $x \in S$, $(\mathbb{Q}_x^n;\ n \geq 1)$ is a consistent family of probability measures on S^n. By Kolmogorov's existence theorem (Theorem 1, Appendix A), we can construct probability measures $(\mathbb{P}_x;\ x \in S)$ on S^∞ such that the restriction of $\mathbb{P}_x$ to S^n is exactly $\mathbb{Q}_x^n$ for each $x \in S$. Now we construct our Markov chain. For any $\omega \in S^\infty$, define

$$X_n(\omega) = \omega^{(n)}, \qquad n \geq 0.$$

It should be recognized that, under the measure $\mathbb{P}_x$, the joint distribution of $X_1, \ldots, X_m$ is given by the measure $\mathbb{Q}_x^m$. Since X satisfies the Chapman–Kolmogorov property, this is clearly equivalent to the Markov property. $\square$

We close this subsection with three examples.

Example 1 Suppose we kill the Markov chain X when it enters a set $A \subset S$. Letting Δ denote the coffin state, we extend the domain of any function $f: S \to \mathbb{R}$ to $S_\Delta = S \cup \{\Delta\}$ by letting $f(\Delta) \equiv 0$. Following the notation of Section 1.3, we will denote this killed process by $Y = (Y_n;\ n \geq 0)$. By Theorem 1.3.1, Y is a Markov chain on S. If $\mathcal{T} = (\mathcal{T}_n;\ n \geq 0)$ designates the transition semigroup of X, that of Y is given by $\overline{\mathcal{T}} = (\overline{\mathcal{T}}_n;\ n \geq 0)$, where for all bounded functions $f: S \to \mathbb{R}$,

$$\overline{\mathcal{T}}_n f(x) = \mathbb{E}_x\left[f(X_n)\mathbb{1}_{(T_A > n)}\right].$$

It is a good exercise to check directly that this is indeed a Markov semigroup; see Supplementary Exercise 8. $\square$

Example 2 Suppose we kill our Markov chain X at rate $\lambda \in\ ]0,1[$ and call this killed process $X^\lambda = (X_n^\lambda;\ n \geq 0)$; cf. Section 1.3 for this notation. By Theorem 1.3.2, X^λ is a Markov chain on S. It is easy to see that its transition operators $(\mathcal{T}_n^\lambda;\ n \geq 0)$ are given by

$$\mathcal{T}_n^\lambda f(x) = \lambda^n \mathcal{T}_n f(x),$$

where $f: S \to \mathbb{R}$ is any bounded function. One can directly check that this is a Markov semigroup; cf. Supplementary Exercise 8. $\square$

Example 3 Our final example for this subsection relates to absorbed Markov chains. Let Z be our Markov chain X, absorbed at $\{a\}$. (The notation is that of Section 1.3.) The transition operators of Z are $\mathcal{T}^a = (\mathcal{T}_n^a; n \geq 0)$, where

$$\mathcal{T}_n^a f(x) = \mathbb{E}_x\big[f(X_n)\mathbb{1}_{(T_a > n)}\big] + f(a)\mathbb{P}_x(T_a \leq n),$$

for any bounded function $f: S \to \mathbb{R}$. These form a Markov semigroup; cf. Supplementary Exercise 8. □

1.5 Resolvents and λ-Potentials

Given a Markov chain X with transition operators $\mathcal{T} = (\mathcal{T}_n; n \geq 0)$ on a denumerable state space S, we define a family of linear operators $\mathcal{R} = (\mathcal{R}_\lambda; \lambda \in \,]0,1[\,)$ by the prescription

$$\mathcal{R}_\lambda f(x) = \sum_{n=0}^\infty \lambda^n \mathcal{T}_n f(x),$$

for all bounded functions $f: S \to \mathbb{R}$ and all $x \in S$. (Clearly, the above sum converges absolutely.) We call $\mathcal{R}_\lambda$ the **λ-potential operator of** X, the function $\mathcal{R}_\lambda f$ is the **λ-potential of** f, and the **resolvent of** X is the entire collection $\mathcal{R}$.

Note that $\lambda \mapsto \mathcal{R}_\lambda f(x)$ is none other then the classical generating function for $n \mapsto \mathcal{T}_n f(x)$. By the dominated convergence theorem, we can reinterpret the above as

$$\mathcal{R}_\lambda f(x) = \mathbb{E}_x\left[\sum_{n=0}^\infty \lambda^n f(X_n)\right]. \qquad (1)$$

In particular, $\mathcal{R}_\lambda \mathbb{1}_A(x)$ is the expected number of visits to $A \subset S$, discounted at rate λ, conditional on $(X_0 = x)$. Example 2 of Section 1.4 provides us with yet another interpretation of these λ-resolvent operators: If $X^\lambda = (X_n^\lambda; n \geq 0)$ is the process X killed at rate λ, then

$$\mathcal{R}_\lambda f(x) = \mathbb{E}_x\left[\sum_{n=0}^\infty f(X_n^\lambda)\right].$$

In particular, $\mathcal{R}_\lambda \mathbb{1}_A(x)$ is the expected number of visits to $A \subset S$ for the killed process, conditional on $(X_0^\lambda = x)$.

The generating function interpretation of λ-potential operators, together with the classical uniqueness theorem for generating functions, shows us that if $\mathcal{R}_\lambda f(x) = \mathcal{R}_\lambda g(x)$ for all $\lambda \in \,]0,1[$ and all $x \in S$, then $\mathcal{T}_n f = \mathcal{T}_n g$ for all $n \geq 0$. This uniqueness assertion can be refined.

278   8. Constructing Markov Processes

Theorem 1.5.1 (The Resolvent Equation) *The resolvent of a Markov chain satisfies the following: For all $\lambda, \gamma \in \,]0,1[$,*

$$\lambda \mathcal{R}_\gamma - \gamma \mathcal{R}_\lambda = (\lambda - \gamma)\mathcal{R}_\lambda \mathcal{R}_\gamma.$$

In particular, suppose there exists a $\lambda \in \,]0,1[$ such that $\mathcal{R}_\lambda f(x) = 0$ for all $x \in S$. Then, for all $\gamma \in \,]0,1[$ and $y \in S$, $\mathcal{R}_\gamma f(y) = 0$.

The second statement shows that if $\mathcal{R}_\lambda f = \mathcal{R}_\lambda g$ for some λ, then $\mathfrak{T}_n f = \mathfrak{T}_n g$ for all $n \geq 0$. This is the improvement over the mentioned uniqueness assertion via generating functions.

Proof By Theorem 1.2.1, for all bounded functions $f : S \to \mathbb{R}$ and all integers $j, n \geq 0$,

$$\mathfrak{T}_j f(X_n) = \mathbb{E}_x[f(X_{j+n}) \mid \mathcal{F}_n], \tag{2}$$

$\mathbb{P}_x$-a.s. for all $x \in S$. Thus, we can use equation (1) to deduce that with $\mathbb{P}_x$ probability one for any $x \in S$,

$$\mathcal{R}_\gamma f(X_n) = \mathbb{E}_{X_n}\left[\sum_{j=0}^\infty \gamma^j f(X_j)\right] = \sum_{j=0}^\infty \gamma^j \mathfrak{T}_j f(X_n)$$

$$= \sum_{j=0}^\infty \gamma^j \mathbb{E}_x[f(X_{j+n}) \mid \mathcal{F}_n] = \gamma^{-n} \mathbb{E}_x\left[\sum_{j=n}^\infty \gamma^j f(X_j) \mid \mathcal{F}_n\right]. \tag{3}$$

Consequently, for all $x \in S$,

$$\mathcal{R}_\lambda \mathcal{R}_\gamma f(x) = \mathbb{E}_x\left[\sum_{n=0}^\infty \lambda^n \mathcal{R}_\gamma f(X_n)\right] = \mathbb{E}_x\left[\sum_{n=0}^\infty \left(\frac{\lambda}{\gamma}\right)^n \sum_{j=n}^\infty \gamma^j f(X_j)\right]$$

$$= \frac{\gamma}{\lambda - \gamma} \mathbb{E}_x\left[\sum_{j=0}^\infty \left\{1 - \left(\frac{\lambda}{\gamma}\right)^{j+1}\right\} \gamma^j f(X_j)\right].$$

The result follows readily from this. □

Equation (3) of the above proof allows us to deduce that for any $\lambda \in \,]0,1[$ and all $x \in S$,

$$\mathbb{E}_x\left[\sum_{j=0}^\infty \lambda^j f(X_j) \,\Big|\, \mathcal{F}_n\right] = \lambda^n \mathcal{R}_\lambda f(X_n) + \sum_{j=0}^{n-1} \lambda^j f(X_j), \quad \mathbb{P}_x\text{-a.s.} \tag{4}$$

Corollary 1.5.1 (Doob–Meyer Decomposition) *For any $\lambda \in \,]0,1[$ and any bounded nonnegative $f : S \to \mathbb{R}_+$, $(\lambda^n \mathcal{R}_\lambda f(X_n); \, n \geq 0)$ is a supermartingale that satisfies the decomposition (4) with respect to $\mathbb{P}_x$ for any $x \in S$.*

1.6 Distribution of Entrance Times

There are deep connections between entrance times and potentials of functions. We will now elaborate on one such connection. Let $X = (X_n; n \geq 0)$ denote a Markov chain on a denumerable space S. Let $\mathcal{F} = (\mathcal{F}_n; n \geq 0)$ designate the underlying filtration with respect to which X is Markov, and as in the previous subsection, we define $\mathcal{R} = (\mathcal{R}_\lambda; \lambda \in \,]0,1[\,)$ to be the resolvent of X. The following computes the generating function of the entrance time as the potential of a function.

Theorem 1.6.1 *Let $T_E = \inf(k \geq 0 : X_k \in E)$ denote the entrance time to $E \subset S$. There exists a function $f_E : S \to [0,1]$ such that for all $x \in S$ and all $\lambda \in \,]0,1[$,*
$$\mathbb{E}_x[\lambda^{T_E}] = \mathcal{R}_\lambda f_E(x).$$

Proof Recall that $\mathbf{g}_\lambda$ is an independent geometric random variable with parameter λ and that $\mathbb{P}(\mathbf{g}_\lambda > n) = \lambda^n$. Let L_E^λ denote the last hitting time of E before $\mathbf{g}_\lambda$. That is,
$$L_E^\lambda = \sup(0 \leq k < \mathbf{g}_\lambda : X_k \in E),$$
where $\sup \varnothing = -1$. The advertised function f_E is given by
$$f_E(x) = \mathbb{P}_x(L_E^\lambda = 0), \qquad x \in S. \tag{1}$$

We now compute directly:

$\mathcal{R}_\lambda f_E(x)$

$= \mathbb{E}_x\left[\sum_{n=0}^\infty \lambda^n \mathbb{P}_{X_n}(L_E^\lambda = 0)\right]$

$= \mathbb{E}_x\left[\sum_{n=0}^\infty \lambda^n \mathbb{P}_{X_n}(X_0 \in E, \; X_\ell \notin E \text{ for all } 1 \leq \ell < \mathbf{g}_\lambda)\right]$

$= (1-\lambda) \sum_{k=1}^\infty \lambda^k \mathbb{E}_x\left[\sum_{n=0}^\infty \lambda^n \mathbb{P}_{X_n}(X_0 \in E, \; X_\ell \notin E \text{ for all } 1 \leq \ell < k)\right],$

by independence. Employing Theorem 1.2.1 leads to

$\mathcal{R}_\lambda f_E(x)$

$= (1-\lambda) \sum_{k=1}^\infty \lambda^k \mathbb{E}_x\left[\sum_{n=0}^\infty \lambda^n \mathbb{P}_x(X_n \in E, \; X_{n+\ell} \notin E \text{ for all } 1 \leq \ell < k \,|\, \mathcal{F}_n)\right]$

$= \mathbb{E}_x\left[\sum_{n=0}^\infty \lambda^n \mathbb{P}_x(X_n \in E, \; X_{n+\ell} \notin E \text{ for all } 1 \leq \ell < \mathbf{g}_\lambda \,|\, \mathcal{F}_n)\right]$

$= \mathbb{E}_x\left[\sum_{n=0}^\infty \lambda^n \mathbb{P}_x(L_E^\lambda = n \,|\, \mathcal{F}_n)\right] = \mathbb{E}_x\left[\sum_{n=0}^\infty \lambda^n \mathbf{1}_{(L_E^\lambda = n)}\right].$

Thus,
$$\mathcal{R}_\lambda f_E(x) = \mathbb{E}_x\left[\sum_{n=0}^{g_\lambda-1} \mathbb{1}_{(L_E^\lambda = n)}\right] = \mathbb{P}_x(T_E < g_\lambda) = \mathbb{E}_x[\lambda^{T_E}].$$

We have used the independence of X and g_λ several times. This completes the proof. □

As a consequence of the proof of the above, we mention the following.

Corollary 1.6.1 *The function f_E above satisfies:*

(i) for all $x \notin E$, $f_E(x) = 0$; and

(ii) for all $x \in E$, $\mathcal{R}_\lambda f_E(x) = 1$.

Exercise 1.6.1 Complete the proof of Corollary 1.6.1. □

The function f_E is called the **equilibrium measure** on E, and $\mathcal{R}_\lambda f_E$ is called the **equilibrium (λ)-potential** of E and has the following interpretation.

Corollary 1.6.2 *For all $x \in S$ and $E \subset S$,*
$$\mathbb{P}_x(T_E < \infty) = \lim_{\lambda \to 1^-} \mathcal{R}_\lambda f_E(x).$$

Exercise 1.6.2 Prove Corollary 1.6.2. □

The equilibrium potentials of Theorem 1.6.1 satisfy the following variational problem.

Theorem 1.6.2 *For any $x \in S$, all $\lambda \in {]0,1[}$, and every $E \subset S$,*
$$\mathcal{R}_\lambda f_E(x) = \inf \mathcal{R}_\lambda f(x),$$
where the infimum is taken over all bounded functions $f : S \to [0,1]$ such that $f = 0$ off of E and $\mathcal{R}_\lambda f \geq 1$ on E.

Proof Fix some bounded function $f : S \to \mathbb{R}$ such that $f = 0$ off of E, and $\mathcal{R}_\lambda f \geq 1$ on E. Fix $x \in S$ and define $M^f = (M_n^f; n \geq 0)$ by
$$M_n^f = \mathbb{E}_x\left[\sum_{j=0}^\infty \lambda^j f(X_j) \,\Big|\, \mathcal{F}_n\right].$$

Clearly, M^f is a bounded martingale with respect to $\mathbb{P}_x$. Moreover, by the Doob–Meyer decomposition (Corollary 1.5.1), with $\mathbb{P}_x$-probability one,
$$M_{T_E}^f \mathbb{1}_{(T_E < \infty)} = \lambda^{T_E} \mathcal{R}_\lambda f(X_{T_E}) \mathbb{1}_{(T_E < \infty)}.$$

Therefore (why?), by the optional stopping theorem (Theorem 1.2.1 of Chapter 1),

$$\mathbb{E}_x[M_0^f] = \lim_{k \to \infty} \mathbb{E}_x[M_{T_E \wedge k}^f] = \mathbb{E}_x[M_{T_E}^f \mathbf{1}_{(T_E < \infty)}]$$
$$= \mathbb{E}_x[\lambda^{T_E} \mathcal{R}_\lambda f(X_{T_E}) \mathbf{1}_{(T_E < \infty)}]$$
$$\geq \mathbb{E}_x[\lambda^{T_E} \mathbf{1}_{(T_E < \infty)}],$$

which equals $\mathbb{E}_x[\lambda^{T_E}]$, since $\lambda \in {]}0,1[$. On the other hand, $\mathbb{E}_x[M_0^f] = \mathcal{R}_\lambda f(x)$. This and Theorem 1.6.1 together complete the proof. □

There are other variational representations of equilibrium potentials. Here is a sample.

***Exercise* 1.6.3** Prove that for all $x \in S$, all $\lambda \in {]}0,1[$, and for every $E \subset S$, $\mathcal{R}_\lambda f_E(x) = \sup \mathcal{R}_\lambda f(x)$, where the supremum is taken over all $f: S \to \mathbb{R}_+$ such that $f \equiv 0$ off of E and $\mathcal{R}_\lambda f(x) \leq 1$ for all $x \in S$. □

We will return to this topic more fully in Chapter 10.

2 Markov Semigroups

At first sight, Markov processes can be thought of as "obvious" extensions of Markov chains. However, upon closer examination, one finds a number of hazardous technical pitfalls in actually carrying out such extensions. To avoid them, one typically focuses on a class of "nice" Markov processes.

In this book we will concern ourselves with one such class of processes known as Feller processes. In order to construct such processes, we need some functional-analytic machinery that will be developed in this section. The following section uses this material to construct and analyze Feller processes. The reader should be *well* familiar with the material of Section 1 before proceeding any further.

We begin our discussion with some elementary notions from functional analysis.

2.1 Bounded Linear Operators

This subsection is a brief review of some relevant facts about bounded (or continuous) linear operators and establishes some useful notation.

Recall that a set X is a **linear space** (or a real **vector space**) if for every $\alpha_1, \alpha_2 \in \mathbb{R}$ and all $f_1, f_2 \in X$, $\alpha_1 f_1 + \alpha_2 f_2 \in X$. The mapping $\| \cdot \|_X : X \to \mathbb{R}_+ \cup \{+\infty\}$ defines a **norm** on X if:

1. for any $f \in X$, $\|f\|_X = 0$ if and only if $f = 0$—the zero element of X;

2. *(The Triangle Inequality)* for all $f_1, f_2 \in X$, $\|f_1 + f_2\|_X \le \|f_1\|_X + \|f_2\|_X$; and

3. for all $\alpha \in \mathbb{R}$ and $f \in X$, $\|\alpha f\|_X = |\alpha| \cdot \|f\|_X$.

Note that any normed linear space X can be metrized: We can simply define the distance between $f, g \in X$ as $\|f - g\|_X$.

If X and Y are linear spaces, recall that a map $\mathcal{L} : X \to Y$ is a **linear operator** (from X in to Y) if for all $\alpha_1, \alpha_2 \in \mathbb{R}$ and all $f_1, f_2 \in X$, $\mathcal{L}(\alpha_1 f_1 + \alpha_2 f_2) = \alpha_1 \mathcal{L}(f_1) + \alpha_2 \mathcal{L}(f_2)$. If X and Y are in fact normed linear spaces (with norms $\|\cdot\|_X$ and $\|\cdot\|_Y$, respectively), we can define

$$\|\mathcal{L}\|_{\mathsf{op}} = \sup_{f \in X:\ \|f\|_X = 1} \|\mathcal{L}(f)\|_Y.$$

We say that the operator $\mathcal{L}$ is a **bounded linear operator** if $\|\mathcal{L}\|_{\mathsf{op}} < \infty$. The collection of all bounded linear operators from X in to Y is denoted by $\mathfrak{B}(X, Y)$.

Lemma 2.1.1 *Suppose X and Y are normed linear spaces. Then endowed with $\|\cdot\|_{\mathsf{op}}$, $\mathfrak{B}(X, Y)$ is a normed linear space. Moreover,*

$$\|\mathcal{L}\|_{\mathsf{op}} = \sup_{f \in X} \frac{\|\mathcal{L}(f)\|_Y}{\|f\|_X}, \tag{1}$$

where $0 \div 0 = 1$.

The norm $\|\mathcal{L}\|_{\mathsf{op}}$ is called the **operator norm** of $\mathcal{L}$.

Corollary 2.1.1 *Suppose X and Y are normed linear spaces with norms $\|\cdots\|_X$ and $\|\cdots\|_Y$, respectively. For any $f_1, f_2 \in X$, and for all $\mathcal{L} \in \mathfrak{B}(X, Y)$,*

$$\|\mathcal{L}(f_1) - \mathcal{L}(f_2)\|_Y \le \|\mathcal{L}\|_{\mathsf{op}} \|f_1 - f_2\|_X.$$

Exercise **2.1.1** Verify Lemma 2.1.1 and Corollary 2.1.1. □

2.2 Markov Semigroups and Resolvents

From now on, let S denote a separable, locally compact space.[3] We are primarily interested in two linear spaces associated with S. The first is $L^\infty(S)$, the collection of all bounded, measurable functions $f : S \to \mathbb{R}$.[4]

---

[3] The space S is **locally compact** at $x \in S$ if for any open neighborhood U of x, we can find an open neighborhood V of x such that $\overline{V} \subset U$ is compact. We say that S is locally compact if it is locally compact at all points $x \in S$.

[4] This is a slight abuse of notation, since, typically, we refer to an L^∞-space as one with a measure structure. This should not cause any confusion, however.

The second linear space of interest is $C_0(S)$. This is the collection of all continuous functions $f : S \to \mathbb{R}$ that vanish at infinity. While this definition is clear when $S = \mathbb{R}^d$, it deserves a line of explanation to cover the general case. We say that $f : S \to \mathbb{R}$ **vanishes at infinity** if for all $\varepsilon > 0$, there exists a compact set $K_\varepsilon \subset S$ such that $\sup_{x \notin K_\varepsilon} |f(x)| \leq \varepsilon$. You should check the following.

***Exercise* 2.2.1** Any function $f \in C_0(S)$ is *uniformly continuous*. □

It can (and should) also be easily checked that $C_0(S) \subset L^\infty(S)$ and that $L^\infty(S)$ and $C_0(S)$ are both normed linear spaces when endowed with the norm
$$\|f\|_\infty = \sup_{x \in S} |f(x)|, \qquad f \in L^\infty(S).$$

A collection $\mathcal{T} = (\mathcal{T}_t; \ t \geq 0)$ is said to be a **Markov semigroup** on S if:

1. For all $t \geq 0$, $\mathcal{T}_t \in \mathfrak{B}(L^\infty(S), L^\infty(S))$.

2. $\mathcal{T}_0$ is the identity operator. That is, for all $f \in L^\infty(S)$, $\mathcal{T}_0(f) = f$.

3. *(Semigroup Property)* For all $t, s \geq 0$, $\mathcal{T}_{t+s} = \mathcal{T}_t \mathcal{T}_s$.

4. *(Nonnegativity-Preserving)* If $f \in L^\infty(S)$ is nonnegative, then for all $t \geq 0$ and all $x \in S$, $\mathcal{T}_t f(x) \geq 0$.

Lemma 2.2.1 *If $\mathcal{T}$ is a Markov semigroup on S, then for all $t \geq 0$, $\|\mathcal{T}_t\|_{\mathsf{op}} = 1$.*

Proof For all $x \in S$, $f(x) - \|f\|_\infty \leq 0$; consequently, $\mathcal{T}_t f(x) \leq \|f\|_\infty \mathcal{T}_t \mathbf{1}(x) = \|f\|_\infty$, pointwise. This shows that $\|\mathcal{T}_t\|_{\mathsf{op}} \leq 1$. On the other hand, $\mathcal{T}_t \mathbf{1}(x) = 1$ for all $t \geq 0$ and all $x \in S$, and our proof is complete. □

To any Markov semigroup we can associate a **resolvent** $\mathcal{R} = (\mathcal{R}_\lambda; \ \lambda > 0)$, defined as follows: For all $f \in L^\infty(S)$, all $\lambda > 0$, and all $x \in S$,
$$\mathcal{R}_\lambda f(x) = \int_0^\infty e^{-\lambda s} \mathcal{T}_s f(x) \, ds.$$

The function $\mathcal{R}_\lambda f$ is said to be the **λ-potential** of $f \in L^\infty(S)$. Note that the above resolvent (in continuous time) is not exactly the same as the discrete-time resolvent of Section 1.5.

Lemma 2.2.2 *For all $\lambda > 0$, $\mathcal{R}_\lambda \in \mathfrak{B}(L^\infty(S), L^\infty(S))$ and $\|\mathcal{R}_\lambda\|_{\mathsf{op}} = \lambda^{-1}$.*

***Exercise* 2.2.2** Prove Lemma 2.2.2. □

The following is the continuous-time analogue of the resolvent equation in discrete time (Theorem 1.5.1) and bears the same name.

Theorem 2.2.1 (The Resolvent Equation) *For all $\lambda, \gamma > 0$,*

$$\mathcal{R}_\gamma - \mathcal{R}_\lambda = (\lambda - \gamma)\mathcal{R}_\lambda \mathcal{R}_\gamma = (\lambda - \gamma)\mathcal{R}_\gamma \mathcal{R}_\lambda.$$

Exercise **2.2.3** Prove Theorem 2.2.1.
(HINT: Model the proof after that of Theorem 1.5.1.) □

Corollary 2.2.1 *Given $f \in L^\infty(S)$, suppose there exists $\lambda > 0$ such that $\mathcal{R}_\lambda f \equiv 0$. Then, for all $\gamma > 0$, $\mathcal{R}_\gamma f \equiv 0$.*

Exercise **2.2.4** Prove Corollary 2.2.1. □

2.3 Transition and Potential Densities

We now discuss a useful method of producing Markov semigroups. Suppose μ is a σ-finite measure on S and $p = (p_t; t \geq 0)$ is a family of functions $p_t : S \times S \to \mathbb{R}_+$ ($t \geq 0$) such that $(t; x, y) \mapsto p_t(x, y)$ is measurable and:

1. for all $t \geq 0$ and $x \in S$, $\int p_t(x, y)\,\mu(dy) = 1$;

2. *(The Chapman–Kolmogorov Property)* for every $t, s \geq 0$ and for all $x, y \in S$, $p_{t+s}(x, y) = \int p_t(x, z) p_s(z, y)\,\mu(dz)$.

The functions p_t are then said to be **transition densities** (with respect to the measure μ). For any $f \in L^\infty(S)$, for all $t \geq 0$, and every $x \in S$, we may define

$$\mathcal{T}_t f(x) = \int p_t(x, y) f(y)\,\mu(dy).$$

Proposition 2.3.1 *The collection $\mathcal{T} = (\mathcal{T}_t; t \geq 0)$ is a Markov semigroup on S.*

Exercise **2.3.1** Verify Proposition 2.3.1. □

Next, let us look at the resolvent $\mathcal{R} = (\mathcal{R}_\lambda; \lambda > 0)$ of $\mathcal{T}$. For any $\lambda > 0$, define $r_\lambda : S \times S \to \mathbb{R}$ by

$$r_\lambda(x, y) = \int_0^\infty e^{-\lambda t} p_t(x, y)\,dt.$$

By the monotone convergence theorem, this integral always exists, although it may be infinite. When it is finite $\mu \times \mu$-almost everywhere, we say that r_λ is the **λ-potential density** of $\mathcal{T}$. This definition is motivated by the following calculation, which is a consequence of Fubini's theorem: For all nonnegative $f \in L^\infty(S)$,

$$\mathcal{R}_\lambda f(x) = \int r_\lambda(x,y) f(y) \, \mu(dy).$$

That is, the measure $A \mapsto \mathcal{R}_\lambda \mathbf{1}_A(x)$ is absolutely continuous with respect to μ and the Radon–Nikodým density (at $y \in S$) is $r_\lambda(x,y)$.

Example Consider $S = \mathbb{R}$, endowed with its Euclidean topology. Define $p_t(x,y) = q_t(y-x)$, where

$$q_t(a) = \frac{1}{\sqrt{2\pi t}} e^{-a^2/2t}, \qquad a \in \mathbb{R}, \ t \geq 0.$$

Clearly, $(t; x, y) \mapsto p_t(x,y)$ is measurable and $\int_{-\infty}^\infty p_t(x,y) dy = 1$, for all $t \geq 0$ and $x \in \mathbb{R}$. We will next show that the collection of p_t's satisfies the Chapman–Kolmogorov condition 2 above with μ being the 1-dimensional Lebesgue measure. In other words, we will check that for all $t, s \geq 0$, $q_{t+s} = q_t \star q_s$, where $\star$ denotes convolution. Let $\widehat{q_t}$ denote the Fourier transform of $a \mapsto q_t(a)$. That is,

$$\widehat{q_t}(\xi) = \int_{-\infty}^\infty e^{i\xi a} q_t(a) \, da.$$

From a direct computation (cf. Exercise 2.3.2 below), we find that for any $t \geq 0$ and all $\xi \in \mathbb{R}$, $\widehat{q_t}(\xi) = e^{-\frac{1}{2} t \xi^2}$. In particular, for all $t, s \geq 0$, $\widehat{q_{t+s}} = \widehat{q_t} \cdot \widehat{q_s}$. Since $\widehat{f \star g} = \widehat{f}\widehat{g}$, we have shown that $q_{t+s} = q_t \star q_s$, as desired.

Define $\mathcal{T} = (\mathcal{T}_t; \ t \geq 0)$ by

$$\mathcal{T}_t f(x) = \int_{-\infty}^\infty p_t(x,y) f(y) \, dy, \qquad t \geq 0, x \in \mathbb{R}, f \in L^\infty(\mathbb{R}).$$

By Proposition 2.3.1, $\mathcal{T}$ is a semigroup on $\mathbb{R}$ and is called the **heat semigroup on** $\mathbb{R}$. Next, we compute the resolvent of the heat semigroup by finding an expression for the λ-potential density r_λ. It is clear that $r_\lambda(x,y) = u_\lambda(y - x)$, where

$$u_\lambda(a) = \int_0^\infty e^{-\lambda r} q_r(a) \, dr, \qquad \lambda > 0, \ a \in \mathbb{R}. \qquad (1)$$

By the inversion theorem for characteristic functions,

$$q_r(a) = \frac{1}{2\pi} \int_{-\infty}^\infty \widehat{q_r}(\xi) e^{-i\xi a} \, d\xi = \frac{1}{2\pi} \mathrm{Re} \int_{-\infty}^\infty e^{-\frac{1}{2} r \xi^2 - i\xi a} \, d\xi$$

$$= \frac{1}{\pi} \int_0^\infty e^{-\frac{1}{2} r \xi^2} \cos(\xi a) \, d\xi.$$

286   8. Constructing Markov Processes

Plugging this into equation (1) and using Fubini's theorem, we arrive at the following:
$$u_\lambda(a) = \frac{2}{\pi} \int_0^\infty \frac{\cos(\xi a)}{\xi^2 + 2\lambda} \, d\xi.$$

The modified Bessel's function, $\mathbf{K}_{\frac{1}{2}}$, of the third kind is defined by
$$\mathbf{K}_{\frac{1}{2}}(y) = \sqrt{\frac{2}{y\pi}} \int_0^\infty \frac{\cos(y\xi)}{\xi^2 + 1} \, d\xi, \qquad y \in \mathbb{R};$$

see Watson (1995). Thus, the λ-resolvent density $r_\lambda(x,y)$ is equal to $u_\lambda(y-x)$, where
$$u_\lambda(a) = \left(\frac{2a^2}{\pi^3}\right)^{\frac{1}{4}} \mathbf{K}_{\frac{1}{2}}\left(a\sqrt{2\lambda}\right), \qquad \lambda > 0, \, a \in \mathbb{R}.$$

This provides the computation of the resolvent of the heat semigroup on $\mathbb{R}^1$. □

Exercise 2.3.2 Verify that in the above example, $\widehat{q_t}(\xi) = e^{-\frac{1}{2}t\xi^2}$, as asserted. Also, check that u_λ is a bounded continuous function. □

Exercise 2.3.3 The heat semigroup on $\mathbb{R}^d$ is defined analogously to the $d = 1$ case as $\mathcal{T}_t f(x) = \int_{\mathbb{R}^d} p_t(x,y) f(y) \, dy$, where
$$p_t(x,y) = (2\pi t)^{-\frac{d}{2}} \exp\left\{-\frac{1}{2t}\|x-y\|^2\right\}.$$

Show that for all $\lambda > 0$, $r_\lambda(x,y) = u_\lambda(y-x)$, where
$$u_\lambda(a) = \frac{2}{(2\pi)^d} \int_{\mathbb{R}^d} \frac{\cos(\xi \cdot a)}{2\lambda + \|\xi\|^2} \, d\xi,$$

for all $a \in \mathbb{R}^d$. Moreover, prove that when $d \geq 3$, even r_0 makes sense and is given by $r_0(x,y) = C\|x-y\|^{2-d}$ for some finite positive constant C. Compute C explicitly. □

Exercise 2.3.4 For the example of this section, prove that
$$u_\lambda(a) = Ce^{-|a|\sqrt{2\lambda}},$$

and compute C. Can you compute, in terms of Bessel functions, the analogous u_λ when $d \geq 2$?
(HINT: Try to show that u_λ, suitably truncated, solves $u'' = 2\lambda u$.) □

2.4 Feller Semigroups

Let S denote a separable, locally compact space and suppose $\mathcal{T} = (\mathcal{T}_t;\ t \geq 0)$ is a Markov semigroup on S. We say that $\mathcal{T}$ is a **Feller semigroup** (**Feller**, briefly) if:

1. for all $t \geq 0$, $\mathcal{T}_t : C_0(S) \to C_0(S)$;
2. *(The Feller Property)* for all $f \in C_0(S)$, $\lim_{t \to 0+} \|\mathcal{T}_t f - f\|_\infty = 0$.

Thus, the Feller property states that the operator-valued map $t \mapsto \mathcal{T}_t$ is right-continuous at 0. In fact, $t \mapsto \mathcal{T}_t$ is uniformly right-continuous at all points $t \geq 0$, as the following shows.

Lemma 2.4.1 *The Feller property is equivalent to the following: For all $f \in C_0(S)$,*
$$\lim_{t \to 0+} \sup_{s \geq 0} \|\mathcal{T}_{s+t} f - \mathcal{T}_s f\|_\infty = 0.$$
Moreover, it implies that for all $f \in C_0(S)$,
$$\lim_{\lambda \to \infty} \|\lambda \mathcal{R}_\lambda f - f\|_\infty = 0.$$

***Exercise* 2.4.1** Prove Lemma 2.4.1. □

Suppose $\mathcal{T}$ has transition densities $p = (p_t;\ t \geq 0)$ with respect to some σ-finite measure μ on S; cf. Section 2.3. It is not hard to find a useful sufficient condition on the transition densities that guarantees that $\mathcal{T}$ is Feller.

Proposition 2.4.1 *Suppose $\mathcal{T}$ is a Markov semigroup on a locally compact, separable metric space S. Suppose further that $\mathcal{T}$ has transition densities p with respect to a σ-finite measure μ on S. Then, $\mathcal{T}$ is a Feller semigroup, provided that for all $\delta > 0$,*
$$\lim_{t \to 0+} \sup_{x \in S} \int_{y \in S:\ d(x,y) \geq \delta} p_t(x,y)\, \mu(dy) = 0,$$
where $d(\bullet, \bullet)$ denotes the metric on S.

Proof For all $f \in C_0(S)$ and for all $t \geq 0$,
$$\mathcal{T}_t f(x) - f(x) = \int p_t(x,y) \{f(y) - f(x)\}\, \mu(dy), \qquad x \in S.$$

We can divide the above integral into two pieces: one where $d(x,y) \leq \delta$, and one where $d(x,y) > \delta$. Since $p_t(x,y) \geq 0$ and $\int p_t(x,z)\, \mu(dz) = 1$ for all $t \geq 0$ and $x, y \in S$,
$$\|\mathcal{T}_t f - f\|_\infty \leq \sup_{x,y \in S:\ d(x,y) \leq \delta} |f(x) - f(y)|$$
$$+ 2\|f\|_\infty \sup_{x \in S} \int_{y \in S:\ d(x,y) \geq \delta} p_t(x,y)\, \mu(dy).$$

Letting $t \to 0+$ and then $\delta \to 0+$, we see that

$$\limsup_{t \to 0+} \|\mathcal{T}_t f - f\|_\infty \le \lim_{\delta \to 0+} \sup_{x,y \in S:\ d(x,y) \le \delta} |f(x) - f(y)|,$$

which is 0 (why?). This completes our proof. □

Example We proceed to use Proposition 2.4.1 to show that the heat semigroup of the example of Section 2.3 is indeed Feller. In the notation of the latter Example, a few lines of calculations show that for any $x \in \mathbb{R}$ and all $\delta > 0$,

$$\int_{y \in \mathbb{R}:\ |x-y| \ge \delta} p_t(x,y)\, dy = \sqrt{\frac{2}{\pi}} \int_{\delta t^{-1/2}}^{\infty} e^{-\frac{1}{2} z^2}\, dz.$$

The Feller property of the heat semigroup follows from Proposition 2.4.1. □

3 Markov Processes

We are now in a position to begin our construction of Markov processes in continuous time. Ignoring the technical difficulties for the moment, a Markov process should be defined as a stochastic process that is conditionally independent of its past, given its present value. See Section 1 on Markov chains for a more precise description in discrete time. In order to carry out such a construction in continuous time, we need additional regularity properties. In this book we content ourselves with a *uniform continuity condition in probability*, which, in more precise terms, translates to the so-called Feller property. Our approach is based on more traditional semigroup methods and is closest to the treatment of Dynkin (1965), Hunt (1956a, 1957, 1958), Itô (1984), and Knight (1981).

3.1 Initial Measures

Suppose $\mathcal{T}$ is a Markov semigroup on a separable, locally compact space S. An S-valued stochastic process $X = (X_t;\ t \ge 0)$ is said to be a **Markov process with initial measure** ν, filtration $\mathcal{X} = (\mathcal{X}_t;\ t \ge 0)$, and **transition operators** $\mathcal{T}$ if there exists a probability measure $\mathbb{P}_\nu$ on our probability space such that:

1. X is adapted to $\mathcal{X}$;

2. for all $s,t \ge 0$ and all $f \in C_0(S)$,

$$\mathbb{E}_\nu[f(X_{t+s}) \mid \mathcal{X}_s] = \mathcal{T}_t f(X_s),$$

$\mathbb{P}_\nu$-a.s., where $\mathbb{E}_\nu$ denotes the expectation operator corresponding to $\mathbb{P}_\nu$; and

3. for all Borel sets $A \subset S$, $\mathbb{P}_\nu(X_0 \in A) = \nu(A)$.

An important choice for an initial measure is the point mass at $x \in S$, which we denote by δ_x. In this case, we write $\mathbb{P}_x$ and $\mathbb{E}_x$ in place of the more cumbersome $\mathbb{P}_{\delta_x}$ and $\mathbb{E}_{\delta_x}$, respectively.

We say that X is a **Markov process** with transition operators $\mathcal{T}$ and filtration $\mathcal{X}$ if it is a Markov process for any initial measure δ_x, $x \in S$. We can associate a collection of probability measures to a Markov process X as follows:

$$\mathcal{T}_t(x, A) = \mathcal{T}_t \mathbb{1}_A(x), \qquad x \in S, A \subset S, \text{ Borel.} \tag{1}$$

It is easy to see that for each $t \geq 0$ and $x \in S$, $\mathcal{T}_t(x, \bullet)$ is a probability measure. In fact, by properties 2 and 3 of Markov processes, $\mathcal{T}_t(x, A) = \mathbb{P}_x(X_t \in A)$. That is, $\mathcal{T}_t(x, \bullet)$ is the the distribution of X_t under the measure $\mathbb{P}_x$. A little thought shows that an equivalent formulation is that $\mathcal{T}_t(x, \bullet)$ is the conditional distribution of X_t, given that $X_0 = x$. However, the somewhat circuitous route that we have taken avoids our having to worry about a suitable choice of regular conditional probabilities.[5]

The measures $(\mathcal{T}_t(x, \bullet); x \in S)$ are called the **transition functions** of X. According to equation (1), we can find the transition functions of a Markov process from its transition operators. On the other hand, by a monotone class argument,

$$\mathcal{T}_t f(x) = \mathbb{E}_x[f(X_t)] = \int f(y)\, \mathcal{T}_t(x, dy), \qquad x \in S,\ f \in C_0(\mathbb{R}). \tag{2}$$

(Why?) That is, we know the transition operators if and only if we know the transition functions. From now on, we will make no distinctions between transition operators and transition functions; they are identified with each other by equations (1) and (2).

Lemma 3.1.1 (The Chapman–Kolmogorov Equation) *Suppose X is a Markov process with filtration $\mathcal{X}$, transition operators $\mathcal{T}$, and initial measure ν. For all $\varphi_0, \ldots, \varphi_k \in L^\infty(S)$ and all $0 = t_0 < t_1 < \cdots < t_k$,*

$$\mathbb{E}_\nu\left[\prod_{j=0}^k \varphi_j(X_{t_j})\right]$$

$$= \int \cdots \int \prod_{j=0}^k \varphi_j(a_j) \nu(da_0) \mathcal{T}_{t_1}(a_0, da_1) \mathcal{T}_{t_2-t_1}(a_1, da_2) \cdots \mathcal{T}_{t_k-t_{k-1}}(a_{k-1}, da_k),$$

---

[5] It is equally possible to use regular conditional probabilities to define such Markov processes. To learn more about regular conditional probabilities, see (Blackwell and Dubins 1975; Walsh 1972).

where $\mathbb{E}_\nu[Y] = \int_\Omega Y(\omega)\,\mathbb{P}_\nu(d\omega)$ denotes the expectation operator corresponding to $\mathbb{P}_\nu$.

Exercise 3.1.1 Prove Lemma 3.1.1 by imitating its discrete analogue, Theorem 1.1.1. □

Lemma 3.1.1 has two important implications. First, it shows that the transition functions completely determine the finite-dimensional distributions of a Markov process. The second implication is the following, which is obtained from Lemma 3.1.1 and a monotone class argument.

Corollary 3.1.1 *Suppose X is a Markov process with transition operators $\mathcal{T}$, filtration $\mathcal{X}$. Let $\mathcal{X}_\infty = \bigvee_{t \geq 0} \mathcal{X}_t$ denote the smallest σ-field generated by the entire process X. For any probability measure ν on $(S, \mathcal{X}_\infty)$ and all bounded, $\mathcal{X}_\infty$-measurable random variables Y,*

$$\mathbb{E}_\nu[Y] = \int_S \mathbb{E}_x[Y]\,\nu(dx), \tag{3}$$

where $\mathbb{E}_x = \mathbb{E}_{\delta_x}$, where δ_x represents the point mass at $x \in S$.

Let us conclude this subsection with two remarks.

- For any $\mathcal{X}_\infty$-measurable random variable $Y \in \bigcup_{x \in S} L^1(\mathbb{P}_x)$, $x \mapsto \mathbb{E}_x[Y]$ is measurable as a function from S into $\mathbb{R}$. This holds thanks to Lemma 3.1.1 and the measurability conditions on Markov semigroups.

- By equation (3), for any Borel set $A \in \mathcal{X}_\infty$, $\mathbb{P}_\nu(A) = \int \mathbb{P}_x(A)\nu(dx)$.

Exercise 3.1.2 Prove that X is a Markov process if and only if it is a Markov process for all choices of the initial measure ν. □

3.2 Augmentation

Aside from the basic definitions, in the previous subsection we studied the elementary properties of a Markov process X with filtration $\mathcal{X}$ and transition functions $\mathcal{T}$ as we considered various initial measures. We now turn our attention to the underlying filtration.

For any probability measure μ on a measure space $(S, \mathcal{X}_\infty)$, define the filtration $\mathcal{X}^\mu = (\mathcal{X}_t^\mu;\ t \geq 0)$ as follows: For all $t \geq 0$, $\mathcal{X}_t^\mu$ is the completion of $\mathcal{X}_t$ with respect to the measure μ.[6] The **complete filtration** $\overline{\mathcal{X}} = (\overline{\mathcal{X}}_t;\ t \geq 0)$ is defined by

$$\overline{\mathcal{X}}_t = \bigcap \mathcal{X}_t^\mu, \qquad t \geq 0,$$

---

[6]Recall that this means that $\mathcal{X}_t^\mu$ is obtained by adding to $\mathcal{X}_t$ all subsets of μ-null sets.

where the intersection is taken over all probability measures μ on $(S, \mathcal{X}_\infty)$. We can extend a probability measure μ on $(S, \mathcal{X}_\infty)$ to a probability measure $\overline{\mu}$ on $(S, \overline{\mathcal{X}}_\infty)$ as usual, by *defining*:

1. $\overline{\mu}(E) \equiv 0$ for all $E \in \overline{\mathcal{X}}_\infty$ such that E is a subset of m-null sets for all probability measures m on $(S, \mathcal{X}_\infty)$; and

2. $\overline{\mu} = \mu$, on $\mathcal{X}_\infty$.

As is customary, we will abuse notation by writing μ in place of $\overline{\mu}$. This should not cause any confusion.

It turns out that, after completing the underlying filtration, we do not alter the Markovian structure of a process.

Lemma 3.2.1 *Suppose X is a Markov process with transition operators $\mathcal{T}$ and filtration $\mathcal{X}$. Then, X is also a Markov process with the same transition functions and with the complete filtration $\overline{\mathcal{X}}$.*

Proof We will show that for all $x \in S$, all $f \in C_0(S)$, and all $s, t \geq 0$,

$$\mathbb{E}_x\big[f(X_{t+s}) \,\big|\, \overline{\mathcal{X}}_s\big] = \mathcal{T}_t f(X_s), \tag{1}$$

$\mathbb{P}_x$-a.s. By the remarks following Corollary 3.1.1, this shows that for any initial measure ν, (1) holds $\mathbb{P}_\nu$-a.s. and proves the desired result. See also Exercise 3.1.2.

Let us choose some $s, t \geq 0$ and hold them fixed. For any $\Lambda \in \overline{\mathcal{X}}_s$, there exists $\Lambda_\star \in \mathcal{X}_s$ such that for all probability measures μ on S, $\mu(\Lambda \triangle \Lambda_\star) = 0$. In particular, for all $x \in S$, $\mathbb{P}_x(\Lambda \triangle \Lambda_\star) = 0$. Hence, for all $f \in C_0(S)$,

$$\mathbb{E}_x\big[f(X_{s+t}) \mathbf{1}_\Lambda\big] = \mathbb{E}_x\big[f(X_{s+t}) \mathbf{1}_{\Lambda_\star}\big] = \mathbb{E}_x\Big[\mathbb{E}_x\big[f(X_{s+t}) \,\big|\, \mathcal{X}_s\big] \mathbf{1}_{\Lambda_\star}\Big].$$

By Property 2 of Markov processes (cf. Section 3.1),

$$\mathbb{E}_x\big[f(X_{s+t}) \mathbf{1}_\Lambda\big] = \mathbb{E}_x\big[\mathcal{T}_t f(X_s) \mathbf{1}_{\Lambda_\star}\big] = \mathbb{E}_x\big[\mathcal{T}_t f(X_s) \mathbf{1}_\Lambda\big].$$

This validates (1) and completes the proof. $\square$

The next step in our program is fairly natural: In order to fully utilize the martingale theory of Chapter 7, we need to extend the Markov property from the complete filtration of Lemma 3.2.1 to a Markov property that holds on the "right-continuous" extension of the mentioned complete filtrations. That is, suppose X is a Markov process with transition functions $\mathcal{T}$ and complete filtration $\mathcal{X}$. Define the filtration $\mathcal{X}^\star = (\mathcal{X}_t^\star;\, t \geq 0\,)$ as in Section 1.4 of Chapter 7, by

$$\mathcal{X}_t^\star = \bigcap_{s > t} \overline{\mathcal{X}}_s, \qquad t \geq 0,$$

and call it the **complete augmented filtration** of $\mathcal{X}$. Our hope is to show that if X is Markov with respect to the filtration $\mathcal{F}$, it is also Markov with respect to $\mathcal{F}^*$. Unfortunately, at this level of generality this is not so![7] However, if the transition operators of X are sufficiently nicely behaved, it is. We will return to this issue in Section 4, where we discuss a class of nice Feller processes.

3.3 Shifts

A collection $\theta = (\theta_t;\ t \geq 0)$ is called a collection of **shift operators** (**shifts**, briefly) for a Markov process X if:

1. for each $t \geq 0$, $\theta_t : \Omega \to \Omega$ is measurable;

2. for all $s, t \geq 0$ and all $\omega \in \Omega$, $X_s \circ \theta_t(\omega) = X_s(\theta_t(\omega)) = X_{t+s}(\omega)$.

The following is the basic existence result in the theory.

Theorem 3.3.1 *Suppose $\mathcal{T}$ is a Markov semigroup on a locally compact, separable metric space S. On an appropriate probability space one can construct a Markov process X with transition functions $\mathcal{T}$, together with shift operators θ.*

Proof For any probability measure ν on S, and for all $0 = t_0 < t_1 < \cdots < t_k$, we define a probability measure $P_{t_0,\ldots,t_k}$ on S^{k+1} as follows: For all Borel sets $A_0, \ldots, A_k \subset S$, $P_{t_0,\ldots,t_k}(A_0, \ldots, A_k)$ equals

$$\int \cdots \int \prod_{j=0}^{k} \mathbb{1}_{A_j}(a_j)\, \nu(da_0)\, \mathcal{T}_{t_1}(a_0, da_1)\, \mathcal{T}_{t_2-t_1}(a_1, da_2) \cdots \mathcal{T}_{t_k-t_{k-1}}(a_{k-1}, da_k);$$

see the Chapman–Kolmogorov equation, Lemma 3.1.1. It is clear that the above measures form a consistent family of probability measures on $S^{\mathbb{R}_+}$ (endowed with the product topology). By Kolmogorov's consistency theorem (Theorem 1, Appendix A), there exists a measure $\mathbb{P}_\nu$ on $S^{\mathbb{R}_+}$ whose cylindrical projections are the $P_{t_0,\ldots,t_k}$'s above. More precisely, for all $\omega \in S^{\mathbb{R}_+}$, define $X_t(\omega)$ as the tth coordinate of ω, i.e., $X_t(\omega) = \omega^{(t)} = \omega(t)$. Then, for all Borel sets $A_0, \ldots, A_k \subset S$,

$$\mathbb{P}_\nu(X_{t_i} \in A_i, \text{ for all } 0 \leq i \leq k) = P_{t_0,\ldots,t_k}(A_0, \ldots, A_k).$$

One now checks directly that $X = (X_t;\ t \geq 0)$ is the desired Markov process. Moreover, we can define a family of shift operators for X as follows:

---

[7] In fact, the existence of such extended Markov properties is closely tied with the strong Markov property. For a glimpse of what is involved, see Supplementary Exercises 11, 12, and 13 below.

For all $t \geq 0$ and $\omega \in S^{\mathbb{R}_+}$, define the function $\theta_t(\omega)$ according to the formula
$$\theta_t(\omega)(s) = \omega(t+s), \qquad s \geq 0.$$
Our proof is now complete. □

When shift operators exist, we can formulate the Markov property in the following elegant manner.

Theorem 3.3.2 (The Markov Property) *Suppose X is a Markov process with respect to a filtration $\mathcal{F}$, and with transition functions $\mathcal{T}$ and shifts θ. Then, for any bounded $\vee_t \mathcal{F}_t$-measurable random variable Y, for all $t \geq 0$ and all $x \in S$,*
$$\mathbb{E}_x[Y \circ \theta_t \,|\, \mathcal{F}_t] = \mathbb{E}_{X_t}[Y], \qquad \mathbb{P}_x\text{-a.s.}$$

Proof Suppose $f : S^m \to \mathbb{R}$ is a bounded, measurable function and hold $0 \leq t_1 < \cdots < t_k$ fixed. Then, for all $t \geq 0$, the following holds $\mathbb{P}_x$-a.s., for all $x \in S$,
$$\mathbb{E}_x\big[f(X_{t_1+t}, \ldots, X_{t_k+t}) \,\big|\, \mathcal{F}_t\big] = \mathbb{E}_{X_t}\big[f(X_{t_1}, \ldots, X_{t_k})\big]. \tag{1}$$

The proof of this assertion is very similar to that of Theorem 1.1.2; see also the Chapman–Kolmogorov equation (Lemma 3.1.1) and Exercise 3.3.1 below for details. Alternatively, we can write the above display as
$$\mathbb{E}_x\big[f(X_{t_1}, \ldots, X_{t_k}) \circ \theta_t \,\big|\, \mathcal{F}_t\big] = \mathbb{E}_{X_t}\big[f(X_{t_1}, \ldots, X_{t_k})\big]. \tag{2}$$

The general result follows from a straightforward monotone class argument. □

***Exercise* 3.3.1** First, verify equation (1) and then, complete the proof of Theorem 3.3.2. □

***Exercise* 3.3.2** If X is a Markov process with respect to some filtration $\mathcal{F}$, it is also a Markov process with respect to its history $\mathcal{H}$. It may help to recall that $\mathcal{H}_t$ denotes the smallest σ-field that makes the random variables $(X_r;\ 0 \leq r \leq t)$ measurable. □

4 Feller Processes

Throughout the remainder of this section we consider a locally compact, separable metric space (S, d).[8] In the previous section we constructed S-valued Markov processes that corresponded to Markov semigroups on S.

---

[8] By Urysohn's metrization theorem, this is the *same* as assuming S to be either a locally compact, separable Hausdorff space or a second countable, locally compact space.

While we have established the existence of such stochastic processes in Theorem 3.3.1, the S-valued random function $t \mapsto X_t$ may well be quite badly behaved. In this section we propose to show that if, in addition, $\mathcal{T}$ is a *Feller* semigroup, then there exists an appropriate modification $\widetilde{X}$ of X that is not only a Markov process with transition functions $\mathcal{T}$, but also has right-continuous paths; i.e., $t \mapsto \widetilde{X}_t$ is right-continuous. Moreover, we will show that this modification is Markov, even with respect to the complete augmented filtration; see the discussion of Section 3.2. In light of Section 1.1 of Chapter 7, we could then use the theory of stopping times to better understand Feller processes.

On the other hand, modifications of stochastic processes often require us to enlarge the state space. For instance, in Section 2.2 of Chapter 5 we showed that $\mathbb{R}$-valued stochastic processes have separable modifications; an inspection of our proofs shows that such separable modifications are quite possibly $\mathbb{R} \cup \{\pm \infty\}$-valued. In order to obtain suitable modifications for Markov processes, we now need to enlarge the state space so that it is compact.

Recall that S is a separable, locally compact space. If S is already compact, there is no need to further compactify it. Otherwise, choose some state $\Delta \notin S$ and let $S_\Delta = S \cup \{\Delta\}$. (We have already encountered such extensions in our study of Markov chains in Section 1.3.) We declare $A \subset S_\Delta$ **open** (in S_Δ) if either:

1. $A \subset S$ is open (in the topology of S); or if

2. there exists a compact $K \subset S$ (K is compact in the topology of S) such that $A = S_\Delta \setminus K$.

The above defines a topology on S_Δ that renders it compact; cf. Supplementary Exercise 1. The space S_Δ is the so-called **one-point compactification of S_Δ**. Henceforth, *any f from S into $\mathbb{R}$ is extended to a function from S_Δ in to $\mathbb{R}$ by defining $f(\Delta) = 0$.* If S is already compact, we define $S_\Delta = S$.

4.1 Feller Processes

Suppose $\mathcal{T}$ is a Markov semigroup on a locally compact, separable metric space (S, d) and let X denote a Markov process with transition functions $\mathcal{T}$. We say that X is a **Feller process** (**Feller**, briefly) if $\mathcal{T}$ is Feller.

The following is the first main result of this subsection.

Theorem 4.1.1 *Suppose X is a Feller process on S with transition functions $\mathcal{T}$. Then, X has a right-continuous, S_Δ-valued modification $\widetilde{X}$ that is itself a Feller process with the same transition functions as X. Moreover, on a suitable probability space we can construct a right-continuous Feller process Y, together with a family of shifts, such that Y has the same transition functions as X.*

Our proof of Theorem 4.1.1 is long and is divided into several steps. We begin with two lemmas, the first of which is the continuous-time analogue of Corollary 1.5.1. Throughout, X denotes our Feller process that has transition functions $\mathcal{T}$ and resolvent $\mathcal{R}$.

Lemma 4.1.1 *For any nonnegative $f \in C_0(S)$ and all $\lambda > 0$, the stochastic process $(e^{-\lambda t} \mathcal{R}_\lambda f(X_t);\ t \geq 0)$ is a supermartingale, under the measure $\mathbb{P}_x$ for any $x \in S$.*

Proof Suppose $f \in C_0(S)$. Since $r \mapsto \mathcal{T}_r f(\cdot)$ is uniformly right-continuous (Lemma 2.4.1), we can apply Fubini's theorem to see that $\mathbb{P}_x$-a.s. for all $x \in S$,

$$\mathbb{E}_x\big[\mathcal{R}_\lambda f(X_{t+s}) \,\big|\, \mathcal{F}_s\big] = \mathbb{E}_x\left[\int_0^\infty e^{-\lambda r} \mathcal{T}_r f(X_{t+s})\, dr \,\bigg|\, \mathcal{F}_s\right]$$

$$= \int_0^\infty e^{-\lambda r} \mathbb{E}_x\big[\mathcal{T}_r f(X_{t+s}) \,\big|\, \mathcal{F}_s\big]\, dr.$$

On the other hand, for all $r, t, s \geq 0$, $\mathbb{E}_x[\mathcal{T}_r f(X_{t+s}) | \mathcal{F}_s] = \mathcal{T}_{r+t} f(X_s)$, $\mathbb{P}_x$-a.s., for all $x \in S$. Using Fubini's theorem once more, we deduce from Theorem 3.3.2 (and equation (2) of Section 3.3 if shifts do not exist) that $\mathbb{P}_x$-a.s. for any $x \in S$,

$$\mathbb{E}_x\big[\mathcal{R}_\lambda f(X_{t+s}) \,\big|\, \mathcal{F}_s\big] = \int_0^\infty e^{-\lambda r} \mathbb{E}_{X_s}\big[\mathcal{T}_r f(X_t)\big]\, dr$$

$$= \int_0^\infty e^{-\lambda r} \mathcal{T}_{r+t} f(X_s)\, dr$$

$$= e^{\lambda t} \int_t^\infty e^{-\lambda r} \mathcal{T}_r f(X_s)\, dr.$$

In other words, for all $x \in S$, the following holds $\mathbb{P}_x$-a.s.:

$$\mathbb{E}_x\big[e^{-\lambda t} \mathcal{R}_\lambda f(X_{t+s}) \,\big|\, \mathcal{F}_s\big] = \mathcal{R}_\lambda f(X_s) - \int_0^t e^{-\lambda r} \mathcal{T}_r f(X_s)\, dr. \qquad (1)$$

If $f(a) \geq 0$ for all $a \subset S$, it follows that for any $x \in S$, $\mathbb{P}_x$-a.s.,

$$\mathbb{E}_x\big[e^{-\lambda(t+s)} \mathcal{R}_\lambda f(X_{t+s}) \,\big|\, \mathcal{F}_s\big] \geq e^{-\lambda s} \mathcal{R}_\lambda f(X_s).$$

This proves the result. □

The second lemma of this subsection is a technical real-variable result. In order to introduce it properly, we need some notation. Let d denote the metric on S and define

$$\overline{B}_d(a;\varepsilon) = \{b \in S:\ d(a,b) \leq \varepsilon\}$$

to be the closed ball of radius ε about a. This is the closure of the open ball $\mathcal{B}_d(a;\varepsilon)$, which appeared earlier in Section 2.1 of Chapter 5. By Urysohn's lemma (Lemma 1.3.1, Chapter 6), for any $a \in S$ and $\varepsilon > 0$, we can find a function $\psi_{a,\varepsilon} \in C_0(S)$ such that:

- for all $y \in S$, $0 \leq \psi_{a,\varepsilon}(y) \leq 1$;
- for all $y \in \overline{\mathcal{B}}_d(a;\varepsilon)$, $\psi_{a,\varepsilon}(y) = 1$; and
- for all $y \notin \mathcal{B}_d(a;2\varepsilon)$, $\psi_{a,\varepsilon}(y) = 0$.

Since S is separable, it contains a dense subset $S_\star$. Define

$$\Psi = \big(\psi_{a,\varepsilon};\ a \in S_\star,\ \varepsilon \in \mathbb{Q}_+\big).$$

The announced technical lemma is the following. It is here that we use the compactness of S_Δ in an essential way.

Lemma 4.1.2 *Suppose the function $x : \mathbb{R}_+ \to S_\Delta$ satisfies the following: For all $f \in \Psi$ and all $t \geq 0$, $\lim_{s \downarrow t:\ s \in \mathbb{Q}_+} f(x(s))$ exists. Then, for all $t \geq 0$, $\lim_{s \downarrow t:\ s \in \mathbb{Q}_+} x(s)$ exists.*

Proof Suppose, to the contrary, that the mentioned limit does not always exist. By compactness, we can find two distinct points $a, a' \in S_\Delta$, $t \in [0,1[$, and $s_1, s_2, \ldots, s_1', s_2', \ldots \in \mathbb{Q}_+$ such that as $n \to \infty$, $s_n, s_n' \downarrow t$ and $a = \lim_{n \to \infty} x(s_n)$, while $a' = \lim_{n \to \infty} x(s_n')$. Suppose $a, a' \in S$. We can then find $b \in S_\star$ and $\varepsilon \in \mathbb{Q}_+$ such that $d(a,b) \leq \varepsilon/2$ and $d(b,a') > 4\varepsilon$. Thus, there exists $n_\star$ such that for all $n \geq n_\star$, $d(x(s_n),b) \leq \varepsilon$ while $d(x(s_n'),b) > 2\varepsilon$. In other words, for all $n \geq n_\star$, $\psi_{b,\varepsilon}(x(s_n)) = 1$ and $\psi_{b,\varepsilon}(x(s_n')) = 0$. In particular,

$$\limsup_{s \downarrow t:\ s \in \mathbb{Q}_+} \psi_{b,\varepsilon}\big(x(s)\big) \geq 1 > 0 \geq \liminf_{s \downarrow t:\ s \in \mathbb{Q}_+} \psi_{b,\varepsilon}\big(x(s)\big).$$

The above still holds true if $a \in S$ but $a' = \Delta$ or vice versa. We have arrived at a contradiction, and the lemma follows. $\square$

We are prepared to demonstrate Theorem 4.1.1.

Proof of Theorem 4.1.1 For all $f \in C_0(S)$ and for every $\lambda > 0$, define $\Lambda_{f,\lambda}$ to be the collection of all $\omega \in \Omega$ such that

$$\lim_{s \downarrow t:\ s \in \mathbb{Q}_+} \mathcal{R}_\lambda f(X_s)(\omega) \text{ exists for all } t \geq 0.$$

Lemma 4.1.1 and Lemma 1.4.1 of Chapter 7 can be combined to show that for all $x \in S$, $\mathbb{P}_x(\Lambda_{f,\lambda}) = 0$. If

$$\Lambda = \bigcap_{f \in \Psi,\ \lambda \in \mathbb{Q}_+} \Lambda_{f,\lambda},$$

then for all $x \in S$, $\mathbb{P}_x(\Lambda^\complement) = 0$. By Lemma 2.4.1, for any $\omega \in \Lambda$,

$$\lim_{s \downarrow t:\ s \in \mathbb{Q}_+} f(X_s)(\omega) \text{ exists for all } t \geq 0 \text{ and all } f \in \Psi.$$

Given $\omega \in \Lambda$, we can apply Lemma 4.1.2 with $x(s) = X_s(\omega)$ to conclude that for all $\omega \in \Lambda$ and all $t \geq 0$, $\lim_{s \downarrow t:\ s \in \mathbb{Q}_+} X_s(\omega)$ exists. Fix some $a \in S$ and define $\widetilde{X} = (\widetilde{X}_t;\ t \geq 0)$ by

$$\widetilde{X}_t(\omega) = \begin{cases} \lim_{s \downarrow t:\ s \in \mathbb{Q}_+} X_s(\omega), & \text{if } \omega \in \Lambda, \\ a, & \text{if } \omega \notin \Lambda. \end{cases}$$

It is clear that $\widetilde{X}$ is right-continuous. We will show that it is a modification of X. Given any $\varphi_1, \varphi_2 \in L^\infty(S)^9$ and any $x \in S$, then for all $t \geq 0$,

$$\begin{aligned}
\mathbb{E}_x\big[\varphi_1(\widetilde{X}_t)\varphi_2(X_t)\big] &= \lim_{s \downarrow t:\ s \in \mathbb{Q}_+} \mathbb{E}_x\big[\varphi_1(X_s)\varphi_2(X_t)\big] \\
&= \lim_{s \downarrow t:\ s \in \mathbb{Q}_+} \mathbb{E}_x\big[\mathbb{E}_x\{\varphi_1(X_s)\,|\,\mathcal{F}_t\}\,\varphi_2(X_t)\big] \\
&= \lim_{s \downarrow t:\ s \in \mathbb{Q}_+} \mathbb{E}_x\big[\mathcal{T}_{s-t}\varphi_1(X_t)\,\varphi_2(X_t)\big] \\
&= \mathbb{E}_x\big[\varphi_1(X_t)\varphi_2(X_t)\big].
\end{aligned}$$

The first equality follows from the dominated convergence theorem, the third from the definition of Markov processes, and the last from the Feller property, together with the dominated convergence theorem. A monotone class argument now shows that for all bounded, continuous $\varphi : S \times S \to \mathbb{R}$, for all $x \in S$, and for every $t \geq 0$,

$$\mathbb{E}_x\big[\varphi(\widetilde{X}_t, X_t)\big] = \mathbb{E}_x\big[\varphi(X_t, X_t)\big].$$

Applying this to $\varphi(a,b) = d(a,b) \wedge 1$, $(a,b \in S)$, we can conclude that for all $x \in S$, $\widetilde{X}$ is a $\mathbb{P}_x$-modification of X. The fact that $\widetilde{X}$ is a Markov process with transitions $\mathcal{T}$ follows immediately from the fact that it is a modification of X.

To conclude, we address the issue of the existence of shifts. The desired probability space is the one appearing in the proof of Theorem 3.3.1, namely, $\Omega = S^{\mathbb{R}_+}$. For all $\omega \in S^{\mathbb{R}_+}$ and all $s, t \geq 0$, let $\widetilde{Y}_t(\omega) = \omega(t)$ and $\theta_t(\omega)(s) = \omega(t+s)$. We recall from the proof of Theorem 3.3.1 that the measures $\mathbb{P}_x$ render $\widetilde{Y} = (\widetilde{Y}_t;\ t \geq 0)$ a Feller process with shift operators θ and transition functions $\mathcal{T}$. We can define Λ^0 in the same way as we did Λ above, except that we replace X by $\widetilde{Y}$ everywhere. Define

$$Y_t(\omega) = \begin{cases} \lim_{s \downarrow t:\ s \in \mathbb{Q}_+} \widetilde{Y}_s(\omega), & \text{if } \omega \in \Lambda^0, \\ \omega(0), & \text{if } \omega \notin \Lambda^0. \end{cases}$$

---

[9] Recall that $\varphi_i(\Delta) = 0$

Finally, let

$$\widetilde{\theta}_t(\omega)(s) = \begin{cases} \lim_{r \downarrow t+s:\ r \in \mathbb{Q}_+} Y_r(\omega), & \text{if } \omega \in \Lambda^0, \\ \omega(0), & \text{if } \omega \notin \Lambda^0. \end{cases}$$

Then, $t \mapsto Y_t(\omega)$ is right-continuous for all $\omega \in S^{\mathbb{R}_+}$, and Y is a Feller process with transition functions $\mathcal{T}$. Finally, a few more lines of calculations show that $\widetilde{\theta} = (\widetilde{\theta}_t;\ t \geq 0)$ are shifts for Y. □

To state the second, and final, result of this subsection, we return to a question that came up at the end of Section 3.2 above. Namely, we now show that if X is Markov with respect to a filtration $\mathcal{F}$, X is also Markov with respect to the complete augmented filtration $\mathcal{F}^\star$, at least if X is Feller.

Theorem 4.1.2 *Suppose X is a right-continuous, S_Δ-valued Feller process with respect to a filtration $\mathcal{F}$. Then, X is also a Feller process with respect to the complete augmented filtration $\mathcal{F}^\star$.*

Proof In light of Lemma 3.2.1, we can assume, without loss of generality, that $\mathcal{F}$ is complete; thus, $\mathcal{F}_t^\star = \cap_{s>t} \mathcal{F}_s$, and we seek to prove the following: For all $f \in L^\infty(S)$, all $x \in S$, and all $s, t > 0$,

$$\mathbb{E}_x\big[f(X_{t+s}) \,\big|\, \mathcal{F}_s^\star\big] = \mathbb{E}_x\big[f(X_{t+s}) \,\big|\, \mathcal{F}_s\big],$$

$\mathbb{P}_x$-a.s. But by the Markov property, for all small $\varepsilon > 0$,

$$\mathbb{E}_x\big[f(X_{t+s}) \,\big|\, \mathcal{F}_{s+\varepsilon}\big] = \mathcal{T}_{t-\varepsilon} f(X_{s+\varepsilon}),$$

$\mathbb{P}_x$-a.s.; cf. Lemma 3.2.1. Let $\varepsilon \downarrow 0$ along a rational sequence to see that the right-hand side converges to $\mathcal{T}_t f(X_s)$, by the Feller property and the right continuity of X. Furthermore, the left-hand side converges to $\mathbb{E}_x[f(X_{t+s}) \,|\, \mathcal{F}_s^\star]$, by Doob's martingale convergence theorem for discrete-parameter martingales; cf. Theorem 1.7.1 of Chapter 1. The result follows. □

Convention Recall that the **history** $\mathcal{F} = (\mathcal{F}_t;\ t \geq 0)$ of the process X is the filtration given by the following: For all $t \geq 0$, $\mathcal{F}_t$ is the smallest σ-field that makes $(X_r;\ 0 \leq r \leq t)$ measurable. In light of Theorems 4.1.1, 4.1.2 and Exercise 3.3.2, when we discuss a Feller process X, we can, and always will, assume, without loss of generality, that X is right-continuous and the underlying filtration is the complete augmented history of X.

4.2 The Strong Markov Property

It is a good time for us to prove that Feller processes satisfy a strong Markov property that is the natural continuous-time extension of the strong Markov property we encountered in the discrete setting; cf. Section 1.2 for the latter.

Suppose (S, d) is a locally compact, separable metric space, $\mathcal{T}$ is a Markov semigroup on S, and X is a Markov process on S with transition functions $\mathcal{T}$. Throughout this subsection, $\mathcal{F} = (\mathcal{F}_t;\ t \geq 0)$ designates the complete augmented history of X. Using our one-point compactification S_Δ, we define $X_\infty(\omega) = \Delta$ and call X a **strong Markov process** if:

1. for all $\mathcal{F}$-stopping times T, $\omega \mapsto X_T(\omega)$ is measurable (as a function from $\Omega \to S_\Delta$); and

2. *(The Strong Markov Property)* for all $\mathcal{F}$-stopping times T, all $f \in L^\infty(S)$, and for every $t \geq 0$,
$$\mathbb{E}_x[f(X_{t+T}) \mid \mathcal{F}_T]\mathbf{1}_{(T<\infty)} = \mathbb{E}_{X_T}[f(X_t)],$$
$\mathbb{P}_x$-a.s., for every $x \in S$.

It is clear that every strong Markov process is a Markov process. However, the converse need not be true, as the following shows.

Exercise **4.2.1** Let T be an exponential random variable with mean 1 and define $X_t = 0$ for all $t \neq T$, whereas $X_T = 1$.

(i) Check that $T = \inf(s \geq 0 : X_s = 1)$ and is a finite stopping time.

(ii) Prove that X is a Markov process that is *not* strong Markov.

(HINT: You can start by showing that a modification of a Markov process is itself a Markov process.)   □

Since $g(\Delta) = 0$ for all $g : S \to \mathbb{R}$, condition 2 holds if and only if for all $x \in S$, the following holds $\mathbb{P}_x$-a.s.:
$$\mathbb{E}_x[f(X_{t+T}) \mid \mathcal{F}_T]\mathbf{1}_{(T<\infty)} = \mathbb{E}_{X_T}[f(X_t)]\mathbf{1}_{(T<\infty)}.$$

The following provides us with an easy-to-use condition for verification of the strong Markov property.

Lemma 4.2.1 *Given condition 1 above, condition 2 is equivalent to the following:*

3. *for all $\mathcal{F}$-stopping times T, $f \in L^\infty(S)$, $x \in S$, and for every $t \geq 0$,*
$$\mathbb{E}_x[f(X_{t+T})] = \mathbb{E}_x\{\mathbb{E}_{X_T}[f(X_t)]\}.$$

Proof If condition 2 holds, then we obtain condition 3 by integration with respect to $\mathbb{P}_x$. We will now show that condition 3 implies condition 2. For any $A \in \mathcal{F}_T$, we define the $\mathbb{R}_+ \cup \{\infty\}$-valued random variable T_A by
$$T_A(\omega) = \begin{cases} T(\omega), & \text{if } \omega \in A, \\ +\infty, & \text{if } \omega \notin A, \end{cases} \quad \omega \in \Omega.$$

For any $t \geq 0$, $(T_A \leq t) = (T \leq t) \cap A \in \mathcal{F}_t$. Thus, T_A is an $\mathcal{F}$-stopping time. By condition 3, for all $x \in S$ and all $t \geq 0$,

$$\mathbb{E}_x[f(X_{t+T_A})] = \mathbb{E}_x\{\mathbb{E}_{X_{T_A}}[f(X_t)]\}.$$

The right-hand side equals $\mathbb{E}_x\{\mathbb{E}_{X_T}[f(X_t)]\mathbf{1}_A\}$, while the left-hand side equals $\mathbb{E}_x[f(X_{t+T})\mathbf{1}_A]$. This proves the lemma. $\square$

Next, we prove the strong version of Theorem 3.3.2 for strong Markov processes.

Theorem 4.2.1 *Suppose X is a strong Markov process with shift operators $\theta = (\theta_t;\ t \geq 0)$. For any $\mathcal{F}$-stopping time T and for all bounded, $\vee_{t \geq 0}\mathcal{F}_t$-measurable random variables Y,*

$$\mathbb{E}_x[Y \circ \theta_T \mid \mathcal{F}_T]\mathbf{1}_{(T<\infty)} = \mathbb{E}_{X_T}[Y], \tag{1}$$

$\mathbb{P}_x$-*a.s., for all $x \in S$.*

Proof Since $t \mapsto \theta_t(\omega)$ is a stochastic process, $\theta_T(\omega) = \theta_{T(\omega)}(\omega)$. To begin with, we need to show that $\theta_T : \Omega \to \Omega$ is measurable. Let $A_1, \ldots, A_m$ be measurable subsets of S and fix $0 \leq t_1 < t_2 < \cdots < t_m$. Note that $A = \cap_{j=1}^m (X_{t_j} \in A_j) \in \vee_{t \geq 0}\mathcal{F}_t$. Moreover,

$$\theta_T^{-1}(A) = (\omega : \theta_T(\omega) \in A) = \bigcap_{j=1}^m (\omega : X_{T+t_j}(\omega) \in A_j),$$

which is in $\mathcal{F}_{T+t_m}$; we have used the simple fact that $T + t_m$ is a stopping time.

By a monotone class theorem, for all $A \in \vee_{t \geq 0}\mathcal{F}_t$, $\theta_T^{-1}(A)$ is a measurable subset of Ω. This shows that $\theta_T : \Omega \to \Omega$ is indeed measurable. Now we proceed to verify the claim of the theorem. By a monotone class theorem, it suffices to prove this result for Y's of the form $\prod_{j=1}^k \varphi_j(X_{t_j})$, where $\varphi_1, \ldots, \varphi_k \in L^\infty(S)$ and $0 \leq t_1 < \cdots < t_k$. Let Φ_m denote the collection of all random variables Y of this form. When $Y \in \Phi_1$, (1) follows from the definition of the strong Markov property. We proceed by induction on m.

Suppose (1) holds true for all $Y \in \Phi_{k-1}$; we will show that it also holds for all $Y \in \Phi_k$. Indeed, if $Y \in \Phi_k$, we can find $\varphi_1, \ldots, \varphi_k \in L^\infty(S)$ and $0 \leq t_1 < \cdots < t_k$ such that $Y = \prod_{j=1}^k \varphi_j(X_{t_j})$. Now, for any $x \in S$, the following holds $\mathbb{P}_x$-a.s.:

$$\mathbb{E}_x[Y \circ \theta_T \mid \mathcal{F}_{T+t_{k-1}}] = \prod_{j=1}^{k-1} \varphi_j(X_{t_j+T})\, \mathbb{E}_x[\varphi_k(X_{T+t_k}) \mid \mathcal{F}_{t_{k-1}+T}]$$

$$= \prod_{j=1}^{k-1} \varphi_j(X_{t_j+T})\, \mathbb{E}_x[\varphi_k(X_{t_k-t_{k-1}}) \circ \theta_{t_{k-1}+T} \mid \mathcal{F}_{t_{k-1}+T}]$$

$$= \prod_{j=1}^{k-1} \varphi_j(X_{t_j+T}) \times \mathbb{E}_{X_k}[\varphi_k(X_{t_k - t_{k-1}})],$$

by the induction hypothesis (on Φ_1). Define

$$\psi(x) = \varphi_{k-1}(x)\,\mathbb{E}_x[\varphi_k(X_{t_k - t_{k-1}})], \qquad x \in S.$$

This is a measurable function from S into $\mathbb{R}$, and we have shown that $\mathbb{P}_x$-a.s. for any $x \in S$,

$$\mathbb{E}_x[Y \circ \theta_T \mid \mathcal{F}_{T+t_{k-1}}] = \prod_{j=1}^{k-2} \varphi_j(X_{t_j+T}) \times \psi(X_{t_{k-1}+T})$$

$$= \left\{ \psi(X_{t_{k-1}}) \times \prod_{j=1}^{k-2} \varphi_j(X_{t_j}) \right\} \circ \theta_T.$$

By the induction hypothesis (on Φ_{k-1}), $\mathbb{P}_x$-a.s. for all $x \in S$,

$$\mathbb{E}_x[Y \circ \theta_T \mid \mathcal{F}_T] = \mathbb{E}_x\left\{ \mathbb{E}_x[Y \circ \theta_T \mid \mathcal{F}_{T+t_{k-1}}] \,\Big|\, \mathcal{F}_T \right\}$$

$$= \mathbb{E}_{X_T}\left[\psi(X_{t_{k-1}}) \times \prod_{j=1}^{k-2} \varphi_j(X_{t_j}) \right]$$

$$= \mathbb{E}_{X_T}\left\{ \prod_{j=1}^{k-1} \varphi_j(X_{t_j})\, \mathbb{E}_{X_{t_{k-1}}}[\varphi_k(X_{t_k - t_{k-1}})] \right\}$$

$$= \mathbb{E}_{X_T}\left[\prod_{j=1}^{k-1} \varphi_j(X_{t_j}) \right].$$

The last line follows from the Markov property of X; cf. Theorem 3.3.2. This completes our proof. $\square$

The main result of this section is the following.

Theorem 4.2.2 *Any right-continuous Feller process on S is a strong Markov process.*

Proof This proof is long, and divided into three steps.

Step 1. Suppose there exists $\alpha \in \mathbb{R}_+$ such that for all $x \in S$, $\mathbb{P}_x(T \in \alpha \mathbb{N}_0 \cup \{\infty\}) = 1$. We will show that condition 3 of the strong Markov property holds for such a stopping time T. Indeed, for all $f \in L^\infty(S)$, all $x \in S$, and all $t \geq 0$,

$$\mathbb{E}_x[f(X_{t+T})] = \sum_{n=0}^{\infty} \mathbb{E}_x[f(X_{t+\alpha n}) \mathbf{1}_{(T=\alpha n)}].$$

Let $\mathcal{F}$ denote the complete augmented history of X. Of course, for all $t \geq 0$, $(T = t) \in \mathcal{F}_t$. Thus,

$$\mathbb{E}_x[f(X_{t+T})] = \sum_{n=0}^{\infty} \mathbb{E}_x\{\mathbb{E}_x[f(X_{t+\alpha n}) \mid \mathcal{F}_{\alpha n}] \mathbf{1}_{(T=\alpha n)}\}$$

$$= \sum_{n=0}^{\infty} \mathbb{E}_x\{\mathbb{E}_{X_{\alpha n}}[f(X_t)] \mathbf{1}_{(T=\alpha n)}\}$$

$$= \mathbb{E}_x\{\mathbb{E}_{X_T}[f(X_t)]\}.$$

The second equality follows from the Markov property; see Theorems 3.3.2 and 4.1.2. This completes Step 1 of our proof.

Step 2. If T is an $\mathcal{F}$-stopping time, we can apply Lemma 1.1.1 of Chapter 7 to construct $\mathcal{F}$-stopping times $T^1, T^2, \ldots$ such that:

- for any $n \geq 1$, $T^n < \infty$ if and only $T < \infty$;
- for any $n \geq 1$, $T^n \geq T^{n+1}$;
- if $T(\omega) < \infty$, then $T^n(\omega) \in 2^{-n}\mathbb{N}$; and
- $\lim_{n \to \infty} T^n = T$.

Applying Step 1 to the T^n's, we see that for all $f \in C_0(S)$, all $x \in S$, every $t \geq 0$, and every $n \geq 0$,

$$\mathbb{E}_x[f(X_{t+T^n})] = \mathbb{E}_x\{\mathbb{E}_{X_{T^n}}[f(X_t)]\}.$$

Letting $n \to \infty$ and using the right continuity of $t \mapsto X_t$, we obtain the following from the dominated convergence theorem: For all $f \in C_0(S)$, all $x \in S$, and every $t \geq 0$,

$$\mathbb{E}_x[f(X_{t+T})] = \mathbb{E}_x\{\mathbb{E}_{X_T}[f(X_t)]\}.$$

That is, condition 3 holds for all $f \in C_0(S)$ and all $\mathcal{F}$-stopping times T.

Step 3. Let $G \subset S$ be an open set. There exists a sequence of open sets $G_1, G_2, \ldots$ and a series of closed sets $F_1, F_2, \ldots$ such that for all $n \geq 1$,

$$G_n \subset G_{n+1} \subset \cdots \subset G \subset \cdots \subset F_{n+1} \subset F_n,$$

and such that $\cap_n F_n = \cup_n G_n = G$. To be concrete, we can consider $G_n = \{y \in G : d(y, G^C) < n^{-1}\}$ and $F_n = \{y \in S : d(y, G) \leq n^{-1}\}$. By Urysohn's lemma (Lemma 1.3.1, Chapter 6), for any $n \geq 1$, we can find functions $\underline{\varphi}_n, \overline{\varphi}_n \in C_0(S)$ such that for all $x \in S$ and all $n \geq 1$, $0 \leq \underline{\varphi}_n(x), \overline{\varphi}_n(x) \leq 1$, and

$$\underline{\varphi}_n(x) = \begin{cases} 1, & \text{if } x \in G_n, \\ 0, & \text{if } x \notin G, \end{cases} \qquad \overline{\varphi}_n(x) = \begin{cases} 1, & \text{if } x \in G, \\ 0, & \text{if } x \notin F_n. \end{cases}$$

A little thought shows that for all $n \geq 1$,

$$\underline{\varphi}_n \leq \underline{\varphi}_{n+1} \leq \mathbf{1}_G \leq \overline{\varphi}_{n+1} \leq \overline{\varphi}_n. \tag{2}$$

Thus, for all $x \in S$, all $\mathcal{F}$-stopping times T, and all $t \geq 0$,

$$\mathbb{E}_x\left[\underline{\varphi}_n(X_{t+T})\right] \leq \mathbb{P}_x(X_{t+T} \in G) \leq \mathbb{E}_x\left[\overline{\varphi}_n(X_{t+T})\right]. \tag{3}$$

By Step 2, the two sides of the above can be identified as $\mathbb{E}_x\{\mathbb{E}_{X_T}[\underline{\varphi}_n(X_t)]\}$ and $\mathbb{E}_x\{\mathbb{E}_{X_T}[\overline{\varphi}_n(X_t)]\}$, respectively. Consequently,

$$\mathbb{E}_x\left\{\mathbb{E}_{X_T}\left[\underline{\varphi}_n(X_t)\right]\right\} \leq \mathbb{P}_x(X_{t+T} \in G) \leq \mathbb{E}_x\left\{\mathbb{E}_{X_T}\left[\overline{\varphi}_n(X_t)\right]\right\}. \tag{4}$$

By equation (2) and the monotone convergence theorem, both sides have finite limits, as $n \to \infty$. We now claim that these two limits are one and the same, and we wish to identify the said limit.

For all $x \in S$, $0 \leq \overline{\varphi}_n(x) - \underline{\varphi}_n(x) \leq \mathbf{1}_{F_n \setminus G_n}(x)$. As $n \to \infty$, the latter converges down to 0. Hence, we can apply the monotone convergence theorem to see that

$$\lim_{n \to \infty} \mathbb{E}_x\{|\overline{\varphi}_n(X_{t+T}) - \underline{\varphi}_n(X_{t+T})|\} = 0.$$

By Step 2, equations (2) and (4), and the monotone convergence theorem,

$$\mathbb{P}_x(X_{t+T} \in G) = \lim_{n \to \infty} \mathbb{E}_x\{\mathbb{E}_{X_T}[\overline{\varphi}_n(X_t)]\} = \mathbb{E}_x\{\mathbb{P}_{X_T}(X_t \in G)\}.$$

The two extreme sides of the above are probability measures as set functions of G. Since the above holds for all open sets $G \subset S$, by a monotone class argument it holds for all measurable $G \subset S$. Another monotone class argument shows that for all $f \in L^\infty(S)$, all $x \in S$, and every $t \geq 0$,

$$\mathbb{E}_x[f(X_{t+T})] = \mathbb{E}_x\{\mathbb{E}_{X_T}[f(X_t)]\}.$$

An appeal to Lemma 4.2.1 finishes our proof. □

4.3 Lévy Processes

It is time for us to see some examples of Feller processes. The examples of this subsection are, in a sense, not much more than continuous-time random walks. With this random-walk model in mind, we say that an $\mathbb{R}^d$-valued stochastic process $X = (X_t; t \geq 0)$ is a **Lévy process** (on $\mathbb{R}^d$) if:

1. $\mathbb{P}(X_0 = 0) = 1$.

2. *(Independence of the Increments)* For all $0 = t_0 \leq t_1 \leq \cdots \leq t_m$, $(X_{t_{i+1}} - X_{t_i}; 0 \leq i \leq m-1)$ is a collection of independent random vectors in $\mathbb{R}^d$.

3. *(Stationarity of the Increments)* For all $0 \leq t_1 \leq t_2$, the distribution of $X_{t_2} - X_{t_1}$ is the same as that of $X_{t_2-t_1}$.

4. The random function $t \mapsto X_t$ is continuous in probability. That is, for all $t \geq 0$ and all $\varepsilon > 0$,
$$\lim_{s \to t} \mathbb{P}(|X_s - X_t| \geq \varepsilon) = 0.$$

Conditions 1–3 ensure that X is a random walk in continuous time, while condition 4 guarantees us some minimal regularity. Barring issues of existence, let us define a family of linear operators $\mathcal{T} = (\mathcal{T}_t;\ t \geq 0)$ by
$$\mathcal{T}_t f(x) = \mathbb{E}[f(X_t + x)], \qquad t \geq 0,\ x \in \mathbb{R}^d,\ f \in L^\infty(\mathbb{R}^d).$$

Lemma 4.3.1 *If the Lévy process X exists, then $\mathcal{T}$, as defined above, is a Feller semigroup on $\mathbb{R}^d$.*

Proof For all $f \in L^\infty(\mathbb{R}^d)$, $x \in \mathbb{R}^d$, and $t \geq 0$,
$$\mathcal{T}_t \mathcal{T}_s f(x) = \mathbb{E}[\mathcal{T}_s f(x + X_t)] = \mathbb{E}\left\{\int_{\mathbb{R}^d} f(x + y + X_t)\, \mathbb{P}(X_s \in dy)\right\}$$
$$= \mathbb{E}\left\{\int_{\mathbb{R}^d} f(x + y + X_t)\, \mathbb{P}(X_{t+s} - X_t \in dy)\right\},$$
by condition 3. Consequently,
$$\mathcal{T}_t \mathcal{T}_s f(x) = \mathbb{E}\{\mathbb{E}[f(x + X_{t+s} - X_t + X_t) \mid X_t]\} = \mathbb{E}\{\mathbb{E}[f(x + X_{t+s}) \mid X_t]\}$$
$$= \mathbb{E}[f(x + X_{t+s})] = \mathcal{T}_{t+s} f(x).$$

That is, $\mathcal{T}$ is a semigroup on $\mathbb{R}^d$. The other properties of Markov semigroups follow readily.

To conclude this proof, note that any $f \in C_0(\mathbb{R}^d)$ is uniformly continuous; cf. Exercise 2.2.1. Moreover,
$$\sup_{x \in \mathbb{R}^d} |\mathcal{T}_t f(x) - f(x)| \leq \mathbb{E}\left[\sup_{x \in \mathbb{R}^d} |f(x + X_t) - f(x)|\right].$$

The bounded convergence theorem implies that it is sufficient to show that as $t \to 0+$, $X_t \to 0$, in probability; this is an immediate consequence of conditions 1 and 4 of Lévy processes. $\square$

Exercise 4.3.1 Show that whenever μ is an infinitely divisible distribution on $\mathbb{R}^d$, there exists an $\mathbb{R}^d$-valued Lévy process $X = (X_t;\ t \geq 0)$ such that the distribution of X_1 is μ. That is, $\mathbb{P}(X_1 \in A) = \mu(A)$. Conversely, show that whenever X is a Lévy process, X_1 is an infinitely divisible random variable.
(HINT: If X is a Lévy process, $(X_t;\ t \in F)$ can be identified with a random walk, as long as F is finite.) $\square$

Based on the transition operators $\mathcal{T}$, we can construct transition functions $\mathfrak{T}$ by the assignment $\mathfrak{T}_t(x, A) = \mathcal{T}_t \mathbf{1}_A(x)$. Note that $A \mapsto \mathfrak{T}_t(x, A)$ is a probability measure on $\mathbb{R}^d$ for each $x \in \mathbb{R}^d$ and for every $t \geq 0$. The following is a prefatory version of the Chapman–Kolmogorov equation and has the very same proof; cf. Lemma 3.1.1.

Lemma 4.3.2 *If the Lévy process X exists, then for all $0 \leq t_1 \leq \cdots \leq t_m$ and all $\varphi_1, \ldots, \varphi_m \in L^\infty(\mathbb{R}^d)$,*

$$\mathbb{E}\left[\prod_{j=1}^m \varphi_j(X_{t_j})\right]$$
$$= \int \cdots \int \prod_{j=1}^m \varphi_j(a_j) \mathfrak{T}_{t_1}(0, da_1) \mathfrak{T}_{t_2-t_1}(a_1, da_2) \cdots \mathfrak{T}_{t_m - t_{m-1}}(d_{m-1}, da_m).$$

If X is a Lévy process, we can view $t \mapsto X_t$ as a random variable that takes its values in the space $S = (\mathbb{R}^d)^{\mathbb{R}_+}$, endowed with the product topology and the corresponding Borel field $\mathcal{S}$.

Remark While S is *not* a metric space, one defines such random variables in the same manner as those of Chapter 5.

For each $x \in \mathbb{R}^d$, we can define the following probability measure on $(S, \mathcal{S})$: $\mathbb{P}_x = \mathbb{P} \circ (X + x)^{-1}$, where $(X + x)_t = X_t + x$. Then, it quickly follows that for any initial measure ν, X is a Feller process on $\mathbb{R}^d$ with respect to the measure $\mathbb{P}_\nu(\bullet) = \int \mathbb{P}_x(\bullet)\nu(dx)$. As such, we can combine Lemma 4.3.1 with Theorems 4.1.1 and 4.2.2 to obtain the following.

Proposition 4.3.1 *After a suitable modification, any Lévy process on $\mathbb{R}^d$ is a right-continuous strong Markov process.*

Example 1 (Multidimensional Brownian Motion) We say that the $\mathbb{R}^d$-valued stochastic process $B = (B_t; t \geq 0)$ is (standard) **Brownian motion** in $\mathbb{R}^d$ if its coordinate processes $B^{(1)}, \ldots, B^{(d)}$ are independent standard Brownian motions, where $B^{(i)} = (B_t^{(i)}; t \geq 0)$, $1 \leq i \leq d$. By Theorem 1.7.1 of Chapter 7, B is a Lévy process, and is, in fact, continuous. Note that for any $t \geq 0$, $B_t \sim \mathcal{N}_d(0, t\mathbf{I}_d)$, where $\mathbf{I}_d$ denotes the identity matrix in $\mathbb{R}^d \times \mathbb{R}^d$. For all $t \geq 0$ and $y \in \mathbb{R}^d$, define

$$q_t(y) = q(t; y) = (2\pi t)^{-\frac{d}{2}} \exp\left(-\frac{\|y\|^2}{2t}\right),$$

where $\|y\|^2 = \sum_{j=1}^d |y^{(j)}|^2$ denotes the square of the Euclidean ℓ^2 norm, as usual. It is then clear that for all $t \geq 0$, $x \in \mathbb{R}^d$, and $f \in L^\infty(\mathbb{R}^d)$,

$$\mathcal{T}_t f(x) = \mathbb{E}[f(B_t + x)] = \int_{\mathbb{R}^d} f(y)\, q(t; y - x)\, dy.$$

In other words, the transition functions of B are given by the following: For all $t \geq 0$, $x \in \mathbb{R}^d$, and $f \in L^\infty(\mathbb{R}^d)$,

$$\mathcal{T}_t f(x) = f \star q_t(x),$$

where $\star$ denotes convolution. Following the discussion of Section 2.3, we conclude that $\mathcal{T}$ has transition densities with respect to Lebesgue's measure, and they are $p_t(x, y) = q_t(y - x)$, $t \geq 0$, $x, y \in \mathbb{R}^d$. □

Example 2 (The Generalized Multidimensional Poisson Process) Let $\mathsf{P} = (\mathsf{P}_t; t \geq 0)$ denote an $\mathbb{R}^d$-valued stochastic process whose ith coordinate process $\mathsf{P}^{(i)} = (\mathsf{P}_t^{(i)}; t \geq 0)$ is a Poisson process with rate $\lambda^{(i)} > 0$, and suppose that the coordinate processes are independent. The process P is called the generalized **Poisson process** on $\mathbb{N}_0^d$ with rate vector λ. According to Section 1.8 of Chapter 7, P is a Lévy process on $\mathbb{R}^d$, and it is not hard to see that it has the following transition densities with respect to the counting measure on $\mathbb{N}_0^d$: For all $t \geq 0$,

$$p_t(x, y) = \begin{cases} \prod_{j=1}^d \dfrac{e^{-\lambda^{(j)} t} (\lambda^{(j)} t)^{y^{(j)} - x^{(j)}}}{(y^{(j)} - x^{(j)})!}, & \text{if } y \succcurlyeq x \text{ and } y - x \in \mathbb{N}_0^d, \\ 0, & \text{otherwise.} \end{cases}$$

□

Example 3 (Isotropic Stable Processes) An $\mathbb{R}^d$-valued stochastic process $X = (X_t; t \geq 0)$ is an **isotropic stable process** of index $\alpha \in]0, 2]$ if it is a Lévy process and if for all $t \geq 0$, the characteristic function of X_t is given by the following formula:

$$\mathbb{E}[e^{i\xi \cdot X_t}] = e^{-\frac{1}{2} t \|\xi\|^\alpha}, \qquad \xi \in \mathbb{R}^d.$$

As usual, $\|\xi\| = (\sum_{j=1}^d |\xi^{(j)}|^2)^{1/2}$ stands for the Euclidean ℓ^2 norm of $\xi \in \mathbb{R}^d$.[10]

When $\alpha = 2$, this is Brownian motion, and existence of such a process has already been verified. When $\alpha = d = 1$, this is called the (symmetric) **Cauchy process**, since the above characteristic function implies that the probability density function of X_t is, in this case, given by

$$\mathbb{P}(X_t \in A) = \frac{2t}{\pi} \int_A \frac{1}{t^2 + 4x^2} \, dx,$$

for all Borel sets $A \in \mathbb{R}$. (Why?) That is, in this case, $t^{-1} X_t$ has a standard Cauchy distribution.

---

[10] If $\alpha \notin]0, 2]$, recall that $\xi \mapsto \exp\{-\frac{1}{2} \|\xi\|^\alpha\}$ is not a characteristic function.

In the general case where $\alpha \in]0, 2]$, such processes always exist. This is worked out in Supplementary Exercise 5, by using the Lévy–Khintchine formula. Moreover, as we will shortly see, their distribution has a density with respect to Lebesgue's measure. But these densities are not explicitly known except when $\alpha = 2$ or $\alpha = d = 1$. Nonetheless, one can obtain much information on the existence and structure of these densities by Fourier-analytic methods. Indeed, by the inversion formula for characteristic functions, X has transition densities with respect to the Lebesgue measure and they are of the form $p_t(x, y) = q_t(y - x)$, where

$$q_t(a) = (2\pi)^{-d} \int_{\mathbb{R}^d} e^{-i\xi \cdot a} e^{-\frac{1}{2} t \|\xi\|^\alpha} \, d\xi.$$

(Check!) We can now compute the transition functions in complete analogy to Example 1: For all $f \in L^\infty(\mathbb{R}^d)$, $\mathcal{T}_t f(x) = f \star q_t(x)$. □

Example 4 (Uniform Motion Along a Ray) Fix $\beta \in \mathbb{R}^d$ and define for all $\omega \in \Omega$ and all $t \geq 0$, $X_t(\omega) = t\beta$. This is a nonrandom stochastic process. However, it is possible to show that it is a Lévy process on $\mathbb{R}^d$ with transition functions $\mathcal{T}_t f(x) = f(x + t\beta)$, $t \geq 0$, $x \in \mathbb{R}^d$, $f \in L^\infty(\mathbb{R}^d)$. □

Exercise **4.3.2** Complete the computations of Example 4 above. □

5 Supplementary Exercises

1. Prove that the one-point compactification of a locally compact, separable metric space is a topology and that it renders S_Δ compact.

2. Let $X = (X_n;\ n \geq 0)$ denote a Markov chain on a denumerable state space S whose k-step transition function is denoted by p_k. Find an explicit formula for p_k in terms of p_{k-1} and p_1 and obtain a recursion.
(HINT: $p_k(x, y)$ is the probability of going from x to y in k steps. This means that in $(k - 1)$ steps, the chain goes from x to some $z \in S$ and in one last step, it goes from z to y.)

3. Let $X = (X_n;\ n \geq 0)$ denote the simple random walk on $\mathbb{Z}^d$. View this as a Markov chain with transition functions p_k.

(i) Prove that for all $x, y \in \mathbb{Z}^d$ and all $n \geq 0$,

$$p_n(x, y) = (2\pi)^{-d} \int_{[-\pi, \pi]^d} e^{-i\xi \cdot y} \left(\frac{1}{d} \sum_{j=1}^{d} \cos(\xi^{(j)}) \right)^n d\xi.$$

(ii) Show that for all positive functions f, for all $x \in \mathbb{Z}^d$, and for all $\lambda \in\,]0,1[$, $\mathcal{R}_\lambda f(x) = \sum_{y \in \mathbb{Z}^d} f(y) u_\lambda(y-x)$, where

$$u_\lambda(a) = (2\pi)^{-d} \int_{[-\pi,\pi]^d} e^{-i\xi \cdot a} \left[1 - \frac{\lambda}{d} \sum_{j=1}^{d} \cos(\xi^{(j)})\right]^{-1} d\xi.$$

When is $\lim_{\lambda \uparrow 1} u_\lambda(a)$ finite? This is related to Section 3.1, Chapter 3.
(HINT: For part (i), use the inversion formula for characteristic functions.)

4. Given a Markov chain $X = (X_n;\ n \geq 0)$ on a denumerable space S, we say that $x_0 \in S$ is **recurrent** if, starting at x_0, S visits x_0 infinitely often with positive probability. Check that $x_0 \in S$ is recurrent if and only if $\sum_{n=0}^{\infty} p_n(x_0, x_0) = +\infty$, where p_n designates the n-step transition function of X. Moreover, whenever $\sum_{n=0}^{\infty} p_n(x_0, x_0) = +\infty$, then $\mathbb{P}_{x_0}$-a.s., $\sum_{n=0}^{\infty} \mathbf{1}_{(X_n = x_0)} = +\infty$.
(HINT: See Proposition 1.3.1, Chapter 3.)

5. Let μ be an infinitely divisible distribution on $\mathbb{R}^d$ whose characteristic function is denoted by φ. Recall the Lévy–Khintchine formula[11] $\varphi(\xi) = \exp\{-\Psi(\xi)\}$, where Ψ has the following representation: There exists $\mathbf{a} \in \mathbb{R}^d$, a symmetric, positive definite matrix $\Sigma \in \mathbb{R}^d \times \mathbb{R}^d$, and a measure L on $\mathbb{R}^d$ with $\mathsf{L}(\{0\}) = 0$ and $\int \{1 \wedge |\tau|^2\} \mathsf{L}(d\tau) < \infty$ such that for all $\xi \in \mathbb{R}^d$,

$$\Psi(\xi) = i\mathbf{a} \cdot \xi + \frac{1}{2}\xi' \Sigma \xi + \int \left\{1 - e^{i\xi \cdot \tau} + \frac{i\xi \cdot \tau}{1 + \|\tau\|^2}\right\} \mathsf{L}(d\tau).$$

(i) Use this to show that for any $\alpha \in\,]0,2[$, there exists an infinitely divisible $\mathbb{R}^d$-valued random variable Y such that $\mathbb{E}[e^{i\xi \cdot Y}] = \exp\{-\frac{1}{2}\|\xi\|^\alpha\}$.

(ii) Conclude that isotropic stable processes exist.

(HINT: For the second part, employ Exercise 4.3.1; it may help to know this fact in $\mathbb{R}^1$.)

6. A strong Markov process $X = (X_t;\ t \geq 0)$ on a separable matric space (S, d) is a **diffusion** if it has a continuous modification. In this exercise we will find a sufficient condition for X to be a diffusion. As usual, S_Δ denotes the one-point compactification of S.

(i) Suppose $f : [0,1] \to S_\Delta$ is a right-continuous function. Show that f is discontinuous if and only if

$$\liminf_{n \to \infty} \max_{0 \leq k \leq n-1} d\left(f\left(\frac{k}{n}\right), f\left(\frac{k+1}{n}\right)\right) > 0.$$

(ii) Use the above to show that X is a diffusion (on S) if for all compact sets $K \subset S$ and for all $\varepsilon > 0$,

$$\lim_{t \to 0^+} \frac{1}{t} \sup_{x \in K} \mathbb{P}_x(X_t \notin \mathcal{B}(x; \varepsilon)) = 0,$$

where $\mathcal{B}(x; \varepsilon) = \{y \in S :\ d(x, y) < \varepsilon\}$.

---

[11] See (Bertoin 1996; Feller 1971; Itô 1984; Sato 1999) and their references for earlier works. The 1-dimensional version appears in most textbooks on probability.

(iii) Provide an alternative proof that Brownian motion has a continuous modification.

7. (Hard. Continued from Supplementary Exercise 6)
Let $X = (X_t; t \geq 0)$ denote an isotropic stable Lévy process of index $\alpha \in]0, 2[$. (Note $\alpha \neq 2$ in this problem!) We intend to prove that X is *not* a diffusion.

   (i) Check that for each fixed $t > 0$, X_t has the same distribution as $t^{1/\alpha} X_1$. This property is called **scaling**.

   (ii) Let $q_t(x)$ denote the density function of X_t. We will need the following estimate, which follows from Proposition 3.3.1 of Chapter 10: There exists a finite constant $C > 1$ such that for all $a \in \mathbb{R}^d$ with $\|a\| > 2$,
   $$C\|a\|^{-(d+\alpha)} \geq q_1(a) \geq C^{-1}\|a\|^{-(d+\alpha)}.$$
   Use this, without proof, to deduce the existence of a finite constant $A > 1$ such that for all $y > 1$,
   $$Ay^{-\alpha} \geq \mathbb{P}(\|X_1\| > y) \geq A^{-1} y^{-\alpha}.$$

   (iii) Show that with positive probability, X is not a diffusion. Use a 0-1 law to demonstrate that X is a.s. not a diffusion. (See Supplementary Exercises 10 and 11.)

   (iv) Check, explicitly, that the condition of Supplementary Exercise 6 fails to hold in this case.

(HINT: For part (iii), apply the Paley–Zygmund lemma; cf. Lemma 1.4.1, Chapter 3.)

8. Prove directly that the operators of Examples 1, 2, and 3 of Section 1.4 form Markov semigroups.

9. (Hard) Suppose $\mathcal{T} = (\mathcal{T}_n; n \geq 0)$ and $\mathcal{T}^\star = (\mathcal{T}_n^\star; n \geq 0)$ are two transition functions on a denumerable state space S. If ν is a probability measure on S, we say that $\mathcal{T}$ and $\mathcal{T}^\star$ are in **duality** with respect to ν if for all $f, g : S \to \mathbb{R}_+$, $\int f(x) \cdot \mathcal{T}_1 g(x) \nu(dx) = \int g(x) \cdot \mathcal{T}_1^\star f(x) \nu(dx)$. Let $X = (X_n; n \geq 0)$ denote the S-valued Markov chain whose transition functions are $\mathcal{T}$, and define $\mathcal{F}_n$ and $\mathcal{G}_n$ to be the σ-fields generated by $(X_0, \ldots, X_n)$ and $(X_n, X_{n+1}, \ldots)$, respectively.

   (i) Prove that for all $n \geq 0$ and all bounded functions $h : S \to \mathbb{R}_+$,
   $$\mathbb{E}_\nu[h(X_{n+1}) \mid \mathcal{F}_1] = \mathcal{T}_n h(X_1), \qquad \mathbb{P}_\nu\text{-a.s.}$$

   (ii) Prove that for all $n \geq 0$ and all bounded functions $h : S \to \mathbb{R}_+$,
   $$\mathbb{E}_\nu[h(X_1) \mid \mathcal{G}_{n+1}] = \mathcal{T}_n^\star h(X_{n+1}), \qquad \mathbb{P}_\nu\text{-a.s.}$$

   (iii) Prove that for all (measurable) functions $f : S \to \mathbb{R}_+$ and for all $p > 1$,
   $$\int_S |\sup_n \mathcal{T}_n \mathcal{T}_n^\star f(x)|^p \nu(dx) \leq \left(\frac{p}{p-1}\right)^p \int_S |f(x)|^p \nu(dx).$$

(iv) Show that when $p = 1$, the above has the following analogue: For all functions $f : S \to \mathbb{R}_+$,

$$\int_S |\sup_n \mathfrak{T}_n \mathfrak{T}_n^\star f(x)| \nu(dx) \le \left(\frac{e}{e-1}\right)\left\{1 + \int f(x) \ln_+ f(x) \nu(dx)\right\}.$$

Moreover, if $\int_S f(x) \ln_+ f(x) \nu(dx) < \infty$, show that $\lim_n \mathfrak{T}_n \mathfrak{T}_n^\star f$ exists ν-a.s. What is this limit?

(HINT: For part (ii), have a look at Exercise 1.1.2.)

10. Suppose $X = (X_t;\ t \ge 0)$ is a Feller process on $\mathbb{R}^d$. For all $t \ge 0$, let $\mathfrak{T}_t$ denote the σ-field generated by the collection of random variables $(X_s;\ s \ge t)$. The **tail σ-field** of X, $\mathfrak{T}$, is defined as

$$\mathfrak{T} = \bigcap_{t \ge 0} \mathfrak{T}_t.$$

(i) Prove that the tail σ-field of any Lévy process is trivial in that whenever $A \in \mathfrak{T}$, then, $\mathbb{P}_x(A) \in \{0,1\}$ for all $x \in \mathbb{R}^d$.

(ii) Construct $\mathbb{R}^d$-valued Markov processes whose tail σ-fields are *not* trivial.

The first portion is a variant of A. N. Kolmogorov's 0-1 law.
(HINT: If X is a Lévy process, $(X_s;\, s \in \mathbb{Q}_+)$ can be approximated by a random walk.)

11. Let $X = (X_t;\ t \ge 0)$ denote a Feller process on a compact separable metric space (S,d). Suppose $\mathcal{F} = (\mathcal{F}_t;\ t \ge 0)$ denotes its complete augmented history of X. Prove that $\mathcal{F}_0$ is trivial in that any $A \in \mathcal{F}_0$ satisfies $\mathbb{P}_x(A) \in \{0,1\}$ for all $x \in S$. This is the celebrated **Blumenthal's 0-1 law**, discovered by Blumenthal (1957).
(HINT: First consider sets A that are measurable with respect to the σ-field generated by X_0. The latter is not quite the same σ-field as $\mathcal{F}_0$. Why?)

12. (Hard. Blumenthal's 0-1 law, continued) Suppose X is a Markov process on a compact metric space (S,d) and with respect to a complete filtration $\mathcal{F} = (\mathcal{F}_t;\ t \ge 0)$. Define the right-continuous filtration $\mathcal{F}^\star = (\mathcal{F}_{t+};\ t \ge 0)$ by $\mathcal{F}_{t+} = \cap_{s>t} \mathcal{F}_s$. We say that X satisfies Blumenthal's 0-1 law if for all $\Lambda \in \mathcal{F}_{0+}$ and all $x \in S$, $\mathbb{P}_x(\Lambda) \in \{0,1\}$; cf. also Supplementary Exercise 11. Prove that X satisfies Blumenthal's 0-1 law if and only if the following holds: For all $t \ge 0$, all $x \in S$, and all $\mathcal{F}_{t+}$-measurable events Λ, there exists an $\mathcal{F}_t$-measurable event Λ_x such that $\mathbb{P}_x(\Lambda = \Lambda_x) = 1$. In words, Blumenthal's 0-1 law holds if and only if $\mathcal{F}_{t+}$ is essentially the same as $\mathcal{F}_t$. Use this and Theorem 4.1.2 to give an alternative proof of the fact that if X is Feller, it satisfies Blumenthal's 0-1 law; cf. Supplementary Exercise 11.
(HINT: Without loss of generality, assume the existence of shifts θ and by considering the finite-dimensional distributions, we can reduce the problem to showing that for all bounded random variables Y, $\mathbb{E}_x[Y \circ \theta_t \mid \mathcal{F}_{t+}] = \mathbb{E}_{X_t}[Y]$, $\mathbb{P}_x$-a.s.)

13. (Hard) In this exercise we construct a Markov process X with respect to a complete filtration $\mathcal{F}$ such that X is *not* Markov with respect to $\mathcal{F}^*$; it also is not a strong Markov process. Let $S_1 = [0,1] \times \{(0,0)\}$ be the interval $[0,1]$ viewed as a subset of $\mathbb{R}^2$; also, let $S_2 = \{(1,0)\} \times \mathbb{R}$ and define $S = S_1 \cup S_2$. Viewed as a subset of $\mathbb{R}^2$, S is a nice metric space and will be the state space of our Markov process.

The dynamics of the process X, in words, are as follows: For any $a \in [0,1[$, on the event that $X_0 = (a,0)$, X moves at unit speed to the right along S_1; when it reaches $(1,0)$, it tosses an independent fair coin and goes up or down along S_2 at unit speed, with probability $\frac{1}{2}$ each. If $X_0 = (1,a)$ for $a > 0$, X moves up along S_2 at unit speed, whereas if $X_0 = (1,-a)$, it moves down along S_2 at unit speed. Finally, if $X_0 = (1,0)$, it tosses an independent fair coin and moves up or down along S_2 at unit speed, with probability $\frac{1}{2}$ each.

(i) Construct X rigorously and prove that it is a Markov process with respect to its complete history $\mathcal{F}$.

(ii) Show that $\mathcal{F}_1$ is trivial, whereas $\mathcal{F}_{1+} = \cap_{s>1} \mathcal{F}_s$ is not. Conclude that X is not Markov with respect to its complete augmented history.

(iii) Prove that X is *not* a strong Markov process.
(HINT: Consider the first time that the second coordinate of X is greater than zero.)

(iv) Check directly that X is not a Feller process.

This is due to J. B. Walsh.

14. Suppose X denotes a Feller process on the one-point compactification of a locally compact metric space (S,d). If $K \subset S$ is compact, $T_K = \inf(s \geq 0 : X_s \notin K)$ is a stopping time. Prove that $t \mapsto X_{t \wedge T_K}$ is a Feller process on K. Identify its transition functions in terms of those of X.
(HINT: See Theorem 1.3.3 of Chapter 8 for a related result.)

6 Notes on Chapter 8

Section 1 The material on denumerable Markov chains is classical and shows only the surface of a rich theory; see Revuz (1984) for a complete treatment, as well as an extensive bibliography of this subject. In relation to Theorem 1.6.1, and in the context of random walks, we also mention Itô and McKean (1960).

Section 2 Some of the many in-depth introductions to operator theory and functional analysis are (Hille and Phillips 1957; Riesz and Sz.-Nagy 1955; Yosida 1995).

Section 3 While our approach is an economical one, the theory of Markov processes is far from being a consequence of semigroup considerations. A more general theory can be found in the combined references Blumenthal and Getoor (1968), Dellacherie and Meyer (1988), Ethier and Kurtz (1986), Fukushima et al. (1994), Getoor (1975, 1990), Rogers and Williams (1987, 1994), and Sharpe (1988). They contain various aspects of a very rich and general theory that is based on the resolvents, rather than on the associated semigroups.

Section 4 Two pedagogical accounts of Lévy processes are (Bertoin 1996; Sato 1999). Even more is known in the case that our Lévy process is Brownian motion. You can start reading on this in (Bass 1995; Föllmer 1984b; Le Gall 1992; Revuz and Yor 1994), as well as Yor (1992, 1997).

Section 5 Supplementary Exercise 7 is quite classical and is a natural introduction to Lévy systems. Revuz and Yor (1994) is an excellent textbook account that is also equipped with a detailed bibliography.

The original form of Supplementary Exercise 9 for $p > 1$ is from Rota (1962); it extends an older result of Burkholder and Chow (1961). When $p = 1$, this was announced in Rota (1969); see Burkholder (1962, footnotes). The proof outlined here is borrowed from Doob (1963). For a multiparameter extension of Rota's theorem, together with a number of results in multiparameter ergodic theory, see Sucheston (1983, Theorem 2.3).

Supplementary Exercise 9 and Exercise 1.1.2 are only the beginnings of the theory of time reversal for Markov processes. Much more on this can be found in (Chung and Walsh 1969; Getoor 1979; Millar 1978; Nagasawa 1964); see also Bertoin (1996) for the special case of Lévy processes. Among many other things, these references will show you that Blumenthal's 0-1 law (Supplementary Exercise 11) holds for all strong Markov processes; this is a deep fact and lies at the heart of much of the general theory.

9
Generation of Markov Processes

In this chapter we briefly discuss a few of the many interactions between Markov processes and integro-differential equations. In particular, we will concentrate on connections involving the so-called infinitesimal generator of the Markov process.

Suppose $\mathcal{T}$ denotes the transition operators of the process in question. Then, the semigroup property $\mathcal{T}_{t+s} = \mathcal{T}_t \mathcal{T}_s$ suggests the existence of an operator $\mathcal{A}$ such that $\mathcal{T}_t = e^{t\mathcal{A}}$, once this operator identity is suitably interpreted. If so, knowing the one operator $\mathcal{A}$ is equivalent to knowing all of the $\mathcal{T}_t$'s, and one expects $\mathcal{A}$ to be the right derivative of $t \mapsto \mathcal{T}_t$ at $t = 0$. The operator $\mathcal{A}$ is the so-called generator of the process, and is the subject of this chapter. The reader who is strictly interested in multiparameter processes may safely skip the material of Section 1.4 and beyond.

1 Generation

In Chapter 8 we saw that nice Markov processes are naturally described by two classes of linear operators: (1) their semigroups; and (2) their resolvents. It is a remarkable fact that there is *one* linear operator that completely describes the process, together with its semigroup and resolvent. This so-called generator often turns out to be an unbounded operator.

In this section we show the existence of generators for Feller processes, while in the next section we shall explicitly compute them for two classes of Feller processes: isotropic Lévy processes such as Brownian motion, and generalized Poisson processes. Supplementary Exercises 1–5 contain further computations. These examples, together with the material of Sections 3 and

4 below, present a glimpse at some of the many deep connections between Markov processes and integro-differential equations.

1.1 Existence

Suppose $\mathcal{T}$ is a Feller semigroup on a locally compact, separable metric space S. Let $\mathcal{R}$ denote the resolvent of $\mathcal{T}$. According to Theorems 3.3.1, 4.1.1, and 4.2.2 of Chapter 8, there exists an associated Feller process X with shift operators θ that is right-continuous and whose transition functions and resolvents are $\mathcal{T}$ and $\mathcal{R}$, respectively. We begin with two technical lemmas.

Lemma 1.1.1 *For all $\lambda > 0$, $\mathcal{R}_\lambda : C_0(S) \to \mathcal{R}_\lambda(C_0(S))$ is a bijection. Thus, its inverse $\mathcal{R}_\lambda^{-1} : \mathcal{R}_\lambda(C_0(S)) \to C_0(S)$ is a linear operator.*

It is important to point out that $\mathcal{R}_\lambda^{-1}$ need *not* be a bounded operator.

Proof If $f \in C_0(S)$ is such that for some $\lambda > 0$, $\mathcal{R}_\lambda f \equiv 0$, we can apply Corollary 2.2.1 of Chapter 8 to the resolvent equation to see that for *all* $\gamma > 0$, $\mathcal{R}_\gamma f \equiv 0$. Applying Lemma 2.4.1 of Chapter 8, we see that $f = \lim_{\gamma \to +\infty} \gamma \mathcal{R}_\gamma f = 0$. In particular, if $f, g \in C_0(S)$ satisfy $\mathcal{R}_\lambda f \equiv \mathcal{R}_\lambda g$, for some $\lambda > 0$, then $f(x) = g(x)$, for all $x \in S$. In other words, $\mathcal{R}_\lambda$ maps $C_0(S)$ bijectively onto its range. Thus, the inverse $\mathcal{R}_\lambda^{-1} : \mathcal{R}_\lambda(C_0(S)) \to C_0(S)$ exists, and for all $f, g \in C_0(S)$ and all $\alpha, \beta \in \mathbb{R}$,

$$\mathcal{R}_\lambda^{-1} \mathcal{R}_\lambda (\alpha f + \beta g) = \alpha \mathcal{R}_\lambda^{-1} \mathcal{R}_\lambda f + \beta \mathcal{R}_\lambda^{-1} \mathcal{R}_\lambda g,$$

as can be seen by applying $\mathcal{R}_\lambda$ to both sides (why is this enough?). This concludes our proof. □

Our second technical lemma is the following consequence of the resolvent equation (Theorem 2.2.1 of Chapter 8).

Lemma 1.1.2 *For any $\lambda, \gamma > 0$, $\mathcal{R}_\lambda(C_0(S)) = \mathcal{R}_\gamma(C_0(S))$.*

Exercise **1.1.1** Prove Lemma 1.1.2. □

Given distinct $\lambda, \gamma > 0$ and $f \in \mathcal{R}_\lambda(C_0(S))$, define $g = \mathcal{R}_\lambda^{-1} \mathcal{R}_\gamma^{-1} f$. By Lemma 1.1.2, this is sensible, and $g \in C_0(S)$. Thus, we can use the resolvent equation (Theorem 2.2.1 of Chapter 8) to conclude that

$$\mathcal{R}_\gamma^{-1} f - \mathcal{R}_\lambda^{-1} f = (\gamma - \lambda) f. \tag{1}$$

(Why?) Equation (1) can be rewritten as $\lambda - \mathcal{R}_\lambda^{-1} = \gamma - \mathcal{R}_\gamma^{-1}$. Thus, we can define a linear operator $\mathcal{A} : \mathcal{R}_\lambda(C_0(S)) \to C_0(S)$ by setting $\mathcal{A} = \lambda - \mathcal{R}_\lambda^{-1}$ for *any* $\lambda > 0$. To be more precise,

$$\mathcal{A} f(x) = \lambda f(x) - \mathcal{R}_\lambda^{-1} f(x), \qquad x \in S, \ f \in \mathcal{D}(\mathcal{A}), \tag{2}$$

where $\mathcal{D}(\mathcal{A}) = \mathcal{R}_\lambda(C_0(S))$ is independent of the choice of λ. The operator $\mathcal{A}$ is interchangeably called the **generator** of $\mathfrak{T}$, $\mathcal{R}$, and the process X. The space $\mathcal{D}(\mathcal{A})$ is the **domain** of the operator $\mathcal{A}$. In other words, $\mathcal{A} : \mathcal{D}(\mathcal{A}) \to C_0(S)$. We re-iterate that neither $\mathcal{A}$ nor its domain depend on $\lambda > 0$.

1.2 Identifying the Domain: The Hille–Yosida Theorem

It is usually very difficult to find the domain $\mathcal{D}(\mathcal{A})$ of the generator $\mathcal{A}$ of a Markov process. Typically, one aims to find a sufficiently large collection $C \subset \mathcal{D}(\mathcal{A})$. The following characterization of $\mathcal{D}(\mathcal{A})$ is sometimes useful; it was found, independently and at around the same time, in (Hille 1958; Yosida 1958). Two nice pedagogical treatments, plus a thorough combined bibliography, can be found in (Hille and Phillips 1957; Yosida 1995).

Theorem 1.2.1 (The Hille–Yosida Theorem) *If $\mathcal{A}$ denotes the generator of a Feller semigroup $\mathfrak{T}$ on a locally compact, separable metric space S, then*

$$\mathcal{D}(\mathcal{A}) = \left\{ \varphi \in C_0(S) : \lim_{t \to 0+} \frac{\mathfrak{T}_t \varphi - \varphi}{t} \text{ exists in } C_0(S) \right\}.$$

Moreover, if $\varphi \in \mathcal{D}(\mathcal{A})$, then

$$\mathcal{A}\varphi = \lim_{t \to 0+} \frac{\mathfrak{T}_t \varphi - \varphi}{t}, \tag{1}$$

where the limit takes place in $C_0(S)$.

Of course, this last statement means that for all $\varphi \in \mathcal{D}(\mathcal{A})$,

$$\lim_{t \to 0+} \left\| \mathcal{A}\varphi - \frac{1}{t}\{\mathfrak{T}_t \varphi - \varphi\} \right\|_\infty = 0.$$

In particular, $\mathcal{A}\varphi(x)$ is the derivative, from the right, of the function $t \mapsto \mathfrak{T}_t\varphi(x)$ at $t = 0$. Informally, $\mathfrak{T}_{t+s} = \mathfrak{T}_s \mathfrak{T}_t$ is an exponential-type property; one may think of $\mathfrak{T}_t$ as $e^{t\mathcal{A}}$. However, this intuitive picture is not necessary for the discussion of this chapter.

The usual treatments of this theorem contain a second half that can be found in Exercise 1.2.1 below.

Proof Throughout this argument we will use the following fact several times: Whenever $\varphi \in \mathcal{D}(\mathcal{A})$, there exist $\lambda > 0$ and $f \in C_0(S)$ such that $\varphi = \mathcal{R}_\lambda f$. In fact, for *any* $\lambda > 0$, such an f can be found.

A direct computation reveals that for all $f \in L^\infty(S)$, all $\lambda > 0$, and all $t \geq 0$,

$$\mathfrak{T}_t \mathcal{R}_\lambda f = e^{\lambda t} \int_t^\infty e^{-\lambda s} \mathfrak{T}_s f \, ds. \tag{2}$$

See the given proof of the resolvent equation (Theorem 2.2.1, Chapter 8). In particular,

$$\mathcal{T}_t \mathcal{R}_\lambda f - \mathcal{R}_\lambda f = e^{-\lambda t} \mathcal{T}_t \mathcal{R}_\lambda f - \mathcal{R}_\lambda f + (1 - e^{-\lambda t}) \mathcal{T}_t \mathcal{R}_\lambda f$$

$$= -\int_0^t e^{-\lambda s} \mathcal{T}_s f \, ds + (1 - e^{-\lambda t}) \mathcal{T}_t \mathcal{R}_\lambda f,$$

by equation (2). Thus,

$$\frac{\mathcal{T}_t(\mathcal{R}_\lambda f) - \mathcal{R}_\lambda f}{t} = -\frac{1}{t} \int_0^t e^{-\lambda s} \mathcal{T}_s f \, ds + \frac{1 - e^{-\lambda t}}{t} \mathcal{T}_t(\mathcal{R}_\lambda f).$$

By the Feller property,

$$\lim_{t \to 0+} \frac{\mathcal{T}_t(\mathcal{R}_\lambda f) - \mathcal{R}_\lambda f}{t} = -f + \lambda \mathcal{R}_\lambda f = \mathcal{A}(\mathcal{R}_\lambda f),$$

where the convergence takes place in $C_0(S)$. Now suppose $\varphi \in \mathcal{D}(\mathcal{A})$. Then, $\varphi = \mathcal{R}_\lambda f$ for some $f \in C_0(S)$ and $\lambda > 0$. This proves equation (1) and that

$$\mathcal{D}(\mathcal{A}) \subset \mathcal{D}^\star = \left\{ \varphi \in C_0(S) : \lim_{t \to 0+} \frac{\mathcal{T}_t \varphi - \varphi}{t} \text{ exists in } C_0(S) \right\}. \quad (3)$$

Conversely, suppose $\varphi \in \mathcal{D}^\star$ and define $\mathcal{A}^\star \varphi = \lim_{t \to 0+} \{\mathcal{T}_t \varphi - \varphi\}/t$, where the limit takes place in $C_0(S)$. It follows that $\mathcal{A}^\star$ is a linear operator with domain $\mathcal{D}(\mathcal{A}^\star) = \mathcal{D}^\star$. Moreover, by equation (1), for any $\lambda > 0$,

$$\mathcal{R}_\lambda(\mathcal{A}^\star \varphi) = \lim_{t \to 0+} \frac{\mathcal{T}_t(\mathcal{R}_\lambda \varphi) - \mathcal{R}_\lambda \varphi}{t} = \mathcal{A}(\mathcal{R}_\lambda \varphi),$$

where the limit takes place in $C_0(S)$. We have used the continuity of $f \mapsto \mathcal{R}_\lambda f$; cf. Corollary 2.1.1 and Lemma 2.2.2 of Chapter 8. Recalling the definition of $\mathcal{A}$, i.e., equation (2) of Section 1.1, we can rewrite the above as follows: For all $\varphi \in \mathcal{D}^\star$,

$$\mathcal{A}^\star(\mathcal{R}_\lambda \varphi) = \mathcal{R}_\lambda(\mathcal{A}^\star \varphi) = \lambda \mathcal{R}_\lambda \varphi - \varphi.$$

Since $\psi = \lambda \varphi - \mathcal{A}^\star \varphi \in C_0(S)$, $\varphi = \mathcal{R}_\lambda \psi \in \mathcal{D}(\mathcal{A})$. We have shown that $\mathcal{D}^\star \subset \mathcal{D}(\mathcal{A})$ and completed the proof. $\square$

Theorem 1.2.1 shows that if we know the semigroup, we can, in principle, find the generator. The converse also holds. In particular, the single operator $\mathcal{A}$ completely determines the entire semigroup $\mathcal{T}$. A more precise statement follows.

Theorem 1.2.2 *Suppose $\mathcal{T}$ and $\mathcal{T}^\star$ are two Feller semigroups on S. Denote their generators by $\mathcal{A}$ and $\mathcal{A}^\star$ and their resolvents by $\mathcal{R}$ and $\mathcal{R}^\star$, respectively. If $\mathcal{D}(\mathcal{A}) = \mathcal{D}(\mathcal{A}^\star) = \mathcal{D}$ and if $\mathcal{A}f = \mathcal{A}^\star f$ on $\mathcal{D}$, then $\mathcal{T}$ and $\mathcal{T}^\star$ are one and the same.*

Proof For all $f \in \mathcal{D}$, $\mathcal{A}f - \mathcal{A}^\star f = 0$. Recalling (2) of Section 1.1 and applying $\mathcal{R}_\lambda$ to the latter equation, we see that for all $\lambda > 0$,

$$0 = \mathcal{R}_\lambda(\mathcal{A}f - \mathcal{A}^\star f) = \mathcal{R}_\lambda(\mathcal{R}_\lambda^\star)^{-1} f - f.$$

By Lemma 1.1.2, for all $\varphi \in C_0(S)$ and all $\lambda > 0$, there exists $f \in \mathcal{D}$ such that $\mathcal{R}_\lambda^\star \varphi = f$. Thus, we have shown that for all $\varphi \in C_0(S)$ and all $\lambda > 0$, $\mathcal{R}_\lambda \varphi = \mathcal{R}_\lambda^\star \varphi$. The present theorem now follows from the uniqueness of Laplace transforms; cf. Theorem 1.1.1 of Appendix B. □

In fact, the cycle $\mathcal{A} \to \mathcal{T} \to \mathcal{R}$ can be completed by showing that the resolvent $\mathcal{R}$ completely determines the generator $\mathcal{A}$.

Exercise 1.2.1 (Theorem 1.2.1. Continued) Suppose $\mathcal{A}$ denotes the generator of a Feller semigroup $\mathcal{T}$ on a locally compact, separable metric space S. Prove that

$$\mathcal{D}(\mathcal{A}) = \left\{ \varphi \in C_0(S) : \lim_{\lambda \to \infty} \lambda[\lambda \mathcal{R}_\lambda \varphi - \varphi] \text{ exists in } C_0(S) \right\}$$

and $\mathcal{A}\varphi = \lim_{\lambda \to \infty} \lambda\{\lambda \mathcal{R}_\lambda \varphi - \varphi\}$, where the limit takes place in $C_0(S)$. □

1.3 The Martingale Problem

There are deep connections between generators and martingales. One such connection is the following.

Theorem 1.3.1 (The Martingale Problem: Sufficiency) *Suppose $\mathcal{A}$ is the generator of a right-continuous Feller process X on a locally compact, separable metric space S. For any $f \in \mathcal{D}(\mathcal{A})$, define the adapted process $M^f = (M_t^f; t \geq 0)$ by*

$$M_t^f = f(X_t) - f(X_0) - \int_0^t \mathcal{A}f(X_s)\,ds.$$

If $\mathcal{F} = (\mathcal{F}_t; t \geq 0)$ denotes the complete augmented history of X, then M^f is a mean-zero martingale with respect to the filtration $\mathcal{F}$, under any of the measures $\mathbb{P}_x$, $x \in S$.

We say that $C \subset \mathcal{D}(\mathcal{A})$ is an **essential core** for $\mathcal{A}$ if $\mathcal{A}(C) \subset C_0(S)$. The previous theorem is essentially sharp as long as the domain of the generator has a sufficiently large essential core, as the following reveals.

Theorem 1.3.2 (The Martingale Problem: Necessity) *Suppose $\mathcal{A}$ is the generator of a Feller process X on a locally compact, separable metric space S. Let C denote an essential core of $\mathcal{A}$. Suppose further that* there exists a linear operator $\mathcal{B}: C \to C_0(S)$ such that for all $f \in C$, $N^f = (N_t^f;\ t \geq 0)$ is an $\mathcal{F}$-martingale $\mathbb{P}_x$-a.s., for all $x \in S$, where

$$N_t^f = f(X_t) - f(X_0) - \int_0^t \mathcal{B}f(X_s)\, ds, \qquad t \geq 0.$$

Then, $\mathcal{B} = \mathcal{A}$ on C.

Roughly speaking, the two parts of the martingale problem together assert that if the generator $\mathcal{A}$ of a Feller process X has a sufficiently large essential core, then $\mathcal{A}$ is determined by knowing all of the martingales adapted to the filtration of X.

The main tool for verifying the sufficiency of the martingale problem is the very important Doob–Meyer decomposition for potentials in continuous time. In the setting of discrete Markov chains, its analogue was shown to be true in Corollary 1.5.1 of Chapter 8.

Theorem 1.3.3 (Doob–Meyer Decomposition of Potentials) *For any $x \in S$, all $\lambda > 0$, and any $\varphi \in C_0(S)$, $\mathbb{P}_x$-a.s.,*

$$e^{-\lambda t}\mathcal{R}_\lambda \varphi(X_t) + \int_0^t e^{-\lambda s}\varphi(X_s)\, ds = \mathbb{E}_x\left[\int_0^\infty e^{-\lambda s}\varphi(X_s)\, ds \,\Big|\, \mathcal{F}_t\right].$$

Proof By Theorem 4.1.1 of Chapter 8, on an appropriate probability space we can construct a Feller process $\overline{X}$ with the same finite-dimensional distributions as X (under $\mathbb{P}_x$, for any $x \in S$) such that $\overline{X}$ also has shift operators. Since the theorem can be reduced to statements about the finite-dimensional distributions of X, it suffices to prove the theorem for $\overline{X}$ replacing X. That is, we can assume, without loss of generality, that X possesses shift operators θ.

For any $x \in S$, $t \geq 0$, and $\lambda > 0$, the following holds $\mathbb{P}_x$-a.s.:

$$\mathcal{R}_\lambda \varphi(X_t) = \int_0^\infty e^{-\lambda r}\mathcal{T}_r\varphi(X_t)\, dr = \int_0^\infty e^{-\lambda r}\mathbb{E}_x[\varphi(X_{r+t})\,|\,\mathcal{F}_t]\, dr$$
$$= e^{\lambda t}\int_t^\infty e^{-\lambda s}\mathbb{E}_x[\varphi(X_s)\,|\,\mathcal{F}_t]\, ds.$$

In fact, we have already seen this calculation in the course of the proof of Lemma 4.1.1 of Chapter 8. What is new, however, is the right continuity of $t \mapsto X_t(\omega)$ for every ω. In particular, $(\omega, t) \mapsto X_t(\omega)$ is measurable. By Fubini's theorem, $\mathbb{P}_x$-a.s. for any $x \in S$,

$$e^{-\lambda t}\mathcal{R}_\lambda \varphi(X_t) = \mathbb{E}_x\left[\int_t^\infty e^{-\lambda s}\varphi(X_s)\, ds \,\Big|\, \mathcal{F}_t\right],$$

from which the theorem ensues. $\square$

In particular, if φ is also assumed to be nonnegative, then $t \mapsto e^{-\lambda t}\mathcal{R}_\lambda \varphi(X_t)$ is a supermartingale. Thus, the Doob–Meyer decomposition of potentials can be viewed as an improvement of Lemma 4.1.1 of Chapter 8. We can now demonstrate Theorem 1.3.1.

Proof of Theorem 1.3.1 Suppose $f \in \mathcal{D}(\mathcal{A})$. By its very definition, this implies that for all $\lambda > 0$, there exists a function $\varphi \in C_0(S)$ such that $f = \mathcal{R}_\lambda \varphi$. Equivalently, $\mathcal{A}f = \lambda \mathcal{R}_\lambda \varphi - \varphi$. Applying the Doob–Meyer decomposition (Theorem 1.3.3), we obtain the following for all $t \geq 0$: For all $x \in S$, $\mathbb{P}_x$-a.s.,

$$e^{-\lambda t}f(X_t) - \int_0^t \mathcal{A}f(X_s)\,ds + \lambda \int_0^t e^{-\lambda s}f(X_s)\,ds = \mathbb{E}_x\left[\int_0^\infty e^{-\lambda s}\varphi(X_s)\,ds \,\Big|\, \mathcal{F}_t\right].$$

In other words, $Z^\lambda = (Z_t^\lambda; t \geq 0)$ is a martingale with respect to $\mathcal{F}$, where

$$Z_t^\lambda = e^{-\lambda t}f(X_t) - \int_0^t \mathcal{A}f(X_s)\,ds + \lambda \int_0^t e^{-\lambda s}f(X_s)\,ds.$$

To address integrability issues, we merely note that thanks to Lemma 2.2.2 of Chapter 8,

$$\|\mathcal{A}f\|_\infty \leq \lambda\|\mathcal{R}_\lambda \varphi\|_\infty + \|\varphi\|_\infty \leq 2\|\varphi\|_\infty.$$

Moreover, by the dominated convergence theorem,

$$\lim_{\lambda \downarrow 0} Z_t^\lambda = f(X_t) - \int_0^t \mathcal{A}f(X_s)\,ds,$$

$\mathbb{P}_x$-a.s. and in $L^1(\mathbb{P}_x)$ for all $x \in S$. By Supplementary Exercise 1 of Chapter 7, $t \mapsto f(X_t) - \int_0^t \mathcal{A}f(X_s)\,ds$ is a martingale, since it is the $L^1(\mathbb{P}_x)$ limit of martingales. The result follows, since $M_0^f = 0$. □

Proof of Theorem 1.3.2 Let $f \in C$ be a fixed function and define $D = (D_t; t \geq 0)$ by

$$D_t = \int_0^t \mathcal{A}f(X_s)\,ds - \int_0^t \mathcal{B}f(X_s)\,ds, \qquad t \geq 0.$$

By our assumptions, D is an $L^2(\mathbb{P}_x)$ continuous martingale of bounded variation, for any $x \in S$. Since $D_0 = 0$, Theorem 3.1.1 of Chapter 7 implies that for any $x \in S$, $\mathbb{P}_x(D_t = 0$ for all $t \geq 0) = 1$. On the other hand, since $t \mapsto X_t$ is right-continuous and $\mathcal{A}f$ and $\mathcal{B}f$ are bounded and continuous, $\mathbb{P}_x$-a.s. for all $x \in S$,

$$\mathcal{A}f(x) = \mathcal{A}f(X_0) = \lim_{t \to 0+} \frac{1}{t}\int_0^t \mathcal{A}f(X_s)\,ds$$

$$= \lim_{t \to 0+} \frac{1}{t}\int_0^t \mathcal{B}f(X_s)\,ds = \mathcal{B}f(X_0) = \mathcal{B}f(x).$$

This completes our derivation. □

2 Explicit Computations

We now study the form of the generator of the Lévy processes that appeared in Examples 1 through 4 of Section 4.3, Chapter 8. The processes of interest are Brownian motion in $\mathbb{R}^d$, isotropic stable processes in $\mathbb{R}^d$, the Poisson process on $\mathbb{R}$, and uniform motion in $\mathbb{R}$.

2.1 Brownian Motion

We have encountered $\mathbb{R}^d$-valued Brownian motion $B = (B_t;\ t \geq 0)$ in Example 1 of Section 4.3, Chapter 8. Recall that it is a continuous Feller process on $\mathbb{R}^d$ whose transition densities (with respect to Lebesgue's measure on $\mathbb{R}^d$) are given by $p_t(x, y) = q_t(y - x) = q(t; y - x)$, where

$$q_t(a) = (2\pi t)^{-\frac{d}{2}} \exp\left\{-\frac{\|a\|^2}{2t}\right\}, \qquad t \geq 0,\ q \in \mathbb{R}^d.$$

As usual, $\|a\|^2 = \sum_{j=1}^d |a^{(j)}|^2$ denotes the square of the ℓ^2-norm of $a \in \mathbb{R}^d$. The transition functions of B are determined as follows: For all $f \in L^\infty(\mathbb{R}^d)$,

$$\mathcal{T}_t f(x) = q_t \star f(x) = \int_{\mathbb{R}^d} q_t(y - x) f(y)\, dy.$$

It is easy to see that q satisfies the classical **heat equation**

$$\dot{q}(t; x) = \frac{1}{2} \Delta q(t; x), \qquad t \geq 0,\ x \in \mathbb{R}^d, \tag{1}$$

where $\dot{q}$ denotes differentiation in the time variable t, and Δ is the usual **Laplace** operator[1] (**Laplacian**, for short), acting on the spatial variable x. That is, for all twice differentiable functions $g : \mathbb{R}^d \to \mathbb{R}$,

$$\Delta g(x) = \sum_{j=1}^d \frac{\partial^2 g}{\left(\partial x^{(i)}\right)^2}(x), \qquad x \in \mathbb{R}^d.$$

In particular, if $f \in L^\infty(\mathbb{R}^d)$ and $\tau_f(t; x) = \mathcal{T}_t f(x)$, then we obtain the following from the dominated convergence theorem:

$$\dot{\tau}_f(t; x) = \frac{1}{2} \Delta \tau_f(t; x), \qquad t \geq 0,\ x \in \mathbb{R}^d. \tag{2}$$

---

[1] In the mathematical physics literature Δ is often written as $\nabla \cdot \nabla$ or ∇^2; in any case, Δ is the trace of the $(d \times d)$ **Hessian** matrix $\mathrm{H} = (D_i D_j)$.

Now let $C_c^2(\mathbb{R}^d)$ denote the collection of all twice continuously differentiable functions from $\mathbb{R}^d$ into $\mathbb{R}$ that vanish outside a compact set. Suppose now that $f \in C_c^2(\mathbb{R}^d)$. Clearly, $\Delta f \in C_0(\mathbb{R}^d)$ and $\tau_f(t;x) = \int_{\mathbb{R}^d} f(x-y) q_t(y)\, dy$. Consequently, equation (2) can be written as follows:

$$\frac{\partial}{\partial t} \mathcal{T}_f(x) = \mathcal{T}_f\!\left(\tfrac{1}{2}\Delta f\right)(x), \qquad t \geq 0,\ x \in \mathbb{R}^d.$$

We can use the identity $\mathcal{T}_0 f = f$, and write the above in integrated form:

$$\frac{\mathcal{T}_t f(x) - f(x)}{t} = \frac{1}{t} \int_0^t \mathcal{T}_s\!\left(\tfrac{1}{2}\Delta f\right)(x)\, ds, \qquad t \geq 0,\ x \in \mathbb{R}^d. \tag{3}$$

Since B is a Feller process, we can use the fact that $\Delta f \in C_0(\mathbb{R}^d)$ once more to observe that $\lim_{s \to 0+} \mathcal{T}_s(\tfrac{1}{2}\Delta f) = \tfrac{1}{2}\Delta f$, in $C_0(\mathbb{R}^d)$. That is, for all $\varepsilon > 0$, there exists $s_0 > 0$ such that whenever $s \in \,]0, s_0[$,

$$\left\| \mathcal{T}_s\!\left(\tfrac{1}{2}\Delta f\right) - \tfrac{1}{2}\Delta f \right\|_\infty \leq \varepsilon.$$

Theorem 1.2.1 together with equation (3) yields the following.

Proposition 2.1.1 *If $\mathcal{A}$ denotes the generator of Brownian motion on $\mathbb{R}^d$, then $C_c^2(\mathbb{R}^d)$ is an essential core for $\mathcal{D}(\mathcal{A})$. Moreover, on $C_c^2(\mathbb{R}^d)$, $\mathcal{A} = \tfrac{1}{2}\Delta$.*

For all intents and purposes this is an adequate result on the generator of Brownian motion. Indeed, by Proposition 2.1.1,

$$C_c^2(\mathbb{R}^d) \subset \mathcal{D}(\mathcal{A}) \subset C_0(\mathbb{R}^d),$$

while it is a standard exercise in analysis to show that $C_c^2(\mathbb{R}^d)$ is dense in $C_0(\mathbb{R}^d)$. However, using ideas from the theory of distributions, one can do a little more in this special setting.

If $F, G : \mathbb{R}^d \to \mathbb{R}$ are two measurable functions such that $FG \in L^1(\mathbb{R}^d)$, define $\langle F, G \rangle = \int_{\mathbb{R}^d} F(x) G(x)\, dx$. We have already seen $\langle \bullet, \bullet \rangle$ as an inner product on $L^2(\mathbb{R}^d)$. However, $\langle F, G \rangle$ makes sense, for example, if $F \in C_0(\mathbb{R}^d)$ and if G vanishes outside a compact set.

Recall equation (2): For all $f \in L^\infty(\mathbb{R}^d)$,

$$\frac{\partial}{\partial t} \mathcal{T}_f(x) = \tfrac{1}{2} \Delta \mathcal{T}_t f(x), \qquad t \geq 0,\ x \in \mathbb{R}^d. \tag{4}$$

Let $C_c^\infty(\mathbb{R}^d)$ denote the collection of all infinitely differentiable functions from $\mathbb{R}^d$ into $\mathbb{R}$ that vanish outside a compact set. We can multiply both sides of (4) by $\psi(x)$ and integrate (dx) to obtain the following for all $t \geq 0$:

$$\left\langle \psi, \frac{\partial}{\partial t} \mathcal{T}_t f \right\rangle = \tfrac{1}{2} \langle \psi, \Delta \mathcal{T}_t f \rangle. \tag{5}$$

Thanks to the dominated convergence theorem, the left-hand side is equal to $(\partial/\partial t)\langle \psi, \mathcal{T}_t f\rangle$. On the other hand, we can integrate the right-hand side of equation (5) by parts, and use the fact that ψ is zero outside a compact set, to see that $\langle \psi, \Delta \mathcal{T}_t f\rangle = \langle \Delta \psi, \mathcal{T}_t f\rangle$. In other words, we have shown that for all $t \geq 0$, all $f \in L^\infty(\mathbb{R}^d)$, and all $\psi \in C_c^\infty(\mathbb{R}^d)$,

$$\frac{\partial}{\partial t}\langle \psi, \mathcal{T}_t f\rangle = \left\langle \frac{1}{2}\Delta\psi, \mathcal{T}_t f\right\rangle.$$

Since $\mathcal{T}$ is Feller, we can integrate the above (in t) from 0 to r and divide by r to see that for all $r > 0$, all $\psi \in C_c^\infty(\mathbb{R}^d)$, and all $f \in C_0(\mathbb{R}^d)$,

$$\left\langle \psi, \frac{\mathcal{T}_r f - f}{r}\right\rangle = \left\langle \frac{1}{2}\Delta\psi, \frac{1}{r}\int_0^r \mathcal{T}_t f\, dt\right\rangle. \tag{6}$$

Now let us restrict $f \in \mathcal{D}(\mathcal{A})$. By the dominated convergence theorem, as $r \to 0+$, the left-hand side converges to $\langle \psi, \mathcal{A}f\rangle$, while the right-hand side converges to $\langle \frac{1}{2}\Delta\psi, f\rangle$. That is,

$$\langle \psi, \mathcal{A}f\rangle = \left\langle \frac{1}{2}\Delta\psi, f\right\rangle, \qquad \psi \in C_c^\infty(\mathbb{R}^d),\ f \in \mathcal{D}(\mathcal{A}).$$

To summarize our efforts thus far, we have established the following:

$$\mathcal{D}(\mathcal{A}) \subset \left\{ f \in C_0(\mathbb{R}^d) : \frac{1}{2}\Delta f \text{ exists in the sense of distributions}\right\},$$

and on its domain, $\mathcal{A} = \frac{1}{2}\Delta$, in the sense of distributions.

On the other hand, Proposition 2.1.1 shows that for all $f \in C_0(\mathbb{R}^d)$ and all $\psi \in C_c^\infty(\mathbb{R}^d)$, $\langle \mathcal{A}\psi, f\rangle = \langle \frac{1}{2}\Delta\psi, f\rangle$. Consequently, we have demonstrated the following theorem.

Theorem 2.1.1 *The domain of the generator of Brownian motion on $\mathbb{R}^d$ is precisely the class of all $f \in C_0(\mathbb{R}^d)$ for which Δf exists, in the sense of distributions. Moreover, for all $f \in \mathcal{D}(\mathcal{A})$, $\mathcal{A}f = \frac{1}{2}\Delta f$, in the sense of distributions.*

2.2 Isotropic Stable Processes

We now compute the form of the generator of the isotropic stable processes of Section 4.3, Chapter 8. Recall that the $\mathbb{R}^d$-valued $X = (X_t;\ t \geq 0)$ is an isotropic stable process with index $\alpha \in\,]0, 2]$ if it is a Lévy process on $\mathbb{R}^d$ with

$$\mathbb{E}[e^{i\xi \cdot X_t}] = \exp(-\tfrac{1}{2}\|\xi\|^\alpha), \qquad \xi \in \mathbb{R}^d,$$

where $\|\bullet\|$ denotes the ℓ^2 Euclidean norm, as usual. Recall, further, that when $\alpha = 2$, this is none other than d-dimensional Brownian motion.

2 Explicit Computations

We begin by recalling some elementary facts from harmonic analysis. Let $\mathbb{C}$ denote the complex plane. For any integrable $f: \mathbb{R}^d \to \mathbb{C}$, $\widehat{f}$ denotes its Fourier transform in the following sense:

$$\widehat{f}(\xi) = \int_{\mathbb{R}^d} e^{i\xi \cdot x} f(x)\, dx, \qquad \xi \in \mathbb{R}^d.$$

We say that $f \in L^1(\mathbb{R}^d)$, if $\mathrm{Re} f$ and $\mathrm{Im} f$ are both in $L^1(\mathbb{R}^d)$. The same remark applies to all other (real) function spaces that we have seen thus far; this includes $L^p(\mathbb{R}^d)$, $C_c^\infty(\mathbb{R}^d)$, etc. In particular, by Fourier inversion, if $\widehat{f} \in L^1(\mathbb{R}^d)$,

$$f(x) = (2\pi)^{-d} \int_{\mathbb{R}^d} e^{-ix \cdot \xi} \widehat{f}(\xi)\, d\xi, \qquad x \in \mathbb{R}^d.$$

Finally, let us recall Parseval's identity: Whenever $f, g \in L^1(\mathbb{R}^d)$ and $f\bar{g} \in L^1(\mathbb{R}^d)$, then $\widehat{f\bar{g}} \in L^1(\mathbb{R}^d)$, and moreover,

$$\int_{\mathbb{R}^d} f(x) \overline{g(x)}\, dx = (2\pi)^{-d} \int_{\mathbb{R}^d} \widehat{f}(\xi) \overline{\widehat{g}(\xi)}\, d\xi.$$

We now begin to compute the form of the generator of X. Let us define

$$q_t(a) = (2\pi)^{-d} \int_{\mathbb{R}^d} e^{-i\xi \cdot a} e^{-\frac{1}{2} t \|\xi\|^\alpha}\, d\xi, \qquad t \geq 0,\ a \in \mathbb{R}^d.$$

The transition densities for X can then be written as $p_t(x, y) = q_t(y - x)$; cf. Example 3, Section 4.3 of Chapter 8. Moreover, the following symmetry relation holds:

$$q_t(a) = q_t(-a), \qquad t \geq 0,\ a \in \mathbb{R}^d.$$

Thus, for all $f \in L^\infty(\mathbb{R}^d)$,

$$\mathcal{T}_t f(x) = \int_{\mathbb{R}^d} q_t(y - x) f(y)\, dy = \int_{\mathbb{R}^d} q_t(x - y) f(y)\, dy$$

$$= (2\pi)^{-d} \int_{\mathbb{R}^d} \int_{\mathbb{R}^d} e^{-i\xi \cdot (x - y)} f(y) e^{-\frac{1}{2} t \|\xi\|^\alpha}\, dy\, d\xi$$

$$= (2\pi)^{-d} \int_{\mathbb{R}^d} \widehat{f}(y) e^{-i\xi \cdot x} e^{-\frac{1}{2} t \|\xi\|^\alpha}\, d\xi,$$

by Fubini's theorem; also consult Example 3, Section 4.3 of Chapter 8. If $f \in L^\infty(\mathbb{R}^d)$ and $\widehat{f} \in L^1(\mathbb{R}^d)$, then by Fourier's inversion theorem,

$$f(x) = (2\pi)^{-d} \int_{\mathbb{R}^d} \widehat{f}(\xi) e^{-i\xi \cdot x}\, d\xi.$$

Combining the last two displays, we conclude that whenever $f \in L^\infty(\mathbb{R}^d)$ and $\widehat{f} \in L^1(\mathbb{R}^d)$,

$$\frac{\mathcal{T}_t f(x) - f(x)}{t} = (2\pi)^{-d} \int_{\mathbb{R}^d} \widehat{f}(\xi) e^{-i\xi \cdot x} \frac{e^{-\frac{1}{2}t\|\xi\|^\alpha} - 1}{2} \, d\xi. \tag{1}$$

Using Taylor's expansion, it is easy to see that for all $\theta \geq 0$, all $\varepsilon_0 > 0$, and all $t \in [0, \varepsilon_0]$,

$$0 \leq \frac{1 - e^{-t\theta}}{\theta} \leq \varepsilon_0 e^{\theta \varepsilon_0}. \tag{2}$$

Define

$$\mathfrak{S}_\alpha = \left\{ f \in L^\infty(\mathbb{R}^d) \,\Big|\, \exists \varepsilon_0 > 0 : \int_{\mathbb{R}^d} |\widehat{f}(\xi)| \cdot \|\xi\|^\alpha \, e^{\frac{1}{2}\varepsilon_0 \|\xi\|^\alpha} \, d\xi < \infty \right\}, \tag{3}$$

where the existence of $\widehat{f}$ implicitly implies that $f \in L^1(\mathbb{R}^d)$ as well. Combining (1) and (2), we can conclude from the dominated convergence theorem that whenever $f \in \mathfrak{S}_\alpha$, then as $t \to 0^+$,

$$\frac{\mathcal{T}_t f(x) - f(x)}{t} \to -\frac{1}{2}(2\pi)^{-d} \int_{\mathbb{R}^d} \widehat{f}(\xi) \cdot \|\xi\|^\alpha \, e^{-i\xi \cdot x} \, d\xi,$$

uniformly in $x \in \mathbb{R}^d$. That is, we have shown that $\mathfrak{S}_\alpha \subset \mathcal{D}(\mathcal{A})$ and

$$\mathcal{A}f(x) = -\frac{1}{2}(2\pi)^{-d} \int_{\mathbb{R}^d} \widehat{f}(\xi) \cdot \|\xi\|^\alpha \, e^{-i\xi \cdot x} \, d\xi, \qquad x \in \mathbb{R}^d, \; f \in \mathfrak{S}_\alpha. \tag{4}$$

Using the dominated convergence theorem once more, it is clear from (1) and (2) that $\mathfrak{S}_\alpha$ is an essential core for $\mathcal{D}(\mathcal{A})$. Thus, we have obtained the following.

Proposition 2.2.1 *Let $\mathcal{A}$ denote the generator of an $\mathbb{R}^d$-valued isotropic stable Lévy process of index $\alpha \in {]}0, 2]$. Then, $\mathfrak{S}_\alpha$ is an essential core for the domain of $\mathcal{A}$, and on $\mathfrak{S}_\alpha$, $\mathcal{A}$ is given by (4).*

In fact, the above description captures the essence of the form of the generator, as the following shows.

Proposition 2.2.2 *Let C denote any essential core of $\mathcal{D}(\mathcal{A})$. Then, for all $f \in C$ such that $\mathcal{A}f \in L^1(\mathbb{R}^d)$, $\mathcal{A}f$ is given by (4).*

Proof Suppose f is in the domain of $\mathcal{A}$. For all $\varphi \in C_c^\infty(\mathbb{R}^d)$,

$$\left\langle \varphi, \frac{\mathcal{T}_t f - f}{t} \right\rangle = (2\pi)^{-d} \int_{\mathbb{R}^d} \widehat{f}(\xi) \, \overline{\widehat{\varphi}(\xi)} \, \frac{e^{-\frac{1}{2}t\|\xi\|^\alpha} - 1}{t} \, d\xi,$$

where $\overline{z}$ denotes the complex conjugate of $z \in \mathbb{C}$. By the dominated convergence theorem, as $t \to 0^+$, the left-hand side converges to $\langle \varphi, \mathcal{A}f \rangle$. Thus, if C denotes an essential core of $\mathcal{A}$,

$$\langle \varphi, \mathcal{A}f \rangle = -\frac{1}{2}(2\pi)^{-d} \int_{\mathbb{R}^d} \widehat{f}(\xi) \, \overline{\widehat{\varphi}(\xi)} \, \|\xi\|^\alpha \, d\xi, \qquad \forall f \in C, \; \varphi \in \mathfrak{S}_\alpha.$$

By Supplementary Exercise 5, $\mathfrak{S}_\alpha$ is dense in $C_0(\mathbb{R}^d)$. Therefore, by Parseval's identity,

$$\widehat{\mathcal{A}f}(\xi) = -\frac{1}{2}\widehat{f}(\xi) \cdot \|\xi\|^\alpha, \qquad \xi \in \mathbb{R}^d, \ f \in C. \tag{5}$$

The proposition now follows from Fourier's inversion theorem. □

When $\alpha = 2$, X is d-dimensional Brownian motion and $\mathcal{A} = -\frac{1}{2}\Delta$, in the sense of distributions; cf. Theorem 2.1.1. In this case, direct computations reveal that for all $f \in C_c^2(\mathbb{R}^d)$,

$$\widehat{\mathcal{A}f}(\xi) = -\frac{1}{2}\widehat{f}(\xi) \cdot \|\xi\|^2, \qquad \xi \in \mathbb{R}^d.$$

Informally, this is written as $\widehat{-\Delta}(\xi) = \|\xi\|^2$. In view of equation (5), when $\mathcal{A}$ denotes the generator of an $\mathbb{R}^d$-valued isotropic stable Lévy process of index $\alpha \in \,]0,2]$, then $\widehat{-2\mathcal{A}}(\xi) = \|\xi\|^\alpha$ (again, informally), and therefore $2\mathcal{A}$ is called the **fractional Laplacian** of power $\frac{\alpha}{2}$. This operator is sometimes written, suggestively, as $\Delta^{\alpha/2}$.

***Exercise* 2.2.1** Verify by direct computations that in the above sense, $\widehat{-\Delta}(\xi) = \|\xi\|^2$ for all $\xi \in \mathbb{R}^d$. □

***Exercise* 2.2.2** Suppose X and Y are independent Lévy processes with generators $\mathcal{A}_X$ and $\mathcal{A}_Y$, respectively. Let $\mathcal{D}(\mathcal{A}_X)$ and $\mathcal{D}(\mathcal{A}_Y)$ designate the respective generator's domain.

(i) Prove that the domain of the generator of the Lévy process Z equals $\mathcal{D}(\mathcal{A}_X) \cap \mathcal{D}(\mathcal{A}_Y)$.

(ii) Verify that the formal generator of Z is $\mathcal{A}_X + \mathcal{A}_Y$,

where Z is the Lévy process given by the assignment $Z_t = X_t + Y_t$. □

2.3 The Poisson Process

Let $\mathsf{P} = (\mathsf{P}_t;\ t \geq 0)$ denote an $\mathbb{R}^d$-valued Poisson process with rate vector $\lambda \in \mathbb{R}_+^d$. (We are following the notation of Example 2, Section 4.3 of Chapter 8.) We can now identify the generator of P, together with its domain.

Proposition 2.3.1 *Let $\mathcal{A}$ denote the generator of P. Then, $\mathcal{D}(\mathcal{A}) = C_0(\mathbb{R}^d)$ and*

$$\mathcal{A}f(x) = \sum_{\ell=1}^d \{f(x+e_j) - f(x)\}\lambda^{(j)}, \qquad f \in C_0(\mathbb{R}^d),\ x \in \mathbb{R}^d,$$

where $e_1, \ldots, e_d$ designate the standard basis vectors of $\mathbb{R}^d$. That is, $e_j^{(i)} = \mathbf{1}_{\{j\}}(i)$, $1 \leq i, j \leq d$.

Exercise 2.3.1 Prove Proposition 2.3.1. □

2.4 The Linear Uniform Motion

We conclude this section by presenting the domain of the generator of the uniform motion in the 1-dimensional case.

Proposition 2.4.1 *Let $X_t = t$ for all $t \geq 0$. Then, $X = (X_t;\ t \geq 0)$ is a Feller process on $\mathbb{R}$ whose generator $\mathcal{A}$ is given by $\mathcal{A}f(x) = f'(x)$, for all $f \in C_0(\mathbb{R})$ whose derivatives exist and are uniformly continuous. The latter collection of f's is precisely the domain of $\mathcal{A}$.*

Exercise 2.4.1 Prove Proposition 2.4.1. □

3 The Feynman–Kac Formula

We now begin in earnest to discuss a few of the connections between Markov processes and some equations of analysis and/or mathematical physics. This section deals with the celebrated Feynman–Kac formula.[2]

Throughout this section let S denote a locally compact, separable metric space with one-point compactification S_Δ. Furthermore, $X = (X_t;\ t \geq 0)$ denotes a right-continuous Feller process on S_Δ whose transition functions and generator are $\mathcal{T} = (\mathcal{T}_t;\ t \geq 0)$ and $\mathcal{A}$, respectively. For the sake of convenience, we will also assume the existence of shift operators $\theta = (\theta_t;\ t \geq 0)$; cf. Section 3.3 of Chapter 8. Finally, we will hold fixed a *nonnegative* function $v \in L^\infty(S)$ and define

$$V_t = \int_0^t v(X_s)\,ds, \qquad t \geq 0.$$

3.1 The Feynman–Kac Semigroup

We begin with a technical lemma.

Lemma 3.1.1 *Suppose that $f, v \in L^\infty(S)$ and that v satisfies*

$$\lim_{t \to 0^+} \|\mathcal{T}_t v - v\|_\infty = 0. \tag{1}$$

---

[2]This was found to various degrees of rigor and generality in (Feynman 1948; Kac 1951). A modern account, together with some applications, can be found in Fitzsimmons and Pitman (1999).

Then, uniformly over all $x \in S$,

$$\lim_{t \to 0} \mathbb{E}_x\left[f(X_t) \cdot \left(\frac{1 - e^{-V_t}}{t}\right)\right] = v(x)f(x).$$

Remark By the Feller property, any $v \in C_0(S)$ satisfies equation (1).

Proof By Taylor's expansion, for all $x \in \mathbb{R}$, $|1 - e^{-x} - x| \leq \sum_{j=2}^{\infty} |x|^j/j!$. Apply this with $x \equiv V_t$ to obtain the following: For all $0 \leq t \leq 1$,

$$\sup_{x \in S} \left| \mathbb{E}_x\left[f(X_t) \cdot \left(\frac{1 - e^{-V_t}}{t}\right)\right] - \mathbb{E}_x\left[f(X_t)\frac{V_t}{t}\right] \right| \leq At, \tag{2}$$

where $A = \|f\|_\infty \cdot \|v\|_\infty^2 e^{\|v\|_\infty}$ (why?). Moreover,

$$\sup_{x \in S} \left| \mathbb{E}_x\left[f(X_t)\frac{V_t}{t}\right] - v(x)\mathcal{T}_t f(x) \right| = \sup_{x \in S} \left| \mathbb{E}_x\left[f(X_t)\frac{V_t}{t}\right] - \mathbb{E}_x[f(X_t)v(x)] \right|$$

$$\leq \|f\|_\infty \cdot \frac{1}{t} \int_0^t \|\mathcal{T}_s v - v\|_\infty \, ds.$$

The lemma follows from this, equation (2), and assumption (1). □

Next, we define a collection of linear operators $\mathcal{T}^v = (\mathcal{T}_t^v;\ t \geq 0)$ as

$$\mathcal{T}_t^v f(x) = \mathbb{E}_x[f(X_t)e^{-V_t}], \qquad f \in C_0(S),\ t \geq 0. \tag{3}$$

Our next result is the first indication that $\mathcal{T}^v$ is an interesting collection of linear operators.

Lemma 3.1.2 *Under the assumption (1), $\mathcal{T}^v$ is a Feller semigroup on S.*

This semigroup is sometimes known as the **Feynman–Kac** semigroup.[3]

Proof Clearly, $\mathcal{T}_0^v f = f$, and $\mathcal{T}_t^v$ is a positive linear operator. Next, we verify the semigroup property. By the Markov property (Theorem 3.3.2, Chapter 8), for all $f \in L^\infty(S)$ and all $x \in S$, the following holds $\mathbb{P}_x$-a.s.:

$$\mathcal{T}_s^v f(X_t) = \mathbb{E}_{X_t}\left[f(X_s)e^{-\int_0^s v(X_r)\,dr}\right] = \mathbb{E}_x\left[f(X_{s+t})e^{-\int_t^{t+s} v(X_r)\,dr}\,\Big|\,\mathcal{F}_t\right],$$

where $\mathcal{F} = (\mathcal{F}_t;\ t \geq 0)$ denotes the complete augmented history of X. On the other hand, for all $f \in L^\infty(S)$ and all $x \in S$,

$$\mathcal{T}_t^v \mathcal{T}_s^v f(x) = \mathbb{E}_x\left[\mathcal{T}_s^v f(X_t)e^{-V_t}\right].$$

---

[3]Oftentimes, this is said to correspond to the so-called potential v. However, we will not use this terminology, since we are already using the term potential for something else.

Whenever $f \in C_0(S)$, $\mathcal{T}_t f \in C_0(S)$ for all $t \geq 0$. By the dominated convergence theorem, $\mathcal{T}_t^v f \in C_0(S)$, also. This implies that $\mathcal{T}_t^v(C_0(S)) \subset C_0(S)$. We obtain the semigroup property promptly.

Let us conclude this argument by verifying the Feller property. For all $f \in C_0(S)$,

$$\|\mathcal{T}_t^v f - f\|_\infty = \sup_{x \in S} \left| \mathbb{E}_x[f(X_t)e^{-V_t}] - f(x) \right|$$

$$\leq \sup_{x \in S} \left| \mathbb{E}_x[f(X_t)(e^{-V_t} - 1)] \right| + \|\mathcal{T}_t f - f\|_\infty.$$

Since X is Feller, Lemma 3.1.1 implies that $\mathcal{T}^v$ is Feller. □

Let $\mathcal{A}^v$ denote the generator of $\mathcal{T}^v$. Our main and final goal in this subsection is to identify $\mathcal{A}^v$ and its domain in terms of $\mathcal{A}$ and its domain.

Theorem 3.1.1 (The Feynman–Kac Formula) *Under the assumption (1), the domains of the generators $\mathcal{A}^v$ and $\mathcal{A}$ are one and the same. Moreover, for all $f \in \mathcal{D}(\mathcal{A})$,*

$$\mathcal{A}^v f(x) = \mathcal{A}f(x) - v(x)f(x), \qquad x \in S.$$

Proof Clearly, for all $f \in L^\infty(S)$ and all $x \in S$,

$$\frac{\mathcal{T}_t^v f(x) - f(x)}{t} = \frac{\mathcal{T}_t f(x) - f(x)}{t} - \mathbb{E}_x\left[f(X_t)\left(\frac{1 - e^{-V_t}}{t}\right)\right].$$

By Lemma 3.1.1, for all $f \in L^\infty(S)$, in particular for all $f \in C_0(S)$,

$$\lim_{t \to 0^+} \sup_{x \in S} \left| \frac{\mathcal{T}_t^v f(x) - f(x)}{t} - \frac{\mathcal{T}_t f(x) - f(x)}{t} + v(x)f(x) \right| = 0.$$

The theorem follows readily. □

3.2 The Doob–Meyer Decomposition

We now prove a Doob–Meyer decomposition that corresponds to the Feynman–Kac semigroup. We will later apply this decomposition to make explicit computations for multidimensional Brownian motion; see the following section.

Theorem 3.2.1 (The Doob–Meyer Decomposition) *Suppose that $v \in L^\infty(S)$ satisfies equation (1) of Section 3.1. For any $f \in \mathcal{D}(\mathcal{A})$, define $M = (M_t;\ t \geq 0)$ by*

$$M_t = e^{-V_t}f(X_t) - f(X_0) - \int_0^t e^{-V_r}\mathcal{A}^v f(X_s)\,ds, \qquad t \geq 0.$$

Then, M is a mean-zero martingale with respect to the complete augmented filtration of X, and under any of the measures $\mathbb{P}_x$, $x \in S$.

Proof Recall that we have assumed the existence of shifts $\theta = (\theta_t;\, t \geq 0)$. If they do not exist, the proof only becomes notationally more cumbersome.

For any $s, t \geq 0$,
$$V_t \circ \theta_s = V_{t+s} - V_s. \tag{1}$$

We now follow the proof of Theorem 1.3.3 closely, only making a few necessary adjustments.

Let $\mathcal{R}^v = (\mathcal{R}^v_\lambda;\, \lambda > 0)$ denote the resolvent corresponding to the generator $\mathcal{A}^v$. For any $\lambda > 0$, all $\varphi \in L^\infty(S)$, and for all $t \geq 0$, we can use the Markov property (Theorem 3.3.2 of Chapter 8) to see that $\mathbb{P}_x$-a.s. for all $x \in S$,
$$\mathcal{R}^v_\lambda \varphi(X_t) = \mathbb{E}_x\!\left[\int_0^\infty e^{-\lambda s - V_s \circ \theta_t}\varphi(X_{s+t})\, ds \,\Big|\, \mathcal{F}_t\right].$$

Since V_t is $\mathcal{F}_t$-measurable, by equation (1), $\mathbb{P}_x$-a.s. for all $x \in S$,
$$e^{-V_t}\mathcal{R}^v_\lambda \varphi(X_t) = e^{\lambda t}\mathbb{E}_x\!\left[\int_0^\infty e^{-\lambda(s+t) - V_{s+t}}\varphi(X_{s+t})\, ds \,\Big|\, \mathcal{F}_t\right]$$
$$= e^{\lambda t}\mathbb{E}_x\!\left[\int_t^\infty e^{-\lambda r - V_r}\varphi(X_r)\, dr \,\Big|\, \mathcal{F}_t\right].$$

Given $f \in \mathcal{D}(\mathcal{A})$, we can use the Feynman–Kac formula (Theorem 3.1.1) to deduce that for any $\lambda > 0$, we can find $\varphi \in C_0(S)$ such that $f = \mathcal{R}^v_\lambda \varphi$. Equivalently, $\mathcal{A}^v f = \lambda f - \varphi$. Thus, we can apply the previous computations to this specific choice of φ to deduce that $\mathbb{P}_x$-a.s. for all $x \in S$,
$$e^{-\lambda t - V_t} f(X_t) = \mathbb{E}_x\!\left[\int_0^\infty e^{-\lambda r - V_r}\varphi(X_r)\, dr \,\Big|\, \mathcal{F}_t\right] - \int_0^t e^{-\lambda r - V_r}\varphi(X_r)\, dr$$
$$= \mathbb{E}_x\!\left[\int_0^\infty e^{-\lambda r - V_r}\varphi(X_r)\, dr \,\Big|\, \mathcal{F}_t\right]$$
$$\quad - \int_0^t e^{-\lambda r - V_r}\{\lambda f(X_r) - \mathcal{A}^v f(X_r)\}\, dr.$$

Define
$$Z^\lambda_t = e^{-\lambda t - V_t} f(X_t) - \int_0^t \mathcal{A}^v f(X_r)\, dr + \lambda \int_0^t e^{-\lambda r - V_r} f(X_r)\, dr, \qquad t \geq 0.$$

Then, we have shown that $Z^\lambda = (Z^\lambda_t;\, t \geq 0)$ is a martingale with respect to $\mathcal{F}$, and under any of the measures $\mathbb{P}_x$, $x \in S$. The result follows, by the dominated convergence theorem, upon letting $\lambda \to 0^+$. $\square$

4 Exit Times and Brownian Motion

Throughout this section we let $B = (B_t;\, t \geq 0)$ denote a d-dimensional Brownian motion with complete augmented filtration $\mathcal{F} = (\mathcal{F}_t;\, t \geq 0)$.

Recall from Chapter 8 that $\mathbb{P}_x$ is the $\mathbb{P}$-distribution of $x + B$, for any $x \in \mathbb{R}^d$. We will write $\mathbb{P}$ for $\mathbb{P}_0$ and use the fact that $t \mapsto B_t$ is continuous. Let us define $\mathcal{B}(a;r)$ as the open ℓ^2-ball of radius $r > 0$ around $a \in \mathbb{R}^d$. That is,

$$\mathcal{B}(a;r) = \{b \in \mathbb{R}^d : \|a - b\| < r\}, \qquad a \in \mathbb{R}^d, \ r > 0.$$

Then, the **exit time** $\zeta(a;r)$ of the ball $\mathcal{B}(a;r)$ is defined as

$$\zeta(a;r) = \inf\left(s > 0 : B_s \notin \mathcal{B}(a;r)\right).$$

Clearly, $\zeta(a;r)$ is an $\mathcal{F}$-stopping time. Among other things, in this section we shall compute the distribution of $\zeta(a;r)$. This will be done by establishing connections between Brownian motion and partial differential equations. The material of this, and the remaining sections of this chapter, are independent of the rest of the book, and can be skipped at first reading.

In order to simplify our exposition, let $\tau(a;r) = T_{\partial \mathcal{B}(a;r)}$ denote the first time to hit the boundary of the ball $\mathcal{B}(a;r)$. If our Brownian motion B starts in the interior of $\mathcal{B}(a;r)$, then $\zeta(a;r) = 0$ (why?). On the other hand, if it start outside of the closure of $\mathcal{B}(a;r)$, the continuity of $t \mapsto B_t$ ensures that $\zeta(a;r) = \tau(a;r)$. Thus, we will concentrate on computing the distribution of $\tau(a;r)$.[4]

4.1 Dimension One

When $d = 1$, it is possible to find the distribution of $\tau(a;r)$ in a simple and elegant way. We will now make this computation without further ado. The remaining calculations, i.e., those for $d \geq 2$, are more delicate, and rely on further analysis that we will perform in the subsequent subsections.

Theorem 4.1.1 (P. Lévy) *If $d = 1$, then for any $x \in \mathbb{R}$ and all $\alpha > 0$,*

$$\mathbb{E}_x[e^{-\alpha \tau(x;r)}] = \operatorname{sech}(r\sqrt{2\alpha}),$$

where sech *denotes the hyperbolic secant.*

Proof Under $\mathbb{P}_x$, $B - x$ has the same distribution as B, under $\mathbb{P}$. Thus,

$$\mathbb{E}_x[e^{-\alpha \tau(x;r)}] = \mathbb{E}[e^{-\alpha(0;r)}].$$

We should recall that $\mathbb{E} = \mathbb{E}_0$. That is, we have reduced the problem for general x to that for $x = 0$.

From now on, we hold $r > 0$ fixed and write $\tau = \tau(0;r)$. The remainder of the proof is a sign of what is to come in higher dimensions: We strive to find a good martingale, and apply the optional stopping theorem.

---

[4] It turns out that if B starts on the boundary of $\mathcal{B}(a;r)$, then $\zeta(a;r)$ is still zero. In this regard, see Supplementary Exercise 6.

Recall from Corollary 1.7.1(*iii*), Chapter 7, that for any $\lambda > 0$, $M = (M_t;\ t \geq 0)$, a mean-one $\mathcal{F}$-martingale, where

$$M_t = \exp\left(\lambda B_t - \frac{\lambda^2}{2}t\right), \qquad t \geq 0.$$

Since $t \mapsto B_t$ is continuous, Corollary 1.7.2 of Chapter 7 shows that τ is a.s. finite. Thus, we can apply the optional stopping theorem (Theorem 1.7.1, Chapter 7) to deduce that for all $t \geq 0$,

$$1 = \mathbb{E}[M_{t\wedge\tau}] = \mathbb{E}\left[e^{-\lambda B_{t\wedge\tau} - \frac{1}{2}\lambda^2(t\wedge\tau)}\right].$$

On the other hand, by path continuity, $\sup_{s<\tau}|B_s| \leq r$, since $\tau < \infty$, $\mathbb{P}$-a.s. Thus, we can apply the dominated convergence theorem to legitimately let $t \to \infty$ and conclude that

$$1 = \mathbb{E}\left[e^{-\lambda B_\tau - \frac{\lambda^2}{2}\tau}\right].$$

Since $\mathbb{P}(B_\tau = \pm r) = 1$,

$$1 = e^{-\lambda r}\mathbb{E}\left[e^{-\frac{\lambda^2}{2}\tau}\mathbf{1}_{(B_\tau=-r)}\right] + e^{\lambda r}\mathbb{E}\left[e^{-\frac{\lambda^2}{2}\tau}\mathbf{1}_{(B_\tau=r)}\right]. \tag{1}$$

The processes B and $-B$ have the same finite-dimensional distributions. Therefore, the random vectors (B_τ, τ) and $(-B_\tau, \tau)$ have the same distributions (why?). As a result, the two expectations in equation (1) are equal; i.e.,

$$\mathbb{E}\left[e^{-\frac{\lambda^2}{2}\tau}\mathbf{1}_{(B_\tau=r)}\right] = \mathbb{E}\left[e^{-\frac{\lambda^2}{2}\tau}\mathbf{1}_{(B_\tau=-r)}\right] = \frac{1}{2}\mathbb{E}\left[e^{-\frac{\lambda^2}{2}\tau}\right].$$

Plugging this into equation (1), we can see that for all $\lambda > 0$,

$$\mathbb{E}\left[e^{-\frac{\lambda^2}{2}\tau}\right] = \frac{2}{e^{-\lambda r} + e^{\lambda r}} = \operatorname{sech}(\lambda r).$$

The theorem follows upon letting $\lambda = \sqrt{2\alpha}$. $\square$

4.2 Some Fundamental Local Martingales

Recall from Section 3.3 of Chapter 7 that a real-valued stochastic process $M = (M_t;\ t \geq 0)$ is a **local martingale** (with respect to $\mathcal{F}$ under all measures $\mathbb{P}_x$, $x \in \mathbb{R}^d$) if:

(a) for all $t \geq 0$, M_t is adapted to $\mathcal{F}_t$;

(b) there exist possibly infinite $\mathcal{F}$-stopping times $\tau_1, \tau_2, \ldots$ such that

(b$_1$) for all $x \in \mathbb{R}^d$, $\mathbb{P}_x(\lim_{k\to\infty} \tau_k = +\infty) = 1$; and (b$_2$) for all $x \in \mathbb{R}^d$ and all $k \geq 1$, $t \mapsto M_{t \wedge \tau_k}$ is a martingale under the measure $\mathbb{P}_x$.

Lemma 3.3.1 of Chapter 7 shows that if we truncate local martingales, we obtain genuine martingales. Conversely, if M is a martingale, so is $t \mapsto M_{t \wedge \tau_k}$ for any $k \geq 1$, where τ_k is the sequence of stopping times given in Lemma 3.3.1, Chapter 7; see the optional stopping theorem (Theorem 1.6.1, Chapter 7).

In this section we will find some useful local martingales. Define

$$u(x) = \begin{cases} -\ln \|x\|, & \text{if } d = 2, \\ \|x\|^{2-d}, & \text{if } d \geq 3, \end{cases} \tag{1}$$

where we interpret $\log \infty = 1 \div 0 = +\infty$. The function u is said to be the **fundamental harmonic function** with pole at the origin. That u has a pole, or singularity, at the origin is clear; the word harmonic comes from the following.

Lemma 4.2.1 *Let $d \geq 2$. The function u given by (1) solves the PDE*

$$\Delta u(x) = 0, \qquad x \in \mathbb{R}^d,$$

where $\Delta = \sum_{j=1}^{d} \partial^2 / \partial (x^{(j)})^2$ denotes the Laplacian.

Exercise 4.2.1 Verify Lemma 4.2.1 directly. □

The partial differential equation $\Delta u = 0$, given above, is the so-called **Dirichlet problem** of mathematical physics; its solutions are called **harmonic** functions. The general solution to this PDE turns out to be $c_1 u(x) + c_2$, where c_1 and c_2 are arbitrary constants (why?). This explains why u is said to be the fundamental harmonic function with pole at $0 \in \mathbb{R}^d$.

The following shows one of the many deep connections between PDEs and Markov processes.

Lemma 4.2.2 *Let $d \geq 2$ and let u denote the fundamental harmonic function with pole at the origin. Then, $u(B) = (u(B_t); \ t \geq 0)$ is a local martingale under the measure $\mathbb{P}_x$, $x \in \mathbb{R}^d \setminus \{0\}$.*

Proof Recall from Theorem 2.1.1 that the Laplacian Δ is the generator of B. Moreover, note that u is *not* in the domain of this generator, since it is not in $C_0(\mathbb{R}^d)$. Before giving an honest proof, let us skirt this issue and formally argue why the result holds true. We will then proceed with a precise demonstration.

By Lemma 4.2.1, $\Delta u \equiv 0$. Therefore, one would expect that (by the martingale problem, cf. Theorem 1.3.1), $t \mapsto u(B_t) - u(x)$ is a mean-zero martingale under the measure $\mathbb{P}_x$, for any $x \in \mathbb{R}^d$. This would imply a strong form of the theorem, i.e., one that asserts a *martingale* property, rather than a local martingale property. However, as it turns out, $u(B)$ is *not* a martingale; cf. Supplementary Exercise 7. Therefore, we need to proceed with more caution.

Now we proceed with our derivation. Let us fix a constant $a > 0$ and define
$$\mathfrak{U}(a) = \{x \in \mathbb{R}^d : |u(x)| < a\}.$$
Clearly, this is an open set. Define
$$\sigma(a) = \inf(s \geq 0 : B_s \notin \overline{\mathfrak{U}(a)}), \qquad a > 0,$$
where, as usual, $\inf \varnothing = \infty$. Note that $\sigma(a)$ is a stopping time (Theorem 1.2.1, Chapter 7). Moreover, it is *not* always the same as $\inf(s > 0 : B_s \notin \mathfrak{U}(a))$.

By Supplementary Exercise 14 of Chapter 8, the absorbed process $B^a = (B_{t \wedge \sigma(a)}; \, t \geq 0)$ is a Feller process on the compact space $\overline{\mathfrak{U}(a)}$. The domain of its generator is described in Supplementary Exercise 8; this domain includes the collection of all continuous functions $f : \overline{\mathfrak{U}(a)} \to \mathbb{R}$ for which Δf exists in the sense of distributions.

Define $u^a(x) = u(x) \mathbf{1}_{\mathfrak{U}(a)}(x)$. Then, u^a is in the domain of the generator of B^a. Furthermore, in the sense of distributions, $\Delta u^a \equiv 0$, since $\Delta u \equiv 0$. Therefore, we can apply the martingale problem (Theorem 1.3.1) to the process B^a, and conclude that $t \mapsto u^a(B^a_t)$ is a *martingale* under any of the measures $\mathbb{P}_x$, $x \in \mathfrak{U}(a)$. On the other hand, for all $t \in \,]0, \sigma(a)]$, $u^a(B^a_t) = u(B_t)$. By the optional stopping theorem, $t \mapsto u(B_{t \wedge \sigma(a)})$ is a martingale under $\mathbb{P}_x$, for any $x \in \mathfrak{U}(a)$. Thus, $(\sigma(k); \, k \geq 1)$ is a localizing sequence, and $u(B)$ is a local martingale under $\mathbb{P}_x$, $x \in \mathfrak{U}(a)$. Since $a > 0$ is arbitrary, we are done. □

Before proceeding with our next lemma, we first need access to an important class of martingales.

***Exercise* 4.2.2** If B denotes $\mathbb{R}^d$-valued Brownian motion, then, $t \mapsto \|B_t\|^2 - t$ is a mean-zero martingale under the measure $\mathbb{P}$. Consequently, $t \mapsto \|B_t - x\|^2 - t$ is a mean-zero martingale under the measure $\mathbb{P}_x$, for any $x \in \mathbb{R}^d$. □

Lemma 4.2.3 *For all $a \in \mathbb{R}^d$, all $r > 0$, and all $x \in \mathcal{B}(a, r)$,*
$$\mathbb{P}_x(\tau(a; r) < \infty) = 1.$$

Proof Combining Exercise 4.2.2 with the optional stopping theorem (Theorem 1.6.1, Chapter 7) yields
$$\mathbb{E}_x\{\|B_{t \wedge \tau(a;r)} - x\|^2\} = \mathbb{E}_x[t \wedge \tau(a; r)]. \qquad (2)$$
On the other hand, $\mathbb{P}_x$-a.s.,
$$\sup_t \|B_{t \wedge \tau(a;r)} - x\| \leq \|x\| + \|a\| + r \leq 2\|a\| + r,$$

which is finite. Thus, we can take $t \to \infty$ in (2), to obtain

$$\mathbb{E}_x\{\|B_{\tau(a;r)} - x\|^2\} = \mathbb{E}_x[\tau(a;r)].$$

You need to envoke the dominated convergence theorem for the left-hand side of (2) and the monotone convergence theorem for the corresponding right-hand side. Since $\|B_{\tau(a;r)} - x\| \leq \|a\| + r + \|x\| \leq 2\|a\| + r$, $\mathbb{P}_x$-a.s.,

$$\mathbb{E}_x[\tau(a;r)] \leq 2\|a\| + r < \infty.$$

The lemma follows readily from this. $\square$

We can now apply the above lemmas to solve a "*gambler's ruin problem*"; it may be helpful to recall that we have already encountered the 1-dimensional gambler's ruin problem in Corollary 1.7.3 of Chapter 7.

Theorem 4.2.1 (Gambler's Ruin) If $d \geq 2$, then for $0 < r < R$ and $x \in \mathbb{R}^d$ with $r \leq \|x\| \leq R$,

$$\mathbb{P}_x(\tau(a;r) < \tau(a;R)) = \begin{cases} \dfrac{\ln(R/\|x\|)}{\ln(R/r)}, & \text{if } d = 2, \\ \dfrac{\|x\|^{2-d} - R^{2-d}}{r^{2-d} - R^{2-d}}, & \text{if } d \geq 3. \end{cases}$$

Figure 9.1 shows the event that the two-dimensional Brownian path starting at $x \in \mathbb{R}^d$ with $\|x\| = \varrho$ satisfies $\tau(a;R) < \tau(a;r)$.

Let us now prove the theorem.

Proof We shall fix x, r, and R as given by the statement of the theorem. Next, let us define $U :\,]0, \infty[\, \to \mathbb{R}$ by

$$U(y) = \begin{cases} -\ln y, & \text{if } d = 2, \\ y^{2-d}, & \text{if } d \geq 3, \end{cases} \quad y > 0.$$

Clearly, $u(x) = U(\|x\|)$, for all $x \in \mathbb{R}^d$. We will show that

$$\mathbb{P}_x(\tau(a;r) < \tau(a;R)) = \frac{U(x) - U(R)}{U(r) - U(R)}, \tag{3}$$

which is the same as the assertion of our theorem.

By Lemma 4.2.2 above, $U(\|B\|) = (U(\|B_t\|); t \geq 0)$ is a local martingale under $\mathbb{P}_x$. On the other hand, with probability one,

$$\sup_{t \geq 0} U\big(\|B_{t \wedge \tau(x,r) \wedge \tau(x,R)}\|\big) \leq |U(R)| \wedge |U(r)|, \tag{4}$$

which is finite. Hence, by the optional stopping theorem (Theorem 1.6.1, Chapter 7), $t \mapsto U(\|B_{t \wedge \tau(x;r) \wedge \tau(x;R)}\|)$ is a martingale. In particular, we can take expectations to deduce

$$U(x) = \mathbb{E}_x\big[U(\|B_{t \wedge \tau(x;r) \wedge \tau(x;R)}\|)\big].$$

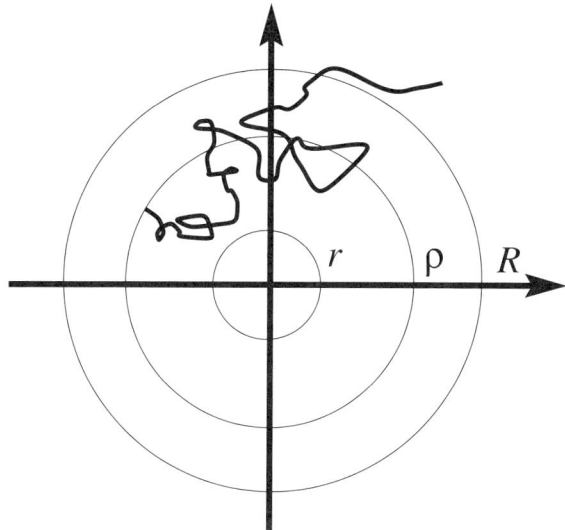

Figure 9.1: Gambler's ruin for 2-dimensional Brownian motion

By combining Lemma 4.2.3, equation (4), and the dominated convergence theorem, we can let $t \to \infty$ to see that

$$\begin{aligned}
U(x) &= \mathbb{E}_x\left[U(\|B_{\tau(x;r) \wedge \tau(x;R)}\|)\right] \\
&= U(r)\mathbb{P}_x\{\tau(a;r) < \tau(a;R)\} + U(R)\mathbb{P}_x\{\tau(a;r) > \tau(a;R)\} \\
&= \left[U(r) - U(R)\right]\mathbb{P}_x\{\tau(a;r) < \tau(a;R)\} + U(R).
\end{aligned}$$

Equation (3) follows. $\square$

4.3 The Distribution of Exit Times

The main result of this subsection is the following computation of the Laplace transform of the distribution of $\tau(a;r)$. At this level of generality, this computation is from Ciesielski and Taylor (1962); see Lévy (1965) for several related results. We follow the elegant exposition of Knight (1981).

Theorem 4.3.1 (The Ciesielski–Taylor Theorem) *For all* $a \in \mathbb{R}^d$, $r > 0$, *every* $x \in \mathcal{B}(a;r)$, *and all* $\alpha > 0$,

$$\mathbb{E}_x[e^{-\alpha \tau(a;r)}] = \frac{\|x-a\|^{1-\frac{d}{2}} \mathbf{I}_{\frac{d}{2}-1}(\sqrt{2\alpha}\|x-a\|)}{r^{1-\frac{d}{2}} \mathbf{I}_{\frac{d}{2}-1}(\sqrt{2\alpha}r)},$$

where $\mathbf{I}_\nu$ *is the modified Bessel function of index* ν.

336   9. Generation of Markov Processes

Recall that when $\nu > -\frac{1}{2}$, the modified Bessel functions $\mathbf{I}_\nu$ and $\mathbf{K}_\nu$ are given by[5]

$$\mathbf{I}_\nu(z) = \frac{z^\nu 2^{1-\nu}}{\Gamma(\nu + \frac{1}{2})\sqrt{\pi}} \int_0^1 e^{-zt}(1-t^2)^{\nu-\frac{1}{2}}\,dt; \tag{1}$$

$$\mathbf{K}_\nu(z) = \frac{z^\nu \sqrt{\pi}}{2^\nu \Gamma(\nu + \frac{1}{2})} \int_1^\infty e^{-zt}(t^2-1)^{\nu-\frac{1}{2}}\,dt. \tag{2}$$

Moreover, when $\nu \leq -\frac{1}{2}$, they are defined by the identities

$$\mathbf{I}_\nu = \mathbf{I}_{-\nu},\ \mathbf{K}_\nu = \mathbf{K}_{-\nu}.$$

(We have already encountered the function $\mathbf{K}_{\frac{1}{2}}$ in Section 2.3 of Chapter 8.)

Now, consider the following Bessel's equation:

$$G''(s) + \frac{1}{s}G'(s) - \left(1 + \frac{\nu^2}{s^2}\right)G(s) = 0, \qquad s \in \mathbb{R}. \tag{3}$$

The general solution to (3) is $G(s) = c_1 \mathbf{I}_\nu(s) + c_2 \mathbf{K}_\nu(s)$, where c_1 and c_2 are arbitrary constants.

Lemma 4.3.1 *Consider the following form of Bessel's equation:*

$$G''(s) + \frac{\theta}{s}G'(s) - \kappa G(s) = 0, \qquad s \in \mathbb{R},$$

where $\theta \in \mathbb{R}$ and $\kappa > 0$. Then, the general solution to the above is of the form

$$G(s) = c_1 s^{\frac{1}{2}(1-\theta)} \mathbf{I}_{\frac{1}{2}(\theta-1)}(\sqrt{\kappa}s) + c_2 s^{\frac{1}{2}(1-\theta)} \mathbf{K}_{\frac{1}{2}(\theta-1)}(\sqrt{\kappa}s),$$

where c_1 and c_2 are arbitrary constants.

***Exercise* 4.3.1** Check the validity of Lemma 4.3.1.   □

Lemma 4.3.2 *For any $r, \alpha > 0$, $x \mapsto \mathbb{E}_x[e^{-\alpha\tau(0;r)}]$ is a radial function.*

Proof For a fixed $r > 0$, we define $v(x) = \mathbb{E}_x[e^{-\alpha\tau(0;r)}]$, $x \in \mathbb{R}^d$, and will show that for any (unitary) rotation matrix $\mathbf{O}$, and for all $x \in \mathbb{R}^d$, $v(\mathbf{O}x) = v(x)$.

It is easy to see that for any $(d \times d)$ (unitary) rotation matrix $\mathbf{O}$, $\mathbf{O}B = (\mathbf{O}B_t;\ t \geq 0)$ is d-dimensional Brownian motion. Indeed, since $\mathbf{O}B$ is a continuous Gaussian process, one does this by simply checking the covariances (check this!). Let $\tau'(0;r)$ denote the entrance time of $\partial \mathcal{B}(0;r)$

---

[5] For a thorough treatment of Bessel functions, see Watson (1995).

for the process $\mathbf{O}B$. It is immediate that the distribution of $\tau'(0;r)$ under $\mathbb{P}_{\mathbf{O}x}$ is the same as that of $\tau(0;r)$ under $\mathbb{P}_x$. The result follows readily from this observation. □

Lemma 4.3.3 *For any $r, \alpha > 0$,*

$$\lim_{\substack{x \in \mathcal{B}(0;r): \\ \|x\| \to r}} \mathbb{E}_x\left[e^{-\alpha \tau(0;r)}\right] = 1.$$

Proof Since this result has to do with distributional properties, we can assume, without loss of generality, that there are shifts $\theta = (\theta_t; \, t \geq 0)$ (why?).

Let us now hold fixed a $y \in \mathcal{B}(0;r)$ and a $u \in\,]\|y\|, r[$. Since $t \mapsto B_t$ is continuous, starting from y, Brownian motion must first exit $\mathcal{B}(0;u)$ before exiting $\mathcal{B}(0;r)$. In other words, by inspecting the paths,

$$\tau(0;r) = \tau(0;u) + \tau(0;r) \circ \theta_{\tau(0;u)}, \qquad \mathbb{P}_y\text{-a.s.}$$

On the other hand, B is a strong Markov process (Example 1 of Section 4.3, Chapter 8). By the strong Markov property (Theorem 4.2.1, Chapter 8),

$$\mathbb{E}_y[e^{-\alpha \tau(0;r)}] = \mathbb{E}_y\left[e^{-\alpha \tau(0;u)} \mathbb{E}_{B_{\tau(0;u)}}\{e^{-\alpha \tau(0;r)}\}\right]. \qquad (4)$$

We have tacitly used Lemma 4.2.3 of the previous subsection. Since $t \mapsto B_t$ is continuous, $B_{\tau(0;u)} \in \partial \mathcal{B}(0;u)$. Hence, Lemma 4.3.2 shows that for *any* $x \in \partial \mathcal{B}(0;u)$,

$$\mathbb{E}_{B_{\tau(0;u)}}\left[e^{-\alpha \tau(0;r)}\right] = \mathbb{E}_x\left[e^{-\alpha \tau(0;r)}\right].$$

Consequently, equation (4) has the following reformulation: For all $x, y \in \mathcal{B}(0;r)$ with $\|x\| > \|y\|$,

$$\mathbb{E}_y\left[e^{-\alpha \tau(0;r)}\right] = \mathbb{E}_y\left[e^{-\alpha \tau(0;\|x\|)}\right] \mathbb{E}_x\left[e^{-\alpha \tau(0,r)}\right].$$

By the continuity of $t \mapsto B_t$, as $\|x\| \to r^-$, $\tau(0; \|x\|) \to \tau(0;r)$, $\mathbb{P}_y$-a.s. The lemma follows immediately from the previous display. □

We can now prove Theorem 4.3.1.

Proof of Theorem 4.3.1 Recall from Theorem 2.1.1 that the generator of B is $\frac{1}{2}\Delta$, i.e., the Laplacian in the sense of distributions. Moreover, its domain, $\mathcal{D}(\Delta)$, is the collection of all $f \in C_0(\mathbb{R}^d)$ for which Δf exists.

Suppose that we could find a *radial* function $f \in \mathcal{D}(\Delta)$ such that

$$\frac{1}{2}\Delta f = \alpha f, \qquad (5)$$

where $\alpha > 0$ is a fixed real number. Let $v(x) \equiv \alpha$ for all $x \in \mathbb{R}^d$, and note that by the Feynman–Kac formula (Theorem 3.1.1), $\mathcal{A}^v f = \frac{1}{2}\Delta f - \alpha f$. That is, equation (5) is *equivalent* to the condition

$$\mathcal{A}^v f(x) = 0, \qquad \forall x \in \mathbb{R}^d.$$

By the Doob–Meyer decomposition (Theorem 3.2.1), for all $x \in \mathcal{B}(0;r)$, $\gamma = (\gamma_t;\ t \geq 0)$ is a continuous, mean-zero martingale, under the measure $\mathbb{P}_x$, where

$$\gamma_t = e^{-\alpha t} f(B_t) - f(x), \qquad t \geq 0.$$

By the optional stopping theorem (Theorem 1.6.1, Chapter 7), for all $t \geq 0$ and all $x \in \mathcal{B}(0;r)$, $\mathbb{E}_x[\gamma_{t \wedge \tau(0;r)}] = 0$. Equivalently,

$$f(x) = \mathbb{E}_x\bigl[e^{-\alpha(t \wedge \tau(0;r))} f(B_{t \wedge \tau(0;r)})\bigr], \qquad x \in \mathcal{B}(0;r).$$

Since $f \in C_0(\mathbb{R}^d)$, we can use the dominated convergence theorem to let $t \to \infty$, and see that

$$f(x) = \mathbb{E}_x\bigl[e^{-\alpha \tau(0;r)} f(B_{\tau(0;r)})\bigr], \qquad x \in \mathcal{B}(0;r).$$

But $B_{\tau(0;r)} \in \partial \mathcal{B}(0;r)$, and f is radial. Therefore,

$$f(x) = f(0,\ldots,0,r) \times \mathbb{E}_x\bigl[e^{-\alpha \tau(0;r)}\bigr], \qquad x \in \mathcal{B}(0;r). \tag{6}$$

That is, in summary, if there is a radial solution to the PDE (5), it must provide the general form of the Laplace transform of the exit time of $\mathcal{B}(0;r)$. Consequently, it suffices to solve equation (5) by a radial function.

Henceforth, let f denote a radial $C_0(\mathbb{R}^d)$ solution to equation (5). We will write $f(x) = F(\|x\|)$, for all $x \in \mathbb{R}^d$, where F is a function on $\mathbb{R}_+$. It is easy to check directly from equation (5) that F must satisfy the differential equation

$$F''(s) + \frac{d-1}{s} F'(s) - 2\alpha F(s) = 0, \qquad s \geq 0.$$

According to Lemma 4.3.1,

$$F(s) = c_1 s^{1-\frac{d}{2}} \mathbf{I}_{\frac{d}{2}-1}(\sqrt{2\alpha}s) + c_2 s^{1-\frac{d}{2}} \mathbf{K}_{\frac{d}{2}-1}(\sqrt{2\alpha}s), \qquad s \geq 0,$$

for some constants c_1 and c_2. On the other hand, $f(x) = F(\|x\|)$ is bounded. Since $z \mapsto z^{-\nu} \mathbf{K}_\nu(z)$ is not bounded near 0 (equation (2)), the constant c_2 above must be 0. That is,

$$f(x) = c_1 \|x\|^{1-\frac{d}{2}} \mathbf{I}_{\frac{d}{2}-1}(\sqrt{2\alpha}\|x\|).$$

Therefore, equation (6) and the above together yield a constant C such that for all $x \in \mathcal{B}(0;r)$,

$$\mathbb{E}_x\bigl[e^{-\alpha \tau(0;r)}\bigr] = C\|x\|^{1-\frac{d}{2}} \mathbf{I}_{\frac{d}{2}-1}(\sqrt{2\alpha}\|x\|).$$

We can let $\|x\| \to r^-$, and apply Lemma 4.3.3 to see that

$$C = \frac{1}{r^{1-\frac{d}{2}}\mathbf{I}_{\frac{d}{2}-1}(\sqrt{2\alpha r})}.$$

In other words, we have proven the theorem when $a = 0$. To prove the general result, note that the distribution of $\tau(a;r)$ under the measure $\mathbb{P}_x$ is the same as the distribution of $\tau(0;r)$ under the measure $\mathbb{P}_{x-a}$. This follows quickly from the definition of $\mathbb{P}_x$ as the distribution of $x + B$. □

5 Supplementary Exercises

1. Use Itô's formula (Theorem 3.8.1, Chapter 7) to give another proof of Proposition 2.1.1 in the case $d = 1$. To complete the proof of Proposition 2.1.1, use the multidimensional Itô's formula of Supplementary Exercise 8, Chapter 7.

2. Suppose $\mathcal{A}$ denotes the generator of a Feller process X on a locally compact, separable metric space S. If $\mathcal{A}$ has an essential core C such that $\mathcal{A}(C)$ is dense in $C_0(S)$, prove that $t \mapsto (t, X_t)$ is a Feller process on $[0, \infty[\times S$ whose generator, $\mathcal{A}^\star$, has the form

$$\mathcal{A}^\star \varphi(t, x) = \frac{\partial \varphi}{\partial t} + \mathcal{A}\varphi(t, x).$$

Find an essential core $C^\star$ for $\mathcal{D}(\mathcal{A}^\star)$ such that $\mathcal{A}^\star(C^\star)$ is dense in $C_0(S)$. The process $t \mapsto (t, X_t)$ is called the **space–time** process corresponding to X, and its generator $\mathcal{A}^\star$ is the **parabolic** operator corresponding to the operator $\mathcal{A}$.

3. Let B denote d-dimensional Brownian motion and define the d-dimensional **Ornstein–Uhlenbeck** $\mathcal{O} = (\mathcal{O}_t;\ t \geq 0)$ by $\mathcal{O}_t = e^{-\frac{t}{2}} B_{e^t}$ $(t \geq 0)$. Prove that $\mathcal{O}$ is a Feller process on $\mathbb{R}^d$ and find the form of its generator. Can you identify the domain of the generator?

4. Use Itô's formula (Supplementary Exercise 8, Chapter 7) to show that when B is d-dimensional Brownian motion and $d \geq 2$, $t \mapsto \|B_t\|$ and $t \mapsto \|B_t\|^2$ are Feller processes. Identify the forms of their respective generators. These processes are the d-dimensional **Bessel process** and the d-dimensional **Bessel squared process**, respectively.

5. Show that when $0 < \alpha < 2$, the collection $\mathfrak{S}_\alpha$, defined by equation (3) of Section 2.2, is dense in $C_0(\mathbb{R}^d)$.
(HINT: It may be easier to prove more: Define $\mathfrak{C}$ to be the collection of all $f \in L^\infty(\mathbb{R}^d) \cap L^1(\mathbb{R}^d)$ such that outside a compact set, $\widehat{f} = 0$, and prove that $\mathfrak{C}$ is dense in $C_0(\mathbb{R}^d)$.)

6. Use Theorem 4.2.1 to show that in the notation of the preamble to Section 4, for all $r > 0$ and all $x \in \partial B(a;r)$, $\mathbb{P}_x(\zeta(a;r) = 0) = 1$.

7. Given a 3-dimensional Brownian motion B, prove that when $x \neq 0$, $t \mapsto \|B_t\|^{-1}$ is a continuous local martingale but is not a continuous martingale. What if $d \geq 4$ and B is d-dimensional Brownian motion?
(HINT: Check that $\lim_{t \to \infty} \mathbb{E}_x[\|B_t\|^{-1}] = 0 \neq \|x\|^{-1}$.)

8. Suppose X is a Feller process on a locally compact, separable metric space (S, d). Fix a compact set $K \subset S$ and let $T_K = \inf(s \geq 0 : X_s \notin K)$. Define Y to be X "killed upon leaving K," i.e., $Y_t = X_{t \wedge T_K}$, $t \geq 0$. Recall from Supplementary Exercise 14 of Chapter 8 that Y is a Feller process on K. If the generators of X and Y are denoted by $\mathcal{A}_X$ and $\mathcal{A}_Y$, respectively, show that for all $\varphi \in \mathcal{D}(\mathcal{A}_X)$ that vanish outside K, $\mathcal{A}_Y \varphi = \mathcal{A}_X \varphi$.

9. Suppose X is a Markov process on S_Δ with generator $\mathcal{A}$.
   (i) Show that whenever $f \in \mathcal{D}(\mathcal{A})$ solves $\mathcal{A}f = 0$, $t \mapsto f(X_t)$ is a local martingale.
   (ii) Suppose $v : [0, \infty[\times S \to \mathbb{R}$ is such that for all $t \geq 0$, $x \mapsto v(t, x)$ is in $\mathcal{D}(\mathcal{A})$ and that
   $$\frac{\partial v}{\partial t} = -\mathcal{A}v.$$
   Prove that $t \mapsto v(t, X_t)$ is a local martingale.

10. Suppose $D \subset \mathbb{R}^d$ is open and has compact closure. The **Dirichlet problem** on D with boundary function g is the following PDE: Find a function v such that $\Delta v(x) = 0$, for all $x \in D$ and $v(x) \equiv g(x)$, for all x on the boundary of D. Prove that whenever g is continuous, and when the Dirichlet problem has a bounded solution v, it has the probabilistic representation $v(x) = \mathbb{E}_x[g(B_S)]$, for all $x \in \mathbb{R}^d$, where B denotes d-dimensional Brownian motion and $S = \inf(s \geq 0 : B_s \in D^\complement)$ is the exit time from D.

11. (Continued from Supplementary Exercise 10) The Poisson equation on D (a bounded, open set in $\mathbb{R}^d$) with boundary function g (a function on ∂D) and potential V is the PDE $\Delta v(x) = V(x)$, for all $x \in D$ and $v(x) = g(x)$, for all $x \in \partial D$. Find a probabilistic representation for the unique bounded solution to this PDE, if one exists.

6 Notes on Chapter 9

Section 1 Our discussion of generators is quite slim. You can learn more about this subject from (Ethier and Kurtz 1986; Fukushima et al. 1994; Yosida 1995). What we call the martingale problem is only half of the martingale problem of D. Stroock and S. R. S. Varadhan; cf. Ethier and Kurtz (1986) for a textbook introduction to this topic.

The Doob–Meyer decomposition has other probabilistic implications, as well as analytical ones; see Getoor and Glover (1984) for a sample.

Section 2 To learn more about the material of this section, you can start with (Bertoin 1996; Knight 1981).

Section 4 In the mathematical literature the connections between Markov processes and partial differential equations date at least as far back as Kakutani (1944b, 1945) and Dynkin (1965). The Ciesielski–Taylor theorem (Theorem 4.3.1) was first found, in part, by Paul Lévy and then completely in Ciesielski and Taylor (1962). See the references in the latter paper. Another proof can be found in Knight (1981, Theorem 4.2.20) that essentially forms the basis for the arguments presented here.

Section 5 Supplementary Exercises 11 and 10 are a starting point in the probabilistic approach to second-order PDEs, and explicitly appear in Kac (1951). To learn more about probabilistic solutions to PDEs and many of their applications, see (Bass 1998; Karatzas and Shreve 1991; Revuz and Yor 1994).

10
Probabilistic Potential Theory

Consider a random subset K of $\mathbb{R}^d$. A basic problem in probabilistic potential theory is the following: *For what nonrandom sets E is $\mathbb{P}(K \cap E \neq \varnothing)$ positive?* The archetypal example of such a set K is the range of a random field. Let $X = (X_t;\ t \in \mathbb{R}_+^N)$ denote an N-parameter stochastic process that takes its values in $\mathbb{R}^d$ and consider the random set $K = \{X_s\ :\ s \in \mathbb{R}_+^N\}$.[1] For this particular random set K, the above question translates to the following: *When does the random function X ever enter a given nonrandom set E with positive probability?* Even though we will study a large class of random fields in the next chapter, the solution to the above problem is sufficiently involved that it is best to start with the easiest one-parameter case, which is the subject of the present chapter. Even in this simpler one-parameter setting, it is not clear, a priori, why such problems are interesting. Thus, our starting point will be the analysis of recurrence phenomena for one-parameter Markov processes that have nice properties. To illustrate the key ideas without having to deal with too many technical issues, our discussion of recurrence concentrates on Lévy processes. The astute reader may recognize Section 1 below as the continuous-time analogue of the results of the first section of Chapter 3.

---

[1]The *set* $\{X_s\ :\ s \in \mathbb{R}_+^N\}$ must not be confused with the *process* $(X_s;\ s \in \mathbb{R}_+^N)$. Since this notation for sets is the typical one in most of mathematics, we shall adopt it with no further comment.

1 Recurrent Lévy Processes

In Section 4.3 of Chapter 8 we saw that Lévy processes are continuous-time analogue of the random walks of Part I. We now address the issues of transience and recurrence, in analogy to the discrete theory of Section 1, Chapter 3.

Throughout this section we let $X = (X_t;\ t \geq 0)$ be an $\mathbb{R}^d$-valued Lévy process. Recall from Section 4.3, Chapter 8, that X is a Feller process and the measures $\mathbb{P}_x$ ($x \in \mathbb{R}^d$) are none other than the distributions of the $(\mathbb{R}^d)^{\mathbb{R}_+}$-valued random variable $X + x$. (Heuristically speaking, under the measure $\mathbb{P}_x$, X is a Lévy process, conditioned on $(X_0 = x)$.) We will write $\mathbb{P}$ for $\mathbb{P}_0$, at no great risk of ambiguity. Finally, $\mathcal{F} = (\mathcal{F}_t;\ t \geq 0)$ denotes the complete augmented history of the process X.

A point $x \in \mathbb{R}^d$ is said to be recurrent (for X) if for all $\varepsilon > 0$, the set

$$\{t \geq 0 : |X_t - x| \leq \varepsilon\}$$

is unbounded, $\mathbb{P}$-almost surely. In order to make this measure-theoretically rigorous, we let $\tau_0 = 0$ and iteratively define

$$\tau_{k+1} = \inf(t > 1 + \tau_k : |X_t - x| \leq \varepsilon),$$

with the usual stipulation that $\inf \varnothing = \infty$. Then, we say that x is **recurrent** (for X) if

$$\mathbb{P}(\tau_k < \infty) = 1, \qquad \forall k \geq 1.$$

If x is not recurrent, it is said to be **transient**. If every $x \in \mathbb{R}^d$ is transient (respectively recurrent), then X is said to be **transient** (respectively **recurrent**). Our immediate goal is to find a useful condition for when 0 is recurrent or transient for a Lévy process X. To this end, we present the methods of Khoshnevisan (1997a).

1.1 Sojourn Times

Recall that $|x|$ denotes the (Euclidean) ℓ^∞ norm of *any* (Euclidean) vector x. For all $a \in \mathbb{R}^d$ and all $r > 0$, define the closed ball

$$\overline{\mathcal{B}(a;r)} = \{b \in \mathbb{R}^d : |a - b| \leq r\}. \tag{1}$$

(This is, geometrically speaking, a cube.) We will use the above notation for the remainder of this section.

Our discussion of recurrence begins with the following technical lemma.

Lemma 1.1.1 *For any $t, \varepsilon > 0$,*

$$\int_0^t \mathbb{P}(|X_s| \leq 2\varepsilon)\, ds \leq 4^d \int_0^t \mathbb{P}(|X_s| \leq \varepsilon)\, ds.$$

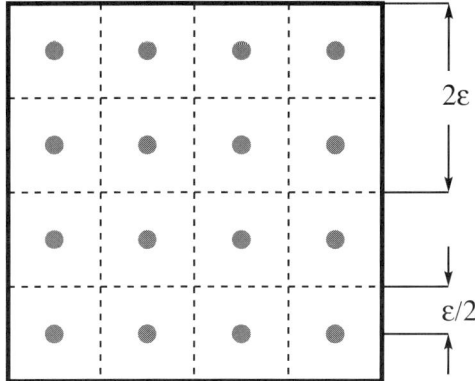

Figure 10.1: Covering $\overline{\mathcal{B}(0;2\varepsilon)}$ with 16 disjoint balls of radius $\frac{1}{2}\varepsilon$; the bullets denote the positions of $a_1, \ldots, a_{16}$.

Proof Since this is a distributional result, we can assume, without loss of generality, that X has shifts $\theta = (\theta_t;\ t \geq 0)$ (why?).

It is simple to see that there exist $a_1, \ldots, a_{4^d} \in \overline{\mathcal{B}(0;2\varepsilon)}$ such that:

- for all $i \neq j$, both in $\{1, \ldots, 4^d\}$, the interiors of $\overline{\mathcal{B}(a_i; \frac{1}{2}\varepsilon)}$ and $\overline{\mathcal{B}(a_j; \frac{1}{2}\varepsilon)}$ are disjoint; and

- $\cup_{i=1}^{4^d} \overline{\mathcal{B}(a_i; \frac{1}{2}\varepsilon)} = \overline{\mathcal{B}(0;2\varepsilon)}$.

Figure 10.1 shows this for $d = 2$. By Fubini's theorem,

$$\int_0^t \mathbb{P}(|X_s| \leq 2\varepsilon)\, ds \leq \sum_{j=1}^{4^d} \mathbb{E}\bigg[\int_0^t \mathbb{1}_{(|X_s - a_j| \leq \frac{1}{2}\varepsilon)}\, ds\bigg]. \qquad (2)$$

Define $\sigma_j = \inf(s > 0:\ |X_s - a_j| \leq \frac{1}{2}\varepsilon)$; by Theorem 1.5.1, Chapter 7, the σ_j's are $\mathcal{F}$-stopping times. Moreover,

$$\int_0^t \mathbb{1}_{(|X_s - a_j| \leq \frac{1}{2}\varepsilon)}\, ds = \int_{\sigma_j}^t \mathbb{1}_{(|X_s - a_j| \leq \frac{1}{2}\varepsilon)}\, ds\ \mathbb{1}_{(\sigma_j \leq t)}$$

$$= \bigg[\int_0^{t-\sigma_j} \mathbb{1}_{(|X_s - a_j| \leq \frac{1}{2}\varepsilon)}\, ds\bigg] \circ \theta_{\sigma_j}\ \mathbb{1}_{(\sigma_j \leq t)}$$

$$\leq \bigg[\int_0^t \mathbb{1}_{(|X_s - a_j| \leq \frac{1}{2}\varepsilon)}\, ds\bigg] \circ \theta_{\sigma_j}\ \mathbb{1}_{(\sigma_j \leq t)}.$$

By the strong Markov property (Theorem 4.2.1, Chapter 8),

$$\mathbb{E}\bigg[\int_0^t \mathbb{1}_{(|X_s - a_j| \leq \frac{1}{2}\varepsilon)}\, ds\bigg] \leq \mathbb{E}\bigg[\mathbb{E}\bigg\{\int_0^t \mathbb{1}_{(|X_s - a_j| \leq \frac{1}{2}\varepsilon)}\, ds\ \bigg|\ \mathcal{F}_{\sigma_j}\bigg\}\ \mathbb{1}_{(\sigma_j \leq t)}\bigg]$$

$$= \mathbb{E}\bigg[\mathbb{E}_{X_{\sigma_j}}\bigg\{\int_0^t \mathbb{1}_{(|X_s - a_j| \leq \frac{1}{2}\varepsilon)}\, ds\bigg\}\ \mathbb{1}_{(\sigma_j \leq t)}\bigg].$$

However, the distribution of $X_s - a_j$ under $\mathbb{P}_x$ is the same as the distribution of X_s under the measure $\mathbb{P}_{x-a_j}$. Thus,

$$\mathbb{E}\left[\int_0^t \mathbf{1}_{(|X_s - a_j| \leq \frac{1}{2}\varepsilon)}\, ds\right] \leq \mathbb{E}\left[\mathbb{E}_{X_{\sigma_j} - a_j}\left\{\int_0^t \mathbf{1}_{(|X_s| \leq \frac{1}{2}\varepsilon)}\, ds\right\} \mathbf{1}_{(\sigma_j \leq t)}\right].$$

By the right continuity of $t \mapsto X_t$, if $\sigma_j < \infty$, then $|X_{\sigma_j} - a_j| \leq \frac{1}{2}\varepsilon$. Therefore,

$$\mathbb{E}\left[\int_0^t \mathbf{1}_{(|X_s - a_j| \leq \frac{1}{2}\varepsilon)}\, ds\right] \leq \mathbb{E}\left[\sup_{x:\, |x| \leq \frac{1}{2}\varepsilon} \mathbb{E}_x\left\{\int_0^t \mathbf{1}_{(|X_s| \leq \frac{1}{2}\varepsilon)}\, ds\right\} \mathbf{1}_{(\sigma_j \leq t)}\right]$$

$$= \mathbb{E}\left[\sup_{x:\, |x| \leq \frac{1}{2}\varepsilon} \mathbb{E}\left\{\int_0^t \mathbf{1}_{(|X_s + x| \leq \frac{1}{2}\varepsilon)}\, ds\right\} \mathbf{1}_{(\sigma_j \leq t)}\right]$$

$$\leq \mathbb{E}\left[\mathbb{E}\left\{\int_0^t \mathbf{1}_{(|X_s| \leq \varepsilon)}\, ds\right\} \mathbf{1}_{(\sigma_j \leq t)}\right]$$

$$\leq \int_0^t \mathbb{P}(|X_s| \leq \varepsilon)\, ds.$$

This, together with equation (2) proves the lemma. □

The following is another useful technical estimate.

Lemma 1.1.2 *For any $b > a > 0$ and for all $\varepsilon > 0$,*

$$\mathbb{E}\left[\left\{\int_a^b \mathbf{1}_{(|X_s| \leq \varepsilon)}\, ds\right\}^2\right] \leq 2 \cdot 4^d \int_a^b \mathbb{P}(|X_s| \leq \varepsilon)\, ds \cdot \int_0^b \mathbb{P}(|X_s| \leq \varepsilon)\, ds.$$

Proof Once more, as in our proof of Lemma 1.1.1, we may assume the existence of shifts $\theta = (\theta_t;\ t \geq 0)$. By Fubini's theorem,

$$\mathbb{E}\left[\left\{\int_a^b \mathbf{1}_{(|X_s| \leq \varepsilon)}\, ds\right\}^2\right] = 2 \int_a^b \int_s^b \mathbb{P}(|X_s| \leq \varepsilon,\, |X_t| \leq \varepsilon)\, dt\, ds.$$

By the definition of Lévy processes, whenever $t \geq s \geq 0$, the random vector $X_t - X_s$ is independent of X_s and has the same distribution as X_{t-s}. Thus,

$$\mathbb{E}\left[\left\{\int_a^b \mathbf{1}_{(|X_s| \leq \varepsilon)}\, ds\right\}^2\right] = 2 \int_a^b \int_s^b \mathbb{P}(|X_s| \leq \varepsilon,\, |X_t - X_s + X_s| \leq \varepsilon)\, dt\, ds$$

$$\leq 2 \int_a^b \int_s^b \mathbb{P}(|X_s| \leq \varepsilon)\, \mathbb{P}(|X_t - X_s| \leq 2\varepsilon)\, ds\, dt$$

$$= 2 \int_a^b \int_s^b \mathbb{P}(|X_s| \leq \varepsilon)\, \mathbb{P}(|X_{t-s}| \leq 2\varepsilon)\, dt\, ds$$

$$\leq 2 \int_a^b \mathbb{P}(|X_s| \leq \varepsilon)\, ds \cdot \int_0^b \mathbb{P}(|X_t| \leq 2\varepsilon)\, dt.$$

Our lemma now follows from Lemma 1.1.1. □

For all $a, \varepsilon > 0$, define

$$A_a(\varepsilon) = \int_0^a \mathbb{1}_{(|X_s| \leq \varepsilon)}\, ds. \tag{3}$$

The process $A(\varepsilon) = (A_a(\varepsilon);\ a > 0)$ is the so-called **sojourn time process** on $\mathcal{B}(0;\varepsilon)$, for X.

Proposition 1.1.1 *The following are equivalent:*

(i) there exists an $\varepsilon > 0$ such that $\mathbb{E}\{A_\infty(\varepsilon)\} = \infty$;

(ii) for all $\varepsilon > 0$, $\mathbb{E}\{A_\infty(\varepsilon)\} = \infty$;

(iii) there exists an $\varepsilon > 0$ such that $\mathbb{P}\{A_\infty(\varepsilon) = \infty\} > 0$;

(iv) for all $\varepsilon > 0$, $\mathbb{P}\{A_\infty(\varepsilon) = \infty\} > 0$; and

(v) for all $\varepsilon > 0$, $\mathbb{P}\{A_\infty(\varepsilon) = \infty\} = 1$.

Proof The equivalence of (i) and (ii) follows from Lemma 1.1.1. Next, we show that $(i) \Rightarrow (iii)$. We apply Lemma 1.1.2 to obtain the following: For all $a, \varepsilon > 0$,

$$\mathbb{E}\big[\{A_a(\varepsilon)\}^2\big] \leq 2 \cdot 4^d \cdot \big\{\mathbb{E}[A_a(\varepsilon)]\big\}^2.$$

Thus, we can combine this with the Paley–Zygmund lemma (Lemma 1.4.1, Chapter 3) to deduce that for all $a, \varepsilon > 0$,

$$\mathbb{P}\{A_\infty(\varepsilon) \geq \tfrac{1}{2}\mathbb{E}[A_a(\varepsilon)]\} \geq \mathbb{P}\{A_a(\varepsilon) \geq \tfrac{1}{2}\mathbb{E}[A_a(\varepsilon)]\} \geq \frac{1}{2 \cdot 4^{1+d}}.$$

Let $a \to \infty$ in the leftmost term to prove that both $(i) \Rightarrow (iii)$ and $(ii) \Rightarrow (iv)$. Since $(iv) \Rightarrow (iii) \Rightarrow (i)$, and since $(i) \Rightarrow (ii)$ has already been established, we can deduce the equivalence of (i) through (iv). It suffices to prove that $(iv) \Rightarrow (v)$ to finish the demonstration. On the other hand, the event $(A_\infty(\varepsilon) = \infty)$ is a tail event for X. By Supplementary Exercise 10 of Chapter 8, the tail σ-field of X is trivial. Therefore, $\mathbb{P}\{A_\infty(\varepsilon) = \infty\}$ is 0 or 1, and our task is finished. □

1.2 Recurrence of the Origin

We are now in a position to decide when the origin $0 \in \mathbb{R}^d$ is recurrent for X. The problem of deciding when other points $x \in \mathbb{R}^d$ are recurrent is deferred to Supplementary Exercise 8.

Theorem 1.2.1 *The following are equivalent:*

(i) there exists $\varepsilon > 0$ such that $\int_0^\infty \mathbb{P}(|X_s| \leq \varepsilon)\,ds = \infty$;

(ii) for all $\varepsilon > 0$, $\int_0^\infty \mathbb{P}(|X_s| \leq \varepsilon)\,ds = \infty$;

(iii) for all $\varepsilon > 0$, $\int_0^\infty \mathbf{1}_{(|X_s| \leq \varepsilon)}\,ds = \infty$, a.s.; and

(iv) $0 \in \mathbb{R}^d$ is recurrent for X.

Our proof of Theorem 1.2.1 rests on one technical lemma.

Lemma 1.2.1 *For all $\varepsilon > 0$ and all $b > a > 0$,*

$$\frac{\int_a^{\frac{1}{2}(a+b)} \mathbb{P}(|X_s| \leq \varepsilon)\,ds}{2 \cdot 4^d \int_0^{\frac{1}{2}(a+b)} \mathbb{P}(|X_s| \leq \varepsilon)\,ds} \leq \mathbb{P}\Big(\inf_{a \leq s \leq \frac{1}{2}(b+a)} |X_s| \leq \varepsilon\Big) \leq \frac{\int_a^b \mathbb{P}(|X_s| \leq 2\varepsilon)\,ds}{\int_0^{\frac{1}{2}(b-a)} \mathbb{P}(|X_s| \leq \varepsilon)\,ds}. \tag{1}$$

Proof Consider the right-continuous bounded martingale $M = (M_t;\ t \geq 0)$ defined by

$$M_t = \mathbb{E}\Big[\int_a^b \mathbf{1}_{(|X_s| \leq \varepsilon)}\,ds \,\Big|\, \mathcal{F}_t\Big], \qquad t \geq 0.$$

For all $0 \leq t \leq \frac{1}{2}(a+b)$,

$$M_t \geq \mathbb{E}\Big[\int_t^b \mathbf{1}_{(|X_s| \leq \varepsilon)}\,ds \,\Big|\, \mathcal{F}_t\Big]\mathbf{1}_{(|X_t| \leq \frac{1}{2}\varepsilon)}$$

$$\geq \mathbb{E}\Big[\int_t^b \mathbf{1}_{(|X_s - X_t| \leq \frac{1}{2}\varepsilon)}\,ds \,\Big|\, \mathcal{F}_t\Big]\mathbf{1}_{(|X_t| \leq \frac{1}{2}\varepsilon)}$$

$$= \mathbb{E}\Big[\int_0^{b-t} \mathbf{1}_{(|X_s| \leq \frac{1}{2}\varepsilon)}\,ds\Big]\mathbf{1}_{(|X_t| \leq \frac{1}{2}\varepsilon)}.$$

We have used the stationarity and the independence of the increments of X in the last line. Since $t \leq \frac{1}{2}(a+b)$, we obtain

$$M_t \geq \int_0^{\frac{1}{2}(b-a)} \mathbb{P}(|X_s| \leq \tfrac{1}{2}\varepsilon)\,ds\,\mathbf{1}_{(|X_t| \leq \frac{1}{2}\varepsilon)}, \qquad \mathbb{P}\text{-a.s.} \tag{2}$$

The above holds $\mathbb{P}$-a.s. for each $t \geq 0$. By the σ-additivity of probability measures, (2) holds $\mathbb{P}$-a.s., simultaneously for all rational $t \geq 0$. On the other hand, M is right-continuous, and so is X; see the last paragraph of Section 1.4, Chapter 7. Thus, equation (2) holds $\mathbb{P}$-a.s., for all $t \geq 0$. (Why?)

Now let $\sigma = \inf(t \in [a, \frac{1}{2}(a+b)] : |X_t| \leq \frac{1}{2}\varepsilon)$, where $\inf \varnothing = \infty$. By Theorem 1.5.1 of Chapter 7, σ is an $\mathcal{F}$-stopping time. Thus,

$$\mathbb{E}[M_\sigma \mathbf{1}_{(\sigma < \infty)}] \geq \int_0^{\frac{1}{2}(b-a)} \mathbb{P}(|X_s| \leq \tfrac{1}{2}\varepsilon)\,ds \cdot \mathbb{P}\Big(\inf_{a \leq t \leq \frac{1}{2}(a+b)} |X_t| \leq \tfrac{1}{2}\varepsilon\Big).$$

Since M is bounded, by the bounded convergence theorem,
$$\mathbb{E}\big[M_\sigma \mathbf{1}_{(\sigma<\infty)}\big] = \lim_{n\to\infty} \mathbb{E}\big[M_{\sigma\wedge n}\big],$$
which, thanks to the optional stopping theorem (Theorem 1.6.1 of Chapter 7), equals $\mathbb{E}[M_0] = \int_a^b \mathbb{P}(|X_s| \leq \varepsilon)\, ds$.

Replacing ε by 2ε, we obtain the first inequality of equation (1). To prove the second inequality of (1), note that
$$\int_a^{\frac{1}{2}(a+b)} \mathbf{1}_{(|X_s|\leq \frac{1}{2}\varepsilon)}\, ds > 0 \implies \inf_{a\leq t\leq \frac{1}{2}(a+b)} |X_t| \leq \varepsilon.$$

Thus,
$$\mathbb{P}\Big(\inf_{a\leq t\leq \frac{1}{2}(a+b)} |X_t| \leq \varepsilon\Big) \geq \mathbb{P}\Big(\int_a^{\frac{1}{2}(a+b)} \mathbf{1}_{(|X_s|\leq\varepsilon)}\, ds > 0\Big)$$
$$\geq \frac{\big\{\mathbb{E}\big[\int_a^{\frac{1}{2}(a+b)} \mathbf{1}_{(|X_s|\leq\varepsilon)}\, ds\big]\big\}^2}{\mathbb{E}\big[\big\{\int_a^{\frac{1}{2}(a+b)} \mathbf{1}_{(|X_s|\leq\varepsilon)}\, ds\big\}^2\big]}.$$

In the last line we have used the Paley–Zygmund lemma; see Lemma 1.4.1 of Chapter 3. Equation (1) follows from this and Lemma 1.1.2. □

Proof of Theorem 1.2.1 Proposition 1.1.1 shows that (i), (ii), and (iii) are equivalent. Next, let us suppose that (ii) fails to hold. By Lemma 1.2.1, for all $a, \varepsilon > 0$,
$$\mathbb{P}\Big(\inf_{t\geq a} |X_t| \leq \varepsilon\Big) \leq \frac{\int_a^\infty \mathbb{P}(|X_s|\leq \frac{1}{2}\varepsilon)\, ds}{\int_0^\infty \mathbb{P}(|X_s|\leq \varepsilon)\, ds}.$$

Consequently, by Lemma 1.1.1,
$$\lim_{a\to\infty} \mathbb{P}\Big(\inf_{t\geq a} |X_t| \leq \varepsilon\Big) = 0.$$

In particular, 0 cannot be recurrent (why?). That is, we have verified $(iv) \Rightarrow (ii)$. Finally, let us suppose that (iv) fails to hold. This means that for any $\varepsilon > 0$, $\mathbb{P}(L_\varepsilon < \infty) = 1$, where
$$L_\varepsilon = \sup(t \geq 0 : |X_t| \leq \varepsilon).$$

(Why?) Since a.s.,
$$\int_0^\infty \mathbf{1}_{(|X_s|\leq\varepsilon)}\, ds \leq L_\varepsilon < \infty,$$

$(iii) \Rightarrow (iv)$ follows. □

Exercise 1.2.1 If 0 is transient, show that

$$\frac{\int_t^\infty \mathbb{P}(|X_s| \leq \varepsilon)\,ds}{2 \cdot 4^d \int_0^\infty \mathbb{P}(|X_s| \leq \varepsilon)\,ds} \leq \mathbb{P}(|X_s| \leq \varepsilon \text{ for some } s \geq t)$$
$$\leq \frac{4^d \int_t^\infty \mathbb{P}(|X_s| \leq \varepsilon)\,ds}{\int_0^\infty \mathbb{P}(|X_s| \leq \varepsilon)\,ds},$$

for all choices of $t, \varepsilon > 0$. □

To put this subsection in the general context of this chapter, define the random set $K = \{X_s : a \leq s \leq b\}$. Lemma 1.2.1 estimates $\mathbb{P}(K \cap E \neq \varnothing)$, where E denotes the closed ball of radius ε about the origin. Recalling the first paragraph of this chapter, we have seen that issues of recurrence can be boiled down to questions about whether K intersects a certain set (here a closed ball) with positive probability.

1.3 Escape Rates for Isotropic Stable Processes

In some specific cases, the method used in our characterization of the recurrence of 0 (for a Lévy process X) has some delicate and intriguing consequences. We now describe one such fact for a special class of Lévy processes.

Let $X = (X_t;\ t \geq 0)$ designate an isotropic stable Lévy process on $\mathbb{R}^d$ with index $\alpha \in\,]0, 2]$; cf. Example 3, Section 4.3 of Chapter 8. Recall also that when $\alpha = 2$, X is d-dimensional Brownian motion.

Define the function $U_\alpha : \mathbb{R}_+ \to \mathbb{R}_+$ by

$$U_\alpha(r) = \begin{cases} 1, & \text{if } \alpha > d, \\ \left[\ln_+\!\left(\tfrac{1}{r}\right)\right]^{-1}, & \text{if } \alpha = d, \\ r^{d-\alpha}, & \text{if } \alpha < d. \end{cases} \qquad (1)$$

The following is the main result of this and the next subsection. When $\alpha = 2$ (i.e., Brownian motion), it was found by Dvoretzky and Erdős (1951) when $d \geq 3$ and by Spitzer (1958) when $d = 2$. At the level of generality described below, the theorem was found in Takeuchi (1964a, 1964b), and Takeuchi and Watanabe (1964); see also Hendricks (1972, 1974, 1979), Sato (1999), and Bertoin (1996, Theorem 6, Chapter VIII.2), and their references for further elaborations. Our proofs will follow the general method of Khoshnevisan (1997a) closely.

Theorem 1.3.1 *Let $\varphi : \mathbb{R}_+ \to \mathbb{R}_+$ be nonincreasing.*

(i) *If $d < \alpha$, then with probability one, the level set $X^{-1}\{0\}$ is unbounded, where*

$$X^{-1}\{0\} = \{t \geq 0 : X_t = 0\}.$$

(ii) If $d \geq \alpha$, then with probability one,

$$\liminf_{t \to \infty} \frac{|X_t|}{t^{\frac{1}{\alpha}} \varphi(t)} = \begin{cases} \infty, & \text{if } \mathfrak{J}(\varphi) < \infty, \\ 0, & \text{if } \mathfrak{J}(\varphi) = \infty, \end{cases}$$

where

$$\mathfrak{J}(\varphi) = \int_1^\infty \frac{U_\alpha(\varphi(t))}{t} \, dt.$$

Theorem 1.3.1 will be demonstrated in this and the next subsection. However, we first examine its content.

Examples

(a) If $d < \alpha$, we see that X is recurrent. Of course, since $\alpha \in]0,2]$, the only possible value for d in this regime is $d = 1$. That is, we conclude that whenever $d = 1$ and $\alpha > 1$, X is recurrent. In fact, one can show the stronger property that when $\alpha > 1$, for any $a \in \mathbb{R}$, there is an a.s. unbounded set of times t such that $X_t = a$; cf. Supplementary Exercise 7. This property is called **point recurrence**. That is, not only does X come within ε of any point $a \in \mathbb{R}$ infinitely often in the long run, but it in fact hits that point infinitely often in the long run. □

(b) If $d > \alpha$, then $U_\alpha(x) = x^{d-\alpha}$. Choose $\varphi(t) = \{\ln_+ t\}^{-\beta}$ for some $\beta > 0$ and note that for this choice of φ,

$$\mathfrak{J}(\varphi) < \infty \iff \beta > \frac{1}{d - \alpha}.$$

In particular, with probability one,

$$\liminf_{t \to \infty} \frac{\{\ln_+ t\}^\beta}{t^{\frac{1}{\alpha}}} |X_t| = \begin{cases} \infty, & \text{if } \beta > (d-\alpha)^{-1}, \\ 0, & \text{if } \beta \leq (d-\alpha)^{-1}. \end{cases}$$

An immediate consequence of this is that $\lim_{t \to \infty} |X_t| = \infty$, a.s. That is, the process X is transient when $d > \alpha$ (why?). □

(c) We now turn to the "critical case" $d = \alpha$. Since $\alpha \in]0,2]$, this means that either $(d, \alpha) = (2, 2)$ or $(d, \alpha) = (1, 1)$. In the latter case, X is 2-dimensional Brownian motion, while in the former case, it is called the 1-dimensional **Cauchy process**; cf. Example 3, Section 4.3 of Chapter 8, for an explanation of this terminology. In either case, $U_\alpha(x) = \{\ln_+(1/x)\}^{-1}$. For any $\beta > 0$ fixed, let us choose $\varphi(x) = \exp(-\{\ln_+ x\}^\beta)$. Then, it is easy to see that

$$\mathfrak{J}(\varphi) < \infty \iff \beta > 1.$$

352   10. Probabilistic Potential Theory

In other words, with probability one,

$$\liminf_{t\to\infty} \frac{|X_t|}{t^{\frac{1}{\alpha}}\exp\{-(\ln t)^\beta\}} = \begin{cases} \infty, & \text{if } \beta > 1, \\ 0, & \text{if } \beta \leq 1. \end{cases}$$

From this we can conclude that a.s., for all t large, $\liminf_{t\to\infty}|X_t| = 0$. In other words, in the case $d = \alpha$, 0 is recurrent. Using Supplementary Exercise 8, this can be strengthened to show that when $d = \alpha$, X is a recurrent process that is not point recurrent. □

We conclude this subsection by starting on our proof of Theorem 1.3.1; the following results are elementary distributional properties of the isotropic stable process X.

Lemma 1.3.1 (Scaling Lemma) *If $c > 0$ is fixed, then the finite-dimensional distributions of $(X_{ct};\ t \geq 0)$ and $(c^{\frac{1}{\alpha}}X_t;\ t \geq 0)$ are the same.*

Proof We will prove this lemma by directly examining the characteristic function of X at various times. Suppose $0 = t_0 < t_1 < \cdots < t_m$ are fixed. We need to show that the distributions of $(X_{ct_1}, \ldots, X_{ct_m})$ and $c^{\frac{1}{\alpha}}(X_{t_1}, \ldots, X_{t_m})$ are one and the same. That is, we will show that for all $\xi_1, \ldots, \xi_m \in \mathbb{R}^d$,

$$\mathbb{E}\left[\exp\left\{i\sum_{j=1}^m \xi_j \cdot X_{ct_j}\right\}\right] = \mathbb{E}\left[\exp\left\{ic^{\frac{1}{\alpha}}\sum_{j=1}^m \xi_j \cdot X_{t_j}\right\}\right].$$

By considering linear combinations, it suffices to show that for all choices of $\zeta_1, \ldots, \zeta_m \in \mathbb{R}^d$,

$$\mathbb{E}\left[\exp\left\{i\sum_{j=1}^m \zeta_j \cdot (X_{ct_j} - X_{ct_{j-1}})\right\}\right] = \mathbb{E}\left[\exp\left\{ic^{\frac{1}{\alpha}}\sum_{j=1}^m \zeta_j \cdot (X_{t_j} - X_{t_{j-1}})\right\}\right].$$

(Why?) On the other hand, by the stationary, independent increments property of Lévy processes,

$$\mathbb{E}\left[\exp\left\{i\sum_{j=1}^m \zeta_j \cdot (X_{ct_j} - X_{ct_{j-1}})\right\}\right] = \prod_{j=1}^m \mathbb{E}\left[\exp\left\{i\zeta_j \cdot X_{c(t_j - t_{j-1})}\right\}\right], \quad (2)$$

while

$$\mathbb{E}\left[\exp\left\{ic^{\frac{1}{\alpha}}\sum_{j=1}^m \zeta_j \cdot (X_{t_j} - X_{t_{j-1}})\right\}\right] = \prod_{j=1}^m \mathbb{E}\left[\exp\{ic^{\frac{1}{\alpha}}\zeta_j \cdot X_{(t_j - t_{j-1})}\}\right]. \quad (3)$$

Recall from Example 3 of Section 4.3, Chapter 8, that the characteristic function of X_t is

$$\mathbb{E}\left[e^{i\xi \cdot X_t}\right] = e^{-\frac{1}{2}\|\xi\|^\alpha}, \qquad \xi \in \mathbb{R}^d.$$

By the form of this characteristic function, the terms in equations (2) and (3) are both equal to

$$\prod_{j=1}^{m} \exp\Big(-\frac{c(t_j - t_{j-1})\|\zeta_j\|^\alpha}{2} \Big),$$

which proves our lemma. □

The following is immediate from the inversion formula for characteristic functions.

Lemma 1.3.2 *For any $t \geq 0$, the random vector X_t has the following (isotropic) probability density function:*

$$q_t(x) = (2\pi)^{-\frac{d}{2}} \int_{\mathbb{R}^d} e^{-i\xi \cdot x} e^{-\frac{1}{2}t\|\xi\|^\alpha} d\xi, \qquad x \in \mathbb{R}^d.$$

This formula has some important consequences, which we list below:

Corollary 1.3.1 *(i) For all $t > 0$, $q_t(0) = \sup_{x \in \mathbb{R}^d} q_t(x) > 0$.*

(ii) $t \mapsto q_t(0)$ is nonincreasing. In fact, $q_t(0) = At^{-\frac{1}{\alpha}}$, where $A = 2^{\frac{1}{\alpha}}(2\pi)^{-\frac{d}{2}} \int_{\mathbb{R}^d} e^{-\|\zeta\|^\alpha} d\zeta$.

(iii) The function $(t, x) \mapsto q_t(x)$ is uniformly continuous on $[\varepsilon, \varepsilon^{-1}] \times \mathbb{R}^d$ for any $\varepsilon > 0$.

(iv) For all $\varepsilon \in\,]0, 1[$, there exist a finite constant $\eta > 0$ and an open set $G \ni 0$ in $\mathbb{R}^d$ such that for all $\varepsilon < t < \varepsilon^{-1}$ and for all $x \in G$, $q_t(x) \geq \eta$.

Exercise **1.3.1** Verify Corollary 1.3.1. □

Exercise **1.3.2** Prove that the above density function q_t is isotropic, i.e., depends on x only through $\|x\|$, and satisfies the following scaling relation: For all $t > 0$ and all $x \in \mathbb{R}^d$, $q_t(x) = t^{-\frac{d}{\alpha}} q_1(xt^{-\frac{1}{\alpha}})$. □

1.4 Hitting Probabilities

The following is the main step in our proof of Theorem 1.3.1; it computes, up to a multiplicative constant, the probability that the process X hits the closed ball $\mathcal{B}(0; \varepsilon)$ at some time between a and b. Recall the function U_α from equation (1) of Section 1.3.

Proposition 1.4.1 *For all $b > a > 0$, there exist finite constants $\varepsilon_0 \in\,]0, 1[$ and $A > 1$ such that for all $\varepsilon \in\,]0, \varepsilon_0[$,*

$$\frac{1}{A} U_\alpha(\varepsilon) \leq \mathbb{P}\Big(\inf_{a \leq s \leq b} |X_s| \leq \varepsilon \Big) \leq A U_\alpha(\varepsilon).$$

10. Probabilistic Potential Theory

The above is proved in two steps, which we present as lemmas. For the sake of notational convenience, throughout our proof of Proposition 1.4.1 $A, A_1, A_2, \ldots$ will denote unimportant constants whose values may change from one lemma to another.

Lemma 1.4.1 *For every $b > a > 0$, there exist $\varepsilon_0 > 0$ and a finite $A > 1$ such that for all $\varepsilon \in]0, \varepsilon_0[$ and all $s \in [a, b]$,*

$$\frac{1}{A}\varepsilon^d \leq \mathbb{P}(|X_s| \leq \varepsilon) \leq A\varepsilon^d.$$

Proof By Lemma 1.3.2 and its corollary (Corollary 1.3.1), for any $s, \varepsilon > 0$,

$$\mathbb{P}(|X_s| \leq \varepsilon) \leq q_s(0)(2\varepsilon)^d.$$

(Recall that $|x|$ is the ℓ^∞ norm of x.) Thus, for all $a \leq s \leq b$,

$$\mathbb{P}(|X_s| \leq \varepsilon) \leq (2\pi)^{-\frac{d}{2}} \int_{\mathbb{R}^d} e^{-\frac{1}{2}s\|\xi\|^\alpha} d\xi \cdot (2\varepsilon)^d \leq A_1 \varepsilon^d,$$

where $A_1 = \left(\frac{2}{\pi}\right)^{\frac{d}{2}} \int_{\mathbb{R}^d} e^{-\frac{1}{2}a\|\xi\|^\alpha} d\xi$. On the other hand,

$$\mathbb{P}(|X_s| \leq \varepsilon) \geq \inf_{x: |x| \leq \varepsilon} q_s(x)(2\varepsilon)^d.$$

By another application of Lemma 1.3.2 and Corollary 1.3.1, there exists $\varepsilon_0 > 0$ small enough and a constant $A_1 > 1$ such that

$$\frac{1}{A_2} = 2^d \inf_{s \in [a,b]} \inf_{x: |x| \leq \varepsilon_0} q_s(x) > 0.$$

The lemma follows with $A = A_1 \vee A_2 \vee 1$. $\square$

Lemma 1.4.2 *For any $c > 0$, there exist finite constants $A > 1$ and $\varepsilon_0 \in]0, 1[$ such that for all $\varepsilon \in]0, \varepsilon_0[$,*

$$\frac{\varepsilon^d}{A U_\alpha(\varepsilon)} \leq \int_0^c \mathbb{P}(|X_s| \leq \varepsilon) \, ds \leq \frac{A \varepsilon^d}{U_\alpha(\varepsilon)}.$$

Proof We will verify the upper bound on the integral; the lower bound is derived similarly. By the scaling lemma (Lemma 1.3.1),

$$\int_0^c \mathbb{P}(|X_s| \leq \varepsilon) \, ds = \int_0^c \mathbb{P}\left(|X_1| \leq \frac{\varepsilon}{s^{\frac{1}{\alpha}}}\right) ds.$$

By Lemma 1.3.2, for all $r > 0$,

$$\mathbb{P}(|X_1| \leq r) \leq \{q_1(0) r^d \wedge 1\} \leq A_1 \{r^d \wedge 1\},$$

where $A_1 = q_1(0) \vee 1$. Thus, for all $\varepsilon \in \,]0, c^{\frac{1}{\alpha}}[$,

$$\int_0^c \mathbb{P}(|X_s| \leq \varepsilon)\, ds \leq A_1 \int_0^c \left\{ \frac{\varepsilon}{s^{\frac{1}{\alpha}}} \wedge 1 \right\}^d ds = A_1 \left\{ \varepsilon^\alpha + \varepsilon^d \int_{\varepsilon^\alpha}^c s^{-\frac{d}{\alpha}}\, ds \right\}.$$

Computing directly, we see that

$$\limsup_{\varepsilon \to 0^+} \frac{U_\alpha(\varepsilon)}{\varepsilon^d} \int_0^c \mathbb{P}(|X_s| \leq \varepsilon)\, ds \leq 2 A_1.$$

The announced upper bound on the integral follows suit. □

Exercise 1.4.1 Verify the lower bound in Lemma 1.4.2. □

Proposition 1.4.1 is now easily seen to follow from Lemmas 1.2.1, 1.4.1, and 1.4.2.

Exercise 1.4.2 When $d > \alpha$, prove there exist two finite positive constants A_1 and A_2 such that for all $T, R > 0$,

$$A_1 \left(\frac{T}{R^\alpha} \right)^{1 - \frac{d}{\alpha}} \leq \mathbb{P}(|X_s| \leq R \text{ for some } s \geq T) \leq A_2 \left(\frac{T}{R^\alpha} \right)^{1 - \frac{d}{\alpha}}$$

as long as $TR^{-\alpha} > 1$.
(HINT: See Exercise 1.2.1.) □

For technical reasons, we will later need the following improvement over the probability upper bound in Proposition 1.4.1.

Lemma 1.4.3 *For all $b > a > 0$, there exist finite constants $\varepsilon_0 \in \,]0, 1[$ and $A > 1$ such that for all $\varepsilon \in \,]0, \varepsilon_0[$,*

$$\sup_{x \in \mathbb{R}^d} \mathbb{P}_x \left(\inf_{a \leq s \leq b} |X_s| \leq \varepsilon \right) \leq A U_\alpha(\varepsilon).$$

Proof Our proof is analogous to that of Lemma 1.2.1. Define

$$M_t = \mathbb{E}_x \left[\int_a^{2b-a} \mathbb{1}_{(|X_s| \leq \varepsilon)}\, ds \,\bigg|\, \mathcal{F}_t \right], \qquad t \geq 0.$$

The process $M = (M_t; t \geq 0)$ is a martingale under $\mathbb{P}_x$. Moreover, $\mathbb{P}_x$-a.s.,

$$M_t \geq \int_0^b \mathbb{P}(|X_s| \leq \tfrac{1}{2}\varepsilon)\, ds\, \mathbb{1}_{(|X_t| \leq \frac{1}{2}\varepsilon)}, \qquad t \geq 0;$$

see the derivation of (2) in Section 1.2 above. Define the $\mathcal{F}$-stopping time $\sigma = \inf(s \in [a, b] : |X_s| \leq \tfrac{1}{2}\varepsilon)$ to see that for any $x \in \mathbb{R}^d$, $\mathbb{P}_x$-a.s.,

$$M_\sigma \mathbb{1}_{(\sigma < \infty)} \geq \int_0^b \mathbb{P}(|X_s| \leq \tfrac{1}{2}\varepsilon)\, ds\, \mathbb{1}_{(\sigma < \infty)}.$$

As in the described proof of Lemma 1.2.1, we can take expectations (this time $\mathbb{E}_x$-expectations), use the optional stopping theorem, and conclude that

$$\mathbb{E}_x[M_0] = \int_0^b \mathbb{P}(|X_s| \leq \tfrac{1}{2}\varepsilon)\,ds \cdot \mathbb{P}_x\bigl(\inf_{a \leq t \leq b} |X_t| \leq \tfrac{1}{2}\varepsilon\bigr). \quad (1)$$

It is very important to note that the right-hand side has one probability term involving $\mathbb{P} = \mathbb{P}_0$ and one involving $\mathbb{P}_x$. But

$$\mathbb{E}_x[M_0] = \int_a^b \mathbb{P}_x(|X_s| \leq \varepsilon)\,ds = \int_a^b \mathbb{P}(|X_s + x| \leq \varepsilon)\,ds$$

$$= \int_a^b \int_{B(0;\varepsilon)} q_s(a+x)\,da\,ds \leq q_a(0)(b-a)(2\varepsilon)^d. \quad (2)$$

In the last line we have used Corollary 1.3.1. By Lemma 1.1.1,

$$\int_0^b \mathbb{P}(|X_s| \leq \tfrac{1}{2}\varepsilon)\,ds \geq 4^{-d} \int_0^b \mathbb{P}(|X_s| \leq \varepsilon)\,ds.$$

Applying Lemma 1.4.2, we obtain finite constants $\varepsilon_0 > 0$ and $A_1 > 1$ such that for all $\varepsilon \in \,]0, \varepsilon_0[$,

$$\int_0^b \mathbb{P}(|X_s| \leq \tfrac{1}{2}\varepsilon)\,ds \geq \frac{A_1 \varepsilon^d}{U_\alpha(\varepsilon)}.$$

Combining this with equations (1) and (2), we deduce the lemma with $A = 8^d q_a(0)(b-a) A_1^{-1}$. $\square$

We will also need the following "converse" to the Borel–Cantelli lemma that is found in (Chung and Erdős 1952; Kochen and Stone 1964); see also Exercise 1.4.3, Chapter 3.

Lemma 1.4.4 (Kochen–Stone Lemma) *Suppose $E_1, E_2, \ldots$ are measurable events. Let $N_n = \sum_{j=1}^n \mathbf{1}_{E_j}$ and assume that:*

(i) $\mathbb{E}[N_\infty] = \infty$; *and*

(ii) $\liminf_{n \to \infty} \mathbb{E}[(N_n)^2]/\{\mathbb{E}[N_n]\}^2 < \infty$.

Then, $\mathbb{P}(E_n \text{ infinitely often}) > 0$.

We now have accumulated the tools necessary for the following.

Proof of Theorem 1.3.1: Part (i) By Proposition 1.4.1, we can find finite constants $\varepsilon_0 > 0$ and $A > 1$ such that for all $\varepsilon \in \,]0, \varepsilon_0[$,

$$\mathbb{P}\bigl(\inf_{1 \leq t \leq e} |X_t| \leq \varepsilon\bigr) \geq A^{-1}.$$

(You should recall the definition of U_α from equation (1) of Section 1.3.) Letting $\varepsilon \to 0$, continuity of probability measures guarantees us that

$$\mathbb{P}(X^{-1}\{0\} \cap [1,e] \neq \varnothing) = \mathbb{P}\left(\inf_{1 \leq t \leq e} |X_t| = 0\right) \geq \frac{1}{A} > 0.$$

On the other hand, by scaling (Lemma 1.3.1), $\inf_{1 \leq t \leq e} |X_t|$ has the same distribution as $e^{-\frac{j}{\alpha}} \inf_{e^j \leq t \leq e^{j+1}} |X_t|$. Thus, for all $j \geq 1$,

$$\mathbb{P}(X^{-1}\{0\} \cap [e^j, e^{j+1}] \neq \varnothing) \geq \frac{1}{A}.$$

Since A is independent of $j \geq 1$,

$$\mathbb{P}\bigl(X^{-1}\{0\} \cap [e^j, e^{j+1}] \neq \varnothing \text{ infinitely often}\bigr) \geq \frac{1}{A} > 0.$$

On the other hand, the tail σ-field of X is trivial (Supplementary Exercise 10, Chapter 8), and the above event is measurable with respect to it. Thus,

$$\mathbb{P}\bigl(X^{-1}\{0\} \cap [e^j, e^{j+1}] \neq \varnothing \text{ infinitely often}\bigr) = 1.$$

This completes our proof of the first part. $\square$

Proof of Theorem 1.3.1: Part (ii) Since the result holds for all such φ, we can assume, without loss of generality, that $\lim_{t \to \infty} \varphi(t) = 0$. (Why?) This assumption holds tacitly from now on.

For all $j \geq 1$ and all $\lambda > 0$, define the measurable events

$$\begin{aligned}\mathsf{E}_j &= \left(\inf_{e^j \leq t \leq e^{j+1}} |X_t| \leq \lambda e^{\frac{j}{\alpha}} \varphi(e^{j+1})\right); \\ \mathsf{F}_j &= \left(\inf_{e^j \leq t \leq e^{j+1}} \frac{|X_t|}{t^{\frac{1}{\alpha}} \varphi(t)} \leq \lambda\right); \quad (3) \\ \mathsf{G}_j &= \left(\inf_{e^j \leq t \leq e^{j+1}} |X_t| \leq \lambda e^{\frac{(j+1)}{\alpha}} \varphi(e^j)\right).\end{aligned}$$

By monotonicity, for all $j \geq 1$,

$$\mathsf{E}_j \subset \mathsf{F}_j \subset \mathsf{G}_j. \quad (4)$$

By scaling (Lemma 1.3.1),

$$\mathbb{P}(\mathsf{E}_j) = \mathbb{P}\left(\inf_{1 \leq t \leq e} |X_t| \leq \lambda \varphi(e^{j+1})\right).$$

Since $\lambda > 0$ is held fixed, by Proposition 1.4.1 and by the form of the function U_α there exist finite constants $J_1 > 1$ and $A_1 \in \,]0,1[$ such that for all $j \geq J_1$,

$$\mathbb{P}(\mathsf{E}_j) \geq A_1 U_\alpha(\varphi(e^{j+1})). \quad (5)$$

(See equation (1) of Section 1.3 for the form of U_α.) Similarly, there exist finite constants $A_2, J_2 > 1$ such that for all $j \geq J_2$,

$$\mathbb{P}(\mathsf{G}_j) \leq A_2 U_\alpha(\varphi(e^j)). \tag{6}$$

Combining the above two estimates with equation (4), we can deduce that

$$\sum_j \mathbb{P}(\mathsf{F}_j) < \infty \iff \sum_j U_\alpha(\varphi(e^j)) < \infty.$$

On the other hand, by monotonicity,

$$\left(\frac{e-1}{e}\right) \sum_{j=1}^{\infty} U_\alpha(\varphi(e^j)) = \sum_{j=0}^{\infty} \int_{e^j}^{e^{j+1}} \frac{U_\alpha(\varphi(e^{j+1}))}{e^{j+1}} \, dt$$

$$\leq \sum_{j=0}^{\infty} \int_{e^j}^{e^{j+1}} \frac{U_\alpha(\varphi(t))}{t} \, dt$$

$$= \int_1^\infty \frac{U_\alpha(\varphi(t))}{t} \, dt \qquad \left(\cdots = \mathfrak{J}(\varphi)\right)$$

$$\leq \sum_{j=0}^{\infty} \int_{e^j}^{e^{j+1}} \frac{U_\alpha(\varphi(e^j))}{e^j} \, dt$$

$$= (e-1) \sum_{j=0}^{\infty} U_\alpha(\varphi(e^j)).$$

Thus far, we have shown that

$$\sum_j \mathbb{P}(\mathsf{F}_j) < \infty \iff \mathfrak{J}(\varphi) < \infty. \tag{7}$$

Now we proceed with our proof. Suppose that $\mathfrak{J}(\varphi) < \infty$. By equation (7) and by the Borel–Cantelli lemma, $\mathbb{P}(\mathsf{F}_j \text{ infinitely often}) = 0$. Recalling equation (3), this means that for any $\lambda > 0$, the following holds a.s.:

$$\inf_{e^j \leq t \leq e^{j+1}} \frac{|X_t|}{t^{\frac{1}{\alpha}} \varphi(t)} \geq \lambda, \text{ eventually.}$$

The above holds a.s., simultaneously over all rational $\lambda > 0$. Thus, we can let $\lambda \to \infty$ along a rational sequence to see that

$$\liminf_{j \to \infty} \inf_{e^j \leq t \leq e^{j+1}} \frac{|X_t|}{t^{\frac{1}{\alpha}} \varphi(t)} = \infty, \qquad \text{a.s.}$$

This proves the theorem in the case $\mathfrak{J}(\varphi) < \infty$.

Conversely, suppose $\mathfrak{J}(\varphi) = \infty$. By equation (7), $\sum_j \mathbb{P}(\mathsf{F}_j) = \infty$. We claim that

$$\liminf_{n \to \infty} \frac{\sum_{i=1}^n \sum_{j=1}^n \mathbb{P}(\mathsf{F}_i \cap \mathsf{F}_j)}{\{\sum_{i=1}^n \mathbb{P}(\mathsf{F}_i)\}^2} < \infty. \tag{8}$$

Having this, we employ the Kochen–Stone lemma (Lemma 1.4.4) to see that $\mathbb{P}(\mathsf{F}_j \text{ infinitely often}) > 0$. The latter event is a tail event, and the tail σ-field for X is trivial; cf. Supplementary Exercise 10 of Chapter 8. Therefore, $\mathbb{P}(\mathsf{F}_j \text{ infinitely often}) = 1$. Now, we can reverse the argument used above (in the summability portion) to argue that if $\mathfrak{J}(\varphi) = \infty$, then

$$\liminf_{j \to \infty} \inf_{e^j \le t \le e^{j+1}} \frac{|X_t|}{t^{\frac{1}{\alpha}} \varphi(t)} = 0, \quad \text{a.s.}$$

Since this would finish our proof of Theorem 1.3.1, we are left to verify the validity of equation (8).

Using equation (4), for all $i, j \ge 1$, $\mathbb{P}(\mathsf{F}_i \cap \mathsf{F}_j) \le \mathbb{P}(\mathsf{G}_i \cap \mathsf{G}_j)$. With (8) in mind, we set out to estimate the latter probability. Since $\mathsf{G}_i \in \mathcal{F}_{e^{i+1}}$, for all $j \ge i + 2$,

$$\mathbb{P}(\mathsf{G}_j \mid \mathcal{F}_{e^{i+1}}) = \mathbb{P}_{X_{e^{i+1}}}\left(\inf_{e^j - e^{i+1} \le t \le e^{j+1} - e^{i+1}} |X_t| \le \lambda e^{\frac{1}{\alpha}} \varphi(e^j)\right)$$
$$\le \sup_{x \in \mathbb{R}^d} \mathbb{P}_x\left(\inf_{e^j - e^{i+1} \le t \le e^{j+1} - e^{i+1}} |X_t| \le \lambda e^{\frac{1}{\alpha}} \varphi(e^j)\right).$$

We have used the Markov property in the first line; cf. Theorem 3.3.2 of Chapter 8. By the Scaling Lemma 1.3.1,

$$\mathbb{P}(\mathsf{G}_j \mid \mathcal{F}_{e^{i+1}}) \le \sup_{x \in \mathbb{R}^d} \mathbb{P}_x\left(\inf_{1 - e^{i-j+1} \le t \le e - e^{i-j+1}} |X_t| \le \lambda e^{\frac{1}{\alpha}} \varphi(e^j)\right)$$
$$\le \sup_{x \in \mathbb{R}^d} \mathbb{P}_x\left(\inf_{\frac{1}{2} \le t \le e} |X_t| \le \lambda e^{\frac{1}{\alpha}} \varphi(e^j)\right),$$

since $j \ge i + 2$. Now we can apply Lemma 1.4.3 and use the form of the function U_α to deduce the existence of finite constants $A_3, J_3 > 1$ such that for all $j \ge J_3$ and all $i \le j - 2$,

$$\mathbb{P}(\mathsf{G}_j \mid \mathcal{F}_{e^{i+1}}) \le A_3 U_\alpha(\varphi(e^j)).$$

Together with (6), the above shows that for all $j \ge i - 2 \ge J_4 = J_3 \vee J_2$,

$$\mathbb{P}(\mathsf{G}_i \cap \mathsf{G}_j) \le A_2 A_3 \, U_\alpha(\varphi(e^i)) \cdot U_\alpha(\varphi(e^j)).$$

Together with equations (5) and (4), this implies that for all $j \ge i - 2 \ge J_5 = J_4 \vee J_1$,

$$\mathbb{P}(\mathsf{G}_i \cap \mathsf{G}_j) \le A_1 A_2 A_3 \, \mathbb{P}(\mathsf{F}_i) \cdot \mathbb{P}(\mathsf{F}_j).$$

By symmetry and by equation (4),

$$\sum_{\substack{i=1 \\ |j-i| \ge J_5 + 2}}^{n} \sum_{j=1}^{n} \mathbb{P}(\mathsf{F}_i \cap \mathsf{F}_j) \le A_1 A_2 A_3 \left(\sum_{i=1}^{n} \mathbb{P}(\mathsf{F}_i)\right)^2. \tag{9}$$

On the other hand,

$$\sum_{\substack{i=1 \\ |j-i|<J_5+2}}^{n}\sum_{j=1}^{n} \mathbb{P}(\mathsf{F}_i \cap \mathsf{F}_j) \leq 2(J_5+2)\sum_{i=1}^{n}\mathbb{P}(\mathsf{F}_i), \qquad (10)$$

using $\mathbb{P}(E \cap F) \leq \mathbb{P}(E)$. Now we can finish our proof. Recall that we have $\mathfrak{J}(\varphi) = \infty$, which, thanks to equation (7), implies $\sum_j \mathbb{P}(\mathsf{F}_j) = \infty$. Moreover, (9) and (10) together yield the following:

$$\sum_{i=1}^{n}\sum_{j=1}^{n} P(\mathsf{F}_i \cap \mathsf{F}_j) \leq A_4\left[\left\{\sum_{j=1}^{n}\mathbb{P}(\mathsf{F}_i)\right\}^2 + \sum_{j=1}^{n}\mathbb{P}(\mathsf{F}_j)\right],$$

where $A_4 = \max\{A_1 A_2 A_3, 2(J_5+2)\}$. In particular, since $\sum_j \mathbb{P}(\mathsf{F}_j) = \infty$,

$$\limsup_{n\to\infty} \frac{\sum_{i=1}^{n}\sum_{j=1}^{n}\mathbb{P}(\mathsf{F}_i \cap \mathsf{F}_j)}{\left\{\sum_{i=1}^{n}\mathbb{P}(\mathsf{F}_i)\right\}^2} \leq A_4,$$

which amply verifies equation (8) and hence the theorem. $\square$

***Exercise* 1.4.3** Suppose X is a d-dimensional isotropic stable process of index $\alpha \in \,]0,2]$, where $d \geq \alpha$. If $\psi : \mathbb{R}_+ \to \mathbb{R}_+$ is monotone, find a necessary and sufficient condition for $\liminf_{t\to 0^+} t^{-\frac{1}{\alpha}}\psi(t)|X_t| = +\infty$.
(HINT: Mimic the demonstration of Theorem 1.3.1 carefully, but instead of using the sequence e^j, use e^{-j}.) $\square$

2 Hitting Probabilities for Feller Processes

The previous section has introduced us to the notion that some of the key properties of Markov processes are based on the determination of "hitting probabilities." (Roughly speaking, the hitting probability of a Borel set E is the probability that the Markov process ever enters the set E.) We now reconsider this problem in the more general setting of some Feller processes on $\mathbb{R}^d$. As it turns out, for many Feller processes of interest, the hitting probability of a compact set E can be described by a certain capacity of the set E. At this point the reader is strongly urged to acquaint him- or herself with the material and terminology of Appendices C and D on capacities before proceeding further.

2.1 Strongly Symmetric Feller Processes

Henceforth, we will focus our attention on a **strongly symmetric Feller process** $X = (X_t; t \geq 0)$ on $\mathbb{R}^d$. This is an $\mathbb{R}^d$-valued Feller process that satisfies the following conditions:

1. the transition functions of X have transition densities $p_t(x, y)$ with respect to some σ-finite measure ν on the Borel subsets of $\mathbb{R}^d$;

2. for each $t > 0$, $(x, y) \mapsto p_t(x, y)$ is symmetric in that $p_t(x, y) = p_t(y, x)$ for all $x, y \in \mathbb{R}^d$; and

3. for every $\lambda > 0$, the corresponding λ-resolvent density r_λ is a gauge function on $\mathbb{R}^d \times \mathbb{R}^d$, where

$$r_\lambda(x, y) = \int_0^\infty e^{-\lambda s} p_s(x, y)\, ds, \qquad x, y \in \mathbb{R}^d,\ \lambda > 0.$$

The measure ν is called the **reference measure** for the process X.

Recall that condition 1 is equivalent to the following: For all bounded measurable $f : \mathbb{R}^d \to \mathbb{R}$,

$$\mathbb{E}_x[f(X_t)] = \int p_t(x, y) f(y)\, \nu(dy), \qquad x \in \mathbb{R}^d,\ t \geq 0;$$

see Section 2.3 of Chapter 8. Moreover, the transition functions of X are given by the relation $\mathcal{T}_t f(x) = \mathbb{E}_x[f(X_t)]$, and the corresponding resolvent is similarly described by $\mathcal{R}_\lambda f(x) = \int r_\lambda(x, y) f(y)\, \nu(dy)$ ($x \in \mathbb{R}^d,\ \lambda > 0$). We also recall from Appendix D that a function f is a **gauge function** on $\mathbb{R}^d \times \mathbb{R}^d$ if for all $\eta > 0$, f is continuous on $O_\eta = \{x \otimes y :\ x, y \in \mathbb{R}^d,\ |x - y| > \eta\}$ and if $f > 0$ on $O_\eta^\complement$, when η is small enough. In words, gauge functions on $\mathbb{R}^d \times \mathbb{R}^d$ are nonnegative functions that are continuous everywhere, except possibly on the "diagonal" of $\mathbb{R}^d \times \mathbb{R}^d$, near which they are strictly positive.

***Exercise* 2.1.1** Any gauge function on $\mathbb{R}^d \times \mathbb{R}^d$ is lower semicontinuous. In fact, if $\mathfrak{g}$ is a gauge function, so is $\mathfrak{g} \wedge M$, for any $M > 0$ sufficiently large. Moreover, if r_λ denotes the λ-resolvent density of a strongly symmetric Feller process, then r_λ is a symmetric function.

You may need to recall that f is lower semicontinuous if there are continuous functions $f_1, f_2, \ldots$ such that as $n \to \infty$, $f_n(x) \uparrow f(x)$, for all x; see also Exercise 1.1.3, Appendix D. □

***Exercise* 2.1.2** Consider an isotropic stable Lévy process X of index $\alpha \in\,]0, 2]$. Verify that X has symmetric, bounded, and continuous transition densities with respect to ν, when ν denotes Lebesgue's measure on the Borel subsets of $\mathbb{R}^d$.

In fact, isotropic stable processes are strongly symmetric. See Section 3 below.

(HINT: Examine Section 4.3 of Chapter 8.) □

Exercise 2.1.3 Show that for all continuous functions $f, g : \mathbb{R}^d \to \mathbb{R}_+$ and for all $\lambda > 0$,

$$\int f(x) \mathcal{R}_\lambda g(x) \, \nu(dx) = \int g(x) \mathcal{R}_\lambda f(x) \, \nu(dx).$$

In words, the symmetry of the transition densities p_t implies that the resolvent operator is self-adjoint as an operator on $L^2(\nu)$. □

The existence of the λ-potential density r_λ and the transition densities p_t give rise to the definitions of potentials and transition operators of measures. Indeed, suppose ϕ is a finite measure on $\mathbb{R}^d$. We can define the **λ-potential of ϕ, $\mathcal{R}_\lambda \phi$**, as

$$\mathcal{R}_\lambda \phi(x) = \int r_\lambda(x, y) \, \phi(dy), \qquad x \in \mathbb{R}^d, \ \lambda > 0.$$

Likewise, we can define

$$\mathcal{T}_t \phi(x) = \int p_t(x, y) \, \phi(dy), \qquad t > 0, \ x \in \mathbb{R}^d.$$

The above make sense due to positivity, even though they may be infinite for some, or even for all, values of x.

Exercise 2.1.4 Check that the above extend our earlier definitions of $\mathcal{R}_\lambda f$ and $\mathcal{T}_t f$, respectively. That is, prove that when $\mu(dx) = f(x)\nu(dx)$, then for all $t, \lambda > 0$, $\mathcal{T}_t \mu = \mathcal{T}_t f$ and $\mathcal{R}_\lambda \mu = \mathcal{R}_\lambda f$. □

Exercise 2.1.5 Check that for any two finite measures ξ and ζ on $\mathbb{R}^d$ and for all $\lambda > 0$,

$$\int \mathcal{R}_\lambda \zeta(x) \, \xi(dx) = \int \mathcal{R}_\lambda \xi(x) \, \zeta(dx).$$

In particular, show that for all probability measures μ on $\mathbb{R}^d$,

$$\int \mathcal{R}_\lambda \mu(x) \, \nu(dx) = \lambda^{-1}.$$

This is the reciprocity theorem of classical potential theory, and by Exercise 2.1.4, it extends Exercise 2.1.3 above. See also Exercise 1.1.4, Appendix D. □

2.2 Balayage

Balayage (French for "*sweeping out*") is a method of approximating potentials of measures by potentials of functions.[2] In the context of probabilis-

---

[2] The roots of this method back to the early works of H. Poincaré and his work on the diffusion equation; see the historical discussions of (Kellogg 1967; Wermer 1981) on this matter.

tic potential theory, balayage addresses the following technical question: *"Given a probability measure μ on $\mathbb{R}^d$, can we approximate $\mathcal{R}_\lambda \mu$ by the λ-potential of some function?"* In light of Exercise 2.1.4, an equivalent goal is to approximate $\mathcal{R}_\lambda \mu$ by the λ-potential of an *absolutely continuous* measure. We shall see, shortly, that one can give an affirmative answer to this question. This answer will turn out to depend on the following unusual-looking result, which is the main theorem of this subsection.

Theorem 2.2.1 (The Balayage Theorem) *Consider a strongly symmetric Feller process $X = (X_t; t \geq 0)$ on $\mathbb{R}^d$ with reference measure ν and resolvent $\mathcal{R} = (\mathcal{R}_\lambda; \lambda > 0)$. Given a compact set $E \subset \mathbb{R}^d$, a bounded open set $F \subset \mathbb{R}^d$ with $E \subset F$, and a $\mu \in \mathcal{P}(E)$,[3] we can find a collection of measures $(\zeta_t; t > 0)$ such that:*

(i) for every $t > 0$, ζ_t is absolutely continuous with respect to ν;

(ii) for all $t, \lambda > 0$ and all $x \in \mathbb{R}^d$, $\mathcal{R}_\lambda \zeta_t(x) \leq \mathcal{R}_\lambda \mu(x)$;

(iii) for every $t > 0$, ζ_t is supported on $\overline{F}$; and

(iv) as $t \to 0^+$, ζ_t converges weakly to μ.

Recall that the measures ζ_t **converge weakly** to μ in the same way that probability measures do: if and only if for all bounded continuous functions $\psi : \mathbb{R}^d \to \mathbb{R}$, $\lim_{t \to 0^+} \int \psi \, d\zeta_t = \int \psi \, d\mu$.

Before proving the balayage theorem, we use it to analyze the λ-potential density of a strongly symmetric Feller process X on $\mathbb{R}^d$, whose λ-resolvent density is denoted by r_λ.

First, we recall from Appendix D that the gauge function r_λ is **proper** if for any compact set $E \subset \mathbb{R}^d$ and all probability measures μ on E, there exist bounded open sets $E_1, E_2, \ldots$ and measures $\mu_1, \mu_2, \ldots$ on $\overline{E_1}, \overline{E_2}, \ldots$, respectively, that are absolutely continuous with respect to ν, and such that:

1. $E_1 \supset E_2 \supset \cdots$;

2. $\cap_n \overline{E_n} = E$; and

3. for all $\varepsilon > 0$, there exist N_0 such that for all $n \geq N_0$,

   (a) $\mu_n(\overline{E_n}) \geq 1 - \varepsilon$; and (b) $\mathcal{R}_\lambda \mu_n(x) \leq \mathcal{R}_\lambda \mu(x)$, for all $x \in \mathbb{R}^d$.

The following is a corollary to Theorem 2.2.1, and captures the salient features of the λ-potential densities of strongly symmetric Feller processes. This is a very important result, and its sole message is that r_λ is proper:

---

[3] Recall that $\mathcal{P}(E)$ denotes the collection of all probability measures on E.

Corollary 2.2.1 *Suppose X is a strongly symmetric Feller process on $\mathbb{R}^d$ with reference measure ν and λ-potential density r_λ. Then, r_λ is a proper, symmetric gauge function on $\mathbb{R}^d \times \mathbb{R}^d$.*

Proof By Exercise 2.1.1, r_λ is symmetric, and it is a gauge function by definition. To argue that r_λ is proper for every λ, consider a compact set $E \subset \mathbb{R}^d$, and let E^n denote its $\frac{1}{n}$-enlargement. That is,

$$E^n = \left\{ x \in \mathbb{R}^d : \mathrm{dist}\{x; E\} < \frac{1}{n} \right\}, \qquad n \geq 1.$$

Fix $n \geq 1$ and apply the balayage theorem (Theorem 2.2.1) to $F = E^n$. This way, we construct measures $(\zeta_t;\ t \geq 0)$ as described. By picking $t = t(n)$ sufficiently small, we can ensure that $\zeta_{t(n)}(\overline{E^n}) \geq 1 - \varepsilon$ for all $n \geq 1$, and our corollary follows. $\square$

The remainder of this section is concerned with proving the balayage theorem (Theorem 2.2.1). This proof is not too difficult, but unless the reader is well familiar with time-reversal, it is somewhat mysterious. Furthermore, our proof relies on facts that are needed only for this argument. Therefore, the rest of this subsection can be safely omitted for the newly initiated student of this subject.

Henceforth, X is a strongly symmetric Feller process on $\mathbb{R}^d$, $E \subset \mathbb{R}^d$ is compact, and $F \subset \mathbb{R}^d$ is bounded, open, and $F \supset E$.

We introduce a sequence of families of bounded linear operators $(\mathcal{T}^n_t;\ t \geq 0)$, $n = 1, 2, \ldots$, by

$$\mathcal{T}^n_t f(x) = \mathbb{E}_x\left[f(X_t) \prod_{j=0}^{2^n} \mathbf{1}_{(X_{j 2^{-n} t} \in F)} \right]. \qquad (1)$$

Lemma 2.2.1 *Fix $n \geq 1$ and $t > 0$. Then:*

(i) *whenever $f : \mathbb{R}^d \to \mathbb{R}_+$ is measurable,*

$$\mathcal{T}_t f(x) \geq \mathcal{T}^n_t f(x) \geq \mathcal{T}^{n+1}_t f(x),$$

for all $x \in \mathbb{R}^d$;

(ii) *there exists a nonnegative, symmetric function p^n_t, supported on $\overline{F} \times \overline{F}$, such that for all measurable $f : \mathbb{R}^d \to \mathbb{R}_+$, and for all $x \in \mathbb{R}^d$,*

$$\mathcal{T}^n_t f(x) = \int p^n_t(x, y) f(y)\, \nu(dy);$$

(iii) *for all measurable $f : \mathbb{R}^d \to \mathbb{R}_+$, and for all $x \notin \overline{F}$, $\mathcal{T}^n_t f(x) = 0$; and*

(iv) *the map $A \mapsto \mathcal{T}^n_t \mathbf{1}_A(x)$ is a subprobability measure on $\overline{F}$, for any $x \in \mathbb{R}^d$.*

2 Hitting Probabilities for Feller Processes   365

Proof Part (i) follows from equation (1) and the elementary observation that dyadic numbers are nested. Part (iii) follows from (ii), since by (ii), $p_t^n(x,y) = 0$, unless x and y are both in $\overline{F}$. Finally, (iv) follows from equation (1). Thus, it remains to prove (ii); this follows from the Chapman–Kolmogorov equation (Lemma 3.1.1, Chapter 8). In fact, we can choose $p_t^n(x,y)$ by the formula

$$\int \cdots \int p_{t2^{-n}}(x, x_1)$$
$$\times \prod_{i=1}^{2^n - 2} p_{t2^{-n}}(x_i, x_{i+1}) \cdot p_{t2^{-n}}(x_{2^n - 1}, y) \, \nu(dx_1) \cdots \nu(dx_{2^n - 1})$$

if $x, y \in \overline{F}$, and $p_t^n(x,y) = 0$ otherwise (check!). The remaining properties follow from this and the symmetry of $(a,b) \mapsto p_{t2^{-n}}(a,b)$. $\square$

Exercise **2.2.1** Check that for all nonnegative $f \in L^\infty(\mathbb{R}^d)$,

$$\lim_{n \to \infty} \mathfrak{T}_t^n f(x) = \mathbb{E}_x\big[f(X_t) \mathbf{1}_{(\tau_F > t)}\big], \quad \forall x \in \mathbb{R}^d, \, t \geq 0,$$

where $\tau_F = \inf(s \geq 0 : X_s \notin \overline{F})$. This exercise is a starting point for investigating deep connections between balayage and killed processes. For some further information, see Supplementary Exercise 6. $\square$

Now that we have a density p_t^n, we can extend the definition of the operator $\mathfrak{T}_t^n$ as we did for $\mathfrak{T}_t$. For all measures ϕ on $\mathbb{R}^d$, define

$$\mathfrak{T}_t^n \phi(x) = \int p_t^n(x,y) \, \phi(dy), \quad x \in \mathbb{R}^d, \, t \geq 0, \, n \geq 1.$$

By Lemma 2.2.1,
$$\mathfrak{T}_t^n \phi = 0, \text{ off of } \overline{F}. \tag{2}$$

As in Exercise 2.1.4, this definition of $\mathfrak{T}_t^n$ is consistent. Namely, if $\phi(dx) = f(x)\nu(dx)$, $\mathfrak{T}_t^n \phi = \mathfrak{T}_t^n f$.

Lemma 2.2.2 *For all $\lambda > 0$, $t \geq 0$, $n \geq 1$, $x \in \mathbb{R}^d$, and for all measures ϕ on $\mathbb{R}^d$,*

$$\mathcal{R}_\lambda \mathfrak{T}_t^n \phi(x) \leq e^{\lambda t} \mathcal{R}_\lambda \phi(x).$$

Proof Since

$$\mathcal{R}_\lambda \mathfrak{T}_t^n \phi(x) = \iint r_\lambda(x,y) p_t^n(y,z) \, \phi(dz) \, \nu(dy), \tag{3}$$

we apply the symmetry of r_λ (Exercise 2.1.1) and p_t^n (Lemma 2.2.1(ii)) to obtain

$$\mathcal{R}_\lambda \mathfrak{T}_t^n \phi(x) = \iint r_\lambda(y,x) p_t^n(z,y) \, \nu(dy) \, \phi(dz) = \int \mathfrak{T}_t^n h_x(z) \, \phi(dz),$$

where $h_x(y) = r_\lambda(y,x)$. By Lemma 2.2.1(i),

$$\mathcal{R}_\lambda \mathcal{T}_t^n \phi(x) \leq \int \mathcal{T}_t h_x(z)\, \phi(dz) = \mathcal{R}_\lambda \mathcal{T}_t \phi(x).$$

In the last equality we have utilized the fact that equation (3) also holds without the superscript of n. Now we finish by computing as follows:

$$\mathcal{R}_\lambda \mathcal{T}_t^n \phi(x) \leq \mathcal{R}_\lambda \mathcal{T}_t \phi(x) = \int_0^\infty e^{-\lambda s} \mathcal{T}_{s+t} \phi(x)\, ds$$
$$= e^{\lambda t} \int_t^\infty e^{-\lambda s} \mathcal{T}_s \phi(x)\, ds,$$

from which the assertion follows. $\square$

We are ready to proceed with our proof of Theorem 2.2.1.

Proof of Theorem 2.2.1 define a sequence of measures ζ_t^n, all on $\mathbb{R}^d$, by

$$\zeta_t^n(dx) = e^{-\lambda t} \mathcal{T}_t^n \mu(x)\, \nu(dx),$$

where μ is given in the statement of Theorem 2.2.1. We list some of the key properties of these measures next:

(P1) each ζ_t^n is absolutely continuous with respect to ν;

(P2) thanks to equation (2), each ζ_t^n is a measure on $\overline{F}$;

(P3) by Lemma 2.2.2,

$$\mathcal{R}_\lambda \zeta_t^n(x) \leq \mathcal{R}_\lambda \mu(x).$$

We claim that for all bounded continuous functions $\psi : \mathbb{R}^d \to \mathbb{R}$,

$$\lim_{t \downarrow 0} \lim_{n \to \infty} \int \psi(x)\, \zeta_t^n(dx) = \int \psi(x)\, \mu(dx). \tag{4}$$

Since we can then relabel the ζ's and finish our proof, it suffices to prove (4). But, thanks to Lemma 2.2.1,

$$\int \psi(x)\, \zeta_t^n(dx) = e^{-\lambda t} \iint \psi(x) p_t^n(x,y)\, \mu(dy)\, \nu(dx)$$
$$= e^{-\lambda t} \iint \psi(x) p_t^n(y,x)\, \mu(dy)\, \nu(dx).$$

Reversing the order of the two integrals yields

$$\int \psi(x)\, \zeta_t^n(dx) = e^{-\lambda t} \int \mathcal{T}_t^n \psi(y)\, \mu(dy)$$
$$= e^{-\lambda t} \int \mathbb{E}_y\!\left[\psi(X_t) \prod_{j=0}^{2^n} \mathbf{1}_{(X_{j2^{-n}t} \in \overline{F})}\right] \mu(dy),$$

owing to (1). By Exercise 2.2.1, and by the bounded convergence theorem,

$$\lim_{n\to\infty} \int \psi(x)\, \zeta_t^n(dx) = e^{-\lambda t} \int \mathbb{E}_y\left[\psi(X_t)\mathbf{1}_{(\tau_F > t)}\right] \mu(dy),$$

where $\tau_F = \inf(s \geq 0 : X_s \notin \overline{F})$. By right continuity, $\lim_{t\downarrow 0} X_t = y$, $\mathbb{P}_y$-a.s. Thus, another application of the bounded convergence theorem yields

$$\lim_{t\to 0^+} \lim_{n\to\infty} \int \psi(x)\, \zeta_t^n(dx) = \int \psi(y) \mathbb{P}_y(\tau_F > 0)\, \mu(dy),$$

which equals $\int \psi\, d\mu$. This uses the facts that F is open, E is compact, $E \subset F$, $t \mapsto X_t$ is right-continuous, and $\mu \in \mathcal{P}(E)$. We have demonstrated equation (4), and hence, the balayage theorem. □

***Exercise* 2.2.2** Show, in complete detail, that in the above proof,

$$\mathbb{P}_y(\tau_F > 0) = 1,$$

for μ-almost all y. □

2.3 Hitting Probabilities and Capacities

Throughout this subsection $X = (X_t;\ t \geq 0)$ denotes a strongly symmetric Feller process on $\mathbb{R}^d$ whose reference measure is ν. We shall write $\mathcal{T}$, $\mathcal{R}$, p_t, and r_λ for the transition functions, resolvent, transition densities, and λ-potential densities of X, respectively. The main result of this section is an estimate for the Laplace transform of the hitting (or entrance) time of a *compact* set $E \subset \mathbb{R}^d$ in terms of capacities whose gauge function is r_λ. Recall from Appendix D that the **energy** (with respect to the gauge function r_λ) of a measure μ is defined as

$$\mathcal{E}_{r_\lambda}(\mu) = \iint r_\lambda(x,y)\, \mu(dy)\, \mu(dy).$$

The **capacity** of a Borel set $E \subset \mathbb{R}^d$ is then given by the principle of minimum energy, viz.,

$$\mathcal{C}_{r_\lambda}(E) = \left[\inf_{\mu \in \mathcal{P}(E)} \mathcal{E}_{r_\lambda}(\mu)\right]^{-1},$$

where $\mathcal{P}(E)$ denotes the collection of all probability measures on E. See Appendix D for further details. Having recalled the definitions of capacity and energy, we are ready to state the main result of this section. In the case that X is Brownian motion, a sharper version appears in Benjamini, Pemantle, and Peres (1995).

Theorem 2.3.1 *Suppose X is a strongly symmetric Feller process as described above. Let $E \subset \mathbb{R}^d$ be a compact set and let $T_E = \inf(s \geq 0 : X_s \in E)$ denote its entrance time. Then, under the above assumptions, for all $x \notin E$ and all $\lambda > 0$,*

$$\frac{\mathcal{I}^2}{2\mathcal{S}} \cdot \mathcal{C}_{r_\lambda}(E) \leq \mathbb{E}_x[e^{-\lambda T_E}] \leq \mathcal{S} \cdot \mathcal{C}_{r_\lambda}(E),$$

where $\mathcal{I} = \inf_{y \in E} r_\lambda(x,y)$ and $\mathcal{S} = \sup_{y \in E} r_\lambda(x,y)$.

As a consequence of Theorem 2.3.1, one can show that, typically, $\mathbb{P}_x(T_E < \infty) > 0$ if and only if there exists a probability measure μ on E that has finite energy with gauge function r_λ. The following states this more precisely.

Corollary 2.3.1 *Suppose $E \subset \mathbb{R}^d$ is compact and let T_E denote its hitting time. If for all $x, y \in \mathbb{R}^d$, $r_1(x,y) > 0$, then for any $x \notin E$,*

$$\mathbb{P}_x(T_E < \infty) > 0 \iff \mathcal{C}_{r_1}(E) > 0.$$

In other words, this states that our Markov process X can hit E, starting from $x \in \mathbb{R}^d$, only if E is a sufficiently large set, in the sense that it carries a probability measure of finite energy with respect to the gauge function r_λ.

We will prove Theorem 2.3.1 in the next subsection.

2.4 Proof of Theorem 2.3.1

Throughout this proof let $T = T_E$, for simplicity. We will prove the upper bound and the lower bound for $\mathbb{E}_x[e^{-\lambda T}]$ separately and in this order. The upper bound is much more difficult to prove. Fortunately, we have already built up all of the necessary ingredients, with one notable exception:

Lemma 2.4.1 *Given a measurable function $\varphi : \mathbb{R}^d \to \mathbb{R}_+$,*

$$\mathcal{R}_\lambda \varphi(x) \geq \mathbb{E}_x[e^{-\lambda T} \mathcal{R}_\lambda \varphi(X_T)],$$

for all $\lambda > 0$ and every $x \in \mathbb{R}^d$.

Proof Recall that our Feller process lives on the one-point compactification $\mathbb{R}^d_\Delta$ of $\mathbb{R}^d$, where Δ is a point outside of $\mathbb{R}^d$. Also recall that $X_\infty = \Delta$ and that the domain of definition for all functions $f : \mathbb{R}^d \to \mathbb{R}$ is extended to $\mathbb{R}^d_\Delta$ upon defining $f(\Delta) = 0$. Thus, $\mathbb{E}_x[e^{-\lambda T} \mathcal{R}_\lambda \varphi(X_T)]$ is *always* well-defined, even if $T \equiv \infty$.

Now we proceed with a proof. By considering $\varphi \wedge k$ and letting $k \to \infty$, we can and will assume that φ is a *bounded measurable function*. By the

Doob–Meyer decomposition (Theorem 1.3.3 of Chapter 9), for all $t \geq 0$ and all $x \in \mathbb{R}^d$,

$$M_t = e^{-\lambda t} \mathcal{R}_\lambda \varphi(X_t) + \int_0^t e^{-\lambda s} \varphi(X_s)\, ds, \qquad \mathbb{P}_x\text{-a.s.,} \tag{1}$$

where

$$M_t = \mathbb{E}_x \Big[\int_0^\infty e^{-\lambda s} \varphi(X_s)\, ds \,\Big|\, \mathcal{F}_t \Big].$$

Since $t \mapsto M_t$ and $t \mapsto X_t$ are both right-continuous, equation (1) holds $\mathbb{P}_x$-a.s., simultaneously for all $t \geq 0$. In particular, we are allowed to plug the random time T for t to see that $\mathbb{P}_x$-a.s. for all $x \in \mathbb{R}^d$,

$$M_T \mathbf{1}_{(T<\infty)} = e^{-\lambda T} \mathcal{R}_\lambda \varphi(X_T) + \int_0^T e^{-\lambda s} \varphi(X_s)\, ds \geq e^{-\lambda T} \mathcal{R}_\lambda \varphi(X_T),$$

since $\varphi(y) \geq 0$ for all $y \in \mathbb{R}^d$. Since $\varphi \in L^\infty(\mathbb{R}^d)$, by Lebesgue's dominated convergence theorem, $\mathbb{E}_x[M_T \mathbf{1}_{(T<\infty)}] = \lim_{n\to\infty} \mathbb{E}_x[M_{T\wedge n}]$. We now apply the optional stopping theorem (Theorem 1.6.1, Chapter 7) to deduce that for all $x \in \mathbb{R}^d$,

$$\mathbb{E}_x[M_0] \geq \mathbb{E}_x\big[e^{-\lambda T} \mathcal{R}_\lambda \varphi(X_T)\big].$$

The result follows from the towering property of conditional expectations (equation (1), Section 1.1 of Chapter 1), together with the definition of M. □

***Exercise* 2.4.1** Suppose that μ is a finite measure that is supported on a compact set E. Check that for all $x \in \mathbb{R}^d$ and all $\lambda > 0$,

$$\mathcal{R}_\lambda \mu(x) = \mathbb{E}_x\big[e^{-\lambda T_E} \mathcal{R}_\lambda \mu(X_{T_E})\big].$$

Thus, Lemma 2.4.1 is best possible. □

We now enlarge the probability space, if need be, to construct a random variable $\mathfrak{e}$ that (i) is independent of $\mathcal{F}_t$ for all $t \geq 0$; and (ii) has a mean 1 exponential distribution. Define

$$\mathfrak{e}(\lambda) = \frac{\mathfrak{e}}{\lambda}, \qquad \lambda > 0.$$

Then, for all $\lambda > 0$:

- $\mathfrak{e}(\lambda)$ is independent of the entire history of X; and
- $\mathfrak{e}(\lambda)$ has an exponential distribution with parameter λ. That is,

$$\mathbb{P}\{\mathfrak{e}(\lambda) > y\} = e^{-\lambda y}, \qquad \text{for all } y \geq 0.$$

Combining the above two facts, it follows that for all $x \in \mathbb{R}^d$ and all $\lambda > 0$,
$$\mathbb{P}_x\{\mathfrak{e}(\lambda) \geq y\} = e^{-\lambda y}, \qquad y \geq 0.$$
In particular, we have the following interpretation of the Laplace transform of the entrance time of E; it follows immediately from the above independence assertion, together with Fubini's theorem.

Lemma 2.4.2 *For all $x \in \mathbb{R}^d$ and all $\lambda > 0$, $\mathbb{E}_x[e^{-\lambda T}] = \mathbb{P}_x(T < \mathfrak{e}(\lambda))$.*

Proof of the Upper Bound in Theorem 2.3.1 We may assume without loss of generality that
$$\mathbb{E}_x[e^{-\lambda T}] = \mathbb{P}_x(T < \mathfrak{e}(\lambda)) > 0.$$
For any measurable set $G \subset \mathbb{R}^d$, define
$$\mu_0(G) = \mathbb{P}_x(X_T \in G \,|\, T < \mathfrak{e}(\lambda)).$$
It is easy to check that $\mu_0 \in \mathcal{P}(E)$, since E is compact and since $t \mapsto X_t$ is right-continuous. We shall pick a bounded open set $F \subset \mathbb{R}^d$ that contains E, and define $(\zeta_t; t > 0)$ to be the collection of measures given by the balayage theorem for μ_0 in place of μ; cf. Theorem 2.2.1. The balayage theorem shows that for all $t, \lambda > 0$ and all $x \in \mathbb{R}^d$,
$$\sup_{y \in E} r_\lambda(x, y) \geq \mathcal{R}_\lambda \mu_0(x) \geq \mathcal{R}_\lambda \zeta_t(x) \geq \mathbb{E}_x[e^{-\lambda T} \mathcal{R}_\lambda \zeta_t(X_T)].$$
We have applied Lemma 2.4.1 to the absolutely continuous measure ζ_t. On the other hand, $e^{-\lambda T}$ is the conditional probability of $(\mathfrak{e}(\lambda) > T)$, given the entire process X. Thus,
$$\sup_{t \in E} r_\lambda(x, y) \geq \mathbb{E}_x[e^{-\lambda T}] \cdot \mathbb{E}_x[\mathcal{R}_\lambda \zeta_t(X_T) \,|\, T < \mathfrak{e}(\lambda)]$$
$$= \mathbb{E}_x[e^{-\lambda T}] \cdot \int \mathcal{R}_\lambda \zeta_t(x) \, \mu_0(dx),$$
by the definition of μ_0. By the balayage theorem, as $t \to 0^+$, ζ_t converges weakly to μ_0. Lemma 2.1.1 of Appendix D then asserts that $\liminf_{t \to 0^+} \mathcal{R}_\lambda \zeta_t(x) \geq \mathcal{R}_\lambda \mu_0(x)$, for all $x \in \mathbb{R}^d$. By Fatou's lemma,
$$\sup_{y \in E} r_\lambda(x, y) \geq \mathbb{E}_x[e^{-\lambda T}] \cdot \int \mathcal{R}_\lambda \mu_0(x) \, \mu_0(dx),$$
which equals $\mathbb{E}_x[e^{-\lambda T}] \mathcal{E}_{r_\lambda}(\mu_0)$; see Exercise 1.1.2, Appendix D. Consequently,
$$\sup_{y \in E} r_\lambda(x, y) \geq \mathbb{E}_x[e^{-\lambda T}] \cdot \mathcal{E}_{r_\lambda}(\mu_0).$$

The fact that r_λ is a gauge function and $x \notin E$ implies that the left-hand side is finite. Since $\mathbb{E}_x[e^{-\lambda T}]$ is assumed to be (strictly) positive, the above energy is finite, and we obtain

$$\mathbb{E}_x[e^{-\lambda T}] \le \frac{\sup_{y \in E} r_\lambda(x,y)}{\mathcal{E}_{r_\lambda}(\mu_0)}.$$

This amply proves the upper bound of Theorem 2.3.1. □

To prove the lower bound in Theorem 2.3.1, we need one more technical lemma.

Lemma 2.4.3 *For all $\varphi \in L^\infty(\mathbb{R}^d)$, all $\lambda > 0$, and all $x \in \mathbb{R}^d$,*

$$\mathbb{E}_x\left[\left\{\int_0^{\mathrm{e}(\lambda)} \varphi(X_s)\,ds\right\}^2\right] \le 2 \iint \mathcal{R}_\lambda \varphi(y) r_\lambda(z,y)\, \varphi(y)\, \nu(dz)\, \nu(dy).$$

In particular, if $\varphi = 0$ off of a compact set E, then

$$\mathbb{E}_x\left[\left\{\int_0^{\mathrm{e}(\lambda)} \varphi(X_s)\,ds\right\}^2\right] \le 2 \sup_{y \in E} r_\lambda(x,y) \mathcal{E}_{r_\lambda}(\phi),$$

where $\phi(dz) = \varphi(z)\nu(dz)$.

Proof We expand the square to see that

$$\mathbb{E}_x\left[\left\{\int_0^{\mathrm{e}(\lambda)} \varphi(X_s)\,ds\right\}^2\right] = 2\mathbb{E}_x\left[\int_0^\infty \int_s^\infty \mathbf{1}_{(\mathrm{e}(\lambda) > t)} \varphi(X_s)\varphi(X_t)\,dt\,ds\right]$$

$$= 2\mathbb{E}_x\left[\int_0^\infty \int_s^\infty e^{-\lambda t} \varphi(X_s)\varphi(X_t)\,dt\,ds\right].$$

Let $\mathcal{T}$ denote the transition functions of X. By the Markov property, whenever $s < t$,

$$\mathbb{E}_x[\varphi(X_s)\varphi(X_t)] = \mathbb{E}_x\Big\{\varphi(X_s)\mathbb{E}_{X_s}[\varphi(X_{t-s})]\Big\} = \mathbb{E}_x\{\varphi(X_s)\mathcal{T}_{t-s}\varphi(X_s)\};$$

cf. Theorem 3.3.2, Chapter 8. Applying Fubini's theorem twice, we obtain

$$\mathbb{E}_x\left[\left\{\int_0^{\mathrm{e}(\lambda)} \varphi(X_s)\,ds\right\}^2\right] = 2\mathbb{E}_x\left[\int_0^\infty \varphi(X_s) \int_s^\infty e^{-\lambda t} \mathcal{T}_{t-s}\varphi(X_s)\,dt\,ds\right]$$

$$= 2\mathbb{E}_x\left[\int_0^\infty e^{-\lambda s} \varphi(X_s) \mathcal{R}_\lambda \varphi(X_s)\,ds\right].$$

The lemma follows immediately from this and a few more lines of direct calculations. □

Exercise 2.4.2 The basic idea behind the estimate of Lemma 2.4.3 can be used to show that r_λ is nonnegative definite. Indeed, recall that $\mathbb{E}_\nu[Z] = \int \mathbb{E}_x[Z]\,\nu(dx)$, and check that for any bounded function $f : \mathbb{R}^d \to \mathbb{R}$, and for all $\lambda > 0$,

$$\mathbb{E}_\nu\left[\left|\int_0^{\mathfrak{e}(\lambda)} f(X_s)\,ds\right|^2\right] = \frac{2}{\lambda}\mathcal{E}_{r_\lambda}(\phi),$$

where $\phi(x) = f(x)\nu(dx)$. Conclude that r_λ is nonnegative definite for any $\lambda > 0$. Also, define as in Appendix D (Section 1.1) the mutual energy $\langle \mu, \nu \rangle_{r_\lambda}$ for any two *signed* measures μ and ν on $\mathbb{R}^d$. Use the above to show that $\langle \bullet, \bullet \rangle_{r_\lambda}$ defines an inner product on the space of *signed* measures whose corresponding norm is $\mathcal{E}_{r_\lambda}$. This improves on Exercise 1.1.1, Appendix D. □

We conclude this subsection by proving the remainder of Theorem 2.3.1.

Proof of the Lower Bound in Theorem 2.3.1 For every $\eta > 0$, let E^η denote the η-enlargement of E described by

$$E^\eta = \{x \in \mathbb{R}^d : \text{dist}\{x; E\} < \eta\}, \qquad \eta > 0.$$

Each E^η is a bounded open set that contains E. Let $T^\eta = \inf(s \geq 0 : X_s \in E^\eta)$ denote the entrance time to E^η. This is a stopping time for every $\eta > 0$ (Theorem 1.2.1, Chapter 7) and $\mathbb{P}_x$-a.s. for all $x \in \mathbb{R}^d$, $T^\eta \downarrow T$, as $\eta \to 0^+$; see Supplementary Exercise 5. We shall first obtain a lower bound for $\mathbb{E}_x[e^{-\lambda T^\eta}]$.

Consider any absolutely continuous probability measure ζ on E^η and let $\varphi(x) = \zeta(dx)/\nu(dx)$ denote its Radon–Nikodým derivative. Note that if and when $\int_0^{\mathfrak{e}(\lambda)} \varphi(X_s)\,ds > 0$, then certainly, $T^\eta < \mathfrak{e}(\lambda)$. That is,

$$\mathbb{E}_x[e^{-\lambda T^\eta}] = \mathbb{P}_x(T^\eta < \mathfrak{e}(\lambda)) \geq \mathbb{P}_x\left(\int_0^{\mathfrak{e}(\lambda)} \varphi(X_s)\,ds > 0\right);$$

cf. Lemma 2.4.2 for the first equality. By the Paley–Zygmund lemma (Lemma 1.4.1, Chapter 3),

$$\mathbb{E}_x[e^{-\lambda T^\eta}] \geq \frac{\{\mathbb{E}_x[\int_0^{\mathfrak{e}(\lambda)} \varphi(X_s)\,ds]\}^2}{\mathbb{E}_x[\{\int_0^{\mathfrak{e}(\lambda)} \varphi(X_s)\,ds\}^2]}.$$

The numerator is the square of

$$\mathbb{E}_x\left[\int_0^\infty e^{-\lambda s}\varphi(X_s)\,ds\right] = \mathcal{R}_\lambda \varphi(x) \geq \inf_{y \in \overline{E^\eta}} r_\lambda(x, y).$$

By Lemma 2.4.3, the denominator is bounded above by $2\sup_{y \in \overline{E^\eta}} r_\lambda(x, y)$ times $\mathcal{E}_{r_\lambda}(\zeta)$. Thus,

$$\mathbb{E}_x[e^{-\lambda T^\eta}] \geq \frac{[\inf_{y \in \overline{E^\eta}} r_\lambda(x, y)]^2}{2\sup_{y \in \overline{E^\eta}} r_\lambda(x, y)}[\mathcal{E}_{r_\lambda}(\zeta)]^{-1}.$$

Since ζ is an arbitrary absolutely continuous probability measure on $\overline{E^\eta}$, we can take the supremum over all such ζ's to obtain the absolutely continuous capacity of $\overline{E^\eta}$ on the right-hand side; see Appendix D. That is,

$$\mathbb{E}_x[e^{-\lambda T^\eta}] \geq \frac{\left[\inf_{y \in \overline{E^\eta}} r_\lambda(x,y)\right]^2}{2\sup_{y \in \overline{E^\eta}} r_\lambda(x,y)} \cdot \mathcal{C}^{\mathrm{ac}}_{r_\lambda}(\overline{E^\eta}) \geq \frac{\left[\inf_{y \in \overline{E^\eta}} r_\lambda(x,y)\right]^2}{2\sup_{y \in \overline{E^\eta}} r_\lambda(x,y)} \cdot \mathcal{C}^{\mathrm{ac}}_{r_\lambda}(E).$$

We now wish to let $\eta \downarrow 0^+$. By Corollary 2.2.1, r_λ is proper. Together with Theorem 2.3.1 of Appendix D, this implies that $\mathcal{C}^{\mathrm{ac}}_{r_\lambda}(E) = \mathcal{C}_{r_\lambda}(E)$. On the other hand, as $\eta \to 0^+$, $\inf_{y \in \overline{E^\eta}} r_\lambda(x,y)$ (respectively $\sup_{y \in \overline{E^\eta}} r_\lambda(x,y)$) converges to $\inf_{y \in E} r_\lambda(x,y)$ (respectively $\sup_{y \in \overline{E^\eta}} r_\lambda(x,y)$), since r_λ is a gauge function and since $x \notin E$. The remainder of this proof follows from the already observed fact that $T^\eta \downarrow T$, $\mathbb{P}_x$-a.s. □

3 Explicit Computations

In order to better understand the general results of the previous section, we now specialize our Markov processes to isotropic stable processes and make some rather explicit computations and/or estimations. While Brownian motion is one of the isotropic stable processes, its analysis is much simpler. Thus, we start with Brownian motion.

3.1 Brownian Motion and Bessel–Riesz Capacities

Let $B = (B_t;\ t \geq 0)$ denote d-dimensional Brownian motion. Our immediate goal is to compute its λ-potential density r_λ. When $d = 1$, this was carried out in the example of Section 2.3, Chapter 8, but in disguise. To begin with, note that for all functions $f \in L^\infty(\mathbb{R}^d)$, all $x \in \mathbb{R}^d$, and all $t \geq 0$,

$$\mathcal{T}_t f(x) = \mathbb{E}_x[f(B_t)] = \mathbb{E}[f(x + B_t)] = \int_{\mathbb{R}^d} f(x+y) q_t(y)\, dy,$$

where $\mathcal{T}$ denotes the semigroup corresponding to B and

$$q_t(a) = (2\pi)^{-\frac{d}{2}} t^{-\frac{d}{2}} \exp\left\{-\frac{\|a\|^2}{2t}\right\}, \qquad t \geq 0,\ a \in \mathbb{R}^d.$$

See Exercise 1.1.1 of Chapter 5. The semigroup $\mathcal{T}$ is called the **heat semigroup** (on $\mathbb{R}^d$), and its corresponding transition density is the so-called **heat kernel**. The latter exists. In fact, the discussion of Section 2.3 of Chapter 8 shows that Brownian motion B has transition densities $p_t(x,y)$ (with respect to Lebesgue's measure) that are given by

$$p_t(x,y) = q_t(y-x), \qquad t \geq 0,\ x,y \in \mathbb{R}^d.$$

Now the λ-potential density r_λ can be computed as follows (why?):
$$r_\lambda(x,y) = u_\lambda(y-x), \qquad \lambda > 0,\ x, y \in \mathbb{R}^d,$$
where
$$u_\lambda(a) = \int_0^\infty e^{-\lambda t} q_t(a)\, dt, \qquad \lambda > 0,\ a \in \mathbb{R}^d.$$

Of course, $\lambda \mapsto u_\lambda(a)$ is the Laplace transform of the *function* (not distribution function) $t \mapsto q_t(a)$. You can find some information on Laplace transforms in Appendix B. In any case, we can assemble the above information in one package to obtain the following formula for the 1-potential density of Brownian motion: For all $x, y \in \mathbb{R}^d$,

$$r_1(x,y) = (2\pi)^{-\frac{d}{2}} \int_0^\infty e^{-t} t^{-\frac{d}{2}} \exp\left\{ -\frac{\|y-x\|^2}{2t} \right\} dt. \qquad (1)$$

Exercise 3.1.1 Compute r_λ for all $\lambda > 0$ and verify that it is indeed a gauge function on $\mathbb{R}^d \times \mathbb{R}^d$.

(i) Conclude that d-dimensional Brownian motion is a strongly symmetric Feller process.

(ii) Check that for all $x, y \in \mathbb{R}^d$ and for all $\lambda > 0$, $r_\lambda(x,y) > 0$.

$\square$

While r_λ is a gauge function on $\mathbb{R}^d \times \mathbb{R}^d$, it may (and in this case, it does when $d \geq 2$) have singularities on the diagonal of $\mathbb{R}^d \times \mathbb{R}^d$. A little thought shows that the nature of such singularities is precisely what makes some sets have positive capacity and others not. Our next two lemmas estimate the nature of this singularity, first in the case $d \geq 3$, and then in the case $d = 2$. See the paragraph below Lemma 3.1.2 for the case $d = 1$.

Lemma 3.1.1 *Given $d \geq 3$ and a finite constant $\varepsilon > 0$, there exist two finite positive constants $A_1 \leq A_2$ such that for all $x, y \in \mathbb{R}^d$ with $\|x - y\| \leq \varepsilon$,*
$$\frac{A_1}{\|y-x\|^{d-2}} \leq r_1(x,y) \leq \frac{A_2}{\|y-x\|^{d-2}}.$$

Proof Recalling that $r_1(x,y) = u_1(y-x)$, we use the formula of equation (1) and change variables to see that
$$u_1(a) = (2\pi)^{-\frac{d}{2}} \|a\|^{2-d} \int_0^\infty e^{-s\|a\|^2} s^{-\frac{d}{2}} e^{-\frac{1}{2s}}\, ds. \qquad (2)$$

Note that
$$\int_0^\infty e^{-s\|a\|^2} s^{-\frac{d}{2}} e^{-\frac{1}{2s}}\, ds \leq \int_0^\infty s^{-\frac{d}{2}} e^{-\frac{1}{2s}}\, ds = 2^{-1+\frac{d}{2}} \Gamma\!\left(\frac{d-2}{2}\right).$$

Of course, we need the condition $d \geq 3$ here to have finite integrals. The upper bound on r_1 follows with

$$A_2 = \frac{1}{2\pi^{\frac{d}{2}}}\Gamma\left(\frac{d-2}{2}\right).$$

Similarly, the lower bound holds with

$$A_1 = (2\pi)^{-\frac{d}{2}}\int_0^\infty e^{-\varepsilon^2 s}s^{-\frac{d}{2}}e^{-\frac{1}{2s}}\,ds,$$

which is positive and finite. □

When $d = 2$, the estimation is only slightly more delicate, as the following shows.

Lemma 3.1.2 *Given $d = 2$ and a positive finite ε, there exist two finite positive constants $A_1 \leq A_2$ such that for all $x, y \in \mathbb{R}^2$ with $\|x - y\| < e^{-\varepsilon}$,*

$$A_1 \ln\left(\frac{1}{\|x-y\|}\right) \leq r_1(x,y) \leq A_2 \ln\left(\frac{1}{\|x-y\|}\right).$$

Proof By equation (2), for all $a \in \mathbb{R}^d$,

$$u_1(a) = (2\pi)^{-1}\int_0^\infty e^{-s\|a\|^2}s^{-1}e^{-\frac{1}{2s}}\,ds.$$

Thus, for all $a \in \mathbb{R}^d$ with $\|a\| < e^{-\varepsilon}$,

$$u_1(a) \geq (2\pi)^{-1}e^{-\frac{3}{2}}\int_1^{\|a\|^{-2}} s^{-1}\,ds = \pi^{-1}e^{-\frac{3}{2}}\ln\left(\frac{1}{\|a\|}\right).$$

Thus, we obtain the lower bound on r_1 with $A_1 = \pi^{-1}e^{-3/2}$. To arrive at the upper bound, we split the integral as follows: For all $a \in \mathbb{R}^d$ with $\|a\| < e^{-\varepsilon}$,

$$u_1(a) = (2\pi)^{-1}\{T_1 + T_2 + T_3\}, \tag{3}$$

where

$$T_1 = \int_0^1 e^{-s\|a\|^2}s^{-1}e^{-\frac{1}{2s}}\,ds,$$

$$T_2 = \int_1^{\|a\|^{-2}} e^{-s\|a\|^2}s^{-1}e^{-\frac{1}{2s}}\,ds,$$

$$T_3 = \int_{\|a\|^{-2}}^\infty e^{-s\|a\|^2}s^{-1}e^{-\frac{1}{2s}}\,ds.$$

Clearly,

$$T_1 \leq \int_0^1 e^{-\frac{1}{2s}} s^{-1} ds = \int_1^\infty e^{-\frac{t}{2}} t^{-1} dt \leq \int_0^\infty e^{-\frac{t}{2}} dt = 2,$$

$$T_2 \leq \int_1^{\|a\|^{-2}} s^{-1} ds = 2\ln\left(\frac{1}{\|a\|^2}\right),$$

$$T_3 \leq \int_{\|a\|^{-2}}^\infty e^{-s\|a\|^2} s^{-1} ds = \int_1^\infty e^{-r} r^{-1} dr \leq \int_0^\infty e^{-r} dr = 1.$$

By equation (3), whenever $\|a\| < e^{-\varepsilon}$,

$$u_1(a) \leq (2\pi)^{-1}\left\{3 + 2\ln\left(\frac{1}{\|a\|}\right)\right\} \leq (2\pi)^{-1}\{3\varepsilon^{-1} + 2\} \ln\left(\frac{1}{\|a\|}\right),$$

proving the result with $A_2 = (2\pi)^{-1}\{2\varepsilon^{-1} + 2\}$. □

The only unresolved case is $d = 1$. Fortunately, this has been dealt with in the example of Section 2.3, Chapter 8, but in disguise. Namely, when $d = 1$,

$$r_1(x, y) = \left(\frac{2|x-y|^2}{\pi^3}\right)^{\frac{1}{4}} \mathbf{K}_{\frac{1}{2}}\left(|x-y|\sqrt{2}\right),$$

where $\mathbf{K}_{\frac{1}{2}}$ is the modified Bessel function of index $\frac{1}{2}$.

Exercise 3.1.2 Show that when $d = 1$, r_1 is continuous on all of $\mathbb{R} \times \mathbb{R}$. Show, moreover, that $r_1(x, y) > 0$ for all $x, y \in \mathbb{R}$. □

Next, we recall the definitions of Bessel–Riesz capacities; cf. Section 2.2 of Appendix C for more details.

For any probability measure μ on Borel subsets of $\mathbb{R}^d$, and for any $\beta > 0$, we define $\mathsf{Energy}_\beta(\mu)$ to be the energy of μ with respect to the gauge function $x \mapsto \|x\|^{-\beta}$. When $\beta = 0$, the gauge function is changed to $x \mapsto \ln_+(1/\|x\|)$. Finally, when $\beta < 0$, we define $\mathsf{Energy}_\beta(\mu)$ to be identically equal to 1. For any Borel set $E \subset \mathbb{R}^d$, we define the **Bessel–Riesz capacity** of E of index $\beta \geq 0$ as

$$\mathsf{Cap}_\beta(E) = \left[\inf_{\mu \in \mathcal{P}(E)} \mathsf{Energy}_\beta(\mu)\right]^{-1}.$$

Exercise 3.1.1 states that X is a strongly symmetric Feller process with a strictly positive λ-potential density. Therefore, we can combine Lemmas 3.1.1 and 3.1.2 with Theorem 2.3.1 in order to obtain the following well-known theorem of S. Kakutani; see Kakutani (1944b, 1945) for the original analysis in two dimensions. The higher-dimensional result can be found in Dvoretzky et al. (1950, Lemma 2).

Theorem 3.1.1 (Kakutani's Theorem) *Suppose $B = (B_t;\ t \geq 0)$ denotes d-dimensional Brownian motion. For any compact set $E \subset \mathbb{R}^d$ and for all $x \notin E$,*

$$\mathbb{P}_x(T_E < \infty) > 0 \iff \mathsf{Cap}_{d-2}(E) > 0.$$

Exercise **3.1.3** Complete the described proof of Kakutani's theorem. □

Remarks (i) When $d = 1$, $\mathbb{P}_x(T_E < \infty) > 0$ for *all* compact sets $E \subset \mathbb{R}$; □

(ii) When $d = 2$, Brownian motion does not hit points in the sense that for all $x \neq y$, $\mathbb{P}_x(T_{\{y\}} < \infty) = 0$. To see this, note that any probability measure on $\{y\}$ is necessarily of the form $c\delta_y$, where δ_y denotes a point mass at the point y and $c > 0$ is a finite constant. On the other hand, $\mathsf{Energy}_0(c\delta_y)$ is clearly infinite. That is, $\mathsf{Cap}_0(\{y\}) = 0$, for all $y \in \mathbb{R}^2$. The same property holds when $d \geq 3$. □

3.2 Stable Densities and Bochner's Subordination

It turns out that it is both possible and interesting to try to extend Kakutani's theorem (Theorem 3.1.1) from Brownian motion to isotropic stable processes. Unfortunately, transition densities of stable processes are not explicitly computable outside a few cases; cf. Example 3, Section 4.3 of Chapter 8. Thus, we need to take a different, more subtle, route.

Let $X = (X_t;\ t \geq 0)$ denote an isotropic stable Lévy process in $\mathbb{R}^d$, whose index[4] is $\alpha \in\ ]0, 2[$ and whose transition functions are denoted by $\mathcal{T}$. Recall from Example 3, Section 4.3 of Chapter 8, that X has transition densities (with respect to Lebesgue's measure on $\mathbb{R}^d$) that can be written as follows: For all $x \in \mathbb{R}^d$, all $t \geq 0$, and all $f \in L^\infty(\mathbb{R}^d)$,

$$\mathcal{T}_t f(x) = \mathbb{E}_x[f(X_t)] = \int_{\mathbb{R}^d} f(y) p_t(x, y)\, dy,$$

with

$$p_t(x, y) = q_t(y - x), \qquad t \geq 0,\ x, y \in \mathbb{R}^d,$$

where

$$q_t(a) = (2\pi)^{-d} \int_{\mathbb{R}^d} e^{-i\xi \cdot a} e^{-\frac{1}{2}\|\xi\|^\alpha}\, d\xi, \qquad t \geq 0,\ a \in \mathbb{R}^d.$$

See Section 1.3 above for further details. While we have studied some of the elementary properties of the function q_t in Lemmas 1.3.1 and 1.3.2 and

---

[4] Since we have already studied the connections between Brownian motion and Bessel–Riesz capacity in the previous section, we can and will restrict attention to the stable processes other than Brownian motion; that is, $\alpha < 2$.

in Corollary 1.3.1, they are not sufficient for our present needs. In this and the next subsection we will study the function q_t in much greater depth. In fact, we will study the behavior of $a \mapsto q_t(a)$, as $\|a\| \to \infty$. By scaling, we can reduce this asymptotic question to one about the behavior of $a \mapsto q_1(a)$ as $\|a\| \to \infty$; cf. Lemma 1.3.1. In order to begin our asymptotic analysis, we relate the function q_1 to the transition densities of Brownian motion. The mechanism for doing so involves a little Laplace transform theory and is called **subordination**; this is a special case of a general method that was proposed in Bochner (1955).

Throughout, for any $\beta > 0$, we define the function $g_\beta : \mathbb{R}_+ \to \mathbb{R}_+$ by

$$g_\beta(\lambda) = \exp\{-(2\lambda)^\beta\}, \qquad \lambda \geq 0.$$

The first important result of this subsection is the following.

Theorem 3.2.1 *If $\beta \in \,]0,1[$, then g_β is the Laplace transform of some probability measure σ_β on $\mathbb{R}_+$.*

In order to prove this, we need the following lemma. Recall from Appendix B that a function $f : \mathbb{R}_+ \to \mathbb{R}_+$ is **completely monotone** if f is infinitely differentiable and satisfies $f \geq 0$, $f' \leq 0$, $f'' \geq 0$, $f''' \leq 0, \ldots$.

Lemma 3.2.1 *Consider two infinitely differentiable functions $f : \mathbb{R}_+ \to \mathbb{R}_+$ and $h : \mathbb{R}_+ \to \mathbb{R}_+$. If f and h' are completely monotone, so is the composition function $f \circ h$.*

Exercise **3.2.1** Prove Lemma 3.2.1. □

Proof of Theorem 3.2.1 Let $f(x) = e^{-x}$ and $h(x) = (2x)^\beta$ and check directly that f and h' are both completely monotone. The theorem follows from Lemma 3.2.1 used in conjunction with Bernstein's theorem; cf. Theorem 1.3.1 of Appendix B. □

When $0 < \beta < 1$, the probability measure σ_β given to us by Theorem 3.2.1 corresponds to the **(completely asymmetric) stable distribution** on $[0, \infty[$ with index β. For the remainder of this subsection we refer to this σ_β as such and write F_β for its distribution function. It is now easy to explain S. Bochner's idea of subordination in our present context.

Recalling that $0 < \alpha < 2$, we let $\beta = \alpha/2$ and notice that $0 < \beta < 1$. In particular, we can construct an a.s. positive random variable $\tau_\beta = \tau_{\alpha/2}$ and an independent $\mathbb{R}^d$-valued random vector $Z \sim \mathcal{N}_d(0, \mathbf{I})$, where $\mathbf{I}$ denotes the identity matrix of dimension $(d \times d)$. Recall that this latter notation merely means that $Z = (Z^{(1)}, \ldots, Z^{(d)})$, where the $Z^{(i)}$'s are i.i.d. standard normal random variables. In our present setting, Bochner's subordination can be phrased as follows.

Theorem 3.2.2 (Bochner's Subordination) *Let $\tau_{\alpha/2}$ be a completely asymmetric stable random variable of index $\alpha/2$, taking its values in $\mathbb{R}_+$, a.s. Let $Z \sim \mathcal{N}_d(0, \mathbf{I})$ be constructed on the same probability space and totally independently of $\tau_{\alpha/2}$. Then, the $\mathbb{R}^d$-valued random vector $\sqrt{\tau_{\alpha/2}}\, Z$ has the same distribution as X_1.*

Proof We check the characteristic functions. By independence and by the form of the characteristic function of Gaussian random vectors (Section 1.1, Chapter 5), for all $\xi \in \mathbb{R}^d$,

$$\mathbb{E}[\exp\{i\xi \cdot \sqrt{\tau_{\alpha/2}}\, Z\}] = \mathbb{E}[\exp\{-\tfrac{1}{2}\tau_{\alpha/2}\|\xi\|^2\}] = g_{\alpha/2}(\tfrac{1}{2}\|\xi\|^2)$$
$$= \exp\{-\tfrac{1}{2}\|\xi\|^\alpha\},$$

which equals $\mathbb{E}[e^{i\xi \cdot X_1}]$. The theorem follows from the uniqueness theorem of characteristic functions. $\square$

Let Z and $\tau_{\alpha/2}$ be given by the above theorem. Conditional on the value of $\tau_{\alpha/2}$, $\sqrt{\tau_{\alpha/2}}\, Z \sim \mathcal{N}_d(0, \tau_{\alpha/2}\mathbf{I})$. Thus, using Bochner's subordination (Theorem 3.2.2), the form of the probability density of Gaussian random vectors, and Fubini's theorem, we see that for all Borel sets $A \subset \mathbb{R}^d$,

$$\mathbb{P}(X_1 \in A) = (2\pi)^{-\frac{d}{2}} \int_A \mathbb{E}\left[\tau_{\alpha/2}^{-\frac{d}{2}} \exp\left\{-\frac{\|x\|^2}{2\tau_{\alpha/2}}\right\}\right] dx$$
$$= (2\pi)^{-\frac{d}{2}} \int_A \int_0^\infty t^{-\frac{d}{2}} \exp\left\{-\frac{\|x\|^2}{2t}\right\} \sigma_{\alpha/2}(dt)\, dx,$$

where σ_β is the measure given by Theorem 3.2.1. But $q_1(a)$ is the density function of X_1 at a (with respect to Lebesgue's measure). Therefore, we have the following. (Why?)

Corollary 3.2.1 *For any $a \in \mathbb{R}^d$,*

$$q_1(a) = (2\pi)^{-\frac{d}{2}} \int_0^\infty t^{-\frac{d}{2}} \exp\left\{-\frac{\|a\|^2}{2t}\right\} \sigma_{\alpha/2}(dt).$$

In particular, $q_1(a) > 0$, for all $a \in \mathbb{R}^d$.

This corollary translates statements about $a \mapsto q_1(a)$ (and its asymptotics) into statements about $x \mapsto F_{\alpha/2}(x)$ (and its asymptotics.) We conclude this subsection with the asymptotics of $x \mapsto F_{\alpha/2}(x)$.

Theorem 3.2.3 *For any $\alpha \in\,]0, 2[$,*

$$\lim_{x \to \infty} x^{\alpha/2} \mathbb{P}(\tau_{\alpha/2} > x) = \lim_{x \to \infty} x^{\alpha/2} \{1 - F_{\alpha/2}(x)\} = 2^{\alpha/2}\left[\Gamma\left(1 - \tfrac{1}{2}\alpha\right)\right]^{-1}.$$

Proof Using Corollary 2.1.2 of Appendix B with $\theta = 1 - \frac{\alpha}{2}$, it suffices to show that

$$\lim_{\lambda \to 0^+} \lambda^{-\alpha/2}\{1 - \widehat{F_{\alpha/2}}(\lambda)\} = 2^{\alpha/2},$$

where $\widetilde{f}$ denotes the Laplace transform of the function f; cf. Appendix B. On the other hand, $\widetilde{F_{\alpha/2}}(\lambda) = g_{\alpha/2}(\lambda) = \exp\{-(2\lambda)^{\alpha/2}\}$, from which the theorem follows immediately. □

3.3 Asymptotics for Stable Densities

Continuing with the setup of the previous subsection, we are now ready to prove the following technical estimate.

Proposition 3.3.1 *There exists a finite constant $A > 1$ such that for all $a \in \mathbb{R}^d$ with $\|a\| \geq 2$,*

$$\frac{1}{A}\|a\|^{-(d+\alpha)} \leq q_1(a) \leq A\|a\|^{-(d+\alpha)}.$$

A slightly better result is possible; cf. Supplementary Exercise 9.

Proof Our proof is performed in several steps. We start by writing

$$q_1(a) = (2\pi)^{-\frac{d}{2}}\{T_1 + T_2\}, \tag{1}$$

where

$$T_1 = \int_0^{\|a\|^2} t^{-\frac{d}{2}} \exp\left\{-\frac{\|a\|^2}{2t}\right\} \sigma_{\alpha/2}(dt),$$

$$T_2 = \int_{\|a\|^2}^{\infty} t^{-\frac{d}{2}} \exp\left\{-\frac{\|a\|^2}{2t}\right\} \sigma_{\alpha/2}(dt).$$

This is justified due to the formula given by Corollary 3.2.1. We now start to obtain upper and lower bounds for T_1 and T_2 in stages. The simplest is the upper bound for T_2. Indeed,

$$T_2 \leq \int_{\|a\|^2}^{\infty} t^{-\frac{d}{2}} \sigma_{\alpha/2}(dt) = \sum_{j=0}^{\infty} \int_{e^j\|a\|^2}^{e^{j+1}\|a\|^2} t^{-\frac{d}{2}} \sigma_{\alpha/2}(dt)$$

$$\leq \|a\|^{-d} \sum_{j=0}^{\infty} e^{-\frac{1}{2}dj} \sigma_{\alpha/2}([e^j\|a\|^2, \infty[)$$

$$= \|a\|^{-d} \sum_{j=0}^{\infty} e^{-\frac{1}{2}dj} \mathbb{P}(\tau_{\alpha/2} > e^j\|a\|^2).$$

By Theorem 3.2.3, we can find a finite constant $A_1 > 1$ such that for all $j \geq 0$ and all $\|a\| > 0$,

$$\mathbb{P}(\tau_{\alpha/2} > e^j\|a\|^2) \leq A_1\|a\|^{-\alpha}e^{-\frac{1}{2}j\alpha},$$

whence
$$T_2 \leq A_2 \|a\|^{-(d+\alpha)}, \qquad (2)$$
where $A_2 = A_1 \sum_{j=0}^{\infty} e^{-\frac{1}{2}(d+\alpha)j}$. The estimation of T_1 starts along similar lines:

$$T_1 = \sum_{j=1}^{\infty} \int_{e^{-j}\|a\|^2}^{e^{-j+1}\|a\|^2} t^{-\frac{d}{2}} \exp\left\{-\frac{\|a\|^2}{2t}\right\} \sigma_{\alpha/2}(dt)$$

$$\leq \|a\|^{-d} \sum_{j=1}^{\infty} \exp\left\{\frac{jd}{2} - \frac{e^j}{2e}\right\} \mathbb{P}(\tau_{\alpha/2} > e^{-j}\|a\|^2).$$

Whenever $e^{-j}\|a\|^2 > 1$, then we can find a finite constant $A_3 > 1$ such that
$$\mathbb{P}(\tau_{\alpha/2} > e^{-j}\|a\|^2) \leq A_3 e^{\frac{1}{2}j\alpha} \|a\|^{-\alpha}.$$
On the other hand, if $e^{-j}\|a\|^2 \leq 1$, an appropriate bound for this above probability is 1. Thus,

$$T_1 \leq A_3 \|a\|^{-(d+\alpha)} \sum_{1 \leq j \leq 2\ln\|a\|} \exp\left\{\frac{j(d+\alpha)}{2} - \frac{e^j}{2d}\right\}$$

$$+ \|a\|^{-d} \sum_{j > 2\ln\|a\|} \exp\left\{\frac{jd}{2} - \frac{e^j}{2e}\right\}$$

$$\leq A_4 \|a\|^{-(d+\alpha)} + \|a\|^{-d} \exp\left\{-\frac{\|a\|^2}{4e}\right\} \sum_{j=1}^{\infty} \exp\left\{\frac{jd}{2} - \frac{e^j}{4e}\right\},$$

where
$$A_4 = A_3 \sum_{j=1}^{\infty} \exp\left\{\frac{j(d+\alpha)}{2} - \frac{e^j}{2e}\right\}.$$

Since $\|a\| \geq 2$, we see that there exists a finite constant $A_5 > 1$ such that $T_1 \leq A_5 \|a\|^{-(d+\alpha)}$. In light of equation (1), we can combine this with equation (2) to deduce the existence of a constant $A_6 = (2\pi)^{-\frac{d}{2}}\{A_5 + A_2\} > 1$ such that that for all $\|a\| \geq 2$,
$$q_1(a) \leq A_6 \|a\|^{-(d+\alpha)}. \qquad (3)$$

It remains to get lower bounds. Using the decomposition of equation (1) one more time,

$$q_1(a) \geq T_2 \geq e^{-\frac{1}{2}} \int_{\|a\|^2}^{\infty} t^{-\frac{d}{2}} \sigma_{\alpha/2}(dt) \geq e^{-\frac{1}{2}} \sum_{j=0}^{\infty} \int_{e^j\|a\|^2}^{e^{j+1}\|a\|^2} t^{-\frac{d}{2}} \sigma_{\alpha/2}(dt)$$

$$\geq e^{-\frac{1}{2}} \|a\|^{-d} \sum_{j=0}^{\infty} e^{-(j+1)\frac{d}{2}} \mathbb{P}\left(e^{j+1}\|a\|^2 \geq \tau_{\alpha/2} > e^j\|a\|^2\right).$$

By Theorem 3.2.3, for all $\varepsilon > 0$, there exists $M > 2$ such that for all $a \in \mathbb{R}^d$ with $\|a\| \geq M$, and for all $j \geq 0$,

$$\mathbb{P}\big(e^{j+1}\|a\|^2 \geq \tau_{\alpha/2} > e^j \|a\|^2\big)$$
$$= \mathbb{P}\big(\tau_{\alpha/2} > e^j \|a\|^2\big) - \mathbb{P}\big(\tau_{\alpha/2} > e^{j+1}\|a\|^2\big)$$
$$\geq (1-\varepsilon)e^{-\frac{1}{2}j\alpha}\|a\|^{-\alpha} - (1+\varepsilon)e^{-\frac{1}{2}(j+1)\alpha}\|a\|^{-\alpha}$$
$$= \|a\|^{-\alpha} e^{-\frac{1}{2}j\alpha}\big\{(1-\varepsilon) + \varepsilon(1+e^{-\frac{\alpha}{2}})\big\}.$$

Pick $\varepsilon = \frac{1}{2}(1-e^{-\alpha/2})(1+e^{-\alpha/2})^{-1}$ to see that whenever $\|a\| \geq M$,

$$\mathbb{P}\big(e^{j+1}\|a\|^2 \geq \tau_{\alpha/2} > e^j\|a\|^2\big) \geq \|a\|^{-\alpha} e^{-\frac{1}{2}j\alpha} \left(\frac{1-e^{-\alpha}}{2}\right) = A_7 \|a\|^{-\alpha}.$$

In particular, whenever $\|a\| \geq M$,

$$q_1(a) \geq A_8 \|a\|^{-(d+\alpha)},$$

where

$$A_8 = e^{-1/2} A_7 \left(\frac{1-e^{-\frac{\alpha}{2}}}{2}\right) \sum_{j=0}^{\infty} e^{-\frac{1}{2}(j+1+\alpha)d}.$$

By the positivity assertion of Corollary 3.2.1, whenever $2 \leq \|a\| < M$, the above lower bound for $q_1(a)$ holds with A_8 replaced by a possibly smaller constant A_9; In light of this and equation (3), we have proved the proposition with an appropriately large choice of A. □

3.4 Stable Processes and Bessel–Riesz Capacity

We are in a position to estimate the hitting probabilities for the process X of this section. Recall that the transition density (with respect to Lebesgue's measure) of X is $p_t(x,y) = q_t(y-x)$, where $q_t(a)$ denotes the density (under $\mathbb{P}$) of the random vector X_t, evaluated at $a \in \mathbb{R}^d$. Thus, (why?) X has a λ-potential density $r_\lambda(x,y)$ that is given by

$$r_\lambda(x,y) = u_\lambda(y-x), \qquad \lambda > 0, \ x,y \in \mathbb{R}^d,$$

where,

$$u_\lambda(a) = \int_0^\infty e^{-\lambda s} q_s(a)\,ds.$$

Note that $u_\lambda(a) \leq u_\lambda(0)$. However, the latter is finite if and only if $d > \alpha$.

Exercise 3.4.1 Verify that $u_\lambda(0) < \infty$ if and only if $d > \alpha$. □

Some analysis shows that whenever $d > \alpha$, r_λ is a gauge function on $\mathbb{R}^d \times \mathbb{R}^d$ and X is a strongly symmetric Feller process with Lebesgue's measure as its reference measure. In fact, this is true even if $d \leq \alpha$, but it is harder to prove. The following addresses such technical issues and provides the analogue of Lemmas 3.1.1 and 3.1.2.

Lemma 3.4.1 *Suppose $d > \alpha$ and $\lambda > 0$ are fixed. Then, there exist two finite positive constants $A_1 \leq A_2$ such that for all $x, y \in \mathbb{R}^d$ with $\|x - y\| \leq 1$,*
$$\frac{A_1}{\|x-y\|^{d-\alpha}} \leq r_\lambda(x,y) \leq \frac{A_2}{\|x-y\|^{d-\alpha}}.$$
In particular, r_λ is a gauge function on $\mathbb{R}^d \times \mathbb{R}^d$ and X is a strongly symmetric Feller process whose reference measure is Lebesgue's measure on $\mathbb{R}^d$.

Proof We will prove the asserted inequalities for the λ-potential density r_λ. The remaining assertions follow from this and are covered in Exercise 3.4.2 below.

By the scaling lemma (Lemma 1.3.1) and by elementary manipulations,
$$q_s(a) = s^{-\frac{d}{\alpha}} q_1(a/s^{\frac{1}{\alpha}}), \qquad s \geq 0, \ a \in \mathbb{R}^d;$$
see Exercise 1.3.2. Thus,
$$u_\lambda(a) = \int_0^\infty s^{-\frac{d}{\alpha}} e^{-\lambda s} q_1(a/s^{\frac{1}{\alpha}}) \, ds.$$
We will split this integral into two parts:
$$u_\lambda(a) = T_1(a) + T_2(a), \tag{1}$$
where
$$T_1(a) = \int_0^{2^{-\alpha}\|a\|^\alpha} s^{-\frac{d}{\alpha}} e^{-\lambda s} q_1(a/s^{\frac{1}{\alpha}}) \, ds,$$
$$T_2(a) = \int_{2^{-\alpha}\|a\|^\alpha}^\infty s^{-\frac{d}{\alpha}} e^{-\lambda s} q_1(a/s^{\frac{1}{\alpha}}) \, ds.$$

By Proposition 3.3.1, there exists a finite constant $A_3 > 1$ such that for all $b \in \mathbb{R}^d$ with $\|b\| \geq 2$, $q_1(b) \leq A_3 \|b\|^{-(d+\alpha)}$. Thus,
$$T_1(a) \leq A_3 \|a\|^{-(d+\alpha)} \int_0^{2^{-\alpha}\|a\|^\alpha} s \, ds = 2^{1-2\alpha} A_3 \, \|a\|^{\alpha-d}. \tag{2}$$

On the other hand, by Corollary 1.3.1, for all $a \in \mathbb{R}^d$, $q_1(a) \leq q_1(0) < \infty$. Consequently,
$$T_2(a) \leq q_1(0) \int_{2^{-\alpha}\|a\|^\alpha}^\infty s^{-\frac{d}{\alpha}} \, ds = A_4 \|a\|^{\alpha-d},$$

where $A_4 = q_1(0) \int_{2-\alpha}^{\infty} t^{-\frac{d}{\alpha}}\, dt$. Together with equations (1) and (2), this proves the upper bound with $A_1 = 2^{1-2\alpha} A_3 + A_4$. To obtain the lower bound, note that whenever $\|a\| \leq 1$,

$$u_\lambda(a) \geq T_1(a) \geq \exp\big\{-\lambda 2^{-\alpha}\big\} \int_0^{2^{-\alpha}\|a\|^\alpha} s^{-\frac{d}{\alpha}} q_1(a/s^{\frac{1}{\alpha}})\, ds.$$

Now we can use the lower bound of Proposition 3.3.1 and argue as above to finish our proof. □

Exercise 3.4.2 Complete the derivation of Lemma 3.4.1 by showing that r_λ is a gauge function on $\mathbb{R}^d \times \mathbb{R}^d$ and that X is a strongly symmetric Feller process whose reference measure is Lebesgue's measure on $\mathbb{R}^d$. □

The "critical case" is when $d = \alpha$. In this section $\alpha \in {]0, 2[}$, which means that the critical case is $d = \alpha = 1$. That is handled by the following estimate.

Lemma 3.4.2 *Suppose $d = \alpha = 1$ and $\lambda > 0$ are fixed. Then, there exist two finite positive constants $A_1 \leq A_2$ such that for all $x, y \in \mathbb{R}$ with $\|x - y\| \leq \frac{1}{2}$,*

$$A_1 \ln\left(\frac{1}{\|x - y\|}\right) \leq r_\lambda(x, y) \leq A_2 \ln\left(\frac{1}{\|x - y\|}\right).$$

In particular, r_λ is a gauge function on $\mathbb{R} \times \mathbb{R}$ and X is a strongly symmetric Feller process with Lebesgue's measure on $\mathbb{R}$ as its reference measure.

Exercise 3.4.3 Prove Lemma 3.4.2.
(HINT: Borrow ideas from Lemmas 3.1.2 and 3.4.1.) □

Exercise 3.4.4 Show that in any case, whenever X is a d-dimensional isotropic stable Lévy process of index $\alpha \in {]0, 2]}$, $r_\lambda(x, y) > 0$ for all $x, y \in \mathbb{R}^d$. □

We are ready to state the following extension of Kakutani's theorem (Theorem 3.1.1).

Theorem 3.4.1 *Suppose $X = (X_t;\ t \geq 0)$ denotes a d-dimensional isotropic stable Lévy process of index $\alpha \in {]0, 2]}$. For any compact set $E \subset \mathbb{R}^d$ and for all $x \notin E$,*

$$\mathbb{P}_x(T_E < \infty) > 0 \iff \mathsf{Cap}_{d-\alpha}(E) > 0,$$

where Cap_β denotes the index-β Bessel–Riesz capacity defined in Section 3.1.

Exercise 3.4.5 Carefully verify Theorem 3.4.1. □

3.5 Relation to Hausdorff Dimension

Let $X = (X_t;\ t \geq 0)$ denote an $\mathbb{R}^d$-valued isotropic stable Lévy process with index $\alpha \in\ ]0, 2]$. Given a compact set $E \subset \mathbb{R}^d$, let $T_E = \inf(s > 0:\ X_s \in E)$ and note that $T_E < \infty$ if and only if

$$X(\mathbb{R}_+) \cap E \neq \varnothing,$$

where $X(T)$ is the **image of T** under the random map $s \mapsto X_s$. That is,

$$X(T) = \{X_s:\ s \in T\}.$$

In particular, $X(\mathbb{R}_+)$ is the **range** of X.

According to Theorem 3.4.1, for all $x \notin E$,

$$\mathbb{P}_x\{X(\mathbb{R}_+) \cap E \neq \varnothing\} > 0 \iff \mathsf{Cap}_{d-\alpha}(E) > 0.$$

We can now combine the above with Frostman's theorem (Theorem 2.2.1, Appendix C) to conclude the following. For a detailed historical account of the development of this result and its kin, see Taylor (1986).

Theorem 3.5.1 *For any compact set $E \subset \mathbb{R}^d$ and for all $x \notin E$,*

$$\dim(E) > d - \alpha \implies \mathbb{P}_x\{X(\mathbb{R}_+) \cap E \neq \varnothing\} > 0,$$
$$\dim(E) < d - \alpha \implies \mathbb{P}_x\{X(\mathbb{R}_+) \cap E \neq \varnothing\} = 0.$$

Remarks (1) The above states that the Hausdorff dimension of a set E essentially determines whether or not E is ever hit by the range of an isotropic stable Lévy process. In the critical case where $d = \alpha$, knowing the Hausdorff dimension alone does not give us enough information on the positivity of the probability of the intersection of the range and the set E; see Carleson (1983, Theorems 4 and 5, Chapter IV), for example.

(2) If the starting point x is inside E, the estimation of hitting probabilities could become either trivial or much more complicated, depending on one's interpretation of hitting probabilities and on how complicated the structure of the set E is. For instance, if $x \in E$, it is trivial to see that for $E \subset \mathbb{R}^d$ compact, $\mathbb{P}_x(\overline{X(\mathbb{R}_+)} \cap E \neq \varnothing) = 1$, since $X_0 = x \in E$, $\mathbb{P}_x$-a.s. However, one can ask about when $\liminf_{x \to E: x \notin E} \mathbb{P}_x(\overline{X(\mathbb{R}_+)} \cap E \neq \varnothing) > 0$. This is another matter entirely and has to do with the development of a theory for so-called regular points; we will not address such issues in this book. However, you may wish to consult Itô and McKean (1974, equation (9), Section 8.1, Chapter 7) under the general heading of Wiener's test for electrostatic (Bessel–) Riesz capacity to appreciate some of the many subtleties involved.

4 Supplementary Exercises

1. Consider an $\mathbb{R}^d$-valued Lévy process $X = (X_t;\ t \geq 0)$. Prove that the following are equivalent: (i) X is transient; (ii) the origin is transient; and (iii) $\lim_{t\to\infty} |X_t| = +\infty$, almost surely.

2. Suppose $\mathcal{R}$ and $\mathcal{T}$ denote the the resolvent and transitions of a strongly symmetric Feller process X on $\mathbb{R}^d$, with reference measure ν.
   (i) If ϕ is a finite measure on $\mathbb{R}^d$, show that as $t \downarrow 0$,
   $$e^{-\lambda t}\mathcal{T}_t\mathcal{R}_\lambda\phi(x) \uparrow \mathcal{R}_\lambda\phi(x)$$
   for each $x \in \mathbb{R}^d$.
   (ii) Given that there are two constants $\lambda, \gamma > 0$ and two probability measures μ_1 and μ_2 such that $\mathcal{R}_\lambda\mu_1(x) = \mathcal{R}_\gamma\mu_2(x)$, for ν-almost all $x \in \mathbb{R}^d$, prove that $\mathcal{R}_\lambda\mu_1(x) = \mathcal{R}_\gamma\mu_2(x)$ for *every* $x \in \mathbb{R}^d$. Conclude that μ_1 equals μ_2.

3. (Hard) Let $X = (X_t;\ t \geq 0)$ denote an $\mathbb{R}^d$-valued isotropic stable Lévy process with index $\alpha \in\ ]0, 2[$. Fix some $R, \varepsilon > 0$ and define the *sausage* X_R^ε to be the random compact set
   $$X_R^\varepsilon = \left\{x \in [-R, R]^d :\ \mathrm{dist}[\{x\}; X([1, 2])] \leq \varepsilon\right\}.$$
   (i) Prove that when $d > \alpha$, there exists a finite constant $A > 0$ that depends on d, α, and R such that the expected value of Lebesgue's measure of X_R^ε is bounded above by $A\varepsilon^{d-\alpha}$.
   (ii) Prove that when $d = \alpha$, there exists a finite constant A' that depends on d, α and R, such that the expected value of Lebesgue's measure of X_R^ε is bounded above by $A/\ln_+(1/\varepsilon)$.
   (iii) Deduce from this that when $d \geq \alpha$, the d-dimensional Lebesgue's measure of the random set $\overline{X(\mathbb{R}_+)}$ is a.s. zero. In particular, this is the case when X is two-dimensional Brownian motion.

   This is essentially due to P. Lévy.
   (HINT: Use Lemma 1.4.3.)

4. Recall from Chapter 8 that when X is a Feller process on $\mathbb{R}^d$, as $\lambda \to 0^+$, $\lambda\mathcal{R}_\lambda f$ converges uniformly to f, for all continuous functions f that vanish at infinity. Suppose further that X is strongly symmetric. Prove that for all finite measures μ on $\mathbb{R}^d$, the measures $\lambda\mathcal{R}_\lambda\mu(x)\nu(dx)$ converge weakly to μ, as $\lambda \to 0^+$.

5. Let E denote a compact set in $\mathbb{R}^d$ and let E^η denote its η-enlargement. That is, $x \in E^\eta$ if and only if the Euclidean distance between x and E is (strictly) less than η. Let $T^\eta = \inf(s \geq 0 :\ X_s \in E^\eta)$ denote the entrance time to E^η and prove that for any $x \in \mathbb{R}^d$, as $\eta \to 0^+$, $T^\eta \downarrow T$, $\mathbb{P}_x$-a.s.
   (HINT: Use the right continuity of $t \mapsto X_t$.)

6. (Hard) Suppose $E \subset \mathbb{R}^d$ is compact, $F \subset \mathbb{R}^d$ is open, and $E \subset F$. If X is a strongly symmetric Feller process, define the operator

$$\mathcal{T}_t^F f(x) = \mathbb{E}_x[f(X_t)\mathbf{1}_{(S_F > t)}],$$

where $S_F = \inf(s \geq 0 : X_s \notin F)$.

(i) Prove that $\mathcal{T}^F = (\mathcal{T}_t^F; t \geq 0)$ is a semigroup.

(ii) Show that $\mathcal{T}^F$ defines a Markov semigroup on the one-point compactification F_Δ of F. Informally, $\mathcal{T}^F$ corresponds to the Markov process X, killed upon leaving the open set F. In this regard, see also Section 1.3, Chapter 8.

(iii) Show that for each $t > 0$, there exists a nonnegative function $p_t^F : \mathbb{R}^d \times \mathbb{R}^d$, where:

  (a) for each $x \in \mathbb{R}^d$, $y \mapsto p_t(x, y)$ is measurable;
  (b) unless x and y are both in $\overline{F}$, $p_t(x, y) = 0$;
  (c) as operators, $\mathcal{T}_t^F \varphi(x) = \int p_t^F(x, y) \varphi(y) \, \nu(dy)$.

(v) Show that for all $\varphi, \psi \in L^\infty(\mathbb{R}^d)$,

$$\int \varphi(x) \mathcal{T}_t^F \psi(x) \, \nu(dx) = \int \psi(x) \mathcal{T}_t^F \varphi(x) \, \nu(dx).$$

Use this to prove that for each $t > 0$, we can choose a version of the density p_t^F with the following further property: There exists a ν-null set $\mathcal{N}_t$ such that for all $x \notin \mathcal{N}_t$ and for *all* $y \in \mathbb{R}^d$, $p_t^F(x, y) = p_t^F(y, x)$.

(vi) One can extend the definition of $\mathcal{T}^F$ to act on a measure ϕ by $\mathcal{T}_t^F \phi(x) = \int p_t^F(x, y) \, \phi(dy)$. Show that for all $\lambda, t > 0$ and all $x \in \mathbb{R}^d$,

$$\mathcal{R}_\lambda \mathcal{T}_t^F \phi(x) \leq e^{\lambda t} \mathcal{R}_\lambda \mathcal{T}_t \phi(x),$$

where $\mathcal{T}$ and $\mathcal{R}$ denote the transitions and the resolvent of X, respectively.

7. Let $X = (X_t; t \geq 0)$ denote an isotropic stable Lévy process on $\mathbb{R}^d$ with index $\alpha \in]0, d[$. Prove that for any given $a \in \mathbb{R}^d$, $\{t \geq 0 : X_t = a\}$ is a.s. unbounded.

8. Let $X = (X_t; t \geq 0)$ denote a Lévy process on $\mathbb{R}^d$ and let $\mathfrak{R}$ denote the collection of all points in $\mathbb{R}^d$ that are recurrent for X.

(i) Prove that $\mathfrak{R}$ is a free abelian subgroup of $\mathbb{R}^d$. In particular, deduce from this that $\mathfrak{R} \neq \varnothing$ if and only if the origin is recurrent.

(ii) A point $x \in \mathbb{R}^d$ is said to be (nearly) possible if there exists $t \geq 0$ such that for all $\varepsilon > 0$, $\mathbb{P}(|X_t - x| \leq \varepsilon) > 0$. Prove that when $\mathfrak{R} \neq \varnothing$, all possible points are recurrent.

(iii) Prove that when $\mathfrak{R} \neq \varnothing$, one can identify X with a recurrent Lévy process on a free abelian group G.

9. Refine Proposition 3.3.1 by showing that for any $\alpha \in\ ]0, 2[$,
$$\lim_{\|x\|\to\infty} \|x\|^{d+\alpha} q_1(x) = \alpha 2^{\alpha-1} \pi^{-\frac{d}{2}-1} \sin\left(\frac{\alpha\pi}{2}\right) \Gamma\left(\frac{d+\alpha}{2}\right) \Gamma\left(\frac{\alpha}{2}\right),$$
where Γ denotes the gamma function. When $d = 1$, this is due to Pólya (1923); the general case is due to Blumenthal and Getoor (1960b); see also (Bendikov 1994; Bergström 1952).
(HINT: Follow the described derivation of Proposition 3.3.1 but pay closer attention to estimating the errors involved. You may need the identity $\beta\Gamma(1-\beta) = \pi\beta/\sin(\pi\beta)$.)

10. Suppose X has transition densities $p_t(x, y)$. Prove that for any finite measure μ, $t \mapsto e^{-\lambda t}\mathcal{R}_\lambda\mu(X_t)$ is a supermartingale. Use this to derive another proof of Lemma 2.4.2.

11. Consider a d-dimensional Brownian motion $B = (B_t;\ t \geq 0)$ and define the process $O = (U_t;\ t \geq 0)$ by $O_t = e^{-\frac{t}{2}} B_{e^t}$, $(t \geq 0)$. This is the Ornstein–Uhlenbeck process also defined earlier in Supplementary Exercise 3, Chapter 9.

   (i) Prove that O is a strongly symmetric Feller process with reference measure $\nu(dx) = e^{\frac{1}{2}x^2}\,dx$.

   (ii) Show that for all compact sets $E \subset \mathbb{R}^d$ and for all $x \notin E$, $\mathbb{P}_x\{O_t \in E \text{ for some } t \in [0, \mathfrak{e}(\lambda)]\} > 0$ if and only if E has positive $(d-2)$-dimensional Bessel–Riesz capacity.

12. (Hard) Suppose X is a strongly symmetric Feller process on $\mathbb{R}^d$, with reference measure ν, and whose λ-potential density is r_λ for any $\lambda > 0$. Prove that r_λ satisfies the maximum principle of Appendix D. In other words, show that for all compactly supported probability measures μ with support E,
$$\sup_{x \in \mathbb{R}^d} \mathcal{R}_\lambda \mu(x) = \sup_{x \in E} \mathcal{R}_\lambda \mu(x).$$
(HINT: First, consider the supremum of $\mathcal{R}_\lambda f$, where f is a function. For this you may need the strong Markov property. Then, apply balayage, together with lower semicontinuity.)

5 Notes on Chapter 10

Section 1 The fact that recurrence/transience and hitting probabilities are quite intertwined has been in the folklore of the subject for a very long time. It seems to have been made explicit in Kakutani (1944a).

Section 2 Our proof of Lemma 2.2.2 involves time-reversal. We come back to this method time and again, since time-reversal is intrinsic to the structure of one-parameter Markov processes. See the Notes in Chapter 8 for some references on time-reversal.

To the generalist, the material on potential theory may seem a little specialized. However, this is mainly due to the nature of the exposition rather than the power of the presented methods. Here are two broad remarks: To go beyond the stated symmetry assumptions, one needs a little bit more on time-reversal and initial measures; to extend the theory to processes on separable, locally compact metric spaces, one needs the general form of Prohorov's theorem. See Section 2.5 of Chapter 6 for the latter and see Chapter 8 for the former. Other variants of Theorem 2.3.1 are also possible; see (Fitzsimmons and Salisbury 1989; Salisbury 1996) for the starting point of many of the works in the subject, as well as (Bauer 1994; Benjamini et al. 1995; Mazziotto 1988; Ren 1990), and the fine series by Hirsch and Song (1994, 1995, 1995b, 1995c, 1995d).

There is a rich theory of 1-parameter Markov processes that requires only a notion of duality, and goes beyond assumptions of symmetry; see Getoor (1990) and its bibliography.

The passing references made to time-reversal have to do with the fact that, in the construction of p_t^n in Lemma 2.2.1, we tacitly reversed time, in that we considered both vectors $(X_0, X_{2^{-n}t}, \ldots, X_{((2^n-1)2^{-n})t}, X_t)$ and $(X_t, X_{((2^n-1)2^{-n})t}, \ldots, X_{2^{-n}t}, X_0)$. Time-reversal was systematically explored in Nagasawa (1964), and leads to a rich theory that takes too long to develop here; see also Millar (1978).

Section 3 Potential theory of Lévy processes is well delineated in (Bertoin 1996; Sato 1999) and their combined bibliography. See Janke (1985), Kanda (1982, 1983), Kanda and Uehara (1981), Hawkes (1979), Orey (1967), Rao (1987), and their references for some of the earlier work. Fukushima et al. (1994) contains a detailed treatment of the potential theory of symmetric Markov processes in the context of the Brelot–Beurling–Cartan–Deny theory.

It is not possible to extend the potential density estimates of Lemmas 3.1.1, 3.1.2, 3.4.1, and 3.4.2 to completely general stable processes; cf. Pruitt and Taylor (1969) for a precise statement.

Theorem 3.5.1 is a starting point for connections between stochastic processes and Hausdorff dimension. We will elaborate on this at greater length in Chapter 11. In the meantime, have a look at the survey article Taylor (1986) for an extensive bibliography.

11
Multiparameter Markov Processes

We can informally interpret Chapter 8's definition of a Markov process $X = (X_t;\ t \geq 0)$ as a (one-parameter) process whose "future" values X_{t+s} depend on the past only through its current value X_t. While this is perfectly intuitively clear (due to the well-ordering of the "time axis"), it is far less clear what a multiparameter Markov process should be. In this chapter we introduce and study a class of random fields called multiparameter Markov processes. The definitions, given early on, are motivated by the potential theory that is developed later in this chapter. We will also see how this multiparameter theory can be used to study intersections of ordinary one-parameter processes.

1 Definitions

Throughout, we let (S, d) denote a separable, locally compact metric space. As in Chapter 8, we may need to compactify it via a one-point compactification and topologize it with its one-point compactification topology. Also as in Chapter 8, we always denote the latter one-point compactification by S_Δ. In agreement with Chapter 8, all measurable functions $f : S \to \mathbb{R}$ are extended to functions from S_Δ into $\mathbb{R}$ via the assignment $f(\Delta) = 0$. This section provides us with a general definition of an N-parameter Markov process that takes its values in the space S_Δ.

1.1 Preliminaries

An N-parameter, S_Δ-valued stochastic process $X = (X_t;\ t \in \mathbb{R}_+^N)$ is said to be a **multiparameter Markov process** if there exists an N-parameter filtration $\mathcal{F} = (\mathcal{F}_t;\ t \in \mathbb{R}_+^N)$ and a family of operators $\mathcal{T} = (\mathcal{T}_t;\ t \in \mathbb{R}_+^N)$ such that for all $x \in S$, there exists a probability measure $\mathbb{P}_x$ for which the following conditions are met:[1]

(i) X is adapted to $\mathcal{F}$;

(ii) $t \mapsto X_t$ is right-continuous $\mathbb{P}_x$-a.s. for all $x \in S$;

(iii) for all $t \in \mathbb{R}_+^N$, $\mathcal{F}_t$ is $\mathbb{P}_x$-complete for all $x \in S$; moreover, $\mathcal{F}$ is a commuting σ-field with respect to all measures $\mathbb{P}_x$, $x \in S$;

(iv) for all $s, t \in \mathbb{R}_+^N$ and all $f \in C_0(S)$, the following holds $\mathbb{P}_x$-a.s. for all $x \in S$:
$$\mathbb{E}_x[f(X_{t+s}) \mid \mathcal{F}_s] = \mathcal{T}_t f(X_s);\ \text{and}$$

(v) For all $x \in S$, $\mathbb{P}_x(X_0 = x) = 1$.

Remarks (1) Part of the assertion of (ii) is that the following event is measurable: $(t \mapsto X_t$ is right-continuous).

(2) From now on, we *assume* the existence of such a process. Later on, we shall see, via diverse examples, that such processes often exist. In a one-parameter setting we have seen that such an assumption can be *proved* to hold under Feller's condition.

(3) Recall that a function $f : \mathbb{R}_+^N \to S$ is right-continuous if for any $s \downarrow t$ (with respect to the partial order $\preccurlyeq$ on $\mathbb{R}_+^N$), $f(s) \to f(t)$.

(4) Suppose ν is a measure on the Borel subsets of S. In complete analogy with the 1-parameter theory, we will write $\mathbb{P}_\nu$ for the measure $\int \mathbb{P}_x(\cdots)\nu(dx)$. Likewise, $\mathbb{E}_\nu$ denotes the operator $\int \mathbb{E}_x(\cdots)\nu(dx)$.[2] Intuitively speaking, $\mathbb{P}_\nu$ denotes the underlying probability measure, given that X_0 has distribution ν. However, this makes honest sense only when ν is itself a probability measure. Nonetheless, the above are both well-defined, irrespective of the total mass of ν; we will call ν the **initial measure** and call $\mathbb{P}_\nu$ and $\mathbb{E}_\nu$ the distribution of the process X and its expectation operator, respectively, with initial measure ν.

---

[1] There are other, equally interesting, notions of Markov property in several dimensions; see Rozanov (1982).

[2] Of course, this discussion makes sense only if $x \mapsto \mathbb{P}_x(\cdots)$ is measurable, a fact that will hold automatically for the multiparameter Markov processes of the remainder of this chapter.

1 Definitions

The $\mathcal{T}_t$'s are called the **transition operators** of the process X. As in Chapter 8, we may identify the *transition operator* $\mathcal{T}_t$ with the *transition function* $\mathcal{T}_t(x, A) = \mathcal{T}_t \mathbf{1}_A(x)$; cf. equation (1), Section 3.1 of Chapter 8. In particular, we are justified in also referring to the $\mathcal{T}_t$'s as the **transition functions** of X. It is extremely important to stress that the underlying filtration *must* be commuting for the forthcoming theory to work. Finally, we say that X is an N-parameter, S_Δ-valued **Feller process** (or simply **Feller**) if:

(i) for all $t \in \mathbb{R}_+^N$, $\mathcal{T}_t : C_0(S) \to C_0(S)$; that is, for all $f \in C_0(S)$, $\mathcal{T}_t f \in C_0(S)$;

(ii) for each $f \in C_0(S)$,
$$\lim_{t \to 0} \|\mathcal{T}_t f - f\|_\infty = 0.$$

Given an N-parameter Markov process with transition functions $\mathcal{T}$ and initial measure ν, we can compute the "one-dimensional marginals" of the process X, i.e., the $\mathbb{P}_\nu$-distribution of X_t for any $t \in \mathbb{R}_+^N$, as follows:

$$\mathbb{P}_\nu(X_t \in A) = \int_S \mathcal{T}_t(x, A)\, \nu(dx), \quad \text{for all measurable } A \subset S,\, t \in \mathbb{R}_+^N.$$

This follows from the fact that $\mathbb{P}_x(X_t \in A) = \mathcal{T}_t(x, A)$.

One can also compute "two-dimensional" marginals, although the expression is slightly more cumbersome:

Lemma 1.1.1 *For all $\varphi_1, \varphi_2 \in L^\infty(S)$, every $x \in S$, and all $s, t \in \mathbb{R}_+^N$,*
$$\mathbb{E}_x\left[\varphi_1(X_s)\varphi_2(X_t)\right] = \mathcal{T}_{s \wedge t}\left[\mathcal{T}_{s-(s \wedge t)}\varphi_1 \cdot \mathcal{T}_{t-(s \wedge t)}\varphi_2\right](x).$$

Proof Clearly, $\varphi(X_t)$ (respectively $\varphi(X_s)$) is measurable with respective to $\mathcal{F}_t$ (respectively $\mathcal{F}_s$). Thus, commutation allows us to write
$$\mathbb{E}_x\left[\varphi_1(X_s)\varphi_2(X_t)\right] = \mathbb{E}_x\left[\mathbb{E}_x\{\varphi_1(X_s)\,|\,\mathcal{F}_{s \wedge t}\} \cdot \mathbb{E}_x\{\varphi_2(X_t)\,|\,\mathcal{F}_{s \wedge t}\}\right].$$

By the multiparameter Markov property,
$$\mathbb{E}_x\left[\varphi_1(X_s)\varphi_2(X_t)\right] = \mathbb{E}_x\left[\mathcal{T}_{s-(s \wedge t)}\varphi_1(X_{s \wedge t}) \cdot \mathcal{T}_{t-(s \wedge t)}\varphi_2(X_{s \wedge t})\right],$$

which has the desired effect. $\square$

Exercise **1.1.1** Suppose $\gamma : [0,1] \to \mathbb{R}_+^N$ is nondecreasing in the sense that whenever $s \le t$, then $\gamma(s) \preccurlyeq \gamma(t)$. Given any such γ, show that the finite-dimensional distributions of the one-parameter process $t \mapsto X_{\gamma(t)}$ are entirely determined by the transition functions $\mathcal{T}$. At the time of writing this book, it is not known whether $\mathcal{T}$ determines the finite-dimensional distributions of the entire process X.

(HINT: You can try showing that if $\widetilde{X}$ is another N-parameter Markov process with transition functions $\mathcal{T}$, $t \mapsto \widetilde{X}_{\gamma(t)}$ has the same finite-dimensional distributions as X.) □

It is time to discuss concrete examples of multiparameter Markov processes.

Example 1 Consider two independent d-dimensional Brownian B^1 and B^2 and define the 2-parameter, d-dimensional **additive Brownian motion** $X = (X_t; t \in \mathbb{R}_+^2)$ as

$$X_t = B^1_{t^{(1)}} + B^2_{t^{(2)}}, \qquad t \in \mathbb{R}_+^2.$$

For all $t \in \mathbb{R}_+^2$, all $x \in \mathbb{R}^d$, and all measurable functions $f : \mathbb{R}^d \to \mathbb{R}_+$, define the bounded linear operator

$$\mathcal{T}_t f(x) = \int q_t(y-x) f(y)\, dy,$$

where $t \succ 0$ is in $\mathbb{R}_+^2$, $x \in \mathbb{R}^d$, and

$$q_t(a) = (2\pi\{t^{(1)} + t^{(2)}\})^{-\frac{d}{2}} \exp\left(-\frac{\|a\|^2}{2\{t^{(1)} + t^{(2)}\}}\right), \qquad a \in \mathbb{R}^d.$$

As usual, $\|a\|$ denotes the ℓ^2-Euclidean norm of $a \in \mathbb{R}^d$. By the elementary properties of Gaussian processes (Exercise 1.1.1, Chapter 5),

$$\mathcal{T}_t f(x) = \mathbb{E}[f(x + X_t)].$$

We shall see later on, in Section 3 (or you can try this at this point), that X is a 2-parameter Feller process on $\mathbb{R}^d$ whose transition operators are precisely $\mathcal{T} = (\mathcal{T}_t; t \in \mathbb{R}_+^2)$. □

Example 2 Continuing with the basic setup of Example 1, define the 2-parameter process $Y = (Y_t; t \in \mathbb{R}_+^2)$ by

$$Y_t = B^1_{t^{(1)}} \otimes B^2_{t^{(2)}}, \qquad t \in \mathbb{R}_+^2.$$

Note that B^1 and B^2 are $\mathbb{R}^d$-valued, while Y is $\mathbb{R}^d \times \mathbb{R}^d = \mathbb{R}^{2d}$-valued. In particular, if $d = 1$, $Y_t = (B^1_{t^{(1)}}, B^2_{t^{(2)}})$. Similar calculations to those in Example 1 can be made to show that Y is a 2-parameter Feller process whose transition operators are

$$\mathcal{T}_t f(x) = (2\pi)^{-d} \{t^{(1)} t^{(2)}\}^{-\frac{d}{2}} \int\!\!\int_{\mathbb{R}^d \times \mathbb{R}^d} f(x + \{u \otimes v\}) \exp\left(-\frac{\|u\|^2}{2t^{(1)}} - \frac{\|v\|^2}{2t^{(2)}}\right) du\, dv,$$

where $t \succ 0$ is in $\mathbb{R}_+^2$, $x \in \mathbb{R}^{2d}$, and $f : \mathbb{R}^{2d} \to \mathbb{R}_+$ is measurable. (Why $t \succ 0$ and not $t \succcurlyeq 0$?) It is a good idea to either verify directly that Y is a 2-parameter Feller process or to peek ahead to Section 3 and peruse the general discussion there. The two-parameter process Y is called **bi-Brownian motion** and is related to multiply harmonic functions; cf. Cairoli and Walsh (1977a, 1977b, 1977c), as well as Walsh (1986b), together with their combined references. □

***Exercise* 1.1.2** The discussion of Example 2 describes the operator $\mathcal{T}_t$ when $t \succ 0$. What does this particular operator look like when $t \in \mathbb{R}_+^2$ but $t \not\succ 0$? □

***Exercise* 1.1.3** Prove that the standard N-parameter Brownian sheet is *not* an N-parameter Markov process. □

1.2 Commutation and Semigroups

Consider an N-parameter, S_Δ-valued Markov process $X = (X_t;\, t \in \mathbb{R}_+^N)$ and let $\mathcal{T} = (\mathcal{T}_t;\, t \in \mathbb{R}_+^N)$ denote its corresponding collection of transition operators. In this subsection we study some of the analytical properties of the family $\mathcal{T}$.

To start, let us note that for all $t \in \mathbb{R}_+^N$, $f \in L^\infty(S)$, and $x \in S$,

$$\mathcal{T}_t f(x) = \mathbb{E}_x[\mathcal{T}_t f(X_0)] = \mathbb{E}_x\big[\mathbb{E}_x\{f(X_t) \mid \mathcal{F}_0\}\big] = \mathbb{E}_x[f(X_t)].$$

Moreover, whenever s is also in $\mathbb{R}_+^N$,

$$\mathcal{T}_t \mathcal{T}_s f(x) = \mathbb{E}_x[\mathcal{T}_s f(X_t)] = \mathbb{E}_x\big\{\mathbb{E}_x[f(X_{t+s}) \mid \mathcal{F}_t]\big\}$$
$$= \mathbb{E}_x[f(X_{t+s})] = \mathcal{T}_{t+s} f(x).$$

That is, $\mathcal{T}$ is a **semigroup** of operators on S. In fact, we have the following result.

Proposition 1.2.1 *The transition functions of an N-parameter, S_Δ-valued Markov process form an N-parameter semigroup of bounded linear operators on S.*

***Exercise* 1.2.1** Complete the proof of Proposition 1.2.1. □

The operators $\mathcal{T}_t$ are interesting in and of themselves, since they describe the local dynamics of the process X. Next, we will show a representation of $\mathcal{T}$ in terms of N one-parameter family of operators.

For any integer $1 \le j \le N$, we can temporarily define $\sigma_j : \mathbb{R}_+ \to \mathbb{R}_+^N$ as follows:

$$\big(\sigma_j(r)\big)^{(\ell)} = \begin{cases} r, & \text{if } \ell = j, \\ 0, & \text{otherwise,} \end{cases} \quad 1 \le \ell \le N.$$

Stated in plain terms, for all $r \geq 0$,

$$\sigma_1(r) = (r, 0, 0, \ldots, 0),$$
$$\sigma_2(r) = (0, r, 0, \cdots, 0),$$
$$\vdots$$
$$\sigma_N(r) = (0, 0, \ldots, 0, r).$$

We can now define N *one-parameter* bounded linear operators $\mathcal{T}^1, \ldots, \mathcal{T}^N$, all on S, by the following prescription: For every $j \in \{1, \ldots, N\}$,

$$\mathcal{T}^j_r f(x) = \mathcal{T}_{\sigma_j(r)} f(x), \qquad f \in L^\infty(S),\ x \in S,\ r \geq 0.$$

The operators $\mathcal{T}^1, \ldots, \mathcal{T}^N$ are called the **marginal semigroups** (equivalently, **marginal transition functions** or **marginal transition operators**) of X. The reason for this terminology is given by the following result.

Proposition 1.2.2 *If $\mathcal{T}$ denotes the transition functions of an N-parameter Markov process, then for all $t \in \mathbb{R}^N_+$,*

$$\mathcal{T}_t = \mathcal{T}^1_{t(1)} \cdots \mathcal{T}^N_{t(N)},$$

where the $\mathcal{T}^i$'s denote the marginal transition functions of $\mathcal{T}$. Furthermore, each $\mathcal{T}^i$ is itself a 1-parameter Markov semigroup. Finally, the $\mathcal{T}^i$'s commute in the following sense: For all $i, j \in \{1, \ldots, N\}$ and all $s, t \geq 0$, $\mathcal{T}^i_s \mathcal{T}^j_t = \mathcal{T}^j_t \mathcal{T}^i_s$.

***Exercise* 1.2.2** Verify Proposition 1.2.2. □

A Notational Remark If $\mathcal{A}$ and $\mathcal{B}$ denote any two bounded linear operators on S, then $\mathcal{AB}$ (itself a bounded linear operator) denotes the composition of the two operators $\mathcal{A}$ and $\mathcal{B}$, in this order. Motivated by this, we sometimes write

$$\mathcal{A}_1 \circ \mathcal{A}_2 \circ \cdots \circ \mathcal{A}_m = \mathcal{A}_1 \cdots \mathcal{A}_m.$$

We will also sometimes write this as $\bigcirc_{j=1}^m \mathcal{A}_j$, all the time noting that $\bigcirc_{j=1}^1 \mathcal{A}_j$ need not equal $\bigcirc_{j=m}^1 \mathcal{A}_j$, since the $\mathcal{A}_j$'s need not commute. However, this is not an issue for the transition operators of a multiparameter Markov process. Indeed, as the above proposition shows, for each $t \in \mathbb{R}^N_+$, $\mathcal{T}_t$ can be written compactly as

$$\mathcal{T}_t = \bigcirc_{j=1}^N \mathcal{T}^{\pi(j)}_{t(j)}, \qquad (1)$$

where $\{\pi(1), \ldots, \pi(N)\}$ denotes an arbitrary permutation of $\{1, \ldots, N\}$.

Let us close this section with a property of Feller processes.

Proposition 1.2.3 *Suppose X is an N-parameter Markov process with transition functions $\mathcal{T}$ and marginal semigroups $\mathcal{T}^1, \ldots, \mathcal{T}^N$. Then, X is Feller if and only if $\mathcal{T}^j$ is a one-parameter Feller semigroup, for every $j \in \{1, \ldots, N\}$.*

***Exercise* 1.2.3** Prove Proposition 1.2.3. □

1.3 Resolvents

Corresponding to X and $\mathcal{T}$ of the previous section, we now define the **resolvents** as the N-parameter family $\mathcal{R} = (\mathcal{R}_\lambda; \lambda \in \mathbb{R}_+^N$ such that $\lambda \succ 0)$, where

$$\mathcal{R}_\lambda = \int_{\mathbb{R}_+^N} e^{-\lambda \cdot s} \mathcal{T}_s \, ds, \qquad \lambda \succ 0 \text{ in } \mathbb{R}_+^N,$$

as bounded linear operators.[3] This is shorthand for the following: For all measurable functions $f : S \to \mathbb{R}_+$,

$$\mathcal{R}_\lambda f(x) = \int_{\mathbb{R}_+^N} e^{-\lambda \cdot s} \mathcal{T}_s f(x) \, ds, \qquad x \in S, \ \lambda \succ 0 \text{ in } \mathbb{R}_+^N.$$

It is often more convenient to use the former operator notation, as we shall see next. Note that as linear operators, for all $\lambda, s \succ 0$ (both in $\mathbb{R}_+^N$),

$$e^{-\lambda \cdot s} \mathcal{T}_s = \bigcirc_{j=1}^N e^{-\lambda^{(j)} s^{(j)}} \mathcal{T}^j_{s^{(j)}}.$$

Thus, viewed once again as bounded linear operators,

$$\int_{\mathbb{R}_+^N} e^{-\lambda \cdot s} \mathcal{T}_s \, ds = \bigcirc_{j=1}^N \int_0^\infty e^{-\lambda^{(j)} r} \mathcal{T}^j_r \, dr.$$

But the right-hand side is $\mathcal{R}^j_{\lambda^{(j)}}$, where $\mathcal{R}^j = (\mathcal{R}^j_\gamma; \gamma > 0)$ denotes the resolvent of the Markov semigroup $\mathcal{T}^j$. That is, we have proved the following:

Proposition 1.3.1 *Consider an N-parameter Markov process X whose transition functions are given by $\mathcal{T}$. If $\mathcal{T}^1, \ldots, \mathcal{T}^N$ and $\mathcal{R}^1, \ldots, \mathcal{R}^N$ denote the associated marginal semigroups and their respective resolvents, then*

$$\mathcal{R}_\lambda = \bigcirc_{j=1}^N \mathcal{R}^j_{\lambda^{(j)}}, \qquad \lambda \succ 0 \text{ (in } \mathbb{R}_+^N).$$

---

[3] As in the earlier chapters, $\lambda \cdot s = \sum_{j=1}^N \lambda^{(j)} s^{(j)}$ denotes the Euclidean inner product between s and λ, both of which are in $\mathbb{R}_+^N$. Also, note that in order for $\mathcal{R}_\lambda$ to be a bounded linear operator, in general, we need to assume that $\lambda \succ 0$ (in $\mathbb{R}_+^N$), since there is no a priori reason for $x \mapsto \int_{\mathbb{R}_+^N} \mathcal{T}_s f(x) \, ds$ to be bounded when f is. See Supplementary Exercise 13.

When X is a Feller process, we have the following N-parameter analogue of Lemma 2.4.1 of Chapter 8.

Lemma 1.3.1 *Suppose X is an N-parameter Feller process with transition functions $\mathcal{T}$ and resolvent $\mathcal{R}$, respectively. Then, for all $f \in C_0(S)$,*

$$\lim_{t \to 0} \sup_{s \in \mathbb{R}_+^N} \|\mathcal{T}_{t+s} f - \mathcal{T}_s f\|_\infty = 0,$$

$$\lim_{\lambda \to 0} \left\| \prod_{j=1}^N \lambda^{(j)} \cdot \mathcal{R}_\lambda f - f \right\|_\infty = 0.$$

***Exercise* 1.3.1** Prove Lemma 1.3.1. $\square$

***Exercise* 1.3.2** Suppose $f, g : S \to \mathbb{R}$ are bounded measurable functions such that for *some* $\gamma \succ 0$ in $\mathbb{R}_+^N$, $\mathcal{R}_\gamma f = \mathcal{R}_\gamma g$. Prove that for *all* $\lambda \succ 0$ in $\mathbb{R}_+^N$, $\mathcal{R}_\lambda f = \mathcal{R}_\lambda g$. $\square$

1.4 Strongly Symmetric Feller Processes

We now focus our attention on a class of N-parameter Markov processes on the Euclidean space $\mathbb{R}^d$.

Let $X = (X_t; \, t \in \mathbb{R}_+^N)$ denote an N-parameter, $\mathbb{R}^d$-valued Markov process with transition operators $\mathcal{T} = (\mathcal{T}_t; \, t \in \mathbb{R}_+^N)$. A measure ν on the Borel subsets of S is a **reference measure** for X if:

1. there exists a measurable function p from $]0, \infty[^N \times \mathbb{R}^d \times \mathbb{R}^d$ into $\mathbb{R}_+$ such that for all measurable $f : S \to \mathbb{R}_+$, all $t \succ 0$ in $\mathbb{R}_+^N$, and all $x \in \mathbb{R}^d$,

$$\mathcal{T}_t f(x) = \int f(y) p_t(x, y) \, \nu(dy); \text{ and}$$

2. for all open sets $G \subset \mathbb{R}^d$, $\nu(G) > 0$.

The second condition is assumed for simplicity, and does not have implications that deeply affect the potential theory to be developed; cf. Exercise 3.4.1 below.

In particular, by choosing $f(x) = \mathbf{1}_A(x)$, the above implies that we can write probabilities such as $\mathbb{P}_x(X_t \in A)$ in terms of the density function p_t. In general, since $f \geq 0$, the above integrals are always well-defined, but they may be infinite. Part of this definition is that either they are both finite, or they are both infinite.

The functions p_t are the **transition densities** of X with respect to ν, while the corresponding **λ-resolvent density** r_λ is defined by the following formula:

$$r_\lambda(x, y) = \int_{\mathbb{R}_+^N} e^{-\lambda \cdot s} p_s(x, y) \, ds,$$

where $\lambda \succ 0$ is in $\mathbb{R}_+^N$ and $x, y \in \mathbb{R}^d$. As in Chapter 10, we can extend the definition of the operators $\mathcal{T}_t$ and $\mathcal{R}_\lambda$ as follows: For all σ-finite measures μ on Borel subsets of $\mathbb{R}^d$, $\mathcal{T}_t \mu(x) = \int p_t(x, y) \, \mu(dy)$ and $\mathcal{R}_\lambda \mu(x) = \int r_\lambda(x, y) \, \mu(dy)$, where $x \in \mathbb{R}^d$ and $\lambda, t \succ 0$ are both in $\mathbb{R}_+^N$.

Exercise 1.4.1 Prove that the above definitions are consistent with the earlier ones. That is, suppose $\mu(dx) = f(x) \nu(dx)$ is a σ-finite measure on $\mathbb{R}^d$. Prove that for all $t, \lambda \succ 0$, both in $\mathbb{R}_+^N$, $\mathcal{R}_\lambda \mu = \mathcal{R}_\lambda f$ and $\mathcal{T}_t \mu = \mathcal{T}_t f$. Moreover, check that r_λ is symmetric if p_t is, i.e., if $p_t(x, y) = p_t(y, x)$. □

We can consistently define $\mathbb{P}_\nu$, *even when ν is not a probability measure*, by

$$\mathbb{P}_\nu(\bullet) = \int \mathbb{P}_x(\bullet) \, \nu(dx),$$

and $\mathbb{E}_\nu(\cdots) = \int \mathbb{E}_x(\cdots) \, \nu(dx)$, in analogy. Recall that when ν is a probability measure, we think of $\mathbb{P}_\nu$ as the distribution of the random function X, when the distribution of X_0 is ν. When ν is not a probability measure, this interpretation breaks down unless we are willing to think of the distribution of X_0 as a (possibly infinite) σ-finite measure ν.

We say that X is a **strongly symmetric Feller process** if:

(i) for every $t \succ 0$ in $\mathbb{R}_+^N$, p_t is a symmetric function on $\mathbb{R}^d \times \mathbb{R}^d$ (that is, $p_t(x, y) = p_t(y, x)$, for all $x, y \in \mathbb{R}^d$); and

(ii) for each $\lambda \succ 0$ in $\mathbb{R}_+^N$, r_λ is a proper gauge function on $\mathbb{R}^d \times \mathbb{R}^d$.

A technical word of caution is in order here. When $N = 1$, we needed to assume only that r_λ is a gauge function. From this and symmetry, it followed that r_λ is proper; cf. the balayage theorem (Theorem 2.2.1, Chapter 10). Our N-parameter processes are sufficiently general, however, that we need to *assume* that r_λ is proper for a nice theory to follow. Nevertheless, we shall see, via a number of examples, that this assumption is, in practice, harmless.

Exercise 1.4.2 Suppose that X is strongly symmetric.

(i) Prove that for all σ-finite measures ξ and ζ on $\mathbb{R}^d$,

$$\int \mathcal{T}_t \zeta(x) \, \xi(dx) = \int \mathcal{T}_t \xi(x) \, \zeta(dx),$$

for all $t \in \mathbb{R}_+^N$. In other words, show that $\mathcal{T}_t$ is self-adjoint on $L^2(\nu)$.

(ii) Prove that for all σ-finite measures ξ and ζ on $\mathbb{R}^d$,

$$\int \mathcal{R}_\lambda \zeta(x) \, \xi(dx) = \int \mathcal{R}_\lambda \xi(x) \, \zeta(dx),$$

whenever $\lambda \succ 0$ is in $\mathbb{R}_+^N$.

This is a multiparameter extension of Exercise 2.1.5, Chapter 10. □

An important property of strongly symmetric Markov processes is that if we allow their initial measure to be their reference measure, then the distribution of X_t becomes independent of t. This property is often referred to as the (weak) **stationarity** of X. More precisely, we state the following result.

Lemma 1.4.1 *Suppose X is a strongly symmetric, N-parameter, $\mathbb{R}^d$-valued Markov process whose reference measure is ν. Then, for all measurable $f : \mathbb{R}^d \to \mathbb{R}_+$,*

$$\mathbb{E}_\nu[f(X_t)] = \int f(x)\,\nu(dx) = \int \mathcal{T}_t f(x)\,\nu(dx).$$

***Exercise* 1.4.3** Prove Lemma 1.4.1. □

In fact, not only does symmetry make it easier to compute the distribution of X_t for a fixed $t \in \mathbb{R}_+^N$, it also often allows for the development of more-or-less simple formulæ for the finite-dimensional distributions of X, as the following extension of Lemma 1.4.1 reveals in a special case.

Lemma 1.4.2 *Let $s, t \in \mathbb{R}_+^N$ and suppose $\varphi_1, \varphi_2 : \mathbb{R}^d \to \mathbb{R}_+$ are measurable functions. Given the hypotheses of Lemma 1.4.1,*

$$\mathbb{E}_\nu[\varphi_1(X_s)\varphi_2(X_t)] = \int \varphi_1(x) \cdot \mathcal{T}_{s+t-2(s\wedge t)}\varphi_2(x)\,\nu(dx).$$

Proof of Lemma 1.4.2 Suppose **1** denotes the function that is identically equal to 1. Since $\mathcal{T}_t \mathbf{1}(x) = \mathbf{1}$, we can apply Exercise 1.4.2, as well as Lemmas 1.1.1 and 1.4.1, to deduce the following:

$$\mathbb{E}_\nu[\varphi_1(X_s)\varphi_2(X_t)] = \int \mathcal{T}_{s\wedge t}\bigl[\mathcal{T}_{s-(s\wedge t)}\varphi_1 \cdot \mathcal{T}_{t-(s\wedge t)}\varphi_2\bigr](x)\,\nu(dx)$$

$$= \int \mathcal{T}_{s-(s\wedge t)}\varphi_1(x) \cdot \mathcal{T}_{t-(s\wedge t)}\varphi_2(x)\,\nu(dx).$$

Another application of Exercise 1.4.2 implies the result. □

To recapitulate, suppose that X is strongly symmetric and that its initial measure is its reference measure ν.[4] Then, the distribution of X_t is independent of t (Lemma 1.4.1), and the joint distribution of X_s and X_t depends only on $s+t-2(s \wedge t)$ (Lemma 1.4.2). Consequently, given a fixed $r \in \mathbb{R}_+^N$, the distribution of (X_s, X_t) is the same as that of (X_{s+r}, X_{t+r}). This property is sometimes called second-order, or L^2, stationarity.

So far, the results of this subsection have relied only on the the symmetry of p_t. However, when X is strongly symmetric, we also know that r_λ is a

---

[4] Heuristically, this means that X_0 is a random variable with "distribution" ν, although this interpretation is sensible only when ν is a probability measure.

proper gauge function on $\mathbb{R}^d \times \mathbb{R}^d$; cf. Sections 1.1 and 2.3 of Appendix D for definitions. These properties will be used later on to connect N-parameter processes to potentials and energy.

2 Examples

We now turn our attention to some examples of multiparameter Markov processes that we shall study in the following two sections. More specifically, we will introduce three large classes of multiparameter Markov processes: product Feller processes, additive Lévy processes, and multiparameter product processes. The first two families share the property that they are both built from N independent one-parameter processes. Earlier in the book we have seen other such constructions in the setting of discrete parameter theory; cf. Section 2.1, Chapter 1, and Section 2 of Chapter 3.

2.1 General Notation

In this section we build examples of multiparameter Markov processes by suitably combining one-parameter Markov processes. The processes that we shall construct have a rich structure, as we shall see in Section 4 below. On the other hand, some cumbersome notation is required in order to carry out our constructions. Thus, before working on the details of the constructions of this section, it is worth our while to spend a few paragraphs and establish some basic notation.

Given N probability spaces $(\Omega_1, \mathcal{G}_1, \mathbb{Q}_1), \ldots, (\Omega_N, \mathcal{G}_N, \mathbb{Q}_N)$, we define the product space Ω, the product σ-field $\mathcal{G}$, and the product probability measure $\mathbb{Q}$ in the usual way, as follows:

$$\Omega = \Omega_1 \times \cdots \times \Omega_N,$$
$$\mathcal{G} = \mathcal{G}_1 \times \cdots \times \mathcal{G}_N,$$
$$\mathbb{Q} = \mathbb{Q}_1 \times \cdots \times \mathbb{Q}_N.$$

Throughout much of this section we will be considering N independent *one-parameter* Feller processes $X^1, \ldots, X^N$, all the time remembering that $X^j = (X_r^j; \, r \geq 0)$ is a one-parameter process defined on the probability space $(\Omega_j, \mathcal{G}_j, \mathbb{Q}_j)$, for $j \in \{1, \ldots, N\}$. Furthermore, for each $j \in \{1, \ldots, N\}$, every X^j is S_j-valued, where $S_1, \ldots, S_N$ are compact spaces; if they are not compact, we will always compactify them, using the one-point compactification.[5]

Define $\mathbb{P}_x^i$ ($x \in S_i$) to be the probability measures in the definition of the Feller process X^i, where $i \in \{1, \ldots, N\}$. As always, $\mathbb{E}_x^i$ ($x \in S_i$) denotes

---

[5] It may help you to consider the special case $S_1 = S_2 = \cdots = S_N$.

the corresponding expectation operator; we also let $\mathsf{T}^i = (\mathsf{T}^i_r;\ r \geq 0)$ and $\mathsf{R}^i = (\mathsf{R}^i_\gamma;\ \gamma > 0)$ denote the transition functions and the resolvent of X^i, respectively. Fixing each $i \in \{1,\ldots,N\}$, we also need some notation for the complete augmented filtration for X^i, which will be written as $\mathcal{F}^i = (\mathcal{F}^i_r;\ r \geq 0)$. Of course, the phrase "complete augmented" refers to the measures $\mathbb{P}^i_x$, $x \in S_i$. In particular, note that for each $i \in \{1,\ldots,N\}$, all bounded measurable functions $f: S_i \to \mathbb{R}$, and for all $u, v \geq 0$,

$$\mathbb{E}^i_x[f(X^i_{u+v}) \mid \mathcal{F}^i_u] = \mathsf{T}^i_v f(X^i_v),$$

$\mathbb{P}^i_x$-a.s. for all $x \in S_i$.

In order to make statements about $X^1,\ldots,X^N$ simultaneously, we let $\mathbb{P}_x = \mathbb{P}^1_{x_1} \times \cdots \times \mathbb{P}^N_{x_N}$ ($x \in S$), as a product measure on $(\Omega, \mathcal{G})$. We have written $x \in S$ in the direct product notation

$$x = x_1 \otimes \cdots \otimes x_N,$$

where $x_i \in S_i$, $1 \leq i \leq N$. To put it another way, whenever $x_i \in S_i$, $1 \leq i \leq N$, $x_1 \otimes \cdots \otimes x_N$ is the element of S with $x^{(i)} = x_i$, for all $1 \leq i \leq N$.[6] As usual, $\mathbb{E}_x$ ($x \in S$) denotes the corresponding expectation operator.

We complete the introduction of our notation by defining the N-parameter filtration $\mathcal{F} = (\mathcal{F}_t;\ t \in \mathbb{R}^N_+)$ as

$$\mathcal{F}_t = \bigvee_{i=1}^{N} \mathcal{F}^i_{t(i)}, \qquad t \in \mathbb{R}^N_+.$$

Observe that $\mathcal{F}$ is the minimal N-parameter filtration whose marginal filtrations are given by $\mathcal{F}^1,\ldots,\mathcal{F}^N$, respectively.

2.2 Product Feller Processes

Let $S = S_1 \times \cdots \times S_N$ denote the corresponding product space endowed with the product topology and the corresponding Borel σ-field. Since S is already compact, we need not one-point compactify it. Consequently, in the notation of Chapter 8, $S_\Delta = S$. Define the N-parameter stochastic process $X = (X_t;\ t \in \mathbb{R}^N_+)$ by defining, in direct product notation,

$$X_t(\omega) = X^1_{t(1)}(\omega_1) \otimes \cdots \otimes X^N_{t(N)}(\omega_N).$$

Let us recall what this means: Whenever $x_i \in S_i$ ($1 \leq i \leq N$), then $x = x_1 \otimes \cdots \otimes x_N$ is the "N-dimensional"[7] point in S whose first coordinate

---

[6] Hence, as a concrete example, consider $(1,2,3,4,5,6) = 1 \otimes (2,3) \otimes (4,5,6)$.

[7] Since S is the product of the N spaces $S_1,\ldots,S_N$, this may justify a relaxed reference to x as being N-dimensional.

is x_1, second is x_2, etc. (Similarly, $\omega \in \Omega$ is written as $\omega_1 \otimes \cdots \otimes \omega_N$, where $\omega_j \in \Omega_j$, $1 \le j \le N$.) The process X is called the **product Feller process** with coordinate processes $X^1, \ldots, X^N$.

We shall next argue that X is an N-parameter, S-valued Feller process. This is achieved by a direct calculation of $\mathbb{E}_x[f(X_t)]$ for all bounded, measurable functions $f : S \to \mathbb{R}$, all $t \in \mathbb{R}_+^N$, and all $x \in S$.

Suppose $f : S \to \mathbb{R}_+$ is measurable and is of the form $f = f_1 \otimes \cdots \otimes f_N$, where $f_i : S_i \to \mathbb{R}_+$. That is, we suppose

$$f(x) = \prod_{j=1}^{N} f_j(x_j), \qquad S \ni x = x_1 \otimes \cdots \otimes x_N, \ x_i \in S_i.$$

Then, for all $x \in S$ of the form $x = x_1 \otimes \cdots \otimes x_N$ with $x_i \in S_i$ for all $1 \le i \le N$,

$$\mathbb{E}_x[f(X_t)] = \prod_{i=1}^{N} \mathbb{E}_{x_i}^i[f_i(X_{t(i)}^i)] = \prod_{j=1}^{N} \mathsf{T}_{t(j)}^j f_j(x_j), \qquad t \in \mathbb{R}_+^N. \tag{1}$$

(Recall that T^j denotes the transition functions of X^j.) Motivated by this calculation, define for all measurable functions $f : S \to \mathbb{R}_+$ of the form $f = f_1 \otimes \cdots \otimes f_N$ ($f_i : S_i \to \mathbb{R}_+$),

$$\mathcal{T}_r^j f(x) = \mathsf{T}_r^j f_j(x_j), \qquad r \ge 0, \ x \in S,$$

where $x \in S$ is (again) written in direct product notation as $x_1 \otimes \cdots \otimes x_N$, $x_i \in S_i$. By the Feller property, we can extend the domain of definition of $f \to \mathcal{T}_r^j f$ to *all* measurable functions $f : S \to \mathbb{R}_+$ in a unique manner:

Exercise **2.2.1** Use a monotone class argument to show that there exists a unique extension of the bounded linear operator $f \mapsto \mathcal{T}_r^j f$ to the collection of all bounded measurable functions $f : \mathbb{R}^d \to \mathbb{R}_+$. Moreover, show that $\mathcal{T}^j$ is a one-parameter semigroup on S (and not always S_j!) and that $\mathcal{T}_t = \bigcirc_{j=1}^{N} \mathcal{T}_{t(j)}^j$, for all $t \in \mathbb{R}_+^N$. □

Having come this far, it should not too difficult to check the following:

Theorem 2.2.1 *The product Feller process X is an N-parameter Feller process whose transition functions are given by $\mathcal{T} = (\mathcal{T}_t;\ t \in \mathbb{R}_+^N)$.*

Exercise **2.2.2** Verify Theorem 2.2.1.
(HINT: Carefully study the proof of Proposition 2.2.1 below.) □

We also mention the following calculation for the resolvent of X in terms of the resolvent R^j of X^j's: For all measurable $f : S \to \mathbb{R}_+$ of the form $f = f_1 \otimes \cdots \otimes f_N$, $f_j : S_j \to \mathbb{R}_+$,

$$\mathcal{R}_\lambda f(x) = \prod_{j=1}^{N} \mathsf{R}_{\lambda(j)}^j f_j(x_j), \tag{2}$$

where $\lambda \in \mathbb{R}_+^N$ and $x = x_1 \otimes \cdots \otimes x_N$, $x_j \in S_j$.

Suppose $X^1, \ldots, X^N$ are strongly symmetric one-parameter Feller processes with reference measures $\nu_1, \ldots, \nu_N$, respectively. Suppose further that each X^j possesses transition densities $(p_r^j(x,y); \, r > 0, \, x, y \in S_j)$, with respect to ν_j. That is, for all bounded measurable functions $f: S_j \to \mathbb{R}$,

$$\mathbb{E}_x^j[f(X_r^j)] = \int_{S_j} f(y) p_r^j(x,y) \, \nu_j(dy).$$

We then have the following.

Proposition 2.2.1 *Suppose that $X^1, \ldots, X^N$ are strongly symmetric Feller processes with reference measures $\nu_1, \ldots, \nu_N$, respectively. Then, the product Feller process X is a strongly symmetric N-parameter Feller process with reference measure $\nu = \nu_1 \times \cdots \times \nu_N$.*

Proof Suppose $f = f_1 \otimes \cdots \otimes f_N$, where $f_i : S_i \to \mathbb{R}$ are bounded and measurable $(1 = 1, \ldots, N)$. Then, for all $x = x_1 \otimes \cdots \otimes x_N$ with $x_i \in S_i$, and for all $t \in \mathbb{R}_+^N$,

$$\mathcal{T}_t f(x) = \mathbb{E}_x[f(X_t)] = \prod_{j=1}^N \int_{S_j} p_{t(j)}^j(x_j, y) f_j(y) \, \nu_j(dy).$$

By a monotone class argument, we can choose the following version of the transition densities for X (why?):

$$p_t(x,y) = \prod_{j=1}^N p_{t(j)}^j(x_j, y_j),$$

where $t \succ 0$ is in $\mathbb{R}_+^N$ and where $x = x_1 \otimes \cdots \otimes x_N$ and $y = y_1 \otimes \cdots \otimes y_N$ are both in S. Note that p_t is symmetric for all $t \succ 0$ in $\mathbb{R}_+^N$, as it should be. Moreover, $\mathcal{R}_\lambda$ has the following density with respect to the reference measure $\nu = \nu_1 \times \cdots \times \nu_N$:

$$r_\lambda(x,y) = \prod_{j=1}^N r_{\lambda(j)}^j(x_j, y_j), \tag{3}$$

where $x = x_1 \otimes \cdots \otimes x_N$, $y = y_1 \otimes \cdots \otimes y_N$ $(x_i, y_i \in S_i)$, $\lambda \succ 0$ is in $\mathbb{R}_+^N$, and $r_\gamma^j(a,b) = \int_0^\infty e^{-\gamma r} p_r^j(a,b) \, dr$ for all $a, b \in S_j$ and all $\gamma > 0$. That is, r_γ^j denotes the γ-potential density of X^j with respect to the measure ν_j. Finally, notice that when r_γ^j is a proper gauge function on $S^j \times S^j$, r_λ is automatically a gauge function on S. (To be consistent with our definitions of gauge functions, we need to assume that the S_j's are Euclidean. It is a good idea to define proper gauge functions on metric spaces by analogy and check that the above remains true, in general.) The proof follows readily from this. □

2.3 Additive Lévy Processes

The second class of multiparameter processes that we are interested in is the family of so-called additive Lévy processes. Suppose $X^1, \ldots, X^N$ of Section 2.1 are in fact (one-parameter) Lévy processes with $S_1 = S_2 = \cdots = S_N = \mathbb{R}^d$. The N-parameter, $\mathbb{R}^d$-valued **additive Lévy process** $X = (X_t; t \in \mathbb{R}_+^N)$ is defined as follows:

$$X_t(\omega) = \sum_{j=1}^N X_{t^{(j)}}^j(\omega_j), \qquad \omega \in \Omega, \ t \in \mathbb{R}_+^N.$$

We sometimes write the process X in direct sum notation as $X = \bigoplus_{j=1}^N X^j$.

For all bounded measurable functions $f: \mathbb{R}^d \to \mathbb{R}$, all $x \in \mathbb{R}^d$, and all $t \in \mathbb{R}_+^N$, define

$$\mathcal{T}_t f(x) = \mathbb{Q}[f(X_t + x)].$$

We recall that $\mathbb{Q}$ denotes both the underlying probability measure and its expectation.

Theorem 2.3.1 *The process X is an N-parameter Feller process whose transition functions are given by $\mathcal{T} = (\mathcal{T}_t; t \in \mathbb{R}_+^N)$.*

***Exercise* 2.3.1** Prove Theorem 2.3.1. □

Let us now turn our attention to the resolvent of X. Suppose $f: \mathbb{R}^d \to \mathbb{R}$ is a bounded, measurable function, $x \in \mathbb{R}^d$, and $\lambda \succ 0$ is in $\mathbb{R}_+^N$. We shall also fix N points $u_1, \ldots, u_N \in \mathbb{R}^d$ such that $\sum_{j=1}^N u_j = x$; otherwise, the choice of $u_1, \ldots, u_N$ is completely arbitrary. After several applications of Fubini's theorem, we obtain

$$\mathcal{R}_\lambda f(x) = \int_{\mathbb{R}_+^N} e^{-\lambda \cdot t} \mathbb{Q}[f(X_t + x)] \, dt = \mathbb{Q}\Big[\int_{\mathbb{R}_+^N} e^{-\lambda \cdot t} f(X_t + x) \, dt\Big]$$

$$= \mathbb{E}_{u_1}^1 \mathbb{E}_{u_2}^2 \cdots \mathbb{E}_{u_N}^N \Big[\int_{\mathbb{R}_+^N} e^{-\lambda \cdot t} f(X_t) \, dt\Big]$$

$$= (\bigcirc_{j=1}^N \mathbb{E}_{u_j}^j) \Big[\int_{\mathbb{R}_+^N} e^{-\lambda \cdot t} f(X_t) \, dt\Big],$$

in operator notation. We can evaluate the above by "peeling the indices back." To be more precise, note that for any $\omega_1 \in \Omega_1, \ldots, \omega_{N-1} \in \Omega_{N-1}$,

$$\mathbb{E}_{u_N}^N \Big[\int_{\mathbb{R}_+^N} e^{-\lambda \cdot t} f(X_t(\omega)) \, dt\Big]$$

$$= \int_{\Omega_N} \int_{\mathbb{R}_+^N} e^{-\lambda \cdot t} f(X_t(\omega)) \, dt \, \mathbb{P}_{u_N}^N(d\omega_N)$$

$$= \int_{\mathbb{R}_+^{N-1}} e^{-\sum_{j=1}^{N-1} \lambda^{(j)} t^{(j)}} \mathsf{R}_{\lambda^{(N)}}^N f\Big(\sum_{j=1}^{N-1} X_{t^{(j)}}^j(\vec{\omega})\Big) \, dt^{(N-1)} \cdots dt^{(1)},$$

where $\vec{\omega} = \omega_1 \otimes \cdots \omega_{N-1}$. By induction (on N),

$$\mathcal{R}_\lambda f(x) = \bigcirc_{j=1}^N \mathsf{R}_{\lambda(j)}^j f(x). \tag{1}$$

We can specialize further by assuming that $X^1, \ldots, X^N$ possess γ-potential densities $r_\gamma^1, \ldots, r_\gamma^N$ with respect to Lebesgue's measure on $\mathbb{R}^d$. That is, for all $\gamma > 0$, for $x \in \mathbb{R}^d$, and all bounded, measurable functions $f : \mathbb{R}^d \to \mathbb{R}$,

$$\mathsf{R}_\gamma^i f(x) = \int f(y) r_\gamma^i(x, y) \, dy.$$

By Example 3, Section 4.3 of Chapter 8, we can always write $r_\gamma^i(x, y) = u_\gamma^i(y - x)$ (why?). Consequently,

$$\mathsf{R}_\gamma^i f(x) = f \star u_\gamma^i(x), \tag{2}$$

where $\star$ denotes ordinary convolution on $\mathbb{R}^d$. For all $\lambda \succ 0$ in $\mathbb{R}_+^N$ and all $a \in \mathbb{R}^d$, define

$$u_\lambda(a) = u_{\lambda(1)}^1 \star \cdots \star u_{\lambda(N)}^N(a), \tag{3}$$

and let $r_\lambda(x, y) = u_\lambda(y - x)$, where $\lambda \succ 0$ is in $\mathbb{R}_+^N$ and $x, y \in \mathbb{R}^d$. By equation (1), r_λ is the density of $\mathcal{R}^\lambda$ with respect to Lebesgue's measure on $\mathbb{R}^d$ in the sense that for all bounded, measurable functions $f : \mathbb{R}^d \to \mathbb{R}$,

$$\mathcal{R}_\lambda f(x) = \int f(y) r_\lambda(x, y) \, dy. \tag{4}$$

We now address the issue of strong symmetry.

Theorem 2.3.2 *Suppose $X^1, \ldots, X^N$ are independent, strongly symmetric $\mathbb{R}^d$-valued Lévy processes whose reference measure is Lebesgue's measure on $\mathbb{R}^d$. Then, $X = \oplus_{j=1}^N X^j$ is a strongly symmetric $\mathbb{R}^d$-valued, N-parameter Feller process whose reference measure is Lebesgue's measure on $\mathbb{R}^d$.*

Proof We begin by reiterating the computations made prior to the statement of Theorem 2.3.2. Suppose that $\lambda > 0$ and that $r_\lambda^1, \ldots, r_\lambda^N$ denote the λ-potential densities of $X^1, \ldots, X^N$, respectively. Then, $r_\lambda^j(x, y) = u_\lambda^i(y - x)$, for $\lambda > 0$ and $y, x \in \mathbb{R}^d$, and $r_\lambda(x, y) = u_\lambda(x, y)$, where u_λ is given by equation (3).

Since the X^i's are strongly symmetric, the r^i's are symmetric functions (Exercise 2.1.1, Chapter 10). Equivalently, $u_\lambda^i(a) = u_\lambda^i(-a)$. Thus, $u_\lambda(a) = u_\lambda(-a)$, and this means that the λ-potential density r_λ of the N-parameter process X is symmetric. Moreover, since continuity is preserved under convolutions, r_λ is a gauge function (why?). That p_t exists and is symmetric is shown by using similar arguments. Thus, to finish, we need to demonstrate that r_λ is proper. For any σ-finite measures μ on $\mathbb{R}^d$, all $\lambda \succ 0$ in $\mathbb{R}_+^N$, and for every $x \in \mathbb{R}^d$, $\mathcal{R}_\lambda \mu(x) = \mathsf{R}_{\lambda(N)}^N \circ \cdots \circ \mathsf{R}_{\lambda(1)}^1 \mu(x)$. Since $r_{\lambda(1)}^1$ is proper by assumption, a little thought shows that so is r_λ. This completes the proof. $\square$

2.4 More General Product Processes

Let N be a positive integer and consider for each $i \in \{1, \ldots, N\}$, N independent multiparameter Markov processes $X^i = (X^i_t;\ t \in \mathbb{R}^{M_i}_+)$, where $M_1, \ldots, M_N$ are positive integers. For the sake of concreteness, we suppose that for each $i \in \{1, \ldots, N\}$, X^i is defined on some probability space $(\Omega_i, \mathcal{G}_i, \mathbb{Q}_i)$ and is S_i-valued, where S_i is a locally compact metric space. (Recall that for our purposes, we can assume, without loss of generality, that each S_i is compact.) In order to keep the notation from becoming overwhelming, you should note at this point that each X^i is an S_i-valued, M_i-parameter stochastic process. As in Section 2.1, we define $(\Omega, \mathcal{G}, \mathbb{Q})$ to be the product probability space, let $S = S_1 \times \cdots \times S_N$, $M = M_1 + \cdots + M_N$, and define the S-valued, M-parameter process $X = X^1 \otimes \cdots \otimes X^N$ by

$$X_t(\omega) = X^1_{t_1}(\omega_1) \otimes \cdots \otimes X^N_{t_N}(\omega_N), \qquad t \in \mathbb{R}^M,$$

where $\omega = \omega_1 \otimes \cdots \otimes \omega_N \in \Omega$ with $\omega_i \in \Omega_i$ and $t = t_1 \otimes \cdots \otimes t_N \in \mathbb{R}^M_+$ with $t_i \in \mathbb{R}^{M_i}_+$. The S-valued, M-parameter process X is the **product process** built with the N processes $X^1, \ldots, X^N$, the ith one of which is S_i-valued and is indexed by M_i parameters. Of course, the underlying probability space is built up as in Section 2.1.

Theorem 2.4.1 *If X^i is an M_i-parameter, S_i-valued Feller process for all $1 \leq i \leq N$, then X is an S-valued, M-parameter Feller process. Moreover, suppose $\mathcal{R}^1, \ldots, \mathcal{R}^N$ and $\mathcal{R}$ denote the resolvents of $X^1, \ldots, X^N$ and X, respectively. Define $f = f_1 \otimes \cdots \otimes f_N$, where $f_i : S_i \to \mathbb{R}_+$ are measurable for each $1 \leq i \leq N$. Then, for all $\lambda \in \mathbb{R}^M_+$ with $\lambda = \lambda_1 \otimes \cdots \otimes \lambda_N$, $\lambda_i \in \mathbb{R}^{M_i}_+$, and all $x \in S$ with $x = x_1 \otimes \cdots \otimes x_N$, $x_i \in S_i$,*

$$\mathcal{R}_\lambda f(x) = \prod_{i=1}^N \mathcal{R}^i_{\lambda_i} f_i(x_i).$$

Finally, if the $X^1, \ldots, X^N$ are all strongly symmetric and if $\nu_1, \ldots, \nu_N$ denote their respective reference measures, X is also strongly symmetric, and its reference measure is $\nu_1 \times \cdots \times \nu_N$.

***Exercise* 2.4.1** Prove Theorem 2.4.1. (WARNING: We have defined strong symmetry for a process only when the process takes its values in a Euclidean space. Theorem 2.4.1, however, remains true for the obvious general definition of strong symmetry. Try some of these details.) □

In particular, the above implies that we can combine the examples of Sections 2.2 and 2.3 in nontrivial ways to construct other interesting multiparameter Markov processes.

3 Potential Theory

We now return to the basic question raised (and answered in a one-parameter setting) in Chapter 10. Namely, *when does the range of a strongly symmetric multiparameter Markov process intersect a given set?*

3.1 The Main Result

Let $X = (X_t;\ t \in \mathbb{R}_+^N)$ denote a strongly symmetric, $\mathbb{R}^d$-valued Markov process with transition functions $\mathcal{T} = (\mathcal{T}_t;\ t \in \mathbb{R}_+^N)$, resolvent $\mathcal{R} = (\mathcal{R}_\lambda;\ \lambda \succ 0 \text{ in } \mathbb{R}_+^N)$, and reference measure ν.

The main result of this section is the following characterization of the positivity of intersection (or hitting) probabilities in terms of capacities. It unifies the results of Hirsch (1995), Hirsch and Song (1995a, 1994, 1995c, 1995d), Fitzsimmons and Salisbury (1989), Ren (1990), and Salisbury (1996) in the setting of multiparameter strongly symmetric Feller process on Euclidean spaces.

Theorem 3.1.1 *Suppose X is a strongly symmetric, $\mathbb{R}^d$-valued, N-parameter Markov process with resolvent $\mathcal{R}$ and reference measure ν. Then, for all compact sets $E \subset \mathbb{R}^d$, and for all $\lambda \succ 0$ in $\mathbb{R}_+^N$,*

$$A_1 \mathcal{C}_{r_\lambda}(E) \le \int_{\mathbb{R}_+^N} e^{-\lambda \cdot t} \mathbb{P}_\nu\{\overline{X([0,t])} \cap E \ne \varnothing\} \, dt \le A_2 \mathcal{C}_{r_{\lambda/2}}(E),$$

where $A_1 = 2^{-N}\{\prod_{j=1}^N \lambda^{(j)}\}^{-1}$ and $A_2 = 8^N\{\prod_{j=1}^N \lambda^{(j)}\}^{-1}$.

Remarks (i) Essentially using the notation of the previous chapter, for any Borel set $G \subset \mathbb{R}_+^N$, we let $X(G)$ denote the image of G under $t \mapsto X_t$. That is,

$$X(G) = \{x \in \mathbb{R}^d : \exists t \in G \text{ such that } X_t = x\}.$$

(ii) Recall that $\mathcal{C}_g(E)$ denotes the capacity of E with respect to any gauge function $g: \mathbb{R}^d \times \mathbb{R}^d \to \mathbb{R}_+ \cup \{\infty\}$; cf. Appendix D.

(iii) Recall that ν need not be a probability measure. See the convention about $\mathbb{P}_\nu$ stated after Exercise 1.4.1.

An important consequence of the above theorem is the following, which we leave as an exercise.

***Exercise* 3.1.1** Prove that under the conditions of Theorem 3.1.1, for any compact set $E \subset \mathbb{R}^d$ and for any $\lambda \succ 0$ in $\mathbb{R}_+^N$,

$$\mathbb{P}_\nu\{\overline{X(\mathbb{R}_+^N)} \cap E \ne \varnothing\} > 0 \iff \mathcal{C}_{r_\lambda}(E) > 0.$$

In words, show that the closed range of the process X hits a compact set E (with positive $\mathbb{P}_\nu$-"probability") if and only if E has positive capacity. (HINT: One can view r_λ as a Laplace transform; cf. also Appendix B.) □

We will prove Theorem 3.1.1 in the next three subsections and close this subsection with an important consequence.

Note that unless ν is a probability (respectively finite) measure, $\mathbb{P}_\nu$ is not a probability (respectively finite) measure. This makes the present form of Theorem 3.1.1 both awkward and difficult to use. Next, we show that under mild conditions, $\mathbb{P}_\nu$ can be replaced with the probability measure $\mathbb{P}_x$. This result is akin to Corollary 2.3.1 of Chapter 10, and its proof is motivated by a positivity condition first mentioned explicitly in Evans (1987a); see Evans (1987b) for related results.

Corollary 3.1.1 *Consider a strongly symmetric, N-parameter, $\mathbb{R}^d$-valued Markov process and assume that there exists $\lambda \succ 0$ in $\mathbb{R}_+^N$ and $x_0 \in \mathbb{R}^d$ such that for all $y \in \mathbb{R}^d$, $r_\lambda(x_0, y) > 0$. Then, for all compact sets $E \subset \mathbb{R}^d$ such that $x_0 \notin E$,*

$$\mathbb{P}_{x_0}\{\overline{X(\mathbb{R}_+^N)} \cap E \neq \varnothing\} > 0 \iff \mathcal{C}_{r_\lambda}(E) > 0.$$

Proof of Corollary 3.1.1 Let x_0 and λ be as in the statement of the corollary, and recall that for all $\lambda \succ 0$ and all $y \in \mathbb{R}^d$,

$$r_\lambda(x_0, y) = \int_{\mathbb{R}_+^N} e^{-\lambda \cdot s} p_s(x_0, y) \, ds.$$

Thus, the condition $r_\lambda(x_0, y) > 0$ is equivalent to the *strict* positivity of $p_s(x_0, y)$ for Lebesgue almost all $s \in \mathbb{R}_+^N$. In particular,

$$\mathbb{P}_\nu\{\overline{X(\mathbb{R}_+^N)} \cap E \neq \varnothing\} > 0$$
$$\iff \int \mathbb{P}_y\{\overline{X(\mathbb{R}_+^N)} \cap E \neq \varnothing\} p_s(x_0, y) \nu(dy) > 0, \quad \text{for almost all } s \in \mathbb{R}_+^N.$$

Now,

$$\int \mathbb{P}_y\{\overline{X(\mathbb{R}_+^N)} \cap E \neq \varnothing\} p_s(x_0, y) \nu(dy) = \mathbb{E}_{x_0}\left[\mathbb{P}_{X_s}\{\overline{X(\mathbb{R}_+^N)} \cap E \neq \varnothing\}\right]$$
$$= \mathbb{P}_{x_0}\{\overline{X([s, \infty[)} \cap E \neq \varnothing\},$$

where $[s, \infty[= \prod_{\ell=1}^N [s^{(\ell)}, \infty[$. In the last step we have used part (iv) of the definition of Markov property in Section 1.1 (why?). In particular, we can combine the above observation with Remark (iii) after Theorem 3.1.1 to deduce that

$$\mathcal{C}_{r_\lambda}(E) > 0 \iff \mathbb{P}_{x_0}\{\overline{X([s, \infty[)} \cap E \neq \varnothing\} > 0, \quad \text{for almost all } s \in \mathbb{R}_+^N.$$

410   11. Multiparameter Markov Processes

To finish, note that by the right continuity of $t \mapsto X_t$,

$$\lim_{s \to 0} \mathbb{P}_{x_0}\{\overline{X([0,s])} \cap E \neq \varnothing\} = 0.$$

Since $s \mapsto \mathbb{P}_{x_0}\{\overline{X([s,\infty[)} \cap E \neq \varnothing\}$ is nonincreasing with respect to the partial order $\preccurlyeq$, this implies the corollary. □

3.2   Three Technical Estimates

Let $\mathfrak{e}_1, \ldots, \mathfrak{e}_N$ denote N independent exponentially distributed random variables whose means are all 1. By further enlarging the underlying probability space, we may, and will, assume without loss of generality that the collection $(\mathfrak{e}_1, \ldots, \mathfrak{e}_N)$ is totally independent of the process X.

For any $\lambda \succ 0$ in $\mathbb{R}_+^N$, define $\mathfrak{e}(\lambda)$ to be the random vector whose jth coordinate is $\mathfrak{e}_j/\lambda^{(j)}$. Throughout this subsection we are interested in obtaining three distributional estimates for $J_\lambda(\varphi)$, where $\varphi : \mathbb{R}^d \to \mathbb{R}_+$ is a measurable function and where

$$J_\lambda(\varphi) = \int_{[0,\mathfrak{e}(\lambda)]} \varphi(X_s)\,ds. \tag{1}$$

We shall often use the self-evident fact that for any $x \in \mathbb{R}^d$ and for all $s \in \mathbb{R}_+^N$,

$$\mathbb{P}_x\{s \preccurlyeq \mathfrak{e}(\lambda)\} = e^{-\lambda \cdot s}. \tag{2}$$

Lemma 3.2.1 *For all $\lambda \succ 0$ in $\mathbb{R}_+^N$ and for every measurable function $\varphi : \mathbb{R}^d \to \mathbb{R}_+$,*

$$\mathbb{E}_\nu[J_\lambda(\varphi)] = \frac{1}{\prod_{j=1}^N \lambda^{(j)}} \cdot \int \varphi(x)\,\nu(dx).$$

Proof By the monotone convergence theorem, we may assume that both sides are finite (why?). Note that for all $x \in \mathbb{R}^d$, $\mathbb{E}_x[J_\lambda(\varphi)] = \mathcal{R}_\lambda \varphi(x)$. Thus,

$$\mathbb{E}_\nu[J_\lambda(\varphi)] = \int_{\mathbb{R}^d} \mathcal{R}_\lambda \varphi(x)\,\nu(dx) = \int_{\mathbb{R}^d} \varphi(x) \cdot \mathcal{R}_\lambda \mathbf{1}(x)\,\nu(dx),$$

where $\mathbf{1}(x) \equiv 1$ for all $x \in \mathbb{R}^d$; we have used Exercise 1.4.2 in this last step. Since $\mathcal{R}_\lambda \mathbf{1}(x) = \int_{\mathbb{R}_+^N} e^{-\lambda \cdot s}\,ds = \left\{ \prod_{j=1}^N \lambda^{(j)} \right\}^{-1}$, the lemma follows. □

Lemma 3.2.2 *For all $\lambda \succ 0$ in $\mathbb{R}_+^N$ and for all measurable functions $\varphi : \mathbb{R}^d \to \mathbb{R}_+$,*

$$\mathbb{E}_\nu[\{J_\lambda(\varphi)\}^2] = \frac{2^N}{\prod_{j=1}^N \lambda^{(j)}} \cdot \int_{\mathbb{R}^d} \varphi(a)\,\mathcal{R}_\lambda \varphi(a)\,\nu(da).$$

In particular,
$$\mathbb{E}_\nu\bigl[\{J_\lambda(\varphi)\}^2\bigr] = \frac{2^N}{\prod_{j=1}^N \lambda^{(j)}} \iint r_\lambda(x,y)\varphi(x)\varphi(y)\,\nu(dx)\,\nu(dy).$$

Proof By the monotone convergence theorem, we can assume that all integrals in question are finite (why?). Directly computing, we obtain the following: For all $a \in \mathbb{R}^d$,
$$\mathbb{E}_a\bigl[\{J_\lambda(\varphi)\}^2\bigr] = \int_{\mathbb{R}_+^N}\int_{\mathbb{R}_+^N} e^{-\lambda\cdot(s\curlyvee t)} \mathbb{E}_a\bigl[\varphi(X_s)\varphi(X_t)\bigr]\,ds\,dt.$$

Thus, we can integrate with respect to $\nu(da)$ to deduce that
$$\mathbb{E}_\nu\bigl[\{J_\lambda(\varphi)\}^2\bigr] = \int_{\mathbb{R}_+^N}\int_{\mathbb{R}_+^N} e^{-\lambda\cdot(s\curlyvee t)} \mathbb{E}_\nu\bigl[\varphi(X_s)\varphi(X_t)\bigr]\,ds\,dt.$$

We now use symmetry in an essential way: By Lemma 1.4.2 and by Fubini's theorem,
$$\mathbb{E}_\nu\bigl[\{J_\lambda(\varphi)\}^2\bigr] = \int_{\mathbb{R}^d}\int_{\mathbb{R}_+^N}\int_{\mathbb{R}_+^N} e^{-\lambda\cdot(s\curlyvee t)}\varphi(a)\,\mathcal{T}_{s+t-2(s\curlywedge t)}\varphi(a)\,ds\,dt\,\nu(da).$$

Clearly, for all $u,v \in \mathbb{R}_+^N$, $u+v-2(u\curlywedge v) = (u\curlyvee v) - (u\curlywedge v)$. Thus,
$$\mathbb{E}_\nu\bigl[\{J_\lambda(\varphi)\}^2\bigr] = \int_{\mathbb{R}^d} \Bigl\{\int_{\mathbb{R}_+^N}\int_{\mathbb{R}_+^N} e^{-\lambda\cdot(s\curlywedge t)}\,\mathcal{T}_{s\curlyvee t - s\curlywedge t}\varphi(a)\,ds\,dt\Bigr\}\varphi(a)\,\nu(da).$$

On the other hand, as operators,
$$\int_{\mathbb{R}_+^N}\int_{\mathbb{R}_+^N} e^{-\lambda\cdot(s\curlywedge t)}\,\mathcal{T}_{s\curlyvee t - s\curlywedge t}\,ds\,dt$$
$$= \bigcirc_{j=1}^N \left(\int_0^\infty \int_0^\infty e^{-\lambda^{(j)}(s^{(j)}\vee t^{(j)})} \mathcal{T}^j_{s^{(j)}\vee t^{(j)} - s^{(j)}\wedge t^{(j)}}\,ds^{(j)}\,dt^{(j)}\right)$$
$$= 2^N \bigcirc_{j=1}^N \left(\int_0^\infty \int_{s^{(j)}}^\infty e^{-\lambda^{(j)} t^{(j)}} \mathcal{T}^j_{t^{(j)} - s^{(j)}}\,dt^{(j)}\,ds^{(j)}\right)$$
$$= 2^N \bigcirc_{j=1}^N \left(\int_0^\infty e^{-\lambda^{(j)} s^{(j)}} \mathcal{R}^j_{\lambda^{(j)}}\,ds^{(j)}\right)$$
$$= \frac{2^N}{\prod_{j=1}^N \lambda^{(j)}} \bigcirc_{j=1}^N \mathcal{R}^j_{\lambda^{(j)}} = \frac{2^N}{\prod_{j=1}^N \lambda^{(j)}} \mathcal{R}_\lambda.$$

We have used Proposition 1.3.1 in the last line. This immediately proves the first assertion of the lemma. The second assertion follows from the first and from Fubini's theorem. $\square$

In order to describe the next technical lemma of this subsection, we finally need to address the case where ν is potentially not a probability measure. While $\mathbb{P}_\nu$ makes sense in any case, the martingale theory that has been developed in this book can be invoked for $\mathbb{P}_\nu$ only if it is a probability measure, i.e., when ν is a probability measure.

Unfortunately, in many of our intended applications ν will be Lebesgue's measure, and this means that we need to approximate ν with probability measures. The precise way to do this is immaterial in that any reasonable approximation serves our purposes.

We describe one such approximation next. For all $k \geq 1$, we define the *probability measure* ν_k as follows: For all Borel sets $G \subset \mathbb{R}^d$,

$$\nu_k(G) = \frac{\nu(G \cap [-k,k]^d)}{\nu([-k,k]^d)}, \qquad k \geq 1, \tag{3}$$

where $1 \div 0 = \infty$. Our definition of reference measures implies that $\nu([-k,k]^d)$ is strictly positive, for all $k \geq 1$. Thus, ν_k is a probability measure on $\mathbb{R}^d$, for all $k \geq 1$, and correspondingly, $\mathbb{P}_{\nu_k}$ is a probability measure (on our underlying probability space) for any and all k large enough.

If $\varphi : \mathbb{R}^d \to \mathbb{R}_+$ is measurable and $\lambda \succ 0$ is in $\mathbb{R}_+^N$, we define $M^k = (M_t^k;\, t \in \mathbb{R}_+^N)$ by

$$M_t^k = \mathbb{E}_{\nu_k}[J_\lambda(\varphi) \mid \mathcal{F}_t], \tag{4}$$

where $J_\lambda(\varphi)$ is defined by equation (1) above. Note that when φ is bounded, for instance, M^k is an N-parameter martingale, which we can always take to be separable. (Otherwise, consider instead the modification of M^k provided by Doob's separability theorem, Theorem 2.2.1 of Chapter 5. Once again, note that we need the probability space to be complete in order to do this.)

Lemma 3.2.3 *Consider a fixed $\lambda \succ 0$ in $\mathbb{R}_+^N$ and a fixed, measurable function $\varphi : \mathbb{R}^d \to \mathbb{R}_+$. Then, for all $t \in \mathbb{R}_+^N$ and all $k \geq 1$,*

$$M_t^k \geq e^{-\lambda \cdot t} \mathcal{R}_\lambda \varphi(X_t), \qquad \mathbb{P}_{\nu_k}\text{-a.s.,}$$

where M^k is defined by (4).

Proof First, we suppose that φ is also bounded. For each fixed $t \in \mathbb{R}_+^N$ and for every $k \geq 1$,

$$\begin{aligned}
M_t^k &= \mathbb{E}_{\nu_k}[J_\lambda(\varphi) \mid \mathcal{F}_t] \\
&\geq \mathbb{E}_{\nu_k}\left[\int_{\mathbb{R}_+^N} \mathbf{1}_{\left(t \preccurlyeq s \preccurlyeq \mathfrak{e}(\lambda)\right)} \varphi(X_s)\, ds \,\Big|\, \mathcal{F}_t\right] \\
&= \int_{s \succcurlyeq t} e^{-\lambda \cdot s}\, \mathbb{E}_{\nu_k}\!\left[\varphi(X_s) \mid \mathcal{F}_t\right] ds \\
&= \int_{s \succcurlyeq t} e^{-\lambda \cdot s}\, \mathcal{T}_{s-t}\varphi(X_t)\, ds
\end{aligned}$$

$$= e^{-\lambda \cdot s} \mathcal{R}_\lambda \varphi(X_t),$$

$\mathbb{P}_{\nu_k}$-a.s. This completes the proof when φ is bounded. The general result follows from Lebesgue's monotone convergence theorem, since we can consider the function $\varphi \wedge n$ instead of φ and then let $n \uparrow \infty$, monotonically. □

3.3 Proof of Theorem 3.1.1: First Half

In this subsection we prove the upper bound on the integral of Theorem 3.1.1. Recalling the definition of $\mathfrak{e}(\lambda)$, $\lambda \succ 0$ in $\mathbb{R}_+^N$, this is equivalent to proving that

$$\mathbb{P}_\nu \left\{ \overline{X([0, \mathfrak{e}(2\lambda)])} \cap E \neq \varnothing \right\} \leq A_2 \mathcal{C}_{r_\lambda}(E). \tag{1}$$

(Why?) This is a natural time to ponder over measurability issues. Define E_n to be the open $\frac{1}{n}$-enlargement of E. That is,

$$E_n = \left\{ x \in \mathbb{R}^d : \operatorname{dist}\{x; E\} < \frac{1}{n} \right\}, \qquad n \geq 1.$$

Of course, n need not be an integer for the above to make sense. Clearly, $\overline{E_1}, \overline{E_2}, \ldots$ are all compact subsets of $\mathbb{R}^d$. Moreover, for any $s \in \mathbb{R}_+^N$ and for all $\omega \in \Omega$ (the underlying sample space),

$$\overline{X([0,s])}(\omega) \cap E \neq \varnothing \iff \forall n \geq 1 : X([0,s])(\omega) \cap E_n \neq \varnothing.$$

We have used the right continuity of $t \mapsto X_t$. Since E is open, this right continuity implies that the collection of all $\omega \in \Omega$ such that $X([0,s])(\omega) \cap E_n \neq \varnothing$ is measurable (why?). Since measurability follows from this, we proceed with the main body of our proof.

Let $\varphi : \mathbb{R}^d \to \mathbb{R}_+$ denote an arbitrary measurable function and recall $J_\lambda(\varphi)$ from equation (1) of Section 3.2. Also recall the probability measures $\nu_1, \nu_2, \ldots$ and the associated N-parameter processes $M^1, M^2, \ldots$ from equations (3) and (4) of Section 3.2, respectively. Note that when φ is bounded (say), the M^k's are N-parameter martingales. (In general, integrability need not hold.)

We shall use these processes in the main step in the proof of equation (1) and show that for every $n \geq 1$ and all $k \geq 1$ sufficiently large,

$$\mathbb{P}_{\nu_k}\left\{ X([0, \mathfrak{e}(2\lambda)]) \cap E_n \neq \varnothing \right\} \leq A_2 \frac{\mathcal{C}_{r_\lambda}(\overline{E_n})}{\nu([-k,k]^d)}. \tag{2}$$

This is the first reduction.

Lemma 3.3.1 *Equation (2) implies equation (1).*

Proof Clearly, for each Borel set G, $\nu([-k,k]^d) \times \mathbb{P}_{\nu_k}(G) \uparrow \mathbb{P}_\nu(G)$, as $k \uparrow \infty$. Thus, equation (2) implies

$$\mathbb{P}_\nu\{X([0,\mathfrak{e}(2\lambda)]) \cap E_n \neq \varnothing\} \leq A_2 \mathcal{C}_{r_\lambda}(\overline{E_n}).$$

Since $E_n \downarrow E$ as $n \to \infty$, by the countable additivity of $\mathbb{P}_\nu$,

$$\lim_{n \to \infty} \mathbb{P}_\nu\{X([0,\mathfrak{e}(2\lambda)]) \cap E_n \neq \varnothing\} = \mathbb{P}_\nu\{\overline{X([0,\mathfrak{e}(2\lambda)])} \cap E \neq \varnothing\},$$

by compactness. It remains to show that $\lim_{n \to \infty} \mathcal{C}_{r_\lambda}(\overline{E_n}) = \mathcal{C}_{r_\lambda}(E)$. But this follows from the outer regularity of capacities; cf. Lemma 2.1.2, Appendix D. □

Before proving equation (2) in earnest, we make a few observations. First, we note that since E_n is open and since ν is a reference measure, $\nu(E_n) > 0$. Using this, together with Lemma 1.4.1, we deduce that for all $t \in \mathbb{R}_+^N$,

$$\mathbb{P}_\nu(X_t \in E_n) = \mathbb{P}_\nu(X_0 \in E_n) = \nu(E_n) > 0. \qquad (3)$$

Our second observation is a mere application of Supplementary Exercise 11. Namely, we can find a $\mathbb{Q}_+^N \cup \{\infty\}$-valued random vector T such that

$$T(\omega) \in \mathbb{Q}_+^N \cap [0,\mathfrak{e}(2\lambda)] \iff \exists t \in \mathbb{Q}_+^N \cap [0,\mathfrak{e}(\lambda)] : X_t \in E_n. \qquad (4)$$

It should be noted that by its very definition, $T = \infty$ if and only if $T \notin \mathbb{Q}_+^N$. For this random vector T, we have the following statement.

Lemma 3.3.2 *For all $k \geq 1$, the following $L^2(\mathbb{P}_{\nu_k})$ estimate holds:*

$$\mathbb{E}_{\nu_k}\left[\sup_{t \in \mathbb{Q}_+} |M_t^k|^2\right] \geq \left[\mathbb{E}_{\nu_k}\{\mathcal{R}_\lambda\varphi(X_T) \mid T \preccurlyeq \mathfrak{e}(2\lambda)\}\right]^2 \cdot \mathbb{P}_{\nu_k}\{T \preccurlyeq \mathfrak{e}(2\lambda)\}.$$

Proof Applying (4), Lemma 3.2.3, and the countable additivity of $\mathbb{P}_{\nu_k}$, all in conjunction, we obtain

$$\sup_{t \in \mathbb{Q}_+^N} M_t^k \geq e^{-\lambda \cdot T} \mathcal{R}_\lambda\varphi(X_T) \mathbf{1}_{(T \in \mathbb{Q}_+^N)},$$

$\mathbb{P}_{\nu_k}$-a.s. (Why?) We can square both sides of the above and take $\mathbb{E}_{\nu_k}$-expectations to see that

$$\mathbb{E}_{\nu_k}\left[\sup_{t \in \mathbb{Q}_+^N} |M_t|^2\right] \geq \mathbb{E}_{\nu_k}\left[e^{-2\lambda \cdot T}\{\mathcal{R}_\lambda\varphi(X_T)\}^2 \mathbf{1}_{(T \neq \infty)}\right]$$

$$= \mathbb{E}_{\nu_k}\left[\mathbf{1}_{(T \preccurlyeq \mathfrak{e}(2\lambda))}\{\mathcal{R}_\lambda\varphi(X_T)\}^2 \mathbf{1}_{(T \neq \infty)}\right]$$

$$= \mathbb{E}_{\nu_k}\left[\mathbf{1}_{(T \preccurlyeq \mathfrak{e}(2\lambda))}\{\mathcal{R}_\lambda\varphi(X_T)\}^2\right]$$

$$= \mathbb{E}_{\nu_k}\left[\{\mathcal{R}_\lambda\varphi(X_T)\}^2 \mid T \preccurlyeq \mathfrak{e}(2\lambda)\right] \cdot \mathbb{P}_{\nu_k}\{T \preccurlyeq \mathfrak{e}(2\lambda)\}.$$

We have used the independence of $\mathfrak{e}$ from the entire process X, together with equation (2) of Section 3.2. By the Cauchy–Schwarz inequality, for all nonnegative random variables Z,

$$\mathbb{E}_{\nu_k}\left[Z^2 \,\middle|\, T \preccurlyeq \mathfrak{e}(2\lambda)\right] \geq \left|\mathbb{E}_{\nu_k}\{Z \mid T \preccurlyeq \mathfrak{e}(2\lambda)\}\right|^2.$$

The result follows from this upon letting $Z = \mathcal{R}_\lambda \varphi(X_T)$. $\square$

For all $k \geq 1$ and all Borel sets $F \subset \mathbb{R}^d$, define

$$\mu_k(F) = \mathbb{P}_{\nu_k}\{X_T \in F \mid T \preccurlyeq \mathfrak{e}(2\lambda)\}. \tag{5}$$

Lemma 3.3.3 *There exists $k_0 \geq 1$ such that for every $k \geq k_0$, the measure μ_k is absolutely continuous with respect to ν_k and hence with respect to ν. Moreover, $\mu_k \in \mathcal{P}(\overline{E_n})$.*

Proof We can see from the right continuity of $t \mapsto X_t$, and from equation (3), that $\mu_k \in \mathcal{P}(\overline{E_n})$ for all $k \geq 1$. Furthermore, we can appeal to equation (4) to see that on $(T \neq \infty)$, $X_T \in \overline{\{X_t : t \in \mathbb{Q}_+^N\}}$. Now suppose there exists a Borel set $G \subset \mathbb{R}^d$ such that $\nu_k(G) = 0$. By invoking equation (4), we obtain

$$\mu_k(G) \leq \sum_{t \in \mathbb{Q}_+^N} \frac{\mathbb{P}_{\nu_k}(X_t \in G)}{\mathbb{P}_{\nu_k}(T \preccurlyeq \mathfrak{e}(2\lambda))} \leq \sum_{t \in \mathbb{Q}_+^N} \frac{\nu_k(G)}{\mathbb{P}_{\nu_k}(T \preccurlyeq \mathfrak{e}(2\lambda))} = 0,$$

as long as $\mathbb{P}_{\nu_k}\{T \preccurlyeq \mathfrak{e}(2\lambda)\} > 0$. However,

$$\lim_{k \to \infty} \nu\bigl([-k,k]^d\bigr) \times \mathbb{P}_{\nu_k}\{T \preccurlyeq \mathfrak{e}(2\lambda)\} = \mathbb{P}_\nu\{T \preccurlyeq \mathfrak{e}(2\lambda)\} > 0,$$

by equation (3). This proves the absolute continuity of μ_k for all k sufficiently large. $\square$

Lemma 3.3.4 *Let μ_k be defined by (5) and let k_0 be the constant given in Lemma 3.3.3. Then, for all $k \geq k_0$, for any measurable $\varphi : \mathbb{R}^d \to \mathbb{R}_+$ and for all $\lambda \succ 0$ in $\mathbb{R}_+^N$,*

$$\frac{8^N}{\prod_{j=1}^N \lambda^{(j)}} \iint r_\lambda(x,y)\varphi(x)\varphi(y)\,\nu(dx)\,\nu(dy)$$
$$\geq \left(\int_{\mathbb{R}^d} \mathcal{R}_\lambda \varphi(y)\, \mu_k(dy)\right)^2 \cdot \mathbb{P}_{\nu_k}\bigl(T \preccurlyeq \mathfrak{e}(2\lambda)\bigr) \cdot \nu\bigl([-k,k]^d\bigr).$$

Proof Truncate φ and apply Lebesgue's monotone convergence theorem to see that, without loss of generality, φ can be assumed to be bounded. In particular, for each $k \geq k_0$, M^k is now an N-parameter martingale!

Combine Lemma 3.3.2 and the definition of μ_k to see that

$$\mathbb{E}_{\nu_k}\left[\sup_{t\in\mathbb{Q}_+^N}\{M_t^k\}^2\right] \geq \left(\mathbb{E}_{\nu_k}[\mathcal{R}_\lambda\varphi(X_T)\mid T\preccurlyeq\mathfrak{e}(2\lambda)]\right)^2 \cdot \mathbb{P}_{\nu_k}(T\preccurlyeq\mathfrak{e}(2\lambda))$$

$$= \left(\int_{\mathbb{R}^d}\mathcal{R}_\lambda\varphi(y)\,\mu_k(dy)\right)^2 \cdot \mathbb{P}_{\nu_k}(T\preccurlyeq\mathfrak{e}(2\lambda)).$$

We now employ Cairoli's second inequality (Theorem 2.3.2, Chapter 7) and deduce that

$$4^N \sup_{t\in\mathbb{R}_+^N}\mathbb{E}_{\nu_k}[|M_t^k|^2] \geq \left(\int_{\mathbb{R}^d}\mathcal{R}_\lambda\varphi(y)\,\mu_k(dy)\right)^2 \cdot \mathbb{P}_{\nu_k}(T\preccurlyeq\mathfrak{e}(2\lambda)).$$

On the other hand, by Lemma 3.2.2,

$$\sup_{t\in\mathbb{R}_+^N}\mathbb{E}_{\nu_k}[M_t^2] \leq \mathbb{E}_{\nu_k}\left[\{J_\lambda(\varphi)\}^2\right]$$

$$\leq \frac{1}{\nu([-k,k]^d)} \cdot \mathbb{E}_\nu\left[|J_\lambda(\varphi)|^2\right]$$

$$= \frac{1}{\nu([-k,k]^d)} \times \frac{2^N}{\prod_{j=1}^N \lambda^{(j)}} \iint r_\lambda(x,y)\varphi(x)\varphi(y)\,\nu(dx)\,\nu(dy).$$

The lemma follows readily from this. $\square$

We are ready to conclude the proof of the first half of Theorem 3.1.1.

Proof of Equation (2) A little thought shows that equation (2) is equivalent to the following: For all $k \geq k_0$,

$$\mathbb{P}_{\nu_k}\{T\preccurlyeq\mathfrak{e}(2\lambda)\} \leq A_2 \frac{\mathcal{C}_{r_\lambda}(\overline{E}_n)}{\nu([-k,k]^d)}. \tag{6}$$

This follows from equation (4) above, used in conjunction with equation (2) of Section 3.2 and the independence of T and $\mathfrak{e}(2\lambda)$. Thus, we may (and will) assume without loss of generality that the above probability is strictly positive, for otherwise, there is nothing to prove.

The previous lemma holds for *any* measurable function φ. We now choose a "good" φ in order to finish. Henceforth, let us fix large finite constants $\ell > 0$ and $k \geq k_0$ and define

$$\varphi(x) = \frac{d\mu_k}{d\nu_k}\mathbf{1}_{\Theta(\ell)}(x), \qquad x\in\mathbb{R}^d,$$

where

$$\Theta(\ell) = \left\{x\in\overline{E}_n:\ \frac{d\mu_k}{d\nu}\leq\ell\right\}.$$

It is very important to note that by picking k_0 large enough (chosen independently of our choice of ℓ), we can ensure that

$$\varphi(x) = \nu\bigl([-k,k]^d\bigr) \frac{d\mu_k}{d\nu}(x) \mathbf{1}_{\Theta(\ell)},$$

since for all $n \geq 1$, $E_n \subset \overline{E_1}$ is bounded. By Lemma 3.3.3, this definition is well-defined. For this choice of φ, the expression in Lemma 3.3.4 can be efficiently estimated, viz.,

$$\int \mathcal{R}_\lambda \varphi(x)\, \mu_k(dx) = \nu\bigl([-k,k]^d\bigr) \int\!\!\int r_\lambda(x,y)\, \varphi(y)\, \mu_k(dx)\,\nu(dy)$$

$$= \nu\bigl([-k,k]^d\bigr) \int_{\Theta(\ell)} \int r_\lambda(x,y)\, \mu_k(dx)\, \mu_k(dy)$$

$$\geq \nu\bigl([-k,k]^d\bigr) \int_{\Theta(\ell)} \int_{\Theta(\ell)} r_\lambda(x,y) \mu_k(dx)\, \mu_k(dy)$$

$$= \nu\bigl([-k,k]^d\bigr) \mathcal{E}_{r_\lambda}\bigl(\mu_k|_{\Theta(\ell)}\bigr),$$

where $\mu_k\bigl|_F(\Gamma) = \mu_k(F \cap \Gamma)$ denotes the restriction of μ_k to any measurable set Γ. Hence, for the above choice of φ, Lemma 3.3.4 becomes

$$\mathbb{P}_{\nu_k}\bigl(T \prec \mathfrak{e}(2\lambda)\bigr) \cdot \bigl[\mathcal{E}_{r_\lambda}(\mu_k|_{\Theta(\ell)})\bigr]^2 \cdot \nu\bigl([-k,k]^d\bigr) \leq \frac{8^N}{\prod_{j=1}^N \lambda^{(j)}} \mathcal{E}_{r_\lambda}(\mu_k|_{\Theta(\ell)}). \quad (7)$$

However, by Exercise 1.4.2,

$$\mathcal{E}_{r_\lambda}(\mu_k|_{\Theta(\ell)}) = \int_{\Theta(\ell)} \varphi(x)\, \mathcal{R}_\lambda \varphi(x)\, \nu_k(dx) \leq \ell \int \mathcal{R}_\lambda \varphi(x)\, \nu_k(dx)$$

$$= \ell \int \varphi(x) \cdot \mathcal{R}_\lambda \mathbf{1}(x)\, \nu_k(dx) = \frac{\ell}{\prod_{j=1}^N \lambda^{(j)}} \int \varphi(x)\, \nu_k(dx),$$

where $\mathbf{1}(x) = 1$ for all $x \in \mathbb{R}^d$. We have used the following calculation, which we also encountered earlier:

$$\mathcal{R}_\lambda \mathbf{1}(x) = \int_{\mathbb{R}_+^N} e^{-\lambda \cdot s}\, ds = \Bigl[\prod_{j=1}^N \lambda^{(j)}\Bigr]^{-1}.$$

Since $\int \varphi(x)\, \nu_k(dx) = \mu_k(\mathbb{R}^d) = 1$, we can conclude that $\mathcal{E}_{r_\lambda}(\mu_k|_{\Theta(\ell)}) < \infty$. We claim that this expression is also strictly positive, as long as ℓ is large enough. Indeed, by Fubini's theorem,

$$\liminf_{\ell \to \infty} \mathcal{E}_{r_\lambda}(\mu_k|_{\Theta(\ell)}) = \liminf_{\ell \to \infty} \int \mathcal{R}_\lambda \mu_k|_{\Theta(\ell)}(x)\, \nu(dx)$$

$$\geq \int \liminf_{\ell \to \infty} \mathcal{R}_\lambda \mu_k|_{\Theta(\ell)}(x)\, \nu(dx)$$

$$= \iint r_\lambda(x,y)\, \mu_k(dx)\, \nu(dy)$$

$$= \frac{1}{\prod_{j=1}^N \lambda^{(j)}} > 0,$$

since $\mu_k(\mathbb{R}^d) = 1$. What all this means is that we can divide both sides of equation (7) by the square of $\mathcal{E}_{r_\lambda}(\mu_k|_{\Theta(\ell)})$ if ℓ is sufficiently large. In other words, for all ℓ large,

$$\nu\big([-k,k]^d\big) \cdot \mathbb{P}_{\nu_k}\big(T \preccurlyeq \mathfrak{e}(2\lambda)\big) \leq \frac{8^N}{\prod_{j=1}^N \lambda^{(j)}} \mathcal{C}_{r_\lambda}^{\mathrm{ac}}\big(\Theta(\ell)\big) \leq \frac{8^N}{\prod_{j=1}^N \lambda^{(j)}} \mathcal{C}_{r_\lambda}^{\mathrm{ac}}\big(\overline{E}_n\big),$$

where $\mathcal{C}_{r_\lambda}^{\mathrm{ac}}(G)$ is the absolutely continuous capacity of G with respect to the gauge function r_λ; cf. Section 2.3 of Appendix D. Since r_λ is a symmetric and proper gauge function, Theorem 2.3.1 of Appendix D reveals that on compact sets, $\mathcal{C}_{r_\lambda}^{\mathrm{ac}} = \mathcal{C}_{r_\lambda}$. This verifies equation (6) and hence, equation (2). The first half of the proof of Theorem 3.1.1 is complete. □

3.4 Proof of Theorem 3.1.1: Second Half

From Section 3.3 we recall the open sets $E_n \downarrow E$, as $n \to \infty$, where E_n denotes the open $\frac{1}{n}$-enlargement of E.

Suppose that $n \geq 1$ is a fixed integer and that $\varphi : \mathbb{R}^d \to \mathbb{R}_+$ is a nontrivial measurable function such that for all $x \notin \overline{E}_n$, $\varphi(x) = 0$. By equation (3) of Section 3.2, for any $\lambda \succ 0$ in $\mathbb{R}_+^N$,

$$\mathbb{P}_\nu(J_\lambda(\varphi) > 0) \leq \mathbb{P}_\nu\{X([0,\mathfrak{e}(\lambda)]) \cap E_n \neq \varnothing\}.$$

Therefore, we can use our proof[8] of the Paley–Zygmund lemma (Lemma 1.4.1, Chapter 3), to obtain

$$\mathbb{P}_\nu\{X([0,\mathfrak{e}(\lambda)]) \cap E_n \neq \varnothing\} \geq \frac{\{\mathbb{E}_\nu[J_\lambda(\varphi)]\}^2}{\mathbb{E}_\nu\big[\{J_\lambda(\varphi)\}^2\big]}.$$

Lemmas 3.2.1 and 3.2.2 together imply that

$$\mathbb{P}_\nu\{X([0,\mathfrak{e}(\lambda)]) \cap E_n \neq \varnothing\} \geq \frac{1}{2^N \prod_{j=1}^N \lambda^{(j)}} \bigg\{\int \varphi(x)\, \nu(dx)\bigg\}^2$$

$$\times \bigg[\iint r_\lambda(x,y)\varphi(x)\varphi(y)\, \nu(dx)\, \nu(dy)\bigg]^{-1}.$$

---

[8] While we can use the proof of the Paley–Zygmund lemma, we cannot invoke it as it appears in Chapter 3, since $\mathbb{P}_\nu$ is not necessarily a probability measure. Alternatively, we can replace ν by ν_k and work with $\mathbb{P}_{\nu_k}$ first. You may want to try this on your own.

We should recognize that this holds for all measurable $\varphi : \mathbb{R}^d \to \mathbb{R}_+$ such that $\varphi(x) = 0$, off of $\overline{E_n}$. Equivalently, suppose ζ is any probability measure on the closure of E_n that is absolutely continuous with respect to ν. Then,

$$\mathbb{P}_\nu\{X([0, \mathfrak{e}(\lambda)]) \cap E_n \neq \varnothing\} \geq \frac{1}{2^N \prod_{j=1}^N \lambda^{(j)}} \left[\mathcal{E}_{r_\lambda}(\zeta)\right]^{-1}.$$

Optimizing over such ζ's, we can conclude that

$$\mathbb{P}_\nu\{X([0, \mathfrak{e}(\lambda)]) \cap E_n \neq \varnothing\} \geq \frac{1}{2^N \prod_{j=1}^N \lambda^{(j)}} \mathcal{C}_{r_\lambda}^{\mathrm{ac}}(\overline{E_n}),$$

where $\mathcal{C}_{r_\lambda}^{\mathrm{ac}}$ denotes the absolutely continuous capacity with respect to the gauge function r_λ; cf. Section 2.3 of Appendix D. Since r_λ is a symmetric, proper gauge function on $\mathbb{R}^d \times \mathbb{R}^d$, Theorem 2.3.1 of Appendix D ensures us that absolutely continuous capacities agree with capacities on compact sets. The proof of the lower bound of Theorem 3.1.1 follows once we observe that $\mathcal{C}_{r_\lambda}(\overline{E_n}) \geq \mathcal{C}_{r_\lambda}(E)$ and let $n \to \infty$ afterwards. $\square$

***Exercise* 3.4.1** Recall that in the N-parameter theory, we *defined* ν to be a reference measure if $\nu(G) > 0$ for all open sets $G \subset \mathbb{R}^d$. This was not needed to establish the 1-parameter potential theory of Chapter 10. Show that it is not needed in the N-parameter case, either. To be more precise, show that whenever $\nu(G) = 0$ for an open set $G \subset \mathbb{R}^d$, the $\mathbb{P}_\nu$-probability that the image of X ever intersects G is zero. Conclude that Theorem 3.1.1 holds without this positivity condition on ν.
(HINT: Use the right continuity of the N-parameter process $t \mapsto X_t$.) $\square$

4 Applications

Having demonstrated Theorem 3.1.1, we are in position to apply it to analyze the fractal structure of a large class of interesting stochastic processes. In this section we show some such applications of Theorem 3.1.1.

4.1 Additive Stable Processes

Consider N independent, $\mathbb{R}^d$-valued, isotropic Lévy processes $X^1, \ldots, X^N$ of index $\alpha \in]0, 2]$, and define the **additive stable process of index** α, $X = (X_t;\ t \in \mathbb{R}_+^N)$, as $X = X^1 \oplus \cdots \oplus X^N$. That is,

$$X_t = \sum_{j=1}^N X_{t^{(j)}}^j, \qquad t \in \mathbb{R}_+^N.$$

Since X is an additive Lévy process, the measure-theoretic details of such a construction have already been worked out in Section 2.3 above. Recall from Section 3.4 that $X^1, \ldots, X^N$ have (the same) γ-potential densities $\{r^1_\gamma(x,y); \gamma > 0, x, y \in \mathbb{R}^d\}$ given by $r^1_\gamma(x,y) = u^1_\gamma(y-x)$, where

$$u^1_\gamma(a) = \int_0^\infty e^{-\gamma s} q_s(a)\, ds, \qquad a \in \mathbb{R}^d,\ \gamma > 0,$$

where $q_s(a)$ denotes the density function of X^1_s at a, with respect to Lebesgue's measure on $\mathbb{R}^d$. By equation (3) of Section 2.3, X has λ-potential densities $(r_\lambda(x,y); \lambda \succ 0$ in $\mathbb{R}^N_+,\ x, y \in \mathbb{R}^d)$ with respect to Lebesgue's measure on $\mathbb{R}^d$. Moreover, $r_\lambda(x,y) = u_\lambda(y-x)$, where

$$u_\lambda(a) = u^1_{\lambda(1)} \star \cdots \star u^1_{\lambda(N)}(a), \qquad a \in \mathbb{R}^d,\ \lambda \succ 0 \text{ in } \mathbb{R}^N_+, \tag{1}$$

where $\star$ denotes convolution. We now estimate this λ-potential density.

Proposition 4.1.1 *Suppose $d > N\alpha$. Then, the λ-potential density u_λ is a symmetric, proper gauge function on $\mathbb{R}^d \times \mathbb{R}^d$. In particular, X is a strongly symmetric Markov process whose reference measure is Lebesgue's measure on $\mathbb{R}^d$. Moreover, for every $\lambda \succ 0$ in $\mathbb{R}^N_+$, there exists a finite constant $A > 1$ such that for all $x, y \in \mathbb{R}^d$ with $\|x-y\| \leq \frac{1}{2}$,*

$$\frac{1}{A} \|x-y\|^{-d+N\alpha} \leq r_\lambda(x,y) \leq A\|x-y\|^{-d+N\alpha}.$$

Proof Throughout, we assume that $d > N\alpha$ and $\lambda \succ 0$ (in $\mathbb{R}^N_+$) are fixed.

The proof of this proposition is divided into two general parts: an upper bound on r_λ and a lower bound on r_λ. The verification of the remaining assertions are deferred to Exercise 4.1.2 below.

For the upper bound, we propose to prove more. Namely, we will show that there exists a finite constant $A_1 > 1$ such that for *all* $a \in \mathbb{R}^d$,

$$u_\lambda(a) \leq A_1 \|a\|^{-d+N\alpha}. \tag{2}$$

When $N = 1$ and $\|a\| \leq 1$, this follows from Lemma 3.4.1 of Chapter 10. We demonstrate equation (2) by induction on N. For all $a \in \mathbb{R}^d$, define $U^0(a) = 1$. Then, we iteratively define for all $n \in \{1, \ldots, N\}$,

$$U^n(a) = U^{n-1} \star u^1_{\lambda(n)}(a), \qquad a \in \mathbb{R}^d.$$

It is important to note that $U^1(a) = u^1_{\lambda(1)}(a)$ and $U^N(a) = u_\lambda(a)$. equation (2) clearly follows once we show that for every $n \in \{1, \ldots, N\}$, there exists a finite constant $A_2 > 1$ such that

$$U^n(a) \leq A_2 \|a\|^{-d+n\alpha}, \qquad a \in \mathbb{R}^d. \tag{3}$$

We have already observed that when $n = 1$, this holds by Lemma 3.4.1 of Chapter 10. Supposing the above holds for n replaced by any $k \leq n-1$, we will verify it for n. The proof (and, in fact, the inequality (3) itself) is very close to that of Lemma 3.3.1 of Chapter 3. Henceforth, we assume that equation (3) holds with n replaced by $n-1$, where $n \in \{1, \ldots, N\}$ and $d > N\alpha \geq n\alpha$. By the definition of a convolution and by the induction hypothesis, for all $a \in \mathbb{R}^d$,

$$U^n(a) = \int_{\mathbb{R}^d} U^{n-1}(b-a) u^1_{\lambda(n)}(b) \, db$$

$$\leq A_3 \int_{\mathbb{R}^d} \|b-a\|^{-d+(n-1)\alpha} \cdot \|b\|^{-d+\alpha} \, db,$$

for some finite constant $A_3 > 1$. We split this integral into three pieces,

$$U^n(a) \leq A_3 \{T_1 + T_2 + T_3\},$$

where

$$T_1 = \int_{b \in \mathbb{R}^d: \, \|b\| > 2\|a\|} \|b-a\|^{-d+(n-1)\alpha} \cdot \|b\|^{-d+\alpha} \, db,$$

$$T_2 = \int_{\substack{b \in \mathbb{R}^d: \|b\| < 2\|a\| \\ \|b-a\| \leq \frac{1}{2}\|a\|}} \|b-a\|^{-d+(n-1)\alpha} \cdot \|b\|^{-d+\alpha} \, db,$$

$$T_3 = \int_{\substack{b \in \mathbb{R}^d: \|b\| < 2\|a\| \\ \|b-a\| \geq \frac{1}{2}\|a\|}} \|b-a\|^{-d+(n-1)\alpha} \cdot \|b\|^{-d+\alpha} \, db,$$

and estimate each term in turn. For the b's in the integral of T_1, $\|b\| \leq \|b-a\| + \|a\| \leq \|b-a\| + \frac{1}{2}\|b\|$. In other words, $\|b-a\| \geq \frac{1}{2}\|b\|$. Thus,

$$T_1 \leq 2^{d-(n-1)\alpha} \int_{b \in \mathbb{R}^d: \, \|b\| > 2\|a\|} \|b\|^{-2d+n\alpha} \, db$$

$$= 2^{d-(n-1)\alpha} \|a\|^{-d+n\alpha} \int_{\xi \in \mathbb{R}^d: \, \|\xi\| > 2} \|\xi\|^{-2d+n\alpha} \, d\xi. \tag{4}$$

By polar coordinates calculations (Supplementary Exercise 7, Chapter 3), the above integral equals a constant times $\int_2^\infty x^{-2+n\alpha-1} \, dx$, which is finite, since $d > n\alpha$, for all $n \leq N$; this estimates T_1.

To estimate T_2, note that for all $b \in \mathbb{R}^d$ with $\|b\| \leq 2\|a\|$ and $\|b-a\| \leq \frac{1}{2}\|a\|$, $\|b\| \geq \|a\| - \|b-a\| \geq \frac{1}{2}\|a\|$. Thus,

$$T_2 \leq 2^{d-\alpha} \|a\|^{-d+\alpha} \int_{b \in \mathbb{R}^d: \, \|b-a\| \leq \|a\|/2} \|b-a\|^{-d+(n-1)\alpha} \, db$$

$$= 2^{d-\alpha} \|a\|^{-d+\alpha} \int_{b \in \mathbb{R}^d: \, \|b\| \leq \|a\|/2} \|b\|^{-d+(n-1)\alpha} \, db$$

$$= 2^{d-\alpha} \|a\|^{-d+n\alpha} \int_{\xi \in \mathbb{R}^d: \, \|\xi\| \leq \frac{1}{2}} \|\xi\|^{-d+(n-1)\alpha} \, d\xi, \tag{5}$$

and the latter integral is finite, as can be seen by once again calculating in polar coordinates; cf. also Lemma 3.1.2 of Chapter 3. Finally, we have

$$T_3 \leq 2^{d-(n-1)\alpha} \|a\|^{-d+(n-1)\alpha} \int_{b \in \mathbb{R}^d: \ \|b\|<2\|a\|} \|b\|^{-d+\alpha} \, db$$

$$= 2^{d-(n-1)\alpha} \|a\|^{-d+n\alpha} \int_{b \in \mathbb{R}^d: \ \|b\|<2} \|b\|^{-d+\alpha} \, db.$$

Since the above integral is finite, we have deduced the existence of a finite constant $A_4 > 1$ such that for all $a \in \mathbb{R}^d$, $U^n(a) \leq A_4 \|a\|^{-d+n\alpha}$. This proves (3), and the asserted upper bound (2) follows by induction.

The last part of our proof is concerned with verifying the lower bound on the λ-potential density. We will show that there exists a finite constant $A_5 > 1$ such that for all $a \in \mathbb{R}^d$ with $\|a\| \leq \frac{1}{2}$,

$$u_\lambda(a) \geq \frac{1}{A_5} \|a\|^{-d+N\alpha}.$$

The proposition follows upon letting $A = \max\{A_1, A_5\}$. We will show that for all $n \in \{1, \ldots, N\}$, there exists a finite constant $A_6 > 1$ such that for every $a \in \mathbb{R}^d$ with $\|a\| \leq \frac{1}{2}$,

$$U^n(a) \geq \frac{1}{A_6} \|a\|^{-d+n\alpha}. \tag{6}$$

This would finish the proof. When $n = 1$, (6) follows from the lower bound in Lemma 3.4.1, Chapter 10. Supposing (6) holds for n replaced by any $k \leq n - 1$, we will show it holds for n. By the induction hypothesis, there exists a finite constant $A_7 > 1$ such that whenever $\|a\| \leq \frac{1}{2}$,

$$U^n(a) = \int U^{n-1}(b-a) u^1_{\lambda(N)}(b) \, db$$

$$\geq \int_{\substack{b \in \mathbb{R}^d: \ \|b-a\| \leq \frac{1}{4}\|a\| \\ \|b\| < 1}} U^{n-1}(b-a) u^1_{\lambda(N)}(b) \, db$$

$$\geq \frac{1}{A_7} \int_{\substack{b \in \mathbb{R}^d: \ \|b-a\| \leq \frac{1}{4}\|a\| \\ \|b\| < 1}} \|b-a\|^{-d+(n-1)\alpha} \cdot \|b\|^{-d+\alpha} \, db.$$

Equation (6), and hence the proposition, follows from similar arguments used in estimating T_2 above; cf. Exercise 4.1.1 for details. □

Exercise 4.1.1 Complete the verification of equation (6). □

***Exercise* 4.1.2** Prove the remaining assertions of Proposition 4.1.1 by showing that X is a strongly symmetric Markov process whose reference measure is Lebesgue's measure on $\mathbb{R}^d$. □

***Exercise* 4.1.3** Prove that in the above setting, $r_\lambda(x,y) > 0$ for all $\lambda \succ 0$ in $\mathbb{R}_+^N$ and all $x, y \in \mathbb{R}^d$. □

In order to present our first application of Theorem 3.1.1, let ν denote Lebesgue's measure on $\mathbb{R}^d$. By Theorem 2.3.2, X is a strongly symmetric Lévy process with reference measure ν. By Exercise 4.1.3, all of the conditions of Corollary 3.1.1 are met. Therefore, we can deduce the following:

Theorem 4.1.1 *Suppose X is an $\mathbb{R}^d$-valued, N-parameter additive stable process with index $\alpha \in]0, 2]$ with $d > N\alpha$. If $E \subset \mathbb{R}^d$ is compact, then for all $x \notin E$,*

$$\mathbb{P}_x\big[\overline{X(\mathbb{R}_+^N)} \cap E \neq \varnothing\big] > 0 \iff \mathsf{Cap}_{d-N\alpha}(E) > 0.$$

In particular, by invoking Frostman's theorem (Theorem 2.2.1, Appendix C), we can deduce the following multiparameter analogue of Theorem 3.5.1, Chapter 10.

Corollary 4.1.1 *If $E \subset \mathbb{R}^d$ is compact, then for all $x \notin E$,*

$$\dim(E) > d - N\alpha \implies \mathbb{P}_x\big\{\overline{X(\mathbb{R}_+^N)} \cap E \neq \varnothing\big\} > 0,$$
$$\dim(E) < d - N\alpha \implies \mathbb{P}_x\big\{\overline{X(\mathbb{R}_+^N)} \cap E \neq \varnothing\big\} = 0.$$

where dim *denotes Hausdorff dimension.*

The proof of Theorem 4.1.1 has another corollary, which we state as the following important exercise.

***Exercise* 4.1.4** (Hard) For all $\beta, M > 0$, there exist constants A_1 and A_2 such that for all compact sets $E \subset [-M, M]^d$,

$$A_1 \mathsf{Cap}_\beta(E) \leq \mathsf{Cap}_\beta^{\mathsf{ac}}(E) \leq A_2 \mathsf{Cap}_\beta(E),$$

where Cap_β denotes the β-dimensional Bessel–Riesz capacity and $\mathsf{Cap}_\beta^{\mathsf{ac}}$ designates the absolutely continuous capacity with respect to the gauge function $x \mapsto \|x\|^{-\beta}$, where absolute continuity is understood to hold with respect to Lebesgue's measure; see Section 2.2, Appendix D, for the definition of absolutely continuous capacities. (In fact, using Fourier analysis, one can show that $A_1 = A_2 = 1$.) Prove also that the above has an analogue when $\beta = 0$.
(HINT: Prove the following refinement of Proposition 4.1.1: For all $M > 0$, there are positive and finite constants A_1 and A_2 such that

$$A_1\Big(1 \vee \|x - y\|^{-d+N\alpha}\Big) \leq r_\lambda(x,y) \leq A_2\Big(1 \vee \|x-y\|^{-d+N\alpha}\Big),$$

for all $x, y \in [-M, M]^d$ and all $\lambda \succ 0$ in $\mathbb{R}_+^N$.) □

4.2 Intersections of Independent Processes

To present another application of Corollary 3.1.1 of Theorem 3.1.1, we investigate the problem of when the trajectories of N independent, strongly symmetric multiparameter Feller processes intersect. To this end, we suppose that $X^1, \ldots, X^N$ are N independent, $\mathbb{R}^d$-valued, strongly symmetric and multiparameter Feller processes. To be concrete, we suppose that for each $1 \leq i \leq N$, X^i has M_i parameters, as well as a λ-potential density r_λ^i for every $\lambda \succ 0$ in $\mathbb{R}_+^{M_i}$. We shall also denote the reference measure of X^i by ν_i.

As in Section 2.4, we define $X = X^1 \otimes \cdots \otimes X^N$ to be the M-parameter, $\mathbb{R}^{Nd}$-valued stochastic process where $M = M_1 + \cdots + M_N$. By Theorem 2.4.1, X is a strongly symmetric, $\mathbb{R}^{Nd}$-valued, M-parameter Feller process. Moreover, for all $\lambda \succ 0$ (in $\mathbb{R}_+^M$), X has a λ-potential density r_λ that is a symmetric, proper gauge function on $\mathbb{R}^d \times \mathbb{R}^d$ and is described by the following:

$$r_\lambda(x, y) = \prod_{j=1}^{N} r_{\lambda_j}^j(x_j, y_j), \qquad x, y \in \mathbb{R}^{Nd}, \; \lambda \succ 0 \text{ in } \mathbb{R}_+^M,$$

where any $x \in \mathbb{R}^{Nd}$ is written in direct product notation as $x_1 \otimes \cdots \otimes x_N$, where $x_i \in \mathbb{R}^d$ for all $1 \leq i \leq N$; similarly, $\lambda \succ 0$ (in $\mathbb{R}_+^M$) is written as $\lambda_1 \otimes \cdots \otimes \lambda_N$, where $\lambda_i \succ 0$ (in $\mathbb{R}_+^{M_i}$) for all $1 \leq i \leq N$. By invoking Theorem 4.1.1, we can conclude that for all compact sets $F \in \mathbb{R}^{Nd}$,

$$\mathbb{P}_\nu\{\overline{X(\mathbb{R}_+^M)} \cap F \neq \varnothing\} > 0 \iff \mathcal{C}_{r_\lambda}(F) > 0.$$

Equivalently, $\mathbb{P}_\nu\{\overline{X(\mathbb{R}_+^M)} \cap F \neq \varnothing\} > 0$ if and only if there exists a probability measure $\sigma \in \mathcal{P}(F)$ such that $\mathcal{E}_{r_\lambda}(\sigma) < \infty$ (why?). Now we turn to the question of intersections. Note that for any compact set $E \subset \mathbb{R}^d$,

$$\bigcap_{i=1}^{N} \overline{X^i(\mathbb{R}_+^{M_i})} \cap E \neq \varnothing \iff \overline{X(\mathbb{R}_+^M)} \cap D_E \neq \varnothing,$$

where $D_E = \{x \otimes \cdots \otimes x : x \in E\}$. In words, the (closed) ranges of the processes $X^1, \ldots, X^N$ intersect in some set $E \subset \mathbb{R}^d$ if and only if the M-parameter process X hits D_E. Hence, we can conclude that $\bigcap_{i=1}^{N} \overline{X^i(\mathbb{R}_+^{M_i})} \cap E$ is nonvoid if and only if there exists a probability measure $\sigma \in \mathcal{P}(D_E)$ such that $\mathcal{E}_{r_\lambda}(\sigma) < \infty$. On the other hand, the set D_E is special; any probability measure σ on D_E *must* be of the form

$$\sigma(dx_1 \otimes \cdots \otimes dx_N) = \mu(dx_1)\delta_{x_1}(dx_2) \cdots \delta_{x_1}(dx_N), \qquad x_i \in \mathbb{R}^d, \; 1 \leq i \leq N,$$

where δ_a denotes a point mass at $a \in \mathbb{R}^d$ and $\mu \in \mathcal{P}(E)$ (why?). It is now easy to check that when σ is of the above form,

$$\mathcal{E}_{r_\lambda}(\sigma) = \iint \prod_{j=1}^N r^j(x,y)\,\mu(dx)\,\mu(dy) = \mathcal{E}_{\prod_{j=1}^N r^j_{\lambda_j}}(\mu),$$

where $\lambda = \lambda_1 \otimes \cdots \otimes \lambda_N$ for $\lambda_i \succ 0$ (in $\mathbb{R}^{M_i}_+$), $1 \leq i \leq N$.

We have proven the following result; see Benjamini et al. (1995), Fitzsimmons and Salisbury (1989), Hirsch and Song (1994, 1995a, 1995c, 1995d), Hirsch (1995), Khoshnevisan (1997b, 1999), Khoshnevisan and Shi (1999), Pemantle et al. (1996), Peres (1996a), Ren (1990), and Salisbury (1992, 1996) for a collection of related results.

Theorem 4.2.1 *Consider N independent, strongly symmetric, $\mathbb{R}^d$-valued Feller processes $X^1, \ldots, X^N$ with potential densities $r^1, \ldots, r^N$. If X^i has M_i parameters and has reference measure ν_i $(1 \leq i \leq N)$, then for any compact set $E \subset \mathbb{R}^d$,*

$$\mathbb{P}_\nu\Big\{\bigcap_{j=1}^N \overline{X^i(\mathbb{R}^{M_i}_+)} \cap E \neq \varnothing\Big\} > 0 \iff \mathcal{C}_g(E) > 0,$$

where $\nu = \nu_1 \times \cdots \times \nu_N$ and $g(x,y) = \prod_{j=1}^N r^j_{\lambda_j}(x,y)$ for any $\lambda_i \succ 0$ (in $\mathbb{R}^{M_i}_+$, $1 \leq i \leq N$) and for all $x, y \in \mathbb{R}^d$.

The following important corollary is proved in similar fashion to Corollary 3.1.1.

Corollary 4.2.1 *Consider the processes of Theorem 4.2.1, and fix some compact set $E \subset \mathbb{R}^{Nd}$. Suppose, in addition, that there exists some $\lambda \succ 0$ in $\prod_{i=1}^N \mathbb{R}^{M_i}$ and some $x_0 \in E^\complement$ such that for all $y \in \mathbb{R}^d$, $r_\lambda(x_0, y) > 0$. Then,*

$$\mathbb{P}_{x_0}\Big(\bigcap_{i=1}^N \overline{X^i(\mathbb{R}^{M_i}_+)} \cap E \neq \varnothing\Big) > 0 \iff \mathcal{C}_g(E) > 0.$$

Here, $g(x,y) = \prod_{j=1}^N r_{\lambda_j}(x,y)$, where $\lambda = \lambda_1 \otimes \cdots \otimes \lambda_N$ with $\lambda_j \succ 0$ in $\mathbb{R}^{M_j}_+$ and $x, y \in \mathbb{R}^d$.

Exercise **4.2.1** Prove Corollary 4.2.1.
(HINT: This is an exercise in Laplace transforms; cf. also Appendix B.) □

In the next subsection we will try to understand this result via a number of examples that are presented to various degrees of generality.

4.3 Dvoretzky–Erdős–Kakutani Theorems

Consider two independent $\mathbb{R}^d$-valued, isotropic Lévy processes X^1 and X^2 with indices α_1 and α_2, respectively. We shall assume that X^1 and X^2 are both transient. That is, we assume that

$$d > \alpha_1 \vee \alpha_2; \tag{1}$$

cf. Theorem 1.3.1, Chapter 10.

Let p^1 and p^2 denote their respective transition densities (with respect to Lebesgue's measure on $\mathbb{R}^d$). That is, for each $\ell = 1, 2$,

$$p_t^\ell(x, y) = (2\pi)^{-d} \int_{\mathbb{R}^d} \exp\left\{ -i\xi \cdot (y - x) - \frac{t\|\xi\|^{\alpha_\ell}}{2} \right\} d\xi, \quad x, y \in \mathbb{R}^d,\ t \geq 0;$$

cf. Example 3, Section 4.3 of Chapter 10. By replacing ξ by $-\xi$ in the above integral(s), we see that for every $t \geq 0$, for all $x, y \in \mathbb{R}^d$, and for $\ell = 1$ or 2, $p_t^\ell(x, y) = p_t^\ell(y, x)$. Consequently, we can appeal to the example of Section 1.4, to see that X^1 and X^2 are strongly symmetric, multiparameter Feller processes and that they both have Lebesgue's measure for their reference measures. Let $X = X^1 \otimes X^2$ be the corresponding product process in the sense of Section 2.2. By Proposition 2.2.1, X is a two-parameter Feller process whose reference measure is Lebesgue's measure on $\mathbb{R}^{2d}$. Combining these observations, together with Corollary 4.2.1 above and with Lemma 3.4.1 of Chapter 10, we can conclude the following:

Theorem 4.3.1 *Let X^1 and X^2 denote two independent, $\mathbb{R}^d$-valued isotropic stable Lévy processes with indices α_1 and α_2, respectively. Under the transience condition (1), for any compact set $E \subset \mathbb{R}^d$, and for all $x = x_1 \otimes x_2$ with distinct $x_1, x_2 \in \mathbb{R}^d$,*

$$\mathbb{P}_x\left[\overline{X^1(\mathbb{R}_+)} \cap \overline{X^2(\mathbb{R}_+)} \cap E \neq \varnothing\right] > 0 \iff \mathsf{Cap}_{2d-\alpha_1-\alpha_2}(E) > 0.$$

***Exercise* 4.3.1** Verify the details of the proof of Theorem 4.3.1. □

As it was mentioned in Chapter 8, $\mathbb{P}_x$ should the thought of as the conditional measure of the process X, given that $X_0^1 = x_1 \in \mathbb{R}^d$ and $X_0^2 = x_2 \in \mathbb{R}^d$, where $x = x_1 \otimes x_2$, in direct product notation. In words, what we have shown, thus far, is that the $(2d - \alpha_1 - \alpha_2)$-dimensional Bessel–Riesz capacity of E is positive if and only if the intersection of the (closed) ranges of X^1 and X^2 intersects E, as long as X starts outside E. We apply Frostman's theorem (Theorem 2.2.1, Appendix C) to Theorem 4.3.1 to deduce the following:

Corollary 4.3.1 *In the setup of Theorem 4.3.1,*

$$\dim(E) > 2d - \alpha_1 - \alpha_2 \implies \mathbb{P}_x\left[\overline{X^1(\mathbb{R}_+)} \cap \overline{X^2(\mathbb{R}_+)} \cap E \neq \varnothing\right] > 0,$$
$$\dim(E) < 2d - \alpha_1 - \alpha_2 \implies \mathbb{P}_x\left[\overline{X^1(\mathbb{R}_+)} \cap \overline{X^2(\mathbb{R}_+)} \cap E \neq \varnothing\right] = 0.$$

A particularly important case is obtained if we let E approximate $\mathbb{R}^d$. In this case, we conclude that if $x = x_1 \otimes x_2$ with distinct $x_1, x_2 \in \mathbb{R}^d$,

$$d < \alpha_1 + \alpha_2 \implies \mathbb{P}_x\left[\overline{X^1(\mathbb{R}_+)} \cap \overline{X^2(\mathbb{R}_+)} \neq \varnothing\right] > 0,$$
$$d > \alpha_1 + \alpha_2 \implies \mathbb{P}_x\left[\overline{X^1(\mathbb{R}_+)} \cap \overline{X^2(\mathbb{R}_+)} \neq \varnothing\right] = 0,$$

with the added proviso that $d > \alpha_1 \vee \alpha_2$; cf. equation (1). In particular, we can take $\alpha_1 = \alpha_2 = 2$ to find that the trajectories of two independent Brownian motions in $\mathbb{R}^d$ intersect if $d = 3$, but not if $d \geq 5$, and this holds whenever their starting points are distinct. This characterization leaves out the cases $d = 2$ and $d = 4$. It also omits $d = 1$, which is the simplest case; cf. Exercise 4.3.3 below for the treatment of the one-dimensional case. We concentrate on $d = 2, 4$ next.

Let us first fix $d = 4$ and use the notation of Theorem 4.3.1. The argument that led to Theorem 4.3.1 shows that for all compact sets $E \subset \mathbb{R}^4$, and all $x = x_1 \otimes x_2$ with distinct $x_1, x_2 \in \mathbb{R}^4$,

$$\mathbb{P}_x\left[\overline{X^1(\mathbb{R}_+)} \cap \overline{X^2(\mathbb{R}_+)} \cap E \neq \varnothing\right] > 0 \iff \mathsf{Cap}_4(E) > 0.$$

Let E approximate $\mathbb{R}^4$ arbitrarily well, and recall from Taylor's theorem that $\mathsf{Cap}_4(\mathbb{R}^4) = 0$; cf. Corollary 2.3.1 of Appendix C. This shows that if they have distinct starting points, the trajectories of two independent Brownian motions in $\mathbb{R}^4$ do not intersect.

To conclude our discussion of trajectorial intersections of two independent Brownian motions, we need to address the case $d = 2$. Let X^1 and X^2 be two independent 2-dimensional Brownian motions. Lemma 3.1.2 of Chapter 10 can be used to show that for any compact set $E \subset \mathbb{R}^2$, the following are equivalent:

- For all $x = x_1 \otimes x_2$ with distinct $x_1, x_2 \in \mathbb{R}^2$,

$$\mathbb{P}_x\left[\overline{X^1(\mathbb{R}_+)} \cap \overline{X^2(\mathbb{R}_+)} \cap E \neq \varnothing\right] > 0.$$

- There exists a probability measure $\mu \in \mathcal{P}(E)$ such that

$$\iint \left|\ln_+\left(\frac{1}{\|a-b\|}\right)\right|^2 \mu(da)\,\mu(db) < \infty.$$

***Exercise* 4.3.2** Prove, in detail, that the above are equivalent. □

We can apply this equivalence to $E = \mathbb{R}^2$ (why? this is not completely obvious, since $\mathbb{R}^2$ is not compact). We note that there exists $\varepsilon_0 \in \,]0,1[$ such that for all $\zeta \in [0, \varepsilon_0]$, $|\ln_+(1/\zeta)|^2 \leq \zeta^{-1}$. Thus, for all $\mu \in \mathcal{P}(\mathbb{R}^2)$,

$$\iint \left|\ln_+\left(\frac{1}{\|a-b\|}\right)\right|^2 \mu(da)\,\mu(db)$$
$$\leq 1 + \iint_{a,b \in \mathbb{R}^2:\ \|a-b\| \leq \varepsilon_0} \|a-b\|^{-1} \mu(da)\,\mu(db)$$
$$\leq 1 + \mathsf{Energy}_1(\mu).$$

On the other hand, since $\mathcal{H}_2(\mathbb{R}^2) = +\infty$, by Frostman's theorem, $\mathsf{Cap}_1(\mathbb{R}^2) > 0$; cf. Theorem 2.2.1 of Appendix C. In particular, there must exist a probability measure μ on $\mathbb{R}^2$ such that $\mathsf{Energy}_1(\mu) < \infty$. By the above estimate, $\iint |\ln_+(1/\|a-b\|)|^2\,\mu(da)\,\mu(db) < \infty$ for such a $\mu \in \mathcal{P}(\mathbb{R}^2)$. That is, whenever they have distinct starting points, the trajectories of two independent Brownian motions in $\mathbb{R}^2$ intersect with positive probability.

Combining the above discussions, we obtain the following result of Dvoretzky et al. (1954).

Theorem 4.3.2 (The Dvoretzky–Erdős–Kakutani Theorem) *If X^1 and X^2 denote two independent d-dimensional Brownian motions with distinct starting points x_1 and x_2, respectively, then*

$$\mathbb{P}_{x_1 \otimes x_2}\left[\overline{X^1(\mathbb{R}_+)} \cap \overline{X^2(\mathbb{R}_+)} \neq \varnothing\right] > 0 \iff d \leq 3.$$

We have not discussed the easiest case in the above theorem, which is when $d = 1$. This is worked out in the following exercise.

***Exercise* 4.3.3** Verify that the trajectories of two real-valued, independent Brownian motions intersect with positive probability. □

Supplementary Exercise 3 discusses other related results of Dvoretzky et al. (1954).

4.4 Intersecting an Additive Stable Process

Let X^1 denote an $\mathbb{R}^d$-valued, isotropic stable Lévy process with index $\alpha_1 \in \,]0,2]$. Consider an independent process X^2, which denotes an $\mathbb{R}^d$-valued, N-parameter additive stable process of index $\alpha_2 \in \,]0,2]$; cf. Section 4.1 for the latter. Throughout this subsection we will assume the following:

$$d > N\alpha_2. \tag{1}$$

Theorem 4.4.1 *Suppose X^1 is an index-α_1 isotropic Lévy process on $\mathbb{R}^d$, and X^2 is an independent, N-parameter, index-α_2 additive stable process.*

If equation (1) holds, then for all compact sets $E \subset \mathbb{R}^d$ and all distinct $x, y \in \mathbb{R}^d$ with $x \in E^\complement$,

$$\mathbb{P}_{x \otimes y}\left[\overline{X^1(\mathbb{R}_+)} \cap \overline{X^2(\mathbb{R}_+^N)} \cap E \neq \varnothing\right] > 0 \iff \mathcal{C}_\kappa(E) > 0,$$

where

$$\kappa(u,v) = \begin{cases} \ln_+\left(\dfrac{1}{\|u-v\|}\right)\|u-v\|^{-d+N\alpha_2}, & \text{if } d = \alpha_1, \\ \|u-v\|^{-2d+\alpha_1+N\alpha_2}, & \text{if } d > \alpha_1. \end{cases}$$

The easier case $d < \alpha_1$ is handled in the following exercise.

Exercise 4.4.1 Show that when $d < \alpha_1$, the above intersection probability is always positive. □

Proof of Theorem 4.4.1 Recall Lemmas 3.4.1 and 3.4.2, both from Chapter 10. In particular, recall that X^1 and X^2 are strongly symmetric and their reference measure is Lebesgue's measure on $\mathbb{R}^d$. Let r^1 and r^2 denote the resolvent densities of X^1 and X^2, respectively. By applying Lemmas 3.1.1 and 3.1.2 of Chapter 10, when $d \geq \alpha_1$, we can find a finite constant $A > 1$ such that whenever $x, y \in \mathbb{R}^d$ satisfy $\|x-y\| \leq \frac{1}{2}$ (say), then for any $\lambda_1 \in \mathbb{R}_+$ and for all $\lambda_2 \succ 0$ in $\mathbb{R}_+^N$,

$$\frac{1}{A}\kappa(x,y) \leq r_{\lambda_1}^1(x,y)\, r_{\lambda_2}^2(x,y) \leq A\kappa(x,y).$$

Consequently, $\mathcal{C}_{r_{\lambda_1}^1 r_{\lambda_2}^2}(E) > 0 \iff \mathcal{C}_\kappa(E) > 0$. Thus, Corollary 4.2.1 implies the result. □

4.5 Hausdorff Dimension of the Range of a Stable Process

Let $X = (X_t;\ t \geq 0)$ denote an $\mathbb{R}^d$-valued isotropic stable Lévy process of index $\alpha \in {]0,2]}$. Our present goal is to study the "size" of the range $\overline{X(\mathbb{R}_+)}$ of the process, viewed as a d-dimensional "random" set.[9] Let $(\Omega, \mathcal{G}, \mathbb{Q})$ denote the underlying probability space and let $\mathbb{P}_x$ $(x \in \mathbb{R}^d)$ denote the distribution of $X + x$, as before. We now introduce an auxilliary probability space $(\Omega', \mathcal{G}', \mathbb{Q}')$ large enough to support the construction of an $\mathbb{R}^d$-valued, N-parameter additive stable process $X' = (X'_t;\ t \in \mathbb{R}_+^N)$ of index $\alpha' \in {]0,2]}$. Let $\mathbb{P}'_x$ $(x \in \mathbb{R}^d)$ denote the distribution of $X' + x$, as before. Now, we combine things to make X and X' independent. That is, define $(\Omega^0, \mathcal{G}^0, \mathbb{Q}^0)$ to be the product probability space given by $\Omega^0 = \Omega \times \Omega'$, $\mathcal{G}^0 = \mathcal{G} \times \mathcal{G}'$ and $\mathbb{Q}^0 = \mathbb{Q} \times \mathbb{Q}'$. For all $x, y \in \mathbb{R}^d$, we can define a measure $\mathbb{P}^0_{x \otimes y}$ on $\mathcal{G}^0$ by $\mathbb{P}^0_{x \otimes y} = \mathbb{P}_x \times \mathbb{P}'_y$ (as a product measure). We shall extend the definition of

---

[9] For a rigorous definition of random sets, see Section 4.7.

the processes X and X' to the probability space $(\Omega^0, \mathcal{G}^0, \mathbb{Q}^0)$ in the natural way. That is, for all $\omega^0 \in \Omega^0$ of the form $\omega^0 = \omega \otimes \omega'$, where $\omega \in \Omega$ and $\omega' \in \Omega'$,

$$X_s(\omega^0) = X_s(\omega), \qquad s \geq 0,$$
$$X'_t(\omega^0) = X_t(\omega'), \qquad t \in \mathbb{R}_+^N.$$

The only thing that the above heavy-handed notation does is to set up some rigorous machinery for the statement that X and X' are two independent additive stable processes on $\mathbb{R}^d$ with 1 and N parameters, respectively, that have indices α and α', respectively. Moreover, for all $x, x' \in \mathbb{R}^d$, X and X' start at x and x', respectively, under the measure $\mathbb{P}^0_{x \otimes x'}$.

Having established the requisite notation, we can employ Theorem 4.4.1 to see that for all compact sets $E \subset \mathbb{R}^d$ and for all distinct $x, x' \in \mathbb{R}^d$ with $x \in E^{\complement}$,

$$\mathbb{P}^0_{x \otimes x'}\left\{\overline{X(\mathbb{R}_+)} \cap \overline{X'(\mathbb{R}_+^N)} \cap E \neq \varnothing\right\} > 0 \iff \mathsf{Cap}_{2d-\alpha-N\alpha'}(E) > 0,$$

as long as $d > \alpha \wedge N\alpha'$. Equivalently, by Fubini's theorem, we can first condition on the entire process X to see that

$$\mathbb{P}'_{x'}\left\{\overline{X'(\mathbb{R}_+^N)} \cap \overline{X(\mathbb{R}_+)} \cap E \neq \varnothing \,\middle|\, X\right\} > 0, \ \mathbb{P}_x\text{-a.s.} \qquad (1)$$
$$\iff \mathsf{Cap}_{2d-\alpha-N\alpha'}(E) > 0.$$

On the other hand,

$$\mathbb{P}'_{x'}\left\{\overline{X'(\mathbb{R}_+^N)} \cap \overline{X(\mathbb{R}_+)} \cap E \neq \varnothing \,\middle|\, X\right\} = \lim_{t \to \infty} \mathbb{P}'_{x'}\left\{\overline{X'(\mathbb{R}_+^N)} \cap F_t \neq \varnothing \,\middle|\, X\right\},$$

$\mathbb{P}_x$-a.s., where $F_t = E \cap \overline{X([0,t])}$ is a compact set in $\mathbb{R}^d$. Furthermore, by Theorem 4.1.1 above,

$$\mathbb{P}'_{x'}\left\{\overline{X'(\mathbb{R}_+^N)} \cap F_t \neq \varnothing \,\middle|\, X\right\} > 0, \ \mathbb{P}_x\text{-a.s.} \iff \mathbb{E}_x\left[\mathsf{Cap}_{d-N\alpha'}(F_t)\right] > 0. \quad (2)$$

We have used the assumption that $x \notin E$. Finally, note that as $t \uparrow \infty$, $F_t \uparrow E \cap \overline{X(\mathbb{R}_+)}$. Thus, by Exercise 1.1.5 of Appendix D, the following holds for all $\omega \in \Omega$:

$$\sup_t \mathsf{Cap}_{d-N\alpha'}(F_t) > 0 \iff \mathsf{Cap}_{d-N\alpha'}\{E \cap \overline{X(\mathbb{R}_+^N)}\} > 0.$$

(Why?) Consequently, we can combine equation (1) with equation (2) to deduce the following:

$$\mathbb{E}_x\left[\mathsf{Cap}_{d-N\alpha'}\{E \cap \overline{X(\mathbb{R}_+)}\}\right] > 0 \iff \mathsf{Cap}_{2d-\alpha-N\alpha'}(E) > 0.$$

(Why is the expression under $\mathbb{E}_x$ measurable?) The neat feature of this formula is the complete absence of the process X'. In summary, equation

(2) holds for all positive integers N and for all choices of $\alpha' \in \,]0,2]$. Note that in this regime, $\beta = d - N\alpha'$ is an arbitrary real number in $]0,d[$. In other words, we have shown that whenever $d > \alpha$ and $x \notin E$, for any $\beta \in \,]0,d[$,

$$\mathbb{E}_x\left[\mathsf{Cap}_\beta\{E \cap \overline{X(\mathbb{R}_+)}\}\right] > 0 \iff \mathsf{Cap}_{d-\alpha+\beta}(E) > 0. \qquad (3)$$

This suggests the validity of the following computation of the size of the range of an isotropic stable process. It was discovered in McKean (1955b).

Theorem 4.5.1 (McKean's Theorem) *Let X denote an $\mathbb{R}^d$-valued, isotropic stable Lévy process of index $\alpha \in \,]0,2]$. For any $x \in \mathbb{R}^d$, $\mathbb{P}_x\{\dim(\overline{X(\mathbb{R}_+)}) = \alpha \wedge d\} = 1$.*

When applied to the special case $\alpha = d = 2$, Theorem 4.5.1 states that the range of 2-dimensional Brownian motion is a (random) set whose Hausdorff dimension is 2, a.s. On the other hand, by Supplementary Exercise 3, Chapter 10, the range of 2-dimensional Brownian motion has zero 2-dimensional Lebesgue's measure and, equivalently, zero 2-dimensional Hausdorff measure; cf. Lemma 1.1.1 of Appendix C.

In summary, we have shown that the $\mathcal{H}_s$-measure of the range of 2-dimensional Brownian motion is 0 when $s = 2$, while it is positive when $s < 2$. In fact, it is possible to show a little more:

Exercise 4.5.1 Show that for all choices of $s < 2$, $\mathcal{H}_s\{B(\mathbb{R}_+)\} = +\infty$, almost surely. □

Figure 11.1 shows a realization of a portion of the range of 2-dimensional Brownian motion.

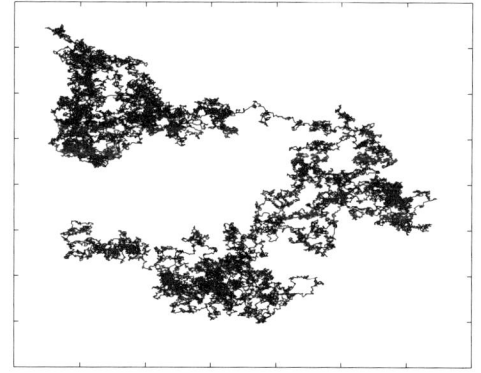

Figure 11.1: $B([0,1])$, where B is 2-dimensional Brownian motion

Proof We will fix some $x \in \mathbb{R}^d$ and first prove the theorem for $d > \alpha$. For all $m \geq 1$ and all $x \in \mathbb{R}^d$, let $\Gamma_m(x)$ denote the following compact set in $\mathbb{R}^d$:

$$\Gamma_m(x) = \{y \in \mathbb{R}^d : m^{-1} \leq |x-y| \leq m\}.$$

Note that $x \notin \Gamma_m(x)$, no matter what the value of $m \geq 1$ is.

Recall that in integer dimensions, Hausdorff's measure and Lebesgue's measure agree (Theorem 1.1.1, Appendix C). Thus, we can apply Frostman's theorem (Theorem 2.2.1, Appendix C) to see that for all $\beta \in \,]\alpha,d[$

and for all $m > 0$ $\mathsf{Cap}_{d-\alpha+\beta}(\Gamma_m(x)) = 0$. By equation (3),

$$\mathbb{E}_x\left[\mathsf{Cap}_\beta\{\Gamma_m(x) \cap \overline{X(\mathbb{R}_+)}\}\right] = 0, \qquad \forall \beta \in \,]\alpha, d[\,.$$

By Exercise 1.1.5 of Appendix D, and by Lebesgue's monotone convergence theorem, we can let $m \uparrow \infty$ to conclude that

$$\mathbb{E}_x\left[\mathsf{Cap}_\beta\{\overline{X(\mathbb{R}_+)}\}\right] = 0, \qquad \forall \beta \in \,]\alpha, d[\,.$$

Consequently,

$$\mathbb{P}_x\left(\mathsf{Cap}_\beta\{\overline{X(\mathbb{R}_+)}\} = 0 \text{ for all } \beta \in \,]\alpha, d[\, \cap \mathbb{Q}_+\right) = 1,$$

from which we get

$$\mathbb{P}_x\left(\dim\{\overline{X(\mathbb{R}_+)}\} \le \alpha\right) = 1. \tag{4}$$

A similar analysis shows that whenever $\beta \in \,]\alpha, d[$,

$$\mathbb{E}_x\left[\mathsf{Cap}_\beta\left(\overline{X(\mathbb{R}_+)}\right)\right] > 0.$$

That is, for any $\beta \in \,]0, \alpha[$, there necessarily exists some $t > 0$ such that

$$\mathbb{E}_x\left[\mathsf{Cap}_\beta\left(\overline{X([t, \infty[)}\right)\right] > 0$$

(Why?) By scaling (Lemma 1.3.1, Chapter 10), $\mathsf{Cap}_\beta\{\overline{X([t, \infty[)}\}$ has the same distribution as $\mathsf{Cap}_\beta\{t^{1/\alpha}\overline{X([1, \infty[)}\}$, which, thanks to Exercise 2.2.2 of Appendix C, equals $t^{-\beta/\alpha}\mathsf{Cap}_\beta\{\overline{X([1, \infty[)}\}$. The above discussion shows that (with positive $\mathbb{P}_x$-probability), $\mathsf{Cap}_\beta\{\overline{X([t, \infty[)}\} > 0$, for some $t > 0$ if and only if $\mathsf{Cap}_\beta\{\overline{X([t, \infty[)}\} > 0$, for all $t > 0$. Moreover, the $\mathbb{P}_x$-probability that $\mathsf{Cap}_\beta\{\overline{X([t, \infty[)}\} > 0$, is independent of the choice of $t > 0$. In particular, we see that for all $\beta \in \,]0, \alpha[$,

$$\lim_{t \to \infty} \mathbb{P}_x\left(\mathsf{Cap}_\beta\{\overline{X([t, \infty[)}\} > 0\right) = \mathbb{P}_x\left(\mathsf{Cap}_\beta\{\overline{X([t, \infty[\,)}\} > 0, \forall t > 0\right) > 0,$$

by monotonicity. The terms inside the second probability above are measurable with respect to the tail σ-field of X. Thus, by Kolmogorov's 0-1 law, the above probability is 1; cf. Supplementary Exercise 10 of Chapter 8. In particular, $\mathbb{P}_x$-a.s.,

$$\mathsf{Cap}_\beta\{\overline{X(\mathbb{R}_+)}\} > 0, \,\forall \beta \in \,]0, \alpha[\, \cap \mathbb{Q}_+.$$

By Frostman's theorem, $\dim\{\overline{X(\mathbb{R}_+)}\} \ge \alpha$, $\mathbb{P}_x$-a.s., which, in light of equation (4), proves the assertion of the theorem when $d > \alpha$.

Next, we prove the theorem when $d = \alpha$. Of course, since d is an integer, this means that X is either 2-dimensional Brownian motion ($\alpha = d = 2$) or a 1-dimensional Cauchy process ($\alpha = d = 1$); cf. Example (c), Section

1.3 of Chapter 10. We will concentrate on the harder case $d = \alpha$. When $d < \alpha$, this follows from the fact that X hits any singleton (proved in Supplementary Exercise 7 of Chapter 10).

When $d = \alpha$, our proof of equation (3) can be used to show that for all $\beta \in \,]0, d[$ and all $z \in E^{\complement}$,

$$\mathbb{E}_z\left[\mathsf{Cap}_\beta(E \cap \overline{X(\mathbb{R}_+)})\right] > 0 \iff \mathcal{C}_\kappa(E) > 0,$$

where $\kappa(x,y) = \|x-y\|^{-\beta}\ln_+\left(1/\|x-y\|\right)$, $x, y \in \mathbb{R}^d$. The remainder of the proof when $d = \alpha$ is carried out in similar fashion to the one in the case $d > \alpha$. □

4.6 Extension to Additive Stable Processes

We now extend the results of Section 4.5 to additive stable processes. Let $X = (X_t;\, t \in \mathbb{R}_+^N)$ denote an N-parameter, $\mathbb{R}^d$-valued additive stable process with index $\alpha \in \,]0, 2]$. When $\alpha = 2$, X is said to be **additive Brownian motion**. Analogously to the proof of (3), we can prove the following: If $d \geq N\alpha$, then for any $\beta \in \,]0, d[$ and for all compact sets $E \subset \mathbb{R}^d$ and all d-dimensional vectors $x \notin E$,

$$\mathbb{E}_x\left[\mathsf{Cap}_\beta\{E \cap \overline{X(\mathbb{R}_+^N)}\}\right] > 0 \iff \mathsf{Cap}_{d-N\alpha+\beta}(E) > 0. \tag{1}$$

Exercise **4.6.1** Fill in the details of the proof that when $d \geq N\alpha$, equation (1) holds. □

Exercise **4.6.2** Prove that when $d < N\alpha$, the $\mathbb{E}_x$-expectation of Lebesgue's measure of the range of X is positive. From this, deduce that in this case, $\mathrm{Leb}\left\{\overline{X(\mathbb{R}_+^N)}\right\} = +\infty$, almost surely. □

Consequently, we can deduce the following N-parameter extension of Theorem 4.5.1:

Theorem 4.6.1 *Let X denote an $\mathbb{R}^d$-valued, N-parameter additive stable process of index $\alpha \in \,]0, 2]$. Then, for any $x \in \mathbb{R}^d$, with $\mathbb{P}_x$-probability one,*

$$\dim\left\{\overline{X(\mathbb{R}_+^N)}\right\} = N\alpha \wedge d.$$

Exercise **4.6.3** Prove Theorem 4.6.1. □

In particular, we can let $\alpha \equiv 2$ to see that whenever $d \leq 2N$, the Hausdorff dimension of the range of additive Brownian motion is a.s. equal to d. Figure 4.6 shows a part of the range of 2-parameter, $\mathbb{R}$-valued additive Brownian motion. To be more precise, it shows the result of a simulation of of $[0,1]^2 \ni s \subset \mathbb{R}_+^2$ versus $X_s \in \mathbb{R}$.

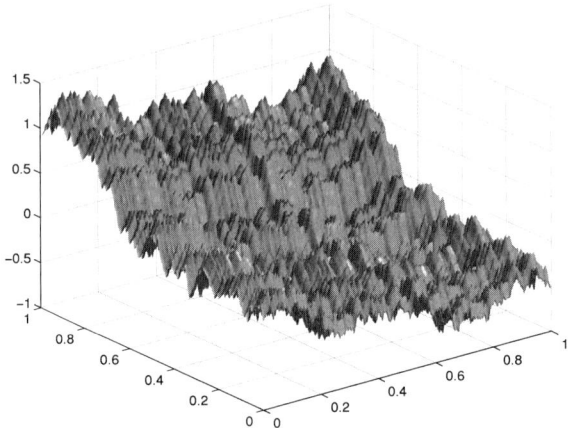

figure 11.2: additive Brownian motion ($N = 2, d = 1$) plotted against time

Exercise 4.6.2 above used Theorem 4.1.1 to conclude that whenever $d < 2N$, then $\mathrm{Leb}\{\overline{X(\mathbb{R}_+^N)}\} = +\infty$, a.s. That is, when $d < 2N$, the range has full d-dimensional Lebesgue measure (and hence, full d-dimensional Hausdorff measure, by Lemma 1.1.1 of Appendix C), whereas we just showed that it has Hausdorff dimension 2. In particular, $\mathcal{H}_2\{\overline{X(\mathbb{R}_+^N)}\} = 0$, a.s., while for any $s \in {]}0,2{[}$, $\mathcal{H}_s\{\overline{X(\mathbb{R}_+^N)}\} = +\infty$, a.s. This discussion leaves out the "critical case," i.e., when $d = 2N$. In this case, the range has zero Lebesgue measure; cf. Supplementary Exercise 12.

We mention another interesting application of (1).

Theorem 4.6.2 *Let X denote an N-parameter, $\mathbb{R}^d$-valued additive stable process of index $\alpha \in {]}0,2]$. Given a compact set $E \subset \mathbb{R}^d$ and a d-dimensional vector $x \notin E$,*

$$\mathbb{P}_x\left\{ \dim\left(\overline{X(\mathbb{R}_+^N)} \cap E\right) \leq \{\dim(E) - d + N\alpha\}^+ \wedge d \right\} = 1.$$

On the other hand,

$$\mathbb{P}_x\left\{ \dim\left(\overline{X(\mathbb{R}_+^N)} \cap E\right) \geq \{\dim(E) - d + N\alpha\}^+ \wedge d \right\} > 0.$$

Exercise **4.6.4** Prove Theorem 4.6.2. □

4.7 Stochastic Codimension

The potential theoretic results of this chapter also apply to other random fields than Markovian ones. In the next two subsections we show how such connections can be established.

A set-valued function K mapping Ω into the collection of all subsets of $\mathbb{R}^d$ is said to be a **random set** if for all nonrandom compact (or σ-compact) sets $E \subset \mathbb{R}^d$, $\mathbf{1}_{K \cap E}$ is a random variable. We have already seen many examples of random sets: The range of an N-parameter, $\mathbb{R}^d$-valued Markov process $X = (X_t; \, t \in \mathbb{R}_+^N)$ is a random set in this sense. Somewhat more generally, suppose $X = (X_t; \, t \in \mathbb{R}_+^N)$ denotes a right-continuous, N-parameter, $\mathbb{R}^d$-valued stochastic process. If $I \subset \mathbb{R}_+^N$ is Borel measurable and if $X(I) = \{X_s; \, s \in I\}$, then the closure of $X(I)$ in $\mathbb{R}^d$ is a random set. In fact, if $t \mapsto X_t$ is continuous, then $X(I)$ is itself a random set (why?).

Throughout this subsection K denotes a random set in $\mathbb{R}^d$. We next define two nonrandom indices for K that we call the upper and the lower (stochastic) codimensions of K, respectively. The **upper (stochastic) codimension** of K (written as $\overline{\mathrm{codim}}(K)$) is the smallest real number $\beta \in [0, d]$ such that $\mathbb{P}(K \cap G \neq \varnothing) > 0$, for all compact sets $G \subset \mathbb{R}^d$ with $\dim(G) > \beta$. If such a β does not exist, we define the upper codimension of K as d. It should be reemphasized that the upper codimension of a random set is always deterministic; i.e., it is *not* a random number.

There is a symmetric definition for the **lower (stochastic) codimension** of a random set: $\underline{\mathrm{codim}}(K)$ is defined as the greatest real number $\beta \in [0, d]$ such that for all compact sets $G \subset \mathbb{R}^d$ with $\dim(K) < \beta$, $\mathbb{P}(K \cap G \neq \varnothing) = 0$. If such a β does not exist, then we define the lower codimension of K to be 0.

In case the two codimensions agree, that is, whenever

$$\overline{\mathrm{codim}}(K) = \underline{\mathrm{codim}}(K),$$

we write $\mathrm{codim}(K)$ for the common value and call it the (stochastic) **codimension** of K.

It may happen that there are several probability measures $\mathbb{P}_x$, $x \in \mathbb{R}^d$ on our probability space. In such a case, we refer to $\overline{\mathrm{codim}}(K)$, $\underline{\mathrm{codim}}(K)$, and $\mathrm{codim}(K)$ as the upper codimension, lower codimension, and the codimension of K, with respect to the measure $\mathbb{P}_x$, when the $\mathbb{P}$ in the definitions of codimensions is replaced by $\mathbb{P}_x$. Let us summarize our efforts thus far.

Lemma 4.7.1 *Suppose K is a random set in $\mathbb{R}^d$. If $G \subset \mathbb{R}^d$ is compact, then*

$$\mathbb{P}(K \cap G \neq \varnothing) \begin{cases} > 0, & \text{whenever } \dim(G) > \overline{\mathrm{codim}}(K), \\ = 0, & \text{whenever } \dim(G) < \underline{\mathrm{codim}}(K). \end{cases}$$

Example 1 Suppose X denotes a d-dimensional, 1-parameter, isotropic stable Lévy process of index $\alpha \in \,]0, 2]$. Theorem 3.5.1 of Chapter 10 can be

recast as follows: With respect to any of the measures $\mathbb{P}_x$, where $x \in \mathbb{R}^d$,
$$\mathrm{codim}\{X(\mathbb{R}_+)\} = (d-\alpha)^+,$$
where $a^+ = \max\{a, 0\}$, as usual. □

Example 2 The previous example can be generalized to a multiparameter setting in various ways. Here is one such method: Let $X = (X_t;\ t \in \mathbb{R}_+^N)$ denote an $\mathbb{R}^d$-valued, N-parameter additive stable process of index $\alpha \in]0, 2]$ and let $\mathbb{P}_x$ denote the distribution of $X + x$. Then, Corollary 4.1.1 is equivalent to the statement that
$$\mathrm{codim}\{X(\mathbb{R}_+^N)\} = (d - N\alpha)^+,$$
under any $\mathbb{P}_x$, where $x \in \mathbb{R}^d$. □

Example 3 Let X^1 and X^2 denote independent, $\mathbb{R}^d$-valued, isotropic stable Lévy processes of indices α_1 and α_2, respectively. If $x, y \in \mathbb{R}^d$ are distinct, Corollary 4.3.1 can be recast as follows: $\mathbb{P}_{x \otimes y}$-a.s.,
$$\mathrm{codim}\{\overline{X^1(\mathbb{R}_+)} \cap \overline{X^2(\mathbb{R}_+)}\} = (2d - \alpha_1 - \alpha_2)^+.$$
In particular, if we choose X^1 and X^2 to both be Brownian motions whose starting points are different, then we see that the codimension of the intersection of the ranges of two independent Brownian motions in $\mathbb{R}^d$ is $2(d-2)^+$. □

The main result of this subsection is the following theorem, which also justifies the use of the term "codimension." In what follows, the assumption that the codimension is *strictly* between 0 and d is indispensable.

Theorem 4.7.1 *Given a random set* K *in* $\mathbb{R}^d$ *whose codimension is strictly between* 0 *and* d,
$$\dim(K) + \mathrm{codim}(K) = d, \qquad \mathbb{P}\text{-}a.s.$$

The main step in our proof is the following result due to Y. Peres; see Peres (1996a, 1996b).

Lemma 4.7.2 (Peres's Lemma) *On a suitable probability space, one can construct a random set* $\Lambda_{\beta,d} \subset \mathbb{R}^d$ *for any* $\beta \in]0, d[$ *such that for all σ-compact sets* $E \subset \mathbb{R}^d$,
$$\mathbb{P}\big(\dim\{\Lambda_{\beta,d} \cap E\} = \dim(E) - \beta\big) = 1,$$
where $\dim(A) < 0$ *means that* $A = \emptyset$, *for any Borel set* $A \subset \mathbb{R}^d$. *Finally, for any Borel set* $E \subset \mathbb{R}^d$ *and all* $\beta \in]0, d[$,
$$\mathbb{P}\big(\Lambda_{\beta,d} \cap E \neq \emptyset\big) \in \{0, 1\}.$$

Proof By enlarging the probability space if necessary, we can assume the existence of independent, identically distributed additive stable processes $X^1, X^2, \ldots$ and choose distinct vectors $x_1, x_2, \ldots \in \mathbb{R}^d$ as their starting points. (That is, the underlying probability measure is taken to be the product measure $\mathbb{P} = \mathbb{P}_{x_1 \otimes x_2 \otimes \cdots} = \prod_{j=1}^{\infty} \mathbb{P}_{x_j}^j$, where $\mathbb{P}_{x_j}^j$ denotes the distribution of $X^j + x_j$.) We shall further assume that these processes have N parameters, are $\mathbb{R}^d$-valued, and have the *same* index α, where the parameters in question satisfy $d - N\alpha = \beta$.

Define the random set

$$\Lambda_{\beta,d}(\omega) = \bigcup_{i=1}^{\infty} \overline{X^i(\mathbb{R}_+^N)}(\omega).$$

(Why is this a random set?) By Exercise 1.2.2 of Appendix C, for any σ-compact set $E \subset \mathbb{R}^d$,

$$\dim\left(\Lambda_{\beta,d} \cap E\right) = \sup_{i \geq 1} \dim\left\{\overline{X^i(\mathbb{R}_+^N)} \cap E\right\}.$$

Applying the Borel–Cantelli lemma together with Theorem 4.6.2, we conclude that

$$\mathbb{P}\big(\dim\{\Lambda_{\beta,d} \cap E\} = \dim(E) - \beta\big) = 1.$$

This is the first half of the desired result. The second half, i.e., the 0-1 law, follows from the Borel–Cantelli lemma and the i.i.d. structure of the X^j's. □

Proof of Theorem 4.7.1 Let $\Lambda_{\beta,d}$ be the random sets of Peres's lemma (Lemma 4.7.2). By further enlarging the probability space, we may assume that all the $\Lambda_{\beta,d}$'s are independent of the random set K. For any $\gamma \in]0, d[$, $\Lambda_{\gamma,d}$ and K are independent. Thus, by Peres's lemma, $\dim(\Lambda_{\gamma,d} \cap \text{K}) = \dim(\text{K}) - \gamma$, almost surely. In particular,

$$\mathbb{P}\big(\Lambda_{\gamma,d} \cap \text{K} \neq \varnothing\big) = \begin{cases} 1, & \text{if } \dim(\text{K}) > \gamma, \\ 0, & \text{if } \dim(\text{K}) < \gamma. \end{cases}$$

Now, suppose there exists $\gamma \in]0, d[$ such that $\mathbb{P}(\dim\{\text{K}\} > \gamma) > 0$. The above, together with the 0-1 law of Peres's lemma, implies that $\mathbb{P}(\Lambda_{\gamma,d} \cap \text{K} \neq \varnothing) = 1$. By Lemma 4.7.1,

$$\mathbb{P}\big(\dim\{\text{K}\} > \gamma\big) > 0 \iff \mathbb{P}\big\{\dim(\Lambda_{\gamma,d}) \geq \text{codim}(\text{K})\big\} > 0.$$

On the other hand, Peres's lemma states that a.s., $\dim(\Lambda_{\gamma,d}) = d - \gamma$. Hence,

$$\mathbb{P}\big(\dim\{\text{K}\} > \gamma\big) > 0 \iff d - \gamma \geq \text{codim}(\text{K}).$$

Equivalently, for all $\gamma > d - \text{codim}(\text{K})$, $\mathbb{P}(\dim\{\text{K}\} > \gamma) = 0$. This proves

$$\mathbb{P}\big\{\dim\{\text{K}\} \leq d - \text{codim}(\text{K})\big\} = 1.$$

Conversely, if $\mathbb{P}(\dim\{K\} < \gamma) > 0$, then $d - \gamma \le \operatorname{codim}(K)$, which shows that
$$\mathbb{P}\{\dim\{K\} \ge d - \operatorname{codim}(K)\} = 1.$$
Our proof is complete. □

5 α-Regular Gaussian Random Fields

The ideas and results introduced in this chapter can be applied, to various degrees, to study other random fields than Markovian ones. We close this chapter with a look at a broad class of Gaussian random fields that (1) are non-Markovian in every reasonable sense; and (2) arise in applications of probability theory to other areas.

Throughout this section $X = (X_t;\ t \in \mathbb{R}^N)$ denotes an $\mathbb{R}^d$-valued Gaussian process that is indexed by all of $\mathbb{R}^N$ and not just $\mathbb{R}^N_+$. We will also assume that for all $t \in \mathbb{R}^N$ and for all $1 \le i \le N$, $\mathbb{E}\bigl[X_t^{(i)}\bigr] = 0$. In order to simplify the exposition, we begin with our study of $\mathbb{R}$-valued, stationary Gaussian random fields.

5.1 One-Dimensional Stationary Gaussian Processes

An N-parameter, real-valued Gaussian process $X = (X_t;\ t \in \mathbb{R}^N)$ is said to be **stationary** if $(s,t) \mapsto \mathbb{E}[X_s X_t]$ is a function of $s - t$. It is said to be **centered** if for all $t \in \mathbb{R}^N$, $\mathbb{E}[X_t] = 0$.

Recall the covariance function of X from Chapter 5, viz.,
$$\Sigma(s,t) = \mathbb{E}[X_s X_t], \qquad s,t \in \mathbb{R}^N.$$

Note that X is stationary if and only if $\Sigma(s,t) = R(s-t)$ for some function $R : \mathbb{R}^N \to \mathbb{R}$. In this case, we can refer to R as the **correlation function** of X. It is abundantly clear that for all $t \in \mathbb{R}^N$, $R(t) = \mathbb{E}[X_t X_0]$. Moreover, from the form of the characteristic function of linear combinations of the X_t's, as t varies, we obtain the following:

Lemma 5.1.1 *If X is a centered, stationary, $\mathbb{R}$-valued, N-parameter Gaussian random field, then for all $s \in \mathbb{R}^N$, $(X_{s+t};\ t \in \mathbb{R}^N)$ has the same finite-dimensional distributions as $(X_t;\ t \in \mathbb{R}^N)$. Moreover, for all $t \in \mathbb{R}^N$, $X_t \sim \mathcal{N}_1(0, R(0))$.*

Exercise **5.1.1** Verify Lemma 5.1.1 by checking covariances. □

In particular, we may assume, without loss of much generality, that
$$R(0) = 1. \tag{1}$$

Indeed, if $R(0) \neq 1$, the results of this section can be applied to the process $[R(0)]^{-1}X$ and then suitably translated.

By Theorem 1.2.1, Chapter 5, Σ is positive definite in the sense that for all $\xi_1, \ldots, \xi_n \in \mathbb{R}$, and for all $s_1, \ldots, s_n \in \mathbb{R}^N$,

$$\sum_{j=1}^{n}\sum_{i=1}^{n} \xi_i \Sigma(s_i, s_j) \xi_j \geq 0.$$

In particular, R is a positive definite function in the classical sense: For all $\xi_1, \ldots, \xi_n \in \mathbb{R}$, and for all $s_1, \ldots, s_n \in \mathbb{R}^N$,

$$\sum_{j=1}^{n}\sum_{i=1}^{n} \xi_i R(s_i - s_j) \xi_j \geq 0.$$

Among its other notable features, we mention that the function R is symmetric in the sense that $R(t) = R(-t)$, for all $t \in \mathbb{R}^N$. Moreover, by the Cauchy–Schwarz inequality, for all $t \in \mathbb{R}^N$,

$$|R(t)| \leq \sqrt{\mathbb{E}[(X_t)^2] \cdot \mathbb{E}[(X_0)^2]} = 1.$$

Let us recall the following form of the so-called **spectral theorem**.

Theorem 5.1.1 (The Spectral Theorem) *Given a symmetric, positive definite function $f : \mathbb{R}^N \to \mathbb{R}$ such that $\sup_{t \in \mathbb{R}^N} |f(t)| = 1$, there exists a probability measure μ, defined on Borel subsets of $\mathbb{R}^N$, such that for all $t \in \mathbb{R}^N$,*

$$f(t) = \int_{\mathbb{R}^N} e^{i\xi \cdot t} \mu_f(d\xi).$$

The probability measure μ_f is the so-called **spectral measure** associated with the function f. We now combine the above ingredients to obtain the following:

Lemma 5.1.2 *Recalling (1), there exists a probability measure μ, on Borel subsets of $\mathbb{R}^N$, such that for all $t \in \mathbb{R}^N$,*

$$R(t) = \int_{\mathbb{R}^N} e^{i\xi \cdot t} \mu(d\xi).$$

Remarks (1) In fact, since R is real-valued, we can write R as the following cosine transform:

$$R(t) = \int_{\mathbb{R}^N} \cos(\xi \cdot t) \mu(d\xi).$$

(2) Conversely, whenever R is the characteristic function of a probability measure on $\mathbb{R}^N$, then R generates a mean-function-zero, stationary, real-valued, N-parameter Gaussian process by the prescription $\Sigma(s, t) = R(s - t)$, for all $s, t \in \mathbb{R}^N$. Clearly, Σ is positive definite in the sense

of Chapter 5. Theorem 1.2.1 of Chapter 5 now constructs the mentioned Gaussian process.

(3) Remark (2) above is often useful in modeling. For instance, it shows that for any $\alpha \in {]}0,2]$, we can construct a centered, stationary, N-parameter, real-valued Gaussian random field whose correlation function is given by the "shape" $R(t) = \exp(-\|t\|^\alpha)$ (why?).

It is also possible to write the natural pseudometric generated by X, in terms of R, viz.,

$$\mathbb{E}\big[(X_s - X_t)^2\big] = 2\{1 - R(s-t)\}, \qquad s, t \in \mathbb{R}^N. \tag{2}$$

We can often use the above, in conjunction with continuity theorems of Chapter 5, to see that X has a continuous modification if $t \mapsto R(t)$ is sufficiently smooth near the origin.

Example By (2) and by Exercise 2.5.1 of Chapter 5, for all $p > 0$, we can find a finite, positive constant $c(p)$, such that

$$\mathbb{E}\big[|X_s - X_t|^p\big] = c(p)\big\{\mathbb{E}[|X_s - X_t|^2]\big\}^{\frac{p}{2}} = 2^{\frac{p}{2}} c(p) \{1 - R(s-t)\}^{\frac{p}{2}}.$$

Suppose there exist finite constants $C > 1$, $\beta \in {]}0,2]$, and $\varepsilon > 0$ such that for all $h \in \mathbb{R}^N$ with $|h| \leq \varepsilon$,

$$1 - R(h) \leq C|h|^\beta.$$

($R(t) = \exp\big(-\|t\|^\beta\big)$ of Remark (2) above is an example of such a correlation function.) Then, for *all* $p > 0$, we can deduce the existence of a finite, positive constant A_p such that for all $s, t \in \mathbb{R}^N$,

$$\mathbb{E}[|X_s - X_t|^p] \leq A_p |s-t|^{\frac{\beta p}{2}}.$$

Choose p so large that $\beta p > 2N$ and use Theorem 2.5.1 of Chapter 5 to conclude that there exists a Hölder continuous modification of X, which we continue to denote by X. One can refine this, since the same theorem shows that X is Hölder continuous of any order $r < \frac{1}{2}\beta$. In fact, we can do even more than this: By Exercise 2.5.1 of Chapter 5, for all $a \preccurlyeq b$, both in $\mathbb{R}^N$, all $0 < \mu < \beta$, and all $p > 0$, there exists a finite constant B_p such that for all $\varepsilon \in {]}0,1[$,

$$\mathbb{E}\bigg[\sup_{s,t \in [a,b]:\ |s-t| \leq \varepsilon} |X_s - X_t|^p\bigg] \leq B_p\, \varepsilon^{\frac{\mu p}{2}}. \tag{3}$$

See Supplementary Exercise 9 for an extension. $\square$

5.2 α-Regular Gaussian Fields

We say that the N-parameter, $\mathbb{R}^d$-valued process X is **α-regular** if $\alpha \in\ ]0,1[$ and if X is a centered, $\mathbb{R}^d$-valued, N-parameter, stationary Gaussian process with i.i.d. coordinate processes whose correlation function R satisfies the following: There exists finite, positive constants $c_1 \leq c_2$ and ε_0 such that for all $h \in \mathbb{R}^N$ with $|h| \leq \varepsilon_0$,

$$c_1|h|^{2\alpha} \leq 1 - R(s-t) \leq c_2|h|^{2\alpha}. \tag{1}$$

Note that for all $1 \leq i \leq N$, $R(t) = \mathbb{E}\bigl[X_t^{(i)} X_0^{(i)}\bigr]$. In particular, each coordinate process $X^{(i)}$ is of the type considered in the example of Section 5.1 with $\beta = 2\alpha$.

The main result of this section is the following codimension calculation for the range of an α-regular process.

Theorem 5.2.1 *Let X denote an N-parameter, $\mathbb{R}^d$-valued, α-regular process, where $\alpha \in\ ]0,1[$. Then,*

$$\mathrm{codim}\bigl(\overline{X(\mathbb{R}^N)}\bigr) = \Bigl[d - \Bigl(\frac{N}{\alpha}\Bigr)\Bigr]^+.$$

In particular, we deduce the following important corollary; it measures the size of the range of the process X by way of Hausdorff dimension.

Corollary 5.2.1 *In Theorem 5.2.1,*

$$\dim\bigl(\overline{X(\mathbb{R}^N)}\bigr) = \Bigl(\frac{N}{\alpha}\Bigr) \wedge d, \qquad a.s.$$

Variants of the above, together with further refinements, can be found in (Adler 1981; Testard 1985; Weber 1983).

Let us conclude this subsection by proving Corollary 5.2.1. The proof of Theorem 5.2.1 is deferred to the next two subsections.

Proof of Corollary 5.2.1 If $N < \alpha d$, Theorem 5.2.1 implies that the range of X has codimension $d - \frac{N}{\alpha}$, which is strictly between 0 and d. Since $K = \overline{X(\mathbb{R}^N)}$ is a random set in the sense of Section 4.7 (why?), we can apply Theorem 4.7.1 to conclude that with probability one, the Hausdorff dimension of the range of X is d minus the codimension; i.e., the Hausdorff dimension is a.s. equal to $\frac{N}{\alpha}$. It remains to prove the corollary when $N \geq \alpha d$. Indeed, it suffices to show that

$$\dim(X([1, 1+\varepsilon_0]^N)) \geq d, \qquad a.s.,$$

where ε_0 is given in the preamble to equation (1) above. (Why?) We shall demonstrate this by bare-hands methods.

Define the random measure μ on Borel subsets of $\mathbb{R}^d$ by assigning to every Borel set $A \subset \mathbb{R}^d$,

$$\mu(A) = \int_{[1,1+\varepsilon_0]^N} \mathbf{1}_A(X_s) \, ds.$$

This is the so-called **occupation measure** for the stochastic process $(X_s; \ s \in [1, 1+\varepsilon_0]^N)$. It should also be clear that $\mu_0 = \varepsilon_0^{-N}\mu$ is a (random) *probability* measure that is supported on the (random, closed) set $X([1,1+\varepsilon_0]^N)$. Moreover, after a change of variables, we can see that for any $\beta \in \mathbb{R}$,

$$\mathsf{Energy}_\beta(\mu_0) = \varepsilon_0^{-2N} \int_{[1,1+\varepsilon_0]^N} \int_{[1,1+\varepsilon_0]^N} \|X_s - X_t\|^{-\beta} \, ds \, dt.$$

In particular, if $\beta \geq 0$,

$$\mathbb{E}\big[\mathsf{Energy}_\beta(\mu_0)\big] = \varepsilon_0^{-2N} \int_{[1,1+\varepsilon_0]^N} \int_{[1,1+\varepsilon_0]^N} \mathbb{E}\big[\|X_s - X_t\|^{-\beta}\big] \, ds \, dt.$$

Since $X_s^{(i)} - X_t^{(i)}$ are i.i.d. (as i varies from 1 to d), for all $s, t \in [1, 1+\varepsilon_0]^N$,

$$\mathbb{E}\big[\|X_s - X_t\|^2\big] = d\mathbb{E}\big[|X_s^{(1)} - X_t^{(1)}|^2\big] = 2d\{1 - R(s-t)\},$$

by equation (2) of Section 5.1. By the definition of α-regularity, for all $s, t \in [1, 1+\varepsilon_0]^N$,

$$2dc_1|s-t|^{2\alpha} \leq \mathbb{E}\big[\|X_s - X_t\|^2\big] \leq 2dc_2|s-t|^{2\alpha}. \tag{2}$$

For any $\beta \in \mathbb{R}$,

$$\mathbb{E}\big[\|X_s - X_t\|^{-\beta}\big] = \big\{\mathbb{E}\big[\|X_s - X_t\|^2\big]\big\}^{-\frac{\beta}{2}} \mathbb{E}[\|Z\|^{-\beta}],$$

where Z is an $\mathbb{R}^d$-valued, Gaussian random variable with mean vector 0 whose covariance matrix is the $(d \times d)$ identity matrix (why?). Furthermore,

$$\mathbb{E}\big[\|Z\|^{-\beta}\big] < \infty \iff \beta < d.$$

(Why? See Exercise 1.1.1 of Chapter 5 for the form of the density.) Now we put things together: Suppose $0 < \beta < d$ and observe by equation (3) that

$$\mathbb{E}\big[\mathsf{Energy}_\beta(\mu_0)\big] \leq A_\beta \int_{[1,1+\varepsilon_0]^N} \int_{[1,1+\varepsilon_0]^N} |s-t|^{-\beta\alpha} \, ds \, dt,$$

where
$$A_\beta = c_2^{-\beta}(2d)^{-\beta}\varepsilon_0^{-2N}C_\beta\mathbb{E}\big[\|Z\|^{-\beta}\big] < \infty.$$

Since $\beta < d \leq \frac{N}{\alpha}$, we can deduce that $\alpha\beta < N$, and in particular, upon changing variable ($w = s - t$), we obtain the following simple bound:

$$\int_{[1,1+\varepsilon_0]^N}\int_{[1,1+\varepsilon_0]^N}|s-t|^{-\beta\alpha}\,ds\,dt \leq \varepsilon_0^N\int_{[-1,1]^N}|w|^{-\beta\alpha}\,dw,$$

which, thanks to Lemma 3.1.2 of Chapter 3, is finite. In summary, we have shown that there exists a probability measure μ_0 on $X([1, 1+\varepsilon_0]^N)$ such that for any $\beta < d$, $\mathsf{Energy}_\beta(\mu_0) < \infty$, a.s. By Frostman's theorem (Theorem 2.2.1, Appendix C), for any $\beta < d$,

$$\dim(X([1, 1+\varepsilon_0]^N)) \geq \beta, \quad \text{a.s.}$$

Letting $\beta \uparrow d$ along a rational sequence, we deduce that the Hausdorff dimension of $X([1, 1+\varepsilon_0]^N)$ is a.s. at least d. This proves the corollary. □

5.3 Proof of Theorem 5.2.1: First Part

The difficulty with handling the processes of this section is that, generally, they are not Markovian. Thus, in our first lemma we are led to examine methods by which one can introduce some conditional independence for Gaussian processes. Throughout, $X = (X_t;\ t \in \mathbb{R}^N)$ denotes an N-parameter, $\mathbb{R}^d$-valued, α-regular process, where $\alpha \in\ ]0, 1[$.

Lemma 5.3.1 *For any fixed $s \in \mathbb{R}^N$, X_s is independent of the entire process $(X_t - R(s-t)X_s;\ t \in \mathbb{R}^N)$. Moreover, the latter is a centered, N-parameter, $\mathbb{R}^d$-valued Gaussian process with i.i.d. coordinates all of which have mean zero and whose covariance is given by*

$$\mathbb{E}\big[\{X_u^{(1)} - R(u-s)X_s^{(1)}\}\{X_v^{(1)} - R(v-s)X_s^{(1)}\}\big] = R(u-v) - R(s-u)R(s-v),$$

for all $u, v \in \mathbb{R}^N$.

Exercise 5.3.1 Verify Lemma 5.3.1 by computing covariances. □

In particular, note that the variance of each coordinate of $X_t - R(s-t)X_s$ equals $1 - \{R(s-t)\}^2$.

Our next technical result is a key step in proving the first half of Theorem 5.2.1.

Lemma 5.3.2 *Suppose $d > \frac{N}{\alpha}$ and let $M > 1$ be fixed. Then, for all $\zeta \in\ ]0,1[$, there exists a finite constant $A > 0$ such that for all $\varepsilon > 0$,*

$$\sup_{x\in[-M,M]^d}\mathbb{P}\bigg(\inf_{t\in[0,1]^N}|X_t - x| \leq \varepsilon\bigg) \leq A\varepsilon^{(1-\zeta)d-\frac{N}{\alpha}}.$$

444   11. Multiparameter Markov Processes

Before commencing with a proof for this lemma, we introduce some notation. Define Γ_n to be the equipartition of $[0,1]^N$ of mesh $\frac{1}{n}$, viz.,

$$\Gamma_n = \left\{ t \in [0,1]^N : \forall 1 \le \ell \le N,\ t^{(\ell)} = \frac{k}{n} \text{ for some } 0 \le k \le n \right\}.$$

Throughout this proof the generic element of Γ_n is distinguished by the letter γ, where the generic elements of $[0,1]^N$ are written as $s, t, \ldots$, as usual.

For each $\gamma \in \Gamma_n$, let $\mathsf{R}_n(\gamma)$ denote the "right-hand box" of side $\frac{1}{n}$ about γ. This is defined, more precisely, as

$$\mathsf{R}_n(\gamma) = \left\{ t \in [0,1]^N : t \succcurlyeq \gamma, |t - \gamma| \le \frac{1}{n} \right\}, \qquad n \ge 1,\ \gamma \in \Gamma_n.$$

(Recall that $|t| = \max_{1 \le \ell \le N} |t^{(\ell)}|$ denotes the ℓ^∞ norm of $t \in \mathbb{R}^N$.) Clearly, for every integer $n \ge 1$,

$$\bigcup_{\gamma \in \Gamma_n} \mathsf{R}_n(\gamma) = \left[0, 1 + \frac{1}{n} \right]^N.$$

Consequently, $\bigcup_{\gamma \in \Gamma_n} \mathsf{R}_n(\gamma) \supset [0,1]^N$.

Proof Let $n \ge 1$ be a fixed integer such that $n \ge \varepsilon_0^{-1}$, where ε_0 is given by the definition of α-regularity. We shall also hold some point $x \in [-M, M]^d$ fixed, where $M > 1$ is a fixed constant, as in the statement of the lemma.

Note that whenever $\inf_{t \in [0,1]^N} |X_t - x| \le n^{-\alpha}$, then there must exist some $\gamma \in \Gamma_n$ such that for some $t \in \mathsf{R}_n(\gamma)$, $|X_t - x| \le n^{-\alpha}$. Consequently,

$$\inf_{t \in [0,1]^N} |X_t - x| \le n^{-\alpha} \implies \exists \gamma \in \Gamma_n : |X_\gamma - x| \le n^{-\alpha} + \sup_{s \in \mathsf{R}_n(\gamma)} |X_s - X_\gamma|.$$

We now utilize Lemma 5.3.1 in order to create some independence. Namely, we note that for any $\gamma \in \Gamma_n$ and for all $s \in \mathsf{R}_n(\gamma)$,

$$|X_s - X_\gamma| \le |X_s - R(s-\gamma)X_\gamma| + |X_\gamma|\{1 - R(s-\gamma)\}$$
$$\le |X_s - R(s-\gamma)X_\gamma| + |X_\gamma - x|\{1 - R(s-\gamma)\}$$
$$+ |x|\{1 - R(s-\gamma)\}.$$

On the other hand, if $s \in \mathsf{R}_n(\gamma)$, then $|s - \gamma| \le n^{-1}$, which is less than or equal to ε_0. Thus, $1 - R(s-\gamma) \le c_2 n^{-2\alpha}$. Consequently, for all $\gamma \in \Gamma_n$ and all $s \in \mathsf{R}_n(\gamma)$,

$$|X_s - X_\gamma| \le |X_s - R(s-\gamma)X_\gamma| + c_2 n^{-2\alpha}|X_\gamma - x| + c_2 n^{-2\alpha} M.$$

Thus, whenever $\inf_{t \in [0,1]^N} |X_t - x| \le n^{-\alpha}$, then

$$\exists \gamma \in \Gamma_n : (1 - c_2 n^{-2\alpha})|X_\gamma - x| \le n^{-\alpha} + c_2 n^{-2\alpha} M + \sup_{s \in \mathsf{R}_n(\gamma)} |X_s - R(s-\gamma)X_\gamma|.$$

The above holds for all integers $n > \varepsilon_0^{-1}$. If n is even larger, we can simplify this further. Indeed, suppose $n > J$, where $J = \max\{\varepsilon_0^{-1}, (c_2 M)^{\frac{1}{\alpha}}, (2c_2)^{\frac{1}{\alpha}}\}$. Then, $1 - c_2 n^{-2\alpha} \leq \frac{1}{2}$ and $c_2 n^{-2\alpha} M \leq n^{-\alpha}$. Thus,

$$\inf_{t \in [0,1]^N} |X_t - x| \leq n^{-\alpha}$$
$$\implies \exists \gamma \in \Gamma_n : \frac{1}{2}|X_\gamma - x| \leq 2n^{-\alpha} + \sup_{s \in \mathsf{R}_n(\gamma)} |X_s - R(s - \gamma) X_\gamma|.$$

Consequently,

$$\mathbb{P}\left(\inf_{t \in [0,1]^N} |X_t - x| \leq n^{-\alpha}\right)$$
$$\leq \sum_{\gamma \in \Gamma_n} \mathbb{P}\left(|X_\gamma - x| \leq 4n^{-\alpha} + 2 \sup_{s \in \mathsf{R}_n(\gamma)} |X_s - R(s - \gamma) X_\gamma|\right).$$

For any $\gamma \in \Gamma_n$, X_γ is a vector of d independent Gaussian random variables, each with mean 0 and variance 1. Thus, the density of X_γ (with respect to Lebesgue's measure on $\mathbb{R}^d$) is bounded above by 1. In particular, for all $z \in \mathbb{R}^d$ and $r > 0$,

$$\mathbb{P}(|X_\gamma - z| \leq r) \leq \mathrm{Leb}([-r,r]^d) = (2r)^d.$$

By Lemma 5.3.1, $|X_\gamma - x|$ and $\sup_{s \in \mathsf{R}_n(\gamma)} |X_s - R(s - \gamma) X_\gamma|$ are independent random variables. Thus, by first conditioning on the latter random variable, we obtain

$$\mathbb{P}\left(\inf_{t \in [0,1]^N} |X_t - x| \leq n^{-\alpha}\right)$$
$$\leq 2^d \sum_{\gamma \in \Gamma_n} \mathbb{E}\left[\left\{4n^{-\alpha} + 2 \sup_{s \in \mathsf{R}_n(\gamma)} |X_s - R(s - \gamma) X_\gamma|\right\}^d\right]$$
$$\leq 4^d \sum_{\gamma \in \Gamma_n} \left\{(4n^{-\alpha})^d + 2^d \mathbb{E}\left[\sup_{s \in \mathsf{R}_n(\gamma)} |X_s - R(s - \gamma) X_\gamma|^d\right]\right\}, \tag{1}$$

since for all $p, q \geq 0$, $(p + q)^d \leq 2^d(p^d + q^d)$. The same inequality shows that

$$\mathbb{E}\left[\sup_{s \in \mathsf{R}_n(\gamma)} |X_s - R(s - \gamma) X_\gamma|^d\right]$$
$$\leq 2^d \mathbb{E}\left[\sup_{s \in \mathsf{R}_n(\gamma)} |X_s - X_\gamma|^d\right] \tag{2}$$
$$+ 2^d \mathbb{E}\left[|X_\gamma|^d\right] \sup_{s \in \mathsf{R}_n(\gamma)} \{1 - R(s - \gamma)\}^d.$$

By equation (3) of the example of Section 5.1, for any $\zeta \in]0, 1[$, there exists a positive finite constant B_1 such that

$$\mathbb{E}\left[\sup_{s \in \mathsf{R}_n(\gamma)} |X_s - X_\gamma|^d\right] \leq B_1 n^{-\alpha(1-\zeta)d}. \tag{3}$$

Moreover, by α-regularity, $\sup_{s \in \mathsf{R}_n(\gamma)}\{1 - R(s-\gamma)\}^d \leq c_2^d n^{-2\alpha d}$, which is less than $n^{-\alpha(1-\varsigma)d}$ for all $n > J$. Combining this with equations (1), (2), and (3), we can conclude that for all $n > J$,

$$\mathbb{P}\left(\inf_{t \in [0,1]^N} |X_t - x| \leq n^{-\alpha}\right)$$
$$\leq 4^d \sum_{\gamma \in \Gamma_n} \left\{4^d n^{-\alpha d} + 4^d B_1 n^{-\alpha(1-\varsigma)d} + 4^d \mathbb{E}[|X_\gamma|^d] n^{-\alpha(1-\varsigma)d}\right\}$$
$$\leq 16^d n^{-\alpha(1-\varsigma)d}\{1 + B_1 + B_2\}\#(\Gamma_n),$$

where $\#(\Gamma_n) = (n+1)^N \leq 2^N n^N$ is a cardinality bound for Γ_n, and for any and all $\gamma \in \Gamma_n$,

$$B_2 = \mathbb{E}[|X_\gamma|^d] = (2\pi)^{-\frac{d}{2}} \int_{\mathbb{R}^d} |u|^d e^{-\frac{1}{2}\|u\|^2}\, du < \infty.$$

We have shown that for all $n > J$,

$$\mathbb{P}\left(\inf_{t \in [0,1]^N} |X_t - x| \leq n^{-\alpha}\right) \leq B_3 n^{-\alpha(1-\varsigma)d+N},$$

where $B_3 = 16^d 2^N \{1 + B_1 + B_2\}$. Now, supposing $0 < \varepsilon < (2J)^{-\alpha}$, we can find an integer $n > J$ such that $(n+1)^{-\alpha} \leq \varepsilon \leq n^{-\alpha}$. Consequently,

$$\mathbb{P}\left(\inf_{t \in [0,1]^N} |X_t - x| \leq \varepsilon\right) \leq \mathbb{P}\left(\inf_{t \in [0,1]^N} |X_t - x| \leq n^{-\alpha}\right) \leq B_3 n^{-\alpha(1-\varsigma)d+N}.$$

Using the fact that $\varepsilon \geq (n+1)^{-\alpha} \geq (2n)^{-\alpha}$, we deduce that for all $0 < \varepsilon < (2J)^{-\alpha}$,

$$\mathbb{P}\left(\inf_{t \in [0,1]^N} |X_t - x| \leq \varepsilon\right) \leq B_4 \varepsilon^{(1-\varsigma)d - \frac{N}{\alpha}},$$

where $B_4 = B_3 2^{(1-\varsigma)d\alpha - N}$. On the other hand, whenever $\varepsilon \geq (2J)^{-\alpha}$,

$$\mathbb{P}\left(\inf_{t \in [0,1]^N} |X_t - x| \leq \varepsilon\right) \leq 1 \leq (2J)^{\alpha(1-\varsigma)d-N} \varepsilon^{(1-\varsigma)d - \frac{N}{\alpha}}.$$

Thus, the lemma follows with $A = B_4 \vee (2J)^{\alpha(1-\varsigma)d-N}$. □

We are ready to prove the first half of Theorem 5.2.1.

Proof of Theorem 5.2.1: First Half In the first half of the proof, we show that
$$\operatorname{codim}\left(\overline{X(\mathbb{R}^N)}\right) \geq \left\{d - \left(\frac{N}{\alpha}\right)\right\}^+.$$

We can assume, without loss of generality, that $d > \frac{N}{\alpha}$. Otherwise, there is nothing to prove.

Suppose $E \subset \mathbb{R}^d$ is a compact set with $0 < \dim(E) < d - \frac{N}{\alpha}$. We are to show that $\mathbb{P}(\overline{X(\mathbb{R}^N)} \cap E \neq \varnothing) = 0$. By countable additivity, it suffices to show that for all $a \preccurlyeq b$, both in $\mathbb{R}^N$,

$$\mathbb{P}(\overline{X([a,b])} \cap E \neq \varnothing) = 0.$$

We will do this for $[a,b] = [0,1]^N$. The somewhat more general case is handled in Exercise 5.3.2 below.

By the definition of Hausdorff dimension (used in conjunction with Lemma 1.1.3(i) of Appendix C), for all $s > \dim(E)$ and for all $\delta \in]0,1[$, there exist cubes $B_1, B_2, \ldots$ of sides $r_1, r_2, \ldots$ such that (1) $\sup_j r_j \le \delta$; (2) $E \subset \cup_j B_j$; and (3)

$$\sum_{j=1}^{\infty} (2r_j)^s \le \delta. \qquad (4)$$

Henceforth, will choose a fixed $\zeta \in]0,1[$ so small that $s = (1-\zeta)d - \frac{N}{\alpha} > \dim(E)$. Since E is compact, there exists some $M > 1$ such that $B_j \subset [-M, M]^d$ for all $j \ge 1$. Hence, by Lemma 5.3.2 above, we can find a finite constant $A > 0$ such that for all $j \ge 1$,

$$\mathbb{P}(X([0,1]^N) \cap B_j \neq \varnothing) \le A r_j^{(1-\zeta)d - \frac{N}{\alpha}} = A_1 (2r_j)^{(1-\zeta)d - \frac{N}{\alpha}},$$

where $A_1 = 2^{-(1-\zeta)d + (N/\alpha)} A$. By (4),

$$\mathbb{P}(X([0,1]^N) \cap E \neq \varnothing) \le \mathbb{P}\left(X([0,1]^N) \cap \bigcup_{j=1}^{\infty} B_j \neq \varnothing\right)$$

$$\le A_1 \sum_{j=1}^{\infty} (2r_j)^{(1-\zeta)d - \frac{N}{\alpha}} \le A_1 \delta.$$

Since $\delta > 0$ and $\zeta \in]0,1[$ are arbitrary, we have shown that whenever $0 < \dim(E) < d - \frac{N}{\alpha}$, then $\mathbb{P}\{X([0,1]^N) \cap E \neq \varnothing\} = 0$. This completes our proof of the first half. □

Exercise 5.3.2 Extend the above proof to cover the general case where $E \subset [a,b]$. □

Exercise 5.3.3 (Hard) Refine a part of Theorem 5.2.1 by showing that for all $M > 0$ and all $a \preccurlyeq b$, both in $\mathbb{R}_+^N$, there exists a positive finite constant C such that for all compact sets $E \subset [-M, M]^d$,

$$\mathbb{P}\{X([a,b]) \cap E \neq \varnothing\} \le C \mathcal{H}_{d - \frac{N}{\alpha}}(E),$$

where $\mathcal{H}_s$ is the s-dimensional Hausdorff measure of Appendix C. □

5.4 Proof of Theorem 5.2.1: Second Part

The second half of the proof of Theorem 5.2.1 relies on the following two technical lemmas.

Lemma 5.4.1 *For all $M > 0$, there exists a finite constant $A > 0$ such that for all measurable functions $f : \mathbb{R}^d \to \mathbb{R}_+$ that are zero outside $[-M, M]^d$,*

$$\mathbb{E}\left[\left\{\int_{[1,1+\varepsilon_0]^N} f(X_s)\,ds\right\}^2\right] \le A \iint \kappa(\|x - y\|)\, f(x) f(y)\, dx\, dy,$$

where ε_0 is given by the definition of α-regularity and $\kappa : \mathbb{R}_+ \to \mathbb{R}_+ \cup \{\infty\}$ is defined by

$$\kappa(r) = \begin{cases} 1, & \text{if } d < N/\alpha, \\ \ln_+(1/r), & \text{if } d = N/\alpha, \\ r^{-d+(N/\alpha)}, & \text{if } d > N/\alpha. \end{cases}$$

Proof Let $\mathbf{I}$ denote the $(d \times d)$ identity matrix and recall that by Lemma 5.3.1, for any $s, t \in \mathbb{R}^N$:

- $X_s \sim \mathcal{N}_d(0, \mathbf{I})$;
- $X_t - R(s - t)X_s \sim \mathcal{N}_d\big(0, \{1 - R^2(s - t)\}\mathbf{I}\big)$; and
- X_s and $X_t - R(s - t)X_s$ are independent.

In particular,

$$\mathbb{E}\big[f(X_s)f(X_t)\big] = (2\pi)^{-\frac{d}{2}} \iint e^{-\frac{1}{2}\|x\|^2} f(x)\, p_{s,t}(y)\, f(R(s - t)x + y)\, dx\, dy,$$

where

$$p_{s,t}(y) = (2\pi)^{-\frac{d}{2}} \{1 - R^2(s - t)\}^{-\frac{d}{2}} \exp\left(-\frac{\|y\|^2}{2\{1 - R^2(s - t)\}}\right).$$

(Why?) If we further assume that $|s - t| \le \varepsilon_0$, then by α-regularity,

$$p_{s,t}(y) \le c_1^{-\frac{d}{2}} |s - t|^{-\alpha d} \exp\left(-\frac{\|y\|^2}{2c_2|s - t|^{2\alpha}}\right).$$

Thus, whenever $|s - t| \le \varepsilon_0$, $\mathbb{E}[f(X_s)f(X_t)]$ is bounded above by

$$c_1^{-\frac{d}{2}} |s - t|^{-\alpha d} \iint f(x) f(R(s-t)x + y) e^{-\frac{\|y\|^2}{2c_2|s-t|^{2\alpha}}}\, dx\, dy$$

$$= c_1^{-\frac{d}{2}} |s - t|^{-\alpha d} \int_{[-M,M]^d} \int_{[-M,M]^d} f(x) f(z) e^{-\frac{\|z + R(s-t)x\|^2}{2c_2|s-t|^{2\alpha}}}\, dx\, dz.$$

5 α-Regular Gaussian Random Fields   449

We have also used the fact that f is supported on $[-M, M]^d$. Since $|R(h)| \leq 1$ for all $g \in \mathbb{R}^N$, by the triangle inequality and by α-regularity, for all $x, z \in [-M, M]^d$ and all $s, t \in \mathbb{R}^N$ with $|s - t| \leq \varepsilon_0$,

$$\|z - x\|^2 \leq 2\Big[\|z - R(s-t)x\|^2 + M^2\{1 - R(s-t)\}^2\Big]$$
$$\leq 2\Big[\|z - R(s-t)x\|^2 + M^2 c_2^2 \varepsilon_0^{2\alpha} |s-t|^{2\alpha}\Big].$$

Thus, whenever $|s - t| \leq \varepsilon_0$, $\mathbb{E}[f(X_s)f(X_t)]$ is bounded above by

$$A_1 |s-t|^{-\alpha d} \iint \exp\Big(-\frac{\|z-x\|^2}{4c_2|s-t|^{2\alpha}}\Big) f(x) f(z) \, dx \, dz,$$

where $A_1 = c_1^{-d/2} \exp(\frac{1}{2} M^2 c_2^2 \varepsilon_0^{2\alpha})$. The remainder of this lemma follows from direct calculations. □

Exercise 5.4.1 Complete the proof of Lemma 5.4.1. □

We are ready to complete our derivation of Theorem 5.2.1.

Proof of Theorem 5.2.1: Second Half We will prove the following stronger result: If $E \subset \mathbb{R}^d$ is compact, then

$$\mathsf{Cap}_{d-\frac{N}{\alpha}}(E) > 0 \implies \mathbb{P}\big(X([1, 1+\varepsilon_0]^N) \cap E \neq \varnothing\big) > 0. \quad (1)$$

An application of Frostman's theorem completes the proof; see Theorem 2.2.1, Appendix C. It remains to demonstrate equation (1).

From now on, we assume that $E \subset [-M, M]^d$ for some fixed finite constant $M > 0$. Suppose $\mu \in \mathcal{P}(E)$ is absolutely continuous with respect to Lebesgue's measure, and let f denote the probability density function $d\mu/dx$. Define

$$\mathfrak{I}(\mu) = \int_{[1, 1+\varepsilon_0]^N} f(X_s) \, ds.$$

Since each $X_s \sim \mathcal{N}_d(0, \mathbf{I})$,

$$\mathbb{E}[\mathfrak{I}(\mu)] = \varepsilon_0^N (2\pi)^{-\frac{d}{2}} \int_{[-M,M]^d} f(x) e^{-\frac{1}{2}\|x\|^2} \, dx \geq \varepsilon_0^N (2\pi)^{-\frac{d}{2}} e^{-\frac{1}{2}M^2}. \quad (2)$$

On the other hand, by Lemma 5.4.1, there exists a finite constant $A > 0$ such that

$$\mathbb{E}[\{\mathfrak{I}(\mu)\}^2] \leq A \cdot \mathsf{Energy}_{d-\frac{N}{\alpha}}(\mu).$$

Combining this and equation (2), together with the Paley–Zygmund inequality (Lemma 1.4.1, Chapter 3), we obtain the following:

$$\mathbb{P}\big(X([1, 1+\varepsilon_0]^N) \cap E \neq \varnothing\big) \geq \mathbb{P}\{\mathfrak{I}(\mu) > 0\} \geq \frac{\{\mathbb{E}[\mathfrak{I}(\mu)]\}^2}{\mathbb{E}[\{\mathfrak{I}(\mu)\}^2]}$$
$$\geq A_1 \big[\mathsf{Energy}_{d-\frac{N}{\alpha}}(\mu)\big]^{-1},$$

where $A_1 = \varepsilon_0^{2N}(2\pi)^{-d}e^{-M^2}/A$. Since the left-hand side and the choice of A_1 are both independent of the choice of μ, we can deduce that

$$\mathbb{P}\big(X([1,1+\varepsilon_0]^N) \cap E \neq \varnothing\big) \geq A_1 \Big[\inf_\mu \mathsf{Energy}_{d-\frac{N}{\alpha}}(\mu)\Big]^{-1},$$

where the infimum is taken over all probability measures $\mu \in \mathcal{P}(E)$ that are absolutely continuous with respect to Lebesgue's measure. Equation (1) now follows from Exercise 4.1.4 above. □

6 Supplementary Exercises

1. (Hard) Suppose $X = (X_t;\ t \in \mathbb{R}_+^N)$ is an N-parameter, S_Δ-valued Markov process with transition operators $\mathcal{T}$ and resolvent $\mathcal{R}$, where S is a separable metric space (say $S = \mathbb{R}^d$) and S_Δ is its one-point compactification. For every $\lambda \succ 0$ in $\mathbb{R}_+^N$, let $\mathfrak{e}(\lambda)$ denote the random vector defined in Section 3.2. Recall that $\mathfrak{e}(\lambda)$ is independent of the entire X process.

(i) Define a new process $X^\lambda = (X_t^\lambda;\ t \in \mathbb{R}_+^N)$ by

$$X_t^\lambda = \begin{cases} X_t, & \text{if } t \prec \mathfrak{e}(\lambda), \\ \Delta, & \text{if } t \not\prec \mathfrak{e}(\lambda). \end{cases}$$

Prove that X^λ is an N-parameter Markov process. Compute its transition operators and its resolvents.
(See also Theorem 1.3.2, Chapter 8.)

(ii) Show that whenever X is a strongly symmetric Markov process on $\mathbb{R}^d$, X^λ is a strongly symmetric Markov process on $\mathbb{R}_\Delta^d$. (What does this mean?)

(iii) By using X^λ, show that the upper bound in Theorem 3.1.1 can be improved to $A_2 \mathcal{C}_{r_\lambda}(E)$.
(HINT: Check that the integral in the latter theorem is equal to $\mathbb{P}_\nu\{\overline{X^\lambda(\mathbb{R}_+^N)} \cap E \neq \varnothing\}$.)

2. Given an N-parameter Markov process on $\mathbb{R}^d$ and a compact set $K \subset \mathbb{R}^d$ whose interior is open, we can define the set

$$\mathfrak{H}_K = \{t \in \mathbb{R}_+^N : X_s \in K \text{ for all } s \preccurlyeq t\}.$$

(i) Prove that $(t \in \mathfrak{H}_K)$ is $\mathcal{F}_t$-measurable, where $\mathcal{F} = (\mathcal{F}_t;\ t \in \mathbb{R}_+^N)$ denotes the complete augmented history of X.

(ii) Define the linear operators $\mathcal{T}^K = (\mathcal{T}_t^K;\ t \in \mathbb{R}_+^N)$ as follows: For all $f \in L^\infty(\mathbb{R}^d)$, all $t \in \mathbb{R}_+^N$, and all $x \in \mathbb{R}^d$,

$$\mathcal{T}_t^K f(x) = \mathbb{E}_x[f(X_t)\mathbf{1}_{\mathfrak{H}_K}(t)].$$

Prove that for any nonnegative function f, all $t, s \in \mathbb{R}_+^N$, and all x, $\mathcal{T}_{t+s}^K f(x) \leq \mathcal{T}_s^K \mathcal{T}_s^K f(x)$. That is, $\mathcal{T}^K$ is a subsemigroup. Prove that when $N=1$, $\mathcal{T}^K$ is a proper Markov semigroup.

6 Supplementary Exercises   451

3. Consider $k \geq 3$ independent, d-dimensional Brownian motions, all starting from distinct points of $\mathbb{R}^d$. Show that their ranges intersect with positive probability if and only if $d = 1$ or $d = 2$. Find an extension of this when the X's are $\mathbb{R}^d$-valued, isotropic stable Lévy processes, all with index $\alpha \in {]0, 2]}$. The first part is due to Dvoretzky et al. (1954).

4. Let X denote an N-parameter Markov process on a compact, separable metric space S. If $\mathcal{R}$ denotes its resolvent, prove that for all $\lambda \succ 0$ in $\mathbb{R}_+^N$ and all $\varphi \in L^\infty(S)$, $t \mapsto e^{-\lambda \cdot t} \mathcal{R}_\lambda \varphi(X_t)$ is an N-parameter supermartingale.

5. (Hard) Consider a one-parameter, isotropic stable process $X = (X_t; \geq 0)$ of index $\alpha \in [0, 2]$, and show that given a compact set $E \subset \mathbb{R}_+$,
$$\mathbb{P}(0 \in \overline{X(E)}) > 0 \iff \mathsf{Cap}_{d/\alpha}(E) > 0.$$
Conclude that no matter the value of α, whenever $d \geq 2$, $\mathrm{Leb}\{\overline{X(E)}\} = 0$, $\mathbb{P}$-a.s. This is from Hawkes (1977b, Theorem 4) and is based on Hawkes (1970, Theorem 3).

6. (Hard) Suppose X is an N-parameter, S-valued stochastic process, where S is a compact separable metric space. Suppose X satisfies all of the conditions for being an N-parameter Feller process, except that $t \mapsto X_t$ need not be right continuous. Prove that it has a modification that is a proper N-parameter Feller process.
(HINT: Use Supplementary Exercise 4 and follow the proof of Theorem 4.1.1, Chapter 8.)

7. Suppose X is an N-parameter Markov process, taking its values in some compact separable metric space S. Suppose further that there exists a measure ν on S and a measurable function p such that
$$\mathcal{T}_t f(x) = \int p_t(x, y) f(y) \, \nu(dy),$$
for all $t \succ 0$ in $\mathbb{R}_+^N$, all $x \in S$, and all $f \in L^\infty(S)$. Supposing that $\int_{\mathbb{R}_+^N} p_t(x, y) \, dt < \infty$, show that for all $\varphi \in L^\infty(S)$, $t \mapsto \mathcal{R}_0 \varphi(X_t)$ is a supermartingale, where $\mathcal{R}_0 \varphi(x) = \int_{\mathbb{R}_+^N} \mathcal{T}_t \varphi(x) \, dt$.

8. (Hard) Consider two independent d-dimensional Brownian motions, B^1 and B^2. Let $P(\varepsilon)$ denote the probability that $B^1([1, 2])$ is within ε of $B^2([1, 2])$. Prove the existence of positive and finite constants A_1 and A_2 such that for all $\varepsilon \in {]0, 1[}$,
$$A_1 U(\varepsilon) \leq P(\varepsilon) \leq A_2 U(\varepsilon),$$
where
$$U(\varepsilon) = \begin{cases} 1, & \text{if } 1 \leq d \leq 3, \\ \{\ln_+(1/\varepsilon)\}^{-1}, & \text{if } d = 4, \\ \varepsilon^{d-4}, & \text{if } d > 4. \end{cases}$$

(i) Show that the above estimate implies Theorem 4.3.2.
(ii) What if B^1 and B^2 were independent d-dimensional isotropic stable Lévy processes with the same index $\alpha \in {]0, 2[}$ each?

This is due to Aizenman (1985).
(HINT: The lower bound on $P(\varepsilon)$ uses the Paley–Zygmund lemma. For the upper bound, consider the 2-parameter martingale $M_t = \mathbb{E}[\int_{[1,3]^2} \mathbf{1}_{(|B_s^1 - B_t^2| \le \varepsilon)}\, ds\, dt \mid \mathcal{F}_t]$, $t \in [1,2]^2$. Use the trivial inequality that for all $t \in [1,2]^2$, $M_t \ge M_t \mathbf{1}_{(|B_t| \le \varepsilon/2)}$ and apply Cairoli's strong $(2,2)$ inequality. This exercise is an extension of Proposition 1.4.1, Chapter 10.)

9. Show that a modification of the process X of the example of Section 5.1 exists such that for all $a \preccurlyeq b$ both in $\mathbb{R}_+^N$,
$$\limsup_{h \to 0^+} \sup_{s,t \in [a,b]:\, |s-t| \le h} \frac{|X_t - X_s|}{h^{\frac{\beta}{2}} \sqrt{\ln_+(1/h)}} < \infty.$$
(HINT: Consult Dudley's theorem; see Theorem 2.7.1, Chapter 5.)

10. Consider a d-dimensional Brownian motion $B = (B_t;\, t \ge 0)$. If $E \subset \mathbb{R}_+$ is a compact set, show that with probability one, $\dim\{B(E)\} = 2\dim(E) \wedge d$. This is due to McKean (1955a).
(HINT: The upper bound on the dimension of $B(E)$ follows readily from Example 2, Section 1.2 of Appendix C. For the lower bound, take a probability measure μ on E and consider the β-dimensional Bessel–Riesz energy of $\mu \circ B^{-1} \in \mathcal{P}(B(E))$. Start your analysis of this half by showing that for any $\beta > 0$, $\mathrm{Energy}_\beta(\mu \circ B^{-1}) = \iint \|B_s - B_t\|^{-\beta}\, \mu(ds)\, \mu(dt)$. Finish by computing the expectation of the latter.)

11. Suppose $Y = (Y_t;\, t \in \mathbb{R}_+^N)$ is an $\mathbb{R}^d$-valued stochastic process with right-continuous trajectories and some open set $E \subset \mathbb{R}^d$. Prove that there is a measurable $T \in \mathbb{Q}_+^N \cup \{\infty\}$ ($\mathbb{R}_+^N \cup \{\infty\}$ denotes the one-point compactification of $\mathbb{R}_+^N$) such that $\exists t \in [0,1]^N$ such that $Y_t(\omega) \in E$ if and only if $T(\omega) \in \mathbb{Q}_+^N \cap [0,1]^N$. Prove that equation (4) of Section 3.3 follows from this.
(HINT: When $N = 1$, you can define $T_n = \inf(j2^{-n} :\ 0 \le j \le 2^n,\ Y_{j2^{-n}} \in E)$ and define $T = \inf_n T_n$. As usual, $\inf \varnothing = +\infty$. When $N > 1$, work one parameter at a time.)

12. (Hard) Let X denote an N-parameter, $\mathbb{R}^d$-valued additive stable process of index $\alpha \in\,]0,2]$. Suppose $d \ge 2N$ and fix $M > 1$. Prove that
$$\lim_{\varepsilon \to 0^+} \sup_{a \in [-M,M]^d} \mathbb{P}\{\overline{X([1,2]^N)} \cap B(a;\varepsilon) \ne \varnothing\} = 0.$$
Conclude that a.s., the closure of $X([1,2]^N)$ has zero Lebesgue measure in $\mathbb{R}^d$.
(HINT: Prove the appropriate variant of Supplementary Exercise 8.)

13. Let X denote N-parameter additive Brownian motion in $\mathbb{R}^d$, where $d \le 2N$. Prove that if $K \subset \mathbb{R}^d$ is compact,
$$\mathbb{E}_0\left[\int_{\mathbb{R}_+^N} e^{-\lambda \cdot s} \mathbf{1}_K(X_s)\, ds\right] = +\infty,$$
whenever $\lambda \in \mathbb{R}_+^N$ has a coordinate that is 0. Conclude that the potential operator $\mathcal{R}_\lambda$ is a bounded linear operator only if $\lambda \succ 0$ ($\in \mathbb{R}_+^N$).

7 Notes on Chapter 11

Section 1 In the literature at large there does not seem to be an agreement on what a multiparameter Markov process should be. The development here is new and is designed to (i) handle the potential theory of many of the interesting processes; and (ii) be sufficiently general without being overburdened with technical details. However, earlier parts of the development of this theory can be found in Cairoli (1966, 1970a), as well as (Hirsch and Song 1995b; Mazziotto 1988; Wong 1989).

Bass and Pyke (1985, 1984a) contain existence and regularity results for a very large class of random fields that include the additive Lévy processes of this chapter. In considering general multiparameter Markov processes, similar questions are largely unanswered at the moment, and much more remains to be done on this topic.

Section 3 The original motivation for the development of Theorem 3.1.1 and its ilk was to decide when the trajectories of k independent Lévy processes intersect. This was the so-called Hendricks–Taylor problem and was solved in the pioneering work of Fitzsimmons and Salisbury (1989). In the background lurks an ingenious result of Fitzsimmons and Maisonneuve (1986); it essentially states that any one-parameter Markov process can be reversed, as long as one starts the process appropriately (that time-reversal again!). In this connection, see also the expository paper Salisbury (1996).

The statement of the Hendricks–Taylor problem, as well as related things, can be found in Hendricks (1972, 1974, 1979) and Taylor (1986, Problem 6).

Hirsch and Song (1994, 1995a, 1995b, 1995c) extend the Fitzsimmons–Salisbury theory, by other methods, to exhibit a very general and elegant potential theory for multiparameter Markov processes. For related works, see Bass et al. (1994), Bass and Khoshnevisan (1993a), Bauer (1994), Dynkin (1980, 1981a, 1981b, 1983, 1984a, 1984b, 1985, 1986, 1987), Geman et al. (1984), Hawkes (1977a, 1978, 1979), Le Gall and Rosen (1991), Le Gall (1987), Peres (1996a, 1996b), Ren (1990), Rosen (1983, 1984), and Werner (1993). Some of the earlier works on the Hendricks–Taylor problem can be found in Evans (1987a, 1987b), Le Gall et al. (1989), and Rogers (1989).

The methods of this section are an adaptation of those of Khoshnevisan and Shi (1999), developed to study the Brownian sheet (which is, unfortunately, not a multiparameter Markov process as we have defined such processes here.)[10]

Section 4 The material of this section is largely motivated by those of (Fitzsimmons and Salisbury 1989; Hirsch 1995; Khoshnevisan and Shi 1999). Technically speaking, Theorem 4.6.1 is new. However, you should read the one-parameter argument McKean (1955b) with care to see that the necessary ideas are all there. Theorem 4.6.2 is a multiparameter extension of some of the results of Hawkes (1971a, 1971b). In fact, much more can be proved. For instance, Blumenthal and Getoor (1960a) compute $\dim\{X(E)\}$ for a compact set E, where X is a one-parameter stable Lévy process on $\mathbb{R}^d$. See Blumenthal and Getoor (1962),

---

[10] As a counterpoint to this last remark, see the Notes in Chapter 12.

Hawkes (1971a), Kahane (1985), Khoshnevisan and Xiao (2002), Le Gall (1992), Pruitt (1970, 1975), and Taylor (1966) for related results and many other references. In other directions, extensions can be found in Taylor (1953, 1955, 1973, 1986). The last two references are survery articles that point to older literature.

In one dimension, stochastic codimension was first formally defined in Khoshnevisan and Shi (2000). However, the real essence of such a notion appears much earlier in the literature. For example, see the proof of Taylor (1966, Theorem 4); see also (Barlow and Perkins 1984; Lyons 1990; Peres 1996b).

Lemma 4.7.2 was first found in Peres (1996b, Lemma 2.2), but with a different proof.

The mentioned bibliography misses three very interesting papers related to the material of this section by way of conditioned processes and time-reversal; see Davis and Salisbury (1988) and Salisbury (1988, 1992).

Section 5 The material from this section follows from the more general treatment of Weber (1983) and Testard (1985, 1986). The proofs presented here, as well as Exercise 5.3.3, are worked out jointly with Z. Shi; they are included here with his permission.

Theorem 5.1.1 is completely classical. In the present probabilistic setting you can find it in Adler (1981, Section 7.2) and in Rozanov (1982, Section 2, Chapter 3).

12
The Brownian Sheet and Potential Theory

The results and discussions of Chapter 11 do not readily extend to cover a multiparameter process such as the Brownian sheet, since the latter is not a multiparameter Markov process according to the definitions of Chapter 11. In this regard, see Exercise 1.1.3, Chapter 11. In the first section of this chapter we extend and refine the arguments of Chapter 11 to estimate intersection probabilities for the range of the Brownian sheet. As in the development of Chapter 11, these estimates yield geometric information about the range of the Brownian sheet by way of Hausdorff dimension calculations.

Section 2 of this chapter is concerned with the evaluation of the Hausdorff dimension of the zero set of the Brownian sheet; here, the zero set of an N-parameter process X is the set of "times" $t \in \mathbb{R}_+^N$ when $X_t = 0$. In order to perform this calculation, we need a way of estimating the probability that the zero set intersects a small set (yet another application of intersection probabilities).

The third, and final, section is a brief introduction to the theory of local times. Roughly speaking, the local time of a process X at zero is the most natural random measure that one can construct on the zero set. The arguments of Section 2 rely on an approximate version of this local time, and this connection will be further developed in Section 3.

Throughout this chapter, $B = (B_t; t \in \mathbb{R}_+^N)$ will denote an N-parameter, d-dimensional Brownian sheet.

1 Polar Sets for the Range of the Brownian Sheet

A (nonrandom) set $E \subset \mathbb{R}^d$ is said to be **polar** for a (random) set $K \subset \mathbb{R}^d$ if $\mathbb{P}(K \cap E \neq \varnothing) = 0$; otherwise, E is **nonpolar** for K. When K is the range of a suitable multiparameter Markov process in $\mathbb{R}^d$, the results of Chapter 11 characterize the collection of all compact, and hence σ-compact, sets in $\mathbb{R}^d$ that are polar for the range of the aforementioned multiparameter Markov process. The primary goal of this chapter is to determine the collection of all polar sets for the range of the Brownian sheet.

1.1 Intersection Probabilities

Recalling the definition of Bessel–Riesz capacities Cap_β from Appendix C, the main result of this chapter is that the range of B intersects a compact set $E \subset \mathbb{R}^d$ if and only if $\mathsf{Cap}_{d-2N}(E) > 0$. This agrees with Kakutani's theorem when $N = 1$; cf. Theorem 3.1.1, Chapter 10.

Before we attempt detailed calculations, note that no matter how small or how thin E is, $\mathbb{P}(\overline{B(\mathbb{R}_+^N)} \cap E \neq \varnothing) = 1$, as long as E intersects the boundary of $\mathbb{R}^d$. This is due to the fact that for all t on the boundary of the parameter space $\mathbb{R}_+^N$, $B_t = 0$. However, this is merely a superficial problem, and there are two ways to avoid it. To explain the first method, note that the portion of E that is away from the axes is polar for the range of B if and only if $E \cap [n^{-1}, n]^d$ is polar for all integers $n \geq 1$. (Why?) Consequently, we need only characterize all compact polar subsets of the interior of $\mathbb{R}^d$, i.e., $\mathbb{R}^d \setminus \partial \mathbb{R}^d$. Alternatively, we can slightly redefine our notion of polarity by thinking of E as polar (for the range of B) if and only if for all $0 \prec a \preccurlyeq b$, $\mathbb{P}(B([a, b]) \cap E \neq \varnothing) > 0$. Following Khoshnevisan and Shi (1999), we take the latter approach, due to its technical simplicity. The former approach is developed in Supplementary Exercise 4.

Theorem 1.1.1 *Let $E \subset \mathbb{R}^d$ denote a compact subset of $\mathbb{R}^d$. For any $0 \prec a \prec b$, both in $\mathbb{R}_+^N$, there exists a finite constant $A_\star \geq 1$ such that*

$$\frac{1}{A_\star}\mathsf{Cap}_{d-2N}(E) \leq \mathbb{P}\big(B([a,b]) \cap E \neq \varnothing\big) \leq A_\star \mathsf{Cap}_{d-2N}(E).$$

The following is an immediate, but important, corollary of this theorem.

Corollary 1.1.1 *Suppose E is a compact subset of $\mathbb{R}^d \setminus \{0\}$. Then, E is polar for the range of the N-parameter Brownian sheet if and only if $\mathsf{Cap}_{d-2N}(E) > 0$.*

In particular, Frostman's theorem implies that the range of the Brownian sheet intersects E with positive probability if $\dim(E) > d - 2N$, while there are no such intersections if $\dim(E) < d - 2N$; cf. Theorem 2.1.1 of Appendix C. Recalling stochastic codimensions from Section 4.7, Chapter 11, we deduce the following as a by-product.

Corollary 1.1.2 *For any $0 \prec a \prec b$, both in $\mathbb{R}_+^N$,*

$$\operatorname{codim}\{B([a,b])\} = \operatorname{codim}\{\overline{B(\mathbb{R}_+^N)}\} = (d - 2N)^+.$$

***Exercise* 1.1.1** Complete the derivation of Corollary 1.1.2. □

According to Theorem 4.7.1 of Chapter 11, we can now deduce the following computation of the Hausdorff dimension of the image of a rectangle $[a, b] \subset \mathbb{R}_+^N$, under the random map B.

Corollary 1.1.3 *For any $0 \prec a \prec b$, both in $\mathbb{R}_+^N$,*

$$\dim\{B([a,b])\} = \dim\{\overline{B(\mathbb{R}_+^N)}\} = 2N \wedge d, \quad \text{a.s.}$$

***Exercise* 1.1.2** Complete the derivation of Corollary 1.1.3. □

In particular, when $2N < d$, the image of $[0, 1]^N$ under the random map B has 0 d-dimensional Lebesgue's measure, while it has Hausdorff dimension $2N$. That is, when $2N < d$, $B([0, 1]^N)$ has irregular—sometimes called fractal—structure in that it has a Hausdorff dimension (here, $2N$) that is different from its Euclidean dimension (here, d). Perhaps this fractal-like behavior can be detected in the (multiparameter) random walk simulations/approximations of Figures 12.1, 12.2, and 12.3. All three figures show a simulation of the same 2-parameter, 1-dimensional Brownian sheet. The figures are plotted such that the (x, y)-plane forms the 2-parameter time space against the z-axis that shows the values of the process. That is, in all three figures, $(x, y, z) = (t^{(1)}, t^{(2)}, B_t)$.

1.2 Proof of Theorem 1.1.1: Lower Bound

Let $E \subset \mathbb{R}^d$ be a fixed compact set. Throughout, M denotes the **outer radius** of E, i.e.,

$$M = \sup\{\|x\| : x \in E\}. \tag{1}$$

We start our proof of Theorem 1.1.1 by defining the following: For all $a \prec b$, both in $\mathbb{R}_+^N$, and for all measurable functions $f : \mathbb{R}^d \to \mathbb{R}_+$ that are supported on $[-M, M]^d$,

$$J_{a,b}(f) = \int_{[a,b]} f(B_s)\, ds. \tag{2}$$

Using the explicit form of the density function of B_s (Exercise 1.1.1, Chapter 5), we can write

$$\mathbb{E}[J_{a,b}(f)] = \int_{[a,b]} \int_{[-M,M]^d} f(y) p_s(y)\, dy\, ds,$$

where for all $s \in \mathbb{R}_+^N$ and all $y \in \mathbb{R}^d$,

$$p_s(y) = (2\pi)^{-\frac{d}{2}} \left(\prod_{\ell=1}^N s^{(\ell)}\right)^{-\frac{d}{2}} \exp\left\{-\frac{\|y\|^2}{2\prod_{\ell=1}^N s^{(\ell)}}\right\}. \tag{3}$$

458   12. The Brownian Sheet and Potential Theory

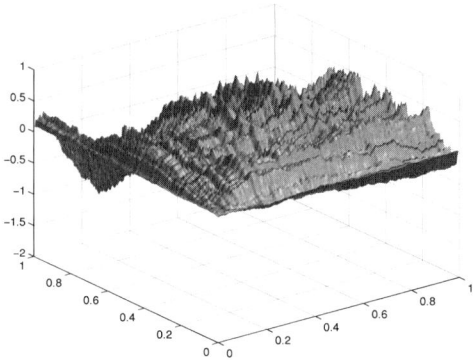

Figure 12.1: Aerial view of an ℝ-valued, 2-parameter Brownian sheet

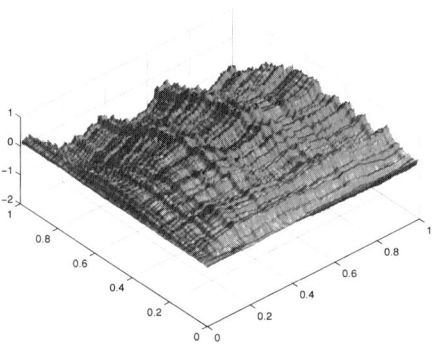

Figure 12.2: Another view of the same Brownian sheet

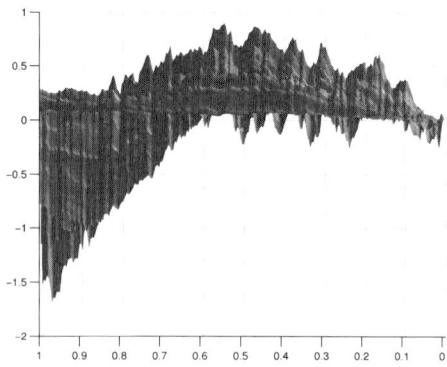

Figure 12.3: A side view of the same Brownian sheet

Note that whenever $s \in [a,b]$ and $y \in [-M,M]^d$,

$$p_s(y) \geq (2\pi)^{-\frac{d}{2}} \Big(\prod_{\ell=1}^{N} b^{(\ell)} \Big)^{-\frac{d}{2}} \exp\Big\{ -\frac{M^2}{2\prod_{\ell=1}^{N} a^{(\ell)}} \Big\} = A_1.$$

Thus, we obtain the following "first moment estimate":

Lemma 1.2.1 *For all $0 \prec a \prec b$, both in $\mathbb{R}_+^N$, there exists a finite constant $A_1 > 0$ such that for all measurable functions $f : \mathbb{R}^d \to \mathbb{R}_+$ that are supported in $[-M,M]^d$,*

$$\mathbb{E}[J_{a,b}(f)] \geq A_1 \cdot \int f(y)\,dy.$$

We shall prove the following "second moment estimate" for $J_{a,b}(f)$ in the next subsection.

Lemma 1.2.2 *For all $0 \prec a \prec b$, both in $\mathbb{R}_+^N$, and all $m > 0$, there exists a finite constant $A_2 > 0$ such that for all measurable functions $f : \mathbb{R}^d \to \mathbb{R}_+$ that are supported in $[-m,m]^d$,*

$$\mathbb{E}\big[\{J_{a,b}(f)\}^2\big] \leq A_2 \cdot \mathsf{Energy}_{d-2N}(f),$$

where $\mathsf{Energy}_\beta(f)$ denotes the β-dimensional Bessel–Riesz energy of the measure $\mu(dx) = f(x)\,dx$; cf. Appendix C.

Armed with this, we can readily demonstrate the lower bound for the probability of Theorem 1.1.1.

Proof of Theorem 1.1.1: Lower Bound Recall the outer radius M of E from equation (1) and hold fixed a probability density function $f : \mathbb{R}^d \to \mathbb{R}_+$ that is supported in E^ε, where $\varepsilon \in\,]0,1[$ is arbitrary and where E^ε denotes the ε-enlargement of E. That is,

$$E^\varepsilon = \{y \in \mathbb{R}^d : \mathrm{dist}\{y;E\} < \varepsilon\}.$$

We note that whenever $J_{a,b}(f)$ is strictly positive, there must exist some $t \in [a,b]$ such that $B_t \in E^\varepsilon$. Equivalently,

$$(J_{a,b}(f) > 0) \subseteq \big(B([a,b]) \cap E^\varepsilon \neq \varnothing\big).$$

In particular,

$$\mathbb{P}\big(B([a,b]) \cap E^\varepsilon \neq \varnothing\big) \geq \mathbb{P}(J_{a,b}(f) > 0) \geq \frac{\{\mathbb{E}[J_{a,b}(f)]\}^2}{\mathbb{E}\big[\{J_{a,b}(f)\}^2\big]}.$$

We have used the Paley–Zygmund lemma in the last line; cf. Lemma 1.4.1 of Chapter 3. Since $\varepsilon \in\,]0,1[$, f is supported in $[-m,m]^d$ with $m = M+1$.

Hence, we can apply Lemmas 1.2.1 and 1.2.2 and deduce the existence of constants A_1 and A_2 such that, *independently of the choice* of f and ε,

$$\mathbb{P}\{B([a,b]) \cap E^\varepsilon \neq \varnothing\} \geq \frac{A_1^2}{A_2 \, \mathsf{Energy}_{d-2N}(f)},$$

where $1 \div \infty = 0$. Optimizing over the choice of all probability density functions f that are supported on E^ε, we see that for all $\varepsilon > 0$,

$$\mathbb{P}\{B([a,b]) \cap E^\varepsilon \neq \varnothing\} \geq \frac{A_1^2}{A_2} \cdot \mathsf{Cap}^{\mathsf{ac}}_{d-2N}(E^\varepsilon),$$

where $\mathsf{Cap}^{\mathsf{ac}}_\beta$ denotes the absolutely continuous capacity corresponding to the gauge function $x \mapsto \|x\|^{-\beta}$ for any $\beta > 0$. By Exercise 4.1.4, Chapter 11, there exists a finite constant $K > 1$ such that

$$K^{-1} \mathsf{Cap}_{d-2N}(E^\varepsilon) \leq \mathsf{Cap}^{\mathsf{ac}}_{d-2N}(E^\varepsilon) \leq K \mathsf{Cap}_{d-2N}(E^\varepsilon).$$

It should be recognized that K depends only on the distance between E and the axes of $\mathbb{R}_+^N$, the outer radius of E, and the parameters N and d.

As $\varepsilon \to 0^+$, $E^\varepsilon \downarrow E$, which is compact. Since $t \mapsto B_t$ is continuous, Lemma 2.1.2 of Appendix D implies that

$$\mathbb{P}\{B([a,b]) \cap E \neq \varnothing\} \geq \frac{A_1^2}{A_2 K} \cdot \mathsf{Cap}_{d-2N}(E).$$

(Why?) This concludes our proof of the lower bound of Theorem 1.1.1, where $A_\star$ is any constant chosen to be larger than the maximum of 1 and $A_2 K / A_1^2$. □

1.3 Proof of Lemma 1.2.2

Our proof of Lemma 1.2.2 is divided into several steps. First, we need a technical real-variable result that estimates the following function: For all $s, t \in \mathbb{R}_+^N$,

$$\sigma^2(s,t) = \frac{\prod_{\ell=1}^N s^{(\ell)} t^{(\ell)} - \prod_{\ell=1}^N (s^{(\ell)} \wedge t^{(\ell)})^2}{\prod_{\ell=1}^N s^{(\ell)} t^{(\ell)}}. \tag{1}$$

Throughout this section we will choose and fix two points $a, b \in \mathbb{R}_+^N$ with $0 \prec a \prec b$. These are the same constants a and b of Lemma 1.2.2 of the previous subsection.

Lemma 1.3.1 *There exists a finite constant $A_1 > 1$ such that for all $s, t \in [a,b]$,*

$$\frac{1}{A_1} \|s-t\| \leq \sigma^2(s,t) \leq A_1 \|s-t\|.$$

That is, as s tends to t, $\sigma(s,t)$ goes to 0, roughly at the rate $\|s-t\| = \{\sum_{\ell=1}^{N} |s^{(\ell)} - t^{(\ell)}|^2\}^{\frac{1}{2}}$. Moreover, this approximation holds uniformly over all $s, t \in [a,b]$. The significance of the function σ will be shown later. Right now, we set out to prove the technical Lemma 1.3.1.

Proof Throughout this proof we define

$$C_\star = \frac{\max_{1 \leq j \leq N} b^{(j)}}{\min_{1 \leq j \leq N} a^{(j)}}.$$

For all $s, t \in [a, b]$,

$$\prod_{j=1}^{N} \left(\frac{s^{(j)} \vee t^{(j)}}{s^{(j)} \wedge t^{(j)}} \right) = \exp\left\{ \sum_{j=1}^{N} \ln\left(\frac{s^{(j)} \vee t^{(j)}}{s^{(j)} \wedge t^{(j)}} \right) \right\}$$

$$= \exp\left\{ \sum_{j=1}^{N} \ln\left(1 + \frac{|s^{(j)} - t^{(j)}|}{s^{(j)} \wedge t^{(j)}} \right) \right\}.$$

On the other hand, since $\ln(1+r) = \int_1^{1+r} u^{-1}\, du$,

$$\frac{r}{1+r} \leq \ln(1+r) \leq r, \qquad r > 0.$$

Consequently, for all $0 \leq r \leq C_\star$,

$$\frac{r}{1+C_\star} \leq \ln(1+r) \leq r.$$

We apply this estimate to $r = |s^{(j)} - t^{(j)}|/(s^{(j)} \wedge t^{(j)})$ and deduce the following:

$$\exp\left\{ \sum_{j=1}^{N} \frac{|s^{(j)} - t^{(j)}|}{(1+C_\star)(s^{(j)} \wedge t^{(j)})} \right\} \leq \prod_{j=1}^{N} \left(\frac{s^{(j)} \vee t^{(j)}}{s^{(j)} \wedge t^{(j)}} \right) \leq \exp\left\{ \sum_{j=1}^{N} \frac{|s^{(j)} - t^{(j)}|}{s^{(j)} \wedge t^{(j)}} \right\}.$$

In the above we have used the simple fact that $a^{(j)} \leq s^{(j)} \wedge t^{(j)} \leq b^{(j)}$ to ensure that our r is indeed in $[0, C_\star]$. Using this simple fact once more, we see that

$$e^{C_1|s-t|} \leq \prod_{j=1}^{N} \left(\frac{s^{(j)} \vee t^{(j)}}{s^{(j)} \wedge t^{(j)}} \right) \leq e^{C_2|s-t|}, \tag{2}$$

where

$$C_1 = \frac{1}{(1+C_\star)\max_{1\leq j \leq N} b^{(j)}},$$

$$C_2 = \frac{N}{\min_{1\leq j \leq N} a^{(j)}}.$$

Now we estimate the above exponentials. Note that for all $x \geq 0$,

$$1 + x \leq e^x \leq 1 + xe^x.$$

This follows immediately from $e^x = 1 + \int_0^x e^u \, du$. Applying equation (2), together with the above estimates, we see that

$$1 + C_1 |s-t| \leq \prod_{j=1}^{N} \left(\frac{s^{(j)} \vee t^{(j)}}{s^{(j)} \wedge t^{(j)}} \right) \leq 1 + C_2 |s-t| e^{C_2 |s-t|}.$$

Since $|s-t| \leq \max_{1 \leq j \leq N} b^{(j)}$, $C_2 |s-t| \leq NC_\star$, and this gives

$$1 + C_1 |s-t| \leq \prod_{j=1}^{N} \left(\frac{s^{(j)} \vee t^{(j)}}{s^{(j)} \wedge t^{(j)}} \right) \leq 1 + C_3 |s-t|, \qquad (3)$$

with $C_3 = C_2 e^{NC_\star}$. Now we finish verifying the lemma. Clearly,

$$\sigma^2(s,t) = \prod_{\ell=1}^{N} \left(\frac{s^{(\ell)} \wedge t^{(\ell)}}{s^{(\ell)} \vee t^{(\ell)}} \right) \cdot \left\{ \prod_{j=1}^{N} \left(\frac{s^{(j)} \vee t^{(j)}}{s^{(j)} \wedge t^{(j)}} \right) - 1 \right\}.$$

Equally clearly,

$$C_\star^{-N} \leq \prod_{\ell=1}^{N} \left(\frac{s^{(\ell)} \wedge t^{(\ell)}}{s^{(\ell)} \vee t^{(\ell)}} \right) \leq C_\star^{N}.$$

Thus, equation (3) implies

$$C_\star^{-N} C_1 \cdot |s-t| \leq \sigma^2(s,t) \leq C_\star^{N} C_3 \cdot |s-t|.$$

By the Cauchy–Schwarz inequality, $N^{-1} \|s-t\| \leq |s-t| \leq \|s-t\|$. Consequently, the lemma follows once we let A_1 be the maximum of 1, $C_\star^N C_3$, and $NC_\star^N C_1^{-1}$. □

For all $s, t \in [a,b]$, define

$$R(s,t) = \frac{\prod_{\ell=1}^{N} (s^{(\ell)} \wedge t^{(\ell)})}{\prod_{\ell=1}^{N} t^{(\ell)}}. \qquad (4)$$

The following is the Brownian sheet analogue of Lemma 5.3.1 of Chapter 11. It also shows the role of the functions $(s,t) \mapsto \sigma^2(s,t)$ via the function $(s,t) \mapsto R(s,t)$.

Lemma 1.3.2 *For any two points $s, t \in [a,b]$, $B_s - R(s,t) B_t$ is independent of B_t. Moreover, $B_s - R(s,t) B_t \sim N_d(0, \sigma^2(s,t) \mathbf{I}_d)$, where $\mathbf{I}_d$ denotes the $(d \times d)$ identity matrix.*

***Exercise* 1.3.1** Prove Lemma 1.3.2. □

Next, we need to estimate the size of $1 - R(s,t)$, when $s \approx t$. The proof is based on similar arguments that were used to prove Lemma 1.3.1.

Lemma 1.3.3 *There exists a finite constant $A_2 \geq 1$ such that for all $s, t \in [a, b]$,*
$$\frac{1}{A_2} \|s - t\| \leq 1 - R(s, t) \leq A_2 \|s - t\|.$$

Proof Later on, we will need only the upper bound on $1 - R(s, t)$. Therefore, we will verify only this part. The lower bound is proved in Exercise 1.3.2 below.

Note that
$$R(s, t) = \exp\left\{ \sum_{\ell=1}^{N} \ln\left(1 - \frac{t^{(\ell)} - s^{(\ell)} \wedge t^{(\ell)}}{t^{(\ell)}}\right)\right\}.$$

For all $r \geq 0$, $\ln(1 - r) = -\int_{1-r}^{1} u^{-1} du \geq -r/(1-r)$. Therefore,
$$R(s, t) \geq \exp\left\{ -\sum_{\ell=1}^{N} \frac{|t^{(\ell)} - s^{(\ell)} \wedge t^{(\ell)}|}{t^{(\ell)}}\right\}$$
$$\geq \exp\left\{ -\frac{1}{\max_{1 \leq j \leq N} b^{(j)}} \cdot \sum_{\ell=1}^{N} |t^{(\ell)} - s^{(\ell)} \wedge t^{(\ell)}|\right\}$$
$$\geq \exp\left\{ -\frac{1}{\max_{1 \leq j \leq N} b^{(j)}} \cdot \sum_{\ell=1}^{N} |t^{(\ell)} - s^{(\ell)}|\right\}$$
$$\geq \exp\left\{ -\frac{\sqrt{N}}{\max_{1 \leq j \leq N} b^{(j)}} \|s - t\|\right\}.$$

We have used the Cauchy–Schwarz inequality in the last line. The upper bound of the lemma follows from the inequality $1 - x \leq e^{-x}$, $x \geq 0$. □

***Exercise* 1.3.2** Verify the lower bound of Lemma 1.3.3. □

We will also need a joint probability estimate.

Lemma 1.3.4 *For all $m > 0$, there exists a finite constant $A_3 > 0$ such that for all $s, t \in [a, b]$ and all measurable functions $f, g : \mathbb{R}^d \to \mathbb{R}_+$ that are supported on $[-m, m]^d$,*
$$\mathbb{E}[f(B_s)g(B_t)] \leq A_3 \|s - t\|^{-\frac{d}{2}} \cdot \iint f(w)g(y) \exp\left\{-\frac{\|w - y\|^2}{4A_1 \|s - t\|}\right\} dw\, dy,$$

where A_1 is the constant of Lemma 1.3.1.

Proof Clearly,

$$\mathbb{E}\big[f(B_s)g(B_t)\big] = \mathbb{E}\Big[f\big(B_s - R(s,t)B_t + R(s,t)B_t\big)\,g(B_t)\Big]$$
$$= \int g(y)\,p_t(y)\,\mathbb{E}\Big[f\big(B_s - R(s,t)B_t + R(s,t)B_t\big)\Big]\,dy,$$

where $p_t(y)$ is given by equation (3) of Section 1.2. Since g is supported on $[-m,m]^d$, it suffices to consider $p_t(y)$ for $t \in [a,b]$ and $y \in [-m,m]^d$. In this case, it follows immediately from (3) of Section 1.2 that $p_t(y) \leq A_4$, where

$$A_4 = (2\pi)^{-\frac{d}{2}}\Big(\prod_{\ell=1}^N a^{(\ell)}\Big)^{-\frac{d}{2}}.$$

Thus,

$$\mathbb{E}\big[f(B_s)g(B_t)\big] \leq A_4 \cdot \int g(y)\,\mathbb{E}\Big[f\big(B_s - R(s,t)B_t + R(s,t)y\big)\Big]\,dy. \quad (5)$$

Now let us use Lemma 1.3.2 and the explicit form of the Gaussian densities to deduce that for all $s,t \in [a,b]$ and all $y \in [-m,m]^d$,

$$\mathbb{E}\Big[f\big(B_s - R(s,t)B_t + R(s,t)y\big)\Big]$$
$$= (2\pi)^{-\frac{d}{2}}\{\sigma(s,t)\}^{-d}\int f(z + R(s,t)y)\exp\Big\{-\frac{\|z\|^2}{2\sigma^2(s,t)}\Big\}\,dz.$$

We can use Lemma 1.3.1 and the fact that $2\pi \geq 1$ (to simplify the constants) and deduce that for all $s,t \in [a,b]$ and for all $y \in [-m,m]^d$,

$$\mathbb{E}\Big[f\big(B_s - R(s,t)B_t + R(s,t)y\big)\Big]$$
$$\leq A_1^{\frac{d}{2}}\|s-t\|^{-\frac{d}{2}}\int f(z+R(s,t)y)\exp\Big\{-\frac{\|z\|^2}{2A_1\|s-t\|}\Big\}\,dz$$
$$= A_1^{\frac{d}{2}}\|s-t\|^{-\frac{d}{2}}\int f(w)\exp\Big\{-\frac{\|w-R(s,t)y\|^2}{2A_1\|s-t\|}\Big\}\,dw, \quad (6)$$

upon changing variables $w = z + R(s,t)y$. On the other hand, it is simple to see that

$$\|w-y\|^2 \leq 2\|w - R(s,t)y\|^2 + 2\|y\|^2 \cdot |1 - R(s,t)|^2.$$

Thus, an application of Lemma 1.3.3 shows us that for all $s,t \in [a,b]$, all $w \in \mathbb{R}^d$, and all $y \in [-m,m]^d$,

$$\|w - R(s,t)y\|^2 \geq \frac{1}{2}\|w-y\|^2 - m^2 A_2^2\|s-t\|^2$$
$$\geq \frac{1}{2}\|w-y\|^2 - 2m^2 A_2^2 N^{\frac{1}{2}}|b|\cdot\|s-t\|.$$

We have used the elementary bound $\|s-t\| \le N^{\frac{1}{2}}|b|$. Plugging this into equation (6), we see that for all $s,t \in [a,b]$ and for all $y \in [-m,m]^d$,

$$\mathbb{E}\Big[f\big(B_s - R(s,t)B_t + R(s,t)y\big)\Big]$$
$$\le A_1^{\frac{d}{2}} e^{m^2 A_2^2 N^{\frac{1}{2}}|b|} \|s-t\|^{-\frac{d}{2}} \int f(w) \exp\left\{-\frac{\|w-y\|^2}{4A_1\|s-t\|}\right\} dw.$$

This and equation (5) together imply the lemma, where the constant A_3 is defined by $A_1^{\frac{d}{2}} A_4 \exp\big(m^2 A_2^2 N^{\frac{1}{2}}|b|\big)$. $\square$

Next, we present a final real-variable lemma. For all $\beta > 0$, define

$$\Phi_\beta(x) = \int_{s \in \mathbb{R}^N:\ \|s\| \le x} \|s\|^{-\beta} e^{-1/\|s\|} ds, \quad x \ge 0. \tag{7}$$

Lemma 1.3.5 *For all $m, \beta > 0$, there exists a finite constant $A_5 \ge 1$ such that for all $x \in [m, \infty[$,*

$$\frac{1}{A_5} U_\beta(x) \le \Phi_\beta(x) \le A_5 U_\beta(x),$$

where

$$U_\beta(x) = \begin{cases} 1, & \text{if } \beta > N, \\ \ln_+(x), & \text{if } \beta = N, \\ x^{-\beta+N}, & \text{if } \beta < N. \end{cases}$$

Remark The function U_β and its variants have already made several appearances in this book. For example, noting the slight change in notation, see equation (1) from Section 1.3 of Chapter 10.

Proof We will prove the upper bound for Φ_β; the lower bound is proved similarly and can be found in Exercise 1.3.3 below.

We can assume, without loss of generality, that $m < 1$. By calculating in polar coordinates (Supplementary Exercise 7, Chapter 3), we can find a finite constant $C_1 > 0$ such that

$$\Phi_\beta(x) = C_1 \int_0^x \lambda^{-\beta+N-1} e^{-1/\lambda} d\lambda.$$

Define $C_2 = C_1 \int_0^1 \lambda^{-\beta+N-1} e^{-1/\lambda} d\lambda$ and note that it is finite. Moreover, for all $m \le x \le 1$,

$$\Phi_\beta(x) \le C_2 \le \frac{C_2}{\inf_{m \le w \le 1} U_\beta(w)} U_\beta(x).$$

Of course, the form of the function U_β makes it apparent that

$$\inf_{m \le w \le 1} U_\beta(w) > 0.$$

Thus, it remains to prove the upper bound of the lemma for $x > 1$. For the remainder of this proof, we assume $x > 1$.

When $\beta > N$,
$$\Phi_\beta(x) \leq C_2 + C_1 \int_1^\infty \lambda^{-\beta+N-1}\, d\lambda = C_3.$$

(Note that $C_3 < \infty$.) When $\beta = N$,
$$\Phi_\beta(x) \leq C_2 + C_1 \int_1^x \lambda^{-1}\, d\lambda \leq (C_2 + C_1)\ln_+(x).$$

Finally, when $\beta < N$,
$$\Phi_\beta(x) \leq C_2 + C_1 \int_1^x \lambda^{-\beta+N-1}\, d\lambda \leq C_2 + C_1(N-\beta)^{-1} x^{-\beta+N}$$
$$\leq (C_2 + C_1) x^{-\beta+N}.$$

The upper bound lemma follows for any choice of A_5 that is greater than the maximum of 1, C_3, $C_1 + C_2$, and $C_2/\inf_{m \leq w \leq 1} U_\beta(x)$. □

Exercise **1.3.3** Verify the lower bound for Φ_β in Lemma 1.3.5. □

We are ready to prove Lemma 1.2.2.

Proof of Lemma 1.2.2 By equation (2) of Section 1.2, Fubini's theorem, and Lemma 1.3.4,
$$\mathbb{E}[\{J_{a,b}(f)\}^2] = \int_{[a,b]} \int_{[a,b]} \mathbb{E}[f(B_s)g(B_t)]\, dt\, ds$$
$$\leq A_3 \cdot \iint f(w)f(y) \int_{[a,b]} \int_{[a,b]} \|s-t\|^{-\frac{d}{2}}$$
$$\times \exp\left\{ -\frac{\|w-y\|^2}{4A_1\|s-t\|} \right\} dt\, ds\, dw\, dy. \quad (8)$$

Let us concentrate on the $ds \times dt$ integrals first. Since f is supported by $[-m,m]^d$, we restrict attention to $y, w \in [-m,m]^d$. Note that whenever $s, t \in [a,b]$, then certainly, $r = s - t \in [-b,b] \subset \{v \in \mathbb{R}^N : \|v\| \leq N^{\frac{1}{2}}|b|\}$. Consequently, for all $y, w \in [-m,m]^d$,
$$\int_{[a,b]} \int_{[a,b]} \|s-t\|^{-\frac{d}{2}} \exp\left\{ -\frac{\|w-y\|^2}{4A_1\|s-t\|} \right\} dt\, ds$$
$$\leq \text{Leb}([a,b]) \cdot \int_{r \in \mathbb{R}^N:\, \|r\| \leq N^{\frac{1}{2}}|b|} \|r\|^{-\frac{d}{2}} \exp\left\{ -\frac{\|w-y\|^2}{4A_1\|r\|} \right\} dr$$
$$\leq (4A_1)^{-(\frac{d}{2})+N} \text{Leb}([a,b]) \cdot \|w-y\|^{-d+2N} \Phi_{\frac{d}{2}}\left(\frac{\|w-y\|^2}{4\sqrt{N}|b|A_1} \right).$$

In the last line we used a change of variables $s = 4A_1 r/\|w-y\|^2$ and used the definition of $\Phi_{\frac{d}{2}}$; see (7). By equation (8),

$$\mathbb{E}[\{J_{a,b}(f)\}^2] \leq A \cdot \iint f(w)f(y)\|w-y\|^{-d+2N} \Phi_{\frac{d}{2}}\left(\frac{\|w-y\|^2}{4\sqrt{N}|b|A_1}\right) dy\, dw,$$

with $A = A_3(4A_1)^{-(\frac{d}{2})+N}\mathrm{Leb}([a,b])$. By Lemma 1.3.5, when $d > 2N$, we can find some finite positive constant A_5 such that $\sup_x \Phi_{\frac{d}{2}}(x) \leq A_5$. Thus, when $d > 2N$,

$$\mathbb{E}[\{J_{a,b}(f)\}^2] \leq A\, A_5 \cdot \iint f(w)f(y)\,\|w-y\|^{-d+2N}\, dw\, dy$$
$$= A\, A_5 \cdot \mathsf{Energy}_{d-2N}(f). \tag{9}$$

Similarly, when $d < 2N$,

$$\mathbb{E}[\{J_{a,b}(f)\}^2] \leq A' \cdot \iint f(w)f(y)\,\|w-y\|^{-d+2N}\, dw\, dy$$
$$= A' \cdot \mathsf{Energy}_{d-2N}(f), \tag{10}$$

where $A' = A\, A_5(4\sqrt{N}|b|A_1)^{\frac{d}{2}-N}$. Finally, a similar analysis shows that when $d = 2N$,

$$\mathbb{E}[\{J_{a,b}(f)\}^2] \leq A\, A_5 \cdot \iint f(w)f(y)\, \ln_+\left(\frac{4\sqrt{N}|b|A_1}{\|w-y\|^2}\right) dw\, dy.$$

Note that for all $w, y \in [-m,m]^d$, $\|y-w\| \leq d^{\frac{1}{2}}M$. Thus,

$$\ln_+\left(\frac{4\sqrt{N}|b|A_1}{\|w-y\|^2}\right) = 2\ln_+\left(\frac{1}{\|w-y\|}\right)\left\{1 + \frac{\ln(2\sqrt{N}|b|A_1)}{\ln_+\left(\frac{1}{\|w-y\|}\right)}\right\}$$
$$\leq 2\ln_+\left(\frac{1}{\|w-y\|}\right)\left\{1 + \frac{\ln(2\sqrt{N}|b|A_1)}{\ln_+\left(\frac{1}{d^{\frac{1}{2}}M}\right)}\right\}$$
$$= A'' \ln_+\left(\frac{1}{\|w-y\|}\right).$$

Thus,

$$\mathbb{E}\left[\{J_{a,b}(f)\}^2\right] \leq A\, A_5\, A'' \cdot \mathsf{Energy}_0(f).$$

Combining this with (9) and (10), we can deduce Lemma 1.2.2 with A_2 there defined to be the maximum of the values of the following parameters from this subsection: $A \cdot A_5$, A', and $A \cdot A_5 \cdot A''$. □

468   12. The Brownian Sheet and Potential Theory

1.4 Proof of Theorem 1.1.1: Upper Bound

Throughout, we let $\mathcal{F} = (\mathcal{F}_t;\ t \in \mathbb{R}_+^N)$ denote the complete augmented history of the process B. We will also hold $0 \prec a \prec b$, both in $\mathbb{R}_+^N$, and a compact set $E \subset \mathbb{R}^d$ fixed whose outer radius is denoted by M; cf. equation (1) of Section 1.2. With the above in mind, for any measurable $f : \mathbb{R}^d \to \mathbb{R}_+$, we can define a multiparameter martingale $M(f) = (M_t(f);\ t \in \mathbb{R}_+^N)$ as

$$M_t(f) = \mathbb{E}\Big[\int_{[a,2b-a]} f(B_s)\,ds \,\Big|\, \mathcal{F}_t\Big], \qquad t \in \mathbb{R}_+^N. \tag{1}$$

We will assume, without loss of generality, that the underlying probability space is complete and hence $M(f)$ can be taken to be separable; cf. Theorem 2.2.1 of Chapter 5. In light of equation (2) of Section 1.2, we note that

$$M_t(f) = \mathbb{E}\big[J_{a,2b-a}(f) \,\big|\, \mathcal{F}_t\big]. \tag{2}$$

Exercise 1.4.1 Demonstrate that for all $s, t \in \mathbb{R}_+^N$, $B_{t+s} - B_t$ is independent of $\mathcal{F}_t$ and compute the mean and covariance matrix of $B_{t+s} - B_t$. □

Lemma 1.4.1 *There exist finite positive constants K_1 and K_2 that depend only on N and d such that for all measurable functions $f : \mathbb{R}^d \to \mathbb{R}_+$ and all $t \in [a, b]$,*

$$M_t(f) \geq K_1 \cdot \int f(y + B_t)\, \|y\|^{-d+2N}\, \Phi_{\frac{d}{2}}\!\left(\frac{\|y\|^2}{K_2}\right) dy, \qquad \text{a.s.,}$$

where $\Phi_{\frac{d}{2}}$ is defined in (7) of Section 1.3.

Proof Since $t \in [a, b]$,

$$M_t(f) \geq \mathbb{E}\Big[\int_{[t,2b-a]} f(B_s)\,ds \,\Big|\, \mathcal{F}_t\Big] \geq \mathbb{E}\Big[\int_{[0,b]} f(B_{s+t})\,ds \,\Big|\, \mathcal{F}_t\Big]$$

$$= \int_{[0,b]} \mathbb{E}\big[f(B_{s+t} - B_t + B_t) \,\big|\, \mathcal{F}_t\big]\, ds. \tag{3}$$

We have used Fubini's theorem in the last line. By Exercise 1.4.1 above, $B_{t+s} - B_t$ is independent of $\mathcal{F}_t$ and $B_{t+s} - B_t \sim \mathcal{N}_d(0, \tau^2 \mathbf{I}_d)$, where $\tau^2 = \text{Leb}([t, t+s])$ and $\mathbf{I}_d$ denotes the $(d \times d)$ identity matrix. By the form of Gaussian densities (Exercise 1.1.1, Chapter 5),

$$\mathbb{E}\big[f(B_{t+s} - B_t + B_t) \,\big|\, \mathcal{F}_t\big] = (2\pi\tau^2)^{-\frac{d}{2}} \int f(y + B_t) \exp\Big\{-\frac{\|y\|^2}{2\tau^2}\Big\} dy.$$

By Lemma 3.1.1 of Chapter 5, we can find two finite positive constants c_1 and c_2 such that for all $s, t \in [a, b]$, $c_1|s| \leq \tau^2 \leq c_2|s|$. Since

$N^{-\frac{1}{2}}\|s\| \leq |s| \leq \|s\|$, we have shown the existence of a finite constant $c_3 \geq 1$ such that for all $s, t \in [a, b]$, $c_3^{-1}\|s\| \leq \tau^2 \leq c_3\|s\|$. Consequently, we obtain

$$\mathbb{E}\big[f(B_{t+s}-B_t+B_t) \,\big|\, \mathcal{F}_t\big] \geq (2\pi c_3)^{-\frac{d}{2}} \|s\|^{-\frac{d}{2}} \cdot \int f(y+B_t) \exp\left\{-\frac{c_3\|y\|^2}{2\|s\|}\right\} dy.$$

Plugging this into equation (3), we deduce the following:

$$M_t(f) \geq (2\pi c_3)^{-\frac{d}{2}} \cdot \int f(y+B_t) \int_{[0,b]} \|s\|^{-\frac{d}{2}} \exp\left\{-\frac{c_3\|y\|^2}{2\|s\|}\right\} ds\, dy$$

$$= c_3^N \pi^{-\frac{d}{2}} 2^{-d} \cdot \int f(y+B_t) \|y\|^{-d+2N} \Phi_{\frac{d}{2}}\left(\frac{c_3\|y\|^2}{2\min_{1\leq\ell\leq N} b^{(\ell)}}\right) dy,$$

by changing variables $\lambda = 2s/(c_3\|y\|^2)$. The lemma follows readily. □

The next ingredient in our derivation of the upper bound in Theorem 1.1.1 is based on Lemma 1.4.1 and Cairoli's maximal inequality (Theorem 2.3.2, Chapter 7): For any measurable function $f : \mathbb{R}^d \to \mathbb{R}_+$,

$$\mathbb{E}\left[\sup_{t\in[a,b]} \{M_t(f)\}^2\right] \leq 4^N \sup_t \mathbb{E}\big[\{M_t(f)\}^2\big] \leq 4^N \mathbb{E}\big[\{J_{2b-a,a}(f)\}^2\big].$$

The second inequality follows from Jensen's inequality for conditional expectations and from equation (2). (An important ingredient in Cairoli's maximal inequality was the commutation of the filtration $\mathcal{F}$; cf. Theorem 2.3.2 of Chapter 7.) Applying Lemma 1.2.2, we have the following result.

Lemma 1.4.2 *For all $m > 0$, there exists a finite constant C_1 that depends only on N, m, and d such that for any measurable function $f : \mathbb{R}^d \to \mathbb{R}_+$ that is supported in $[-m, m]^d$,*

$$\mathbb{E}\left[\sup_{t\in[a,b]} \{M_t(f)\}^2\right] \leq C_1 \cdot \text{Energy}_{d-2N}(f).$$

For any $\varepsilon \in\,]0, 1[$, let E^ε denote the ε-enlargement of E. That is,

$$E^\varepsilon = \Big\{x \in \mathbb{R}^d : \text{dist}\{x; E\} < \varepsilon\Big\}.$$

Note that E^ε is open. As such,

$$\mathbb{P}\big(B([a, b]) \cap E^\varepsilon \neq \varnothing\big) \geq \mathbb{P}(B_1 \in E^\varepsilon) > 0, \tag{4}$$

thanks to the form of the Gaussian density. In particular, by Supplementary Exercise 11 of Chapter 11, we can find a $\mathbb{Q}_+^N \cup \{+\infty\}$-valued random variable T^ε such that $\mathrm{T}^\varepsilon \neq +\infty$ if and only if there exists $t \in \mathbb{Q}_+^N$ such that $B_t \in E^\varepsilon$. Since $t \mapsto B_t$ is continuous and E^ε is open,

$$\mathrm{T}^\varepsilon \neq +\infty \iff B([a, b]) \cap E^\varepsilon \neq \varnothing. \tag{5}$$

Now we can define a set function μ^ε as follows: For all Borel sets $F \subset \mathbb{R}^d$,

$$\mu^\varepsilon(F) = \mathbb{P}\bigl(B_{T^\varepsilon} \in F \mid T^\varepsilon \neq +\infty\bigr).$$

Lemma 1.4.3 *For all $\varepsilon > 0$, μ^ε is an absolutely continuous probability measure on E^ε.*

Proof By (4), for all Borel sets $F \subset \mathbb{R}^d$, $\mu^\varepsilon(F)$ is a properly defined conditional probability. Consequently, μ^ε is a probability measure. By the continuity of $t \mapsto B_t$ and since E^ε is open, conditional on $(T^\varepsilon \neq +\infty)$, $B_{T^\varepsilon} \in E^\varepsilon$. Thus, $\mu^\varepsilon \in \mathcal{P}(E^\varepsilon)$, as desired. To conclude, suppose $G \subset \mathbb{R}^d$ has zero Lebesgue measure. We wish to show that $\mu^\varepsilon(G) = 0$. To argue this, recall that conditional on $(T^\varepsilon \neq +\infty)$, $T^\varepsilon \in \mathbb{Q}_+^N$. In particular,

$$\mu^\varepsilon(G) \leq \sum_{t \in \mathbb{Q}_+^N} \mathbb{P}\bigl(B_t \in G \mid T^\varepsilon \neq +\infty\bigr),$$

which is 0, thanks to the fact that $\mathrm{Leb}(G) = 0$ and to equation (4). This concludes our proof. $\square$

We are ready to prove Theorem 1.1.1.

Proof of Theorem 1.1.1 Owing to Lemma 1.4.3, we can define the probability density

$$f^\varepsilon(x) = \frac{\mu^\varepsilon(dx)}{dx}, \qquad x \in \mathbb{R}^d.$$

In fact, by Lemma 1.4.3, f^ε is a probability density on E^ε. In particular, for any choice of $\varepsilon \in {]0,1[}$, f^ε is supported on $[-m,m]^d$, where $m = M + 1$. By Lemma 1.4.2, for all $\varepsilon \in {]0,1[}$, there exists a finite constant $C_2 > 0$ such that

$$\mathbb{E}\Bigl[\sup_{t \in [a,b]} \{M_t(f^\varepsilon)\}^2\Bigr] \leq C_2 \cdot \mathsf{Energy}_{d-2N}(f^\varepsilon). \tag{6}$$

On the other hand, we can apply Lemma 1.4.1 with f replaced by f^ε and deduce the existence of finite constants C_3 and C_4 such that for all $\varepsilon \in [0,1[$,

$$\sup_{t \in [a,b]} \{M_t(f^\varepsilon)\}^2 \tag{7}$$
$$\geq C_3 \Bigl\{\int f^\varepsilon(y + B_{T^\varepsilon}) \, \|y\|^{-d+2N} \, \Phi_{\frac{d}{2}}\Bigl(\frac{\|y\|^2}{C_4}\Bigr) dy\Bigr\}^2 \mathbf{1}_{(T^\varepsilon \neq +\infty)}.$$

The expectation of the right-hand side of (7) equals

$$\mathbb{P}(T^\varepsilon \neq +\infty) \cdot \int \Bigl\{\int f^\varepsilon(y+w) \, \|y\|^{-d+2N} \, \Phi_{\frac{d}{2}}\Bigl(\frac{\|y\|^2}{C_4}\Bigr) dy\Bigr\}^2 f^\varepsilon(w)\, dw$$
$$\geq \mathbb{P}(T^\varepsilon \neq +\infty) \cdot \Bigl\{\int\!\!\int f^\varepsilon(y+w) \, \|y\|^{-d+2N} \, \Phi_{\frac{d}{2}}\Bigl(\frac{\|y\|^2}{C_4}\Bigr) dy \, f^\varepsilon(w)\, dw\Bigr\}^2.$$

The last line follows from the Cauchy–Schwarz inequality. Lemma 1.3.5 and a few lines of calculations reveal the existence of a finite constant $C_5 > 0$ such that the above is bounded below by

$$C_5\, \mathbb{P}(T^\varepsilon \neq +\infty) \cdot \{\mathsf{Energy}_{d-2N}(f^\varepsilon)\}^2.$$

This is a good lower bound for the expectation of the right-hand side of (7). By Lemma 1.4.2, the expectation of the left-hand side of (7) is bounded above by $C_1 \cdot \mathsf{Energy}_{d-2N}(f^\varepsilon)$, for some constant C_1 that is independent of the choice of $\varepsilon \in]0,1[$. If we knew that $\mathsf{Energy}_{d-2N}(f^\varepsilon)$ were finite, this development would show us that

$$\mathbb{P}(T^\varepsilon \neq +\infty) \leq \frac{C_1}{C_5\, \mathsf{Energy}_{d-2N}(\mu^\varepsilon)}, \tag{8}$$

which is a key inequality. (Recall that the energy $\mathsf{Energy}_{d-2N}(f^\varepsilon)$ of the function f^ε was *defined* to be the energy $\mathsf{Energy}_{d-2N}(\mu^\varepsilon)$ of the measure μ^ε; cf. equation (5) and the subsequent display.) Unfortunately, we do not a priori know that f^ε has finite energy. To get around this difficulty we will use a truncation argument to prove (8).

For all $q > 0$ and all $\varepsilon \in]0,1[$, we define the function f_q^ε by

$$f_q^\varepsilon(x) = f^\varepsilon(x)\, \mathbb{1}_{[0,q]}(f^\varepsilon(x)), \qquad x \in \mathbb{R}^d.$$

Since f^ε is supported on E^ε, so is f_q^ε. Moreover, the latter is a subprobability density function that is bounded above by q. Note that $\mathsf{Energy}_{d-2N}(f_q^\varepsilon) < \infty$ (why?). Exactly the same argument that led to (7) shows us that

$$\sup_{t \in [a,b]} \{M_t(f_q^\varepsilon)\}^2$$
$$\geq C_3 \left\{ \int f_q^\varepsilon(y + B_{T^\varepsilon})\, \|y\|^{-d+2N}\, \Phi_{\frac{d}{2}}\!\left(\frac{\|y\|^2}{C_4}\right) dy \right\}^2 \mathbb{1}_{(T^\varepsilon \neq +\infty)}.$$

Taking expectations, we can see that

$$\mathbb{E}\!\left[\sup_{t \in [a,b]} \{M_t(f_q^\varepsilon)\}^2 \right]$$
$$\geq C_3 \int \left\{ \int f_q^\varepsilon(y+w)\, \|y\|^{-d+2N}\, \Phi_{\frac{d}{2}}\!\left(\frac{\|y\|^2}{C_4}\right) dy \right\}^2 f^\varepsilon(w)\, dw$$
$$\quad \times \mathbb{P}(T^\varepsilon \neq +\infty)$$
$$\geq C_3 \left\{ \int \int f_q^\varepsilon(y+w)\, \|y\|^{-d+2N}\, \Phi_{\frac{d}{2}}\!\left(\frac{\|y\|^2}{C_4}\right) dy\, f^\varepsilon(w)\, dw \right\}^2$$
$$\quad \times \mathbb{P}(T^\varepsilon \neq +\infty),$$

by the Cauchy–Schwarz inequality. By Lemma 1.4.2, the left-hand side is bounded above by $C_1\, \mathsf{Energy}_{d-2N}(f_q^\varepsilon)$. We have already pointed out that this is finite. Thus, by Lemma 1.3.5 and a few lines of calculations, there exists a finite constant $C_5 > 0$ such that

$$\mathbb{E}\left[\sup_{t \in [a,b]} \{M_t(f_q^\varepsilon)\}^2\right]$$
$$\geq C_5 \left\{ \int \int f_q^\varepsilon(y+w)\, f^\varepsilon(w)\, \kappa(y)\, dw \right\}^2 \cdot \mathbb{P}(T^\varepsilon \neq +\infty) \quad (9)$$
$$\geq C_5 \left\{\mathsf{Energy}_{d-2N}(f_q^\varepsilon)\right\}^2 \cdot \mathbb{P}(T^\varepsilon \neq +\infty),$$

where

$$\kappa(y) = \begin{cases} 1, & \text{if } d < 2N, \\ \ln_+(1/\|y\|), & \text{if } d = 2N, \\ \|y\|^{-d+2N}, & \text{if } d > 2N. \end{cases}$$

(Since it depends only on M, d, and N, the above is the same constant C_5 that we found in equation (8).) By Lemma 1.4.2, the left-hand side of (9) is bounded above by $C_1 \cdot \mathsf{Energy}_{d-2N}(f_q^\varepsilon)$. On the other hand, we have already observed that this energy is finite. Thus,

$$\mathbb{P}(T^\varepsilon \neq +\infty) \leq \frac{C_1}{C_5\, \mathsf{Energy}_{d-2N}(f_q^\varepsilon)}.$$

Now we let $q \uparrow +\infty$ and use the monotone convergence theorem to obtain (8).

The remainder of our proof is simple. By equations (8) and (5),

$$\mathbb{P}\big(B([a,b]) \cap E^\varepsilon \neq \varnothing\big) \leq \frac{C_1}{C_5\, \mathsf{Energy}_{d-2N}(\mu^\varepsilon)},$$

and recall that $\mu^\varepsilon \in \mathcal{P}(E^\varepsilon)$. In particular,

$$\mathbb{P}\big(B([a,b]) \cap E^\varepsilon \neq \varnothing\big) \leq \frac{C_1}{C_5}\, \mathsf{Cap}_{d-2N}(\overline{E^\varepsilon}).$$

The upper bound in Theorem 1.1.1 now follows from the continuity of $t \mapsto B_t$ and from Lemma 2.1.2 of Appendix D. $\square$

2 The Codimension of the Level Sets

Theorem 1.1.1 of Section 1 and its corollaries have shown us that the range of the Brownian sheet has quite an intricate structure. See also Figures 12.1, 12.2, and 12.3 of Section 1 for some simulations. In this section we study the level sets (or contours) of the Brownian sheet; we shall see that they, too, have interesting geometric and/or measure-theoretic properties.

2.1 The Main Calculation

As before, we will let $B = (B_t; \ t \in \mathbb{R}_+^N)$ denote an N-parameter, $\mathbb{R}^d$-valued Brownian sheet. For any $a \in \mathbb{R}^d$, the **level set** $B^{-1}(a)$ of B at a is defined by the following N-dimensional "temporal" set:

$$B^{-1}(a) = \{t \in \mathbb{R}_+^N : \ B_t = a\}.$$

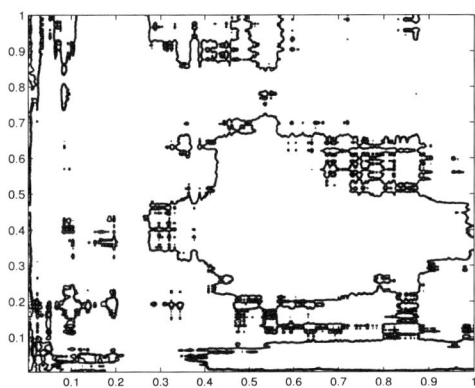

Figure 12.4: The zero set of the $\mathbb{R}$-valued, 2-parameter Brownian sheet

In this section we shall concern ourselves with the analysis of the **zero set** of B, which is $B^{-1}(0)$. This is the level set of B at $0 \in \mathbb{R}^d$, and the other level sets can be similarly studied; cf. Supplementary Exercise 1. Figure 12.4 shows a simulation of the portion of the zero set of B that lies in $[0,1]^2$; here, $d = 1$ and $N = 2$.

In order to analyze the zero set of B, we need to first ask when $B^{-1}(0)$ is nonempty. To begin, note that for all t on the axes of the parameter space $\mathbb{R}_+^N$, $B_t = 0$ (why?). Thus, we always have $\partial \mathbb{R}_+^N \subset B^{-1}(0)$. We say that $B^{-1}(0)$ is **trivial** if $B^{-1}(0) = \partial \mathbb{R}_+^N$. That is, the zero set of B is trivial if and only if the only t's for which $B_t = 0$ are the ones on the axes.

By Theorem 1.1.1, $B^{-1}(0)$ is nontrivial if and only if $\mathsf{Cap}_{d-2N}(\{0\}) > 0$. (Why?) This is *equivalent* to the following, more useful, result.

Corollary 2.1.1 *The zero set of the N-parameter, $\mathbb{R}^d$-valued Brownian sheet is nontrivial if and only if $d < 2N$.*

***Exercise* 2.1.1** Prove Corollary 2.1.1. □

Thus, in all of our subsequent discussions of the zero set of B we are naturally led to assume that $d < 2N$. The main result of this section is the following computation of the codimension of the zero set.

Theorem 2.1.1 *When $d < 2N$,*

$$\mathrm{codim}\{B^{-1}(0) \setminus \partial \mathbb{R}_+^N\} = \frac{d}{2}.$$

Together with Theorem 4.7.1 of Chapter 11, Theorem 2.1.1 implies the following computation of the size of the zero set of the Brownian sheet.

Corollary 2.1.2 *When $d < 2N$,*

$$\dim\{B^{-1}(0) \setminus \partial\mathbb{R}_+^N\} = N - \frac{d}{2}, \quad a.s.$$

***Exercise* 2.1.2** Complete this proof of Corollary 2.1.2. □

Figure 12.4 shows $B^{-1}(0)$ for $N = 2$ and $d = 1$; this is a (random) set in $\mathbb{R}_+^2$ whose Hausdorff dimension is (almost surely) equal to $\frac{3}{2}$.

2.2 Proof of Theorem 2.1.1: The Lower Bound

Proving the "lower bound" in Theorem 2.1.1 consists in showing that the lower stochastic codimension of $B^{-1}(0)$ is at least $\frac{d}{2}$. That is, we wish to verify that whenever $K \subset \mathbb{R}_+^N$ is a compact set whose Hausdorff dimension is strictly less than $\frac{d}{2}$, then $B^{-1}(0) \cap K$ is empty, almost surely; cf. Lemma 4.7.1 of Chapter 11.

We do this by appealing to a counting principle as we did in our proof of Theorem 5.2.1 of Chapter 11. Namely, we derive an estimate for the probability that $B^{-1}(0)$ intersects a small ball as follows:

Lemma 2.2.1 *Suppose $d < 2N$, $M > 1$, and $\eta \in {]0, \frac{d}{2}[}$ are fixed. Then, there exists a finite constant $A > 0$ such that for all $a \in [M^{-1}, M]^N$ and all $\varepsilon \in {]0, 1[}$,*

$$\mathbb{P}\bigl(B^{-1}(0) \cap [a, a + p(\varepsilon)] \neq \varnothing\bigr) \leq A\varepsilon^\eta,$$

where $p(\varepsilon)$ denotes the point in $\mathbb{R}_+^N$ all of whose coordinates are equal to ε.

The above is a useful estimate for small values of ε and can be shown to be essentially sharp, as the following shows.

***Exercise* 2.2.1** Suppose $M > 1$ is fixed. Then, there exists a finite constant $A' > 0$ such that for all $\varepsilon \in {]0, 1[}$,

$$\mathbb{P}\bigl(B^{-1}(0) \cap [a, a + p(\varepsilon)] \neq \varnothing\bigr) \geq A'\varepsilon^{\frac{d}{2}},$$

where $p(\varepsilon)$ is defined in Lemma 2.2.1. □

Proof We begin with an aside, which is a simple calculation involving Gaussian densities: For any $a \in [M^{-1}, M]^N$ and for all $z > 0$,

$$\mathbb{P}(|B_a| \leq z) = \left\{2\pi \prod_{\ell=1}^N a^{(\ell)}\right\}^{-\frac{d}{2}} \cdot \prod_{j=1}^d \int_{-z}^z \exp\left(-\frac{w^2}{2\prod_{k=1}^N a^{(k)}}\right) dw$$

$$\leq \left(\frac{2}{\pi}\right)^{\frac{d}{2}} M^{\frac{Nd}{2}} z^d. \qquad (1)$$

We have merely used the fact that $e^{-u} \leq 1$ for $u \geq 0$.

From here on, the proof of this lemma is similar in spirit to that of Lemma 5.3.2, Chapter 11. Note that $B^{-1}(0) \cap [a, a+p(\varepsilon)]$ is nonempty if and only if there exists a point $\tau \in [a, a+p(\varepsilon)]$ such that $B_\tau = 0$. In particular, for this choice of τ,

$$|B_a| \le |B_a - B_\tau| \le \sup_{a \preccurlyeq t \preccurlyeq a+p(\varepsilon)} |B_a - B_t| = \sup_{0 \preccurlyeq t \preccurlyeq p(\varepsilon)} |B_{a+t} - B_a|.$$

Therefore,

$$\mathbb{P}\bigl(B^{-1}(0) \cap [a, a+p(\varepsilon)] \ne \varnothing\bigr) \le \mathbb{P}\Bigl(|B_a| \le \sup_{0 \preccurlyeq t \preccurlyeq p(\varepsilon)} |B_{a+t} - B_a|\Bigr).$$

On the other hand, B_a is independent of the entire process $(B_{a+t} - B_a;\ t \in \mathbb{R}_+^N)$ (why?). In particular, B_a is independent of $\sup_{0 \preccurlyeq t \preccurlyeq p(\varepsilon)} |B_{a+t} - B_a|$, and equation (1) and Fubini's theorem together imply

$$\mathbb{P}\bigl(B^{-1}(0) \cap [a, a+p(\varepsilon)] \ne \varnothing\bigr) \le A_1 \cdot \mathbb{E}\Bigl[\sup_{0 \preccurlyeq t \preccurlyeq p(\varepsilon)} |B_{a+t} - B_a|^d\Bigr],$$

where $A_1 = (2M^N/\pi)^{\frac{d}{2}}$. On the other hand, metric entropy considerations imply that for any fixed $\eta \in\,]0, \frac{d}{2}[$, there exists $A_2 > 0$ such that for all $\varepsilon \in\,]0, 1[$,

$$\mathbb{E}\Bigl[\sup_{0 \preccurlyeq t \preccurlyeq p(\varepsilon)} |B_{a+t} - B_a|^d\Bigr] \le \varepsilon^\eta; \tag{2}$$

see Kolmogorov's continuity theorem (in particular, Exercise 2.5.1 of Chapter 5). This proves our lemma. □

Exercise **2.2.2** Verify (2) in detail.
(HINT: Consult Lemma 3.1.1, Chapter 5.) □

A more ambitious exercise is to refine Lemma 2.2.1 as follows.

Exercise **2.2.3** Prove that for all $a \in\,]0, \infty[^d$ fixed, there exists $C_1 > 0$ such that for all $\varepsilon \in\,]0, 1[$,

$$\mathbb{E}\Bigl[\sup_{0 \preccurlyeq t \preccurlyeq p(\varepsilon)} |B_{a+t} - B_a|^d\Bigr] \le C_1 \varepsilon^{\frac{d}{2}}.$$

Conclude that there exists $C_2 > 0$ such that for all $\varepsilon \in\,]0, 1[$,

$$\mathbb{P}\bigl(B^{-1}(0) \cap [a, a+p(\varepsilon)] \ne \varnothing\bigr) \le C_2 \varepsilon^{\frac{d}{2}},$$

thus showing that Exercise 2.2.2 is sharp, up to a multiplicative constant. □

We conclude this subsection by proving half of Theorem 2.1.1.

Proof of the Lower Bound in Theorem 2.1.1 We are to show that for all compact sets $K \subset \mathbb{R}_+^N$, if $\dim(K) < \frac{d}{2}$, then $B^{-1}(0) \cap K$ is a.s. empty. In fact, it suffices to assume that $K \subset [-M^{-1}, M]^N$, where $M > 1$ is chosen arbitrarily large (why?). Throughout, we hold fixed some $\eta \in \,]0, \frac{d}{2}[$.

By Lemma 1.1.3 of Appendix C, for any $n > 0$, we can find cubes $C_{j,n}$ of side $r_{j,n}$ such that $\cup_{j=1}^\infty C_{j,n} \supset K$ and

$$\sum_{j=1}^\infty (2r_{j,n})^\eta \leq \frac{1}{n}.$$

Clearly, whenever $B^{-1}(0) \cap K$ is nonempty, there must exist some $j \geq 1$ such that $B^{-1}(0) \cap C_{j,n}$ is nonempty. Consequently, by Lemma 2.2.1 above, there exists a finite constant $A > 0$ such that

$$\mathbb{P}\big(B^{-1}(0) \cap K \neq \varnothing\big) \leq \mathbb{E}\Big[\sum_{j=1}^\infty \mathbf{1}_{(B^{-1}(0) \cap C_{j,n} \neq \varnothing)}\Big] \leq A \cdot \sum_{j=1}^\infty (2r_{j,n})^\eta \leq \frac{A}{n}.$$

Since $n \geq 1$ is arbitrary, this shows that $B^{-1}(0) \cap K = \varnothing$, a.s., and completes our proof. □

2.3 Proof of Theorem 2.1.1: The Upper Bound

To complete our derivation of Theorem 2.1.1, it suffices to show that for all $M > 1$ and all compact sets $K \subset \,]0, M]^N$ with $\dim(K) > \frac{d}{2}$, $\mathbb{P}(B^{-1}(0) \cap K \neq \varnothing) > 0$ (why?).

Frostman's theorem implies the existence of a probability measure μ on K such that $\mathsf{Energy}_{\frac{d}{2}}(\mu) < \infty$; cf. Theorem 2.2.1, Appendix C. Fix such a μ and consider

$$J_\varepsilon = \varepsilon^{-d} \cdot \int \mathbf{1}_{(|B_s| \leq \varepsilon)}\, \mu(ds).$$

By Fatou's lemma and by the form of the Gaussian density function,

$$\liminf_{\varepsilon \to 0^+} \mathbb{E}[J_\varepsilon] \geq (2\pi)^{-\frac{d}{2}} \int \Big\{\prod_{j=1}^N s^{(j)}\Big\}^{-\frac{d}{2}} \mu(ds) \geq (2\pi)^{-\frac{d}{2}} M^{-\frac{Nd}{2}}. \quad (1)$$

On the other hand, by Lemma 1.3.4, for all $s, t \in [-M, M]^N$, we can find a finite constant A such that for all $\varepsilon > 0$,

$$\mathbb{P}\big(|B_s| \leq \varepsilon,\, |B_t| \leq \varepsilon\big) \leq A \|s - t\|^{-\frac{d}{2}} \int_{|y| \leq \varepsilon} \int_{|w| \leq \varepsilon} dw\, dy$$

$$\leq 4^d A \cdot \|s - t\|^{-\frac{d}{2}} \varepsilon^{2d}.$$

Thus, for all $\varepsilon > 0$,
$$\mathbb{E}[J_\varepsilon^2] \leq 4^d A \cdot \mathsf{Energy}_{\frac{d}{2}}(\mu). \tag{2}$$

By the Paley–Zygmund lemma (Lemma 1.4.1, Chapter 3), used in conjunction with equations (1) and (2),

$$\liminf_{\varepsilon \to 0^+} \mathbb{P}\big(K \cap \{s \in \mathbb{R}_+^N : |B_s| \leq \varepsilon\} \neq \varnothing\big) \geq \liminf_{\varepsilon \to 0^+} \frac{\{\mathbb{E}[J_\varepsilon]\}^2}{\mathbb{E}[J_\varepsilon^2]}$$
$$\geq A' \big\{\mathsf{Energy}_{\frac{d}{2}}(\mu)\big\}^{-1},$$

where $A' = (8\pi)^{-d} A^{-1} M^{-Nd}$. We should recognize that the above is strictly positive. By the compactness of K and by the continuity of $t \mapsto B_t$, the left-hand side equals the probability that $B^{-1}(0)$ intersects K. This completes our proof. $\square$

In fact, the above proof implies the following.

Corollary 2.3.1 *Suppose $E \subset \mathbb{R}_+^N$ is compact and $\mathsf{Cap}_{\frac{d}{2}}(E) > 0$. Then, with positive probability, $B^{-1}(0) \cap E \neq \varnothing$.*

One can prove the following refinement, using similar arguments.

Corollary 2.3.2 *Suppose $E \subset \mathbb{R}_+^N$ is compact and $\mathsf{Cap}_{\frac{d}{2}}(E) > 0$. Then, for any $x \in \mathbb{R}^d$, with positive probability, $B^{-1}(x) \cap E \neq \varnothing$.*

***Exercise* 2.3.1** Verify Corollaries 2.3.1 and 2.3.2. $\square$

3 Local Times as Frostman's Measures

We conclude this chapter with a glimpse of further refinements in the analysis of the level sets of random fields. For concreteness, we concentrate on the zero set $B^{-1}(0)$ of an N-parameter, $\mathbb{R}^d$-valued Brownian sheet B that has already been subject to our studies of Sections 1 and 2 above.

Thus far, we have seen that $B^{-1}(0)$ is nontrivial if and only if $d < 2N$ (Corollary 2.1.1). Moreover, in the nontrivial case, the Hausdorff dimension of $B^{-1}(0)$ is a.s. equal to $N - \frac{d}{2}$; see Corollary 2.1.2. Before we use this observation, note that for any compact set $K \subset \mathbb{R}_+^N$, $B^{-1}(0) \cap K$ is compact. Accordingly, the aforementioned dimension result, together with Frostman's lemma (Theorem 2.1.1, Appendix C), guarantees the existence of a (random) measure L on $B^{-1}(0)$ such that a.s., for all $0 < s < N - \frac{d}{2}$,

$$\limsup_{r \to 0^+} r^{-s} \sup_{x \in K} L\{\mathcal{B}(x; r)\} < \infty. \tag{1}$$

Furthermore, with probability one, for all $s > N - \frac{d}{2}$,

$$\limsup_{r \to 0^+} r^{-s} \sup_{x \in K} L\{\mathcal{B}(x;r)\} = \infty. \tag{2}$$

(Why?) In this section we show how one can explicitly construct this random measure L—called the **local time of** B at 0. Of course, the choice of L is not unique. For instance, replace the above L by $2L$. It, too, is a measure on $B^{-1}(0)$, and it, too, satisfies the smoothness properties given by equations (1) and (2). Thus, we will be concerned with finding one "natural" construction. It turns out that our "natural" construction has attractive "uniqueness" properties, which will be explored further in Supplementary Exercise 3 below.

3.1 Construction

What does Frostman's measure of $B^{-1}(0)$ look like? Inspecting our proof of Frostman's theorem, we see that, to a great degree, Frostman's measure of a possibly irregular fractal-like set is the flattest, most uniform, measure that one can construct on the given set.

Now let us consider the zero set of a discrete-time random field $Z = (Z_t; \ t \in \mathbb{N}_0^N)$. Intuitively speaking, if Z looks the same everywhere, then the flattest measure that one can construct on $Z^{-1}(0)$ is the following: For all $A \subset \mathbb{N}_0^N$,

$$\mu(A) = \sum_{t \in A} \mathbb{1}_{(Z_t = 0)}.$$

In the continuous-time setting of the Brownian sheet, we should be trying to define a flat measure L on $B^{-1}(0)$ as

$$L(A) = \int_A \delta_0(B_t) \, dt,$$

for all Borel sets $A \subset \mathbb{R}_+^N$, where δ_0 denotes the delta function, or point mass, at $\{0\}$. Of course, δ_0 is not a function but a measure, and the above definition is an improper one. However, it does imply that if we take a sequence of proper functions φ_ε that converge in some reasonable sense to δ_0 as $\varepsilon \to 0$, one might expect that the measures $\int_\bullet \varphi_\varepsilon(B_s) \, ds$ should converge to $L(\bullet)$. Moreover, one can try to *define* L this way.

As it turns out, the choice of φ_ε is more or less immaterial, as long as φ_ε looks like δ_0 for small ε. An effective method for constructing such φ's is as follows: Consider a probability density function $\varphi_1 : \mathbb{R}_+^N \to \mathbb{R}_+$ and let

$$\varphi_\varepsilon(x) = \varepsilon^{-N} \varphi_1\left(\frac{x}{\varepsilon}\right), \qquad x \in \mathbb{R}_+^N.$$

Such a sequence is called an **approximation to the identity** (or a collection of **mollifiers**), and it is easy to check that φ_ε is a probability density function for every $\varepsilon > 0$. To see what happens as $\varepsilon \to 0^+$, let us temporarily concentrate on the specific φ_1 given by $\varphi_1(x) = 2^{-N}\mathbf{1}_{[-1,1]^N}(x)$, ($x \in \mathbb{R}$). Then, for all $x \in \mathbb{R}$, $\varphi_\varepsilon(x) = (2\varepsilon)^{-N}\mathbf{1}_{[-\varepsilon,\varepsilon]}(x)$. In particular, for small values of ε, $\varphi_\varepsilon(x) = 0$ for essentially all x but $x \approx 0$. In the latter case, $\varphi_\varepsilon(0) \approx \infty$. That is, for small values of ε, the function φ_ε looks like the delta function δ_0, as planned.

Henceforth, we shall use a different approximation to the identity that is easier to use. Namely, for all $x \in \mathbb{R}^N$ and for all $\varepsilon > 0$, define

$$\varphi_\varepsilon(x) = (2\pi\varepsilon)^{-\frac{N}{2}} \exp\left\{-\frac{\|x\|^2}{2\varepsilon}\right\}. \tag{1}$$

It should be recognized that the above is also an approximation to the identity.

Based on the preceding, we now *define* **approximate local times** L_ε as follows: For all Borel sets $A \subset \mathbb{R}_+^N$,

$$L_\varepsilon(A) = \int_A \varphi_\varepsilon(B_s)\,ds. \tag{2}$$

That is, $L_\varepsilon(\bullet) = \int_\bullet \varphi_\varepsilon(B_s)\,ds$, where $\varphi_1(x) = (2\pi)^{-\frac{N}{2}}\exp\{-\frac{1}{2}\|x\|^2\}$. It is very important to try to recognize $L_\varepsilon(\bullet)$ as a variant of J_ε of Section 2.3 and φ_1 as the heat kernel. As such, we have already used such objects to calculate the Hausdorff dimension of $B^{-1}(0)$. In brief, we are now taking a second, deeper, look at the J_ε of Section 2.3, together with its ilk.

Based on the preliminary discussions above, we should expect the following result, which is the main result of this section.

Theorem 3.1.1 *Suppose $d < 2N$. Then, there exists a random measure L on $B^{-1}(0)$ such that:*

(i) *Equations (1) and (2) of the preamble to this subsection hold for this L; and*

(ii) *for any Borel set $A \subset \mathbb{R}_+^N$,*

$$\lim_{\varepsilon \to 0^+} L_\varepsilon(A) = L(A),$$

where the convergence takes place in $L^p(\mathbb{P})$ for every $p > 0$.

The measure L is sometimes called the **local time** of the Brownian sheet (at 0). Once more, we emphasize that the local time (at 0) can be viewed as an explicit construction of Frostman's measure of the (random) level set of B (at 0).

We will derive Theorem 3.1.1 in detail in Sections 3.3 and 3.4 below. However, in order to understand some of the essential features of local times, it is best (and by far the easiest) to start with the one-parameter case of Brownian motion. Recall that in this case $N = 1$, and the condition $d < 2N$ forces us to assume $d = 1$.

3.2 Warmup: Linear Brownian Motion

Throughout this subsection let $B = (B_t;\ t \geq 0)$ denote linear (i.e., one-dimensional) Brownian motion. We begin our investigation of the local times of B with the following analytical estimate:

Proposition 3.2.1 *For all $f \in L^1(\mathbb{R})$, all $t, h \geq 0$, and all integers $k \geq 1$,*

$$\mathbb{E}\bigg[\bigg\{\int_t^{t+h} f(B_s)\, ds\bigg\}^k\bigg] \leq \bigg(\frac{2}{\pi}\bigg)^k k!\, \|f\|_1^k\, h^{\frac{k}{2}}.$$

The constant factors can be slightly improved; cf. Supplementary Exercise 2. We obtain the following as a consequence.

Corollary 3.2.1 *Given $f \in L^1(\mathbb{R})$ and $t \geq 0$, for $0 < \lambda < \frac{\pi}{2} t^{-\frac{1}{2}} \|f\|_1^{-1}$, $\mathbb{E}[e^{\lambda \int_0^t |f(B_s)|\, ds}] < \infty$.*

***Exercise* 3.2.1** Prove Corollary 3.2.1. □

Proof of Proposition 3.2.1 It suffices to prove the result for nonnegative functions f. Otherwise, replace f by $|f|$ everywhere.

Let $\mathcal{T}$ denote the heat semigroup that forms the transition operators of B; cf. Example 1, Section 4.3 of Chapter 8. We will shortly reduce the study of the kth moment of $\int_t^{t+h} f(B_s)\, ds$ to the study of integrals of $\mathcal{T}_s f$.

First, let us note that

$$\mathbb{E}\bigg[\bigg\{\int_t^{t+h} f(B_s)\, ds\bigg\}^k\bigg]$$

$$= \mathbb{E}\bigg[\int_t^{t+h}\cdots\int_t^{t+h} f(B_{s_1})\cdots f(B_{s_k})\, ds_1 \cdots ds_k\bigg]$$

$$= k! \int\cdots\int_{t \leq s_1 \leq \cdots \leq s_k \leq t+h} \mathbb{E}\bigg[\prod_{\ell=1}^k f(B_{s_\ell})\bigg]\, ds_1 \cdots ds_k. \tag{1}$$

Fix any $t = s_0 \leq s_1 \leq \cdots \leq s_k \leq t+h$ and consider $\mathbb{E}[f(B_{s_k})\,|\,\mathcal{F}_{s_{k-1}}]$, where $\mathcal{F} = (\mathcal{F}_s;\ s \geq 0)$ denotes the natural filtration of B. Clearly,

$$\mathbb{E}[f(B_{s_k})\,|\,\mathcal{F}_{s_{k-1}}] = \mathbb{E}[f(B_{s_k} - B_{s_{k-1}} + B_{s_{k-1}})\,|\,\mathcal{F}_{s_{k-1}}].$$

Since the increments of B are stationary and independent, $B_{s_k} - B_{s_{k-1}}$ is independent of $\mathcal{F}_{s_{k-1}}$. It also has the same distribution as $B_{s_k - s_{k-1}}$. Thus, recalling that $f \geq 0$,

$$\mathbb{E}[f(B_{s_k})\,|\,\mathcal{F}_{s_{k-1}}] \leq \sup_{a \in \mathbb{R}} \mathbb{E}[f(B_{s_k - s_{k-1}} + a)] = \|\mathcal{T}_{s_k - s_{k-1}} f\|_\infty.$$

Iterating this and remembering that $s_0 = t$, we see that equation (1) yields

$$\mathbb{E}\Big[\int_0^t \cdots \int_0^t f(B_{s_1}) \cdots f(B_{s_k}) \, ds_1 \cdots ds_k\Big]$$
$$\leq k! \int \cdots \int_{t \leq s_1 \leq \cdots \leq s_k \leq t+h} \prod_{\ell=1}^k \|\mathcal{T}_{s_\ell - s_{\ell-1}} f\|_\infty \, ds_1 \cdots ds_k. \quad (2)$$

On the other hand, for any $r \geq 0$ and for all $x \in \mathbb{R}$,

$$\mathcal{T}_r f(x) = (2\pi r)^{-\frac{1}{2}} \int_{-\infty}^\infty f(y) e^{-|x-y|^2/2r} \, dy \leq (2\pi r)^{-\frac{1}{2}} \|f\|_1.$$

Thus, (2) implies that the kth moment of $\int_t^{t+h} f(B_s) \, ds$ is bounded above by

$$k! \, (2\pi)^{-\frac{k}{2}} \|f\|_1^k \int \cdots \int_{t \leq s_1 \leq \cdots \leq s_k \leq t+h} \prod_{\ell=1}^k (s_\ell - s_{\ell-1})^{-\frac{1}{2}} \, ds_1 \cdots ds_k.$$

Changing variables $r_\ell = s_\ell - s_{\ell-1}$, it is easy to see that

$$\int \cdots \int_{t \leq s_1 \leq \cdots \leq s_k \leq t+h} \prod_{\ell=1}^k (s_\ell - s_{\ell-1})^{-\frac{1}{2}} \, ds_1 \ldots ds_k$$
$$\leq \Big\{\int_0^h r^{-\frac{1}{2}} \, dr\Big\}^{k-1} \cdot \int_t^{t+h} s_1^{-\frac{1}{2}} \, ds_1 \leq 2^{k-1} h^{\frac{k}{2}}.$$

The proposition follows from this. □

Recalling equation (2) of Section 3.1, we begin by showing that for each Borel set $A \subset \mathbb{R}_+$, $\{L_\varepsilon(A); \, \varepsilon > 0\}$ is a Cauchy sequence in $L^2(\mathbb{P})$.

Lemma 3.2.1 *For each $T > 0$,*

$$\lim_{|\delta - \varepsilon| \to 0} \sup_{A \subset [0,T], \, \text{Borel}} \mathbb{E}\big[|L_\varepsilon(A) - L_\delta(A)|^2\big] = 0.$$

In particular, $L(A) = \lim_{\varepsilon \to 0^+} L_\varepsilon(A)$ exists, where the limit holds in $L^2(\mathbb{P})$.

In fact, it is possible to prove the following extension; see Exercise 3.2.2 below.

Corollary 3.2.2 *For all $T > 0$, there exists a finite $C > 0$ such that for all Borel sets $A \subset [0,T]$ and all $\varepsilon, \delta > 0$,*

$$\mathbb{E}\big[|L_\varepsilon(A) - L_\delta(A)|^2\big] \leq C|\delta - \varepsilon|.$$

Proof Once we show that the expectation goes to 0, the existence of the limit follows from the completeness of $L^2(\mathbb{P})$.

By (2) and by the inversion theorem for characteristic functions,

$$L_\varepsilon(A) - L_\delta(A) = \int_A [\varphi_\varepsilon(B_s) - \varphi_\delta(B_s)]\, ds$$
$$= \frac{1}{2\pi} \int_A \int_{-\infty}^\infty e^{-i\xi B_s}\left[e^{-\frac{1}{2}\varepsilon\xi^2} - e^{-\frac{1}{2}\delta\xi^2}\right] d\xi\, ds.$$

Since the left-hand side is real-valued, its square is the same as its square modulus in the sense of complex numbers. That is,

$$\mathbb{E}\big[\{L_\varepsilon(A) - L_\delta(A)\}^2\big]$$
$$= \frac{1}{4\pi^2} \int_A \int_A \int_{-\infty}^\infty \int_{-\infty}^\infty \left[e^{-\frac{1}{2}\varepsilon\xi^2} - e^{-\frac{1}{2}\delta\xi^2}\right]$$
$$\times \left\{e^{-\frac{1}{2}\varepsilon\zeta^2} - e^{-\frac{1}{2}\delta\zeta^2}\right\} \mathbb{E}\big[e^{-i\xi B_s + i\zeta B_r}\big]\, d\xi\, d\zeta\, ds\, dr.$$

Suppose $\delta \geq \varepsilon > 0$ (say). Then,

$$e^{-\frac{1}{2}\varepsilon\xi^2} - e^{-\frac{1}{2}\delta\xi^2} = e^{-\frac{1}{2}\varepsilon\xi^2}\left(1 - e^{-\frac{1}{2}(\delta-\varepsilon)\xi^2}\right) \leq \left(\frac{\delta-\varepsilon}{2}\right)\xi^2 \wedge 1.$$

We have used the inequality $1 - x \leq e^{-x}$ for $x \geq 0$. Furthermore, since B is a Gaussian process, $\mathbb{E}\big[e^{-i\xi B_s + i\zeta B_r}\big] = e^{-\frac{1}{2}\mathbb{E}[\{\xi B_s - \zeta B_r\}^2]}$, which is nonnegative. Thus, for all Borel sets $A \subset [0,T]$,

$$\mathbb{E}\big[\{L_\varepsilon(A) - L_\delta(A)\}^2\big]$$
$$\leq \frac{1}{4\pi^2} \int_{[0,T]} \int_{[0,T]} \int_{-\infty}^\infty \int_{-\infty}^\infty \left\{\left(\frac{\delta-\varepsilon}{2}\right)\xi^2 \wedge 1\right\}$$
$$\times \left\{\left(\frac{\delta-\varepsilon}{2}\right)\zeta^2 \wedge 1\right\} e^{-\frac{1}{2}\mathbb{E}[\{\xi B_s - \zeta B_r\}^2]}\, d\xi\, d\zeta\, ds\, dr$$
$$= \frac{1}{2\pi^2} \iint_{0 \leq s \leq r \leq T} \int_{-\infty}^\infty \int_{-\infty}^\infty \left\{\left(\frac{\delta-\varepsilon}{2}\right)\xi^2 \wedge 1\right\}$$
$$\times \left\{\left(\frac{\delta-\varepsilon}{2}\right)\zeta^2 \wedge 1\right\} e^{-\frac{1}{2}\mathbb{E}[\{\xi B_s - \zeta B_r\}^2]}\, d\xi\, d\zeta\, ds\, dr.$$

If $0 \leq s \leq r \leq T$, $\mathbb{E}[\{\xi B_s - \zeta B_r\}^2] = (r-s)\zeta^2 + s|\xi - \zeta|^2$. (Why?) Accordingly,

$$\mathbb{E}\big[\{L_\varepsilon(A) - L_\delta(A)\}^2\big]$$
$$\leq \frac{1}{2\pi^2} \iint_{0 \leq s \leq r \leq T} \int_{-\infty}^\infty \int_{-\infty}^\infty \left\{\left(\frac{\delta-\varepsilon}{2}\right)^2 \xi^2\zeta^2 \wedge 1\right\}$$
$$\times e^{-\frac{1}{2}(r-s)\zeta^2 + \frac{1}{2}s|\xi-\zeta|^2}\, d\xi\, d\zeta\, ds\, dr.$$

Clearly, $\int_s^T e^{-\frac{1}{2}(r-s)\xi^2}\,dr \leq 2(T \wedge \xi^{-2})$. Thus (why?),

$$\iint_{0 \leq s \leq r \leq T} e^{-\frac{1}{2}(r-s)\zeta^2 + \frac{1}{2}s|\xi-\zeta|^2}\,dr\,ds \leq 4(T \wedge \xi^{-2}) \times (T \wedge |\xi-\zeta|^{-2}).$$

In particular,

$$\mathbb{E}\bigl[\{L_\varepsilon(A) - L_\delta(A)\}^2\bigr] \leq \frac{2}{\pi^2} \int_{-\infty}^\infty \int_{-\infty}^\infty \left\{\left(\frac{\delta-\varepsilon}{2}\right)^2 \xi^2 \zeta^2 \wedge 1\right\} \times (T \wedge \xi^{-2})$$
$$\times (T \wedge |\xi-\zeta|^{-2})\,d\xi\,d\zeta$$
$$\leq \frac{2}{\pi^2} \int_{-\infty}^\infty \int_{-\infty}^\infty \left\{\left(\frac{\delta-\varepsilon}{2}\right)^2 \xi^2 |\xi+\zeta|^2 \wedge 1\right\}$$
$$\times (T \wedge \xi^{-2}) \times (T \wedge \zeta^{-2})\,d\xi\,d\zeta.$$

Since $\iint_{\mathbb{R}^2} (T \wedge \xi^{-2})(T \wedge \zeta^{-2})\,d\xi\,d\zeta$ is finite, the result follows from Lebesgue's dominated convergence theorem. $\square$

Exercise **3.2.2** Prove Corollary 3.2.2. $\square$

We are ready to prove Theorem 3.1.1 in the case $N = d = 1$.

Proof of Theorem 3.1.1 for Linear Brownian Motion By Proposition 3.2.1, for all $k \geq 1$ and all intervals $A \subset \mathbb{R}$,

$$\mathbb{E}\bigl[\{L_\varepsilon(A)\}^k\bigr] \leq \left(\frac{2}{\pi}\right)^k k!\,|\mathrm{Leb}(A)|^{\frac{k}{2}}, \tag{3}$$

where Leb denotes Lebesgue's measure. In fact, a monotone class argument reveals that (3) holds for all Borel sets $A \subset \mathbb{R}_+$. In particular, for each Borel set $A \subset \mathbb{R}_+$, $\{L_\varepsilon(A);\ \varepsilon > 0\}$ is bounded in $L^k(\mathbb{P})$ for all $k \geq 1$. By Lemma 3.2.1 and by uniform integrability, we see that for each Borel set $A \subset \mathbb{R}_+$, $L(A) = \lim_{\varepsilon \to 0^+} L_\varepsilon(A)$ in $L^k(\mathbb{P})$ for all $k \geq 0$. Consequently, we can take $\varepsilon \to 0$ in equation (3) and deduce that for all $k \geq 0$ and for all Borel sets $A \subset \mathbb{R}_+$,

$$\mathbb{E}\bigl[\{L(A)\}^k\bigr] \leq \left(\frac{2}{\pi}\right)^k k!\{\mathrm{Leb}(A)\}^{k/2}. \tag{4}$$

A priori we do not know that this L is a measure on $B^{-1}(0)$, since for each A, our construction of $L(A)$ comes with a number of null sets, and as A varies, so do these uncountably many null sets. To circumvent this problem, we will construct a suitable modification $\widetilde{L}$ of L and prove Theorem 3.1.1 for this modification instead.

Define $\ell_t = L([0,t])$ and think of ℓ as the distribution function of L. Equation (4) implies that for all $t, h \geq 0$ and for all integers $k \geq 1$,

$$\mathbb{E}\bigl[|\ell_{t+h} - \ell_t|^k\bigr] \leq \left(\frac{2}{\pi}\right)^k k!\,h^{\frac{k}{2}}.$$

By Kolmogorov's continuity theorem, $\ell = (\ell_t; t \geq 0)$ has a Hölder modification of any order $0 < q < \frac{1}{2}$; cf. Theorem 2.5.1, Chapter 5. Let $\widetilde{\ell} = (\widetilde{\ell}_t; t \geq 0)$ denote this modification. By Exercise 3.2.4 below, $t \mapsto \widetilde{\ell}_t$ is continuous and increasing on $\mathbb{R}_+$ and satisfies $\widetilde{\ell}_0 = 0$. Thus, $\widetilde{\ell}$ is a random distribution function. Let $\widetilde{L}$ denote the corresponding measure. We shall now argue that this $\widetilde{L}$ is the L of Theorem 3.1.1.

Since ℓ and $\widetilde{\ell}$ are modifications of one another, so are L and $\widetilde{L}$. In particular, for each Borel set $A \subset \mathbb{R}_+$, $\widetilde{L}(A) = \lim_{\varepsilon \to 0^+} L_\varepsilon(A)$, where the convergence takes place in $L^k(\mathbb{P})$ for any $k \geq 1$. Thus (why?), for all $\eta > 0$,

$$\widetilde{L}(A)\mathbf{1}_{(\exists s \in A:\ |B_s| > \eta)} = \lim_{\varepsilon \to 0^+} L_\varepsilon(A)\mathbf{1}_{(\exists s \in A:\ |B_s| > \eta)},$$

where the convergence takes place in $L^k(\mathbb{P})$ for any $k \geq 1$. However, by Exercise 3.2.5 below, the above is 0. Equivalently, for all $\eta > 0$, $\widetilde{L}(A)\mathbf{1}_{(\exists s \in A:\ |B_s| > \eta)} = 0$, almost surely. In particular, the following holds with probability one:

$\forall \eta > 0$, rational, $\forall$ intervals $A \subset \mathbb{R}_+$ with rational endpoints:
$$\widetilde{L}(A)\mathbf{1}_{(\exists s \in A:\ |B_s| > \eta)} = 0.$$

Since $\widetilde{L}$ is a measure and since $B^{-1}(0) = \bigcap_{\eta > 0}\{s:\ |B_s| \leq \eta\}$, thanks to the continuity of $t \mapsto B_t$, the above implies that $\widetilde{L}$ is in fact a measure on $B^{-1}(0)$. Since $t \mapsto \ell_t$ is Hölder continuous of any order strictly less than $\frac{1}{2}$, for all $s < \frac{1}{2}$ and all $T > 0$,

$$\lim_{h \to 0^+} \sup_{0 \leq t \leq T} \frac{|\widetilde{\ell}_{t+h} - \widetilde{\ell}_t|}{h^s} < +\infty.$$

Equivalently, for all $s < \frac{1}{2}$ and all $T > 0$,

$$\lim_{h \to 0^+} \sup_{0 \leq t \leq T} \frac{\widetilde{L}\{\mathcal{B}(t;h)\}}{h^s} < \infty.$$

Thus, equation (1) of the preamble of this section holds for $\widetilde{L}$. To finish this proof, it suffices to show that equation (2) of the preamble of this section holds for $\widetilde{L}$. This follows from Frostman's theorem; cf. the following Exercise 3.2.3. □

Exercise 3.2.3 Demonstrate equation (2) of the preamble to this section for the local time constructed above.
(HINT: You will need to show that $\widetilde{L}$ is not a trivial measure, i.e., a.s., $\widetilde{L}(\mathbb{R}_+) = +\infty$.) □

Exercise 3.2.4 Prove that in our proof of Theorem 3.1.1, $t \mapsto \widetilde{\ell}_t$ is indeed continuous and increasing on $\mathbb{R}_+$. □

Exercise 3.2.5 Complete the present derivation of Theorem 3.1.1 by showing that with probability one, for all $\eta > 0$, there exists $\varepsilon_0 > 0$ such that for all $\varepsilon \in]0, \varepsilon_0[$, $L_\varepsilon(A) \mathbf{1}_{(\exists s \in A:\, |B_s| > \eta)} = 0$. □

3.3 A Variance Estimate

In order to carry out our analysis of the Brownian sheet local times in the more interesting $N > 1$ case, we shall need a technical variance bound that is the focus of this subsection. Roughly speaking, what we need is a careful estimate for the variance of $\sum_{j=1}^{m} \xi_j \cdot B_{t_j}$, where B denotes the N-parameter, d-dimensional Brownian sheet, $t_1, \ldots, t_m \in \mathbb{R}_+^N$, and $\xi_1, \ldots, \xi_m \in \mathbb{R}^d$. The reason for needing this estimate has already arisen within our proof of Lemma 3.2.1, where we needed the quantity $\mathbb{E}[e^{-i\xi B_s + i\zeta B_r}]$ (in the notation of Section 3.2). A similar but more intricate problem arises in the multiparameter setting.

Once again, we start with 1-dimensional Brownian motion. That is, where $d = N = 1$.

Lemma 3.3.1 *Suppose $Z = (Z_t;\ t \geq 0)$ denotes standard Brownian motion. For all $0 \leq r_1 \leq \cdots \leq r_m$ and for all $\zeta_1, \ldots, \zeta_m \in \mathbb{R}$,*

$$\mathbb{E}\Big[\Big(\sum_{j=1}^{m} \zeta_j Z_{r_j}\Big)^2\Big] = \sum_{k=1}^{m} \Big(\sum_{j=k}^{m} \zeta_j\Big)^2 \cdot (r_k - r_{k-1}),$$

where $r_0 = 0$.

Proof To better understand this calculation, note that

$$\mathbb{E}\Big[\Big(\sum_{j=1}^{m} \zeta_j Z_{r_j}\Big)^2\Big] = \sum_{j=1}^{m} \sum_{i=1}^{m} \zeta_j \zeta_i \cdot \{r_i \wedge r_j\},$$

since $\mathbb{E}[B_u B_v] = u \wedge v$. Our lemma above gives us what turns out to be a useful reordering of this sum. Let us now proceed with a proof in earnest. Note that

$$\sum_{j=1}^{m} \zeta_j Z_{r_j} = \sum_{j=1}^{m} \zeta_j \cdot \Big\{\sum_{k=1}^{k}(Z_{r_k} - Z_{r_{k-1}})\Big\} = \sum_{k=1}^{m}\Big\{\sum_{j=k}^{m} \zeta_j\Big\} \cdot \{Z_{r_k} - Z_{r_{k-1}}\},$$

all the time remembering that $r_0 = Z_{r_0} = 0$. On the other hand, for all integers $1 \leq k, k' \leq m$,

$$\mathbb{E}\big[\{Z_{r_k} - Z_{r_{k-1}}\} \cdot \{Z_{r_{k'}} - Z_{r_{k'-1}}\}\big] = \begin{cases} r_k - r_{k-1}, & \text{if } k = j, \\ 0, & \text{otherwise}. \end{cases}$$

That is, the differences $Z_{r_k} - Z_{r_{k-1}}$ are uncorrelated and, in fact, independent. The result follows readily from this. $\square$

Next, we describe a comparison result that relates the variance mentioned for the Brownian sheet to a similar object, but for additive Brownian motion. Recall that $X = (X_t;\ t \in \mathbb{R}_+^N)$ is N-parameter, d-dimensional additive Brownian if for all $t \in \mathbb{R}_+^N$, $X_t = \sum_{j=1}^N X_{t^{(j)}}^j$, where $X^1, \ldots, X^N$ are independent, d-dimensional Brownian motions.

Lemma 3.3.2 *Suppose $a \in \mathbb{R}_+^N$ satisfies the strict inequality $0 \prec a$. Then, for all $t_1, \ldots, t_m \in \mathbb{R}_+^N$ and for all $\xi_1, \ldots, \xi_m \in \mathbb{R}^d$,*

$$\mathbb{E}\left[\left(\sum_{j=1}^m \xi_j \cdot B_{t_j+a}\right)^2\right] \geq \eta^{N-1}\, \mathbb{E}\left[\left(\sum_{j=1}^m \xi_j \cdot X_{t_j}\right)^2\right],$$

where $\eta = \min_{1 \leq j \leq N} a^{(j)}$ and $X = (X_t;\ t \in \mathbb{R}_+^N)$ denotes N-parameter, d-dimensional additive Brownian motion.

Proof We begin by defining N one-parameter, d-dimensional Brownian motions $\beta^1, \ldots, \beta^N$:

$$\beta_r^i = B_{q_r^i} - B_a, \qquad r \geq 0,$$

where q_r^i is a vector in $\mathbb{R}_+^N$ all of whose coordinates agree with the corresponding ones in a, except in the ith coordinate, where $(q_r^i)^{(i)}$ equals r. For instance, when $N = 2$, $\beta_r^1 = B_{r,a^{(2)}} - B_a$ and $\beta_r^2 = B_{a^{(1)},r} - B_a$.

For all $t \in \mathbb{R}_+^N$ such that $a \preccurlyeq t$, we can write

$$B_t = C_t + \sum_{j=1}^N \beta_{t^{(j)}-a^{(j)}}^j, \qquad t \succcurlyeq a.$$

Applying Čentsov's representation (Theorem 1.5.1, Chapter 5), one can check that the processes $C = (C_t;\ t \succcurlyeq a)$, $(\beta_{t^{(1)}-a^{(1)}}^1;\ t^{(1)} \geq a^{(1)}), \ldots$, and $(\beta_{t^{(N)}-a^{(N)}}^N;\ t^{(N)} \geq a^{(N)})$ are all independent, centered, and Gaussian; cf. Exercise 3.3.1 below. Moreover, for all $j = 1, \ldots, N$, $Z^j = (Z_r^j;\ r \geq 0)$ is a d-dimensional Brownian motion, where

$$Z_r^j = \left\{\prod_{\substack{\ell=1 \\ \ell \neq j}}^N a^{(\ell)}\right\}^{-\frac{1}{2}} \beta_r^j, \qquad r \geq 0.$$

Finally, $\mathbb{E}[C_t^{(i)} Z_r^{(j)}] \geq 0$ for all $1 \leq i, j \leq N$, $t \in \mathbb{R}_+^N$ and $r \geq 0$. Hence,

$$\mathbb{E}\left[\left\{\sum_{j=1}^m \xi_j \cdot B_{t_j}\right\}^2\right] \geq \mathbb{E}\left[\left\{\sum_{j=1}^m \sum_{\ell=1}^N \xi_j \cdot \beta_{t_j^{(\ell)}-a^{(\ell)}}^\ell\right\}^2\right]$$

$$= \mathbb{E}\Bigg[\Big\{\sum_{j=1}^{m}\sum_{i=1}^{N}\prod_{\substack{1\leq \ell \leq N \\ \ell \neq j}}\sqrt{a^{(\ell)}}\,\xi_j \cdot Z^i_{t_j^{(i)}-a^{(i)}}\Big\}^2\Bigg]$$

$$\geq \eta^{N-1}\,\mathbb{E}\Bigg[\Big\{\sum_{j=1}^{m}\sum_{i=1}^{N}\xi_j \cdot Z^i_{t_j^{(i)}-a^{(i)}}\Big\}^2\Bigg].$$

(Why?) The result follows, since $\sum_{i=1}^{N} Z_\bullet^{(i)}$ is a d-dimensional, N-parameter additive Brownian motion. □

Now we combine Lemmas 3.3.1 and 3.3.2 to obtain our desired variance estimate, by way of a lower bound. To describe it, we need some messy notation. Suppose $t_1,\ldots,t_m, a \in \mathbb{R}_+^N$ are held fixed, and for some $\eta > 0$, $t_j^{(k)} \geq a^{(k)} \geq \eta$, for all $k = 1,\ldots,N$ and $j = 1,\ldots,m$. Let $\pi_\ell(1),\ldots,\pi_\ell(m)$ denote the indices that order $t_1^{(\ell)},\ldots,t_m^{(\ell)}$. More precisely, for all $\ell = 1,\ldots,N$, π_ℓ is the m-vector defined by

$$t_{\pi_\ell(1)}^{(\ell)} \leq \cdots \leq t_{\pi_\ell(m)}^{(\ell)}.$$

We will also define $\pi_\ell(0) \equiv 0$. Then, by Lemma 3.3.2 above, for all $\xi_1,\ldots,\xi_m \in \mathbb{R}^d$,

$$\mathbb{E}\Bigg[\Big\{\sum_{j=1}^{m}\xi_j \cdot B_{t_j}\Big\}^2\Bigg] \geq \eta^{N-1}\,\mathbb{E}\Bigg[\Big\{\sum_{j=1}^{m}\xi_j \cdot X_{t_j-a}\Big\}^2\Bigg],$$

where X is a d-dimensional, N-parameter additive Brownian motion. Note that we can write $X_t = X^1_{t^{(1)}} + \cdots + X^N_{t^{(N)}}$, where $X^1,\ldots,X^N$ are independent Brownian motions. Thus,

$$\mathbb{E}\Bigg[\Big\{\sum_{j=1}^{m}\xi_j \cdot B_{t_j}\Big\}^2\Bigg] \geq \eta^{N-1}\sum_{\ell=1}^{N}\sum_{p=1}^{d}\mathbb{E}\Bigg[\Big\{\sum_{j=1}^{m}\xi_{\pi_\ell(j)}^{(p)}\,\big(X^\ell_{t_{\pi_\ell(j)}-a^{(\ell)}}\big)^{(p)}\Big\}^2\Bigg].$$

To simplify the notation somewhat, let Z denote a standard (one-dimensional) Brownian motion and note that we have

$$\mathbb{E}\Bigg[\Big\{\sum_{j=1}^{m}\xi_j \cdot B_{t_j}\Big\}^2\Bigg] \geq \eta^{N-1}\sum_{\ell=1}^{N}\sum_{p=1}^{d}\mathbb{E}\Bigg[\Big\{\sum_{j=1}^{m}\xi_{\pi_\ell(j)}^{(p)}\,Z_{t_{\pi_\ell(j)}-a^{(\ell)}}\Big\}^2\Bigg].$$

According to Lemma 3.3.1, we have proved the following:

Proposition 3.3.1 *Suppose $a, t_1,\ldots,t_m$ are in $\mathbb{R}_+^N$, and satisfy $t_j \succcurlyeq a$, and $\min_{1\leq j \leq N} a^{(j)} = \eta > 0$. For all $\ell = 1,\ldots,N$, let $\pi_\ell(1),\ldots,\pi_\ell(m)$ denote the indices that order $t_1^{(\ell)},\ldots,t_m^{(\ell)}$. Then,*

$$\mathbb{E}\Bigg[\Big\{\sum_{j=1}^{m}\xi_j \cdot B_{t_j}\Big\}^2\Bigg] \geq \eta^{N-1}\sum_{\ell=1}^{N}\sum_{p=1}^{d}\sum_{k=1}^{m}\Big(\sum_{j=k}^{m}\xi_{\pi_\ell(j)}^{(p)}\Big)^2 \cdot \big(t_{\pi_\ell(k)}^{(\ell)} - t_{\pi_\ell(k-1)}^{(\ell)}\big),$$

where $\pi_\ell(0) = 0$ for all $\ell = 1, \ldots, N$.

Exercise 3.3.1 Prove that in our proof of Lemma 3.3.2, $(C_t;\ t \succcurlyeq a)$, $(\beta^1_{t^{(1)} - a^{(1)}};\ t^{(1)} \geq a^{(1)}), \ldots, (\beta^N_{t^{(N)} - a^{(N)}};\ t^{(N)} \geq a^{(N)})$ are independent Gaussian processes. Compute their mean and covariance functions. □

3.4 Proof of Theorem 3.1.1: General Case

The following technical proposition is the key step in our proof of Theorem 3.1.1; it is the general N analogue of Proposition 3.2.1. With this underway, Theorem 3.1.1 follows by refining and extending the one-parameter method of Section 3.2. We defer the remainder of this derivation of Theorem 3.1.1 to Exercise 3.4.1 below and state and prove the hardest portion of the argument, which is the following.

Proposition 3.4.1 *Suppose $\eta, h > 0$ are fixed and that $d < 2N$. For all integers $m \geq 1$, there exists a finite constant $\Gamma > 0$ that only depends on η, N, and d such that for every Borel set $A \subset [\eta, \infty[^N$ of diameter bounded above by h, and for all functions $f \in L^1(\mathbb{R}^d)$,*

$$\mathbb{E}\Big[\Big\{\int_A f(B_s)\, ds\Big\}^m\Big] \leq \Gamma^m \, \|f\|_1^m (m!)^N \, h^{\frac{1}{2}(N-2d)m}.$$

Proof Without loss of generality, we can assume that $f(x) \geq 0$ for all $x \in \mathbb{R}^d$. Clearly,

$$\mathbb{E}\Big[\Big\{\int_A f(B_s)\, ds\Big\}^m\Big] = \mathbb{E}\Big[\int_A \cdots \int_A f(B_{s_1}) \cdots f(B_{s_m})\, ds_1 \cdots ds_m\Big].$$

We can denote this mth moment by $\mathfrak{M}_m$ and observe that by Fubini's theorem,

$$\mathfrak{M}_m = \int_A \cdots \int_A \mathbb{E}\Big[\prod_{j=1}^m f(B_{s_j})\Big]\, ds_1 \cdots ds_m.$$

On the other hand, by the inversion theorem for characteristic functions,

$$\mathbb{E}\Big[\prod_{j=1}^m f(B_{s_j})\Big] = (2\pi)^{-dm} \int_{\mathbb{R}^d}\cdots\int_{\mathbb{R}^d} \times \int_{\mathbb{R}^d}\cdots\int_{\mathbb{R}^d} \prod_{j=1}^m \{e^{-i\xi_j \cdot \lambda_j} f(\lambda_j)\}$$
$$\times \mathbb{E}[e^{i\sum_{j=1}^m \xi_j \cdot B_{s_j}}]\, d\lambda_1 \cdots d\lambda_m \times d\xi_1 \cdots d\xi_m,$$

which equals

$$(2\pi)^{-dm} \int_{\mathbb{R}^d}\cdots\int_{\mathbb{R}^d} \times \int_{\mathbb{R}^d}\cdots\int_{\mathbb{R}^d} \prod_{j=1}^m \{e^{-i\xi_j \cdot \lambda_j} f(\lambda_j)\} e^{-\frac{1}{2}\mathbb{E}[\{\sum_{j=1}^m \xi_j \cdot B_{s_j}\}^2]}$$
$$\times d\lambda_1 \cdots d\lambda_m \times d\xi_1 \cdots d\xi_m,$$

since B is a Gaussian process. Moreover, we can use the inequality $|e^{i\theta}| \le 1$ and the fact that all else is nonnegative to see that $\mathbb{E}[\prod_{j=1}^{m} f(B_{s_j})]$ is bounded above by

$$(2\pi)^{-dm} \|f\|_1^m \int_{\mathbb{R}^d} \cdots \int_{\mathbb{R}^d} e^{-\frac{1}{2} \mathbb{E}[\{\sum_{j=1}^{m} \xi_j \cdot B_{s_j}\}^2]} d\xi_1 \cdots d\xi_m.$$

On the other hand, the above exponential can be bounded by Proposition 3.3.1. This way, we can bound $\mathfrak{M}_m$ from above by

$$(2\pi)^{-dm} \|f\|_1^m \int_A \cdots \int_A \times \int_{\mathbb{R}^d} \cdots \int_{\mathbb{R}^d}$$

$$\times \exp\Bigg(-\frac{\eta^{N-1}}{2} \sum_{\ell=1}^{N} \sum_{p=1}^{d} \sum_{k=1}^{m} \Big| \sum_{j=k}^{m} \xi_{\pi_\ell(j)}^{(p)} \Big|^2 \cdot \{s_{\pi_\ell(k)}^{(\ell)} - s_{\pi_\ell(k-1)}^{(\ell)}\} \Bigg)$$

$$\times d\xi_1 \cdots d\xi_m \times ds_1 \cdots ds_m,$$

where $\pi_\ell(\bullet)$ orders $s_1^{(\ell)}, \ldots, s_m^{(\ell)}$ in the sense of the previous subsection. This is, of course, a slight abuse of notation, since π_ℓ depends on s. Our subsequent changes of variables need to be made with care, all the time keeping this in mind. Nonetheless, it is fortunate that there are many symmetries in the above manyfold integral. For instance, we can change variables: For all $1 \le k \le m$ and $1 \le p \le d$, let $\zeta_k^{(p)} = \sum_{j=k}^{m} \xi_{\pi_\ell(j)}^{(p)}$. Since the absolute value of the Jacobian of this map is 1, $\prod_{j=1}^{m} d\xi_j = \prod_{j=1}^{m} d\zeta_j$. Thus, $\mathfrak{M}_m$ is bounded above by

$$(2\pi)^{-dm} \|f\|_1^m \int_A \cdots \int_A \times \int_{\mathbb{R}^d} \cdots \int_{\mathbb{R}^d}$$

$$\times \exp\Bigg(-\frac{\eta^{N-1}}{2} \sum_{\ell=1}^{N} \sum_{p=1}^{d} \sum_{k=1}^{m} |\zeta_k^{(p)}|^2 \cdot \{s_{\pi_\ell(k)}^{(\ell)} - s_{\pi_\ell(k-1)}^{(\ell)}\} \Bigg)$$

$$\times d\zeta_1 \cdots d\zeta_m \times ds_1 \cdots ds_m,$$

which can be simplified to

$$(2\pi)^{-dm} \|f\|_1^m \int_A \cdots \int_A \times \int_{\mathbb{R}^d} \cdots \int_{\mathbb{R}^d}$$

$$\times \exp\Bigg(-\frac{\eta^{N-1}}{2} \sum_{\ell=1}^{N} \sum_{k=1}^{m} \|\zeta_k\|^2 \cdot \{s_{\pi_\ell(k)}^{(\ell)} - s_{\pi_\ell(k-1)}^{(\ell)}\} \Bigg)$$

$$\times d\zeta_1 \cdots d\zeta_m \times ds_1 \cdots ds_m.$$

At this juncture we choose to integrate over the $ds_1 \cdots ds_m$ integral, first. Note that $\int_A \cdots \int_A \exp\{\cdots\} ds_1 \cdots ds_m$ equals

$$(m!)^N \int \cdots \int_{\substack{s_1,\ldots,s_m \in A: \\ \forall 1 \le \ell \le N: s_1^{(\ell)} \le \cdots \le s_m^{(\ell)}}} e^{-\frac{1}{2} \eta^{N-1} \sum_{\ell=1}^{N} \sum_{k=1}^{m} \|\zeta_k\|^2 \cdot (s_k^{(\ell)} - s_{k-1}^{(\ell)})} ds_1 \cdots ds_m,$$

where $s_0 = 0$. Since the diameter of A is bounded above by $h > 0$, we can deduce the following estimate:

$$\int_A \cdots \int_A \exp\{\cdots\} ds_1 \cdots ds_m$$
$$\leq (m!)^N \int_{[0,h]^N} \cdots \int_{[0,h]^N} e^{-\frac{1}{2}\eta^{N-1} \sum_{\ell=1}^N \sum_{k=1}^m \|\zeta_k\|^2 \cdot r_k^{(\ell)}} dr_1 \cdots dr_m$$
$$= (m!)^N \prod_{k=1}^m \int_{[0,h]^N} e^{-\frac{1}{2}\eta^{N-1} \sum_{\ell=1}^N r^{(\ell)} \cdot \|\zeta_k\|^2} dr.$$

In particular, $\mathfrak{M}_m$ is bounded above by

$$(2\pi)^{-dm} \|f\|_1^m (m!)^N \int_{\mathbb{R}^d} \cdots \int_{\mathbb{R}^d}$$
$$\times \prod_{k=1}^m \int_{[0,h]^N} e^{-\frac{1}{2}\eta^{N-1} \sum_{\ell=1}^N r^{(\ell)} \cdot \|\zeta_k\|^2} dr \, d\zeta_1 \cdots d\zeta_m.$$

Once more, we choose to change the order of integration. Note that

$$\int_{\mathbb{R}^d} e^{-\frac{1}{2}\eta^{N-1} \sum_{\ell=1}^N r^{(\ell)} \cdot \|\zeta_k\|^2} d\zeta_k$$
$$= 2^{\frac{d}{2}} \eta^{-\frac{1}{2}d(N-1)} \Big\{ \sum_{\ell=1}^N r^{(\ell)} \Big\}^{-\frac{d}{2}} \int_{\mathbb{R}^d} e^{-\|\zeta\|^2} d\zeta$$
$$\leq C_1 \|r\|^{-\frac{d}{2}},$$

for some finite constant C_1 that depends only on N, η, and d. Thus, $\mathfrak{M}_m$ is bounded above by

$$(2\pi)^{-dm} \|f\|_1^m (m!)^N C_1^m \left(\int_{[0,h]^N} \|r\|^{-\frac{d}{2}} dr \right)^m$$
$$\leq C_2^m \|f\|_1^m (m!)^N \left(\int_0^h r^{-\frac{d}{2}+N-1} dr \right)^m,$$

for some finite constant C_2 that depends only on N, η, and d. This follows from integration in polar coordinates; cf. Supplementary Exercise 7, Chapter 3. The proposition follows from this upon letting $\Gamma = C_2$. □

Exercise 3.4.1 (Hard) Complete the presented proof of Theorem 3.1.1. This is from Ehm (1981).
(HINT: Imitate the argument given in the Brownian case, but use Propositions 3.3.1 and 3.4.1 in place of their 1-parameter counterparts.) □

4 Supplementary Exercises

1. Let $B^{-1}(a) = \{s \in \mathbb{R}_+^N : B_s = a\}$ denote the level set at $a \in \mathbb{R}^d$ of the N-parameter, $\mathbb{R}^d$-valued Brownian sheet B. Prove that irrespective of the choice of $a \in \mathbb{R}^d$, $B^{-1}(a)$ is nontrivial if and only if $d < 2N$. In the case $d < 2N$, show that it has codimension $(\frac{d}{2})$ and Hausdorff dimension $N - (\frac{d}{2})$.

2. Improve Proposition 3.2.1 by showing the existence of a finite constant $C > 0$ such that
$$\mathbb{E}\Big[\Big\{\int_t^{t+h} f(B_s)\,ds\Big\}^{2k}\Big] \leq C^k \frac{(2k)!}{k!}\|f\|_1^{2k} h^k.$$
(HINT: Reconsider the k-dimensional multiple integral that arose in the course of our proof of the mentioned proposition.)

3. Let B denote the N-parameter, $\mathbb{R}^d$-valued Brownian sheet and consider the **occupation measure** μ of $(B_t;\ t \in [1,2]^N)$: For all Borel sets $A \subset \mathbb{R}^d$,
$$\mu(A) = \int_{[1,2]^N} \mathbf{1}_A(B_s)\,ds.$$
Let $\widehat{\mu}$ denote the Fourier transform of μ. That is, for all $\xi \in \mathbb{R}^d$, $\widehat{\mu}(\xi) = \int e^{i\xi \cdot a}\mu(da)$.

   (i) Prove that when $d < 2N$, $\mathbb{E}[\int_{\mathbb{R}^d} |\widehat{\mu}(a)|^2\,da\,] < \infty$.

   (ii) Conclude that when $d < 2N$, there exists a stochastic process $(\mathcal{L}(x);\ x \in \mathbb{R}^d)$ such that:

   (a) $\mathcal{L}(x) \geq 0$, a.s. for all $x \in \mathbb{R}^d$ and $\mathbb{E}[\int_{\mathbb{R}^d}\{\mathcal{L}(x)\}^2\,dx] < \infty$; and

   (b) for all $f \in L^\infty(\mathbb{R}^d)$, $\int_{[1,2]^N} f(B_s)\,ds = \int f(x)\mathcal{L}(x)\,dx$, a.s.

   (iii) It can be shown that $x \mapsto \mathcal{L}(x)$ has a continuous modification that can be denoted by $\mathcal{L}$. Using this, show that $\mathcal{L}(0) = L([1,2]^N)$, a.s., where L denotes the local times of Section 3.2.

Various parts of this are due to P. Lévy, H. Trotter, S. M. Berman, and W. Ehm. (HINT: First, prove Plancherel's theorem in the following form: If ζ is a finite measure on $\mathbb{R}^d$ whose Fourier transform is in $L^2(\mathbb{R}^d)$, then ζ is absolutely continuous with respect to Lebesgue's measure, and the Radon–Nikodým derivative is in $L^2(\mathbb{R}^d)$.)

4. (Hard) For any compact set $E \subset \mathbb{R}^d$, define $\rho_i(E)$ and $\rho_o(E)$ to be the inner and outer radii of E, respectively. More precisely, $\rho_i(E) = \inf\{|x| : x \in E\}$ and $\rho_o(E) = \sup\{|x| : x \in E\}$. Prove that if B denotes the N-parameter Brownian sheet and if $\eta \in\,]0, 1[$ is fixed, there exists a finite constant $K_\eta > 1$ such that for all compact sets E with $\rho_i(E) > \eta$ and $\rho_o(E) < \eta^{-1}$,
$$K_\eta^{-1}\mathsf{Cap}_{d-2N}(E) \leq \mathbb{P}(B([0,1]^N) \cap E \neq \emptyset) \leq K_\eta \mathsf{Cap}_{d-2N}(E).$$
(HINT: Adapt the given proof of Theorem 1.1.1.)

5. Suppose $X = (X_t;\ t \in \mathbb{R}_+^N)$ denotes N-parameter, $\mathbb{R}^d$-valued additive Brownian motion. Prove that for all $M > 1$, there exists a finite constant $K_M > 1$ such that for every compact set E whose outer radius is no more than M,

$$K_M^{-1}\mathsf{Cap}_{d-2N}(E) \leq \mathbb{P}(X([1,2]^N) \cap E \neq \varnothing) \leq K_M \mathsf{Cap}_{d-2N}(E).$$

Improve this by replacing $[1,2]^N$ by $[a,b]$ for any $0 \prec a \prec b$ both in $\mathbb{R}_+^N$. (The constant K will then depend on a and b as well.)

6. Suppose $X = (X_t;\ t \in \mathbb{R}_+^N)$ denotes N-parameter, $\mathbb{R}^d$-valued additive Brownian motion. In the case $d < 2N$, construct local times for X and prove that Theorem 3.1.1 holds true for the process X.
In a slightly different form, this is due to E. B. Dynkin.
(When $N = 2$, the 2-parameter process $\mathbb{R}_+^2 \ni (s,t) \mapsto B_s^1 - B_t^2$ is an additive Brownian motion (why?). The corresponding local times are traditionally called the **intersection local times** between the two d-dimensional Brownian motions B^1 and B^2.)

7. (Hard) Let $B^1, \ldots, B^k$ denote k independent $\mathbb{R}^d$-valued Brownian sheets with $N^1, \ldots, N^k$ parameters, respectively. Find a necessary and sufficient condition on $N^1, \ldots, N^k$ and d for $B^1([1,2]^{N^1}) \cap \cdots \cap B^k([1,2]^{N^k})$ to be nonempty. That is, when do the trajectories of k independent Brownian sheets intersect?

8. Suppose B denotes an $\mathbb{R}^d$-valued, N-parameter Brownian sheet.

(i) Prove that for all $a \prec b$ both in $\mathbb{R}_+^N$ and for all $M > 1$, there exist positive and finite constants A_1 and A_2 such that for all $\varepsilon \in]0,1[$ and for all $x \in [-M, M]^d$,

$$A_1 U_{d-2N}(\varepsilon) \leq \mathbb{P}(B([a,b]) \cap \mathcal{B}(x;\varepsilon) \neq \varnothing) \leq A_2 U_{d-2N}(\varepsilon),$$

where $U_\beta(y) = 1$ if $\beta < 0$, $U_\beta(x) = \{\ln_+(1/x)\}^{-1}$ if $\beta = 0$, and $U_\beta(x) = x^\beta$ if $\beta > 0$. Moreover, $\mathcal{B}(x;\varepsilon) = \{z \in \mathbb{R}^d : |z - x| < \varepsilon\}$.

(ii) Conclude that when $d \geq 2N$, with probability one, $\text{Leb}\{\overline{B(\mathbb{R}_+^N)}\} = 0$.

9. (Hard) Consider a standard (1-dimensional) Brownian motion $B = (B_t;\ t \geq 0)$.

(i) Prove that for all $x \in \mathbb{R}$ and all $t > 0$, $\lim_{\varepsilon \to 0+} \int_0^t \varphi_\varepsilon(B_s - x)\,ds$ exists in $L^2(\mathbb{P})$, where φ_ε is defined by equation (1) of Section 3.1. Let L_t^x denote this limit.

(ii) Show that for all $T > 0$ and all integers $k \geq 1$, there exists a finite constant $C_k > 0$ such that for all $x \in \mathbb{R}$, all integers $k \geq 1$, and all $s, t \in [0, T]$,

$$\mathbb{E}\{|L_t^x - L_s^x|^{2k}\} \leq C_k |t - s|^k.$$

(iii) Show that for all $T > 0$ and all integers $k \geq 1$, there exists a finite constant $D_k > 0$ such that for all $x, y \in \mathbb{R}$, and all $t > 0$,

$$\mathbb{E}\{|L_t^x - L_t^y|^{2k}\} \leq D_k |x - y|^k.$$

(iv) Conclude that there exists a continuous modification of $\mathbb{R} \times [0, \infty[\ni (x, t) \mapsto L_t^x$. Let us continue to write this modification as L_t^x.

(v) Prove that there exists one null set outside which the following holds for all $f \in L^\infty(\mathbb{R})$ and all $t \geq 0$:

$$\int_0^t f(B_s)\, ds = \int_{-\infty}^\infty f(x) L_t^x \, dx.$$

Parts (i) and (iv) are essentially due to P. Lévy; part (iii) is essentially due to H. Trotter and, in part, to Berman (1983). Part (iv) is called the **occupation density formula**; it represents the local times L_t^x as the density of the occupation measure of B with respect to Lebesgue's measure.
(HINT: For part (iii), use the fact that $|e^{i\xi x} - e^{i\xi y}| \leq |\xi| \cdot |x - y| \wedge 1$.)

5 Notes on Chapter 12

Preamble That the Brownian sheet is not a multiparameter Markov process in the sense of Chapter 11 can only be interpreted as such, since the Brownian sheet does have a number of Markovian properties. For instance, consider an $\mathbb{R}^d$-valued, N-parameter Brownian sheet B. If we view $t^{(i)} \mapsto B_t$ as a one-parameter process taking its values in the space of continuous functions from $[0, 1]^{N-1}$ into $\mathbb{R}^d$, we are observing an infinite-dimensional Feller process. Other Markovian properties, far more intricate than the one mentioned here, can be found in Dalang and Walsh (1992), Dorea (1983, 1982), and Zhang (1985).

Section 1 Theorem 1.1.1 is from Khoshnevisan and Shi (1999) and resolves an old problem that was partially settled in Orey and Pruitt (1973). Corollary 1.1.3 is a consequence of the general theory of Adler (1977, 1981) and Weber (1983); see also Chen (1997). Recent applications, as well as nontrivial extensions of Theorem 1.1.1 to the potential theory of stochastic partial differential equations, can be found in Nualart (2001a, 2001b).

Section 2 In the context of random fields, the codimension approach that we have taken is new. However, see the Notes on Chapter 11 for earlier works on codimension. Theorem 1.1.1 is a consequence of the general theory of Adler (1981); see also Kahane (1985, Chapters 17,18).

When $N = 2$, a deeper understanding of Theorem 2.1.1 is available; cf. Khoshnevisan (1999). Khoshnevisan and Xiao (2002) study the level sets of N-parameter additive Lévy processes via capacities.

It is a strange fact that the level sets of various random processes are intimately connected to the size of the range of those very processes! This observation is at the heart of the extension of Theorem 2.1.1, given in Khoshnevisan (1999), and appears earlier on in the literature in Fitzsimmons and Port (1990), Kahane (1982, 1983), and Port (1988). For a pedagogical discussion, see Kahane (1985, Chapters 17,18).

Section 3 The introduction of local times as the natural measures that live on the level sets of processes is not new and goes back to Paul Lévy; see Itô and McKean (1974, Section 2.8, Chapter 1).

Proposition 3.2.1 is a consequence of M. Kac's moment formula; cf. Fitzsimmons and Pitman (1999). Our proof of Theorem 3.1.1 is essentially a refinement of Ehm (1981, equation 1.6) for $m = 2$, using the notation of the mentioned paper.

The literature on local times is monstrously large, and it is simply impossible to provide a comprehensive list of all of the appropriate references here. Suffice it to say that three good starting points are the following, together with their combined references:

- There are deep connections between local times and the theory of Markov processes. To learn more about these, you can start by reading (Dellacherie and Meyer 1988; Blumenthal and Getoor 1968; Fukushima et al. 1994; Sharpe 1988). Depending on which reference you look at, you may need to look at "continuous additive functionals" and/or "homogeneous random measures" first.

- Local times of nice processes can also be studied by appealing to function-theoretic and Fourier-analytic methods. This aspect is very well documented in Geman and Horowitz (1980); and

- There are still deeper connections to stochastic calculus, for which you can start by reading (Chung and Williams 1990; Karatzas and Shreve 1991; Revuz and Yor 1994; Rogers and Williams 1994).

As a representative bibliography on local times of processes related to the Brownian sheet, we mention only Adler (1977, 1980), Dozzi (1988), Imkeller (1984, 1986), Lacey (1990), and Vares (1983). The combined bibliography of (Adler 1981; Geman and Horowitz 1980) contains further references to this subject.

Section 4 Supplementary Exercise 9 started as a problem of P. Lévy that was subsequently solved by H. Trotter, who proved that Brownian local times are continuous. The problem of when a Lévy process possesses (jointly) continuous local times, however, remained open until recently. It was finally settled in Barlow (1988), using earlier metric entropy ideas from Barlow and Hawkes (1985). See Bass and Khoshnevisan (1992b) and Bertoin (1996, Section 3, Chapter V) for simpler proofs of the sufficiency half of Barlow's theorem. This has, in turn, generated further recent interest. In particular, see the impressive results of Marcus and Rosen (1992). Interestingly enough, the local times of one-parameter processes are related to the range of the Brownian sheet and other random fields; see Csáki et al. (1988, 1989, 1992), Eisenbaum (1995, 1997), Rosen (1991), Weinryb (1986), Weinryb and Yor (1988, 1993), and Yor (1983).

Epstein (1989) extends the connections between the Brownian sheet and local times, and more general additive functionals, in new and innovative directions.

In the past decade our understanding of the geometry of the Brownian sheet has been rapidly developing. It is a shame that this book has to end before I can describe some of these exciting results. You can start off, where this book ends, by reading Dalang and Mountford (1996, 1997, 2000, 2001), Dalang and Walsh (1992,

1993, 1996), Kendall (1980), Khoshnevisan (1995), Mountford (1993, 2002), and Walsh (1982).

Part III

Appendices

Appendix A
Kolmogorov's Consistency Theorem

The Daniell–Kolmogorov existence theorem, or Kolmogorov's consistency theorem, describes precisely when a given stochastic process exists.

Given an arbitrary set T, we can define $\mathbb{R}^T$ as the collection of all functions $f : T \to \mathbb{R}$. For any finite set $F \subset T$, let $\pi_F : \mathbb{R}^T \to \mathbb{R}^F$ denote the projection onto $\mathbb{R}^F$. Of course, $\mathbb{R}^F$ is the same thing as the finite-dimensional Euclidean space $\mathbb{R}^{\#F}$, up to an identification of terms. Recall that the product topology on $\mathbb{R}^T$ is the smallest topology that makes π_F continuous for all finite $F \subset T$.

Suppose for each finite F, we are given a probability measure μ_F that lives on the Borel subsets of the finite-dimensional Euclidean space $\mathbb{R}^F$. Then, we can define a probability measure $\mu_F \circ \pi_F^{-1}$ on the Borel subsets of $\mathbb{R}^T$ (with the product topology). Recall that this means that for all Borel sets $A \subset \mathbb{R}^T$, $\mu_F \circ \pi_F^{-1}(A) = \mu_F\{\pi_F(A)\}$.

For instance, suppose $T = \{1, 2\}$ and $F = \{1\}$. Then, $\mathbb{R}^T = \mathbb{R}^2$, and π_F is the projection $\pi_F(x) = x^{(1)}$ for all $x \in \mathbb{R}^2$. As another, somewhat more interesting, example, consider $T = \mathbb{N}$ and $F = \{1, 2\}$. Then, for all $x \in \mathbb{R}^T$, $\pi_F(x)$ is the vector $(x^{(1)}, x^{(2)})$. In this latter case, for any finite $F \subset \mathbb{N}$ of the form $F = \{t_1, \ldots, t_k\}$, π_F maps the vector $x = (x^{(1)}, x^{(2)}, \ldots) \in \mathbb{R}^\mathbb{N}$ to $\pi_F(x)$, which is the vector $(x^{(t_1)}, \ldots, x^{(t_k)}) \in \mathbb{R}^{\#F} = \mathbb{R}^F$.

Let us first consider a measure μ on $\mathbb{R}^\mathbb{N}$, which is usually written as $\mathbb{R}^\omega = \mathbb{R} \times \mathbb{R} \times \cdots$. What do probability measures on $\mathbb{R}^\omega$ look like?

Recall that $\mathbb{R}^\omega$ is endowed with the product topology. Moreover, any $A \in \mathbb{R}^\omega$ is of the form $A = A_1 \times A_2 \times \cdots$, where $A_i \in \mathbb{R}$. Such a set is a **cylinder set** if all but a finite number of the A_j's are equal to $\mathbb{R}$. We can define a very natural partial order $\overset{*}{\subset}$ on all cylinder sets. Suppose $A = A_1 \times A_2 \times \cdots$ and $B = B_1 \times B_2 \times \cdots$ are cylinder sets. Let $\mathcal{A}$ denote $\{i \geq 1 : A_i \neq \mathbb{R}\}$ and similarly define $\mathcal{B}$ for B. Then, $A \overset{*}{\supset} B$ if and only if $\mathcal{B} \supset \mathcal{A}$. That is, whenever a coordinate of B is $\mathbb{R}$, the corresponding coordinate of A is also $\mathbb{R}$. Note that any probability measure μ on the Borel subsets of $\mathbb{R}^\omega$ is defined consistently on all cylinder sets. Formally speaking, whenever A and B are cylinder sets with $A \overset{*}{\supset} B$, $\mu(\pi_\mathcal{B}(A)) = \mu(B)$. Note that $\mu \circ \pi_F$ is a probability measure on the Borel subsets of $\mathbb{R}^F$, and consistency is equivalent to the following: If $F \subset G$ are both finite, $\mu_G \circ \pi_F^{-1} = \mu_F$. (Why?)

The above motivates the following definition. Suppose that for each finite $F \subset T$, μ_F is a probability measure on $\mathbb{R}^F$. We say that this collection of μ_F's is **consistent** if

$$F \subset G \text{ finite} \iff \mu_F = \mu_G \circ \pi_F^{-1}.$$

You should check that when $T = \mathbb{N}$, this is the same as the previous notion of consistency.

Kolmogorov's consistency theorem asserts that any collection of consistent probability measures actually comes from one measure on $\mathbb{R}^T$.

Theorem 1 (Kolmogorov's Consistency Theorem, I) *Given a consistent collection $(\mu_F : F \subset T, \text{finite})$ of probability measures on $(\mathbb{R}^F; F \subset T, \text{finite})$, respectively, there exists a unique probability measure $\mathbb{P}$ on the Borel subsets of $\mathbb{R}^T$ such that $\mathbb{P} \circ \pi_F^{-1} = \mu_F$, for all finite $F \subset T$.*

The measure $\mathbb{P}$ is called **Kolmogorov's extension** of the μ_F's.
The above is *equivalent* to the following.

Theorem 2 (Kolmogorov's Consistency Theorem, II) *Given a consistent collection $(\mu_F; F \subset T, \text{finite})$ of probability measures on $(\mathbb{R}^F; F \subset T, \text{finite})$, respectively, let $\mathbb{P}$ denote the extension of the μ_F's to all of $\mathbb{R}^T$. Then, there exists a probability space $(\Omega, \mathcal{F}, \mathbb{P})$ on which we construct a stochastic process $X = (X_t; t \in T)$ whose finite-dimensional distributions are the μ_F's. That is, for all finite $F \subset T$ of the form $F = \{t_1, \ldots, t_k\}$, and for all Borel sets $A_1, \ldots, A_k \in \mathbb{R}$,*

$$\mathbb{P}(X_{t_1} \in A_1, \ldots, X_{t_k} \in A_k) = \mu_F(A_1 \times \cdots \times A_k).$$

Theorems 1 and 2 are proved in detail in (Bass 1995; Billingsley 1968; Dudley 1989), for instance.

Appendix B
Laplace Transforms

In this appendix we collect some important facts about Laplace transforms.[1] Throughout, μ denotes a σ-finite measure on $\mathbb{R}_+$ (endowed with its Borel sets). The **distribution function** F of the measure μ is defined as

$$F(s) = \mu\{[0, s]\}, \qquad s \geq 0.$$

Suppose $h : \mathbb{R}_+ \to \mathbb{R}_+$ is Borel measurable. Then,

$$\int_0^\infty h(s)\, dF(s) = \int_0^\infty h(s)\, \mu(ds).$$

That is, the abstract Lebesgue integral with respect to μ can be completely identified with the Stieltjes integral with respect to F. With this in mind, we can define the **Laplace transform** $\tilde{\mu}$ of μ by

$$\tilde{\mu}(\lambda) = \tilde{F}(\lambda) = \int_0^\infty e^{-\lambda s} \mu(ds) = \int_0^\infty e^{-\lambda s}\, dF(s), \qquad \lambda \geq 0.$$

1 Uniqueness and Convergence Theorems

The first two important results in this theory are the uniqueness and continuity theorems.

---

[1] The material of this chapter, and much more, can be found in Feller (1971).

1.1 The Uniqueness Theorem

Roughly speaking, the uniqueness theorem for Laplace transforms states that whenever two measures—or equivalently, two distribution functions—have the same Laplace transform, they are indeed one and the same. This is reminiscent of the uniqueness theorem for characteristic functions. One main difference is that in order to obtain uniqueness, it is sufficient that the two Laplace transforms agree on an infinite half-line. On the other hand, there are distinct measures whose Fourier transforms agree on an infinite half-line.

Theorem 1.1.1 (The Uniqueness Theorem) *Suppose μ and ν are two σ-finite measures on $\mathbb{R}_+$ such that for some $\lambda_0 > 0$, $\widetilde{\mu}(\lambda_0)$ and $\widetilde{\nu}(\lambda_0)$ are both finite. Then the following are equivalent:*

(i) $\mu = \nu$;

(ii) for all $\lambda > \lambda_0$, $\widetilde{\mu}(\lambda) = \widetilde{\nu}(\lambda)$.

The above is an immediate consequence of the following *inversion formula*:

Theorem 1.1.2 (Widder's Inversion Formula) *Let μ denote a measure on $\mathbb{R}_+$ such that for some $\lambda_0 > 0$, $\widetilde{\mu}(\lambda_0) < \infty$. Then, for all $s > 0$ that are points of continuity of F,*

$$\lim_{\lambda \to \infty} \sum_{j \leq \lambda s} \frac{(-\lambda)^j}{j!} \frac{\partial^j \widetilde{\mu}}{\partial \lambda^j}(\lambda) = F(s).$$

Proof To begin with, note that for all $\lambda > \lambda_0$, $\widetilde{\mu}(\lambda) < \infty$, and by the dominated convergence theorem, the following derivatives exist and are finite:

$$\frac{\partial^j \widetilde{\mu}}{\partial \lambda^j}(\lambda) = (-1)^j \int_0^\infty e^{-\lambda s} s^j \, \mu(ds), \qquad j \geq 0, \ \lambda > \lambda_0.$$

Consequently, by Fubini's theorem, we see that for all $s > 0$,

$$\sum_{j \leq \lambda s} \frac{(-\lambda)^j}{j!} \frac{\partial^j \widetilde{\mu}}{\partial \lambda^j}(\lambda) = \int_0^\infty \mathbb{P}(X_{\lambda t} \leq \lambda s) \, \mu(dt),$$

where X_α denotes a Poisson random variable with mean α, for any $\alpha \geq 0$.[2] By the (weak) law of large numbers, for all $t \geq 0$ and all $s > 0$,

$$\lim_{\lambda \to \infty} \mathbb{P}(X_{\lambda t} \leq \lambda s) = \begin{cases} 0, & \text{if } s < t, \\ 1, & \text{if } s > t. \end{cases} \qquad (1)$$

---

[2] A Poisson random variable with mean 0 is identically zero.

On the other hand, for all $\varepsilon, s > 0$ and all $\lambda > \lambda_0$,

$$\sum_{j \leq \lambda s} \frac{(-\lambda)^j}{j!} \frac{\partial^j \widetilde{\mu}}{\partial \lambda^j}(\lambda) = \int_0^{(1-\varepsilon)s} \mathbb{P}(X_{\lambda t} \leq \lambda s) \, \mu(dt)$$

$$+ \int_{(1+\varepsilon)s}^{\infty} \mathbb{P}(X_{\lambda t} \leq \lambda s) \, \mu(dt)$$

$$+ \int_{(1-\varepsilon)s}^{(1+\varepsilon)s} \mathbb{P}(X_{\lambda t} \leq \lambda s) \, \mu(dt)$$

$$= T_1(\lambda) + T_2(\lambda) + T_3(\lambda).$$

By equation (1) and by the dominated convergence theorem,

$$\lim_{\lambda \to \infty} T_1(\lambda) = \mu\{[0, (1-\varepsilon)s]\} = F\{(1-\varepsilon)s\},$$

$$\lim_{\lambda \to \infty} T_2(\lambda) = 0,$$

$$\sup_{\lambda \geq 0} T_3(\lambda) \leq F\{(1+\varepsilon)s\} - F\{(1-\varepsilon)s\}.$$

Whenever s is a continuity point for F, $\lim_{\varepsilon \to 0+} \sup_{\lambda \geq 0} T_3(\lambda) = 0$ and $\lim_{\varepsilon \to 0+} F\{(1-\varepsilon)s\} = F(s)$. This proves the theorem. □

1.2 The Convergence Theorem

The convergence theorem of Laplace transforms parallels the convergence theorem of Fourier transforms, but is simpler.

Theorem 1.2.1 (The Convergence Theorem) *Consider a collection of measures $(\mu_n; 1 \leq n \leq \infty)$, all on $\mathbb{R}_+$.*

(i) *Suppose that μ_n converges weakly to μ_∞ and that $\sup_n \widetilde{\mu}_n(\lambda) < \infty$, for all λ greater than some $\lambda_0 \geq 0$. Then $\lim_{n \to \infty} \widetilde{\mu}_n(\lambda)$ exists.*

(ii) *Conversely, suppose there exists $\lambda_0 \geq 0$ such that for each and every $\lambda \geq \lambda_0$, $L(\lambda) = \lim_{n \to \infty} \widetilde{\mu}_n(\lambda)$ exists. Then, L is the Laplace transform of a measure ν on $\mathbb{R}_+$, and μ_n converges weakly to this measure ν.*

In particular, by combining (i) and (ii), we see that if μ_n converges weakly to some μ_∞, then under the boundedness condition, $\lim_n \widetilde{\mu}_n(\lambda) = \widetilde{\mu}_\infty(\lambda)$, for all $\lambda > \lambda_0$.

Proof We first prove the theorem in the case $\mu_1, \mu_2, \ldots$ are all *probability measures*. Of course, in this case, we always have $\sup_n \widetilde{\mu}_n(\lambda) \leq 1$, for all $\lambda > 0$.

(i) Since μ_n converges weakly to μ_∞, then μ_∞ is necessarily a subprobability measure, and this part follows from the definition of weak convergence.

(ii) Recall Helly's selection theorem: Any subsequence $(n_k)_{k\geq 1}$ has a further subsequence $(n_{k_j})_{j\geq 1}$ such that as $j \to \infty$, $\mu_{n_{k_j}}$ converges weakly to some subprobability measure σ. By the part of the theorem that we verified earlier, for all $\lambda > 0$, $\lim_{j\to\infty} \widetilde{\mu_{n_{k_j}}}(\lambda) = \widetilde{\sigma}(\lambda)$. That is, $L = \widetilde{\sigma}$. Since this is independent of the choice of the subsequence (n_k), the theorem follows in the case that $\mu_1, \mu_2, \ldots$ are all probability measures.

We now proceed with the general case.

(i) The first part is similar to the probability case discussed above. Namely, note that by the boundedness assumption and by the dominated convergence theorem,

$$\lim_{T\to\infty} \sup_{n\geq 1} \int_T^\infty e^{-\lambda t} \mu_n(dt) = 0, \qquad \lambda > \lambda_0.$$

Equivalently, for any $\varepsilon > 0$ and for all $\lambda > \lambda_0$, there exists $T_0 > 0$ such that

$$\sup_{n\geq 1} \int_{T_0}^\infty e^{-\lambda t} \mu_n(dt) \leq \varepsilon.$$

On the other hand, by properties of weak convergence, for all $\lambda > \lambda_0$,

$$\lim_{n\to\infty} \int_0^{T_0} e^{-\lambda s} \mu_n(ds) = \int_0^{T_0} e^{-\lambda s} \mu_\infty(ds).$$

In particular, for all $\lambda > \lambda_0$,

$$\lim_{n,m\to\infty} \left| \int_0^\infty e^{-\lambda s} \mu_n(ds) - \int_0^\infty e^{-\lambda s} \mu_m(ds) \right| \leq 2\varepsilon.$$

Since $\varepsilon > 0$ is arbitrary, we see that $\lim_{n\to\infty} \widetilde{\mu_n}(\lambda)$ exists for all $\lambda > \lambda_0$.

(ii) Let us fix some $\lambda_1 > \lambda_0$ and define

$$\nu_n(dt) = \frac{e^{-\lambda_1 t}}{\widetilde{\mu_n}(\lambda_1)} \mu_n(dt), \qquad t \geq 0.$$

(Why can we assume the strict positivity of $\widetilde{\mu_n}(\lambda_1)$?) It follows immediately that $\nu_1, \nu_2, \ldots$ are *probability measures* on $\mathbb{R}_+$. Moreover, for all $\lambda \geq 0$,

$$\widetilde{\nu_n}(\lambda) = \frac{\widetilde{\mu_n}(\lambda + \lambda_1)}{\widetilde{\mu_n}(\lambda_1)}.$$

In particular,

$$\lim_{n\to\infty} \widetilde{\nu_n}(\lambda) = \frac{L(\lambda + \lambda_1)}{L(\lambda_1)}, \qquad \lambda \geq 0. \tag{1}$$

By what we have shown about probability measures, there must exist a subprobability measure ν_∞ on $\mathbb{R}_+$ such that ν_n converges weakly to ν_∞ and
$$\widetilde{\nu_\infty}(\lambda) = \frac{L(\lambda + \lambda_1)}{L(\lambda_1)}, \qquad \lambda \geq 0.$$

Define the measure
$$\sigma(dt) = L(\lambda_1) e^{\lambda_1 t} \nu_\infty(dt), \qquad t \geq 0.$$

Since $\lim_{n\to\infty} \widetilde{\mu_n}(\lambda_1) = L(\lambda_1)$, the definition of ν_n shows that μ_n converges weakly to σ. To finish, note that for all $\lambda \geq \lambda_1$,
$$\widetilde{\sigma}(\lambda) = L(\lambda_1) \int_0^\infty e^{-(\lambda - \lambda_1)t} \nu_\infty(dt) = L(\lambda_1) \lim_{n\to\infty} \widetilde{\nu_n}(\lambda - \lambda_1) = L(\lambda).$$

The last step uses (1). Since $\lambda_1 > \lambda_0$ is arbitrary, the result follows. □

1.3 Bernstein's Theorem

Suppose $f : \mathbb{R}_+ \to \mathbb{R}_+$ is measurable. S. N. Bernstein's theorem addresses the following interesting question: *When is f the Laplace transform of a measure?* That is, we seek conditions under which for all $\lambda \geq 0$, $f(\lambda) = \int_0^\infty e^{-\lambda t} \mu(dt)$ for some measure μ on $\mathbb{R}_+$. The answer to this question revolves around the notion of complete monotonicity.

A function $f : \mathbb{R}_+ \to \mathbb{R}_+$ is said to be **completely monotone** if it is infinitely differentiable and
$$(-1)^n \frac{d^n f}{d\lambda^n}(\lambda) \geq 0, \qquad \lambda > 0.$$

Theorem 1.3.1 (Bernstein's Theorem) *A function $f : \mathbb{R}_+ \to \mathbb{R}_+$ is the Laplace transform of a measure if and only if it is completely monotone. It is the Laplace transform of a probability measure if and only if it is completely monotone and $f(0^+) = 1$.*

Proof Suppose $f = \widetilde{\mu}$ for some measure μ. By the dominated convergence theorem, f is completely monotone. In fact,
$$(-1)^n \frac{d^n f}{d\lambda^n}(\lambda) = \int_0^\infty s^n e^{-\lambda s} \mu(ds) \geq 0.$$

Conversely, let us suppose that f is completely monotone. We fix some number $r > 0$ and define $F : [0, 1[\to \mathbb{R}_+$ by
$$F(s) = f(r - rs), \qquad 0 \leq s < 1.$$

Computing directly, we see that for all $n \geq 0$,
$$\frac{d^n F}{ds^n}(s) = (-r)^n \frac{d^n f}{d\lambda^n}(r - rs), \qquad 0 \leq s < 1.$$
Since this is positive, we can apply Taylor's expansion to see that
$$F(s) = \sum_{n=0}^{\infty} \frac{d^n f}{d\lambda^n}(r) \frac{(-r)^n s^n}{n!}, \qquad 0 \leq s < 1.$$
In particular, we apply the above to $s = e^{-\frac{\lambda}{r}}$, where $\lambda > 0$, and see that
$$f(r - re^{-\frac{\lambda}{r}}) = \sum_{n=0}^{\infty} \frac{d^n f}{d\lambda^n}(r) \frac{(-r)^n}{n!} e^{-\frac{1}{r}n\lambda}.$$
That is, for any fixed $r > 0$, $f(r - re^{-\frac{\lambda}{r}}) = \widetilde{\mu_r}(\lambda)$, where μ_r is the purely atomic nonnegative measure that assigns nonnegative mass $(-r)^n f^{(n)}(r)/n!$ to every point of the form $\frac{n}{r}$, $n = 0, 1, \ldots$. (Temporarily, we have written $f^{(n)}$ for the nth derivative of f.) Since $\lim_{r \to \infty}(r - re^{-\lambda/r}) = \lambda$,
$$f(\lambda) = \lim_{r \to \infty} \widetilde{\mu_r}(\lambda), \qquad \lambda > 0.$$
By the convergence theorem (Theorem 1.2.1), f is the Laplace transform of some measure. At this point it should be clear that f is also the Laplace transform of a probability measure if and only if $f(0+) = 1$. □

2 A Tauberian Theorem

A **Tauberian** theorem is one that states that the asymptotic behavior of the distribution function $F(t)$, as $t \to \infty$, can be read from the behavior of $\widetilde{F}(\lambda)$, as $\lambda \to 0^+$. An **Abelian** theorem is one that states the converse. Following Feller (1971), we use the term Tauberian theorem loosely to stand for both kinds of results.

Theorem 2.1.1 (Feller's Tauberian Theorem) *Consider a σ-finite measure μ with distribution function F such that $\widetilde{\mu}(\lambda) < \infty$ for all $\lambda > 0$. The following are equivalent: There exists $\theta > -1$ such that:*

(i) as $t \to \infty$,
$$\frac{F(tx)}{F(t)} \to x^\theta, \qquad \forall x > 0; \text{ and}$$

(ii) as $s \to 0^+$,
$$\frac{\widetilde{\mu}(s\lambda)}{\widetilde{\mu}(s)} \to \lambda^{-\theta}, \qquad \forall \lambda > 0.$$

Moreover, either of the above two conditions above implies

$$\lim_{t\to\infty} \frac{\widetilde{\mu}(\frac{1}{t})}{F(t)} = \Gamma(\theta+1). \tag{1}$$

In particular, we obtain the following very important corollary.

Corollary 2.1.1 *Suppose μ is given by Theorem 2.1.1. If there exist two finite constants $C > 0$ and $\theta > -1$ such that as $s \to 0^+$, $s^\theta \widetilde{\mu}(s) \to C$, then, as $t \to \infty$, $t^{-\theta} F(t) \to D$, where $D = C/\Gamma(\theta+1)$. The converse also holds.*

Proof of Theorem 2.1.1 First, we show that $(i) \Rightarrow (1)$. For all $x \geq 0$, define

$$G(x) = x^\theta/\Gamma(\theta+1),$$

where $1 \div 0 = \infty$. Viewing G as a distribution function, we immediately see that

$$\widetilde{G}(\lambda) = \frac{1}{\Gamma(\theta+1)} \int_0^\infty e^{-\lambda x} \theta x^{\theta-1}\, dx = \lambda^{-\theta}. \tag{2}$$

For each $t > 0$, define the distribution function

$$H_t(x) = \frac{F(tx)}{F(t)}, \qquad x \geq 0.$$

It is easy to check that for all $t \geq 0$,

$$\widetilde{H}_t(\lambda) = \frac{\widetilde{\mu}(\frac{\lambda}{t})}{F(t)}, \qquad \lambda > 0.$$

Assertion (i) states as $t \to \infty$, H_t converges weakly to the distribution function $\Gamma(\theta+1)G$. Suppose we could show that for all $\lambda > 0$,

$$\sup_t \widetilde{H}_t(\lambda) < \infty. \tag{3}$$

Then, by equation (2) and by the convergence theorem for Laplace transforms (Theorem 1.2.1), we would have

$$\lim_{t\to\infty} \frac{\widetilde{\mu}(\frac{\lambda}{t})}{F(t)} = \lim_{t\to\infty} \widetilde{H}_t(\lambda) = \Gamma(\theta+1)\widetilde{G}(\lambda) = \Gamma(\theta+1)\lambda^{-\theta}.$$

Applying this with $\lambda = 1$, we obtain (1). It remains to demonstrate the validity of (3). By (i), there exists $t_0 > 0$ such that for all $t \geq t_0$, $F(te) \leq e^{1+\theta} F(t)$. By iterating this, we see that for all $t \geq t_0$,

$$F(te^k) \leq e^{k(1+\theta)} F(t), \qquad k \geq 0. \tag{4}$$

Thus, by monotonicity, for all $t \geq t_0$,

$$\tilde{\mu}(\tfrac{1}{t}) = \int_0^t e^{-\frac{s}{t}}\mu(ds) + \sum_{j=0}^{\infty}\int_{te^j}^{te^{j+1}} e^{-\frac{s}{t}}\mu(ds)$$

$$\leq F(t) + \sum_{j=0}^{\infty}\exp(-e^j)F(te^{j+1})$$

$$\leq F(t)\Big[1 + \sum_{j=0}^{\infty}\exp\{-e^j + (j+1)(1+\theta)\}\Big].$$

We have used equation (4) in the last line. This proves (3) and that $(i) \Rightarrow (1)$. Combining (1) with the assertion of (i) itself, we obtain (ii) readily.

Next, we prove that $(ii) \Rightarrow (1)$. For each $t > 0$, define the distribution function

$$F_t(x) = \frac{F(tx)}{\tilde{\mu}(\tfrac{1}{t})}, \qquad x \geq 0.$$

It is easy to check that

$$\widetilde{F}_t(\lambda) = \frac{\tilde{\mu}(\tfrac{\lambda}{t})}{\tilde{\mu}(\tfrac{1}{t})}, \qquad \lambda > 0.$$

In light of equation (2), assertion (ii) of the theorem can be recast as follows:

$$\lim_{t\to\infty} \widetilde{F}_t(\lambda) = \widetilde{G}(\lambda), \qquad \lambda > 0.$$

By the convergence theorem (Theorem 1.2.1), as $t \to \infty$, F_t converges weakly to G. Since G is continuous,

$$\lim_{t\to\infty} F_t(x) = G(x), \qquad \forall x \geq 0.$$

In particular, $\lim_{t\to\infty} F_t(1) = G(1)$, which is (1) written in shorthand. That is, we have shown that $(ii) \Rightarrow (1)$. Together with (ii) itself, (1) implies (i). □

We conclude this appendix with an application to the computation of the asymptotic rate of probability distribution functions. Now let F denote a probability distribution function on $\mathbb{R}_+$. Of course, $\lim_{x\to\infty} F(x) = 1$. Here is an instructive exercise.

Exercise 2.1.1 Show that

$$\frac{1}{\lambda}\{1 - \widetilde{F}(\lambda)\} = \int_0^{\infty} e^{-\lambda x}\{1 - F(x)\}\,dx,$$

whenever $\lambda > 0$. □

Let $\mu_0(dx) = \{1 - F(x)\}\,dx$. Its distribution function is, of course, F_0, where
$$F_0(x) = \int_0^x \{1 - F(y)\}\,dy, \qquad x \geq 0.$$
Moreover, $\widetilde{F_0}(\lambda) = \lambda^{-1}\{1 - \widetilde{F}(\lambda)\}$, $\lambda > 0$. Applying Corollary 2.1.1 to μ_0 and F_0, we can deduce that for some positive finite C and some $\theta > -1$, $x^{-\theta} F_0(x) \to C$ (as $x \to \infty$) if and only if $\lambda^{\theta-1}\{1-\widetilde{F}(\lambda)\} \to D$ (as $\lambda \to 0^+$), where $D = C/\Gamma(1+\theta)$. By applying L'Hôpital's rule of elementary calculus, we obtain the following:

Corollary 2.1.2 *Suppose F is a probability distribution function on $\mathbb{R}_+$. The following are equivalent: For some $\theta > -1$ and some constant $C > 0$:*

(i) as $x \to \infty$, $x^{1-\theta}\{1 - F(x)\} \to C$; and

(ii) as $\lambda \to 0^+$, $\lambda^{\theta-1}\{1 - \widetilde{F}(\lambda)\} \to D$, where $D = C\,\Gamma(\theta + 1)$.

Clearly, this is useful and sensible only when $\theta \in\,]-1, 1[$.

Appendix C
Hausdorff Dimensions and Measures

Hausdorff measures can be thought of as extensions of Lebesgue's measure. While a rather abstract treatment is possible, we restrict our attention to such measures on Euclidean spaces.

1 Preliminaries

Let (S, d) denote a metric space and recall that a set function $\mu : S \to \mathbb{R}_+$ is a **Carathéodory outer measure** (or outer measure, for brevity) if:

(i) $\mu(\varnothing) = 0$;

(ii) [$\boldsymbol{\sigma}$-**subadditivity**] for all $E_1, E_2, \ldots \subset S$, $\mu\left(\bigcup_{i=1}^\infty E_i\right) \leq \sum_{i=1}^\infty \mu(E_i)$; and

(iii) [**Monotonicity**] If $E_1 \subset E_2$ are both in S, then $\mu(E_1) \leq \mu(E_2)$.

The essential difference between outer measures and ordinary measures is countable additivity.

1.1 Definition

Throughout, we restrict attention to $S = \mathbb{R}^d$, endowed with its Euclidean metric topology and its Borel σ-field.

Appendix C. Hausdorff Dimensions and Measures

Given fixed numbers $s \geq 0$ and $\varepsilon > 0$ and given a set $E \subset \mathbb{R}^d$, we define

$$\mathcal{H}_s^\varepsilon(E) = \inf \left\{ \sum_{j=1}^\infty (2r_j)^s : E \subset \bigcup_{k=1}^\infty \overline{\mathcal{B}(x_k; r_k)}, \; \sup_\ell r_\ell \leq \varepsilon \right\},$$

where $\mathcal{B}(x; r)$ denotes, as usual, the open ℓ^∞-ball of radius $r > 0$ about $x \in \mathbb{R}^d$, and $\overline{E}$ the (Euclidean) closure of $E \subset \mathbb{R}^d$.

It is not hard to deduce the following.

***Exercise* 1.1.1** Show that $\mathcal{H}_s^\varepsilon$ is an outer measure on the subsets of $\mathbb{R}^d$. □

Moreover, $\varepsilon \mapsto \mathcal{H}_s^\varepsilon(E)$ is nonincreasing. Therefore, we can unambiguously define

$$\mathcal{H}_s(E) = \lim_{\varepsilon \to 0^+} \mathcal{H}_s^\varepsilon(E), \qquad E \subset \mathbb{R}^d.$$

It immediately follows that $\mathcal{H}_s$ is an outer measure. In fact, it is much more than that. Recall that a set $E \subset \mathbb{R}^d$ is **measurable** for an outer measure μ if

$$\mu(B) = \mu(B \cap E) + \mu(B \setminus E), \qquad \forall B \subseteq \mathbb{R}^d.$$

Recall that the collection of all measurable sets of μ is a σ-field.

Theorem 1.1.1 *For any $s \geq 0$, $\mathcal{H}_s$ is a measure on its measurable sets. Moreover, all Borel sets are measurable for any and all of the outer measures $\mathcal{H}_s$.*

We can, therefore, view $\mathcal{H}_s$ as a measure on the Borel σ-field of $\mathbb{R}^d$. It is called the d-**dimensional Hausdorff measure** on $\mathbb{R}^d$.

Proof Throughout, $s \geq 0$ is held fixed.

In light of Exercise 1.1.1, that $\mathcal{H}_s$ is a measure on its measurable sets is a consequence of Carathéodory's extension theorem of measure theory. We finish the proof by showing that Borel sets are measurable for $\mathcal{H}_s$. That is, we need to show that for any Borel set $E \subset \mathbb{R}^d$,

$$\mathcal{H}_s(B) = \mathcal{H}_s(B \cap E) + \mathcal{H}_s(B \setminus E), \qquad \forall B \subset \mathbb{R}^d.$$

Since the collection of all measurable subsets of $\mathbb{R}^d$ is a σ-field, it suffices to prove that the above holds for all *closed* sets $E \subset \mathbb{R}^d$ (why?).

Thanks to Exercise 1.1.1, it suffices to show that for all closed sets $E \subset \mathbb{R}^d$ and for all $B \subset \mathbb{R}^d$,

$$\mathcal{H}_s(B) \geq \mathcal{H}_s(B \cap E) + \mathcal{H}_s(B \setminus E). \tag{1}$$

With this in mind, let us hold fixed a closed set $E \subset \mathbb{R}^d$ and an arbitrary set $B \subset \mathbb{R}^d$. If either $E \subset B$ or $B \subset E$, then (1) holds trivially. Therefore, we may assume that $E \triangle B \neq \emptyset$.[1] We may also assume, without loss of any generality, that $\mathcal{H}_s(B) < \infty$, for otherwise there is nothing left to prove.

For any integer $j \geq 1$, let B_j denote the collection of all points x in B such that the distance between x and E is at least j^{-1}. Of course, B_j and $B \cap E$ are disjoint. Thus, by the definition of Hausdorff measure and by covering $B \cap E$ and B_j separately,

$$\mathcal{H}_s(B \cap E) + \mathcal{H}_s(B_j) \leq \mathcal{H}_s(B). \tag{2}$$

(Why?) There are two cases to consider at this point. First, suppose there exists j such that $B_j = B$. In this case, equation (1) follows immediately from equation (2). Next, let us suppose that B_j is a proper subset of B for all $j \geq 1$. This means that for all $j \geq 1$, B_j is a proper subset of B_{j+1} (why?). Consequently, we may write $B \setminus E$ as the following disjoint union:

$$B \setminus E = B_j \cup \bigcup_{k=j}^{\infty} (B_{k+1} \setminus B_k).$$

By the first part of the theorem,

$$\mathcal{H}_s(B \setminus E) = \mathcal{H}_s(B_j) + \sum_{k=j}^{\infty} \mathcal{H}_s(B_{k+1} \setminus B_k).$$

In particular, $\lim_{j \to \infty} \mathcal{H}_s(B_j) = \mathcal{H}_s(B \setminus E)$. Equation (1) now follows from equation (2), and our proof is complete. □

In order to better understand these measures, let us first consider Hausdorff measures of integral dimensions.

Lemma 1.1.1 $\mathcal{H}_0$ *is counting measure on Borel subsets of* $\mathbb{R}^d$, *and* $\mathcal{H}_d$ *is d-dimensional Lebesgue's measure on the subsets of* $\mathbb{R}^d$.

Proof The remark about $\mathcal{H}_0$ is immediate. To prove the remainder of the lemma, write Leb, $\mathcal{H}^\varepsilon$, and $\mathcal{H}$ for Leb_d, $\mathcal{H}_d^\varepsilon$, and $\mathcal{H}_d$, respectively. Clearly, for any Borel set $E \subset \mathbb{R}^d$,

$$\mathcal{H}^\varepsilon(E) = \inf \left\{ \sum_{j=1}^{\infty} \text{Leb}\left[\overline{\mathcal{B}(x_j; r_j)}\right] : E \subset \bigcup_{k=1}^{\infty} \overline{\mathcal{B}(x_k; r_k)}, \sup_k r_k \leq \varepsilon \right\}.$$

As $\varepsilon \to 0^+$, the right-hand side converges to $\text{Leb}(E)$, by the definition of Lebesgue's measure, and the left-hand side to $\mathcal{H}(E)$, by the definition of Hausdorff's measure. The result follows. □

---

[1] Recall that $E \triangle B = (E \setminus B) \cup (B \setminus E)$.

Exercise 1.1.2 Suppose we defined $\mathcal{H}_s$ based on ℓ^p balls, not ℓ^∞ balls, and where $1 \leq p < \infty$. Show that when $1 \leq s \leq d$ is integral, for all E compact, $\mathcal{H}_s(E) = c_p \operatorname{Leb}(E)$, where c_p is a constant. Compute c_p explicitly. □

In fact, for nonintegral values of s, we always have the following.

Lemma 1.1.2 *For all Borel sets $E \subset \mathbb{R}^d$, and for all real numbers $s \geq 0$,*

$$\mathcal{H}_s(E) \geq \operatorname{Leb}_s(E).$$

Proof Let us fix $\varepsilon > 0$ and find arbitrary closed ℓ^∞-balls $B_1, B_2, \ldots$ whose radii $r_1, r_2, \ldots$ are all less than or equal to ε. Whenever $E \subset \cup_i B_i$, then

$$\operatorname{Leb}_s(E) \leq \sum_j \operatorname{Leb}_s(B_i) = \sum_j (2r_j)^s.$$

Taking the infimum over all such balls, we conclude that $\operatorname{Leb}_s(E) \leq \mathcal{H}_s^\varepsilon(E)$. Let $\varepsilon \to 0^+$ to obtain the lemma. □

What if $s > d$? In this case, the following identifies $\mathcal{H}_s$ as the trivial measure.

Lemma 1.1.3 *Given a Borel set $E \subset \mathbb{R}^d$:*

(i) Whenever $s < t$ and $\mathcal{H}_s(E) < \infty$, then $\mathcal{H}_t(E) = 0$; and

(ii) whenever $s > t$ and $\mathcal{H}_s(E) > 0$, then $\mathcal{H}_t(E) = \infty$.

Exercise 1.1.3 Prove Lemma 1.1.3. □

Next, we state an invariance property of Hausdorff measures. In light of Lemma 1.1.1, the following extends the usual invariance properties of Lebesgue's measure on $\mathbb{R}^d$ under the action of Poincaré motions.[2]

Lemma 1.1.4 *Given numbers $r, s \geq 0$, a point $x \in \mathbb{R}^d$, and a Borel set $E \subset \mathbb{R}^d$,*

$$\mathcal{H}_s(rE + x) = r^s \mathcal{H}_s(E),$$

where $0^0 = 0$ and $rE + x = \{y \in \mathbb{R}^d : y = rz + x \text{ for some } z \in E\}$. Moreover, for any $(d \times d)$ unitary rotation matrix O, $\mathcal{H}_s(OE) = \mathcal{H}_s(E)$.

Exercise 1.1.4 Prove Lemma 1.1.4. □

---

[2] A function $f : \mathbb{R}^d \to \mathbb{R}^d$ is a **Poincaré motion** if there exists a $(d \times d)$ unitary rotation matrix O and a vector $\beta \in \mathbb{R}^d$ such that for all $x \in \mathbb{R}^d$, $f(x) = Ox + \beta$.

Finally, we mention that there is a relationship between Hausdorff measures and metric entropy of Section 2.1, Chapter 5. Let d denote both the dimension of $\mathbb{R}^d$ and the Euclidean metric induced by the ℓ^∞ norm on $\mathbb{R}^d$ (while this is abusing our notation, no confusion should arise). Recall that for any totally bounded set $E \subset \mathbb{R}^d$, $D(\varepsilon; E) = D(\varepsilon; E, d)$ denotes the minimum number of balls of radius at most ε required to cover E. This immediately gives the following.

Lemma 1.1.5 *For any totally bounded $E \subset \mathbb{R}^d$ and for all $s \geq 0$,*

$$\mathcal{H}_s(E) \leq \liminf_{\varepsilon \to 0^+} (2\varepsilon)^s D(\varepsilon; E).$$

1.2 Hausdorff Dimension

To any Borel set $E \subset \mathbb{R}^d$ we associate a number $\dim(E)$ as follows:

$$\dim(E) = \sup\{s > 0 : \mathcal{H}_s(E) = \infty\},$$

where $\sup \varnothing = 0$. This is the **Hausdorff dimension** of the set E. By Lemma 1.1.3, the above is well-defined. Furthermore (why?),

$$\dim(E) = \inf\{s > 0 : \mathcal{H}_s(E) = 0\}.$$

A simple but important property of Hausdorff dimension is monotonicity.

Lemma 1.2.1 (Monotonicity) *If $E \subset B$ are Borel subsets of $\mathbb{R}^d$, then $\dim(E) \leq \dim(B)$.*

Exercise **1.2.1** Prove the monotonicity lemma, Lemma 1.2.1. □

In light of Lemma 1.1.5, it is easy to obtain upper bounds for the Hausdorff dimension of a Borel set $E \subset \mathbb{R}^d$:

Lemma 1.2.2 *Suppose $E \subset \mathbb{R}^d$ is a Borel set with the following property: There exists a sequence $\varepsilon_k \downarrow 0$ for which there are N_k closed ℓ^∞-balls of radius $\varepsilon_k > 0$ that cover E. Then*

$$\dim(E) \leq \inf\{s > 0 : \liminf_{k \to \infty} \varepsilon_k^s N_k = 0\}.$$

We apply this to two examples.

Example 1 [Cantor's Tertiary Set]

Let $C_0 = [0,1]$, $C_1 = [0, \frac{1}{2}] \cup [\frac{2}{3}, 1]$, $C_2 = [0, \frac{1}{2}] \cup [\frac{2}{9}, \frac{1}{3}] \cup [\frac{2}{3}, \frac{7}{9}] \cup [\frac{8}{9}, 1]$, etc. In general, we obtain C_{k+1} by removing the middle third of every interval subset of C_k; cf. Figure C.1. **Cantor's tertiary set** is simply defined by $C = \cap_{k=0}^\infty C_k$. This is clearly a nonempty compact set. On the other hand, for any $k \geq 0$, C_k is made up of 2^k disjoint intervals of length 3^{-k}.

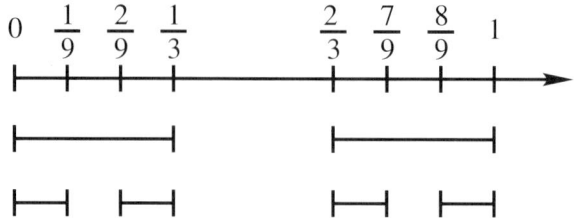

Figure C.1: The first 3 stages in the construction of Cantor's tertiary set

Thus, $\mathrm{Leb}(C_k) = (\frac{2}{3})^k$, which goes to 0 as $k \to \infty$. Since $C_k \downarrow C$, this implies that $\mathrm{Leb}(C) = 0$. Let $\varepsilon_k = 3^{-k}$ and note that C_k is made up of $N_k = 2^k$ ℓ^∞-balls of radius ε_k. Since $C \subset C_k$, by Lemma 1.2.2,

$$\dim(C) \leq \inf\left\{s > 0 : \liminf_{k \to \infty} 2^k 3^{-ks} = 0\right\} = \frac{\ln 2}{\ln 3}.$$

This turns out to be the correct value of the Hausdorff dimension of the tertiary Cantor set, as we shall see later on in Exercise 2.1.1.

Example 2 Recall from Section 2.5, Chapter 5, that a function $f : [0,1]^d \to \mathbb{R}^k$ is said to be Hölder continuous of order $\alpha > 0$ whenever there exists a finite positive constant Γ such that for all $s, t \in [0,1]^d$, $|f(t) - f(s)| \leq \Gamma |t - s|^\alpha$. By Taylor's expansion, the only cases of interest are $\alpha \in\,]0,1]$. If f is such a function, then for all Borel sets $E \subset [0,1]^d$,

$$\dim\{f(E)\} \leq \frac{1}{\alpha} \dim(E) \wedge k. \tag{1}$$

Indeed, let us hold fixed a constant $\gamma > \alpha^{-1} \dim(E)$. By Lemma 1.1.3, we can find a collection of closed ℓ^∞-balls $B_{1,\varepsilon}, B_{2,\varepsilon}, \ldots$ of radii $r_{1,\varepsilon}, r_{2,\varepsilon}, \ldots$ such that:

- $\sup_j r_{j,\varepsilon} \leq \varepsilon$;
- $\cup_{j=1}^\infty B_{j,\varepsilon} \supset E$; and
- $\limsup_{\varepsilon \to 0^+} \sum_{j=1}^\infty (2r_{j,\varepsilon}^\alpha)^\gamma < \infty$.

By the Hölder condition on f, each $f(B_{j,\varepsilon})$ is inside a closed ℓ^∞-ball $X_{j,\varepsilon}$ of radius $\Gamma r_{j,\varepsilon}^\alpha \leq \delta = \Gamma \varepsilon^\alpha$. Since $f(E) \subset \cup_{j=1}^\infty X_{j,\varepsilon}$,

$$\mathcal{H}_\gamma^\delta \{f(E)\} \leq \sum_{j=1}^\infty (2\Gamma r_{j,\varepsilon}^\alpha)^\gamma,$$

which remains bounded as $\delta \to 0^+$. Thus, $\mathcal{H}_\gamma\{f(E)\} < \infty$, which implies that $\dim\{f(E)\} \leq \gamma$. Since this holds for every $\gamma > \alpha^{-1} \dim(E)$, we can deduce that $\dim\{f(E)\} \leq \alpha^{-1} \dim(E)$. On the other hand, since $f(E) \subset \mathbb{R}^k$, by the monotonicity lemma (Lemma 1.2.1), $\dim\{f(E)\} \leq k$. The claim (1) follows.

Exercise 1.2.2 Show that for all compact sets $E_1, E_2, \ldots$ in $\mathbb{R}^d$,

$$\dim\left\{\bigcup_{i=1}^{\infty} E_i\right\} = \sup_i \dim(E_i).$$

This is the **inner regularity** of Hausdorff dimensions. □

2 Frostman's Theorems

In order to compute Hausdorff dimensions, we need a way of obtaining good lower bounds. This is the gist of two theorems due to O. Frostman.

2.1 Frostman's Lemma

The following is the first theorem of Frostman, which, for historical reasons, is called Frostman's lemma.

Theorem 2.1.1 (Frostman's Lemma) *For a compact set $E \subset \mathbb{R}^d$, the following are equivalent. For $s \geq 0$:*

(i) $\mathcal{H}_s(E) > 0$;

(ii) there exists a probability measure μ on E such that

$$\limsup_{r \to 0^+} r^{-s} \sup_{x \in \mathbb{R}^d} \mu\{\mathcal{B}(x; r)\} < \infty.$$

Proof Interestingly enough, the more useful half of this theorem is the easier one to prove. We begin with this case, which is $(ii) \Rightarrow (i)$. Since $\mu \in \mathcal{P}(E)$, (ii) is *equivalent* to the following:

$$A = \sup_{r > 0} \sup_{x \in \mathbb{R}^d} \frac{\mu\{\mathcal{B}(x; r)\}}{r^s} < \infty.$$

Now let $B_1, B_2, \ldots$ denote any sequence of closed ℓ^∞-balls of radii $r_1, r_2, \ldots$ all bounded above by ε. Then,

$$\sum_{j=1}^{\infty} (2r_j)^s \geq \frac{1}{2^s A} \sum_{j=1}^{\infty} \mu\{\overline{\mathcal{B}(x_j; r_j)}\} \geq \frac{1}{2^s A} \mu\left\{\bigcup_{j=1}^{\infty} \mathcal{B}(x_j; r_j)\right\}.$$

If, in addition, $B_1, B_2, \ldots$ cover E, then the right-hand side is greater than or equal to $2^{-s} A^{-1} \mu(E) = 2^{-s} A^{-1}$. That is, we have shown that for all

$\varepsilon > 0$, $\mathcal{H}_s^\varepsilon(E) \geq 2^{-s}A^{-1}$. Let $\varepsilon \to 0$ to obtain (i). Note that in this part of the proof we did not need compactness.

Suppose (i) holds; we are to verify the truth of (ii). Since E is compact, there exists a finite constant $M > 0$ such that $E \subset [-M, M]^d$. Equivalently,

$$\frac{1}{2M}E + 1 \subset [0,1]^d.$$

By the invariance properties of Hausdorff measures (Lemma 1.1.4), we may and will assume that $E \subset [0,1]^d$; otherwise, apply the result for compact subsets of $[0,1]^d$ to the compact set $(2M)^{-1}E + 1$.

From now on, suppose $E \subset [0,1]^d$ is compact and $\mathcal{H}_s(E) > 0$. The latter positivity condition implies that $\mathbf{i} := \inf_{\varepsilon > 0} \mathcal{H}_s^\varepsilon(E) > 0$ (why?). That is, for *any* collection of closed ℓ^∞-balls $B_1, B_2, \ldots$ whose radii are $r_1, r_2, \ldots$, respectively,

$$E \subset \bigcup_{j=1}^\infty B_j \implies \sum_{j=1}^\infty (2r_j)^s \geq \mathbf{i}. \tag{1}$$

Let $\mathcal{D}_n$ denote the collection of all closed, dyadic ℓ^∞-balls of side 2^{-n}:

$$I \in \mathcal{D}_n \iff \exists k \in \mathbb{Z}^d : I = \prod_{j=1}^d [k^{(j)}2^{-n}, (k^{(j)}+1)2^{-n}].$$

Since $E \subset [0,1]^d$, there must exist a maximal $\mathbf{n} \geq 0$ and some $I \in \mathcal{D}_\mathbf{n}$ such that $E \subset I$. We shall need this $\mathbf{n}$ throughout this proof.

For all integers $n > \mathbf{n}$ we can define a measure $\mu_{n,n}$ on $\mathbb{R}^d$ as follows: For each $I \in \mathcal{D}_n$, if $I \cap E \neq \varnothing$, then $\mu_{n,n}$ uniformly assigns mass 2^{-ns} to I; otherwise, $\mu_{n,n}(I) \equiv 0$. More precisely, if $\mu|_F$ denotes the restriction of the measure μ to the set F, for every integer $n > \mathbf{n}$ and all $I \in \mathcal{D}_n$, we define

$$\mu_{n,n}|_I = \begin{cases} 2^{-ns} \dfrac{\text{Leb}|_I}{\text{Leb}(I)}, & \text{if } E \cap I \neq \varnothing, \\ 0, & \text{if } I \cap E = \varnothing, \end{cases}$$

where Leb denotes Lebesgue's measure on $\mathbb{R}^d$. We would like to let $n \to \infty$ and obtain a limiting measure. In order to ensure that $\mu_{n,n}$ remains nontrivial, it needs to be slightly modified. For all $I \in \mathcal{D}_{n-1}$, define

$$\mu_{n,n-1}|_I = \begin{cases} \mu_{n,n}|_I, & \text{if } \mu_{n,n}(I) \leq 2^{-(n-1)s}, \\ 2^{-(n-1)s} \dfrac{\mu_{n,n}|_I}{\mu_{n,n}(I)}, & \text{if } \mu_{n,n}(I) > 2^{-(n-1)s}. \end{cases}$$

We continue this way to define a measure $\mu_{n,n-j-1}$ from $\mu_{n,n-j}$, for all $0 \leq j < n - \mathbf{n}$. That is, for each $0 \leq j < n - \mathbf{n}$ and for all $I \in \mathcal{D}_{n-j-1}$,

$$\mu_{n,n-j-1}|_I = \begin{cases} \mu_{n,n-j}|_I, & \text{if } \mu_{n,n-j}(I) \leq 2^{-(n-j-1)s}, \\ 2^{-(n-j-1)s} \dfrac{\mu_{n,n-j}|_I}{\mu_{n,n-j}(I)}, & \text{if } \mu_{n,n-j}(I) > 2^{-(n-j-1)s}. \end{cases}$$

2 Frostman's Theorems

At the end of this construction you should verify that we have a measure $\mu_{n,n-\mathbf{n}}$ that has the following properties:

P1 $\mu_{n,n-\mathbf{n}}$ is a nonatomic measure on E.

P2 For any $j \in \{0, 1, \ldots, n\}$ and all $I \in \mathcal{D}_{n-j}$, $\mu_{n,n-\mathbf{n}}(I) \leq 2^{-(n-j)s}$.

P3 For each $x \in E$, there exists an integer $j \in \{0, \ldots, n\}$ and some $I \in \mathcal{D}_{n-j}$ such that $x \in I$ and $\mu_{n,n-\mathbf{n}}(I) = 2^{-(n-j)s}$. Equivalently, $\mu_{n,n-\mathbf{n}}(I) = \{2 \operatorname{rad}(I)\}^s$, where $\operatorname{rad}(I)$ denotes the side, i.e., the ℓ^∞ radius, of the box I.

By **P3**, we can find closed ℓ^∞-balls $I_1, \ldots, I_m$ such that (i) $E \subset \cup_{j=1}^m I_j$; (ii) the interiors of $I_1, \ldots, I_m$ are disjoint; and (iii) $\mu_{n,n-\mathbf{n}}(I_j) = \{2 \operatorname{rad}(I_j)\}^s$. Since $\mu_{n,n-\mathbf{n}}$ is nonatomic (**P1**), for each $n > \mathbf{n}$,

$$\mu_{n,n-\mathbf{n}}(E) = \sum_{j=1}^m \mu_{n,n-\mathbf{n}}(I_j) = \sum_{j=1}^m \{2 \operatorname{rad}(I_j)\}^s.$$

By (1), for all $n > \mathbf{n}$,

$$\mu_{n,n-\mathbf{n}}(E) \geq \mathbf{i}. \tag{2}$$

Define, for each $n > \mathbf{n}$,

$$\mu_n = \frac{\mu_{n,n-\mathbf{n}}}{\mu_{n,n-\mathbf{n}}(E)}.$$

Thus, $\{\mu_n; n > \mathbf{n}\}$ is a collection of probability measures on E. We can apply (2) and **P2** to deduce that for every $I \in \mathcal{D}_{n-j}$ $(0 \leq j < n - \mathbf{n})$,

$$\mu_n(I) \leq \frac{1}{\mathbf{i}} \{2 \operatorname{rad}(I)\}^s.$$

For *any* closed ℓ^∞-ball B of radius $2^{-(n-j)}$ $(0 \leq j < n - \mathbf{n})$, we can find (not necessarily distinct) $J_1, \ldots, J_{2^d} \in \mathcal{D}_{n-j}$ such that $B \subset \cup_{\ell=1}^{2^d} J_\ell$; cf. Figure 10.1 of Chapter 10. Thus,

$$\mu_n(B) \leq \sum_{\ell=1}^{2^d} \mu_n(J_\ell) \leq \frac{2^d}{\mathbf{i}} \{2 \operatorname{rad}(B)\}^s.$$

For any ℓ^∞-ball B of radius at most $2^{-\mathbf{n}}$, there exists a ball $B' \supset B$ such that $\operatorname{rad}(B') = 2^{-(n-j)} \leq 2 \operatorname{rad}(B)$, for some $0 \leq j < n - \mathbf{n}$. Thus, for *any* closed ℓ^∞-ball B of radius at most $2^{-\mathbf{n}}$, $\mu_n(B) \leq 2^{d+1+2s} \mathbf{i}^{-1} \{\operatorname{rad}(B)\}^s$. On the other hand, if $\operatorname{rad}(B) > 2^{-\mathbf{n}}$, $\mu_n(B) \leq 1 \leq 2^{\mathbf{n}s} \{\operatorname{rad}(B)\}^s$. Thus, for *all* closed ℓ^∞-balls B,

$$\mu_n(B) \leq A \{\operatorname{rad}(B)\}^s, \tag{3}$$

where $A = \max\{2^{d+1+2s} \mathbf{i}^{-1}, 2^{\mathbf{n}s}\}$. Thus, we have constructed a sequence of probability measures $\{\mu_n; n > \mathbf{n}\}$, all on the compact set E, all of which satisfy (3). Let μ denote any one of the subsequential limits. Since E is compact, the μ_n's form a tight sequence. Consequently, μ is a probability measure on E. By (3), this is the desired probability measure. Indeed, for this measure μ,

$$\limsup_{r \to 0^+} \sup_{x \in \mathbb{R}^d} \frac{\mu\{\overline{B(x;r)}\}}{r^s} \leq A.$$

This completes the proof. $\square$

Frostman's lemma can be used to complete the calculation of Example 1, as the following shows.

Exercise 2.1.1 Use Frostman's lemma to show that the Hausdorff dimension of Cantor's tertiary set is equal to $\ln 2/\ln 3$.
(HINT: In the notation of Example 1 of Section 1.2, construct a measure on C_n by putting equal mass on all the intervals that form C_n. (a) If this measure is denoted by μ_n, estimate the μ_n measure of a small ball, uniformly in n. (b) Show that the μ_n's have a subsequential weak limit μ_∞ that is necessarily a probability measure on Cantor's tertiary set. (c) Estimate the μ_∞ measure of a small ball.) $\square$

Corollary 2.1.1 *Let C denote Cantor's tertiary set in $[0,1]$ and define $s = \ln 2/\ln 3$. Then, $\mathcal{H}_s(C) > 0$. In particular, $\dim(C) = \ln 2/\ln 3$.*

2.2 Frostman's Theorem and Bessel–Riesz Capacities

Given a Borel set $E \subset \mathbb{R}^d$, we write $\mathcal{P}(E)$ for the collection of all probability measures on E. That is, $\mu \in \mathcal{P}(E)$ if and only if $\mu(U^\complement) = 0$, for all open sets U in the complement of E. When E is compact, $\mathcal{P}(E)$ is exactly the collection of all probability measures whose closed support is E. This is an important property and makes the measure theory of compact sets particularly nice. Much of the time we shall concentrate on compacta for this very reason.

For any $\beta > 0$ and any probability measure μ on $\mathbb{R}^d$, we define the β-dimensional (Bessel–Riesz) **energy** of μ as

$$\mathsf{Energy}_\beta(\mu) = \iint \|x-y\|^{-\beta}\, \mu(dy)\, \mu(dx).$$

The 0-dimensional Bessel–Riesz **energy** of μ is given by

$$\mathsf{Energy}_0(\mu) = \iint \ln_+\left(\frac{1}{\|x-y\|}\right) \mu(dy)\, \mu(dx).$$

Finally, when $\beta < 0$, we define $\mathsf{Energy}_\beta(\mu) = 1$, for any probability measure μ.

For all $\beta \in \mathbb{R}$, the β-dimensional Bessel–Riesz **capacity** of a Borel set $E \subset \mathbb{R}^d$ can now be defined by the **principle of minimum energy** as follows:

$$\mathsf{Cap}_\beta(E) = \left\{ \inf_{\mu \in \mathcal{P}(E)} \mathsf{Energy}_\beta(\mu) \right\}^{-1}.$$

Frostman's second theorem—henceforth Frostman's theorem—is about the following beautiful connection between these capacities and measure-theoretic objects such as Hausdorff measures.

Theorem 2.2.1 (Frostman's Theorem) *Let $E \subset \mathbb{R}^d$ be a compact set. Whenever $s > \beta > 0$,*

$$\mathsf{Cap}_s(E) > 0 \implies \mathcal{H}_s(E) > 0 \implies \mathsf{Cap}_\beta(E) > 0.$$

In particular,

$$\dim(E) = \sup\{\beta > 0 : \mathsf{Cap}_\beta(E) > 0\} = \inf\{\beta > 0 : \mathsf{Cap}_\beta(E) = 0\}.$$

The proof of this result requires a simple lemma that is stated as an exercise.

***Exercise* 2.2.1** Show that for all compacts $E \subset \mathbb{R}^d$,

$$\sup\{\beta > 0 : \mathsf{Cap}_\beta(E) > 0\} = \inf\{\beta > 0 : \mathsf{Cap}_\beta(E) = 0\}.$$

Thus, Frostman's theorem identifies this number with the Hausdorff dimension of E. □

Proof Suppose there exists an $s > 0$ such that $\mathcal{H}_s(E) > 0$. By Lemmas 1.1.1 and 1.1.3, this means implicitly that $0 < s \le d$. By Frostman's lemma (Theorem 2.1.1) we can find a probability measure μ on E and a finite constant $A > 0$ such that

$$\sup_{x \in \mathbb{R}^d} \int_{y: |x-y| \le r} \mu(dy) \le Ar^s, \qquad \forall r > 0.$$

Since $\|x - y\| \ge |x - y|$,

$$\sup_{x \in \mathbb{R}^d} \int_{y: \|x-y\| \le r} \mu(dy) \le Ar^s, \qquad \forall r > 0. \qquad (1)$$

Let D denote any finite number greater than the diameter of E. Then, for all $0 < \beta < s$,

$$\mathsf{Energy}_\beta(\mu) = \int\int \|x-y\|^{-\beta} \mu(dx)\,\mu(dy)$$

$$= \sum_{j=0}^\infty \int\int_{2^{-j-1}D \le \|x-y\| < 2^{-j}D} \|x-y\|^{-\beta} \mu(dx)\,\mu(dy)$$

$$\le \sum_{j=0}^\infty 2^{(j+1)\beta} D^{-\beta} \sup_{x \in \mathbb{R}^d} \int_{y: \|x-y\| \le 2^{-j}D} \mu(dy).$$

We have used the fact that $\mu(E) = 1$. Equation (1) can now be invoked to show that

$$\text{Energy}_\beta(\mu) \le AD^{s-\beta} \sum_{j=0}^\infty 2^{-js+(j+1)\beta},$$

which is finite. That is, we have shown that

$$\exists s > 0 : \mathcal{H}_s(E) > 0 \implies \forall \beta \in \,]0,s[: \text{Cap}_\beta(E) > 0.$$

In particular, we obtain the following immediately.

$$\sup\{\beta > 0 : \text{Cap}_\beta(E) > 0\} \ge \dim(E).$$

This proves one half of the result.

Next, let us suppose there exists $s \in \,]0,d]$, such that $\text{Cap}_s(E) > 0$. There must exist $\mu \in \mathcal{P}(E)$ such that $\text{Energy}_s(\mu) < \infty$. The key estimate in this half is the following *maximal inequality*:

$$\mu\left\{x \in E : \sup_{r>0} \frac{\mu\{\mathcal{B}(x;r)\}}{r^s} \ge \lambda\right\} \le \frac{2^{\frac{sd}{2}}}{\lambda} \text{Energy}_s(\mu), \qquad \lambda > 0. \quad (2)$$

Let us prove this first. Since $|x - y| \le 2^{\frac{d}{2}} \|x - y\|$, for any $x \in \mathbb{R}^d$ and for all $r > 0$,

$$\mu\{\mathcal{B}(x;r)\} \le \int_{y : \|y-x\| \le 2^{\frac{d}{2}} r} \mu(dy) \le 2^{\frac{sd}{2}} r^s \int \|x-y\|^{-s} \mu(dy).$$

Equation (2) follows immediately from this and Chebyshev's inequality. Now we conclude the proof. Let $\lambda = 2^{1+\frac{sd}{2}} \text{Energy}_s(\mu)$ and

$$F = \left\{x \in E : \sup_{r>0} \frac{\mu\{\mathcal{B}(x;r)\}}{r^s} \le \lambda\right\}.$$

Thus, using (2), we can conclude that $F \subset E$ and $\mu(F) \ge \frac{1}{2}$. Let us cover F with closed ℓ^∞-balls $B_1, B_2, \ldots$ with radii $r_1, r_2, \ldots$, respectively. We can always arrange things so that for all $k \ge 1$, $B_k \cap F \ne \emptyset$; the centers of the B_k's lie in F; and $\sup_i r_i \le \varepsilon$, where $\varepsilon > 0$ is a preassigned, arbitrary number. Note that for our current choice, $\mu(B_j) \le \lambda r_j^s$ for all $j \ge 1$. Thus,

$$\frac{1}{2} \le \mu(F) \le \sum_{j=1}^\infty \mu(B_j) \le 2^{-s} \lambda \sum_{j=1}^\infty (2r_j)^s.$$

We have used (2) in the last inequality; in particular, we have used the fact that the centers of the B_j's lie in F. Taking the infimum over all such choices of $B_1, B_2, \ldots$, we can deduce that $\frac{1}{2} \le 2^{-s} \lambda \mathcal{H}_s^\varepsilon(F)$. Let $\varepsilon \to 0^+$ and invoke monotonicity (Lemma 1.2.1) to conclude that whenever $\mathsf{Cap}_s(E) > 0$, $\mathcal{H}_s(E) \geq 2^{s-1}\lambda^{-1} > 0$. In particular,

$$\dim(E) \geq \inf\{s > 0 : \mathsf{Cap}_s(E) > 0\}.$$

By Exercise 2.2.1,

$$\inf\{s > 0 : \mathsf{Cap}_s(E) > 0\} = \sup\{s > 0 : \mathsf{Cap}_s(E) = 0\}.$$

This completes our proof. □

***Exercise* 2.2.2** Supposing that $\beta > 0$ and $E \subset \mathbb{R}^d$ is compact, verify that for all finite constants $c > 0$, $\mathsf{Cap}_\beta(cE) = c^{-\beta}\mathsf{Cap}_\beta(E)$, where $cE = \{cy : y \in E\}$. □

2.3 Taylor's Theorem

The first half of Frostman's theorem (Theorem 2.2.1) states that if a compact set $E \subset \mathbb{R}^d$ has positive s-dimensional Bessel–Riesz capacity for some $s > 0$, then it has positive s-dimensional Hausdorff measure. In fact, one can prove slightly more. While the following is only a consequence of a theorem of S. J. Taylor, we will refer to it as Taylor's theorem, nonetheless.

Theorem 2.3.1 (Taylor's Theorem) *Let $E \subset \mathbb{R}^d$ be a compact set such that $\mathsf{Cap}_s(E) > 0$, for some $s > 0$. Then, $\mathcal{H}_s(E) = +\infty$.*

We first need a technical lemma.

Lemma 2.3.1 *Suppose μ is a probability measure on a compact set $E \subset \mathbb{R}^d$ and $s > 0$. Define D to be the outradius of E. That is, $D = \sup\{\|x\| : x \in E\}$. Then,*

$$D^s \mathfrak{S} \leq \mathsf{Energy}_s(E) \leq (2D)^s \mathfrak{S},$$

where

$$\mathfrak{S} = \sum_{j=0}^{\infty} 2^{js}\, \mu \times \mu\{(a,b) \in E \times E : 2^{-j-1}D < \|a-b\| \leq 2^{-j}D\}.$$

***Exercise* 2.3.1** Prove Lemma 2.3.1. □

Our proof of Taylor's theorem will also require an elementary technical result about real numbers. It, too, is stated as an exercise.

***Exercise* 2.3.2** Suppose $a_1, a_2, \ldots$ is a summable sequence of real numbers. Then, there exists a sequence $b_1, b_2, \ldots$ such that $\lim_{n\to\infty} b_n = +\infty$ and yet $\sum_n a_n b_n < \infty$. □

***Exercise* 2.3.3** Suppose μ is a finite measure on some locally compact, separable metric space S. Let $\mathcal{D}$ denote the diagonal of $S \times S$ (endowed with the product topology). That is, $\mathcal{D} = \{x \otimes x : x \in S\}$. Prove that $\mu \times \mu(\mathcal{D}) = 0$. What if μ is σ-finite? $\square$

Proof of Theorem 2.3.1 If $\operatorname{Cap}_s(E) > 0$, there must exist some probability measure $\mu \in \mathcal{P}(E)$ that has finite s-dimensional Bessel–Riesz energy. By Lemma 2.3.1, this is equivalent to the condition

$$\sum_{j=0}^{\infty} 2^{js} \, \mu \times \mu\{(a,b) \in E \times E : 2^{-j-1}D < \|a - b\| \leq 2^{-j}D\} < \infty,$$

where D denotes the outradius of E. By Exercise 2.3.2, we can find a nonincreasing function $h : \mathbb{R}_+ \to \mathbb{R}_+ \cup \{+\infty\}$ such that $\lim_{r \to 0^+} h(r) = +\infty$ and

$$\sum_{j=0}^{\infty} 2^{js} h(2^{-j-1}) \, \mu \times \mu\{(a,b) \in E \times E : 2^{-j-1}D < \|a - b\| \leq 2^{-j}D\} < \infty.$$

Define $\kappa(x) = x^s h(x)$ for all $x \in \mathbb{R}_+$. Clearly, $\lim_{x \to 0^+} \kappa(x)/|x|^s = +\infty$. Equivalently,

$$\forall M > 0, \ \exists \varepsilon_0 > 0 : \forall x \in \,]0, \varepsilon_0], \ \kappa(x) \geq M x^s. \tag{1}$$

Moreover, the method of proof of Lemma 2.3.1 can be used to see that the finiteness of the above double sum is *equivalent* to the condition $\iint \kappa(a - b) \, \mu(da) \, \mu(db) < +\infty$. An argument similar to the proof of equation (2) can now be used to verify the existence of a finite constant $A > 1$ such that

$$\mu\left(x \in E : \sup_{r > 0} \frac{\mu\{B(x;r)b\}}{\kappa(r)} \geq \lambda\right) \leq \frac{A}{\lambda} \iint \kappa(a-b) \, \mu(da) \, \mu(db), \qquad \lambda > 0.$$

We now continue our application of the proof of Frostman's theorem (Theorem 2.2.1) to conclude that $\mathcal{H}_\kappa(E) > 0$, where $\mathcal{H}_\kappa$ is defined exactly like $\mathcal{H}_s(E)$, except that the function $x \mapsto x^s$ is replaced by $x \mapsto \kappa(x)$ there. To be more precise, we can define $\mathcal{H}_\kappa(E) = \lim_{\varepsilon \to 0^+} \mathcal{H}_\kappa^\varepsilon(E)$, where

$$\mathcal{H}_\kappa^\varepsilon(E) = \inf\left\{\sum_{j=1}^{\infty} \kappa(2 r_j) : E \subset \bigcup_{j=1}^{\infty} \overline{B(x_j; r_j)}, \ \sup_\ell r_\ell \leq \varepsilon\right\}.$$

Since $\mathcal{H}_\kappa(E) > 0$, there exists $\delta > 0$ such that for all $\varepsilon > 0$, $\mathcal{H}_\kappa^\varepsilon(E) \geq \delta$. By equation (1), for all $M > 0$, there exists $\varepsilon_0 > 0$ such that for all $\varepsilon \in \,]0, \varepsilon_0]$, $\mathcal{H}_s^\varepsilon(E) \geq M \mathcal{H}_\kappa^\varepsilon(E) \geq M \delta$. Let $\varepsilon \to 0^+$ and $M \to \infty$, in this order, to deduce Theorem 2.3.1. $\square$

Theorem 2.3.1 has the following important consequence.

Corollary 2.3.1 *For all integers $d \geq 1$, $\mathsf{Cap}_d(\mathbb{R}^d) = 0$.*

Proof Suppose, to the contrary, that for some $d \geq 1$, $\mathsf{Cap}_d(\mathbb{R}^d) > 0$. Equivalently, there exists a probability measure $\mu \in \mathcal{P}(\mathbb{R}^d)$ such that $\mathsf{Energy}_d(\mu) < \infty$. We will derive a contradiction from this.

We first claim that there must exist a compact set $E \subset \mathbb{R}^d$ such that $\mathsf{Cap}_d(E) > 0$. For all Borel sets $G \subset \mathbb{R}^d$ and for all compact sets $E \subset \mathbb{R}^d$, define $\mu_E(G) = \mu(E \cap G)/\mu(E)$, where $0 \div 0 = 0$. Since $\mu \in \mathcal{P}(\mathbb{R}^d)$, one can always find a compact set $E \subset \mathbb{R}^d$ such that $\mu(E) > 0$, in which case $\mu_E \in \mathcal{P}(E)$. Moreover,

$$\begin{aligned}\mathsf{Energy}_d(\mu_E) &= \left(\frac{1}{\mu(E)}\right)^2 \int_E \int_E \|a-b\|^{-d}\, \mu(da)\, \mu(db) \\ &\leq \left(\frac{1}{\mu(E)}\right)^2 \mathsf{Energy}_d(\mu) < \infty,\end{aligned}$$

which implies that $\mathsf{Cap}_d(E) > 0$. By Taylor's theorem (Theorem 2.3.1), $\mathcal{H}_d(E) = +\infty$. Lemma 1.1.1 now implies that $\mathsf{Leb}(E) = +\infty$, where Leb denotes Lebesgue's measure on $\mathbb{R}^d$. On the other hand, Leb is a σ-finite measure. Equivalently, the Lebesgue measure of a compact set must be finite. This provides us with the desired contradiction. □

3 Notes on Appendix C

Section 1 This is standard material and can be found, in excellent textbook forms, in (Kahane 1985; Mattila 1995); see also Taylor (1986).

Section 2 Taylor's theorem can be found within (Carleson 1958; Taylor 1961).

Appendix D
Energy and Capacity

Appendix C contains a discussion of Hausdorff measure and dimension that naturally leads to Bessel–Riesz energy and capacity. We now study some of the fundamental properties of rather general energy forms and capacities. This appendix also includes a brief physical discussion of the role of energy and capacity in classical electrostatics.

1 Preliminaries

In this section we aim to introduce some important notions such as energy, potentials, and capacity (Section 1.1). Deeper properties will be studied in the remainder of this appendix. Our definitions of energy and capacity can be viewed as abstractions of the Bessel–Riesz capacities introduced in Appendix C.

There is also a heuristic discussion on classical electrostatics, where many of these objects first arose (Section 1.2). At the very least, Section 1.2 will partly explain the usage of the terms *capacity*, *energy*, etc. Section 1.2 can be skipped at first reading.

1.1 General Definitions

We say that a function $\mathfrak{g} : \mathbb{R}^d \times \mathbb{R}^d \to \mathbb{R}_+$ is a **gauge function** (on $\mathbb{R}^d \times \mathbb{R}^d$) if:

- for all $\eta > 0$, $(x, y) \mapsto \mathfrak{g}(x, y)$ is continuous (and finite) on O_η, where

$$O_\eta = \{x \otimes y \in \mathbb{R}^d \times \mathbb{R}^d : |x - y| > \eta\}, \tag{1}$$

and where $|\bullet|$ denotes the ℓ^∞-norm of a Euclidean vector; and

- there exists $\eta > 0$ such that on $O_\eta^\complement$, $\mathfrak{g} > 0$.

Recall that a gauge function $\mathfrak{g}$ is **symmetric** if $\mathfrak{g}(x, y) = \mathfrak{g}(y, x)$.

To every gauge function $\mathfrak{g}$ on $\mathbb{R}^d \times \mathbb{R}^d$ we associate a bilinear form, called mutual energy, defined next. Given two measures μ and ν, both on $\mathbb{R}^d$, the **mutual energy** between μ and ν, corresponding to the gauge function $\mathfrak{g}$, is defined as

$$\langle \mu, \nu \rangle_\mathfrak{g} = \frac{1}{2}\left\{\iint \mathfrak{g}(x, y)\, \mu(dx)\, \nu(dy) + \iint \mathfrak{g}(x, y)\, \nu(dx)\, \mu(dy)\right\}.$$

This is always defined, since $\mathfrak{g}$ is nonnegative; it may, however, be infinite.

The **energy** of a finite positive measure μ with respect to the gauge function $\mathfrak{g}$ is defined by the relation $\mathcal{E}_\mathfrak{g}(\mu) = \langle \mu, \mu \rangle_\mathfrak{g}$. Equivalently,

$$\mathcal{E}_\mathfrak{g}(\mu) = \iint \mathfrak{g}(x, y)\, \mu(dx)\, \mu(dy).$$

Note that as long as the total variation measure $|\mu|$ has finite energy, $\mathcal{E}_\mathfrak{g}(\mu)$ can be defined for all signed measures μ as well.

***Exercise* 1.1.1** Show that mutual energy can be used to define an inner product on the space of probability measures. Moreover, the seminorm corresponding to this inner product is energy. In particular, conclude that for any two probability measures ν and μ, both of finite energy,

$$\langle \mu, \nu \rangle_\mathfrak{g}^2 \leq \mathcal{E}_\mathfrak{g}(\mu) \cdot \mathcal{E}_\mathfrak{g}(\nu).$$

There is a dramatic improvement to this result when $\mathfrak{g}$ is nonnegative definite; consult Exercise 2.4.2 of Chapter 10, which is the beginning of the Beurling–Brelot–Cartan–Deny theory. □

Finally, the corresponding **capacity** $\mathcal{C}_\mathfrak{g}(A)$ of a Borel set $A \subset \mathbb{R}^d$ is defined by the principle of minimum energy of Appendix C. Namely,

$$\mathcal{C}_\mathfrak{g}(A) = \left[\inf_{\mu \in \mathcal{P}(A)} \mathcal{E}_\mathfrak{g}(\mu)\right]^{-1};$$

we recall that $\mathcal{P}(E)$ designates the collection of all probability measures on E. In this book we will be interested only in capacities of compact sets, and may refer to $\mathcal{E}_\mathfrak{g}(\mu)$ and $\mathcal{C}_\mathfrak{g}(E)$ as the **g-energy** of μ and the **g-capacity** of E, respectively.

Remarks (a) The function $\mathfrak{g}(x,y) = \|x-y\|^{-\beta}$ is the gauge function that corresponds to the Bessel–Riesz energy and capacity of Appendix C. □

(b) It is important to observe that E has positive $\mathfrak{g}$-capacity if and only if there exists a probability measure μ, on E, that has finite $\mathfrak{g}$-energy. Depending on the "shape" of the kernel $\mathfrak{g}$, this gives some information on how large, or thick, the set in question is. For instance, the singleton $\{y\}$ has positive $\mathfrak{g}$-capacity if and only if $\mathfrak{g}(y,y) < \infty$. Thus, if $\mathfrak{g}(x,y)$ is proportional to $|x-y|^{-\beta}$ or $\ln_+(1/|x-y|)$, singletons are thin sets in gauge $\mathfrak{g}$. The latter kernels are discussed in some depth in Appendix C. □

Given a measure μ on $\mathbb{R}^d$, we can define the **potential** of μ (with respect to $\mathfrak{g}$) as the function $\mathfrak{G}\mu$, where

$$\mathfrak{G}\mu(x) = \int \mathfrak{g}(x,y)\,\mu(dy), \qquad x \in \mathbb{R}^d.$$

See Section 1.2 below for a heuristic physical interpretation of this for the gauge function $\mathfrak{g}(x,y) = c\|x-y\|^{-1}$.

The following exercises show important connections among energies, capacities, and potentials.

***Exercise* 1.1.2** Verify that for any gauge function $\mathfrak{g}$ on $\mathbb{R}^d \times \mathbb{R}^d$, and for all measures μ and ν on $\mathbb{R}^d$,

$$\langle \mu, \nu \rangle_\mathfrak{g} = \frac{1}{2}\left\{ \int \mathfrak{G}\mu(x)\,\nu(dx) + \int \mathfrak{G}\nu(x)\,\mu(dx) \right\}.$$

In particular, check that $\mathcal{E}_\mathfrak{g}(\mu) = \int \mathfrak{G}\mu(x)\,\mu(dx)$. □

***Exercise* 1.1.3** Suppose $\mathfrak{g}$ is a gauge function on $\mathbb{R}^d$, and μ is a measure on $\mathbb{R}^d$ of finite $\mathfrak{g}$-energy.

(i) Check that $\mathfrak{G}\mu < +\infty$, μ-a.e.

(ii) Prove that $x \mapsto \mathfrak{G}\mu(x)$ is lower semicontinuous.

(HINT: Estimate the μ-measure of the set $\{x \in \mathbb{R}^d : \mathfrak{G}\mu(x) \geq \lambda\}$.) □

We say that a gauge function $\mathfrak{g}$ is **symmetric** if $\mathfrak{g}(x,y) = \mathfrak{g}(y,x)$.

***Exercise* 1.1.4** Verify that for any symmetric gauge function $\mathfrak{g}$ on $\mathbb{R}^d \times \mathbb{R}^d$,

$$\langle \mu, \nu \rangle_\mathfrak{g} = \int \mathfrak{G}\mu(x)\,\nu(dx) = \int \mathfrak{G}\nu(x)\,\mu(dx),$$

where μ and ν are two arbitrary (nonnegative) measures on the Borel subsets of $\mathbb{R}^d$.

(This is the **reciprocity theorem** of classical potential theory.) □

Exercise 1.1.5 As the preceding Remark(a) implies, one often wants to know whether or not a given set has positive $\mathfrak{g}$-capacity. Along these lines, suppose $\mathfrak{g}$ is a gauge function on $\mathbb{R}^d$, and show that for any compact sets $\Lambda_1, \Lambda_2, \ldots,$

$$\sup_{i \geq 1} \mathcal{C}_{\mathfrak{g}}(\Lambda_i) = 0 \iff \mathcal{C}_{\mathfrak{g}}\left(\bigcup_{i \geq 1} \Lambda_i\right) = 0.$$

Show that there are no obvious extensions for uncountably many Λ's. More precisely, construct a gauge $\mathfrak{g}$ on $\mathbb{R}^d$, and uncountably many sets $(\Lambda_\alpha; \alpha \in A)$, all of which have zero $\mathfrak{g}$-capacity, such that $\cup_{\alpha \in A} \Lambda_\alpha$ has positive $\mathfrak{g}$-capacity. □

1.2 Physical Interpretations

It has long been known that a positively charged particle in $\mathbb{R}^3$ is subject to a certain amount of force if there is a charged body in the vicinity. Moreover, this force depends on the position of the particle and is proportional to the amount of its charge. Thus, we can define the force as $qF(x)$, if $q > 0$ denotes the charge of the particle and if the particle is at position $x \in \mathbb{R}^3$. The function $F : \mathbb{R}^3 \to \mathbb{R}^3$ is called the **electrical field** and is due to the effect of the charged body that is in the vicinity.

It is also known that electrical fields are *additive* in the following sense: Given n positively charged particles at $x_1, x_2, \ldots, x_n \in \mathbb{R}^3$, with charges $q_1, q_2, \ldots, q_n$ and with electrical fields $F_1, \ldots, F_n$, respectively, the force on the ith particle is $q_i \cdot \sum_{\ell=1}^{N} F_\ell(x_\ell)$.

In order to simplify things, we now suppose that the charged body is concentrated at some point x_0 and has charge $-q_0$. In this case, Coulomb's law of electrostatics states that the electrical field of the ith particle (which is $F_i(x_i)$) is a three-dimensional vector that points from x_i to x_0 and whose magnitude is proportional to the inverse of the distance between x_0 and x_i. In fact, the constant of proportionality is precisely the charge of the body, and thus $\|F_i(x_i)\| = q_0 / \|x_0 - x_i\|^2$. That is,

$$F_i(x_i) = \frac{q_0}{\|x_i - x_0\|^2} \cdot \frac{x_i - x_0}{\|x_i - x_0\|} = q_0 \cdot \frac{x_i - x_0}{\|x_i - x_0\|^3}.$$

Therefore, if we put a charge of $q(x)$ at some point $y \in \mathbb{R}^3$, the induced electrical field at $x \in \mathbb{R}^3$ is $F(x) = q(x)(x-y)/\|x-y\|^3$. By the additivity property of electrical fields, if we have n charges of amounts $q(y_1), \ldots, q(y_n)$ at $y_1, \ldots, y_n \in \mathbb{R}^3$, respectively, for any $x \in \mathbb{R}^3$, the force exerted at point $x \in \mathbb{R}^3$ is precisely

$$F(x) = \sum_{\ell=1}^{n} \frac{q(y_\ell)}{\|x - y_\ell\|^3} (x - y_\ell).$$

1 Preliminaries

Somewhat more generally, suppose μ is a measure on $\mathbb{R}^3$ that describes the distribution of charge in space. For example, in the previous paragraph, μ puts weight $q_1,\ldots,q_n$ at $x_1,\ldots,x_n$, respectively. Then, the amount of force that this charged field exerts at $x \in \mathbb{R}^3$ should be $F(x) = \int (x-y)\|x-y\|^{-3}\,\mu(dy)$. We can formally interchange the order of differentiation and integration, and make a line or two of calculations, to deduce that[1] F solves the partial differential equation $F = -\nabla \mathcal{R}\mu$, where ∇ denotes the gradient operator whose ith coordinate is $(\partial/\partial x^{(i)})$ $(i=1,2,3)$ and

$$\mathcal{R}\mu(x) = \int \|x-y\|^{-1}\,\mu(dy).$$

The function $\mathcal{R}\mu$ is the **potential functions** of the force field F. In this three-dimensional setting, this potential function is called the **Coulomb potential** of μ.

Suppose, once more, that we place n positively charged particles in space. Let $q_1,\ldots,q_n$ denote the respective charges and $x_1,\ldots,x_n$ the positions. We have already seen that the nth particle is subject to a certain amount of force. In fact, this force is $q_n \nabla p_n(x_n)$, where $p_n(x) = \sum_{\ell=1}^{n-1} q_\ell \|x - x_\ell\|^{-1}$ is the potential function for the nth particle. Consequently, particle n will be indefinitely pushed away from the other $n-1$ particles. This holds simultaneously for all particles.

We now focus our attention on the nth particle. Let $\tau : [0,\infty] \to \mathbb{R}^3$ denote a parametrization of the trajectory of this nth particle such that $\tau(0) = x_n$ and $\tau(\infty) = +\infty$. The amount of **work** done by the nth particle is the sum total of all of the forces on it along its path. That is,

$$W_n = \int F(\tau)d\tau = \int_0^\infty -q_n\, \nabla p_n(\tau(s)) \cdot \tau'(s)\,ds$$
$$= -q_n \int_0^\infty \frac{dp_n(\tau(s))}{ds}\,ds = -q_n\{p_n(\infty) - p_n(x_n)\} = q_n p_n(x_n)$$
$$= \sum_{\ell=1}^{n-1} \frac{q_n q_\ell}{\|x_n - x_\ell\|},$$

where, in the second equation, "$\cdot$" denotes the Euclidean inner product. Particle n is pushed away from the others, and W_n is the total work performed by it. We have now $n-1$ particles and view the $(n-1)$st as it is being repelled from the remaining $n-2$ particles. The total amount of work performed by the $(n-1)$st particle as it is being pushed away is

$$W_{n-1} = \sum_{k=1}^{n-2} \frac{q_{n-1} q_k}{\|x_{n-1} - x_k\|}.$$

---

[1] This was first observed by J.-L. Lagrange. See Wermer (1981), where the discussion of this section is carried through with more care and in much greater detail.

Now repeat this procedure to compute the work W_i done by the ith particle as it is being pushed away. The principle of conservation of work shows that the total amount of work done by this system of n particles is $W = \sum_{\ell=1}^{n} W_\ell$, which we have just computed as

$$W = \sum_{\ell=1}^{n} \sum_{k=1}^{\ell-1} \frac{q_\ell q_k}{\|x_\ell - x_k\|}.$$

This is half of $\sum_{k \neq \ell} q_k q_\ell / \|x_k - x_\ell\|$.

More generally, if we spread a charge distribution μ in space, the total amount of work expended by this (possibly infinite) system of particles is

$$W = \frac{1}{2} \iint \|x - y\|^{-1} \mu(dx) \mu(dy),$$

provided that we can ignore the effect of the diagonal terms in the above double integral. In other words, total work equals one-half of the 1-dimensional Bessel–Riesz energy $\mathsf{Energy}_1(\mu)$ of the charge distribution, and this is more commonly known as the **Coulomb energy** of μ in this three-dimensional setting.

The physical interpretation of the corresponding capacity $\mathsf{Cap}_1(E)$ of a set $E \subset \mathbb{R}^3$ follows from the construction of a so-called capacitor.[2] Imagine a conductor C that is surrounded by a grounded body G. To be more concrete, let us think of C in the shape of a unit ball in $\mathbb{R}^3$ and of G in the shape of the boundary of the concentric ball of radius $r > 1$, also in $\mathbb{R}^3$. That is, in symbols, $G = \partial(rC)$.

If we put q units of charge in C, it will redistribute itself on ∂C and reach an equilibrium state. Moreover, in this equilibrium state, the charge distribution is as uniform as possible. This, in turn, produces an electric field F on $\mathbb{R}^3 \setminus C$. Since G is grounded, the only area that is subject to this electric field is $\{y \in \mathbb{R}^3 : 1 \leq \|y\| < r\}$.

Next, suppose the electric field F comes from a potential function of the form ∇f, where f is a reasonable function. Since $\nabla f = 0$ on ∂C, f equals some constant K on ∂C. One can show that this observation alone implies that K is *necessarily* proportional to q. The capacity of the "condenser" (or capacitor) created by adjoining C and G is precisely this constant of proportionality. In other words, in physical terms, the total amount of charge in the capacitor equals the potential difference across the capacitor multiplied by the capacity of the capacitor.

C.-F. Gauss showed that capacity can also be obtained by minimum energy considerations. This is a cornerstone in axiomatic potential theory and partial differential equations, and is developed in Theorem 2.1.2 below, as well as in Exercise 2.1.3 below. However, for the purposes of our

---

[2] When E is a three-dimensional set, $\mathsf{Cap}_1(E)$ is commonly referred to as the **Newtonian capacity** of E.

present physical discussion, the principle of requiring minimum energy at equilibrium should seem a physically sound one.

2 Choquet Capacities

Let Ω denote a topological space. A nonnegative set function C, defined on compact subsets of Ω, is said to be a **Choquet capacity** (or a **natural capacity**) if it satisfies the following criteria:[3]

(a) (*Outer Regularity*) Consider $K_1 \supset K_2 \supset \cdots$, all compact subsets of Ω, and the compact set $K = \cap_n K_n$. Then, $\lim_{n\to\infty} C(K_n) = C(K)$.

(b) (*Subadditivity*) For all compact sets E and F, $C(E \cup F) \leq C(E) + C(F)$.

We plan to show, among other things, that if $\mathfrak{g}$ is a symmetric gauge function on $\mathbb{R}^d \times \mathbb{R}^d$, $\mathcal{C}_{\mathfrak{g}}$ is a Choquet capacity on compact subsets of $\mathbb{R}^d$. In the meantime, we note that any probability measure is a Choquet capacity.

2.1 Maximum Principle and Natural Capacities

We say that a gauge function $\mathfrak{g}$ satisfies a **maximum principle** if for all probability measures μ with compact support $E \subset \mathbb{R}^d$,

$$\sup_{x \in \mathbb{R}^d} \mathfrak{G}\mu(x) = \sup_{x \in E} \mathfrak{G}\mu(x),$$

where $\mathfrak{G}\mu$ denotes the $\mathfrak{g}$-potential of μ. Our next theorem is the main result of this subsection.

Theorem 2.1.1 *If $\mathfrak{g}$ is a gauge function on $\mathbb{R}^d$, then $\mathcal{C}_{\mathfrak{g}}$ is outer regular. Furthermore, if $\mathfrak{g}$ is a symmetric gauge function that satisfies the maximum principle, $\mathcal{C}_{\mathfrak{g}}$ is a Choquet capacity on compact subsets of $\mathbb{R}^d$.*

The first portion of this theorem, i.e., the outer regularity claim, is the only part that is used in this book. However, the second portion is essential if one wants to go from capacities on compacts to more general sets, such as measurable, or even analytic, sets. Some of this development can be found in the first chapter of Dellacherie and Meyer (1978).

It is easiest to prove Theorem 2.1.1 in a few steps, each of which is stated separately. Throughout, the notation of Theorem 2.1.1 is used and enforced.

---

[3] In axiomatic potential theory this is actually a Choquet outer capacity on compacts, and is then proved to agree with another object called a Choquet capacity. We do not make such distinctions here.

Lemma 2.1.1 (Lower Semicontinuity of Potentials) *Consider probability measures μ_n ($1 \leq n \leq \infty$), all defined on Borel subsets of $\mathbb{R}^d$. If μ_n converges weakly to μ_∞,*

$$\liminf_{n \to \infty} \mathcal{E}_{\mathfrak{g}}(\mu_n) \geq \mathcal{E}_{\mathfrak{g}}(\mu_\infty).$$

Proof Properties of gauge functions dictate that $\mathfrak{g} \wedge \lambda$ is a continuous function for all λ sufficiently large (why?). For such $\lambda > 0$,

$$\mathcal{E}_{\mathfrak{g}}(\mu_n) \geq \mathcal{E}_{\mathfrak{g} \wedge \lambda}(\mu_n) \to \mathcal{E}_{\mathfrak{g} \wedge \lambda}(\mu_\infty),$$

as $n \to \infty$. We have applied the continuous mapping theorem of weak convergence to the probability measures $\mu_n \times \mu_n$; cf. Theorem 2.2.1 of Chapter 6, for example. By Lebesgue's monotone convergence theorem, $\lim_{\lambda \to \infty} \mathcal{E}_{\mathfrak{g} \wedge \lambda}(\mu_\infty) = \mathcal{E}_{\mathfrak{g}}(\mu_\infty)$, which proves the result. □

Lemma 2.1.2 (Outer Regularity) *If $K_1 \supset K_2 \supset \cdots$ are compact, and if $K = \cap_n K_n$, then, as $n \to \infty$, $\mathcal{C}_{\mathfrak{g}}(K_n) \downarrow \mathcal{C}_{\mathfrak{g}}(K)$.*

Proof The set function $\mathcal{C}_{\mathfrak{g}}$ is obviously monotone. That is, whenever $E \subset F$, $\mathcal{C}_{\mathfrak{g}}(E) \leq \mathcal{C}_{\mathfrak{g}}(F)$. Hence, it suffices to prove

$$\limsup_{n \to \infty} \mathcal{C}_{\mathfrak{g}}(K_n) \leq \mathcal{C}_{\mathfrak{g}}(K). \tag{1}$$

Unless $\mathcal{C}_{\mathfrak{g}}(K_n) > 0$ for all but finitely many n's, this is a trivial matter. Hence, without loss of generality, we may assume that for all n large, the $\mathfrak{g}$-capacity of K_n is positive. This means that for all n large, we can find $\mu_n \in \mathcal{P}(K_n)$ such that $\mathcal{E}_{\mathfrak{g}}(\mu_n) < \infty$. Moreover, we can arrange things so that for all n large and for these very μ_n's,

$$\left[\mathcal{E}_{\mathfrak{g}}(\mu_n)\right]^{-1} \geq \mathcal{C}_{\mathfrak{g}}(K_n) + \frac{1}{n}. \tag{2}$$

Since K_1 includes all the K_n's, as well as K, and since K_1 is compact, there exists a probability measure μ_∞ and a subsequence n_j going to ∞ such that as $j \to \infty$, μ_{n_j} converges weakly to μ_∞. Moreover, $K_n \downarrow K$ shows that $\mu \in \mathcal{P}(K)$ (why?). By Lemma 2.1.1,

$$\liminf_{j \to \infty} \mathcal{E}_{\mathfrak{g}}(\mu_{n_j}) \geq \mathcal{E}_{\mathfrak{g}}(\mu_\infty) \geq \inf_{\mu \in \mathcal{P}(K)} \mathcal{E}_{\mathfrak{g}}(\mu).$$

Take reciprocals of this inequality and apply (2) to complete this proof. □

The above proves the first assertion of Theorem 2.1.1. Its proof can be used to complete the following important exercise.

2 Choquet Capacities

***Exercise* 2.1.1** Lemma 2.1.1 has the following compactness consequence: If $\mathcal{C}_{\mathfrak{g}}(E) > 0$ for some compact set E, there exists $\mu \in \mathcal{P}(E)$ such that $\mathcal{C}_{\mathfrak{g}}(E) = 1/\mathcal{E}_{\mathfrak{g}}(\mu)$. In particular, any compact $F \subset E$ of zero $\mathfrak{g}$-capacity is μ-null. □

***Exercise* 2.1.2** If μ is a measure of finite energy, prove that for all $\lambda > 0$,

$$\mathcal{C}_{\mathfrak{g}}(\{x : \mathfrak{G}\mu(x) \geq \lambda\}) \leq \frac{1}{\lambda^2} \mathcal{E}_{\mathfrak{g}}(\mu).$$

In particular, conclude that outside a set of zero $\mathfrak{g}$-capacity, $\mathfrak{G}\mu < +\infty$. Show that this, together with Exercise 2.1.1, improves on Exercise 1.1.3. (HINT: Exercise 1.1.1.) □

In order to prove Theorem 2.1.1, we will need to better understand the measure μ of minimum energy, as supplied to us by the preceding exercise. The discussion of Section 1.2 suggests that, in some cases, the potential of μ should equal a constant on E. In general, this is almost the case, as the following shows.

Lemma 2.1.3 *Suppose μ is a compactly supported probability measure on $\mathbb{R}^d$ that has finite $\mathfrak{g}$-energy. Then, outside a set of zero $\mathfrak{g}$-capacity, $\mathfrak{G}\mu \geq \mathcal{E}_{\mathfrak{g}}(\mu)$, while for all x in the support of μ, $\mathfrak{G}\mu(x) \leq \mathcal{E}_{\mathfrak{g}}(\mu)$.*

Proof Define for all $\alpha > 0$,

$$\Lambda_\alpha = \{x \in E : \mathfrak{G}\mu(x) \leq \alpha \mathcal{E}_{\mathfrak{g}}(\mu)\}.$$

By lower semicontinuity, Λ_α is compact for all $\alpha > 0$; cf. Exercise 1.1.3. We claim that whenever $\alpha \in \,]0,1[$, $\mathcal{C}_{\mathfrak{g}}(\Lambda_\alpha) = 0$. Indeed, if it were not the case, one could find $\alpha \in \,]0,1[$ and $\nu \in \mathcal{P}(\Lambda_\alpha)$ such that $\mathcal{E}_{\mathfrak{g}}(\nu) < +\infty$. For any $\eta \in \,]0,1[$, define $\zeta_\eta = \eta\nu + (1-\eta)\mu$, and note that $\zeta_\eta \in \mathcal{P}(E)$. We can write

$$\zeta_\eta = \mu + \eta(\nu - \mu),$$

and compute its $\mathfrak{g}$-energy, to be

$$\mathcal{E}_{\mathfrak{g}}(\zeta_\eta) = \langle \zeta_\eta, \zeta_\eta \rangle_{\mathfrak{g}} = \mathcal{E}_{\mathfrak{g}}(\mu) + \eta^2 \mathcal{E}_{\mathfrak{g}}(\nu - \mu) + 2\eta \langle \mu, \nu \rangle_{\mathfrak{g}} - 2\eta \mathcal{E}_{\mathfrak{g}}(\mu).$$

(Check!) On the other hand, μ is a probability measure on E that has minimum energy. In particular, $\mathcal{E}_{\mathfrak{g}}(\mu) < \mathcal{E}_{\mathfrak{g}}(\zeta_\eta)$, for all $\eta \in \,]0,1[$. Using the displayed expression for $\mathcal{E}_{\mathfrak{g}}(\zeta_\eta)$ leads to the inequality

$$\mathcal{E}_{\mathfrak{g}}(\mu) \leq \frac{\eta}{2} \mathcal{E}_{\mathfrak{g}}(\nu - \mu) + \langle \mu, \nu \rangle_{\mathfrak{g}}.$$

We can let $\eta \downarrow 0$ to deduce that

$$\mathcal{E}_{\mathfrak{g}}(\mu) \leq \langle \mu, \nu \rangle_{\mathfrak{g}}.$$

Since $\mathfrak{g}$ is symmetric, $\langle \mu, \nu \rangle_\mathfrak{g} = \int \mathfrak{G}\mu \, d\nu$. Moreover, since $\nu \in \mathcal{P}(\Lambda_\alpha)$,

$$\langle \mu, \nu \rangle_\mathfrak{g} \leq \alpha \mathcal{E}_\mathfrak{g}(\mu).$$

(Why? This is not entirely trivial when μ is not purely atomic.) This implies the contradiction $\mathcal{E}_\mathfrak{g}(\mu) \leq \alpha \mathcal{E}_\mathfrak{g}(\mu)$. Thus, as asserted, Λ_α has zero $\mathfrak{g}$-capacity. By Exercise 1.1.5, $\mathcal{C}_\mathfrak{g}(\Lambda) = 0$, where $\Lambda = \{x \in E : \mathfrak{G}\mu(x) < \mathcal{E}_\mathfrak{g}(\mu)\}$. This is best seen by noticing that Λ is the countable union of Λ_α, as α ranges over $]0,1[\cap \mathbb{Q}$. We have shown that outside a set of zero capacity, $\mathfrak{G}\mu \geq \mathcal{E}_\mathfrak{g}(\mu)$; this proves the first half of our theorem. Moreover, by Exercise 2.1.1, if a set in E has zero $\mathfrak{g}$-capacity, it has zero μ-measure, i.e.,

$$\mathfrak{G}\mu \geq \mathcal{E}_\mathfrak{g}(\mu), \qquad \mu\text{-almost everywhere.} \tag{3}$$

This proves half of our lemma.

For the second half, define for all $\varepsilon \in]0,1[$,

$$A_\varepsilon = \{x \in E : \mathfrak{G}\mu(x) > (1+\varepsilon)\mathcal{E}_\mathfrak{g}(\mu)\}.$$

If there exists some x_0 in the support of μ such that $\mathfrak{G}\mu(x_0) > \mathcal{E}_\mathfrak{g}(\mu)$, by lower semicontinuity (Exercise 1.1.3) we can find $\varepsilon \in]0,1[$ such that A_ε contains a relatively open neighborhood (in E) of x_0 that has positive μ measure for some $\varepsilon \in]0,1[$; this follows from the definition of the support of a measure. In particular, there exists $\varepsilon \in]0,1[$ such that $\mu(A_\varepsilon) > 0$. However, according to equation (3),

$$\begin{aligned}
\mathcal{E}_\mathfrak{g}(\mu) &= \int_{A_\varepsilon} \mathfrak{G}\mu(x)\,\mu(dx) + \int_{A_\varepsilon^\complement} \mathfrak{G}\mu(x)\,\mu(dx) \\
&\geq \mathcal{E}_\mathfrak{g}(\mu)\mu(A_\varepsilon^\complement) + (1+\varepsilon)\mu(A_\varepsilon)\mathcal{E}_\mathfrak{g}(\mu) \\
&= \mathcal{E}_\mathfrak{g}(\mu)\{1 + \varepsilon\mu(A_\varepsilon)\},
\end{aligned}$$

which is the desired contradiction. $\square$

In light of Lemma 2.1.2, Theorem 2.1.1 follows, once we show that $\mathcal{C}_\mathfrak{g}$ is subadditive. We prove the latter property by first deriving an alternative characterization of capacities.

Theorem 2.1.2 *Suppose $\mathfrak{g}$ is a symmetric gauge function and $E \subset \mathbb{R}^d$ is a compact set of positive $\mathfrak{g}$-capacity. Then,*

$$\mathcal{C}_\mathfrak{g}(E) = \inf \sigma(E),$$

where the infimum is taken over all measures σ of finite energy on E such that $\{x \in E : \mathfrak{G}\sigma(x) < 1\}$ has zero $\mathfrak{g}$-capacity.

Proof By Exercise 2.1.1, we can find such a probability measure $\mu \in \mathcal{P}(E)$ that satisfies $\mathcal{C}_\mathfrak{g}(E) = 1/\mathcal{E}_\mathfrak{g}(\mu)$. We also know, from Lemma 2.1.3, that

μ has a potential $\mathfrak{G}\mu$ that is bounded below by $\mathcal{E}_\mathfrak{g}(\mu)$, off a set of zero $\mathfrak{g}$-capacity. Since $\mathfrak{g}$ is a gauge function, $\mathcal{E}_\mathfrak{g}(\mu) > 0$ (why?). Hence, we can define

$$\sigma_0(\bullet) = \frac{\mu(\bullet)}{\mathcal{E}_\mathfrak{g}(\mu)},$$

and check that the following hold:

- $\mathcal{E}_\mathfrak{g}(\sigma_0) = \mathcal{C}_\mathfrak{g}(E)$;
- $\sigma_0(E) = \mathcal{C}_\mathfrak{g}(E)$; and
- $\mathcal{C}_\mathfrak{g}(\{x \in E : \mathfrak{G}\sigma_0(x) < 1\}) = \mu(\{x \in E : \mathfrak{G}\sigma_0(x) < 1\}) = 0$.

The gauge function properties of $\mathfrak{g}$ imply that $\mathcal{C}_\mathfrak{g}(E) < +\infty$. Thus, we have shown that $\mathcal{C}_\mathfrak{g}(E) \geq \inf \sigma(E)$, where the infimum is taken as in the statement of the theorem.

For the converse, suppose σ is a measure on E such that (a) it has finite energy; and (b) outside a μ-null set, $\mathfrak{G}\sigma \geq 1$. In this way, we obtain

$$1 \leq \int \mathfrak{G}\sigma(x)\,\mu(dx) = \int \mathfrak{G}\mu(dx)\,\sigma(dx) \leq \sup_{x \in \mathbb{R}^d} \mathfrak{G}\mu(x)\,\sigma(E).$$

We have used the reciprocity theorem (Exercise 1.1.4). On the other hand, Lemma 2.1.3 and the maximum principle together imply that $\sup_{x \in \mathbb{R}^d} \mathfrak{G}\mu(x) \leq \mathcal{E}_\mathfrak{g}(\mu)$. Thus, $1 \leq \mathcal{E}_\mathfrak{g}(\mu)\sigma(E)$. Divide by $\mathcal{E}_\mathfrak{g}(\mu)$ to deduce that $\mathcal{C}_\mathfrak{g}(E) \leq \inf \sigma(E)$, where the inf is as in the statement of the theorem. □

***Exercise* 2.1.3** If E is compact and has positive $\mathfrak{g}$-capacity, where $\mathfrak{g}$ is a symmetric gauge function on $\mathbb{R}^d$, show that $\mathcal{C}_\mathfrak{g}(E) = \sup \sigma(E)$, where the sup is taken over all measures on E of finite energy with $\sup_{x \in E} \mathfrak{G}\sigma(x) \leq 1$. □

With Theorem 2.1.2 under way, it is possible to prove Theorem 2.1.1, and conclude this subsection.

***Exercise* 2.1.4** Complete the proof of Theorem 2.1.1. □

2.2 Absolutely Continuous Capacities

Let $\mathfrak{g}$ denote a gauge function on $\mathbb{R}^d$, and suppose we are interested in minimizing the energy

$$\mathcal{E}_\mathfrak{g}(\mu) = \iint \mathfrak{g}(x,y)\,\mu(dx)\,\mu(dy),$$

as μ varies over all probability measures on some set E. Equivalently, suppose we are interested in computing the reciprocal of the $\mathfrak{g}$-capacity of E. In principle, there are many such probability measures μ which tends to make the above optimization problem difficult. However, sometimes one can reduce this problem to the case where $\mu(dx)$ is approximable by $f(x)dx$ for some reasonable nonnegative function f on $\mathbb{R}^d$. In such a case, the energy of μ has the following simpler form:

$$\mathcal{E}_{\mathfrak{g}}(\mu) = \iint \mathfrak{g}(x,y) f(x) f(y)\, dx\, dy.$$

We now study the problem of when it is enough to study such energies, in the more general setting where "dx" is replaced by a general measure.

Throughout, ν denotes a fixed measure on measurable subsets of $\mathbb{R}^d$, and we will say that a measure is **absolutely continuous** if it is absolutely continuous with respect to this measure ν.

Consider a gauge function $\mathfrak{g}$ on $\mathbb{R}^d \times \mathbb{R}^d$, together with its induced capacity $\mathcal{C}_{\mathfrak{g}}$. We can modify this capacity a little, by defining a set function $\mathcal{C}_{\mathfrak{g}}^0$ on measurable subsets of $\mathbb{R}^d$ as follows:

$$\mathcal{C}_{\mathfrak{g}}^0(E) = \left[\inf_{\substack{\mu \in \mathcal{P}(E):\\ \text{absolutely continuous}}} \mathcal{E}_{\mathfrak{g}}(\mu) \right]^{-1}.$$

As usual, inf of anything over the empty set is defined to be $+\infty$.

On the one hand, $\mathcal{C}_{\mathfrak{g}}^0$ is a nicer set function than $\mathcal{C}_{\mathfrak{g}}$ in the sense that the measures involved are absolutely continuous. On the other hand, $\mathcal{C}_{\mathfrak{g}}^0$ has some undesirable properties, one of which is that $\nu(E) = 0$ implies $\mathcal{C}_{\mathfrak{g}}^0(E) = 0$. The second, even more serious, problem is that, generally, $\mathcal{C}_{\mathfrak{g}}^0$ is not outer regular on compacts. We try to address both of these issues by using, instead, a regularization $\mathcal{C}_{\mathfrak{g}}^{ac}$ of $\mathcal{C}_{\mathfrak{g}}^0$. Namely, for all bounded Borel sets $E \subset \mathbb{R}^d$, we define

$$\mathcal{C}_{\mathfrak{g}}^{ac}(E) = \inf\left\{ \mathcal{C}_{\mathfrak{g}}^0(\overline{F}) :\ F \supset E \text{ is bounded and open}\right\}.$$

We can extend this definition to any Borel set $E \subset \mathbb{R}^d$ by

$$\mathcal{C}_{\mathfrak{g}}^{ac}(E) = \sup\left\{ \mathcal{C}_{\mathfrak{g}}^{ac}(G) :\ G \subset E \text{ is a bounded Borel set}\right\}.$$

However, since we are interested in $\mathcal{C}_{\mathfrak{g}}^{ac}(E)$ only when E is compact, the above extension is superfluous (for such E's), as the following shows.

***Exercise* 2.2.1** Show that $\mathcal{C}_{\mathfrak{g}}^{ac}$ is outer regular on compacts. Moreover, show that $\mathcal{C}_{\mathfrak{g}}^{ac}$ is a Choquet capacity whenever $\mathcal{C}_{\mathfrak{g}}$ is. Finally, check that for all compact E, $\mathcal{C}_{\mathfrak{g}}(E) \geq \mathcal{C}_{\mathfrak{g}}^{ac}(E)$. □

In words, $\mathcal{C}_{\mathfrak{g}}^{ac}$ is the "best" possible construction of a $\mathfrak{g}$-based capacity that solely uses absolutely continuous measures in its definition. As such, we are justified in calling it the **absolutely continuous capacity** based on the gauge $\mathfrak{g}$.

2.3 Proper Gauge Functions and Balayage

Consider a gauge function $\mathfrak{g}$ on $\mathbb{R}^d \times \mathbb{R}^d$. A question of paramount importance to us is, *when do the capacities and the absolutely continuous capacities based on $\mathfrak{g}$ agree on compact sets?* where absolute continuity is meant to hold with respect to some underlying measure ν on $\mathbb{R}^d$. Equivalently, when is it the case that for all compact sets $E \subset \mathbb{R}^d$, $\mathcal{C}_\mathfrak{g}(E) = \mathcal{C}^{ac}_\mathfrak{g}(E)$? In general, a satisfactory answer to this question does not seem to exist. However, we will see below that there is a complete answer when $\mathfrak{g}$ is "nice."

We say that the gauge function $\mathfrak{g}$ is **proper** if for all compact sets $E \subset \mathbb{R}^d$ and all $\mu \in \mathcal{P}(E)$, there exist bounded open sets $E_1, E_2, \ldots$ such that:

1. $E_1 \supset E_2 \supset \cdots$;

2. $\cap_n \overline{E_n} = E$; and

3. for all $n \geq 1$ large, there exist absolutely continuous measures μ_n on $\overline{E_n}$ such that for all $\varepsilon > 0$, there exists N_0 such that for all $n \geq N_0$,

    (a) $\mu_n(\overline{E_n}) \geq 1 - \varepsilon$; and (b) $\mathfrak{G}\mu_n(x) \leq \mathfrak{G}\mu(x)$ for all $x \in \mathbb{R}^d$.

Roughly speaking, $\mathfrak{g}$ is a proper gauge function if the potential of a probability measure on a compact set can be approximated from *below* by the potential of an absolutely continuous measure on a fattening of E. Moreover, the latter measure can be taken to be close to a probability measure. Following J. H. Poincaré, we can think of $\mathfrak{G}\mu_n$ as a smoothing out of the potential $\mathfrak{G}\mu$ and refer to $\mathfrak{G}\mu_n$ as a **balayage** for $\mathfrak{G}\mu$.

Theorem 2.3.1 *Suppose ν is a measure on $\mathbb{R}^d$ and suppose that $\mathfrak{g}$ is a proper, symmetric gauge function on $\mathbb{R}^d \times \mathbb{R}^d$. Then, for all compact sets $E \subset \mathbb{R}^d$, $\mathcal{C}_\mathfrak{g}(E) = \mathcal{C}^{ac}_\mathfrak{g}(E)$.*

Proof According to Exercise 2.2.1, it suffices to show that for all compact sets $E \subset \mathbb{R}^d$, $\mathcal{C}_\mathfrak{g}(E) \leq \mathcal{C}^{ac}_\mathfrak{g}(E)$. For any measure $\mu \in \mathcal{P}(E)$, we can construct bounded open sets E_n such that as $n \to \infty$, $\overline{E_n} \downarrow E$. We can also find absolutely continuous measures μ_n on $\overline{E_n}$ such that for all n large, $\mu_n(\overline{E_n}) \leq 1 + \varepsilon$ and $\mathfrak{G}\mu_n \leq \mathfrak{G}\mu$, pointwise. We can apply Exercise 1.1.2 and the reciprocity theorem (Exercise 1.1.4) as follows:

$$\mathcal{E}_\mathfrak{g}(\mu) = \int \mathfrak{G}\mu(x)\,\mu(dx) \geq \int \mathfrak{G}\mu_n(x)\,\mu(dx)$$
$$= \int \mathfrak{G}\mu(x)\,\mu_n(dx) \geq \int \mathfrak{G}\mu_n(x)\,\mu_n(dx) = \mathcal{E}_\mathfrak{g}(\mu_n).$$

Define $\mu_n^\star(\bullet) = \mu_n(\bullet)/\mu_n(\overline{E_n})$ and note that $\mu_n^\star \in \mathcal{P}(\overline{E_n})$. Also note that

$$\mathcal{E}_\mathfrak{g}(\mu_n^\star) = \frac{1}{|\mu(\overline{E_n})|^2}\mathcal{E}_\mathfrak{g}(\mu_n),$$

which is less than or equal to $(1-\varepsilon)^{-2}\mathcal{E}_\mathfrak{g}(\mu_n)$, for all n large. Thus, we have shown that for all n large,

$$\mathcal{E}_\mathfrak{g}(\mu) \geq (1-\varepsilon)^2 \inf_\phi \mathcal{E}_\mathfrak{g}(\phi),$$

where the infimum is taken over all absolutely continuous $\phi \in \mathcal{P}(\overline{E_n})$. Since the above holds for any $\mu \in \mathcal{P}(E)$,

$$\inf_{\mu \in \mathcal{P}(E)} \mathcal{E}_\mathfrak{g}(\mu) \geq (1-\varepsilon)^2 \inf_\phi \mathcal{E}_\mathfrak{g}(\phi).$$

We can now invert this and see that for all n large, $\mathcal{C}_\mathfrak{g}(E)$ is bounded above by $(1-\varepsilon)^{-2}\mathcal{C}_\mathfrak{g}^{ac}(\overline{E_n})$. Consequently, Exercise 2.2.1 implies that $\mathcal{C}_\mathfrak{g}(E) \leq (1-\varepsilon)^{-2}\mathcal{C}_\mathfrak{g}^{ac}(E)$, which proves the theorem, since $\varepsilon > 0$ is arbitrary. □

3 Notes on Appendix D

Section 1 Two useful references are (Kellogg 1967; Wermer 1981); these treatments are not only rigorous and thorough, but they also contain lively discussion of the physical aspects of classical potential theory. While the brief discussions of this appendix form a minute portion of the history of the subject, it would be a shame to leave it out altogether. For two wonderful general references, see (Helms 1975; Landkof 1972).

Section 2 Capacities can be treated more generally and with less focus on compact sets. In particular, we have omitted the capacitability theorem of Gustav Choquet. In the present context it states that when $\mathfrak{g}$ is a symmetric gauge function, it is also inner regular. That is, for all measurable $E \subset \mathbb{R}^d$,

$$\mathcal{C}_\mathfrak{g}(E) = \sup\{\mathcal{C}_\mathfrak{g}(K) : K \subset E \text{ is compact}\};$$

see (Bass 1995; Dellacherie and Meyer 1978).

Some of the essential ideas behind this chapter's extremal characterizations of capacity (Theorem 2.1.2 and Exercise 2.1.3) date back to C.-F. Gauss. See Carleson (1983, Theorem 6, Section III, Chapter 1) and Itô and McKean (1974, Section 7.9, Chapter 7) for special cases, together with references to the older literature.

While maximum principles are powerful tools in probability and analysis, we have discussed them only briefly. The reason for this is that our sole intended application of maximum principles is in proving Theorem 2.1.1, which itself is never used in this book.

Carleson (1983) contains a delightful introduction to a class of energies and capacities, with applications to diverse areas in analysis, such as the boundary theory for differential equations, nontriviality of certain H^p spaces, and singular points for harmonic functions.

I have learned the terminology *gauge functions* from Yuval Peres.

Since capacities are introduced via mutual energy in this book, the notion of balayage becomes a somewhat different matter and naturally leads to the introduction of proper gauge functions and absolutely continuous capacities. A probabilist may find some appeal in this approach, due to its measure-theoretic (as opposed to potential-theoretic) nature.

References

Adler, R. J. (1977). Hausdorff dimension and Gaussian fields. *Ann. Probability* 5(1), 145–151.

Adler, R. J. (1980). A Hölder condition for the local time of the Brownian sheet. *Indiana Univ. Math. J.* 29(5), 793–798.

Adler, R. J. (1981). *The Geometry of Random Fields*. Chichester: John Wiley & Sons Ltd. Wiley Series in Probability and Mathematical Statistics.

Adler, R. J. (1990). *An Introduction to Continuity, Extrema, and Related Topics for General Gaussian Processes*. Hayward, CA: Institute of Mathematical Statistics.

Adler, R. J. and R. Pyke (1997). Scanning Brownian processes. *Adv. in Appl. Probab.* 29(2), 295–326.

Aizenman, M. (1985). The intersection of Brownian paths as a case study of a renormalization group method for quantum field theory. *Comm. Math. Phys.* 97(1-2), 91–110.

Bakry, D. (1979). Sur la régularité des trajectoires des martingales à deux indices. *Z. Wahrsch. Verw. Gebiete* 50(2), 149–157.

Bakry, D. (1981a). Limites "quadrantales" des martingales. In *Two-index random processes (Paris, 1980)*, pp. 40–49. Berlin: Springer.

Bakry, D. (1981b). Théorèmes de section et de projection pour les processus à deux indices. *Z. Wahrsch. Verw. Gebiete* 55(1), 55–71.

Bakry, D. (1982). Semimartingales à deux indices. *Ann. Sci. Univ. Clermont-Ferrand II Math.* (20), 53–54.

Barlow, M. T. (1988). Necessary and sufficient conditions for the continuity of local time of Lévy processes. *Ann. Probab.* 16(4), 1389–1427.

Barlow, M. T. and J. Hawkes (1985). Application de l'entropie métrique à la continuité des temps locaux des processus de Lévy. *C. R. Acad. Sci. Paris Sér. I Math. 301*(5), 237–239.

Barlow, M. T. and E. Perkins (1984). Levels at which every Brownian excursion is exceptional. In *Seminar on probability, XVIII*, pp. 1–28. Berlin: Springer.

Bass, R. (1987). L_p inequalities for functionals of Brownian motion. In *Séminaire de Probabilités, XXI*, pp. 206–217. Berlin: Springer.

Bass, R. F. (1985). Law of the iterated logarithm for set-indexed partial sum processes with finite variance. *Z. Wahrsch. Verw. Gebiete 70*(4), 591–608.

Bass, R. F. (1995). *Probabilistic Techniques in Analysis*. New York: Springer-Verlag.

Bass, R. F. (1998). *Diffusions and Elliptic Operators*. New York: Springer-Verlag.

Bass, R. F., K. Burdzy, and D. Khoshnevisan (1994). Intersection local time for points of infinite multiplicity. *Ann. Probab. 22*(2), 566–625.

Bass, R. F. and D. Khoshnevisan (1992a). Local times on curves and uniform invariance principles. *Probab. Theory Related Fields 92*(4), 465–492.

Bass, R. F. and D. Khoshnevisan (1992b). Stochastic calculus and the continuity of local times of Lévy processes. In *Séminaire de Probabilités, XXVI*, pp. 1–10. Berlin: Springer.

Bass, R. F. and D. Khoshnevisan (1993a). Intersection local times and Tanaka formulas. *Ann. Inst. H. Poincaré Probab. Statist. 29*(3), 419–451.

Bass, R. F. and D. Khoshnevisan (1993b). Rates of convergence to Brownian local time. *Stochastic Process. Appl. 47*(2), 197–213.

Bass, R. F. and D. Khoshnevisan (1993c). Strong approximations to Brownian local time. In *Seminar on Stochastic Processes, 1992 (Seattle, WA, 1992)*, pp. 43–65. Boston, MA: Birkhäuser Boston.

Bass, R. F. and D. Khoshnevisan (1995). Laws of the iterated logarithm for local times of the empirical process. *Ann. Probab. 23*(1), 388–399.

Bass, R. F. and R. Pyke (1984a). The existence of set-indexed Lévy processes. *Z. Wahrsch. Verw. Gebiete 66*(2), 157–172.

Bass, R. F. and R. Pyke (1984b). Functional law of the iterated logarithm and uniform central limit theorem for partial-sum processes indexed by sets. *Ann. Probab. 12*(1), 13–34.

Bass, R. F. and R. Pyke (1984c). A strong law of large numbers for partial-sum processes indexed by sets. *Ann. Probab. 12*(1), 268–271.

Bass, R. F. and R. Pyke (1985). The space $\mathcal{D}(A)$ and weak convergence for set-indexed processes. *Ann. Probab. 13*(3), 860–884.

Bauer, J. (1994). Multiparameter processes associated with Ornstein-Uhlenbeck semigroups. In *Classical and Modern Potential Theory and Applications (Chateau de Bonas, 1993)*, pp. 41–55. Dordrecht: Kluwer Acad. Publ.

Bendikov, A. (1994). Asymptotic formulas for symmetric stable semigroups. *Exposition. Math. 12*(4), 381–384.

Benjamini, I., R. Pemantle, and Y. Peres (1995). Martin capacity for Markov chains. *Ann. Probab. 23*(3), 1332–1346.

Bergström, H. (1952). On some expansions of stable distribution functions. *Ark. Mat. 2*, 375–378.

Berman, S. M. (1983). Local nondeterminism and local times of general stochastic processes. *Ann. Inst. H. Poincaré Sect. B (N.S.) 19*(2), 189–207.

Bertoin, J. (1996). *Lévy Processes.* Cambridge: Cambridge University Press.

Bickel, P. J. and M. J. Wichura (1971). Convergence criteria for multiparameter stochastic processes and some applications. *Ann. Math. Statist. 42*, 1656–1670.

Billingsley, P. (1968). *Convergence of Probability Measures.* New York: John Wiley & Sons Inc.

Billingsley, P. (1995). *Probability and Measure* (Third ed.). New York: John Wiley & Sons Inc. A Wiley-Interscience Publication.

Blackwell, D. and L. Dubins (1962). Merging of opinions with increasing information. *Ann. Math. Statist. 33*, 882–886.

Blackwell, D. and L. E. Dubins (1975). On existence and non-existence of proper, regular, conditional distributions. *Ann. Probability 3*(5), 741–752.

Blumenthal, R. M. (1957). An extended Markov property. *Trans. Amer. Math. Soc. 85*, 52–72.

Blumenthal, R. M. and R. K. Getoor (1960a). A dimension theorem for sample functions of stable processes. *Illinois J. Math. 4*, 370–375.

Blumenthal, R. M. and R. K. Getoor (1960b). Some theorems on stable processes. *Trans. Amer. Math. Soc. 95*, 263–273.

Blumenthal, R. M. and R. K. Getoor (1962). The dimension of the set of zeros and the graph of a symmetric stable process. *Illinois J. Math. 6*, 308–316.

Blumenthal, R. M. and R. K. Getoor (1968). *Markov Processes and Potential Theory.* New York: Academic Press. Pure and Applied Mathematics, Vol. 29.

Bochner, S. (1955). *Harmonic Analysis and the Theory of Probability.* Berkeley and Los Angeles: University of California Press.

Borodin, A. N. (1986). On the character of convergence to Brownian local time. I, II. *Probab. Theory Relat. Fields 72*(2), 231–250, 251–277.

Borodin, A. N. (1988). On the weak convergence to Brownian local time. In *Probability theory and mathematical statistics (Kyoto, 1986)*, pp. 55–63. Berlin: Springer.

Burkholder, D. L. (1962). Successive conditional expectations of an integrable function. *Ann. Math. Statist. 33*, 887–893.

Burkholder, D. L. (1964). Maximal inequalities as necessary conditions for almost everywhere convergence. *Z. Wahrscheinlichkeitstheorie und Verw. Gebiete 3*, 75–88 (1964).

Burkholder, D. L. (1973). Distribution function inequalities for martingales. *Ann. Probability 1*, 19–42.

Burkholder, D. L. (1975). One-sided maximal functions and H^p. *J. Functional Analysis 18*, 429–454.

Burkholder, D. L. and Y. S. Chow (1961). Iterates of conditional expectation operators. *Proc. Amer. Math. Soc. 12*, 490–495.

Burkholder, D. L., B. J. Davis, and R. F. Gundy (1972). Integral inequalities for convex functions of operators on martingales. In *Proceedings of the Sixth Berkeley Symposium on Mathematical Statistics and Probability (Univ. California, Berkeley, Calif., 1970/1971), Vol. II: Probability theory*, pp. 223–240. Berkeley, Calif.: Univ. California Press.

Burkholder, D. L. and R. F. Gundy (1972). Distribution function inequalities for the area integral. *Studia Math. 44*, 527–544. Collection of articles honoring the completion by Antoni Zygmund of 50 years of scientific activity, VI.

Cabaña, E. (1991). The Markov property of the Brownian sheet associated with its wave components. In *Mathématiques appliquées aux sciences de l'ingénieur (Santiago, 1989)*, pp. 103–120. Toulouse: Cépaduès.

Cairoli, R. (1966). Produits de semi-groupes de transition et produits de processus. *Publ. Inst. Statist. Univ. Paris 15*, 311–384.

Cairoli, R. (1969). Un théorème de convergence pour martingales à indices multiples. *C. R. Acad. Sci. Paris Sér. A-B 269*, A587–A589.

Cairoli, R. (1970a). Processus croissant naturel associé à une classe de processus à indices doubles. *C. R. Acad. Sci. Paris Sér. A-B 270*, A1604–A1606.

Cairoli, R. (1970b). Une inégalité pour martingales à indices multiples et ses applications. In *Séminaire de Probabilités, IV (Univ. Strasbourg, 1968/1969)*, pp. 1–27. Lecture Notes in Mathematics, Vol. 124. Springer, Berlin.

Cairoli, R. (1971). Décomposition de processus à indices doubles. In *Séminaire de Probabilités, V (Univ. Strasbourg, année universitaire 1969-1970)*, pp. 37–57. Lecture Notes in Math., Vol. 191. Berlin: Springer.

Cairoli, R. (1979). Sur la convergence des martingales indexées par $\mathbf{n} \times \mathbf{n}$. In *Séminaire de Probabilités, XIII (Univ. Strasbourg, Strasbourg, 1977/1978)*, pp. 162–173. Berlin: Springer.

Cairoli, R. and R. C. Dalang (1996). *Sequential Stochastic Optimization*. New York: John Wiley & Sons Inc. A Wiley-Interscience Publication.

Cairoli, R. and J.-P. Gabriel (1979). Arrêt de certaines suites multiples de variables aléatoires indépendantes. In *Séminaire de Probabilités, XIII (Univ. Strasbourg, Strasbourg, 1977/1978)*, pp. 174–198. Berlin: Springer.

Cairoli, R. and J. B. Walsh (1975). Stochastic integrals in the plane. *Acta Math. 134*, 111–183.

Cairoli, R. and J. B. Walsh (1977a). Martingale representations and holomorphic processes. *Ann. Probability 5*(4), 511–521.

Cairoli, R. and J. B. Walsh (1977b). Prolongement de processus holomorphes. Cas "carré intégrable". In *Séminaire de Probabilités, XI (Univ. Strasbourg, Strasbourg, 1975/1976)*, pp. 327–339. Lecture Notes in Math., Vol. 581. Berlin: Springer.

Cairoli, R. and J. B. Walsh (1977c). Some examples of holomorphic processes. In *Séminaire de Probabilités, XI (Univ. Strasbourg, Strasbourg, 1975/1976)*, pp. 340–348. Lecture Notes in Math., Vol. 581. Berlin: Springer.

Cairoli, R. and J. B. Walsh (1978). Régions d'arrêt, localisations et prolongements de martingales. *Z. Wahrsch. Verw. Gebiete 44*(4), 279–306.

Calderón, A. P. and A. Zygmund (1952). On the existence of certain singular integrals. *Acta Math. 88*, 85–139.

Cao, J. and K. Worsley (1999). The geometry of correlation fields with an application to functional connectivity of the brain. *Ann. Appl. Probab. 9*(4), 1021–1057.

Carleson, L. (1958). On the connection between Hausdorff measures and capacity. *Ark. Mat. 3*, 403–406.

Carleson, L. (1983). *Selected Problems on Exceptional Sets*. Belmont, CA: Wadsworth. Selected reprints.

Čentsov, N. N. (1956). Wiener random fields depending on several parameters. *Dokl. Akad. Nauk. S.S.S.R. (NS) 106*, 607–609.

Chatterji, S. D. (1967). Comments on the martingale convergence theorem. In *Symposium on Probability Methods in Analysis (Loutraki, 1966)*, pp. 55–61. Berlin: Springer.

Chatterji, S. D. (1968). Martingale convergence and the Radon-Nikodym theorem in Banach spaces. *Math. Scand. 22*, 21–41.

Chen, Z. L. (1997). Properties of the polar sets of Brownian sheets. *J. Math. (Wuhan) 17*(3), 373–378.

Chow, Y. S. and H. Teicher (1997). *Probability theory* (Third ed.). New York: Springer-Verlag. Independence, Interchangeability, Martingales.

Chung, K. L. (1948). On the maximum partial sums of sequences of independent random variables. *Trans. Amer. Math. Soc. 64*, 205–233.

Chung, K. L. (1974). *A Course in Probability Theory* (Second ed.). Academic Press, New York-London. Probability and Mathematical Statistics, Vol. 21.

Chung, K. L. and P. Erdős (1952). On the application of the Borel-Cantelli lemma. *Trans. Amer. Math. Soc. 72*, 179–186.

Chung, K. L. and W. H. J. Fuchs (1951). On the distribution of values of sums of random variables. *Mem. Amer. Math. Soc. 1951*(6), 12.

Chung, K. L. and G. A. Hunt (1949). On the zeros of $\sum_1^n \pm 1$. *Ann. of Math. (2) 50*, 385–400.

Chung, K. L. and D. Ornstein (1962). On the recurrence of sums of random variables. *Bull. Amer. Math. Soc. 68*, 30–32.

Chung, K. L. and J. B. Walsh (1969). To reverse a Markov process. *Acta Math. 123*, 225–251.

Chung, K. L. and R. J. Williams (1990). *Introduction to Stochastic Integration* (Second ed.). Boston, MA: Birkhäuser Boston Inc.

Ciesielski, Z. (1959). On Haar functions and on the Schauder basis of the space $C_{\langle 0, 1 \rangle}$. *Bull. Acad. Polon. Sci. Sér. Sci. Math. Astronom. Phys. 7*, 227–232.

Ciesielski, Z. (1961). Hölder conditions for realizations of Gaussian processes. *Trans. Amer. Math. Soc. 99*, 403–413.

Ciesielski, Z. and J. Musielak (1959). On absolute convergence of Haar series. *Colloq. Math. 7*, 61–65.

Ciesielski, Z. and S. J. Taylor (1962). First passage times and sojourn times for Brownian motion in space and the exact Hausdorff measure of the sample path. *Trans. Amer. Math. Soc. 103*, 434–450.

Csáki, E., M. Csörgő, A. Földes, and P. Révész (1989). Brownian local time approximated by a Wiener sheet. *Ann. Probab. 17*(2), 516–537.

Csáki, E., M. Csörgő, A. Földes, and P. Révész (1992). Strong approximation of additive functionals. *J. Theoret. Probab. 5*(4), 679–706.

Csáki, E., A. Földes, and Y. Kasahara (1988). Around Yor's theorem on the Brownian sheet and local time. *J. Math. Kyoto Univ. 28*(2), 373–381.

Csáki, E. and P. Révész (1983). Strong invariance for local times. *Z. Wahrsch. Verw. Gebiete 62*(2), 263–278.

Csörgő, M. and P. Révész (1978). How big are the increments of a multiparameter Wiener process? *Z. Wahrsch. Verw. Gebiete 42*(1), 1–12.

Csörgő, M. and P. Révész (1981). *Strong Approximations in Probability and Statistics*. New York: Academic Press Inc. [Harcourt Brace Jovanovich Publishers].

Csörgő, M. and P. Révész (1984). Three strong approximations of the local time of a Wiener process and their applications to invariance. In *Limit theorems in probability and statistics, Vol. I, II (Veszprém, 1982)*, pp. 223–254. Amsterdam: North-Holland.

Csörgő, M. and P. Révész (1985). On the stability of the local time of a symmetric random walk. *Acta Sci. Math. (Szeged) 48*(1-4), 85–96.

Csörgő, M. and P. Révész (1986). Mesure du voisinage and occupation density. *Probab. Theory Relat. Fields 73*(2), 211–226.

Dalang, R. C. and T. Mountford (1996). Nondifferentiability of curves on the Brownian sheet. *Ann. Probab. 24*(1), 182–195.

Dalang, R. C. and T. Mountford (1997). Points of increase of the Brownian sheet. *Probab. Theory Related Fields 108*(1), 1–27.

Dalang, R. C. and T. Mountford (2001). Jordan curves in the level sets of additive Brownian motion. *Trans. Amer. Math. Soc. 353*(9), 3531–3545 (electronic).

Dalang, R. C. and T. S. Mountford (2000). Level sets, bubbles and excursions of a Brownian sheet. In *Infinite dimensional stochastic analysis (Amsterdam, 1999)*, pp. 117–128. R. Neth. Acad. Arts Sci., Amsterdam.

Dalang, R. C. and J. B. Walsh (1992). The sharp Markov property of the Brownian sheet and related processes. *Acta Math. 168*(3-4), 153–218.

Dalang, R. C. and J. B. Walsh (1993). Geography of the level sets of the Brownian sheet. *Probab. Theory Related Fields 96*(2), 153–176.

Dalang, R. C. and J. B. Walsh (1996). Local structure of level sets of the Brownian sheet. In *Stochastic analysis: random fields and measure-valued processes (Ramat Gan, 1993/1995)*, pp. 57–64. Ramat Gan: Bar-Ilan Univ.

Davis, B. and T. S. Salisbury (1988). Connecting Brownian paths. *Ann. Probab. 16*(4), 1428–1457.

de Acosta, A. (1983). A new proof of the Hartman–Wintner law of the iterated logarithm. *Ann. Probab. 11*(2), 270–276.

de Acosta, A. and J. Kuelbs (1981). Some new results on the cluster set $c(\{S_n/a_n\})$ and the LIL. *Ann. Prob. 11*, 102–122.

Dellacherie, C. and P.-A. Meyer (1978). *Probabilities and Potential*. Amsterdam: North-Holland Publishing Co.

Dellacherie, C. and P.-A. Meyer (1982). *Probabilities and Potential. B.* Amsterdam: North-Holland Publishing Co. Theory of martingales, Translated from the French by J. P. Wilson.

Dellacherie, C. and P.-A. Meyer (1988). *Probabilities and Potential. C.* Amsterdam: North-Holland Publishing Co. Potential theory for discrete and continuous semigroups, Translated from the French by J. Norris.

Dembo, A. and O. Zeitouni (1998). *Large Deviations Techniques and Applications* (second ed.). Berlin: Springer.

Donsker, M. D. (1952). Justification and extension of Doob's heuristic approach to the Kolmogorov-Smirnov theorems. *Ann. Math. Statistics 23*, 277–281.

Doob, J. L. (1962/1963). A ratio operator limit theorem. *Z. Wahrscheinlichkeitstheorie und Verw. Gebiete 1*, 288–294.

Doob, J. L. (1990). *Stochastic Processes*. New York: John Wiley & Sons Inc. Reprint of the 1953 original, A Wiley-Interscience Publication.

Dorea, C. C. Y. (1982). A characterization of the multiparameter Wiener process and an application. *Proc. Amer. Math. Soc. 85*(2), 267–271.

Dorea, C. C. Y. (1983). A semigroup characterization of the multiparameter Wiener process. *Semigroup Forum 26*(3-4), 287–293.

Dozzi, M. (1988). On the local time of the multiparameter Wiener process and the asymptotic behaviour of an associated integral. *Stochastics 25*(3), 155–169.

Dozzi, M. (1989). *Stochastic Processes with a Multidimensional Parameter*. Harlow: Longman Scientific & Technical.

Dozzi, M. (1991). Two-parameter stochastic processes. In *Stochastic Processes and Related Topics (Georgenthal, 1990)*, pp. 17–43. Berlin: Akademie-Verlag.

Dubins, L. E. and J. Pitman (1980). A divergent, two-parameter, bounded martingale. *Proc. Amer. Math. Soc. 78*(3), 414–416.

Dudley, R. M. (1973). Sample functions of the Gaussian process. *Ann. Probability 1*(1), 66–103.

Dudley, R. M. (1984). A Course on Empirical Processes. In *École d'été de probabilités de Saint-Flour, XII—1982*, pp. 1–142. Berlin: Springer.

Dudley, R. M. (1989). *Real Analysis and Probability*. Pacific Grove, CA: Wadsworth & Brooks/Cole Advanced Books & Software.

Durrett, R. (1991). *Probability*. Pacific Grove, CA: Wadsworth & Brooks/Cole Advanced Books & Software. Theory and Examples.

Dvoretzky, A. and P. Erdős (1951). Some problems on random walk in space. In *Proceedings of the Second Berkeley Symposium on Mathematical Statistics and Probability, 1950.*, Berkeley and Los Angeles, pp. 353–367. University of California Press.

Dvoretzky, A., P. Erdős, and S. Kakutani (1950). Double points of paths of Brownian motion in n-space. *Acta Sci. Math. Szeged 12*(Leopoldo Fejer et Frederico Riesz LXX annos natis dedicatus, Pars B), 75–81.

Dvoretzky, A., P. Erdős, and S. Kakutani (1954). Multiple points of paths of Brownian motion in the plane. *Bull. Res. Council Israel 3*, 364–371.

Dvoretzky, A., P. Erdős, and S. Kakutani (1958). Points of multiplicity $\mathfrak{c}$ of plane Brownian paths. *Bull. Res. Council Israel Sect. F 7F*, 175–180 (1958).

Dvoretzky, A., P. Erdős, S. Kakutani, and S. J. Taylor (1957). Triple points of Brownian paths in 3-space. *Proc. Cambridge Philos. Soc. 53*, 856–862.

Dynkin, E. B. (1965). *Markov Processes. Vols. I, II*. Publishers, New York: Academic Press Inc. Translated with the authorization and assistance of the author by J. Fabius, V. Greenberg, A. Maitra, G. Majone. Die Grundlehren der Mathematischen Wissenschaften, Bände 121, 122.

Dynkin, E. B. (1980). Markov processes and random fields. *Bull. Amer. Math. Soc. (N.S.) 3*(3), 975–999.

Dynkin, E. B. (1981a). Additive functionals of several time-reversible Markov processes. *J. Funct. Anal. 42*(1), 64–101.

Dynkin, E. B. (1981b). Harmonic functions associated with several Markov processes. *Adv. in Appl. Math. 2*(3), 260–283.

Dynkin, E. B. (1983). Markov processes as a tool in field theory. *J. Funct. Anal. 50*(2), 167–187.

Dynkin, E. B. (1984a). Local times and quantum fields. In *Seminar on stochastic processes, 1983 (Gainesville, Fla., 1983)*, pp. 69–83. Boston, Mass.: Birkhäuser Boston.

Dynkin, E. B. (1984b). Polynomials of the occupation field and related random fields. *J. Funct. Anal. 58*(1), 20–52.

Dynkin, E. B. (1985). Random fields associated with multiple points of the Brownian motion. *J. Funct. Anal. 62*(3), 397–434.

Dynkin, E. B. (1986). Generalized random fields related to self-intersections of the Brownian motion. *Proc. Nat. Acad. Sci. U.S.A. 83*(11), 3575–3576.

Dynkin, E. B. (1987). Self-intersection local times, occupation fields, and stochastic integrals. *Adv. in Math. 65*(3), 254–271.

Dynkin, E. B. (1988). Self-intersection gauge for random walks and for Brownian motion. *Ann. Probab. 16*(1), 1–57.

Edgar, G. A. and L. Sucheston (1992). *Stopping Times and Directed Processes.* Cambridge: Cambridge University Press.

Ehm, W. (1981). Sample function properties of multiparameter stable processes. *Z. Wahrsch. Verw. Gebiete 56*(2), 195–228.

Eisenbaum, N. (1995). Une version sans conditionnement du théorème d'isomorphisms de Dynkin. In *Séminaire de Probabilités, XXIX*, pp. 266–289. Berlin: Springer.

Eisenbaum, N. (1997). Théorèmes limites pour les temps locaux d'un processus stable symétrique. In *Séminaire de Probabilités, XXXI*, pp. 216–224. Berlin: Springer.

Epstein, R. (1989). Some limit theorems for functionals of the Brownian sheet. *Ann. Probab. 17*(2), 538–558.

Erdős, P. (1942). On the law of the iterated logarithm. *Ann. Math. 43*, 419–436.

Erdős, P. and S. J. Taylor (1960a). Some intersection properties of random walk paths. *Acta Math. Acad. Sci. Hungar. 11*, 231–248.

Erdős, P. and S. J. Taylor (1960b). Some problems concerning the structure of random walk paths. *Acta Math. Acad. Sci. Hungar 11*, 137–162. (unbound insert).

Esquível, M. L. (1996). Points of rapid oscillation for the Brownian sheet via Fourier-Schauder series representation. In *Interaction between Functional Analysis, Harmonic Analysis, and Probability (Columbia, MO, 1994)*, pp. 153–162. New York: Dekker.

Etemadi, N. (1977). Collision problems of random walks in two-dimensional time. *J. Multivariate Anal. 7*(2), 249–264.

Etemadi, N. (1991). Maximal inequalities for partial sums of independent random vectors with multi-dimensional time parameters. *Comm. Statist. Theory Methods 20*(12), 3909–3923.

Ethier, S. N. (1998). An optional stopping theorem for nonadapted martingales. *Statist. Probab. Lett. 39*(3), 283–288.

Ethier, S. N. and T. G. Kurtz (1986). *Markov Processes Characterization and Convergence.* New York: John Wiley & Sons Inc.

Evans, S. N. (1987a). Multiple points in the sample paths of a Lévy process. *Probab. Theory Related Fields 76*(3), 359–367.

Evans, S. N. (1987b). Potential theory for a family of several Markov processes. *Ann. Inst. H. Poincaré Probab. Statist. 23*(3), 499–530.

Feller, W. (1968). *An Introduction to Probability Theory and Its Applications. Vol. I* (Third ed.). New York: John Wiley & Sons Inc.

Feller, W. (1968/1969). An extension of the law of the iterated logarithm to variables without variance. *J. Math. Mech. 18*, 343–355.

Feller, W. (1971). *An Introduction to Probability Theory and Its Applications. Vol. II.* (Second ed.). New York: John Wiley & Sons Inc.

Feynman, R. J. (1948). Space-time approach to nonrelativistic quantuum mechanics. *Rev. Mod. Phys. 20*, 367–387.

Fitzsimmons, P. J. and B. Maisonneuve (1986). Excessive measures and Markov processes with random birth and death. *Probab. Theory Relat. Fields 72*(3), 319–336.

Fitzsimmons, P. J. and J. Pitman (1999). Kac's moment formula and the Feynman-Kac formula for additive functionals of a Markov process. *Stochastic Process. Appl. 79*(1), 117–134.

Fitzsimmons, P. J. and S. C. Port (1990). Local times, occupation times, and the Lebesgue measure of the range of a Lévy process. In *Seminar on Stochastic Processes, 1989 (San Diego, CA, 1989)*, pp. 59–73. Boston, MA: Birkhäuser Boston.

Fitzsimmons, P. J. and T. S. Salisbury (1989). Capacity and energy for multiparameter Markov processes. *Ann. Inst. H. Poincaré Probab. Statist. 25*(3), 325–350.

Föllmer, H. (1984a). Almost sure convergence of multiparameter martingales for Markov random fields. *Ann. Probab. 12*(1), 133–140.

Föllmer, H. (1984b). Von der Brownschen Bewegung zum Brownschen Blatt: einige neuere Richtungen in der Theorie der stochastischen Prozesse. In *Perspectives in mathematics*, pp. 159–190. Basel: Birkhäuser.

Fouque, J.-P., K. J. Hochberg, and E. Merzbach (Eds.) (1996). *Stochastic Analysis: Random Fields and Measure-Valued Processes*. Ramat Gan: Bar-Ilan University Gelbart Research Institute for Mathematical Sciences. Papers from the Binational France-Israel Symposium on the Brownian Sheet, held September 1993, and the Conference on Measure-valued Branching and Superprocesses, held May 1995, at Bar-Ilan University, Ramat Gan.

Frangos, N. E. and L. Sucheston (1986). On multiparameter ergodic and martingale theorems in infinite measure spaces. *Probab. Theory Relat. Fields 71*(4), 477–490.

Fukushima, M., Y. Ōshima, and M. Takeda (1994). *Dirichlet Forms and Symmetric Markov Processes*. Berlin: Walter de Gruyter & Co.

Gabriel, J.-P. (1977). Martingales with a countable filtering index set. *Ann. Probability 5*(6), 888–898.

Gänssler, P. (1983). *Empirical Processes*. Hayward, Calif.: Institute of Mathematical Statistics.

Garsia, A. M. (1970). *Topics in Almost Everywhere Convergence*. Markham Publishing Co., Chicago, Ill. Lectures in Advanced Mathematics, 4.

Garsia, A. M. (1973). *Martingale Inequalities: Seminar Notes on Recent Progress*. W. A. Benjamin, Inc., Reading, Mass.–London–Amsterdam. Mathematics Lecture Notes Series.

Garsia, A. M., E. Rodemich, and H. Rumsey, Jr. (1970/1971). A real variable lemma and the continuity of paths of some Gaussian processes. *Indiana Univ. Math. J. 20*, 565–578.

Geman, D. and J. Horowitz (1980). Occupation densities. *Ann. Probab.* 8(1), 1–67.

Geman, D., J. Horowitz, and J. Rosen (1984). A local time analysis of intersections of Brownian paths in the plane. *Ann. Probab.* 12(1), 86–107.

Getoor, R. K. (1975). *Markov Processes: Ray Processes and Right Processes.* Berlin: Springer-Verlag. Lecture Notes in Mathematics, Vol. 440.

Getoor, R. K. (1979). Splitting times and shift functionals. *Z. Wahrsch. Verw. Gebiete* 47(1), 69–81.

Getoor, R. K. (1990). *Excessive Measures.* Boston, MA: Birkhäuser Boston, Inc.

Getoor, R. K. and J. Glover (1984). Riesz decompositions in Markov process theory. *Trans. Amer. Math. Soc.* 285(1), 107–132.

Griffin, P. and J. Kuelbs (1991). Some extensions of the LIL via self-normalizations. *Ann. Probab.* 19(1), 380–395.

Gundy, R. F. (1969). On the class $L \log L$, martingales, and singular integrals. *Studia Math.* 33, 109–118.

Gut, A. (1978/1979). Moments of the maximum of normed partial sums of random variables with multidimensional indices. *Z. Wahrsch. Verw. Gebiete* 46(2), 205–220.

Hall, P. and C. C. Heyde (1980). *Martingale Limit Theory and Its Application.* New York: Academic Press Inc. [Harcourt Brace Jovanovich Publishers]. Probability and Mathematical Statistics.

Hawkes, J. (1970). Polar sets, regular points and recurrent sets for the symmetric and increasing stable processes. *Bull. London Math. Soc.* 2, 53–59.

Hawkes, J. (1970/1971b). Some dimension theorems for the sample functions of stable processes. *Indiana Univ. Math. J.* 20, 733–738.

Hawkes, J. (1971a). On the Hausdorff dimension of the intersection of the range of a stable process with a Borel set. *Z. Wahrscheinlichkeitstheorie und Verw. Gebiete* 19, 90–102.

Hawkes, J. (1976/1977a). Intersections of Markov random sets. *Z. Wahrscheinlichkeitstheorie und Verw. Gebiete* 37(3), 243–251.

Hawkes, J. (1977b). Local properties of some Gaussian processes. *Z. Wahrscheinlichkeitstheorie und Verw. Gebiete* 40(4), 309–315.

Hawkes, J. (1978). Multiple points for symmetric Lévy processes. *Math. Proc. Cambridge Philos. Soc.* 83(1), 83–90.

Hawkes, J. (1979). Potential theory of Lévy processes. *Proc. London Math. Soc. (3)* 38(2), 335–352.

Helms, L. L. (1975). *Introduction to Potential Theory.* Robert E. Krieger Publishing Co., Huntington, N.Y. Reprint of the 1969 edition, Pure and Applied Mathematics, Vol. XXII.

Hendricks, W. J. (1972). Hausdorff dimension in a process with stable components—an interesting counterexample. *Ann. Math. Statist.* 43(2), 690–694.

Hendricks, W. J. (1973/1974). Multiple points for a process in R^2 with stable components. *Z. Wahrscheinlichkeitstheorie und Verw. Gebiete* 28, 113–128.

Hendricks, W. J. (1979). Multiple points for transient symmetric Lévy processes in $\mathbf{R}^d$. *Z. Wahrsch. Verw. Gebiete* 49(1), 13–21.

Hewitt, E. and L. J. Savage (1955). Symmetric measures on Cartesian products. *Trans. Amer. Math. Soc.* 80, 470–501.

Hille, E. (1958). On roots and logarithms of elements of a complex Banach algebra. *Math. Ann.* 136, 46–57.

Hille, E. and R. S. Phillips (1957). *Functional Analysis and Semi-Groups*. Providence, R. I.: American Mathematical Society. rev. ed, American Mathematical Society Colloquium Publications, vol. 31.

Hirsch, F. (1995). Potential theory related to some multiparameter processes. *Potential Anal.* 4(3), 245–267.

Hirsch, F. and S. Song (1995a). Markov properties of multiparameter processes and capacities. *Probab. Theory Related Fields* 103(1), 45–71.

Hirsch, F. and S. Song (1995b). Symmetric Skorohod topology on n-variable functions and hierarchical Markov properties of n-parameter processes. *Probab. Theory Related Fields* 103(1), 25–43.

Hirsch, F. and S. Song (1995c). Une inégalité maximale pour certains processus de Markov à plusieurs paramètres. I. *C. R. Acad. Sci. Paris Sér. I Math.* 320(6), 719–722.

Hirsch, F. and S. Song (1995d). Une inégalité maximale pour certains processus de Markov à plusieurs paramètres. II. *C. R. Acad. Sci. Paris Sér. I Math.* 320(7), 867–870.

Hirsch, F. and S. Q. Song (1994). Propriétés de Markov des processus à plusieurs paramètres et capacités. *C. R. Acad. Sci. Paris Sér. I Math.* 319(5), 483–488.

Hoeffding, W. (1960). The strong law of large numbers for U-statistics. *University of North Carolina Institute of Statistics, Mimeo. Series*.

Hunt, G. A. (1956a). Markoff processes and potentials. *Proc. Nat. Acad. Sci. U.S.A.* 42, 414–418.

Hunt, G. A. (1956b). Semi-groups of measures on Lie groups. *Trans. Amer. Math. Soc.* 81, 264–293.

Hunt, G. A. (1957). Markoff processes and potentials. I, II. *Illinois J. Math.* 1, 44–93, 316–369.

Hunt, G. A. (1958). Markoff processes and potentials. III. *Illinois J. Math.* 2, 151–213.

Hunt, G. A. (1966). *Martingales et Processus de Markov*. Paris: Dunod. Monographies de la société Mathématique de France, No. 1.

Hürzeler, H. E. (1985). The optional sampling theorem for processes indexed by a partially ordered set. *Ann. Probab.* 13(4), 1224–1235.

Imkeller, P. (1984). Local times for a class of multiparameter processes. *Stochastics* 12(2), 143–157.

Imkeller, P. (1985). A stochastic calculus for continuous N-parameter strong martingales. *Stochastic Process. Appl.* 20(1), 1–40.

Imkeller, P. (1986). Local times of continuous N-parameter strong martingales. *J. Multivariate Anal.* 19(2), 348–365.

Imkeller, P. (1988). *Two-Parameter Martingales and Their Quadratic Variation.* Berlin: Springer-Verlag.

Itô, K. (1944). Stochastic integral. *Proc. Imp. Acad. Tokyo 20*, 519–524.

Itô, K. (1984). *Lectures on Stochastic Processes* (Second ed.). Distributed for the Tata Institute of Fundamental Research, Bombay. Notes by K. Muralidhara Rao.

Itô, K. and J. McKean, H. P. (1960). Potentials and the random walk. *Illinois J. Math. 4*, 119–132.

Itô, K. and J. McKean, Henry P. (1974). *Diffusion Processes and Their Sample Paths.* Berlin: Springer-Verlag. Second printing, corrected, Die Grundlehren der mathematischen Wissenschaften, Band 125.

Ivanoff, G. and E. Merzbach (2000). *Set-Indexed Martingales.* Chapman & Hall/CRC, Boca Raton, FL.

Ivanova, B. G. and E. Mertsbakh (1992). Set-indexed stochastic processes and predictability. *Teor. Veroyatnost. i Primenen.* 37(1), 57–63.

Jacod, J. (1998). Rates of convergence to the local time of a diffusion. *Ann. Inst. H. Poincaré Probab. Statist.* 34(4), 505–544.

Janke, S. J. (1985). Recurrent sets for transient Lévy processes with bounded kernels. *Ann. Probab.* 13(4), 1204–1218.

Janson, S. (1997). *Gaussian Hilbert Spaces.* Cambridge: Cambridge University Press.

Kac, M. (1949). On deviations between theoretical and empirical distributions. *Proceedings of the National Academy of Sciences, U.S.A.* 35, 252–257.

Kac, M. (1951). On some connections between probability theory and differential and integral equations. In *Proceedings of the Second Berkeley Symposium on Mathematical Statistics and Probability, 1950*, Berkeley and Los Angeles, pp. 189–215. University of California Press.

Kahane, J.-P. (1982). Points multiples du mouvement brownien et des processus de Lévy symétriques, restreints à un ensemble compact de valeurs du temps. *C. R. Acad. Sci. Paris Sér. I Math.* 295(9), 531–534.

Kahane, J.-P. (1983). Points multiples des processus de Lévy symétriques stables restreints à un ensemble de valeurs du temps. In *Seminar on Harmonic Analysis, 1981–1982*, pp. 74–105. Orsay: Univ. Paris XI.

Kahane, J.-P. (1985). *Some Random Series of Functions* (Second ed.). Cambridge: Cambridge University Press.

Kakutani, S. (1944a). On Brownian motions in n-space. *Proc. Imp. Acad. Tokyo 20*, 648–652.

Kakutani, S. (1944b). Two-dimensional Brownian motion and harmonic functions. *Proc. Imp. Acad. Tokyo 20*, 706–714.

Kakutani, S. (1945). Markoff process and the Dirichlet problem. *Proc. Japan Acad. 21*, 227–233 (1949).

Kanda, M. (1982). Notes on polar sets for Lévy processes on the line. In *Functional analysis in Markov processes (Katata/Kyoto, 1981)*, pp. 227–234. Berlin: Springer.

Kanda, M. (1983). On the class of polar sets for a certain class of Lévy processes on the line. *J. Math. Soc. Japan 35*(2), 221–242.

Kanda, M. and M. Uehara (1981). On the class of polar sets for symmetric Lévy processes on the line. *Z. Wahrsch. Verw. Gebiete 58*(1), 55–67.

Karatsuba, A. A. (1993). *Basic Analytic Number Theory*. Berlin: Springer-Verlag. Translated from the second (1983) Russian edition and with a preface by Melvyn B. Nathanson.

Karatzas, I. and S. E. Shreve (1991). *Brownian Motion and Stochastic Calculus* (Second ed.). New York: Springer-Verlag.

Kargapolov, M. I. and J. I. Merzljakov (1979). *Fundamentals of the Theory of Groups*. New York: Springer-Verlag. Translated from the second Russian edition by Robert G. Burns.

Kellogg, O. D. (1967). *Foundations of Potential Theory*. Berlin: Springer-Verlag. Reprint from the first edition of 1929. Die Grundlehren der Mathematischen Wissenschaften, Band 31.

Kendall, W. S. (1980). Contours of Brownian processes with several-dimensional times. *Z. Wahrsch. Verw. Gebiete 52*(3), 267–276.

Kesten, H. and F. Spitzer (1979). A limit theorem related to a new class of self-similar processes. *Z. Wahrsch. Verw. Gebiete 50*(1), 5–25.

Khoshnevisan, D. (1992). Level crossings of the empirical process. *Stochastic Process. Appl. 43*(2), 331–343.

Khoshnevisan, D. (1993). An embedding of compensated compound Poisson processes with applications to local times. *Ann. Probab. 21*(1), 340–361.

Khoshnevisan, D. (1994). A discrete fractal in $\mathbb{Z}_+^1$. *Proc. Amer. Math. Soc. 120*(2), 577–584.

Khoshnevisan, D. (1995). On the distribution of bubbles of the Brownian sheet. *Ann. Probab. 23*(2), 786–805.

Khoshnevisan, D. (1997a). Escape rates for Lévy processes. *Studia Sci. Math. Hungar. 33*(1-3), 177–183.

Khoshnevisan, D. (1997b). Some polar sets for the Brownian sheet. In *Séminaire de Probabilités, XXXI*, pp. 190–197. Berlin: Springer.

Khoshnevisan, D. (1999). Brownian sheet images and Bessel–Riesz capacity. *Trans. Amer. Math. Soc. 351*(7), 2607–2622.

Khoshnevisan, D. (2000). On sums of i.i.d. random variables indexed by N parameters. In *Séminaire de Probabilités, XXXIV*, pp. 151–156. Berlin: Springer.

Khoshnevisan, D. and T. M. Lewis (1998). A law of the iterated logarithm for stable processes in random scenery. *Stochastic Process. Appl. 74*(1), 89–121.

Khoshnevisan, D. and Z. Shi (1999). Brownian sheet and capacity. *Ann. Probab.* *27*(3), 1135–1159.

Khoshnevisan, D. and Z. Shi (2000). Fast sets and points for fractional Brownian motion. In *Séminaire de Probabilités, XXXIV*, pp. 393–416. Berlin: Springer.

Khoshnevisan, D. and Y. Xiao (2002). Level sets of additive Lévy processes. *Ann. Probab. (To appear)*.

Kinney, J. R. (1953). Continuity properties of sample functions of Markov processes. *Trans. Amer. Math. Soc.* *74*, 280–302.

Kitagawa, T. (1951). Analysis of variance applied to function spaces. *Mem. Fac. Sci. Kyūsyū Univ. A.* *6*, 41–53.

Knight, F. B. (1981). *Essentials of Brownian motion and Diffusion*. Providence, R.I.: American Mathematical Society.

Kochen, S. and C. Stone (1964). A note on the Borel–Cantelli lemma. *Illinois J. Math.* *8*, 248–251.

Körezlioğlu, H., G. Mazziotto, and J. Szpirglas (Eds.) (1981). *Processus Aléatoires À Deux Indices*. Berlin: Springer. Papers from the E.N.S.T.-C.N.E.T. Colloquium held in Paris, June 30–July 1, 1980.

Krengel, U. and R. Pyke (1987). Uniform pointwise ergodic theorems for classes of averaging sets and multiparameter subadditive processes. *Stochastic Process. Appl.* *26*(2), 289–296.

Krickeberg, K. (1963). *Wahrscheinlichkeitstheorie*. B. G. Teubner Verlagsgesellschaft, Stuttgart.

Krickeberg, K. (1965). *Probability Theory*. Addison-Wesley Publishing Co., Inc., Reading, Mass.–London.

Krickeberg, K. and C. Pauc (1963). Martingales et dérivation. *Bull. Soc. Math. France* *91*, 455–543.

Kunita, H. and S. Watanabe (1967). On square integrable martingales. *Nagoya Math. J.* *30*, 209–245.

Kuroda, K. and H. Manaka (1987). The interface of the Ising model and the Brownian sheet. In *Proceedings of the symposium on statistical mechanics of phase transitions—mathematical and physical aspects (Trebon, 1986)*, Volume 47, pp. 979–984.

Kuroda, K. and H. Manaka (1998). Limit theorem related to an interface of three-dimensional Ising model. *Kobe J. Math.* *15*(1), 17–39.

Kuroda, K. and H. Tanemura (1988). Interacting particle system and Brownian sheet. *Keio Sci. Tech. Rep.* *41*(1), 1–16.

Kwon, J. S. (1994). The law of large numbers for product partial sum processes indexed by sets. *J. Multivariate Anal.* *49*(1), 76–86.

Lacey, M. T. (1990). Limit laws for local times of the Brownian sheet. *Probab. Theory Related Fields* *86*(1), 63–85.

Lachout, P. (1988). Billingsley-type tightness criteria for multiparameter stochastic processes. *Kybernetika (Prague)* *24*(5), 363–371.

Lamb, C. W. (1973). A short proof of the martingale convergence theorem. *Proc. Amer. Math. Soc. 38*, 215–217.

Landkof, N. S. (1972). *Foundations of Modern Potential Theory*. New York: Springer-Verlag. Translated from the Russian by A. P. Doohovskoy, Die Grundlehren der mathematischen Wissenschaften, Band 180.

Lawler, G. F. (1991). *Intersections of Random Walks*. Boston, MA: Birkhäuser Boston, Inc.

Le Gall, J.-F. (1987). Temps locaux d'intersection et points multiples des processus de Lévy. In *Séminaire de Probabilités, XXI*, pp. 341–374. Berlin: Springer.

Le Gall, J.-F. (1992). Some Properties of Planar Brownian Motion. In *École d'Été de Probabilités de Saint-Flour XX—1990*, pp. 111–235. Berlin: Springer.

Le Gall, J.-F. and J. Rosen (1991). The range of stable random walks. *Ann. Probab. 19*(2), 650–705.

Le Gall, J.-F., J. S. Rosen, and N.-R. Shieh (1989). Multiple points of Lévy processes. *Ann. Probab. 17*(2), 503–515.

LeCam, L. (1957). Convergence in distribution of stochastic processes. *University of California Pub. Stat. 2*(2).

Ledoux, M. (1981). Classe $L \log L$ et martingales fortes à paramètre bidimensionnel. *Ann. Inst. H. Poincaré Sect. B (N.S.) 17*(3), 275–280.

Ledoux, M. (1996). *Lectures on Probability Theory and Statistics*. Berlin: Springer-Verlag. Lectures from the 24th Saint-Flour Summer School held July 7–23, 1994, Edited by P. Bernard (with Dobrushin, R. and Groeneboom, P.).

Ledoux, M. and M. Talagrand (1991). *Probability in Banach Spaces*. Berlin: Springer-Verlag.

Lévy, P. (1965). *Processus stochastiques et mouvement brownien*. Gauthier-Villars & Cie, Paris. Suivi d'une note de M. Loève. Deuxième édition revue et augmentée.

Li, D. L. and Z. Q. Wu (1989). The law of the iterated logarithm for B-valued random variables with multidimensional indices. *Ann. Probab. 17*(2), 760–774.

Lyons, R. (1990). Random walks and percolation on trees. *Ann. Probab. 18*(3), 931–958.

Madras, N. and G. Slade (1993). *The Self-Avoiding Walk*. Boston, MA: Birkhäuser Boston Inc.

Mandelbrot, B. B. (1982). *The Fractal Geometry of Nature*. San Francisco, Calif.: W. H. Freeman and Co. Schriftenreihe für den Referenten. [Series for the Referee].

Marcus, M. B. and J. Rosen (1992). Sample path properties of the local times of strongly symmetric Markov processes via Gaussian processes. *Ann. Probab. 20*(4), 1603–1684.

Mattila, P. (1995). *Geometry of Sets and Measures in Euclidean Spaces*. Cambridge: Cambridge University Press.

Mazziotto, G. (1988). Two-parameter Hunt processes and a potential theory. *Ann. Probab. 16*(2), 600–619.

Mazziotto, G. and E. Merzbach (1985). Regularity and decomposition of two-parameter supermartingales. *J. Multivariate Anal. 17*(1), 38–55.

Mazziotto, G. and J. Szpirglas (1981). Un exemple de processus à deux indices sans l'hypothèse F4. In *Seminar on Probability, XV (Univ. Strasbourg, Strasbourg, 1979/1980) (French)*, pp. 673–688. Berlin: Springer.

Mazziotto, G. and J. Szpirglas (1982). Optimal stopping for two-parameter processes. In *Advances in filtering and optimal stochastic control (Cocoyoc, 1982)*, pp. 239–245. Berlin: Springer.

Mazziotto, G. and J. Szpirglas (1983). Arrêt optimal sur le plan. *Z. Wahrsch. Verw. Gebiete 62*(2), 215–233.

McKean, Henry P., J. (1955a). Hausdorff–Besicovitch dimension of Brownian motion paths. *Duke Math. J. 22*, 229–234.

McKean, Henry P., J. (1955b). Sample functions of stable processes. *Ann. of Math. (2) 61*, 564–579.

Merzbach, E. and D. Nualart (1985). Different kinds of two-parameter martingales. *Israel J. Math. 52*(3), 193–208.

Millar, P. W. (1978). A path decomposition for Markov processes. *Ann. Probability 6*(2), 345–348.

Mountford, T. S. (1993). Estimates of the Hausdorff dimension of the boundary of positive Brownian sheet components. In *Séminaire de Probabilités, XXVII*, pp. 233–255. Berlin: Springer.

Mountford, T. S. (2002). Brownian bubbles and the local time of the Brownian sheet. (In progress).

Munkres, J. R. (1975). *Topology: A First Course*. Englewood Cliffs, N.J.: Prentice-Hall Inc.

Muroga, S. (1949). On the capacity of a discrete channel. *J. Phys. Soc. Japan 8*, 484–494.

Nagasawa, M. (1964). Time reversions of Markov processes. *Nagoya Math. J. 24*, 177–204.

Neveu, J. (1975). *Discrete-Parameter Martingales* (Revised ed.). Amsterdam: North-Holland Publishing Co. Translated from the French by T. P. Speed, North-Holland Mathematical Library, Vol. 10.

Nualart, D. (1985). Variations quadratiques et inégalités pour les martingales à deux indices. *Stochastics 15*(1), 51–63.

Nualart, D. (1995). *The Malliavin Calculus and Related Topics*. New York: Springer-Verlag.

Nualart, E. (2001a). Ph. D. thesis (in preparation). *Ecole Polytechnique Fèderále de Lausanne*.

Nualart, E. (2001b). Potential theory for hyperbolic SPDEs. *Preprint*.

Orey, S. (1967). Polar sets for processes with stationary independent increments. In *Markov Processes and Potential Theory*, pp. 117–126. New York: Wiley. Proc. Sympos. Math. Res. Center, Madison, Wis., 1967.

Orey, S. and W. E. Pruitt (1973). Sample functions of the N-parameter Wiener process. *Ann. Probability* 1(1), 138–163.

Ornstein, D. S. (1969). Random walks. I, II. *Trans. Amer. Math. Soc. 138 (1969), 1–43; ibid. 138*, 45–60.

Oxtoby, J. C. and S. Ulam (1939). On the existence of a measure invariant under a transformation. *Ann. Math. 40*(2), 560–566.

Paley, R. E. A. C. and A. Zygmund (1932). A note on analytic functions in the unit circle. *Proc. Camb. Phil. Soc. 28*, 366–272.

Paranjape, S. R. and C. Park (1973). Laws of iterated logarithm of multiparameter Wiener processes. *J. Multivariate Anal. 3*, 132–136.

Park, W. J. (1975). The law of the iterated logarithm for Brownian sheets. *J. Appl. Probability 12*(4), 840–844.

Pemantle, R., Y. Peres, and J. W. Shapiro (1996). The trace of spatial Brownian motion is capacity-equivalent to the unit square. *Probab. Theory Related Fields 106*(3), 379–399.

Peres, Y. (1996a). Intersection-equivalence of Brownian paths and certain branching processes. *Comm. Math. Phys. 177*(2), 417–434.

Peres, Y. (1996b). Remarks on intersection-equivalence and capacity-equivalence. *Ann. Inst. H. Poincaré Phys. Théor. 64*(3), 339–347.

Perkins, E. (1982). Weak invariance principles for local time. *Z. Wahrsch. Verw. Gebiete 60*(4), 437–451.

Petrov, V. V. (1995). *Limit Theorems of Probability*. Oxford: Oxford Univ. Press.

Pollard, D. (1984). *Convergence of Stochastic Processes*. New York: Springer-Verlag.

Pólya, G. (1921). Über eine Aufgabe der Wahrscheinlichkeitsrechnung betreffend der Irrfahrt im Straßennetz. *Math. Annalen 84*, 149–160.

Pólya, G. (1923). On the zeros of an integral function represented by Fourier's intergal. *Mess. of Math. 52*, 185–188.

Port, S. C. (1988). Occupation time and the Lebesgue measure of the range for a Lévy process. *Proc. Amer. Math. Soc. 103*(4), 1241–1248.

Port, S. C. and C. J. Stone (1971a). Infinitely divisible processes and their potential theory, I. *Ann. Inst. Fourier (Grenoble) 21*(2), 157–275.

Port, S. C. and C. J. Stone (1971b). Infinitely divisible processes and their potential theory, II. *Ann. Inst. Fourier (Grenoble) 21*(4), 179–265.

Pruitt, W. E. (1969/1970). The Hausdorff dimension of the range of a process with stationary independent increments. *J. Math. Mech. 19*, 371–378.

Pruitt, W. E. (1975). Some dimension results for processes with independent increments. In *Stochastic processes and related topics (Proc. Summer Res. Inst. on Statist. Inference for Stochastic Processes, Indiana Univ., Bloomington, Ind., 1974, Vol. 1; dedicated to Jerzy Neyman)*, New York, pp. 133–165. Academic Press.

Pruitt, W. E. and S. J. Taylor (1969). The potential kernel and hitting probabilities for the general stable process in $\mathbf{R}^N$. *Trans. Amer. Math. Soc. 146*, 299–321.

Pyke, R. (1973). Partial sums of matrix arrays, and Brownian sheets. In *Stochastic analysis (a tribute to the memory of Rollo Davidson)*, pp. 331–348. London: Wiley.

Pyke, R. (1985). Opportunities for set-indexed empirical and quantile processes in inference. In *Proceedings of the 45th session of the International Statistical Institute, Vol. 4 (Amsterdam, 1985)*, Volume 51, pp. No. 25.2, 11.

Rao, M. (1987). On polar sets for Lévy processes. *J. London Math. Soc. (2) 35*(3), 569–576.

Ren, J. G. (1990). Topologie *p*-fine sur l'espace de Wiener et théorème des fonctions implicites. *Bull. Sci. Math. (2) 114*(2), 99–114.

Révész, P. (1981). A strong invariance principle of the local time of RVs with continuous distribution. *Studia Sci. Math. Hungar. 16*(1-2), 219–228.

Révész, P. (1990). *Random Walk in Random and Nonrandom Environments*. Teaneck, NJ: World Scientific Publishing Co. Inc.

Revuz, D. (1984). *Markov Chains* (Second ed.). Amsterdam: North-Holland Publishing Co.

Revuz, D. and M. Yor (1994). *Continuous Martingales and Brownian Motion* (Second ed.). Berlin: Springer-Verlag.

Ricci, F. and E. M. Stein (1992). Multiparameter singular integrals and maximal functions. *Ann. Inst. Fourier (Grenoble) 42*(3), 637–670.

Riesz, F. and B. Sz.-Nagy (1955). *Functional Analysis*. New York: Fredrick Ungar Publishing Company. Seventh Printing, 1978. Translated from the second French edition by Leo F. Boron.

Rogers, C. A. and S. J. Taylor (1962). On the law of the iterated logarithm. *J. London Math. Soc. 37*, 145–151.

Rogers, L. C. G. (1989). Multiple points of Markov processes in a complete metric space. In *Séminaire de Probabilités, XXIII*, pp. 186–197. Berlin: Springer.

Rogers, L. C. G. and D. Williams (1987). *Diffusions, Markov Processes, and Martingales. Vol. 2*. New York: John Wiley & Sons Inc. Itô calculus.

Rogers, L. C. G. and D. Williams (1994). *Diffusions, Markov Processes, and Martingales. Vol. 1* (Second ed.). Chichester: John Wiley & Sons Ltd. Foundations.

Rosen, J. (1983). A local time approach to the self-intersections of Brownian paths in space. *Comm. Math. Phys. 88*(3), 327–338.

Rosen, J. (1984). Self-intersections of random fields. *Ann. Probab.* 12(1), 108–119.

Rosen, J. (1991). Second order limit laws for the local times of stable processes. In *Séminaire de Probabilités, XXV*, pp. 407–424. Berlin: Springer.

Rosen, J. (1993). Uniform invariance principles for intersection local times. In *Seminar on Stochastic Processes, 1992 (Seattle, WA, 1992)*, pp. 241–247. Boston, MA: Birkhäuser Boston.

Rota, G.-C. (1962). An "Alternierende Verfahren" for general positive operators. *Bull. Amer. Math. Soc.* 68, 95–102.

Rota, G.-C. (1969). Syposium on Ergodic Theory, Tulane University. Presented in October 1969.

Royden, H. L. (1968). *Real Analysis* (Second ed.). New York: Macmillan Publishing Company.

Rozanov, Yu. A. (1982). *Markov Random Fields*. New York: Springer-Verlag. Translated from the Russian by Constance M. Elson.

Rudin, W. (1973). *Functional Analysis* (First ed.). New York: McGraw-Hill Inc.

Rudin, W. (1974). *Real and complex analysis* (Third ed.). New York: McGraw-Hill Book Co.

Salisbury, T. S. (1988). Brownian bitransforms. In *Seminar on Stochastic Processes, 1987 (Princeton, NJ, 1987)*, pp. 249–263. Boston, MA: Birkhäuser Boston.

Salisbury, T. S. (1992). A low intensity maximum principle for bi-Brownian motion. *Illinois J. Math.* 36(1), 1–14.

Salisbury, T. S. (1996). Energy, and intersections of Markov chains. In *Random discrete structures (Minneapolis, MN, 1993)*, pp. 213–225. New York: Springer.

Sato, K. (1999). *Lévy Processes and Infinitely Divisble Processes*. Cambridge: Cambridge University Press.

Segal, I. E. (1954). Abstract probability spaces and a theorem of Kolmogoroff. *Amer. J. Math.* 76, 721–732.

Serfling, R. J. (1980). *Approximation Theorems of Mathematical Statistics*. New York: John Wiley & Sons Inc. Wiley Series in Probability and Mathematical Statistics.

Shannon, C. E. (1948). A mathematical theory of communication. *Bell System Tech. J.* 27, 379–423, 623–656.

Shannon, C. E. and W. Weaver (1949). *A Mathematical Theory of Communication*. Urbana, Ill.: University of Illinois Press.

Sharpe, M. (1988). *General Theory of Markov Processes*. Boston, MA: Academic Press Inc.

Shieh, N. R. (1982). Strong differentiation and martingales in product spaces. *Math. Rep. Toyama Univ.* 5, 29–36.

Shorack, G. R. and R. T. Smythe (1976). Inequalities for max $\mid S_{\mathbf{k}} \mid /b_{\mathbf{k}}$ where $\mathbf{k} \in \mathbf{N}^r$. *Proc. Amer. Math. Soc.* 54, 331–336.

Slepian, D. (1962). The one-sided barrier problem for Gaussian noise. *Bell System Tech. J. 41*, 463–501.

Smythe, R. T. (1973). Strong laws of large numbers for r-dimensional arrays of random variables. *Ann. Probability 1*(1), 164–170.

Smythe, R. T. (1974a). Convergence de sommes de variables aléatoires indicées par des ensembles partiellement ordonnés. *Ann. Sci. Univ. Clermont 51*(9), 43–46. Colloque Consacré au Calcul des Probabilités (Univ. Clermont, Clermont-Ferrand, 1973).

Smythe, R. T. (1974b). Sums of independent random variables on partially ordered sets. *Ann. Probability 2*, 906–917.

Song, R. G. (1988). Optimal stopping for general stochastic processes indexed by a lattice-ordered set. *Acta Math. Sci. (English Ed.) 8*(3), 293–306.

Spitzer, F. (1958). Some theorems concerning 2-dimensional Brownian motion. *Trans. Amer. Math. Soc. 87*, 187–197.

Spitzer, F. (1964). *Principles of Random Walk*. D. Van Nostrand Co., Inc., Princeton, N.J.–Toronto–London. The University Series in Higher Mathematics.

Stein, E. M. (1961). On the maximal ergodic theorem. *Proc. Nat. Acad. Sci. U.S.A. 47*, 1894–1897.

Stein, E. M. (1970). *Singular Integrals and Differentiability Properties of Functions*. Princeton, N.J.: Princeton University Press. Princeton Mathematical Series, No. 30.

Stein, E. M. (1993). *Harmonic Analysis: Real-Variable Methods, Orthogonality, and Oscillatory Integrals*. Princeton, NJ: Princeton University Press. With the assistance of Timothy S. Murphy, Monographs in Harmonic Analysis, III.

Stein, E. M. and G. Weiss (1971). *Introduction to Fourier Analysis on Euclidean Spaces*. Princeton, N.J.: Princeton University Press. Princeton Mathematical Series, No. 32.

Stoll, A. (1987). Self-repellent random walks and polymer measures in two dimensions. In *Stochastic processes—mathematics and physics, II (Bielefeld, 1985)*, pp. 298–318. Berlin: Springer.

Stoll, A. (1989). Invariance principles for Brownian intersection local time and polymer measures. *Math. Scand. 64*(1), 133–160.

Stone, C. J. (1969). On the potential operator for one-dimensional recurrent random walks. *Trans. Amer. Math. Soc. 136*, 413–426.

Stout, W. F. (1974). *Almost Sure Convergence*. Academic Press [a subsidiary of Harcourt Brace Jovanovich, Publishers], New York-London. Probability and Mathematical Statistics, Vol. 24.

Strassen, V. (1965/1966). A converse to the law of the iterated logarithm. *Z. Wahrscheinlichkeitstheorie und Verw. Gebiete 4*, 265–268.

Stratonovich, R. L. (1966). A new representation for stochastic integrals and equations. *SIAM J. of Control 4*, 362–371.

Stroock, D. W. (1993). *Probability Theory, an Analytic View*. Cambridge: Cambridge University Press.

Sucheston, L. (1983). On one-parameter proofs of almost sure convergence of multiparameter processes. *Z. Wahrsch. Verw. Gebiete 63*(1), 43–49.

Takeuchi, J. (1964a). A local asymptotic law for the transient stable process. *Proc. Japan Acad. 40*, 141–144.

Takeuchi, J. (1964b). On the sample paths of the symmetric stable processes in spaces. *J. Math. Soc. Japan 16*, 109–127.

Takeuchi, J. and S. Watanabe (1964). Spitzer's test for the Cauchy process on the line. *Z. Wahrscheinlichkeitstheorie und Verw. Gebiete 3*, 204–210 (1964).

Taylor, S. J. (1953). The Hausdorff α-dimensional measure of Brownian paths in n-space. *Proc. Cambridge Philos. Soc. 49*, 31–39.

Taylor, S. J. (1955). The α-dimensional measure of the graph and set of zeros of a Brownian path. *Proc. Cambridge Philos. Soc. 51*, 265–274.

Taylor, S. J. (1961). On the connexion between Hausdorff measures and generalized capacity. *Proc. Cambridge Philos. Soc. 57*, 524–531.

Taylor, S. J. (1966). Multiple points for the sample paths of the symmetric stable process. *Z. Wahrscheinlichkeitstheorie und Verw. Gebiete 5*, 247–264.

Taylor, S. J. (1973). Sample path properties of processes with stationary independent increments. In *Stochastic analysis (a tribute to the memory of Rollo Davidson)*, pp. 387–414. London: Wiley.

Taylor, S. J. (1986). The measure theory of random fractals. *Math. Proc. Cambridge Philos. Soc. 100*(3), 383–406.

Testard, F. (1985). Quelques propriétés géométriques de certains processus gaussiens. *C. R. Acad. Sci. Paris Sér. I Math. 300*(14), 497–500.

Testard, F. (1986). Dimension asymétrique et ensembles doublement non polaires. *C. R. Acad. Sci. Paris Sér. I Math. 303*(12), 579–581.

Tihomirov, V. M. (1963). The works of A. N. Kolmogorov on ε-entropy of function classes and superpositions of functions. *Uspehi Mat. Nauk 18*(5 (113)), 55–92.

Vares, M. E. (1983). Local times for two-parameter Lévy processes. *Stochastic Process. Appl. 15*(1), 59–82.

Walsh, J. B. (1972). Transition functions of Markov processes. In *Séminaire de Probabilités, VI (Univ. Strasbourg, année universitaire 1970–1971)*, pp. 215–232. Lecture Notes in Math., Vol. 258. Berlin: Springer.

Walsh, J. B. (1978/1979). Convergence and regularity of multiparameter strong martingales. *Z. Wahrsch. Verw. Gebiete 46*(2), 177–192.

Walsh, J. B. (1981). Optional increasing paths. In *Two-Index Random Processes (Paris, 1980)*, pp. 172–201. Berlin: Springer.

Walsh, J. B. (1982). Propagation of singularities in the Brownian sheet. *Ann. Probab. 10*(2), 279–288.

Walsh, J. B. (1986a). An introduction to stochastic partial differential equations. In *École d'été de probabilités de Saint-Flour, XIV—1984*, pp. 265–439. Berlin: Springer.

Walsh, J. B. (1986b). Martingales with a multidimensional parameter and stochastic integrals in the plane. In *Lectures in probability and statistics (Santiago de Chile, 1986)*, pp. 329–491. Berlin: Springer.

Watson, G. N. (1995). *A Treatise on the Theory of Bessel Functions*. Cambridge: Cambridge University Press. Reprint of the second (1944) edition.

Weber, M. (1983). Polar sets of some Gaussian processes. In *Probability in Banach Spaces, IV (Oberwolfach, 1982)*, pp. 204–214. Berlin: Springer.

Weinryb, S. (1986). Étude asymptotique par des mesures de r^3 de saucisses de Wiener localisées. *Probab. Theory Relat. Fields* 73(1), 135–148.

Weinryb, S. and M. Yor (1988). Le mouvement brownien de Lévy indexé par R^3 comme limite centrale de temps locaux d'intersection. In *Séminaire de Probabilités, XXII*, pp. 225–248. Berlin: Springer.

Weinryb, S. and M. Yor (1993). Théorème central limite pour l'intersection de deux saucisses de Wiener indépendantes. *Probab. Theory Related Fields* 97(3), 383–401.

Wermer, J. (1981). *Potential Theory* (Second ed.). Berlin: Springer.

Werner, W. (1993). Sur les singularités des temps locaux d'intersection du mouvement brownien plan. *Ann. Inst. H. Poincaré Probab. Statist.* 29(3), 391–418.

Wichura, M. J. (1973). Some Strassen-type laws of the iterated logarithm for multiparameter stochastic processes with independent increments. *Ann. Probab. 1*, 272–296.

Williams, D. (1970). Decomposing the Brownian path. *Bull. Amer. Math. Soc. 76*, 871–873.

Williams, D. (1974). Path decomposition and continuity of local time for one-dimensional diffusions. I. *Proc. London Math. Soc. (3) 28*, 738–768.

Wong, E. (1989). Multiparameter martingale and Markov process. In *Stochastic Differential Systems (Bad Honnef, 1988)*, pp. 329–336. Berlin: Springer.

Yor, M. (1983). Le drap brownien comme limite en loi de temps locaux linéaires. In *Seminar on probability, XVII*, pp. 89–105. Berlin: Springer.

Yor, M. (1992). *Some Aspects of Brownian Motion. Part I*. Basel: Birkhäuser Verlag. Some special functionals.

Yor, M. (1997). *Some Aspects of Brownian Motion. Part II*. Basel: Birkhäuser Verlag. Some recent martingale problems.

Yosida, K. (1958). On the differentiability of semigroups of linear operators. *Proc. Japan Acad. 34*, 337–340.

Yosida, K. (1995). *Functional Analysis*. Berlin: Springer-Verlag. Reprint of the sixth (1980) edition.

Zhang, R. C. (1985). Markov properties of the generalized Brownian sheet and extended OUP_2. *Sci. Sinica Ser. A* 28(8), 814–825.

Zygmund, A. (1988). *Trigonometric Series. Vol. I, II*. Cambridge: Cambridge University Press. Reprint of the 1979 edition.

Name Index

A:
Abel, N. H., 103, 387, 506
Adler, R. J., 158, 170, 178, 179, 441, 454, 493, 494
Aizenman, M., 452
Alaoglu, L., 191
André, D., 133
Arzelá, C., 194, 196, 198
Ascoli, R., 194, 196, 198

B:
Bakry, D., 45, 266
Banach, S., 191
Barlow, M. T., 454, 494
Bass, R. F., iv, 103, 135, 226, 239, 266, 312, 341, 453, 494, 500, 540
Bauer, J., 389, 453
Beebe, N. H. F., iv
Bendikov, A., 388
Benjamini, I., 102, 367, 389, 425
Bergström, H., 388
Berman, S. M., 491, 493
Bernstein, F., 122, 127
Bernstein, S. N., 505

Bertoin, J., iv, 308, 312, 341, 350, 389, 494
Bessel, F. W., 286, 339, 376, 377, 384, 385, 423, 426, 452, 456, 459, 520, 523, 524, 527, 529, 532
Beurling, A. C.-A., 389, 528
Bickel, P. J., 202
Billingsley, P., 44, 104, 200, 202, 212, 213, 500
Blackwell, D., 38, 45, 289
Blumenthal, R. M., 272, 310–312, 388, 453, 454
Bochner, S., 378, 379
Borel, É., 6, 35, 42, 48, 62, 66, 78, 101, 112, 115, 116, 128, 131, 143, 144, 147, 148, 166, 170, 181, 182, 188, 195, 210, 220, 221, 226, 262, 263, 270, 289, 305, 306, 356, 358, 361, 367, 376, 379, 392, 398, 399, 402, 408, 412, 415, 435–437, 442, 470, 478, 479, 481–484, 488, 491,

499, 501, 511–516, 520,
521, 525, 528, 533
Borodin, A. N., 103
Bowman, P., iv
Brelot, M., 389, 528
Burdzy, K., iv, 453
Burkholder, D. L., 44, 134, 253,
256, 261, 266, 312

C:

Cabaña, E., 266
Cairoli, R., iv, 16, 19, 26, 30, 38,
44–46, 63, 86, 134, 135,
237, 266, 395, 416, 452,
453, 469
Calderón, A. P., 62
Cantelli, F. P., 35, 78, 101, 112,
115, 116, 128, 131, 166,
170, 212, 262, 356, 358,
437
Cantor, G. F. L. P., 192, 515
Carathéodory, C., 511
Carleson, L., 385, 525, 540
Cartan, E. J., 389, 528
Cauchy, A.-L., 72, 134, 146, 182,
192, 261, 306, 351, 415,
432, 439, 462, 463, 471,
481
Čentsov, N. N., 147, 173
Chapman, S., 270, 276, 284, 285,
289, 292, 293, 305, 365
Chatterji, S. D., 12, 40, 44, 45
Chebyshev, P., 3, 12, 122, 124, 130,
145, 158, 162, 163, 522
Chen, Z. L., 493
Chernoff, H., 115
Choquet, G., 533, 538, 540
Chow, Y. S., 44
Chung, K. L., 72, 73, 102, 133, 239,
266, 312, 356, 494
Ciesielski, Z., 63, 335, 341
Coulomb, C. A., de, 530
Cramér, H., 206
Csáki, E., 103, 494
Csörgő, M., 103, 179, 494

D:

D'Alembert, J. L., 257

Dalang, R. C., iv, 45, 46, 134, 135,
493, 495
Daniell, P. J., 499
Davis, B. J., 253, 256, 261, 266, 454
de Acosta, A., 117, 136
de Moivre, A. P., 100
Dellacherie, C., 8, 44, 45, 226, 239,
266, 311, 533, 540
Dembo, A., 136
Deny, J., 528
Dirichlet, J. P. G. L., 118, 121, 125,
127, 332, 340
Donsker, M. D., 202
Doob, J. L., 7–9, 11, 12, 18–20, 24,
34, 36, 38, 40, 41, 44, 45,
75, 155, 178, 237, 245,
248, 250, 253, 254, 278,
280, 298, 312, 318, 319,
328, 338, 340, 368, 412
Doob, K. L., iii, iv
Dorea, C. C. Y., 493
Dozzi, M., 239, 266, 494
Dubins, L. E., 33, 35, 37, 38, 45,
289
Dudley, R. M., 44, 150, 151, 158,
171, 172, 174, 178, 179,
202, 212, 452, 500
Durrett, R., 44, 91
Dvoretzky, A., 104, 350, 376, 428,
451
Dynkin, E. B., 104, 288, 341, 453

E:

Edgar, G. A., 45, 135
Ehm, W., 491, 494
Eisenbaum, N., 494
Epstein, R., 494
Erdős, P., 73, 101, 102, 104, 135,
350, 356, 376, 428, 451
Esquível, M. L., 178
Etemadi, N., 104, 136
Ethier, S. N., iv, 45, 202, 311, 340
Euclid, 285, 305, 322, 344, 398, 404,
407, 408, 457, 511, 528,
531
Evans, S. N., 409, 453

F:

Fatou, P., 13, 26

NAME INDEX

Feller, W., 133, 267, 281, 287, 288, 292, 294, 295, 297, 301, 303, 305, 308, 310, 311, 313–318, 320–322, 326–328, 339, 340, 344, 363, 367, 368, 383, 384, 386–388, 392, 394–396, 398, 399, 401, 403–407, 424–426, 451, 493, 501, 506
Fernique, X., 179
Feynman, R. J., 326–329, 338
Fisk, D. L., 250
Fitzsimmons, P. J., 326, 389, 408, 425, 453, 493, 494
Földes, A., 494
Föllmer, H., 33, 45, 312
Fouque, J.-P., 46
Fourier, J. B. J., 177, 285, 307, 323, 325, 423, 491, 502, 503
Frangos, N. E., 135
Frostman, O., 385, 423, 426, 431, 432, 449, 456, 517, 520, 521, 523, 524
Fubini, G., 97, 285, 286, 295, 323, 345, 346, 370, 371, 379, 405, 411, 417, 430, 468, 502
Fuchs, W. H. J., 72
Fukushima, M., 239, 311, 340, 389

G:
Gabriel, J.-P., 16, 134
Gänssler, P., 212
Garsia, A. M., 44, 179
Gauss, C.-F., 100, 103, 115, 116, 127, 129, 135, 137, 139, 141, 142, 146, 147, 170–172, 176, 178, 181, 194, 201, 206, 209, 211, 212, 265, 379, 438, 439, 442, 443, 445, 464, 469, 532, 540
Geman, D. N., 453, 494
Getoor, R. K., 311, 312, 340, 388, 389, 453, 454
Glivenko, V. I., 212
Glover, J., 340
Griffin, P., 134

Gundy, R. F., 63, 253, 256, 261, 266
Gut, A., 134

H:
Haar, A., 47–49, 51, 53–55, 61–63
Hall, P., 44
Hardy, G. H., 57
Hausdorff, F., 293, 385, 389, 423, 431, 433, 434, 441, 443, 447, 455, 457, 512, 515, 517, 520, 521, 523, 527
Hawkes, J., 389, 451, 453, 454, 494
Helly, E., 504
Helms, L. L., 540
Hendricks, W. J., 350, 453
Hesse, L. O., 320
Hewitt, E., 14, 82
Heyde, C. C., 44
Hilbert, D., 137, 176
Hille, E., 311, 315
Hirsch, F., 389, 408, 425, 453
Hochberg, K. J., 46
Hoeffding, W., 45
Hölder, O. L., 165, 166, 174, 176, 254, 263, 265, 440, 484, 516
Horowitz, J., 453, 494
Horváth, L., iv
Hunt, G. A., 7, 8, 45, 226, 239, 272, 288
Hürzeler, H. E., 46

I:
Imkeller, P., 239, 266, 494
Itô, K., 178, 251, 252, 254, 264–266, 288, 308, 311, 339, 385, 494, 540
Ivanoff, B. G., 46
Ivanova, B. G., *see* Ivanoff

J:
Jacod, J., 103
Janke, S. J., 389
Janson, S., 35, 45, 178
Jensen, J., 4, 6, 9, 18, 19, 21, 28, 29, 31, 59, 75, 235, 244
Jessen, B., 54, 58

K:
Kac, M., 232, 326–329, 338, 494

Kahane, J.-P., 454, 493, 525
Kakutani, S., 104, 341, 376, 377, 388, 428, 451, 456
Kanda, M., 389
Karatsuba, A. A., 119, 127
Karatzas, I., 239, 266, 341, 494
Kargapolov, M. I., 78
Kasahara, Y., 494
Kellogg, O. D., 362, 540
Kendall, W. S., 495
Kesten, H., 103
Khintchine, A. I., 116, 117, 307
Khoshnevisan, D., 46, 102–104, 135, 213, 344, 350, 425, 453, 454, 493–495
Kinney, J. R., 272
Kitagawa, T., 178
Knight, F. B., 103, 288, 335, 341
Kochen, S., 73, 356, 359
Kolmogorov, A. N., 3, 14, 111, 123, 142, 150, 151, 159, 160, 162, 164–166, 170, 176, 179, 270, 276, 284, 285, 289, 292, 293, 305, 310, 365, 432, 475, 484, 499, 500
Krengel, U., 135
Krickeberg, K., 45
Krone, S., iv
Kronecker, L., 117, 121, 134
Kuelbs, J., 134, 136
Kunita, H., 251, 266
Kuroda, K., 178
Kurtz, T. G., 202, 311, 340
Kwon, J. S., 135
Körezlioğlu, H., 266

L:

L'Hôpital, G. F. A. M., de, 114, 138, 171, 174, 509
Lévêque, O., iv
Lacey, M. T., 494
Lachout, P., 200
Lagrange, J.-L., 531
Lamb, C. W., 12, 40, 46
Landkof, N. S., 540
Laplace, P.-S., 100, 102, 103, 230, 264, 317, 320, 332, 337, 338, 367, 370, 374, 378, 380, 409, 425, 501–507
Lawler, G., 102, 104
Le Gall, J.-F., 104, 312, 453, 454
Lebesgue, H., 12, 47, 49, 54, 60, 61, 100, 134, 138, 142, 143, 146, 148, 152, 165, 177, 203, 218, 244, 251, 285, 306, 307, 320, 361, 369, 373, 377, 379, 382–384, 386, 406, 409, 412, 413, 415, 420, 423, 426, 431–434, 445, 449, 450, 452, 457, 483, 491, 501, 525, 534
LeCam, L., 213
Ledoux, M., 16, 135, 158, 170, 178, 179
Lévy, P., 63, 113, 114, 123, 175, 178, 303–305, 307, 309, 310, 312, 313, 320, 322, 325, 330, 335, 341, 343, 344, 346, 350, 352, 377, 385–387, 389, 405, 423, 428, 435, 451, 453, 491, 493, 494
Lewis, T. M., iv, 213
Li, D., 135
Liouville, J., 62
Lipschitz, R. O. S., 260, 263, 265
Littlewood, J. E., 57
Lyons, R., 454

M:

Madras, N., 104
Manaka, H., 178
Mandelbrot, B., 42
Marcinkiewicz, J., 54, 58, 266
Marcus, M. B., 494
Markov, A. A., 66, 67, 69, 73, 75, 76, 78, 83, 84, 86, 101, 114, 239, 267–313, 315, 327, 332, 341, 343, 345, 359, 360, 371, 373, 387–389, 391, 393, 395–398, 400, 401, 407–409, 420, 423, 435, 438, 443, 450, 451, 453, 455, 456, 493, 494

Mattila, P., 525
Mazziotto, G., 46, 266, 389, 453
McKean, H. P., 178, 255, 385, 431, 452, 453, 494
Mertsbakh, E., *see* Merzbach
Merzbach, E., 46, 266
Merzljakov, I. J., 78
Meyer, P.-A., 8, 44, 45, 226, 239, 278, 280, 311, 318, 319, 328, 338, 340, 368, 533, 540
Millar, P. W., 312, 389
Milton, G. W., iv
Mountford, T. S., iv, 495
Munkres, J. R., 150, 184, 191, 193
Muroga, S., 179

N:
Nagasawa, M., 312, 389
Neveu, J., 44, 45, 63
Newton, I., 532
Nikodým, O. M., 62, 285, 372, 491
Nualart, D., 266
Nualart, E., iv, 493

O:
Orey, S., 178, 179, 389, 493
Orlicz, W., 45
Ornstein, D. S., 79, 103, 339, 388
Ōshima, Y., 239, 311, 340, 389
Ottaviani, G., 123
Oxtoby, J. C., 213

P:
Paley, R. E. A. C., 72, 75, 82, 85, 309, 347, 349, 372, 418, 449, 452, 459
Paranjape, S. R., 135
Park, C., 135
Parseval, M.-A., des Chênes, 323, 325
Pauc, C., 45
Pemantle, R., 102, 367, 389, 425
Peres, Y., 102, 367, 389, 425, 436, 453, 454, 541
Perkins, E., 103, 454
Petrov, V. V., 136
Phillips, R. S., 311, 315
Picard, C. É., 261

Pitman, J. W., iv, 33, 35, 37, 45, 326, 494
Plancherel, M., 491
Pólya, G., 72, 74, 388
Poincaré, J. H., 40, 362, 539
Poisson, S. D., 340, 502
Pollard, D., 178, 212
Port, S. C., 79, 103, 493
Prohorov, Yu. V., 190, 193, 210, 213, 389
Pruitt, W. E., 178, 389, 454, 493
Pyke, R., 123, 135, 453

R:
Rademacher, H., 134
Radon, J., 62, 285, 372, 491
Rao, K. M., 389
Révész, P., 102, 103, 179, 494
Ren, J. G., 389, 408, 425, 453
Revuz, D., 103, 239, 255, 266, 311, 312, 341, 494
Ricci, F., 135
Riemann, G. F. B., 206
Riesz, F., 62, 193, 311, 376, 377, 384, 385, 423, 426, 452, 456, 459, 520, 523, 524, 527, 529, 532
Rogers, C. A., 135
Rogers, L. C. G., 239, 311, 453, 494
Rosen, J. S., 104, 453, 494
Rota, G.-C., 312
Royden, H. L., 192
Rozanov, Yu. A., 39, 46, 392, 454
Rudin, W., 193, 213

S:
Salisbury, T. S., 389, 408, 425, 453, 454
Sato, K., 308, 312, 350, 389
Savage, L. J., 14, 82
Schwarz, K. H. A., 72, 261, 415, 439, 462, 463, 471
Segall, I. E., 178
Serfling, R. J., 45
Shannon, C. E., 150, 179
Sharpe, M., 239, 311
Shi, Z., iv, 425, 453, 454
Shieh, N.-R., 63, 104
Shorack, G. R., 123, 135

Shreve, S. E., 239, 266, 341, 494
Slade, G., 104
Slepian, D., 176, 178
Smythe, R. T., 110, 112, 117, 134, 135, 178
Song, R. G., 46
Song, S., 389, 408, 425, 453
Spitzer, F., 103, 350
Stein, E. M., 62, 63, 135, 266
Stieltjes, T. J., 203, 264, 501
Stoll, A., 104
Stone, C. J., 73, 79, 103, 356, 359
Stone, M. H., 192
Stout, W. F., 135
Strassen, V., 133
Stratonovich, R. L., 264
Stroock, D. W., 44, 340
Sucheston, L., 10, 45, 135, 312
Sz.-Nagy, B., 62, 311
Szpirglas, J., 46, 266

T:
Takeda, M., 239, 311, 340, 389
Takeuchi, J., 350
Talagrand, M., 135, 158, 170, 178, 179
Tanemura, H., 178
Tauber, A., 102, 506
Taylor, B., 90, 121, 122, 204, 205, 252, 324, 327, 506, 516
Taylor, S. J., 101, 102, 104, 135, 335, 341, 385, 389, 427, 453, 454, 523, 525
Teicher, H., 44
Testard, F., 441, 454
Tihomirov, V. M., 150
Trotter, H., 491, 493, 494

U:
Uhlenbeck, G. E., 339, 388
Ulam, S., 213
Urysohn, P. S., 184, 186, 191, 293, 296, 302

V:
Varadhan, S. R. S., 340
Vares, M. E., 494

W:
Wald, A., 206

Walsh, J. B., iv, 16, 43–46, 63, 135, 237, 239, 266, 289, 311, 312, 395, 493, 495
Watanabe, S., 251, 266, 350
Watson, G. N., 286
Weaver, W., 179
Weber, M., 441, 454, 493
Weierstrass, K. T. W., 192
Weinryb, S., 494
Weiss, G., 62
Wermer, J., 362, 531, 540
Werner, W., 453
Wichura, M. J., 114, 117, 123, 135, 202
Widder, D. V., 502
Wiener, N., 144, 178, 385
Williams, D., 70, 239, 311, 453, 494
Williams, R. J., 239, 266, 494
Wong, E., 250, 453
Wu, Z., 135

X:
Xiao, Y., iv, 454, 493

Y:
Yor, M., 239, 255, 266, 312, 341, 494
Yosida, K., 311, 315, 340

Z:
Zakai, M., 250
Zeitouni, P., 136
Zermelo, E., 15
Zhang, R. C., 493
Zygmund, A., 54, 58, 61–63, 72, 75, 82, 85, 309, 347, 349, 372, 418, 449, 452, 459

Subject Index

Symbols:
ε-enlargement, **372**, **413**

A:
Abelian theorems, *see* Tauberian theorems
adapted, *see* stochastic process
additive Brownian motion, 394, *see also* Lévy processes, *see also* additive stable processes, 492
    hitting probabilities of, 492
additive stable process
    Hausdorff dimension of the range, 433
    Lebesgue's measure of the range, 433
additive stable processes, 419, *see also* Lévy process, additive
    codimension of the range, 423
    hitting probabilities of, 423
    Lebesgue measure of range, 452
    potential density of, 420
Arzelá–Ascoli theorem, 194, **194**

B:
$\mathcal{B}(a;r)$
    Euclidean ball/cube, 344
$\mathcal{B}_d(t;r)$
    metric balls, **149**
Bakry's regularity theorem, 266
balayage, 362, **539**
balayage theorem
    for strongly symmetric Feller processes, **363**
Bernstein's theorem, 505
Bessel functions, 286
    modified, *see* $\mathbf{K}_\nu$, 338
Bessel processes, 339
Bessel's equation, 338
bi-Brownian motion, 395
Borel–Cantelli lemma, 35, 78, 101, 112, 116, 131, 166, 170, 262, 356, 358, 437
branching process, 5, 13, 42
Brownian bridge, 212
Brownian motion, 63
    additive, **433**, *see* additive Brownian motion
    and the heat semigroup, **373**
    as a Lévy process, **229**

Hausdorff dimension of images of, 452
Hausdorff measure of the range, 431
intersection of 3, 451
martingales related to, 229, *see also* harmonic functions and Brownian martingales
multidimensional, **147**, 305
as a Lévy process, 305
one-dimensional, **147**
potential density of, **374**
estimates, **374, 375**
standard, 174
and martingales, 228
Brownian sheet, *see also* stochastic integrals, 455, 462
Čentsov's representation, **148**
and commutation, *see* The Cairoli–Walsh commutation theorem
and its natural pseudometric, 172
as an infinite-dimensional Feller process, 493
as mixed derivative of white noise, 148
SPDE interpretation, 259
codimension of the range, 456
Hölder continuity of, 174, 175
Hausdorff dimension of range, 457
hitting probabilities of, **456**, 491
intersections of, 492
Lebesgue measure of the range, 492
martingales related to, 236
mean and covariance functions, 468
modulus of continuity of, 178
multidimensional, **147**
one-dimensional, **147**
pinned, 212, *see also* Brownian bridge
polar sets for, 456
represented by a random series, 178

standard, 174
Burkholder–Davis–Gundy inequality, *see* inequalities

C:

$C_0(S)$, **283**
$C_b(S)$, **184**
Cairoli's
convergence theorems, *see* orthosmartingales, convergence theorem
inequality, *see* maximal inequalities
maximal inequality, *see* discrete, multiparameter maximal inequalities, *see also* maximal inequalities, multiparameter, continuous
Cantor's tertiary set, **515**
Hausdorff dimension of, 516, **520**
Hausdorff measure of, 520
capacitance, *see* Kolmogorov capacitance
capacities, *see also* Hausdorff dimension, **528**
absolutely continuous, 460, **538**
and Markov processes, *see* Markov processes, multiparameter, hitting probabilities, *see* Markov processes *and* intersections
Bessel–Riesz, **376**, 377, *see* Markov processes *and* intersections *and* Lévy processes, 456, 523, 527, **529**
β-dimensional, **520**
connection to Hausdorff measures, *see also* Frostman's theorem, **521**
vs. absolutely continuous, 423
Choquet, 533

Newtonian, *see* capacities, Bessel–Riesz, 532
  outer
    Choquet, 533
Carathéodory outer measure, *see* outer measure
Cauchy process, 306, 351, 432
Cauchy random walks, 134
cemetery, 272, *see also* Markov chains, killed *and* Markov processes, killed
Čentsov's representation, 173
Chapman–Kolmogorov equation, **270**, 276, 292, 293
  for Markov processes, **289**
  for Markov semigroups, **284**
characteristic functions, 134, 352
  and spectral theorem, 439, *see also* spectral theorem
  and subordination, 379
  inversion theorem, 91, 285, 307, 308
codimension, **435**, 493
  and α-regular Gaussian processes, 441, *see also* Gaussian random variables
  and Hausdorff dimension, 436
coffin state, *see* cemetery
commutation, 35, **35**, 37, 42, **233**
  and conditional independence, 39, 233
  and marginal filtrations, **36**
  and multiparameter random walks, 107
  and orthosmartingales, 37, 38
completely monotone functions, **378**, 505
conditional independence, *see* independence
consistency, **500**
continuity in probability, **158**
continuous mapping theorem, the, **188**
convergence theorems
  for multiparameter martingales, 38
coordinate processes, **80**
correlation function, 438

and pseudometrics, 440
Coulomb's law of electrostatics, 530
Cramér–Wald device, **206**
cumulative distribution function, **188**
  and uniqueness of measures, 188
  and weak convergence in $\mathbb{R}^d$, 189
cylinder sets, 499

D:
$\mathcal{D}$
  domain of a generator, **315**
differentiation theorems, *see also* Lebesgue
  of Jessen, Marcinkiewicz and Zygmund, **63**
  of Jessen, Marcinkiewicz, and Zygmund, **58**
diffusion, **308**
direct products, 80, 402, 403
Dirichlet problem, 332
  and Brownian motion, 340
Dirichlet's divisor lemma, 118, **118**, 121, 125, 127
Donsker's theorem, 202, *see also* random walks, multiparameter, weak convergence
Doob's
  decomposition, **8**, 38, 41, 45
  inequality, *see* maximal inequalities
  martingale convergence theorem, *see* martingales
  separability theorem, **155**
Doob–Meyer decomposition, 280, 340, 368
  for the Feynman–Kac semigroup, 328, 338
  of potentials
    for Feller processes, **318**, 319
    for Markov chains, **278**
downcrossings, *see* upcrossings
duality, 309
Dudley's theorem, **171**, 172, 174

Dvoretzky, Erdős, Kakutani
        theorem, 428
dyadic filtrations, *see also* Haar
        systems and binary
        expansions, *see* filtrations

E:
elementary functions, 144
elementary process
        multiparameter, 265
elementary processes, 246
embedding proposition, **18**, 21
energy, **528**
        Bessel–Riesz, **529**, 532
            β-dimensional, 459, **520**
            0-dimensional, **520**
        Coulomb, 532
        logarithmic, *see* energy,
            0-dimensional
            Bessel–Riesz
        mutual, **528**
        Newtonian, *see* energy,
            β-dimensional
            Bessel–Riesz
enlargement of a set, *see*
        ε-enlargement
entrance times, **221**
        measurability of, 221, **226**
        of Markov chains
            distribution of, **279**
equilibrium measure, 280
equilibrium potential, *see*
        equilibrium measure
essential core, **317**
essential supremum, 48
exchangeability, *see* zero–one laws,
        Hewitt–Savage
exchangeable σ-field, **14**
excursions, 78

F:
F4, *see* commutation
Feller processes, **294**
        characterization via resolvents,
            386
        infinite-dimensional, 493
        multiparameter, 392
        product, 403
        reference measures for, 361

resolvents of, **361**
right continuity of, **294**
strong Markov property of,
    301
strongly symmetric, **360**, 399
Feller semigroup, **294**
Feynman–Kac formula, the, **328**,
    329, 338
Feynman–Kac semigroup, **327**
filtrations
        augmented, **290**
        complete, **290**
        dyadic, 55, 59
            commutation of, **52**
            martingales associated to,
                52
            multidimensional, **52**
            one-dimensional, **49**
        marginal, **32**
            continuous, 233
        multiparameter, 31
            continuous, 233
        one-parameter
            continuous, **218**
            continuous, anomalous
                behavior, 218–220
            discrete, **4**
            reversed, discrete, **30**
            right-continuous extension
                of, **223**
finite-dimensional distributions, **66**,
        68, 69, 154, 194
        and modifications, 154
        not determining weak
            convergence in C, **197**
flow
        stochastic, *see* stochastic flow
Fourier transform
        and the generator of stable
            processes, 323
Fourier transforms, 177, 285, 491,
        see also characteristic
        functions, 502, 503
        inversion theorem, 177, 323,
            325
fractal percolation, 41
fractals, *see* Hausdorff measures
        and Hausdorff dimensions

fractional Laplacian, 325, *see also* stable processes, isotropic, generator of
Frostman's lemma, **517**, 520, 521
Frostman's theorem, 385, 456, **521**, 523, 524
Fubini's theorem, 468

G:
gambler's ruin problem
    for Brownian motion, 229
    for simple walks, 263
gauge function, 361
gauge functions, 460, *see also* energy *and* capacities, **527**, 533
    proper, 363, **539**
Gaussian distributions
    and independence, 139
    characterization of, 139
    covariance, 138
    density function of, 114, **138**, 265, 464, 468, 469
    mean, 138
    tails, **135**
Gaussian processes, *see* Gaussian random variables
Gaussian random variables, **137**, **140**
    and α-regular process, 441
    and martingales, 143
    and stationarity, 438
    and their natural pseudometric, 170
    centered, 438
    covariance, **140**
    existence of, 141
    Hausdorff dimension of the range, 441
    mean, **140**
generators, **315**
    and integro-differential equations, 340
    existence, *see* Hille–Yosida theorem
global helix, 153
Gronwall's lemma, 260, 262, 263
groups
    free abelian, **78**, 103
    and Lévy processes, 387
    and random walks, 78
    the free abelian group theorem, **78**

H:
Hölder continuity, **165**, 516
Haar functions, *see also* Haar systems
    martingales associated to, 50
Haar systems
    and binary expansions, 49
    and dyadic filtrations, *see* Haar systems and binary expansions
    in Lebesgue's differentiation theorem, 55, 61
    multidimensional, 51
        as a basis for $L^1[0,1]^N$, **53**
        as a basis for $L^p[0,1]^N$, 53
    one-dimensional, **48**
        as a basis for $L^1[0,1]$, 51
        as a basis for $L^p[0,1]$, 51
    orthonormality of, 49
harmonic functions, 62
    and Brownian martingales, 332
Hausdorff dimension, 455, 457, **515**, 517
    and capacity, **521**
    and metric entropy, 515
    outer regularity, 517
Hausdorff measures, **512**, 521, *see also* capacity *and* Frostman's theorem, 523
    and metric entropy, 515
    as extensions of Lebesgue's measure, 513
    as outer measures, 512
    invariance properties, 514
heat equation
    and the transition density of Brownian motion, **320**
heat kernel, 373
heat semigroup, **285**
    Feller property of, **288**
    potential density of, **285**
    transition density of, **285**
Helly's selection theorem, 504

Hessian, 320
Hewitt–Savage 0–1 law, *see*
    zero–one laws
Hille–Yosida theorem, **315**
history, *see* orthohistories, 69, 228, **234**, 298
hitting times, *see* entrance times
Hölder continuity, 516
Hölder continuous, 166, 176
    Gaussian processes
        via correlations, 440
Hunt's lemma, **38**, 45
    continuous case, **227**

I:
image, **385**
inclusion–exclusion formula, 38, 40, **107**, 134, 238
increments of, 106
independence
    conditional, **38**, *see also* commutation
    and Markov property, 38, 39, 69
inequalities, *see also* maximal inequalities
    Bernstein, 122
    Burkholder–Davis–Gundy, **253**, 261, 266
        multiparameter, **257**
    Cauchy–Schwarz, 72, 261, 415, 439, 462, 463, 471
    Chebyshev, 3, 12, 124, 130, 145, 158, 162, 163, 522
    Hölder, 254
    Jensen, 4, 6, 9, 18, 19, 21, 28, 29, 31, 59, 75, 235, 244
    Kolmogorov, 3
    maximal, *see* maximal inequalities
    Paley–Zygmund, **72**, 75, 82, 85, 309, 347, 349, 372, 449
        for σ-finite measures, 418
    Slepian, 176, 178
intersection local times, *see* local times
invariance principle, 202

isonormal process, **146**, 148, 176, 238
    existence, 176
    properties, 146
Itô's formula, **251**, 252, 254, 264, **264**, 265
    and the generator of a Bessel process, 339
    and the generator of Brownian motion, 339
Itô's lemma, *see* Itô's formula

K:
$\mathbf{K}_\nu(x)$, 286
Kakutani's theorem, 376, 456
Kochen–Stone lemma, *see also* inequalities,
    Paley–Zygmund, **356**, 359
Kolmogorov capacitance, *see also* metric entropy, **150**
Kolmogorov's consistency theorem, *see* Kolmogorov's existence theorem
Kolmogorov's continuity theorem, 158, **158**, 162, 166, 170
Kolmogorov's existence theorem, 142, 276, 292, 499, **500**
Kronecker's lemma, 117
    multiparameter, **117**

L:
$L^2(\mathbb{R}^N)$, 144
$L^2_{loc}(\mathbb{R}^N)$, 147
$L^\infty(S)$, **282**
L^p bounded
    martingales, *see* martingales
$L^p[0,1]^N$, 48
Laplace operator, *see* Laplacian
Laplace transforms, 102, 103, 230, 338, 370, 378, 425, 501, **501**
    and completely asymmetric stable distributions, 378
    and resolvents, 317
    convergence theorem, 503
    inversion theorem, 502
    of hitting times, 368

resolvents and transition densities, 374
uniqueness theorem, 502
Laplacian, 264, **264**, 320, 332, 337, *see also* fractional Laplacian
law of large numbers
    Kolmogorov's, 3, 111
    moment conditions, 134
    Smythe's, 110, 112, 117, 134, 178
    L^p version, 134
    weak, 502
law of the iterated logarithm, 113
    Chung's, 133
    converse to, 133
    for one-parameter Gaussian walks, **113**
    Khintchine's, 116
    moment conditions, 134
    multiparameter, **117**
    self-normalized, 134
Lebesgue
    differentiation theorem of, 47, **54**, 58, 62
    enhanced, **61**
    monotone convergence theorem of, 12
Lévy processes, *see also* Brownian motion, **303**
    additive, 405
        and codimension of intersection, stable case, 436
        and codimension of range, stable case, 436
        as Feller processes, 405
        reference measure, 406
    and codimension of range, stable case, 436
    and Hausdorff dimension, stable case, 431
    as Feller processes, **304**
    intersections, *see also* Dvoretzky, Erdős, Kakutani theorem
    intersections of, 426
    intersections of, stable case, 428

Liouville's theorem, 62
Lipschitz condition, 265
local central limit theorem, 91, 100
local times, 455, 494
    approximate, 479
    for additive Brownian motion, 492
    for Brownian motion, 491, 493
    intersection, 104
    inverse of, 103
localization, 240, 242, 246, *see also* localizing sequence, **246**, 249, 254, 255
localizing sequence, 246
locally compact space, **282**
lower stochastic codimension, *see* codimension

M:
marginal filtrations
    and commutation, **36**
Markov chains, 78, **267**
    k-step transition functions of, **269**
    absorbed, 274
        transition operators of, 277
    homogeneous, associated to inhomogeneous Markov chains, 268
    initial distribution, 269
    killed, 273
        transition operators of, 276
    Markov property, 86, *see also* Chapman–Kolmogorov equation
    recurrent, **308**
    strong Markov property, **272**
    time-homogeneous, 268
Markov processes, **288**
    capacities associated to, **367**
    generator of, *see* generators
    initial measures of, **288**
    intersections of, 425
    killed
        generator of, 340
    multiparameter, **392**, 455, 456, 493
        hitting probabilities, 408
        initial measures, **392**

reference measures, 398
product, 407
    Feller property of, 407
stationary, **400**
Markov property, 75, 83, 268, **293**,
    see also Markov processes
    and random walks, **67**, 73, 76
    strong, **69**, 78, 114, *see also*
        Feller processes, **299**
        and Lévy's maximal
            inequality, 114
        and random walks, 69, 71
Markov semigroups, 275, 283
    and Feller semigroups, **287**
martingales, *see also* Gaussian
        random variables
    L^p bounded, **12**
    additive, **17**
    and quadratic variation, *see*
        quadratic variation
    continuous
        and unbounded variation,
            239
        convergence theorem
            $L^2(\mathbb{P})$ case, 42
            multiparameter, lack of, 33,
                35, 37
            one-parameter, 36, 37, 41,
                51, 144
            one-parameter, discrete, **12**
            one-parameter, reversed
                case, 41
        convergence theorem, discrete
            one-parameter, 237
        convergence theorem,
            one-parameter, 12, 34
        indexed by directed sets, 63
        local, 245, 331
        multiparameter, *see also*
            orthomartingales, **31**
            and quadratic variation,
                256
            continuous, 234
            strong, 43, 44
        multiplicative, 17
        one-parameter
            continuous, **222**
            discrete, **4**

existence of a regular
    modification, **223**
reversed, discrete, **30**, 227
one-parameter convergence
    theorem, discrete, 41
that are local but not
    martingales, 340
the martingale problem, **317**
upcrossings, *see* upcrossings
maximal inequalities, 522
    and multiparameter walks,
        123, 124
    application of upcrossing
        inequalities, 11
    for continuous smartingales,
        222, 245, 248, 250, 253,
        254, 266
        strong L^p inequality, **235**,
            469
        weak $L\{\ln_+ L\}^{N-1}$
            inequality, **235**
    for multiparameter
        martingales, 38
    in differentiation theory, 55, 59
    in potential theory, 416
    in the LIL, 113
    Kolmogorov, *see* inequalities
    Lévy's, 113, 114
    multiparameter, *see also*
        maximal inequalities for
        continuous smartingales
        martingales, lack of, 33
    of Hardy and Littlewood type,
        57, 59
    one-parameter
        Doob's inequality, discrete,
            8
        strong (p,q) inequality,
            discrete, **8**
        strong $L\ln_+ L$ inequality,
            discrete, **9**, 10
        strong L^p inequality,
            continuous, *see* maximal
            inequalities for
            continuous smartingales
            and maximal inequalities,
            multiparameter,
            continuous

strong L^p inequality,
    discrete, 9, **9**, 12, 40, 75
weak $(1,1)$ inequality,
    discrete, **8**, 18
strong (p,p) inequality for
    orthosubmartingales
    discrete, **20**, 86
strong $L\{\ln_+ L\}^p$ inequality
    for orthosubmartingales
    discrete, 20
weak $(1, L\{\ln_+ L\}^{N-1})$
    inequality for
    orthosubmartingales
    discrete, 22
maximal operator, 55, 58
maximum principle, 388, **533**
McKean's theorem, *see* Lévy
    processes, and Hausdorff
    dimension
method of characteristics, 257
metric entropy, **150**, 515, *see also*
    chaining
    for global helices, 153
    of contractions on $[0,1]$, 151
    relation to Kolmogorov
        capacitance, 150
Mill's ratios, 138
mixing, **100**
modification
    of a stochastic process, **154**
modulus of continuity, 163, **165**, 194
    in probability, **167**
monotonicity argument, **166**, 170
multiparameter martingales, *see*
    martingales

N:

Normal distributions, *see* Gaussian
    distributions

O:

occupation density, *see* local times
    formula, the, 493
occupation measure, 442
one-point compactification, 42,
    294, 391, 401
optional increasing paths, **43**
optional stopping theorem

multiparameter
    and stopping domains, **43**, 44
    and stopping points, **42**
    for strong martingales, 44
one-parameter
    continuous, **226**, 253, 331,
        332, 338, 349, 356, 369
    discrete, **7**, 8, 12, 40, 226,
        281
Orlicz norms, 45
Ornstein–Uhlenbeck process, 339
    Feller property of, 388
orthohistories
    discrete, **22**
orthomartingales, *see also*
    martingales
    and commutation, **36**
    discrete, **16**
    reversed
        convergence theorem,
            discrete, 30
        discrete, **30**
orthosmartingales, *see also*
    orthomartingale
    convergence theorem
        discrete, **26**, 44, 53
        Lamb's method, 44
    discrete, **16**
orthosubmartingales
    discrete, *see also*
        orthomartingale, **16**
orthosupermartingales, *see also*
    orthomartingale, **16**
outer measure, 511
outer radius, 457, 459, 460

P:

$\mathcal{P}(E)$, **520**
Paley–Zygmund lemma, *see*
    inequalities,
    Paley–Zygmund, 459
parabolic operator, 339
Parseval's identity, 323, 325
patching argument, **160**, 166, 174
Peres's lemma, 436
φ-mixing, **100**
Plancherel's theorem, 491
Pólya's criterion, **72**

point recurrence, 351
Poisson processes, **230**
    a construction of, 232
    generator of, **325**
Poisson's equation, 340
polar coordinates, 90, 100, 465
polar set, 455
polarization, 247
Portmanteau theorem, **186**
portmanteau theorem, the, 187
possible points, **76**
    and recurrence, 77
    as a semigroup, 76
    as a subgroup, 76
potential density
    for Markov semigroups, **284**
potential functions, **283**, 531
potentials, *see also* entrance times, *see also* balayage
    of a measure, 529
pretightness, **200**
Prohorov's theorem, **190**, 210
pseudometric, **149**
pseudometric spaces, 149
    totally bounded, 149

Q:
quadratic variation, 213, 242

R:
Rademacher random variables, 134
Radon–Nikodým theorem, 62, 491
random processes, *see* stochastic processes
random set, **435**
random variables, **182**, 194
    distributions of, **183**, 195, *see also* finite-dimensional distributions
    existence of, 183, *see also* Urysohn's lemma
random walks, 5, 65, **66**, 69
    and martingales, 110
    increments of, 40, **66**, *see also* inclusion exclusion formula
    intersection of two independent, **85**
    intersection probabilities, 93
    intersections of, 80, *see also* random walks, simple
    intersections of several, 87, **89**
    Markov property of, 270
    multiparameter, 17, **106**
        and martingales, **108**
        and simulation of the Brownian sheet, 457
        weak convergence to the Brownian sheet, **204**
    nearest-neighborhood, **89**
    simple, 69, **89**
        and weak convergence, 211
        as Markov chains, 308
        intersection of four or more, **97**
        intersection of three, **93**
        intersections of two, **90**, 91
    symmetric, 76, **113**
range, **385**
reciprocity theorem, 529
recurrence
    for Lévy process
        probabilistic criterion, **347**
    for Lévy processes, **344**
    for random walks, **70**, 71, *see also* Pólya's criterion
    and possible points, **77**
recurrence–transience dichotomy, **78**, 79
recurrent points
    as a subgroup, 76
reference measure, *see* Markov processes, multiparameter
reflection principle
    André's, 133, 265
regular conditional probabilities, 289
resolvent
    density
        for stopped random walk, **100**
    multiparameter, **397**
resolvent density, *see* potential density
resolvent equation
    for Markov chains, 278
    for Markov semigroups, **284**

SUBJECT INDEX

resolvents, *see also* resolvent
 equation
 and supermartingales, 295
 corresponding to Markov
  semigroups, **283**
 density
  for multiparameter Markov
   processes, 398
 of Markov chains, **277**
 of Markov process and
  supermartingales, 388
 of Markov processes
  and supermartingales, 451
reversed martingales, *see*
  martingales
reversed orthomartingales, *see*
  orthomartingales
right continuity
  multiparameter, 236

S:

S_Δ, 391
sector condition, **84**
sectorial limits
 discrete, **24**
semigroups, **67**, *see also* Markov
  semigroups, *see also*
  transition operators
 and Markov chains, 275
 and random walks, 67
 Feller, **287**
 marginal
  associated to
   multiparameter
   semigroups, 396
 multiparameter, **395**
 one-parameter representation
  of multiparameter, 403
separability, *see* stochastic
  processes
shift operators (*or* shifts), **292**
simple functions, 144
simple process
 multiparameter, 265
simple processes, 247
smartingales, *see* martingales
sojourn times, 347, *see also*
  occupation measure
space–time process, 339

SPDEs
 hyperbolic
  and Picard's iteration, 261
  Hölder continuity of
   solutions, 263
  motivation, 259, 266
spectral theorem, 439
splitting fields, 46
stable processes, *see* Lévy processes
 completely asymmetric, 378
  asymptotics of the
   distribution, 379
 isotropic, **306**, *see also*
  Bessel–Riesz capacities
  and capacities, 384
  and Hausdorff dimension,
   385
  are not diffusions, **309**
  asymptotics of the density
   function, 380
  escape rates for, **350**
  existence of, 308
  generator of, **324**
  Lebesgue's measure of
   range, 386
  reference measure of, 384
  resolvent density, 382
  resolvent density estimates,
   384
  scaling properties of, **352**
  strong symmetry of, 361
  transition density, 382
stationarity, *see* Markov processes,
  see and Gaussian random
  variables
stick breaking, 5, 13
stochastic codimension, *see*
  codimension
stochastic flow, 265
stochastic integrals
 against martingales
  continuous adapted
   integrands, 248
  elementary integrands, 246
  quadratic variation, 249
  simple integrands, 247
 against the Brownian sheet,
  148
 of Stratonovich, 264

stochastic partial differential
    equation, *see* SPDEs
stochastic process
    adapted, **234**
stochastic processes
    adapted, 4, **218**
    existence, *see* Kolmogorov's
        existence theorem
    modification
        continuous, **168**
    modification, continuous, *see*
        Kolmogorov's continuity
        theorem *and* metric
        entropy *and* Dudley's
        theorem
    one-parameter
        discrete, 4
    separable, *see also* Doob's
        separability theorem, **155**
stopping domains, **43**
stopping points, **42**
stopping times
    continuous, **218**
        approximation by discrete,
            220
    discrete, **5**, 69
strong law of large numbers, *see*
    law of large numbers
strong Markov property, *see*
    Markov property, *see also*
    Markov chains
strong martingales, *see* martingales
strong symmetry, *see* Feller
    processes
submartingales
    multiparameter, *see also*
        multiparameter
        martingales, **31**
    continuous, 234
    one-parameter
        continuous, **222**
    discrete, 4
subordination, 378, **378**
summation by parts, **118**
supermartingales
    multiparameter, 31
        continuous, 234
    one-parameter
        continuous, **222**

discrete, **4**
symmetry, *see also* random walks
    and Gaussian random
        variables, 140
    and positive definiteness, 140
    and resolvent densities, 399
    and the law of the iterated
        logarithm, 134
    for random variables, 113
    of gauge functions, 528, 529
    strong, *see* Feller processes

T:
tail σ-field, **310**
Tauberian theorems, 102, 506
    Feller's, **506**
Taylor's theorem, 523, **523**, 525
The Cairoli–Walsh commutation
    theorem, 237
tightness, **189**
    in C, **200**
time-reversal, 271, 388, 453, 454
total boundedness, **149**
    relation to compactness and
        completeness, 150
towering property
    of conditional expectations, **7**,
        23, 32, 36, 39, 369
trajectories, 67, **84**
transience, *see also* recurrence
    for Lévy processes, *see*
        recurrence for Lévy
        processes
    for random walks, 74
transition densities
    of Markov semigroups, **284**
transition functions, *see* transition
    operators
transition operators, 68, **275**
    and Markov semigroups, 275
    for Markov processes, **288**
    for multiparameter Markov
        processes, **393**
    for random walks, **67**
    marginal, 396

U:
U-Statistics, 45

upcrossing times, discrete, 10, *see also* upcrossings
upcrossings
    inequality
        application to continuous-time, 224
    multiparameter
        inequality, **44**
    one-parameter
        inequality, discrete, 11, **11**, 12
        inequality,discrete, 11
upper stochastic codimension, *see* codimension
Urysohn's lemma, **184**, 296
    and existence of random variables, 186
Urysohn's metrization theorem, 293
usual conditions, **225**, 248

V:

vibrating string problem, *see* wave equation

W:

wave equation, 258
weak convergence, **185**, *see also* Portmanteau theorem
    in C, *see also* Donsker's theorem
weak convergence in C, **198**
white noise, **142**, 148
    is not a measure, 175
    properties, 142
Wiener's test, 385
Williams' path decomposition, 70

Z:

zero set, 455
    of random walks, 102
zero–one laws
    Blumenthal, **310**
    for tail fields, *see* tail σ-field
    Hewitt–Savage, 14, 82
    Kolmogorov, **14**, 310, 432